American Romanian Academy of Arts and Sciences

MIRCEA IFRIM

GHEORGHE NICULESCU

CRIS PRECUP

LIANA MOȘ

ZSOLT GYORI

CASIANA BORU

CSONGOR TOTH

OVIDIU BULZAN

HUMAN ATLAS OF TOPOGRAPHICAL AND FUNCTIONAL ANATOMY

THE NERVOUS SYSTEM AND ANALYZERS

ISBN: 978-1-935924-22-7

Published by:
ARA Publisher: http://www.AmericanRomanianAcademy.org

Printed in the United States

CONTENT

Preface

The extraordinary performances of the brain reflect the unique character of its organization. The human brain, made up of 200 billions of cells which communicate with one another through approximately 100 trillions of contacts, is nowadays the object of an increasingly extensive multi-disciplinary research. It suffices to mention that during the last three decades 15 Nobel Prizes have bean awarded for discoveries in the field of the nervous system. Nevertheless, in spite of the enormous advances achieved, beginning with I. P. Pavlov great discoveries, the unknown field remains, like in all the other sciences, infinitely larger than the known one. In the absence of the demonstration of devices with a deterministic functioning, probabilistic - undoubtedly interesting - theories on the brain functioning were emitted. The concepts of aleatory organization, „*Markovian machine*"(Ross Ashby), fuzziness, applied in neurophysiology, illustrate this tendency. The failure of the excessive localizationism has lead to antilocalizationism and even to the equipotentialism of the cortex *(K. Lashley)*. I do not believe that these trends would contribute significantly to the understanding of the brain. Recent researches on the visual cortex *(Rubel and Wiesel)* plead, on the contrary, for a perfect organization of the cerebral structures. They demonstrate the existence of neuron aggregates in the form of vertical columns arranged in the visual cortex and making up true morphofunctional units of the latter. The connections between the neurons forming these columns are achieved according to a precise scheme, which repeats itself identically in all the columns. The same principle of organization is remarked also in the somatosensitive, as well as in the motor cortex. Nothing seems „*aleatory*", all appears to be built up according to a determined plan, the miraculous transposition of which in neuronal networks is studied in the framework of developmental neurobiology.

The volume dealing with the anatomy of the central nervous system, edited by *Professor Mircea Ifrim*, offers, in a condensed form, a remarkable amount of information reflecting the current stage of knowledge in the field of neuroanatomy. The general conception of the work is that of *Professor Fr. Rainer* concerning the unity between function and structure. Of the vast bulk of literature data, the author eliminates judiciously all those which are not relevant for the understanding of the function of the nervous system and for application in the practice of the neurologic clinic. On the other hand, the description of the connections between structures, as well as the hodology - the description of nervous pathways - are presented exhaustively.

The work is particularly useful for those who want to understand the function of the brain, as well as for those who want to establish the diagnosis of a neurologic syndrome. The establishment of the topographic diagnosis as a compulsory stage preceding the establishment of the etiologic diagnosis has not in the least lost the importance which the classical authors - *Dejerine, Levy-Valensi, Purves Stewart* -had attached to it. It is not a paradox if I state that, by the discovery of the modern laboratory investigation methods, the clinician's role is not at all diminished, but - on the contrary- increased. Only an irreproachable anatomical diagnosis will permit to opt between a myelography and a cranial investigation etc. Among the ever increasing number of possible investigation — some, of which are not devoid of risks —, the neurologist will choose those that are more relevant and will decide the sequence in which to perform them. Avoiding stereotypy in investigation, an individualized strategy will be adopted, allowing to spare time and material means.

As regards the anatomical structures, the work shows the latest discoveries in the field of physiology — neuropeptides, functions of the rhinencephalon, functions of the autonomic nervous system etc. A series of medicosurgical applications, explained by the anatomical data, confer to the work an original character.

The main contribution of the Romanian School of Anatomy and Neurology is pointed out. The clearness, precision and good systematization of the material represent remarkable qualities of the work. We appreciate the systematic utilization of the International anatomic nomenclature and we would be glad if it could be introduced in the current use of clinicians.

The very valuable work of *Professor Mircea Ifrim* will be consulted with a particular utility by the students and the physicians of all specialities, as well as by all those who are interested in the understanding of cerebral structures and functions. It has the merit of being the first work of wide extent in this field appeared in the Romanian medical literature.

Prof. Dr. V. VOICULESCU,
Director of the Institute of
Neurology, Psychiatry and Neurosurgery
of the Academy of Medical Sciences,
Romania

The nervous system

Among the various functions or properties of living matter, one of the most important is excitability or irritability, present in all living beings, from the simplest, unicellular, to the most highly organized ones. Whereas the protozoa react adequately to any excitation of any point of their surface, the metazoa differentiate some portions of their body to accomplish this function. Vertebrates, from the simplest to man, who is the most highly organized living being, are endowed with a special, highly differentiated system, the nervous system, which receives by means of peripheral receptors the excitations arriving from the environment, processes them and gives the appropriate response.

According to its topography, the nervous system may be divided into two parts: *the central nervous system* and *the peripheral nervous system*, forming life's relationship system, through which the organism is able to receive any excitation coming from the environment and to give the appropriate response. The excitations or stimuli are received by the various nerve endings or peripheral sense organs which, after a rough analysis, convert them into a nervous inflow and transmit them, via the peripheral nervous system which is made up of conductors (the nerves) to the central nervous system, where the fine analysis of inflow is performed and the necessary response is conveyed towards the periphery, again via the nerves.

The central nervous system consists of two segments which differ in shape and site, but forming, with the peripheral nervous system, a functional whole. The two segments are: *the encephalon*, called so because it is placed within the head (the cranial cavity) (in Greek: *en* = in; *cephalon* = head), and *the spinal cord (medulla spinalis)*, sited in the vertebral or spinal canal, after which it was named.

The encephalon develops from the cranial portion of the neural plate, passing through a number of cerebral vesicle stages.

Thus, from the stage of the three cerebral vesicles, *the prosencephalon* (forebrain), *the mesencephalon* (midbrain) and *the rhombencephalon* (hindbrain), the stage of five cerebral vesicles is reached through the division of the first and the last cerebral vesicle. These vesicles are: *the telencephalon, the diencephalon, the mesencephalon, the metencephalon (epencephalon)* and the *myelencephalon*.

Each of these cerebral vesicles will give rise, by its development to parts of the encephalon.

The telencephalon, the secondary forebrain or hemispheric brain, is the bulkiest part of the encephalon and is divided into two symmetric parts, the cerebral hemispheres, through a longitudinal fissure of the brain or deep scissure *(scissura interhemispherica)*. The hemispheres are connected through the great interhemispheric commissure *(corpus callosum* and *trigonum)*, each presenting a cavity (the lateral ventricles) which communicates with the middle ventricle *(diencephalon)* through Monro's foramina. Each hemisphere presents, at the contact area with *the diencephalon*, a bulky mass of central grey nuclei, the striate body *(corpus striatum)*.

The diencephalon, also called betweenbrain, is made up of a large ganglionic mass, containing a ventricular cavity (the middle ventricle or third ventricle). The walls of the ventricle, six in number, are: two lateral walls (the thalamus and the suboptic or subthalamic region); the superior wall or ceiling, constituted of *the choroid plexus* of the third ventricle, covered with *the fornix* and *the corpus callosum*; the posterior wall (the pineal body or gland and the tectorial membrane or *membrana tectoria*); the postero-inferior wall (the dorsal *cribrosa lamina*, the mamillary bodies or tubercles, *the tuber cinereum*, the hypophysis, the optic chiasma); the anterior wall (the anterior white commisoure, the terminal lamina).

The mesencephalon or midbrain comprises the cerebral peduncles, the quadrigeminal bodies or *corpora quadrigemina* and the mesencephalic tectum and is crossed by *the aqueduct of Sylvius* or central aqueduct.

The metencephalon is constituted of *the pons* and *the cerebellum*.

The mylencephalon is made up of *the medulla oblongata*.

The dilated cavity of the ependymal canal is sited between the three above mentioned formations, constituting the fourth ventricle.

Functionally, the nervous system is formed of *the cerebro-spinal nervous system* and *the autonomic or vegetative nervous system*. The first one has its name from the two component segments: *the cerebrum* (brain) and the spinal cord, and is also termed the nervous system of relation of the organism with the environment. By its function it converts the stimuli coming from the environment into movements, either of defence or of adaptation, according to the nature and intensity of excitations. In man it also accomplishes higher functions, such as storing of engrams, which are results of excitations, their memorization, learning and education, representing also the substratum of his individuality and personality.

The autonomic nervous system or the system of the internal, *vegetative* life regulates and monitors the activity of internal organs (nutrition, respiration, circulation, excretion etc.) and is constituted, too, of two components, which are functionally antagonistic: *the sympathetic* (or orthosympathetic) and *parasympathetic* divisions, the first being consumptive and the second reparative.

Actually, the term „autonomic" is not a very happy one, since the autonomy of the system is relative, its function being controlled by the higher segments of the central nervous system, so that the nervous system constitutes also from this point of view a unitary whole.

Development of the nervous system
(Fig. 1 - 8)

The nervous system, as the most highly differentiated part of the organism, can be understood only in its development and in close relation with the external and internal environment.

Considered *phylogenetically*, the nervous system appears in the division of the invertebrates, in *the coelenterata*, under the form of an asynaptic diffuse network, in which the intensity of the nervous inflow diminishes gradually from the centre towards the periphery, spreading in all senses. The „nerve" cells are situated in the nodal points of the network and are continuous with each other. This manner of conduction of the nervous inflow represents a progress in comparison with the neuroid conduction, characteristic to sponges in which, owing to the absence of the nervous system, the excitations are transmitted "de proche en proche", i.e. all the cells are excited under the form of the movement of the vibratile cilia.

The sympathetic nervous system appears in worms and arthropods, being formed of nerve cells, the projections of which are united with each other by contact (Ramon I. Cajal), assuming the aspect of neuron chains. Simultaneously appears also a ganglionated cord, sited anteriorly to the digestive tube; at the cephalic extremity it is continuous with a perioesophageal ring and with cerebroid ganglia. As this stage of development we may speak about the existence of an anatomical substratum, necessary for the segmentary reflexes.

In vertebrates, the nervous system - of a tubular aspect - is situated dorsally to the digestive tube and differentiates into an intracranial portion *(the encephalon)* and one which lies in the vertebral canal (the cord).

The encephalon is developed in three phylogenetic stages, at each stage new nervous centres and pathways being added to the already existent formations.

1. During the first stage, *the tectal (archeogenic)* brain - *the archicerebrum* - appears in cyclostomes (fishes with the simplest organization, leading a parasitic life). As a consequence of the necessity of achieving a balance and performing locomotion appears *the tectum opticum* (the equivalent of *the superior colliculi in the mesencephalon*). At the same time differentiates the median cerebellum, which may be considered equivalent to the vermis.

2. The second stage is marked by the appearance of *the basal brain*. We find it in fishes, Amphilia, reptiles and birds. At this stage the basal ganglion appears, under the form of a group of nuclear masses representing the equivalent of *the corpus striatum*.

8

The paleothalamus develops now, too. All these formations make up the paleogenic brain *(palaeocerebrum)* (ancient brain).

3. During the third stage, to the tectal (archaic) and basal (paleogenic) brain is added the cerebral cortex *(cortex pallium)*, forming the neogenic brain *(neocerebrum)* or cortical brain. It appears under a rudimentary form in reptiles and is much more developed in mammals and man. At this stage appears also the *neothalamus* which forms, with the cortex, *the thalamocortical brain*.

The tectal and the basal brain form together *the paleoencephalon*, whereas the cortex represents *the neoencephalon*. The component parts of *the paleoencephalon* assume the automatic and co-ordination functions, making possible to the cortex to remain the site of sensation induction and of movement and action impulse formation. In this way is achieved the relationship between the external and the internal environment, which harmonize in a single whole.

The cerebral cortex, in turn, also develops on the animal scale by three stages:

- first stage: *the archicortex* (the hippocampus area);
- second stage; *the paleocortex* (the olfactory bulb and the fornix);
- third stage: *the neocortex* (the remaider of the cortex).

Generally, there are described:

a) the allocortex, represented by *the rhinencephalon*, with a variable structure according to the species and;

b) the isocortex, of a similar constitution in all the mammals.

Concomitantly, *the cerebellum* is also developing, with the role of dosage and coordination of the movements governed by the cortex. Corresponding to the tectal and to the basal brain, the archicerebellum *(archaeocerebellum)* appears, represented by the lingula, the flocculonodular formations and the vermis, which is predominantly connected with the vestibular system. Phylogenetically, to this portion belongs also the fastigial nucleus. At *the paleocerebrum* stage develop the central lobe, termed culmen, the quadrangular lobe, the pyramid and the uvula, with the emboliform and globose nuclei, that belong to the paleocerebellum *(palaeocerebellum)*. At the cortical brain stage *the neocerebellum* appears, represented by the cerebellar hemispheres and the dentate nucleus. The neocerebellum acquires significant connections with the cerebral hemispheres and the olivary nuclei. Simultaneously with the neocerebellum appears the pons, a formation characteristic of mammals. The pons is crossed by new connections, established between the cerebral cortex and *the neocerebellum* (the corticoponto-cerebellar pathway or *tractus corticoponto-cerebellaris*).

Ontogenetically, the nervous system derives from the ectoblast of the posteromedian region of the embryo, situated in front of Hensen's knot and posteriorly to the chordal bud, At this level, the ectoblast thickens and *the neural plate* appears. The ectoblastic cells forming the neural plate are uniform in size. In the course of its development, the neural plate invaginates, changing into the neural groove. Afterwards, its edges thicken becoming neural crests. They gradually approach the median line, fuse with each other and, as a result, the neural plate changes into a tube, *the neural tube*, which soon separates from the ectoblast from which it arose. The neural tube is flatened transversally and is made up of: two thick lateral walls, a thin anterior wall (basal plate) and a posterior wall (dorsal plate). In its centre lies the ependymal canal. From this moment onwards, the ectoblast regenerates above the neural tube and develops into the epiblast (fig. 1).

During the closure process of the neural tube, some cells sited in the neighbourhood of the neural crests separate completely from the neural tube and form the epiblast, forming on the right and on the left the ganglionic crests. After closure of the neural groove, the ganglionic crests fuse for a short time with each other on the median line and interpose between the neural tube and the epiblast, in a single formation. They separate again, forming longitudinal bands, situated symmetrically along the neural tube.

Then, through a fragmentation process, they will give rise to cell masses arranged metamerically, elements which constitute the sensory ganglia of the cranial and spinal nerves, as well as the ganglia of the sympathetic system.

The closure of the neural tube begins in the median part, advancing towards the two ends of the tube. In the cephalic area, before the closure of the neural tube, occurs the flexion of the head, which brings about the delimitation between the portion of the neural plate which is thicker at this level and the remainder of the neural plate. From the first portion will arise the encephalon, whereas from the second, which has a uniform, tubular calibre, will derive the spinal cord.

After its closure, the neural tube has a double communication with the outside: an anterior one, through an opening situated in the cephalic area, *the anterior neuropore*, and a posterior one, through *the posterior neuropore*. The communication in the posterior area between the neural and the endoblast is temporary and is termed *neurenteric canal*. The two neuropores will close gradually.

The wall of the neural tube is initially made up of elements of an epithelial type, which are disposed around the central canal. Afterwards they increase rapidly in number and arrange themselves in the following layers:

- the germinative layer, constituted of the cells which have proliferated, adjacent to the lumen of the neural tube, and

- the paleal layer, formed of cells also resulted from the initial proliferation, but sited outside the first layer.

The cells originating from the germinative layer differentiate into neuroblasts and spongioblasts. The neuroblasts loose their epithelial appearance and send processes; at first differentiates the axon and then the dendrites, which transform into neurons, From the spongioblasts arise the supporting, glial cells (Fig. 2).

The basal plate and the dorsal plate remain thin; from the basal plate develops the anterior white and the anterior grey commissure and from the dorsal plate, the posterior median septum.

The spinal cord
(Medulla spinalis)

During the development of the grey matter, the lumen of the central canal of the neural tube is at first oval-shaped on cross-section, but through the development of the alar and basal lamina, which are the anterior and posterior portions of the walls of the medullary tube, it assumes a rhombic form on section, to change afterwards into a canal with a round lumen, *the ependymal canal*. In the third month of intrauterine life, the architecture of the spinal cord is roughly set up. From the fourth month starts the myelination of fibres emitted by the neuroblasts, which ends only at the beginning of th second year of life; the last pathway which undergoes myelination is the pyramidal tract, fact which is in close relation with its function: walking, locomotion. After the third month of intrauterine life, the spinal cord lags behind the development of the spinal canal, which has the following consequences:

- the spinal cord does not occupy the whole length of the spinal canal (the lower end of the spinal cord does no more correspond to that of the vertebral column);

- the roots of the lumbar, sacral and coccygeal spinal nerves have an oblique downwards course up to *the intervertebral foramina*, instead of their initial horizontal disposition;

- the arrest in growth of the lower medullary segment, which becomes the terminal filament *(filum terminale)* as a result of the ascension of the spinal cord and of the traction exerted on its lower portion.

The neural tube has initially a constant size, but after the formation of buds, from which the lower and upper limbs will arise, appear the two enlargements of the spinal cord; *the intumescentia cervicalis* and *the intumescentia lumbalis*.

The encephalon
(Encephalon)

The encephalon develops from the cranial part of the neural tube, where the neural plate is thicker. By joining of its margins, the encephalic tube will have a larger volume than that portion of the neural tube from which the medulla develops. In this way forms a cerebral vesicle, *the protoencephalon*, separated into two portions: *the preencephalon* or *archencephalon* (corresponding to the part situated in front of the dorsal cord) and *the postencephalon* or *deuterencephalon* (corresponding to the cordal part). These two vesicles are separated by a faintly represented groove. The preencephalon divides, in the course of its development, into two other vesicles and thus appear the three cerebral vesicles: *the prosencephalon* (forebrain), *the mesencephalon* (midbrain) and *the rhombencephalon* (hindbrain), which is continuous with the spinal cord (fig. 4).

At the same time, changes occur at the level of the encephalon, bringing about the unequal development of the walls of the vesicles (the posterior walls grow to a greater extent), so that the three cerebral vesicles do not remain disposed in a straight line, continuous with the spinal cord. The rhombencephalon forms a nearly right angle with the spinal cord, resulting in the nuchal flexure; the prosencephalon bends too, forming an angle with the mesencephalon.

The development continues and the first and third of the three vesicles divide, the stage of five vesicles being thus reached. From the hindbrain or rhombencephalon arise two formations: *the myelencephalon*, continuous with the spinal cord, and *the metencephalon*. *The mesencephalon* remains undivided, whereas from *the prosencephalon* derive the *diencephalon* and *the telencephalon* (the most rostral, remote, cerebral vesicle) (fig. 5).

The flexures, merely outlined in the three vesicle stage, become more marked and thus the following three flexures appear:

- the nuchal or cervical flexure, between the spinal cord and the mesencephalon, with the concavity directed anteriorly;

- the pontine flexure, between the myelencephalon and the metencephalon , with the concavity directed posteriorly;

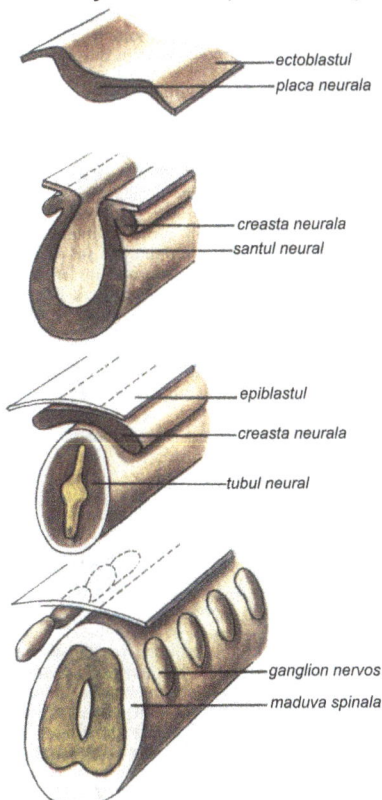

ectoblastul
placa neurala
creasta neurala
santul neural
epiblastul
creasta neurala
tubul neural
ganglion nervos
maduva spinala

Fig. 1 Development of the ectoblast; formation of the neural tube.

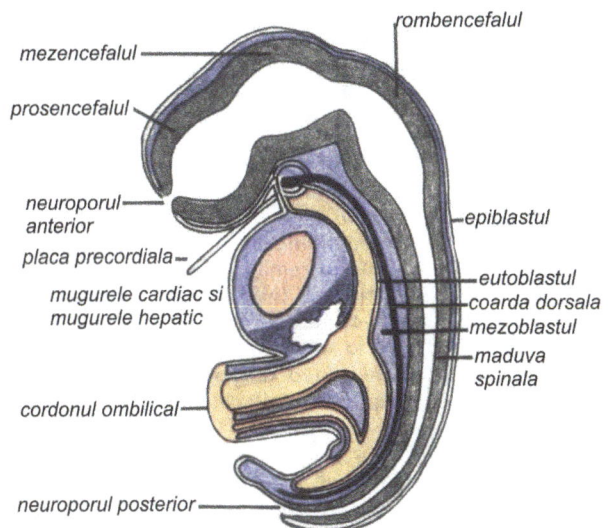

rombencefalul
mezencefalul
prosencefalul
neuroporul anterior
placa precordiala
mugurele cardiac si mugurele hepatic
cordonul ombilical
neuroporul posterior
epiblastul
eutoblastul
coarda dorsala
mezoblastul
maduva spinala

Fig. 3 Embryonic aspect of the end of the neurolation.

11

- the apical (vertex or cephalic) flexure, corresponding to the mesencephalon with the concavity directed anteriorly.

From the five cerebral vesicles the various segments of the encephalon will arise (fig. 8). *The myelencephalon* (medullary brain) forms *the medulla oblongota. The metencephalon* (statoacoustic brain) is the site of origin of *the cerebellum* and *the pons. The mesencephalon* (midbrain) generates *the cerebral peduncles* and *the tectum. The diencephalon* (interbrain) forms *the thalamus* and the tectorial membrane of the third ventricle. *The telencephalon* develops into *the cerebral cortex* and the basal nuclei. Like for the spinal cord, from the alar lamina develop sensory formations and form the basal lamina, motor cell elements. From the intermediate portions arise the vegetative medullary centres.

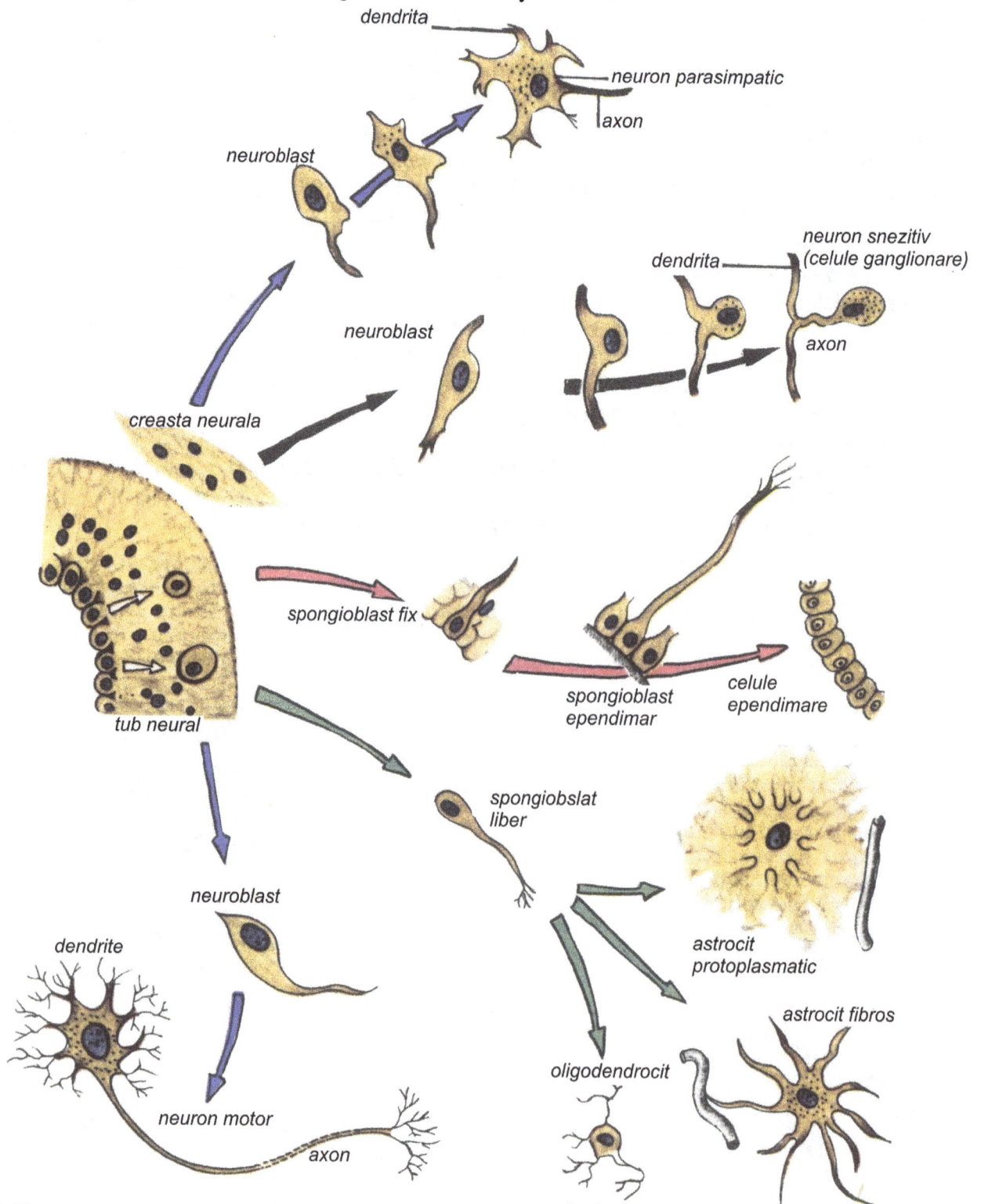

Fig. 2 Histogenesis of the nervous tissue

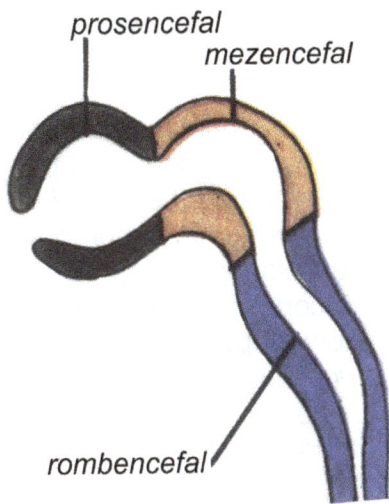

Fig. 4 Three-cerebral-vesicle stage

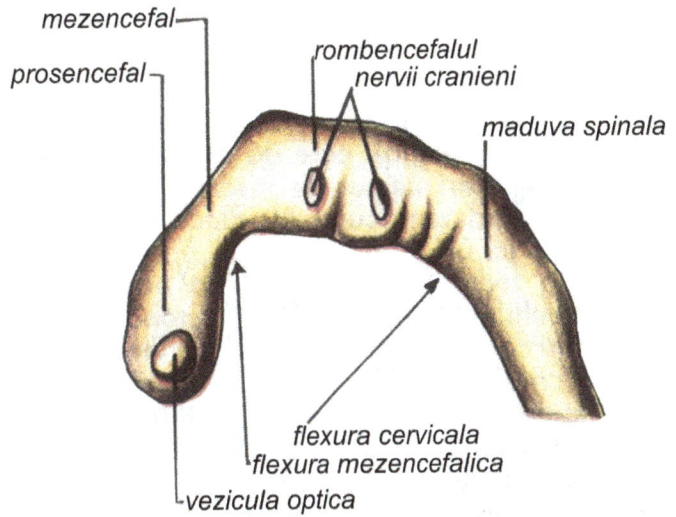

Fig. 5 28 day human embryo – external view

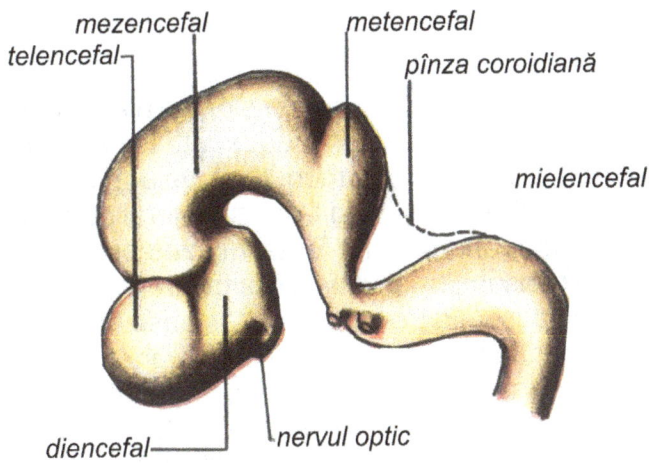

Fig. 6 38 day human embryo – external view

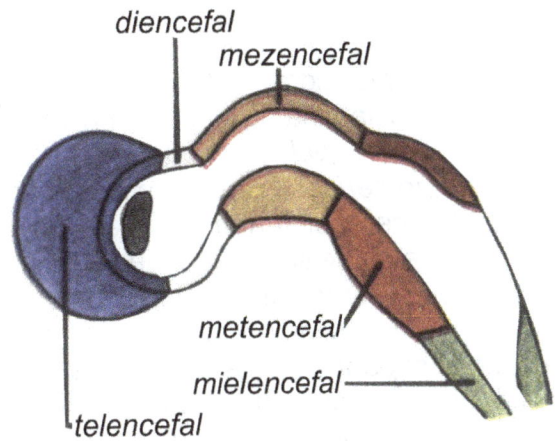

Fig. 7 Five-cerebral-vesicle-stage

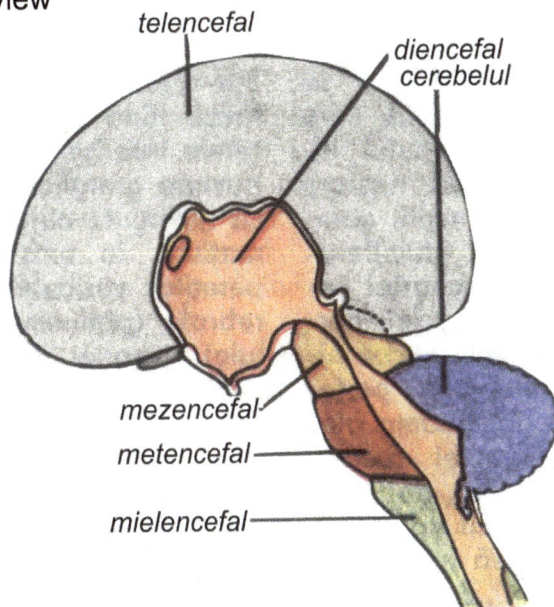

Fig. 8 Development of the cerebral vesicles.

13

The myelencephalon
(Mielencephalon)

The portion of the neural tube from which develops the myelencephalon undergoes rather important changes. Its cavity enlarges, forming the fourth ventricle. The dorsal plate becomes thinner and reduces itself to a layer of ependymal cells, becoming *the tectorial* or *obturator membrane* (Kölliker's membrane) of the fourth ventricle. *The tectorial membrane* together with the adjacent mesenchyme in which neoformation vessels develop, makes up *the tela choroidea*. At the point where the tectorial membrane is continuous with the alar lamina appears a thickening called *lingula* and at the level of the lower apex of the ventricle develops another thickening termed obex. The alar and *basal laminae*, separated by the *sulcus limitans*, are disposed in the same transverse frontal plane. The basal plate is included in the basal lamina and, by a process similar to the "opening of a book" arises *the medulla oblongata* with its structures. From the basal lamina arises the hypoglossal trigone *(trigonum hypoglossy)*, in the depth of which is sited the motor nucleus of this nerve, and from the alar lamina develop the sensory formations of *the medulla oblongata* (*nucleus gracilis, nucleus cuneatus* and *nuclei terminations* of the vestibula nerves). From the alar lamina derives also the oliva. From the intermediate piece develop the visceromotor nuclei (towards the basal lamina) and the viscerosensory nuclei (towards the alar lamina).

The metencephalon
(Metencephalon)

From this segment the cerebellum and the pons develop. The roots of the alar lamina and the dorsal lamina give rise to the cerebellar lamina, which is continuous with the anterior medullar velum *(velum medullare anterior)* and the posterior medullary velum *(velum medullare posterius)*. From the central portion arises the vermis and from the lateral portions develop the cerebellar hemispheres. During the fourth and the fifth foetal months occures a fast increase of the surface of the cerebellum, the consequence of which is the development of numerous fissures and of the lobules. From the basal and the lateral plates arise the pons and the inferior, middle and superior cerebellar peduncles *(pedunculi cerebellary inferius, medius et superius)*, by a process resembling the development of *the medulla oblongata*, contributing in association with the latter to the formation of the rhomboid fossa *(fossa rhomboidea)*. Moreover, like at the level of *the medulla oblongata*, are present motor nuclei *(eminentia medialis)*, developed from the basal lamina, and sensory nuclei *(area vestibularis)* derived from the alar lamina. From the intermediate piece arise the viscerosensory areas.

The mesencephalon
(Mesencephalon)

This cerebral vesicle undergoes smaller changes in comparison with those of the vesicle previously dealt with. Its cavity remains narrow, becoming the cerebral aqueduct *(aqueductus cerebri or aqueductus Sylvii)*, which connects the fourth and the third ventricles. The basal lamina, including the basal plate, gives origin to the cerebral peduncles (motor area). The cerebral peduncles (crus) represent a formation specific to mammals and appear as the result of the significant role of the cerebral cortex in motion. The mesencephalon contains the nuclei of *the trochlear* and *oculomotor nerves,* the red nucleus *(nucleus ruber)* and *the locus niger*. The basal plate takes part in the formation of the posterior perforated substance *(substantia perforata posterior)*. The alar lamina, including the dorsal plate, gives rise to the tectal lamina *(lamina tecti)* (during its development appear at first two colliculi, separated by a anteroposterior groove, after which, owing to the appearance of a furrow perpendicular on the first, arise the four colliculi, two superior and two inferior, in relation with the visual and respectively auditory pathways).

14

From the lateral walls of the mesencephalon the medial geniculate bodies also develop. The vegetative formations are not grouped, as usually, in the area of *the sulcus limitans*. Actually, *the sulcus limitans* is visible only at the surface.

The diencephalon
(Diencephalon)

The diencephalon results from the division of the prosencephalon. A long time before the occurrence of this division, the portion that will become the diencephalon gives rise to the optic vesicles. The diencephalon has an axis represented by the third ventricle. The epithalamus develops from the dorsal lamina and from the posterior portion of the alar plate. During the second embryonic month, this area extends, gets narrower and becomes *the lamina tectoria* of the third ventricle. In the seventh week, arises from the dorsal lamina, posteriorly, also the glove-finger shaped epiphysis, its cavity communicating with the cavity of the ventricle. Afterwards, its lumen disappears, as a result of the thickening of the walls. Concomitantly, at the junction between the alar lamina and the dorsal plate appear *the habenulae* which are part of the olfactory pathways.

The lateral walls of the diencephalon develop and thicken considerably; from the alar lamina arises a sensory nucleus, *the dorsal thalamus*. From the alar lamina develops also the metathalamus (the lateral geniculate bodies). In the basal lamina originates the so-called suboptic region or *ventral thalamus*, which has afferent and efferent connections with the striate bodies (nuclei of Luys, *zona incerta*). From the basal lamina *the globus pallidus (paleostriatum)* also arises. As a result of the exagerated development of the lateral walls, the diencephalon flattens laterally and lengthens anteroposteriorly and its cavity becomes the third ventricle. At the level of *the sulcus limitans*, from the lateral wall of this ventricle develop a number of vegetative formations, which will shift anteriorly, giving rise to *the hypothalamus*. From the floor of the third ventricle arises a diverticulum, the extremity of which, surrounded by Rathke's pounch, generates the nervous lobe of the hypophysis. The evaginated wall forms *the tuber cinereum* and the cavity of the diverticulum is called *infundibulum*. Behind the hypophysis lie two elevations, the mamillary tubercles *(corpora mamillaria)*, separated by a median furrow.

The telencephalon
(Telencephalon)

It is the most rostral part of the neural tube and appears through the division of the prosencephalon, at first as an unpaired formation. It will soon give rise to a median and the lateral portion, termed telencephalic vesicles. The median portion is *the lamina terminalis*, representing in the initial stages - the only connection between the two telencephalic vesicles. Here develop the commissural tracts. The cavities of the two telencephalic vesicles communicate, in the embryon, with the third ventricle through two wide openings, called *interventricular foramina* (Monro's foramina).

At the level of the telencephalon, the separation into an alar lamina and a basal lamina is no longer observed, the telencephalic vesicles being considered as derived from the alar lamina. The inferior wall of the telencephalic vesicles thickens, giving rise to the basal ganglia (encountered in fishes, reptiles and birds, as elements characteristic of the paleoencephalon). All these formations make up the striate body *(corpus striatum)*. The remaining walls of the telencephalic vesicles give rise to the cerebral cortex (*pallium*, the superficial grey highly cellular layer) and to the area of white matter of the hemispheres *(Flechsig's centrum ovale)*. Phylogenetically it is the striate body which appears at first followed by the cortex: initially the olfactory cortex or *archipallium* containing the formations of *the rhinencephalon*, and then the nonolfactory cortex or *neopallium*. They will be studied below in this order.

15

The striate body, coordinator of some complex muscular activities, develops from the anterolateral region of the telencephalon. The striate body is made up of the caudate nucleus *(nucleus caudatus)* and the lentiform nucleus *(nucleus lentiformis)*. The caudate nucleus *(nucleus caudatus)* appears in the second month, in the floor of the interventricular foramen, and bends antero-inferiorly above the diencephalon. Laterally to the caudate nucleus develop a part of the lentiform nucleus, *the putamen* representing *the neostriatum*. Initially the putamen is not well delimited, but later appear fibrons tracts which demarcate it, maintainig nevertheless, the connection with the caudate nucleus. The striate body lies in that portion of the telencephalon which covers the diencephalon. The union between the diencephalon and the telencephalon is achieved by fusion of the lateral surface of the thalamus with the internal surface of the basal ganglia. At the site of contact between the telencephalon and the diencephalon fibres which connect the cerebral cortex with the centres situated at the base of the brain (internal capsule) develop.

The rhinencephalon (olfactory cortex) develops early both phylogenetically and embryopically. In the sixth week, on the anterior surface of each telencephalic vesicle appears the olfactory bulb under the form of a swelling. At first its cavity communicates with the ventricular cavity, then it is obliterated. This bud grows distally, forming the olfactory tract , and from the root of the bud derives the olfactory trigone *(trigonum olfactorium)*. Below the olfactory bulbs is outlined the anterior perforated substance *(substantia perforata anterior)*. *The hippocampus* develops assuming an arched form, owing to the expansion of the telencephalic lobes. The same arched disposition has also the fornix, another portion of the olfactory tract.

The non-olfactory cortex *(neopallium)*, in mammals and man, overlaps completely, *the rhinencephalon*, which is most developed part of nervous system in lower mammals. In primates and humans, as the importance of the olfactory sense diminishes to a great extent, it is the neopallium which develops considerably. It has initially a smooth appearance *(lissencephalon)*. As a result of the continuous development of the surface of the encephalon, the latter as constrained to grow within a limited space, which leads to the appearance on its surface of numerous *gyri* (convolutions) and *sulci* (furrows). Already from the third month, on each lateral surface of the future hemispheres appears a depression, called fosa of Sylvius *(forsa lateralis cerebri)*. In the meantime, each hemisphere develops under on arched form, covering the diencephalon and the mesencephalon. The cortex is more evident around the fossa of Sylvius, giving rise to the rudiments of lobes called at this stage opercula: frontal, parietal and temporal. The floor of the fossa of Sylvius grows slower, becoming *the insula*, sited laterally to the striate body. The three opercula, approximating to each other, cover completely *the insula* and delineate, on the lateral surface of each hemisphere the lateral cerebral sulcus *(sulcus cerebri lateralis Sylvii)*. Both in the course of the phylogenetic and ontogenetic development, the frontal lobe grows later than the others. In the sixth month of intrauterine life are outlined the central incisure *(incisura centralis)*, then the parieto-occipital and the calcarine incisures, delimited by *gyri* of an utmost importance.

Hystogenesis of the cerebral cortex. At the level of the cerebral hemispheres there are the same substances (gray and white matter) as in the spinal cord. The area of the insula is the first that shows a differentiation, which afterwards also extends to the adjacent areas. In the third month, a large number of neuroblasts migrate towards the periphery constituting the primordial grey cortex. From the fourth and fifth month onwards, these neuroblasts will differentiate and the characteristic *six-layered structure* of the cortex will appear. The connective tissue, which gets in contact with the nervous substance, will give rise to *the pia mater*.

The commissural system. At the initial stage, *the lamina terminalis* represents the only connection between the two telencephalic vesicles. Fibres condense around it, forming various transverse commissure which unite the hemispherical buds. The first which appears, both on the phylogenetic and ontogenetic scale, is the anterior white commissure *(commissura alba anterior)*, that interconnects the olfactory centres.

16

In relation with the olfactory sense , during the third month, the fornix also appears. The last commissural system which develops is *the corpus callosum*, a missive system of fibres, connecting the non-olfactory areas and showing the maximum development in man. Through its growth it pushes downwards and backwards *the hippocampus*, on both sides. Between the central trigon and *the corpus callosum* - which has an arciform disposition - lies a narrow zone of the hemispherical wall, termed *septum pellucidum* (transparent septum) which tapers as *the corpus callosum* develops. The two *laminae* of *the septum pellucidum* may be separated by the cavity of the septum pellucidum *(cavum septi pellucidi)*.

The cerebral ventricles appear as a result of the change of the primary of the neural tube. In the spinal cord arises the ependymal canal (central canal of the spinal cord).

The cavities of the mesencephalon and metencephalon respectively take part in the formation of the fourth ventricle. The cavity of the mesencephalon forms the cerebral aqueduct of Sylvius *(apeductus cerebri)*, while the cavity of the diencephalon and the medial portions of the telencephalic cavities constitute the third ventricle. The cavities of the telencephalic vesicles will become the lateral (first and second) ventricles. Concomitantly with the growth of the hemispheres, the appearance of the lateral ventricles changes, owing to the projections which they send into the frontal, temporal and occipital lobes. In some areas, along the ventricles (lateral third and fourth ventricles) their walls remain thin, made up of a layer of ependymal cells disposed on a basement membrane. At this level, the neoformation blood vessels give rise to the choroid plexus. The capillaries push the ependymal membrane, forming fringes which bulge into the ventricular lumen. On this way, a secretory mechanism is formed, in which the ependymal cells extract fluid from vascular plexus and pass the product of their secretion into the lumen of the ventricles.

Structure of the nervous system
(Fig. 9-15)

Generalities. The nervous system has the ability to receive the various excitations, from outside and inside the body, to convert them into certain forms of inflows, to process them and to transmit them into various parts of the organism, achieving thus the unity of the organism with the environment.

The substances which elicit signals for the receptors are either produced by the respective organism (neurotransmitters, neuromodulators, hormones, antigens, toxins, etc), either synthesized, like drugs. Consequently, their actions are extremely diversified: growth control, setting the organism in "rest" or "activity" conditions, regulation of the cardiovascular system, control of movements of the skeletal musculature, transfer of signals and informations to the central nervous system, as well as pharmacological modelling of all these precesses.

As a matter of fact, a substance -whether a neurotransmitter, a hormone or a drug - is "recognized" selectively and this recognition triggers a specific reaction, a stimulus, which acts on the effector system via the various links of the nervous system.

Among these substances are those produced by the human organism itself, such as neurotransmitter, neuromodulators, hormones, antigens, toxic substances, as well as synthesis products, such as drugs, regulators of growth processes, of the menstrual cycle, of the cardiovascular system, etc.

It is considered that a significant role in the carrying out of these processes is played, at the level of the receptors, by the adenyl cyclase, which is activated by a number of substances (hormones, beta-adrenoceptors, etc) into cyclic adenosine monophosphate (cAMP), that in turn, elicits series of physiological processes.

The first researches which demonstrated this type of chemoreceptors were performed on the motor plate of striated muscle, which is an extremely significant system, controlled by receptors that regulate with a very high precision and at a very fast rate the contraction of the muscle fibre.

Thus, the neuromotor plate is made up of the axon of the cholinergic neuron, which expands considerably, forming the presynaptic membrane, where acetylcholine vesicles accumulate. On the postsynaptic aspect acetylcholine sensitive chemoreceptors are disposed. The binding of acetylcholine (Ach) on the receptors brings about a flux of ions which changes the electrical potential of the motor plate, causing muscle contractions.

When the receptors are blocked by another substance, curare for example, the contractions does no more occur. Recent researches show that these receptors of the neuromotor plate vary in size from 45 to 55,000 L and are grouped by five in rosettes, with an ionic channel in the centre host frequently two such rosettes are coupled. Moreover, it was observed that such neuroreceptors for acetylcholine are present also in the central nervous system, Ach being also a central neurotransmitter. Likewise, the amino acid sequence in various receptors was determined and such receptors were synthesized by genetic engineering. The isolation of receptors made possible the obtention of antireceptor antibodies and led to the conclusion that myasthenia is a severe disease of autoimmune type, in which the organism produces antireceptor antibodies.

Other categories of receptors were isolated from other tissues than the neuromotor plate. Thus the beta-adrenergic receptor were isolated, which were divided, according to the amino acid sequence, into beta -1 and beta - 2 receptors; in the case of these receptors, in ionic channels were identified, but a permanent connection with the adenyl cyclase was detected which, through cAMP, may elicit and modulate the physiological processes. In addition, molecule units, which recognize the benzdiazepines and which in this way, can control various physiological processes, were isolated.

The existence of these receptors in different part of the nervous system was demonstrated, with an increased frequency concomitantly with the introduction in current practice of the use of radioactive substances, which have a differentiated selectivity and affinity for the various categories of receptors, representing thus true radioligands. In this way, tissues taken from various parts of the central nervous system or from various peripheral organs are brought into contact with these radioligands and the manner in which a differentiated binding occurs is followed-up. The number of these receptors and the affinity of the radioligand for the receptor are established on the basis of the saturation curves. The affinity of non-radioactive substances (termed cold substances) for the receptor is determined by inhibition of the fixation of the radioactive substance by various concentrations of cold substance.

The method has furnished numerous informations on the density and distribution of these receptors in the central or peripheral nervous system, as well as in the interaction of various substances with the receptors.

Moreover, the autohistoradiographic method permitted the photographic demostration of the same receptors together with the radioligands. Thus receptor for opiates were identified in the central nervous system and it was observed that their concentration was increased in areas related to emotivity and pain integration.

The problem was raised whether these receptors for vegetal opiates are not actuated also by endogenous opiates. Starting from this question, the existence of endogenous morphine derivatives (enkephalins) was demonstrated, which were isolated and proved to be constituted of five amino acids. In addition, anti-enkephalin antibodies were prepared by the immunohistochemical method. By analogy with the system of neurotransmitters, the existence of an endogenous opiate system was proven, which plays a significant role in acupuncture and in the induction of emotional and stress states. It was also shown that the benzodiazepine receptors have a topographical distribution similar to that of the opiate receptors (fig.9).

At the same time, the electrophysilogical researches show the existence of an interaction between the benzodiazepine and the GABA receptors (receptors for the gamma - aminobutyric acid). The gamma aminobutyric acid is an inhibitory transmitter, which diminishes the excitability of the neurons.

18

The GABA receptors modulate the ionic channel for the Cl ions whereas the Ach receptors modulate the ionic channel for Na and K ions. The penetration of the Cl ions into the neurons lowers their excitability. The benzodiazepine receptors are directly coupled with the GABA receptors, this structure being similar. In this way may be explained the anticonvulsant, anxiolytic and myorelaxing projecties of benzodiazepines. The substances with an effect contrary to that of benzodiazepines prevent the passage of the Cl ions, occupy the GABA receptors and disinhibit the neurons. The existence of the benzodiazepine receptors and the appearance of anxiety states through the administration of substances antagonistic to benzodiazepines have led to the conclusion, according to recent researches, that in the organism beside the system of endomorphines, there is also a system of endogenous benzodiazepines. It has been also proved that the cells innervated by the sympathetic system are present beta type receptors, respectively beta-1 type receptors prevalent in the heart and beta-2 type receptors predominant in the lungs. Starting from this finding, a series of beta-blocking substances were discovered, which have a favourable effect, for example, in various types of coronary diseases, such as angor pectoris etc. The range of pharmaceutical preparations of the type has considerably increased: among beta-blockers may be mentioned Propranolol, Inderal etc. selective for various categories of beta receptors (fig. 10-14).

Moreover, we mention that the number of receptors in an organ is variable and strictly individualized. In addition it was demonstrated that functionally their number can increase or diminish in dependence on the various conditions existent. If the receptors are for a long time in contact with synergistic substances, they diminish quantitatively to a great extent, but if, on the contrary they are not-demanded or if they are blocked, their number increases. This explains the reverse effect of sensitization to sympathomimetic agents. Likewise quantitatively, the neuoreceptors undergo nycthemeral and seasonal variations and are subjected to a circadian rhythm which is regulated as it results from the most recent neurological researches, by the suprachiasmatic nucleus of the diencephalon.

Recently, the identification and purification of adrenoreceptors were facilitated by the production of agonistic and antagonistic synthetic substances with a high affinity for receptors (radioligands). The following types of adrenoreceptors were separated by means of this method; alpha from the hepatic tissue, beta from erythrocytes beta 1 from the heart and beta 2 from the lung.

The beta receptors in the heart are polypeptides bound to the cell membrane; their molecular weight exceeds 60,000 and they are coupled with an enzymatic system which stimulates the formation of adenyl cyclase. A high variability is observed at the level of the heart with respect to the rate of beta 1 and beta 2 subtypes. Thus, the atria contain in equal amounts the beta 1 and beta 2 subtypes, whereas in the ventricle prevail the beta 1 receptors.

As regards to the above - mentioned inhibitors of the adrenoceptors; we underline that until 1958, date of appearance of the dichloroisoproterenol (Powell) - the first beta blocker -, only alpha-blockers were available. Afterwards a series of propranolol type beta blocking substances appeared, which are characterized by the following features:

- they are competitive antagonists of the endogenous or exogenous catecholamines (isoprenaline);

- they diminish the sensitivity of organs to catecholamines;

- pharmacologically, their effect can be potentiated by the administration of a supplementary dose of agonists.

At present, a wide range of anatagonists of the beta adrenoceptors (beta-blocking agents) is available to the clinicians, such as; alprenolol, atenolol, metoprolol, oxprenolol, pindolol, propanolol, timolol, between which there are slight differences with aspect to the selectivity and intensity to their action that permit by a rigorous posology to obtain a blockade leading to the disappearance of the sympathetic tonus. It was observed that some beta-blocking agents are weakly liposoluble and have a slower renal eliminations, while others are highly liposoluble (lipophilic) and have a rapid renal elimination and a fast metabolisation at the level of the liver.

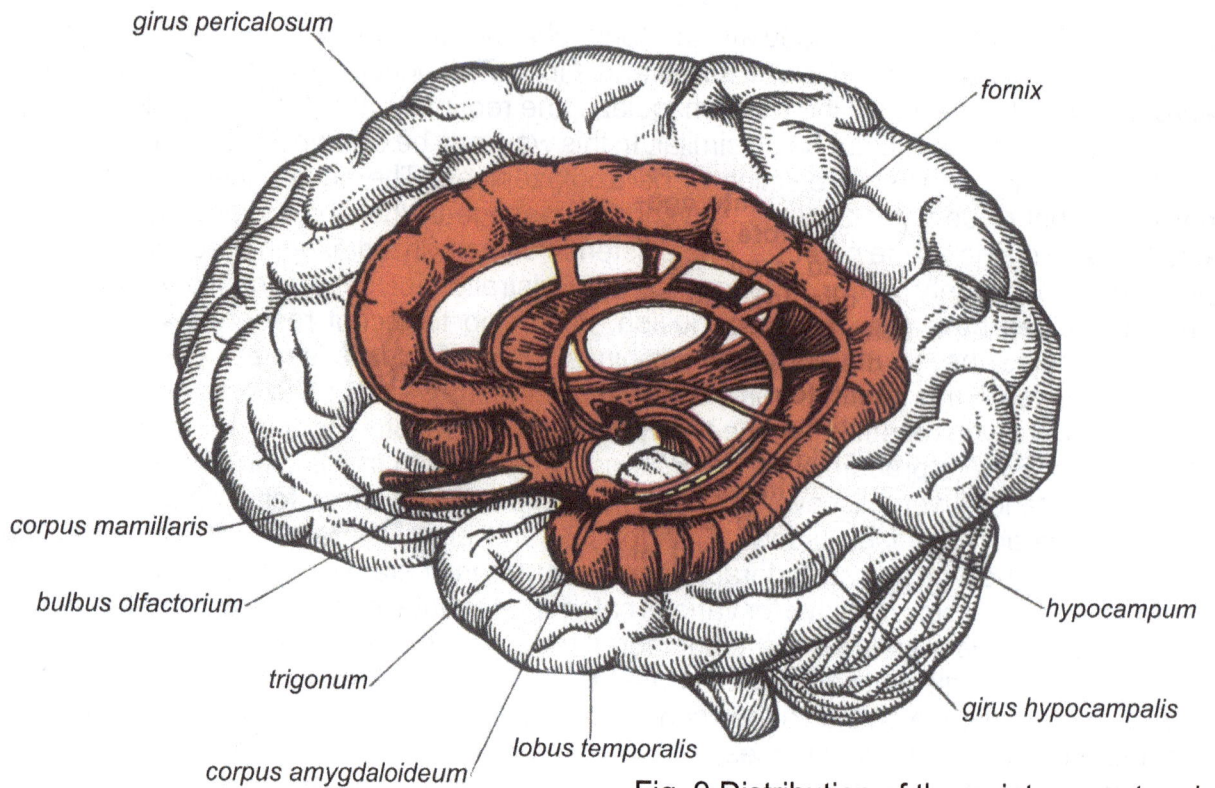

girus pericalosum

fornix

corpus mamillaris

bulbus olfactorium

hypocampum

trigonum

girus hypocampalis

corpus amygdaloideum

lobus temporalis

Fig. 9 Distribution of the opiate receptors in the human brain (the highest concentrations are found in the limbic system, which is involved in the perception of emotions, fielings and pain).

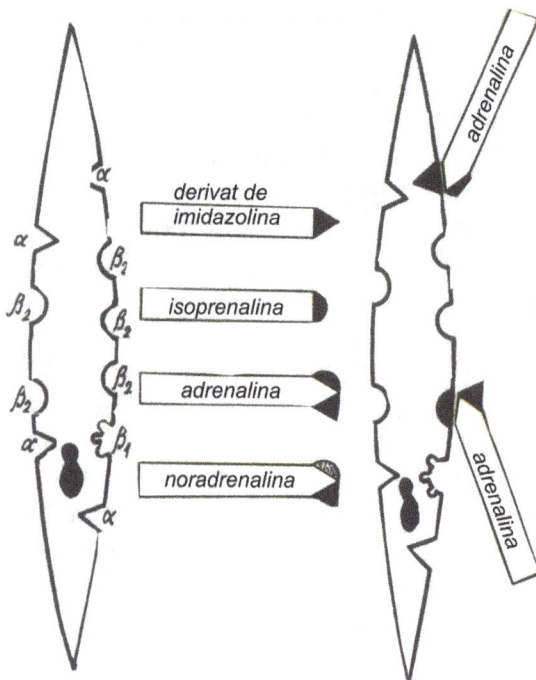

α

derivat de imidazolina

α

β₂

β₂

β₂

isoprenalina

β₂

β₂

adrenalina

α

β₁

noradrenalina

α

adrenalina

adrenalina

Fig. 10 Schematic representation of membranous alpha- and beta- receptors of a smooth muscle cell in the vascular wall.

Fig. 11 Schematic representation of the stimulation and inhibition relationships of the receptors of a smooth muscle cell in the vascular wall.

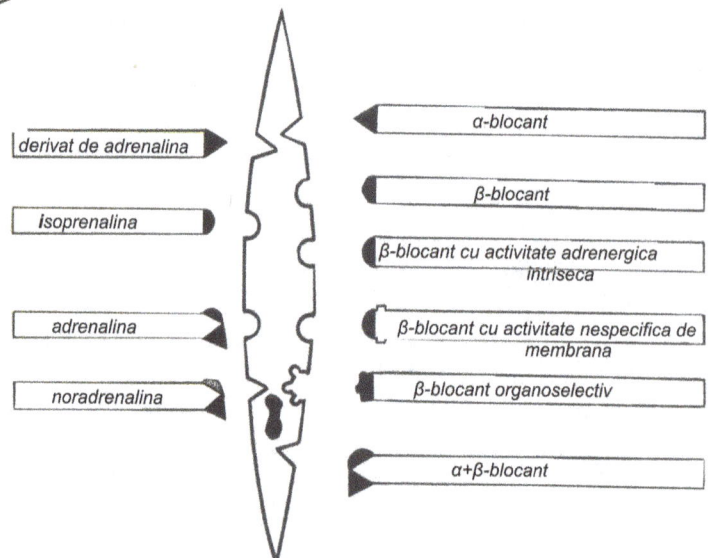

derivat de adrenalina

α-blocant

isoprenalina

β-blocant

β-blocant cu activitate adrenergica intriseca

β-blocant cu activitate nespecifica de membrana

adrenalina

β-blocant organoselectiv

noradrenalina

α+β-blocant

20

Moreover, we mention that as recent researches have proved to the activity of these neuroceptors is related also the appearance of some psychic diseases. Thus, for example, in schizophrenia, disturbances occur in the formation of dopaminergic and alpha type adrenergic receptors. The dopaminergic receptors were identified and localized topographically in the human brain by means of the positron emission tomography (Wagner). There are researches which demonstrate the structural similarity between the peripheral and the central neuroceptors, like, for example, in the case of alpha 2 . adrenergic neuroreceptors, although the mechanism of the central noradrenergic activity is far from being elucidated.

At the level of the central receptors extremely complex interactions take place. At the same time, the neuropeptide concentration represents a homeostatic element of an utmost importance, which in some diseases, undergoes serious impairments. Thus, in Huntington's chorea and in schizophrenia occur concentration variations of neurotensin which increases in the caudate nucleus, of somatostatins, which increas three times in the arcuate nucleus, as well as of thyroliberin in the amygdeloid nucleus, on the contrary, in cortical frontal areas the somatostatin decreases.

The specific base of these processes is the nervous tissue, made up of:

- nerve cells, with their projections and endings, to the study of which has also contributed, to a great extent, the Romanian neurologist Gh. Marinescu;

- supporting (or sustentacular) cells (neuroglia and microglia).

The elements of the nervous system penetrate, through their endings *(reticulum terminale)*, into all the structures of the organism. In this way, the whole organism lies under the influence of the nervous system.

The nervous tissue exhibits some peculiarities in comparison with the other tissues:

- its cells divide only until the seventh month of intrauterine life;

- the cells which are destroyed beyond this age cannot be replaced. However, under favourable conditions, the regeneration of projections is possible, but only in the presence of the cell.

The radioactive thymidine autoradiography has shown that in the subependymal layer of the mature telencephalon arise astrocytes and oligodendrocytes. Under the influence of a diffusible substance, produced by the nervous system, the migrations from the non-proliferative to the proliferative compartment (Patel, 1980) occurs. The mitotic activity of the astrocytes is intense the third day following the cerebral accident, after which it diminishes. The oligodendroylic is not involved in the mitosis process. The proliferation of the ependymal and the epithelial cells forming the choroid plexures was also observed (Chauchan and Lewis,1979).

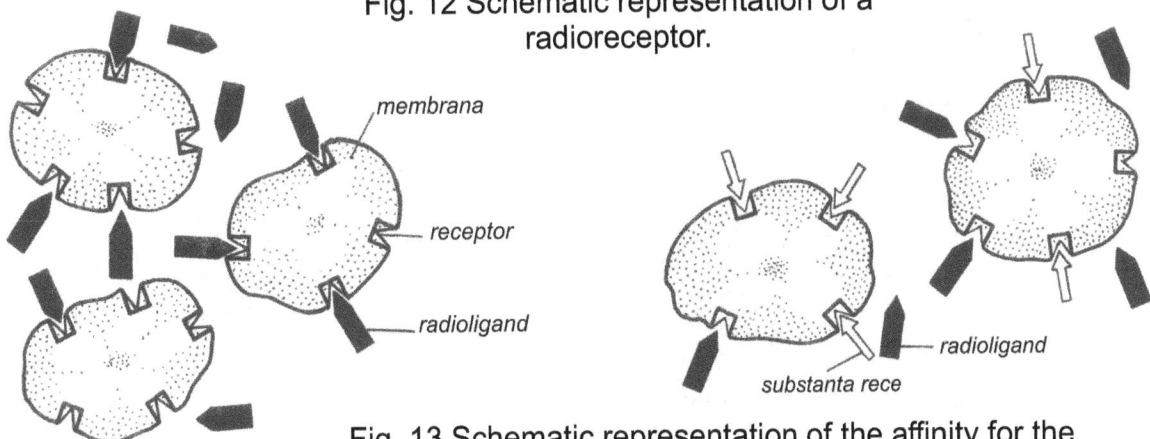

Fig. 12 Schematic representation of a radioreceptor.

Fig. 13 Schematic representation of the affinity for the receptor of a neutral substance. The neutral substance inhibits the fixation of the radioligand (the concentration at which the inhibition is of 50 % expresses the affinity of the substance for the receptor).

21

A complete morphological and functional regeneration of monominergic fibres; of non-myelinated cholinergic axons and of neurosecretory fibres (Berry, 1979).

The remyelination of the central nervous system is imperfect. The regeneration of oligodendrocytes is limited and the neoformed cells are unable to myelinate normally (Blakemore, 1978). As a result, the re-establishment of the nervous conduction is very limited.

The nervous system is characterized by its compartmentation, in its structure existing various metabolic sections separate through membranes at cellular and subcellular level. The neurological and psychic functions are related to the neuronal and glial membranes. Likewise, the axoplasmatic transport of the various substances, transport which may be of a slow (1-5 mm/24 h) or of a rapid type (400mm/24h), plays a significant role in the achievement of the basic functions of the nervous system.

Microscopic aspect of the nervous system. The component elements of the nervous tissue are demonstrated by special staining methods. By impregnations with silver salts, the nervous cells and the neurofibrils are clearly identified; the myelin sheath is stained with osmic acid. We mention that not all the details can be visualized on a single preparation.

Lately, by means of the selective radioactive substance the distribution and density of receptors in various parts of the central nervous system were identified ("*organum sensitorium*" of Ferrara Marquez).

The method permitted also to obtain important data regarding the receptors - endogenous or exogenous substances interaction.

In addition, the autoradiography of the encephalon allowed the drawing-up of a true "atlas" referring to cerebral receptors. The distribution of opiate receptors in the human brain is taken as example. These receptors are present at a higher concentration in the regions where pain and emotions are perceived with an increased intensity.

Starting from these data, the endogenous morphine derivatives were investigated by later studies, leading to the isolation of enkephalins.

Owing to immuno-histochemical researches, the pathways of encephalic neurons were detected and it was established that analogously to other systems of neurotransmitters, there is an endogenous system of opiates, which could explain the mechanism of acupuncture and the reactions in various emotional or stress reactions.

These enkephalins could be released in painful states and during intense and prolonged physical efforts. Beside endomorphines, the existence in the human organism of benzodiazepines, experting anticonvulsant, anxiolytic, sedative and myorelaxing properties.

Fig. 14 The GABA benzodiazepine complex with representation of the channel for chlorine.

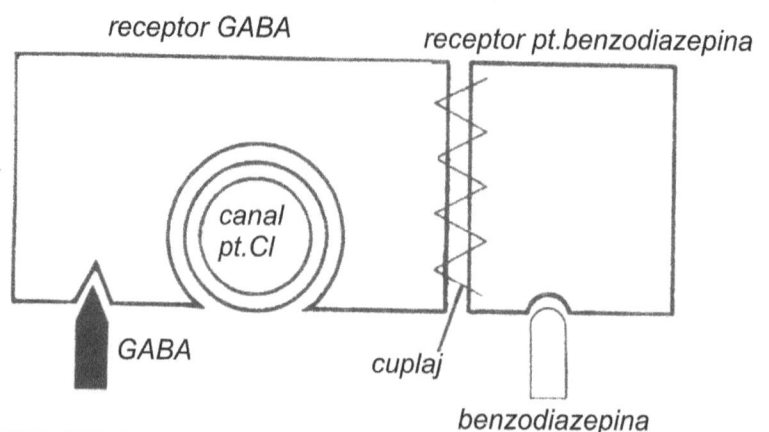

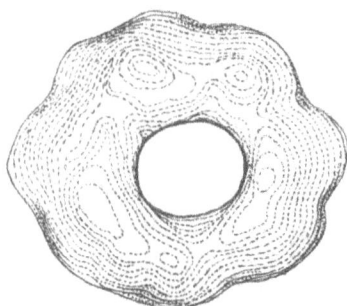

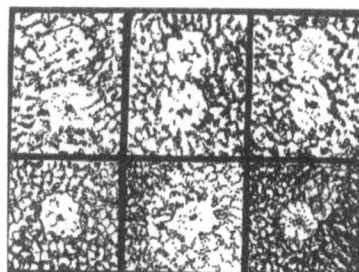

receptor GABA

receptor pt.benzodiazepina

canal pt.Cl

GABA

cuplaj

benzodiazepina

Fig. 15 a) Electron microscopic aspect of some receptor complexes;
b) Reconstruction of a receptor complex by mean of the computer.

b

a

The neuron

The neuron represents a morphological, trophic and functional unit (see also fig.322).

The nuclei of the neurons are usually single, large, spherical and centrally sited . They are strongly hypochromic; the nucleoplasm is faintly coloured and the chromatin is finaly scattered, their enchromatic character demonstrating their intense activity. They are bounded by a nuclear membrane and centred by one or two well visible nucleoli. The cell body is represented by the cytoplasm (neuroplasm), delimited by the cell membrane.

The neuroplasm contains the common and the specific organelles. Common organelles are the mitochondria, the Golgi apparatus well represented close to the nucleus, the lysosomes and the cell centre present in young cells (neuroblasts).

The specific organelles are represented by the Nissl tigroid substance, neurofibrils and neurotubuli.

The Nissl tigroid substance consists of granular elements, scattered throughout the cytoplasm but lacking at the cone of emergence of the axon. Their structure varies under excitation and inhibition conditions (Gh. Marinescu).

The neurofibrils are slender filaments present in the cytoplasm and in axons and dendrites under the form of parallel bundles. They are of a proteic nature with a phospholipidic component and their physiological role is not yet completely elucidated. They represent support and conduction structures.

The neurotubules are electron-microscopic structures described by Palay present in the cytoplasm, the dendrites and the axons. They are involved in the electric excitability process of the axon; their rhythmical contractions modify the shape of the axonal segment, as well as the permeability of the cell membrane at this level, modifying the membrane potential. Moreover, the movement of the proteins, synthesized at the perikaryon, along the neuronic processes, occurs through the neurotubules.

Functionally, the neurons may be motor, sensory, vegetative and intercalary or of association.

The motor neurons transmit the nervous impulse from the centre to the periphery, to the muscle fibre; they are efferent neurons.

The sensory neurons convey excitations from the periphery to the nerve centres; they are afferent neurons.

The vegetative secretory and vasomotor neurons convey the stimuly from the centre to the periphery (glands and blood vessels).

The association neurons or intercalary or internuncial neurons, called also interneurons, are interposed between the motor and the sensory neurons.

A reversal of the conduction direction to the same neuron is exluded (law of dynamic polarization).

From the viewpoint of their location, the neurons are central (in the encephalon) and peripheral (the cell bodies are sited in the spinal cord or in the ganglia, while their projections enter in the constitution of the peripheral nerves).

The motor efferent pathway is made up of two neurons; a central neuron and a peripheral one.

The sensory, afferent pathway is formed of three neurons: one peripheral and two central neurons. The peripheral motor neuron differs from the sensory neuron both functionally and morphologically.

The peripheral motor neuron. The cells of the peripheral motor neuron are stellate-shaped, 80-100 microns in diameter, with 5-8 projections, processes, and are called multipolar nerve cells. Such cells are present in the anterior horn of the spinal cord and in the motor nuclei of the brain stern. The nucleus of the cell is spherical and poor in chromatin. The cytoplasm (neuroplasm) is traversed by a dense reticulum of neurofibrils and contains granules that stain with basic dyes and which are called Nissl bodies or tigroid substance or chromatophil substance, with a significant role in the cell metabolism.

The cell processes are of two types:
- dendrites (protoplasmic processes) and
- axon (cylindraxis or neurite).

Dendrites are rather short and branch in proximity of the cell; they can be also multiple. The neurite is single and attains sometimes a considerable length, as it is constituted of neurofibrils and neuroplasm (axoplasm). Sometimes, perpendicularly to its length are detailed a few short side branches termed collaterals, which behave similarly to the dendrites. It ends in the form of a motor plate in the striated muscle. Outside, the neurite is enveloped in the sheath of Schwann or neurilemma and the space between the axons and the sheath of Schwann is filled with the myelin sheath. The sheath of Shwann is made up of Schwann's cells, provided with nuclei. According to their origin, these cells are considered peripheral glia cells.

The myelin, secreted by Schwann's cells, is constituted of a lipoid complex soluble in pyridine, to which other substances are also associated. It is responsible for the glistening white colour of the nerve fibre. It has the role of isolating the axon and is also involved in the transmission of the nerve influx. At intervals (0.5-1mm) along the axon occur ring-shaped constrictions, called Ranvier's nodes or constrictions, at the level of which myelin is absent and the axon is covered only with neurilemma. The distance between two constrictions is termed internode. Within an internode, the myelin sheath is interrupted here and there by conical spaces, termed Schmidt-Lantermann's incisures.

As we have mentioned above, the peripheral motor neuron ends in the striated muscle, in the form of a motor plate.

The motor plate is a very important functional system, adapted for reception, which regulates "with a diabolical speed and precision" (N. Bittinger, 1984) the muscle fibre contraction (fig. 16). It is made up of the axon of the cholinergic neuron, which leans upon the teloglia (a palisade-like arranged derivative of sarcoplasm) and widens at its extremity, assuming an oval, elongated shape, 50 microns in the large diameter, in which the acetylcholine vesicles accumulate. This is the place where the stimulus passes from the nerve to the muscle. Before its penetration into the muscle fibres, the nerve fibre loses its myelin sheath, the axon remaining covered only with the sheath of Schwann. After its penetration into the sarcolemma, the axon loses also the sheath of Schwann and the neurofibrils are scattered in a fine reticulum, within the muscle fibre. The impulses coming from the axon to the synapse release the acetylcholine into the synaptic cleft. The fixation of acetylcholine on the receptors brings about an ion flux which is objectified in the electric signal - the motor plate potential.

The transmission of the nerve influx to the periphery occurs in the form of depolarization waves. At the passage of the wave, the muscle fibre responds even to subliminal stimuli; this is the so-called „exaltation" period, which is followed by the refractory period, characterized by a decrease of the response even to supraliminal stimuli.

We consider necessary to remind the definition of two notions which we encounter in these situations:
- the rheobase represents the weakest stimulus able to elicit a muscle response and
- the chronaxy is the minimum duration necessary to obtain an excitation with an electric current having the strength equal to twice the value of the rheobase.

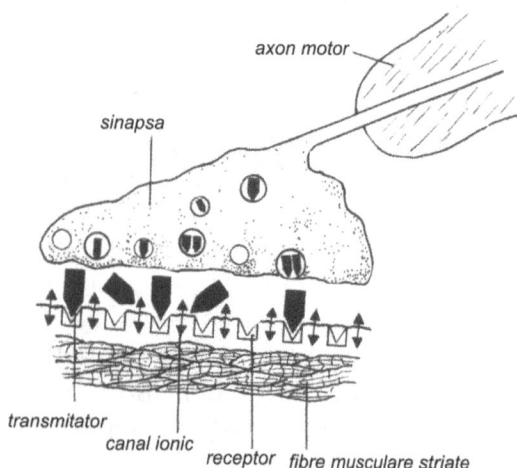

axon motor

sinapsa

transmitator
canal ionic
receptor fibre musculare striate

Fig. 16 Scheme of a motor plate

If curare is used, the respective potential is no longer formed at the level of the motor plate, although the plate contains a sufficiently large number of receptors.

Detailed studies of the motor plate have allowed to isolate 45,000-50,000 Ĺ - sized *subunits*, rosette-like arranged in groups of five units each, which a central „hole" representing the ion channel. They are often grouped into two rosettes.

Likewise, small motor plate particles - microsacs - containing only a small number of ion channels, were isolated.

The striated muscle fibre is, through the motor plate, in a trophic dependence on the afferent neuron. After the destruction of the nerve, which may be due to various causes, the muscle is paralysed and undergoes a progressive atrophy. Degenerative changes occur which, after six weeks, become irreversible, the motor plate disappearing. The reinnervation is achieved by *penetration* of the new nerve fibres into the muscle, either through the place of the former motor plate, or through other sites adjacent to the muscle.

Curare acts, as it was shown, especially on the motor plates, which it disconnects, without modifying the nerve conductivity and the muscle contractility.

In the muscles performing movements of high precision (muscles of the eye and of the larynx, lumbrical muscles), one or several muscle fibres are supplied by a single motor neuron, whereas in predominantly force muscles (*latissimus dorsi* muscle), a single motor nerve, cell supplies up to 150 muscle fibres.

The peripheral sensory neuron. The cell body of the peripheral sensory neurons is spherical and has 150 microns in diameter.

These cells have a process which, after a certain course, undergo a V-shaped division and are consequently named bipolar or pseudounipolar ganglionic cells. These cells are not present in the central nervous system, but outside it, in the spinal ganglia and in the sensory ganglia of the cranial nerves.

The bipolar cells are enveloped in a layer of flattened glial cells, which have a role in their protection and nutrition. The nucleus of the cell, poor in chromatin is provided with a nucleolus. Nissl bodies are scattered in the form of small granules throughout the neuroplasm.

The cell body is disproportionately large in comparison with its processes, so that it is rarely observed in histological preparations. The neurite and the dendrite cannot be differentiated as both are enveloped in a myelin sheath. Moreover, the peripheral sensory fibre has the same structure as the peripheral motor fibre. The differentiation of sensory neurons as sensory receptors occurs under various forms, often specific to a certain excitation type (fig. 17,18):

1. *Free nerve endings.* They are present in epithelia or in tissues and consist of an arborization of neurofibrils. They are pain receptors.

2. *Tactile corpuscles.* They are fine networks of neurofibrils inside the epithelial cells; they are found in the tongue and the epidermis.

3. *Sensitive terminal corpuscles.* They are cell groups included in the connective tissue, in which myelin nervous fibers enter, their neurofibrils forming a network outside and inside of the connective cells.

4. *Krause terminal corpuscles.* These are 100 micron-sized spherical or oval corpuscles and serve to the reception of cold excitations. They are scattered in the skin and some mucous membranes.

5. *Ruffini's corpuscles.* They are spindle-shaped and located at the boundary between the skin and the subcutaneous tissue, as well as in some mucosae. They are heat receptors.

6. *Meissner's tactile corpuscles.* They are made up of overlapping, pine-cone-like arranged cells, scattered in the papillae the corium.

7. *Vater-Pacini's corpuscles* or *lamellated corpuscles.* They are oval-elongated, up to 2 mm in length and composed of 30 or more concentrically arranged lamellae of connective tissue. Between the lamellae is a fluid rich in proteins. The centre of the corpuscle is formed of neurofibrils (Timofeev-Dogiel). They occur in the skin, between the skin and the subcutaneous cellular tissue, but also in the wall of the aorta.

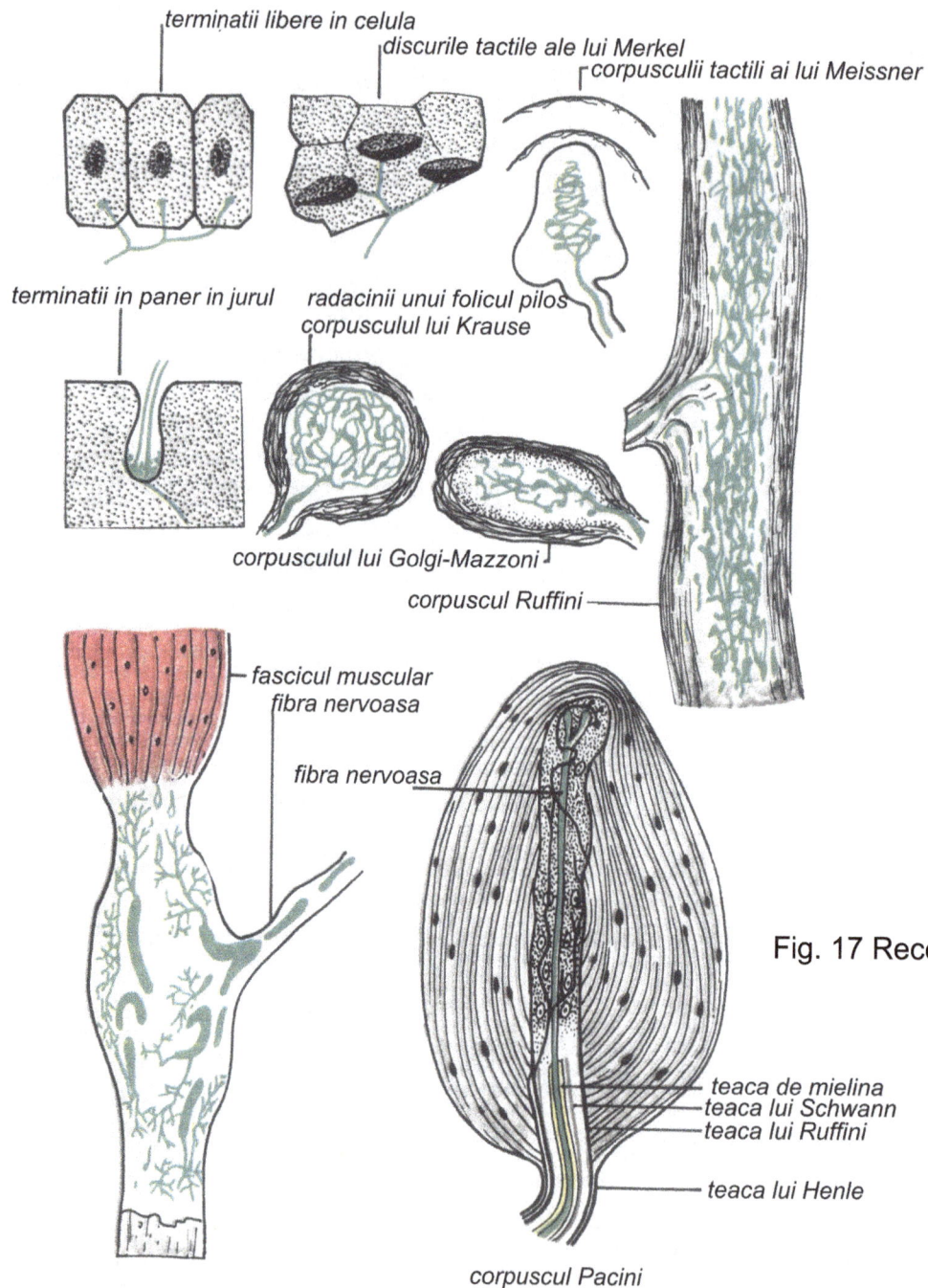

terminatii libere in celula
discurile tactile ale lui Merkel
corpusculii tactili ai lui Meissner
terminatii in paner in jurul
radacinii unui folicul pilos
corpusculul lui Krause
corpusculul lui Golgi-Mazzoni
corpuscul Ruffini
fascicul muscular
fibra nervoasa
fibra nervoasa
teaca de mielina
teaca lui Schwann
teaca lui Ruffini
teaca lui Henle
corpuscul Pacini

Fig. 17 Receptors.

They are receptors of the pressure excitation.

8. *Deep sensitivity reception elements.* By deep sensitivity should be understood the muscle and tendon sensivity which informs about the position of the organism in the space. This information is completed by the stimuli received by the optic and vestibular apparatuses.

Deep sensitivity receptors are the muscular and tendinous spindles:

a) *The muscular spindles* are located inside the muscle; they are oval-elongated, 1-3 mm - sized and enclosed in a capsule of connective tissue. Inside the muscle are fine striated muscle fibres, enveloped by a dense reticulum of sensory fibres. The tension degree of the muscle fibre is converted into a nerve influx and conducted into the nerve centres.

b) In a similar way functions also *the tendinous spindle* of the articular capsules and tendons, which receive the state of tension of tendon fibres.

By training, the muscular and tendinous sensibility may be improved to such a degree, that it can replace the activity of the eyes. This explains the fact that we can type on a typewriter without regarding the letters or that we can play piano without looking at the keyboard.

26

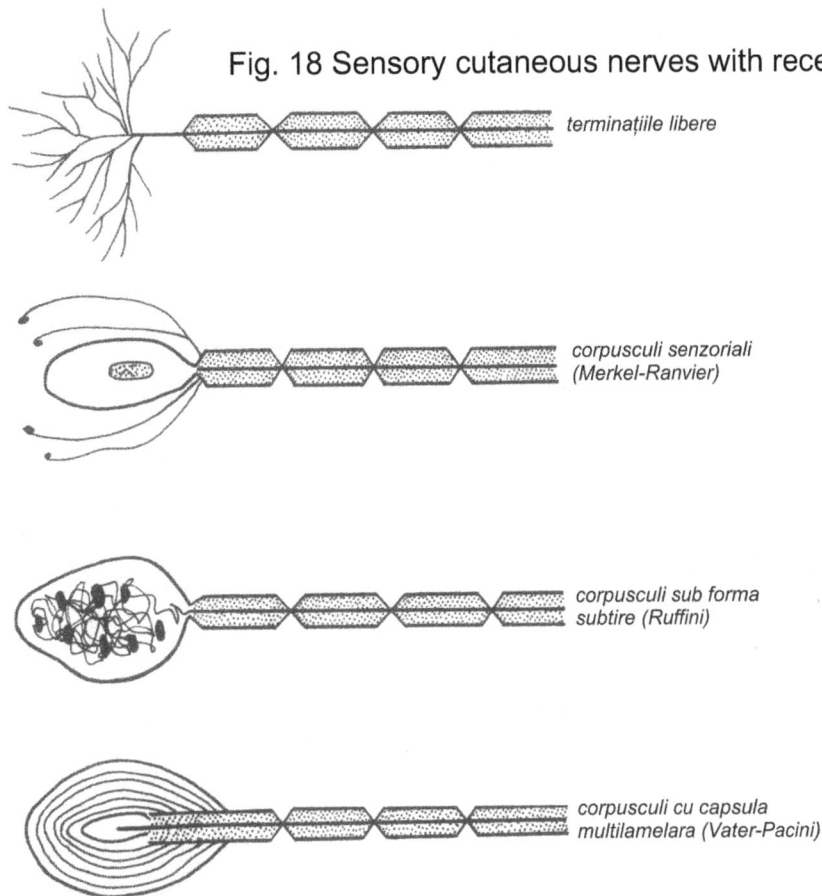
Fig. 18 Sensory cutaneous nerves with receptor endings.

terminaţiile libere

corpusculi senzoriali (Merkel-Ranvier)

corpusculi sub forma subtire (Ruffini)

corpusculi cu capsula multilamelara (Vater-Pacini)

The vegetative neurons are classified into four groups:

a) *The neurons in the tuber cinereum*, are middle-sized, have a circular or piriform outline, sometimes with a single process, but most often they are bipolar. Their chromatophilic substance is accumulated at the periphery, forming blocks and acquiring a hyperchromatic appearance to basic dyes. The central part of the cytoplasm is clear and contains the yellow pigment, the amount of which increases with the age. The nuclei are globulous and shifted towards the periphery.

b. *The neurons in the reticulated bulbo-ponto-peduncular nuclei* are hyperchromic and multipolar. The nucleus is shifted towards the periphery and the cytoplasm is overloaded with yellow pigment.

c. *The neurons in the vegetative ganglia* are comparatively large-sized with a pericellular capsule, and are provided with several processes. They often possess several nuclei.

d. *The vegative neurons in the walls of viscera* are middle - or small - sized and have several processes.

In silver salt impregnations it may be seen that the vegetative neurons are connected with the adjacent cells through their processes. These processes may be rich în myelin, poor in myelin or unmyelinated. The unmyelinated fibres have as a single cover the sheath of Schwann (gray Remak's fibres). The peripheral parts of the autonomic nervous system form a more or less dense plexus, with an extensive reticulum of fibres. These plexuses follow, in most cases, the course of vessels. The visceromotor endings of the vegetative fibres course either towards the smooth musculature, forming terminal arborizations, or towards the glands, becoming secretory fibres.

The vegetative fibres end finally in a bundle *(reticulum terminale)*, which is scattered within the cells of the various tissues.

The association neurons are small-sized, generally multipolar neurons, with short porcesses, which are present in all the grey structures of the neuraxis. The characteristic type of the association neuron is represented by the cells of the second layer of the cortex.

27

The improvements of the higher cortical reflective activity are dependent on the morphophysiology of these association neurons.

Nerve fibre
(neurite)

It is the protoplasmic extension of the neuron (axon or dendrite). Macroscopically, eight categories of myelinated and unmyelinated fibres are distinguished as follows:
- myelinated nerve fibres;
- devoid of the sheath of Schwann:
    - in the white matter of the brain
    - in the spinal cord
- with the sheath of Schwann
    - in cranial and spinal nerves
    - in vegetative nerves
- unmyelinated nerve fibres:
    - with the sheath of Schwann:
    - "grey" or Remak's fibres
    - in autonomic nerves
- devoid of the sheath of Schwann.

The most adequate classification seems to be that based on the physiological criterion, including four categories:
- motor fibre
- sensory fibre
- associative fibre
- autonomic fibre (receptor and motor).

However, this classification is less used, since it is not based upon precise facts.

The fibres derived from the neurons of the anterior and intermedio-lateral-medullar columns, as well as from the spinal ganglia, make up t*he peripheral nerves*, whose axons or dendrites make up fascicles, true bundles enveloped by a connective muff - *the epineurium*. Each fascicle is enclosed in a proper connective sheath - *the perineurium*, formed of flattened cells in an arcuate arrangement and of collagen fibres, disposed longitudinally. The perineurium consists of an *internal layer,* which communicates with the rachidian spaces *(neurothelium)*, and an *external layer*, made up of supporting fibres. *The endoneurium* is represented by the loose tissue containing blood capillaries arranged in a fine reticulum.

The bulky myelinated nerve fibres are enveloped by two sheaths rich in collagen: an *internal sheath*, the latticed layer or *internal endoneural sheath of Plenk-Laidlam*, and an *external sheath* or *the external endoneural sheath of Key-Retzius*.

The connective sheaths have a nutritive, trophic and protective role.

As a consequence of lesions, the internal mucoid degeneration of the endoneurium leads to the appearance of Renaut's bodies, which represent fragments of a fibrillary reticulum with a hyaline halo, assuming an onion bulb aspect.

The Schwann's cell, which is defined as a cell enclosing a nerve fibre invaginated in the cytoplasm (Causey, 1960), has macrophagic properties, its cytoplasm containing myelin detritus.

The unmyelinated fibre, or grey fibre, called also *Remak's fibre*, contains a small amount of myelin and is made up of bundles of axons surrounded by the sheath of Schwann and by their basement membrane.

Classically, the nerve fibre (neurite) is made up of fibrils (neurofibrils), contained in an axial protoplasm (axoplasm) which has at the periphery a denser structure (Mauthner's sheath). Around this axon are arranged as three successive sheats; *the myelin sheath, the sheath of Schwann and the Henlé's sheath* (of ectodermic origin).

28

The myelin sheath extends, with the nerve fibre, up to the entrance into the muscle, where the neurilemma fuses with the sarcolemma, and the Schwann cells form the roof of the groove in which lies the axonal ending.

The terminal axonal branchings, surrounded by sarcoplasm, make up the motor end-plate.

Synapses

Synapses are specialized structures which achieve the contact (contiguity) between neurons, on the one hand, and between neurons and effector cells, on the other hand (see also Fig. 322-332)

There are the following types of synapses:

- axo-somatic synapses, which achieve the immediate contact between the axon terminals of a neuron and the perikaryon of another neuron;

- axo-dendritic synapses, which achieve the junction between an axon and a dendrite.

At the level of synapses takes place the transmission of the nerve influx through chemical transmitters.

Synapses are made up of two membranes: presynaptic and postsynaptic, separated by a space. The axonal presynaptic membrane has the shape of a knob bounded by axolemma and formed of axoplasm, which contains proximally protoneurofibrils and distally, numerous mitochondria and vesicles, called synaptic vesicles containing chemical transmitters. The synaptic vesicles are formed in the perikaryon, from which they migrate distally.

The postsynaptic membrane (dendrite or cell body) is not provided with synaptic vesicles.

Hence, the central nervous system has the appearance of a compact tissue, in which *the extracellular space* forms a true *"labyrinth"* made up of submicroscopic interstitial clefts, the so-called Nissl's grey spaces.

Neuroglia

The neuroglia has, like the nervous tissue, an ectodermal origin, with the exception of the microglia, which derives from the mesoderm. It has the role of supporting, nourishing and isolating the nerve cells and fibres which are components of the control nervous system. Moreover, the neuroglia secretes myelin from the central nerve fibres, while Schwann cells secrete myelin from the peripheral fibres (actually, they have a common origin from spongioblasts). The glial cells have various shapes; the most important are the astrocyte and the microglia

The astrocytes, the main component of the macroglia, has numerous processes, by which the connection between the cells is established. The number of astrocytes varies in humans from 15 to30/0.1 mm3; they are homogeneously distributed in the different segments of the central nervous system.

The examination under the light microscope shows that from a cell body representing the neuroglia body start in all the directions 10-15 protoplasmic processes. Their branching degrees and the length are variable. Astrocytes with short processes and with long processes are distinguished. Astrocytes with short processes are spherical and their processes are arranged radially. They are situated in the grey matter of the brain and spinal cord. The astrocytes with long processes are elongated and their processes arise from the extremities, under the form of bundles. They make up the glia of the white matter. Generally, astrocytes form a tridimensional network, composed of cell bodies and protoplasmic processes spreading in all directions, which are in a reciprocal relationship through the very permeable contact points.

The whole surface of the cell bodies and of the processes is covered by *digitiform diverticula*, representing 50% of the bulk and 60-80% of the surface of astroglia.

In the turgescense phase, the astroglial processes modify their shape and the number of diverticula diminishes, changes which have a significant pathophysiological role (they explain the cerebral edema).

The processes of the glial cells attach to the wall of the capillaries by giving off „glial filaments" (Cajal's tentacles), surrounding the capillary under the form of a perivascular glial limiting membrane and being sited between capillaries and neurons. Thus, the whole metabolism of the nerve cell is conditioned by the existence of the glia which retains some substances from the blood, preventing their penetration into the central nervous system (the blood-brain harrier) (see also Figs. 322-332).

The glial cells with few processes are termed *oligodendroglial cells*. They are situated in immediate proximity to the nerve cells.

The microglial cells have many processes. This kind of cells may lose their processes, become spherical and store lipids, proteins and mineral substances. In case of lesions of the central nervous system, these cells multiply, engulp the destroyed substance and, with the astrocytes, repair the defect (glial cicatrix).

The ependyma is formed, too, of cylindrical or arboid glial cells lining the cerebral ventricles and ependymal canal (the central canal of the spinal cord), under the form of a single cell layer.

The cerebrospinal nervous system
(The nervous system of relationship life)

The nervous system of relationship life (cerebrospinal nervous system) is made up of the central nervous system and the peripheral nervous system.

The central nervous system is represented by the nervous segments situated in the cranium, which, on the whole, constitute the encephalon, and by a single nervous segment lodged in the vertebral or spinal canal - the spinal cord.

The encephalon is made up of the medulla oblongata, the pons, the mesencephalon (known also under the term of brainstem), cerebellum, diencephalon and cerebral hemispheres.

The peripheral nervous system, in close anatomical and functional relationship with the central nervous system, is represented by the spinal rachidian) and cranial (cerebral) nerves.

The spinal cord
(Medulla spinalis)
Fig. 19-38

The spinal cord *(medulla spinalis)* is the segment of the central nervous system contained in the vertebral canal, which it does not fill completely.

The vertebral canal in its cervical portion is triangular and tapers from above downward, beginning from C3. Thus, the mean values of the diameter are: atlas 17-30 mm; C2 14-22 mm and C3 12-21 mm. Its enlarged shape in the upper cervical segment has a great practical significance, since the medullar compression is avoided, even if the intervertebral disc herniates.

The vertebral canal is the narrowest at the level of the C6 vertebra (17mm), just in the area which is the most mobile and the most frequently exposed to injuries or degenerative lesions. The spinal cord adapts to the physiological and even to the pathological curvatures of the vertebral column.

External configuration

Shape and boundaries. The spinal cord has the shape of a slightly anteroposteriorly flattened cylinder and a length of 45 cm in the male and 43 cm in the female. On the transverse section it is oval in shape and 10-16 mm in diameter.

Above it is continuous, without an obvious delineation, with the medulla oblongata at the level of the atlanto-occipital joint. Below it diminishes in thickness and at the level of the L1-L2 vertebrae it is continuous with the terminal cone *(conus medullaris)*, the projection of which, callled *filum terminale*, has functionally no significance, as it consists of glial cells. *The filum terminale* traverses the dura matter and attaches to the second coccygeal vertebra.

The lower extremity of the spinal cord corresponds to the second lumbar vertebra, which is an important surgical landmark for the carrying out of spinal punctures. *The filum terminale* is surrounded by the last spinal nerves, forming *the cauda equina*, which ends at the level of the second coccygeal vertebra.

The spinal cord has not the same thickness (on the average 12mm) through its length. In the cervical region, between the C3 and T2 vertebrae and in the lumbar region, between the T10 and the L1-L2 vertebrae, the spinal cord presents an enlargement, called *intumescentia cervicalis* and respectnively *intumescentia lumbalis*, which narrow considerably the subarachoid space. The diameter of t*he intumescentia cervicalis* (cervical enlargement) attains its maximum size at the level of the C5 vertebra (18mm) and that of t*he intumescentia lumbalis* at the T12 vertebra (13-15mm). The occurrence of these intumescences on the phylogenetic scale is related to the development of limbs. The more important is the role of limbs in locomotion, the layer are the intumescences. In snakes, which are devoid of limbs, the spinal cord is uniformly calibrated throughout its length.

Topographically, the spinal cord is formed of the succession of 31 segments known under the name of *neuromeres*.

Surface of the spinal cord. The spinal cord has two surfaces - anterior and posterior - at the level of which are present several furrows (Fig. 19, 20).

On *the anterior surface* of the spinal cord, on the median line is situated the anterior medium fissure *(fissura mediana anterior)*, the bottom of which is formed of a lamina of white matter, named anterior white commissure, into which penetrates a process of the pia mater. It is of about 2-4 mm in depth.

On *the posterior surface* lies the posterior median sulcus *(sulcus medianus posterior)*, which ends toward the cord with the posterior median septum, visible on a transverse section of the cord.

A mid-sagittal plane passing through the two above-mentioned landmarks divides the cord into two relatively symmetrical halves. On each half, anterolaterally, lies an anterior lateral sulcus *(sulcus lateralis anterior)*, through which emerge the anterior roots *(radix ventralis)* of the spinal nerves, while posterolaterally is situated a posterior lateral sulcus *(sulcus lateralis posterior)*, the entrance site of the posterior roots *(radix dorsalis)* of the spinal nerves. Between the posterior median sulcus and the posterolateral sulcus lies, only in the upper thoracic and in the cervical regions of the cord, the posterior intermediate sulcus *(sulcus intermedius posterior)*, delineating the Goll's fasciculus *(fasciculus gracilis)* from the Burdach's fasciculus *(fasciculus cuneatus)*.

Internal configuration
Macroscopic structure

A transverse section of the spinal cord demonstrates the arrangement of the white and grey matter *(substantia alba et grisea)*. The grey matter consists especially of nerve cells and nerve fibres, while the white matter contains exclusively myelinated fibres. The white matter surrounds the grey matter in such a manner that the latter does not reach in any place the surface of the spinal cord.

Transverse sections of the spinal cord, performed at various levels, show significant variations in the shape, the size and the ratio between the grey and the white matter (Fig. 22-24)

Grey matter
(Substantia grisea)

It is more abundant at the level of the cervical and lumbar intumescens and, comparatively, its amount is greater at the cervical than at the lumbar intumescence (owing to the fact that the thoracic limb possesses grater movement possibilities). In the thoracic portion, the amount of grey matter is somewhat lesser (limited movements at the level of intercostal and abdominal muscles). On the transverse section the shape of the grey matter resembles the letter "H", with the sagittal lines very thickened and the transverse line relatively narrow.

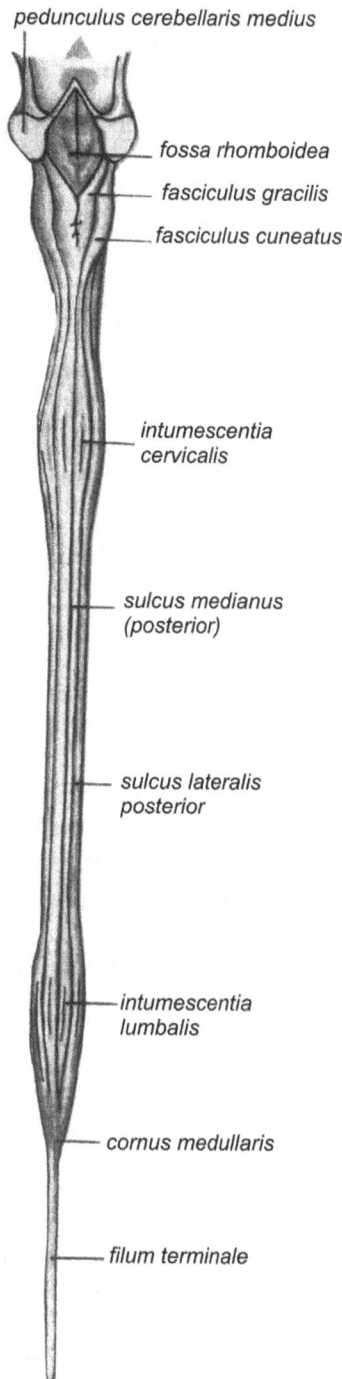

Fig. 20 Spinal cord – anterior view.

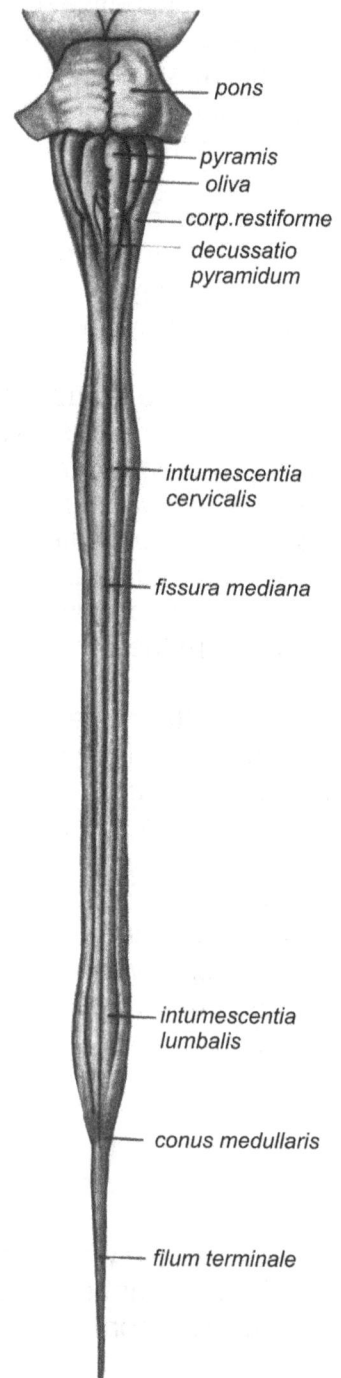

Fig. 19 Spinal cord – posterior view.

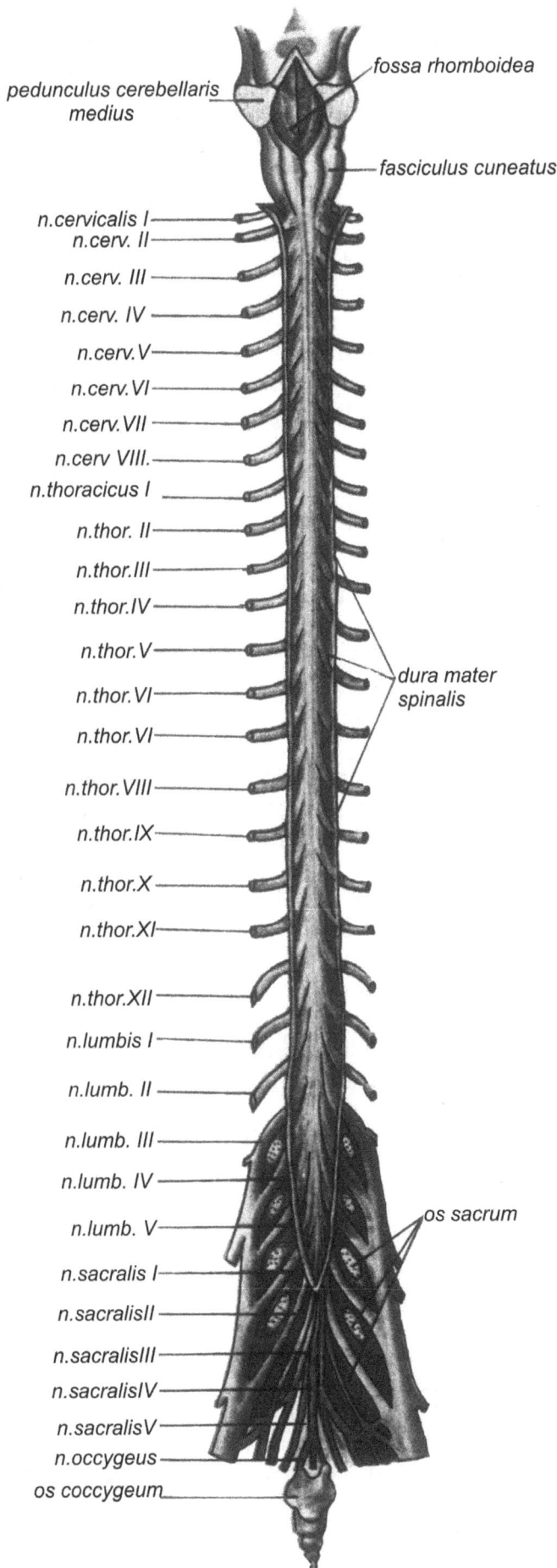

pedunculus cerebellaris medius

fossa rhomboidea

fasciculus cuneatus

n.cervicalis I
n.cerv. II
n.cerv. III
n.cerv. IV
n.cerv. V
n.cerv. VI
n.cerv. VII
n.cerv VIII.
n.thoracicus I
n.thor. II
n.thor.III
n.thor.IV
n.thor.V
n.thor.VI
n.thor.VI
n.thor.VIII
n.thor.IX
n.thor.X
n.thor.XI
n.thor.XII
n.lumbis I
n.lumb. II
n.lumb. III
n.lumb. IV
n.lumb. V
n.sacralis I
n.sacralisII
n.sacralisIII
n.sacralisIV
n.sacralisV
n.occygeus
os coccygeum

dura mater spinalis

os sacrum

Fig. 21 Spinal cord – posterior view, spinal nerves.

The two extremities of the (anteroposterior) sagittal lines are called *anterior horn (cornu anterius* - more developed) and respectively *posterior horn (cornu posterius* - less developed). The anterior horn is also named *anterior column (columna anterior)*, since the grey matter traverses the spinal cord through its whole length as a column; the posterior horn is also termed *posterior column (columna posterior)*. The horizontal branch of the letter H, which unites transversally the two sagittal branches, is named *grey commissure (commissura grisea)*, inside which is situated the *ependymal canal (canalis centralis)*.

The motor anterior horn *(cornu anterius)* is broader and has a festooned outline. Its anterior extremity is separated from the surface of the cord by a layer of white matter. The anterior horn is made up of two portions: the head and the base. In the upper half of the spinal cord, from the external surface of the base of the anterior horn starts a projection of grey matter (of the form of a triangle on the transverse section), termed lateral horn or lateral column *(columna lateralis)*.

The posterior horn *(cornu posterius)*, thinner and more elongated, extends toward the posterolateral sulcus. It is made up of a head (the most posterior part), a base (the most anterior part) and a narrower portion, named neck, situated between the head and the base. In the basal, medial region of the posterior thoracic and lumbar horns, from C7 to L2, extends the thoracic Clarke -Stilling's column, to which is ascribed also the intermediate lateral substance of the lateral column.

Between the head of the posterior horn and the surface of the cord is situated a portion of white matter, called Lissaner's marginal zone *(zona terminalis)*. The head of the posterior horn is covered by a gelatinous substance called Rolando's gelatinous substance *(substantia gelatinosa posterior)*; between the gelatinous substance and the marginal zone lies the spongy substance *(substantia spongiosa)*.

The **lateral horns** *(cornu laterale seu columna intermediolateralis)* contain sympathetic and parasympathetic vegetative formations.

Between the lateral horn and the base of the posterior horn may be seen,. especially in the cervical spinal cord, a number of fibres which penetrate into the lateral funiculus, forming a net of small extension, with irregular meshes, containing a part of the white matter of the lateral funiculus. This zone is termed *reticular formation* (Dieters).

The **ependymal** *(canalis centralis)*, situated inside the grey commissure, is lined with ciliated ependymal cells and contains cerebrospinal fluid *(liquidus cerebrospinalis)*. The ependymal canal communicates above with the fourth ventricle and ends below in a dilation situated in the terminal cone, termed terminal ventricle (Krause).

The shape of the lumen varies in the different segments of the spinal cord. On the transverse section, it is oval in the cervical region, almost round in the thoracic region and laterally flattened in the lumbar and sacral regions. In the cervical and thoracic portions it is often obliterated. It represents an embryological remnant of the central canal of the neural tube.

White matter
(Substantia alba)

Owing to the H-shaped arrangement of the grey matter, the white matter is divided into three funiculi:
- an anterior funiculus *(funiculus anterior)*
- a lateral funiculus *(funiculus lateralis)* and
- a posterior funiculus *(funiculus posterior)*.

These funiculi are disposed symmetrically

The **anterior funiculus** is situated between the anterior median fissure and the anterior collateral furrow and is sometimes divided by a secondary furrow - the anterior paramedian farrow *(sulcus intermedius anterior)* - into two secondary paired fasciculi or tracts: *the pyramidal fasciculus (tract) and the anterior basic fasciculus (tract)*. The direct pyramidal fasciculus crosses its pair at the level of the spinal cord, at each segment, forming the anterior white commissure *(commissura alba anterior)*, a lamina of white matter situated between the median fissure and the grey commissure.

The **posterior funiculi** are separated by the median septum. A posterior intermediate septum *(sulcus intermedius posterior)* existent between C1 and T5, starts from the posterior intermediate sulcus and, without reaching the grey matter, divides the white matter of the posterior funiculi into the Goll's fasciculus *(fasciculus gracilis or pars medialis)* and the Burdach's fascicles *(fasciculus cuneatus or pars lateralis)*.

The **lateral funiculus** is situated between the anterior collateral sulcus and the posterior sulcus.

Spinal fasciculi (tracts)

The spinal cord is traversed by fibres which convey the nerve influxes to and from the brain and are organized in paired tracts (Fig. 26,27).

The **fasciculus gracilis** (Goll's fasciculus) and t**he fasciculus cuneatus** (Burdach's fasciculus) conduct stimuli of the conscious proprioceptive sensibility, as well as of the epicritic tactile sensibility from muscles, tendons and articulations. They are situated between the posteromedian and the posterolateral sulci. In the cervical and the thoracic region these fasciculi are separated by a septum, which at its lower extremity is continuous with the tract of Schultze, as they are situated in the posterior funiculi.

The **lateral spinothalamic tract or fasciculus** *(tractus spinothalamicus lateralis)* conveys the pain, thermal and, partially, also the tactile stimuli of the protopathic sensibility and is situated in the lateral funiculus.

34

At the level of the anterior commissure it crosses its pair and then ascends the lateral funiculus toward the thalamus (posterior central nucleus).

The anterior spinothalamic tract or fasciculus *(tractus spinothalamicus anterior)* transmits the tactile protopathic stimuli and is situated in the anterolateral funiculus. After traversing the anterior commissure towards the opposite side, it ascends the anterior funiculus toward the thalamus (near the Reil band).

The posterior spinocerebellar tract or fasciculus, called also Flechsig's direct spinocerebellar tract *(tractus spinocerebellaris posterior)* transmits the stimuli from the muscles of the pelvic members and of the trunk (C6-L2), i.e. the uncoscious deep sensibility; it is situated on the lateral surface, in front of the posterolateral sulcus, ascending toward the restiform body and the cerebellum.

Gower's crossed spinocerebellar tract or fasciculus *(tractus spinocerebellaris anterior)* is situated at the periphery of the anterolateral surface of the lateral funiculus of the spinal cord and transmits the deep unconscious sensibility to the cerebellum through the pons and the anterior medullary velum.

The tectospinal tract or fasciculus *(tractus tectospinalis)* transmits the cephalogyric auditory and optical reflexes. It arises in the homolateral quadrigeminal superior and inferior colliculi; it crosses the similar tract on the opposite side (Meynert's decussation) and descends into the anterior funiculus, ending at the level of the motor cells of the anterior grey column (horn).

The rubrospinal tract or fasciculus *(tractus rubrospinalis)* conveys impulses from the red nucleus and, after crossing in Forel's decussation, descends into the lateral funiculus of the spinal cord, in front of the crossed corticospinal tract, ending in the anterior horn, at the level of the gamma motor neurons.

The olivospinal tract called also Helweg's bundle *(tractus olivospinalis)*, **reticulospinal tract** *(tractus reticulospinalis)* and **vestibulospinal tract** *(tractus vestibulospinalis)* are also extrapyramidal pathways (involuntary movements), also situated in the lateral funiculus.

The anterior and lateral corticospinal tracts or direct and crossed pyramidal tracts *(tractus corticospinalis anterior lateralis)* transmit the impulses of involuntary movements. The crossed pyramidal tract has its origin in Betz's pyramidal cells (the first neuron) of the precentral gyrus and, after the pyramidal decussation in the medulla oblongata, enters in the spinal lateral funiculus; it is situated between the posterior spinocerebellar tract and the posterior horn. The direct pyramidal tract is much thinner and originates in the central motor area. It descends along the same side and, at the level of the anterior commissure of the spinal cord, it crosses its pair on the opposite side, ending at various levels by synapses with the cells of the anterior horn. Both end either directly or through interneurons at the alpha motor cells and are pathways specific to primates.

Microscopic structure

Histologically, the grey matter is made up of the body of nerve cells and of their unmyelinated processes, while *the white matter* is almost completely formed of myelinated fibres. The spinal grey matter contains both nervous tissue and supporting tissue. The nervous elements of the grey matter are represented by pigment cells and unmyelinated fibres.

Nerve cells

They are divided into three categories.

The radicular neurons are provided with a long axon (neurite), which penetrates into the anterior roots of the spinal cord. These neurons are of two types: *vegetative*, situated at the level of the lateral horn, from where the preganglionic vegetative fibres arise, and *somatic*, situated in the body of the anterior horn.

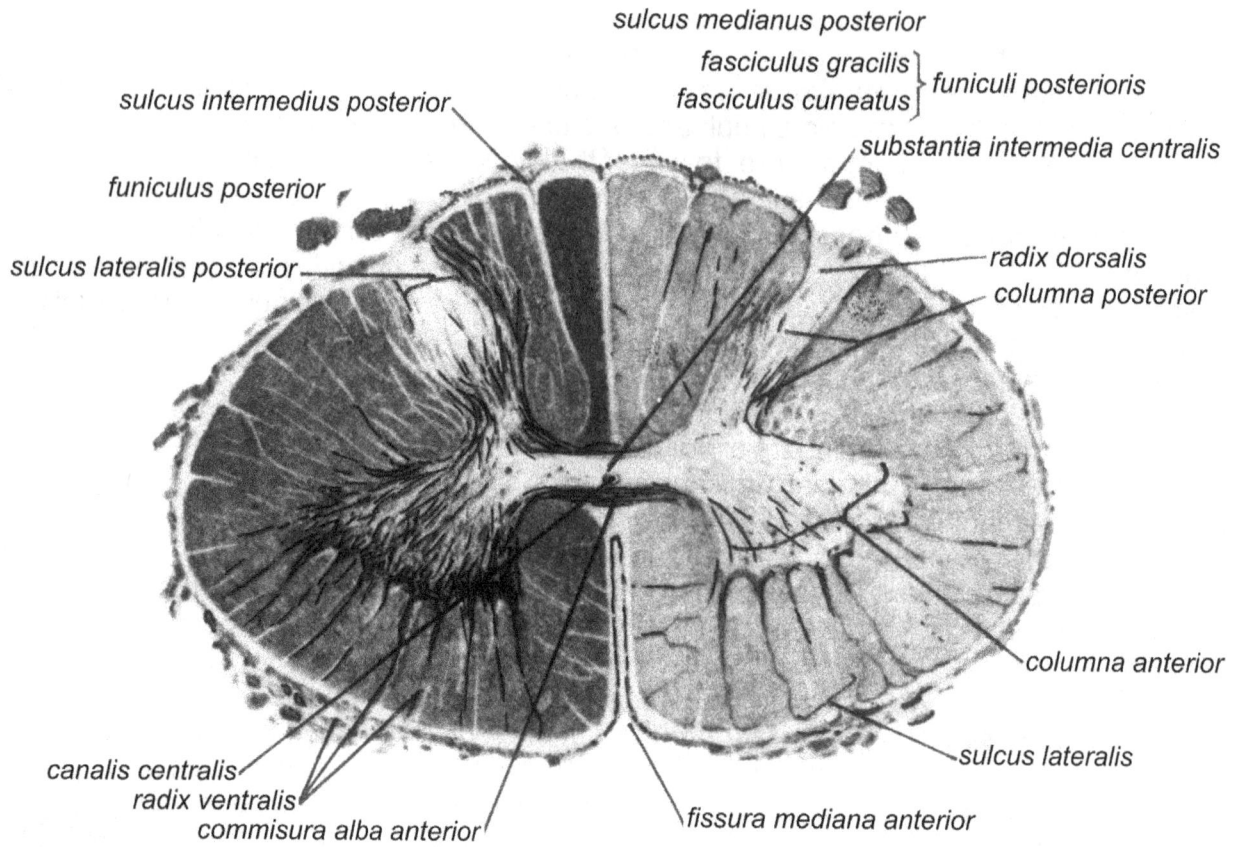

Fig. 22 Spinal cord – horizontal section, cervical region.

Labels on Fig. 22:
- sulcus medianus posterior
- fasciculus gracilis
- fasciculus cuneatus
- funiculi posterioris
- substantia intermedia centralis
- sulcus intermedius posterior
- funiculus posterior
- sulcus lateralis posterior
- radix dorsalis
- columna posterior
- columna anterior
- sulcus lateralis
- canalis centralis
- radix ventralis
- commisura alba anterior
- fissura mediana anterior

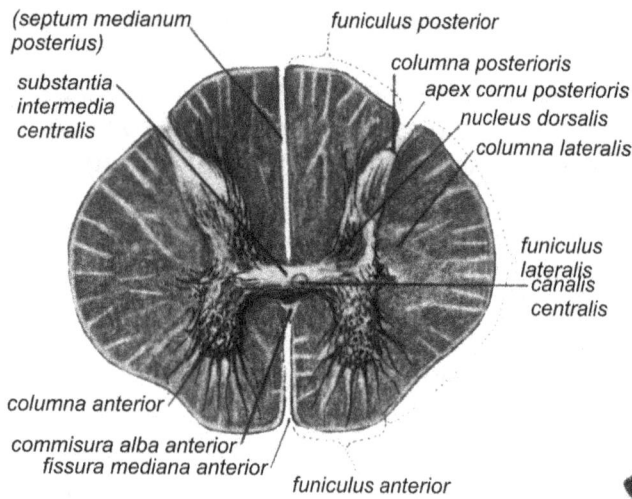

Fig. 23 Spinal cord – horizontal section, thoracic region.

Labels on Fig. 23:
- (septum medianum posterius)
- funiculus posterior
- columna posterioris
- apex cornu posterioris
- nucleus dorsalis
- columna lateralis
- substantia intermedia centralis
- funiculus lateralis
- canalis centralis
- columna anterior
- commisura alba anterior
- fissura mediana anterior
- funiculus anterior

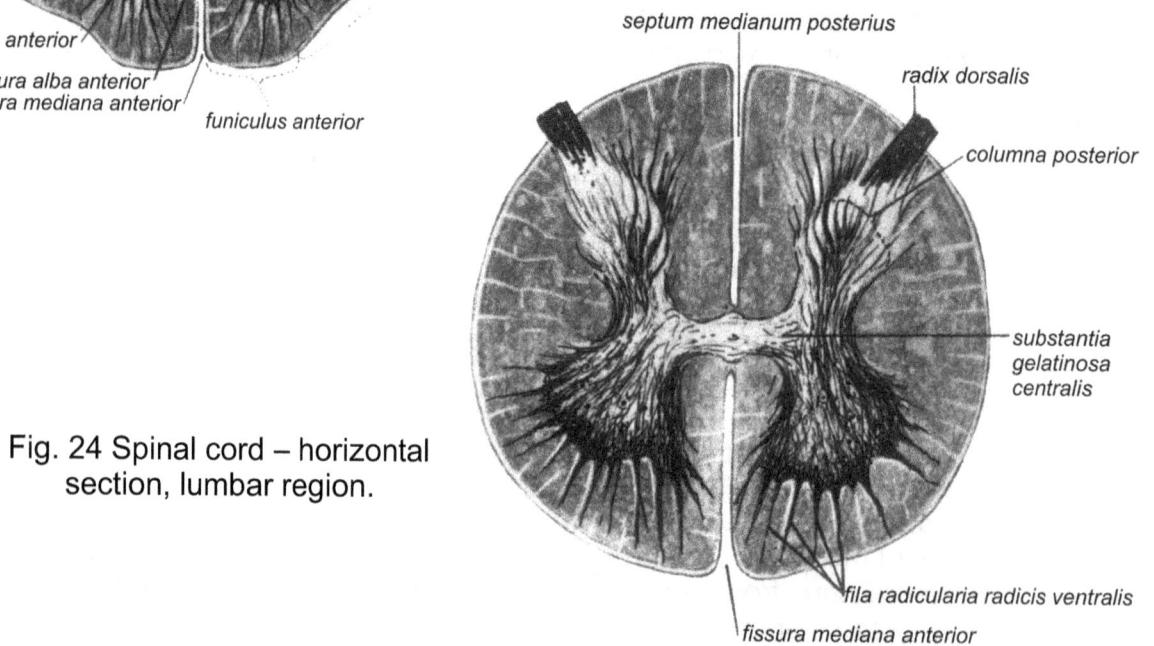

Fig. 24 Spinal cord – horizontal section, lumbar region.

Labels on Fig. 24:
- septum medianum posterius
- radix dorsalis
- columna posterior
- substantia gelatinosa centralis
- fila radicularia radicis ventralis
- fissura mediana anterior

36

The somatic neurons are of three types:

a) *alpha*, which supply the skeletal musculature; the alpha neurons may be divided into *alpha tonic neurons*, supplying the slow muscle fibre, and *alpha phasic neurons* supplying the fast muscle fibre; they are chiefly stimulated by fibres deriving from the neuromuscular spindles and, to a lesser extent, by the corticospinal and vestibulospinal tracts;

b) *beta* neurons, suplying the slow muscle fibre and the intrafusal muscle fibre;

c) *gamma* neurons, which supply the neuromuscular spindles and are of two categories: *gamma dynamic neurons*, which play a role in the control of the muscle contraction degree, and *static gamma neurons*, which have a function in the stages preceding the muscle contraction. By the contraction of intrafusal fibres, these neurons influence the neuromuscular bundle sensibility.

Moreover, there are also Renshaw'a neurons, which play a role of motor inhibition; they are interneurons with a feed back type inhibition mechanism acting especially on alpha cells.

The cordonal neurons, the processes of which connect the various levels of the spinal cord, pass through the white matter of the funiculi. When they connect two levels on the same side, they are called *homolateral cordonal neurons*, while when they connect a level on one side of the spinal cord with a level on the opposite side they are termed *heterolateral cordonal neurons*. There are also *bilateral cordonal neurons*, which behave both as homolateral and as heterolateral neurons.

Fig. 25
Medullovertebral
topography.

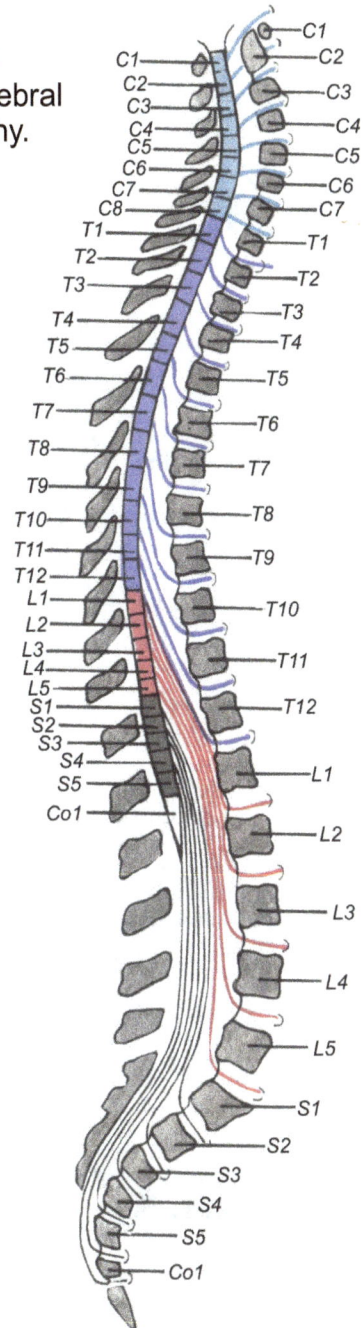

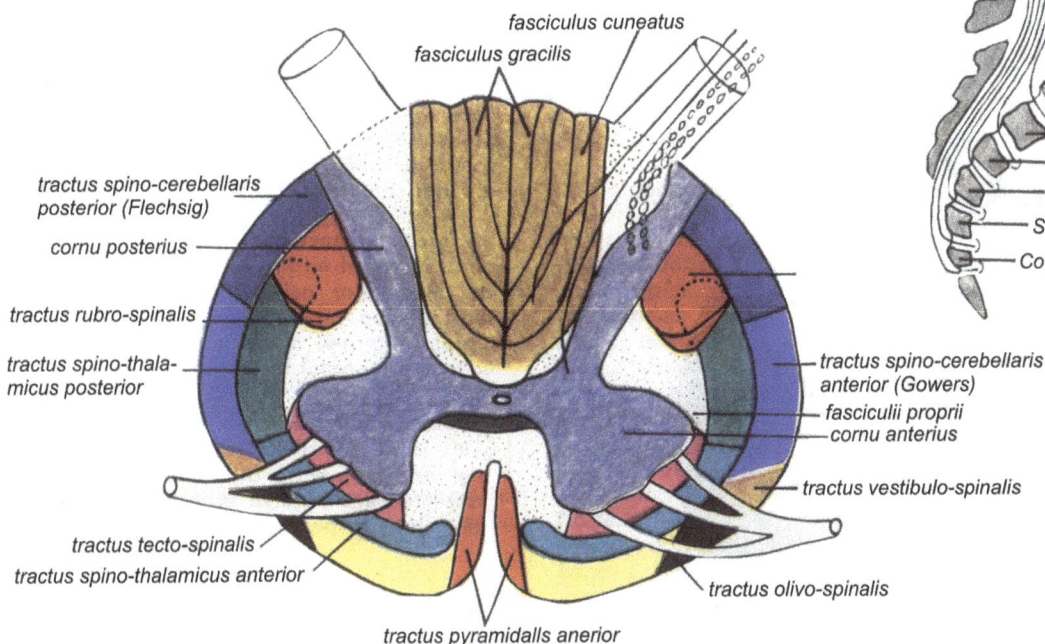

Fig. 26 Spinal cord – topography of nerve fascicles.

37

fasciculus gracilis(Goll)
fasciculus cuneatus (Burdach)

tractus spinocerebellaris anterior (Gowers)

tractus spinocerebellaris posterior (Flechsig)

tractus corticospinalis (pyramidalis) anterior (piramidal direct)

sulcus medianus posterior

fasciculul lui Hoche (descendent posterior)

fasciculus semilunaris (Schultze)

tractus corticospinalis lateralis (piramidal încrucişat)

tractus rubrospinalis (Monakow)

zona terminalis
zona spongiosa
substantia gelatinosa (Rolando)
nucleus proprius

nucleus dorsalis (Clarke-Stilling)

columna intermedia lateralis

n.dorsolateralis

n.ventrolateralis
n.centralis

n.dorsomedialis
n.ventromedialis

tractus corticospinalis lateralis

tractus rubrospinalis (Monakow)

tractus spinothalamicus et spinotectalis

fasciculus longitudinalis medialis
fissura mediana anterior

tractus tectospinalis
tractus corticospinalis anterior (piramidal direct)
tractus reticulospinalis et vestibulospinalis
tractus vestibulospinalis
tractus olivospinalis et spinoolivaris

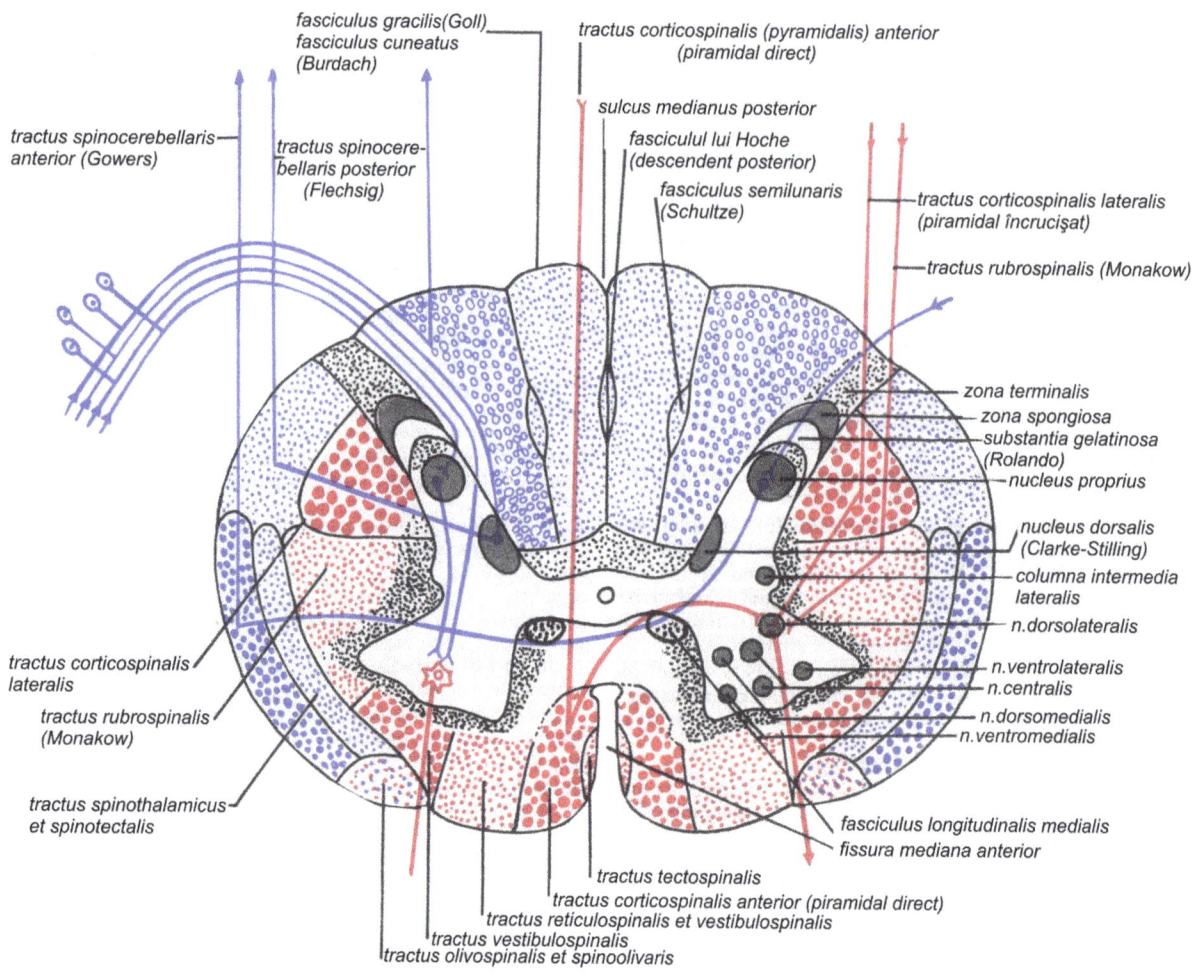

Fig. 27 Section at the level of the spinal cord – scheme after prof. R. Robacki.

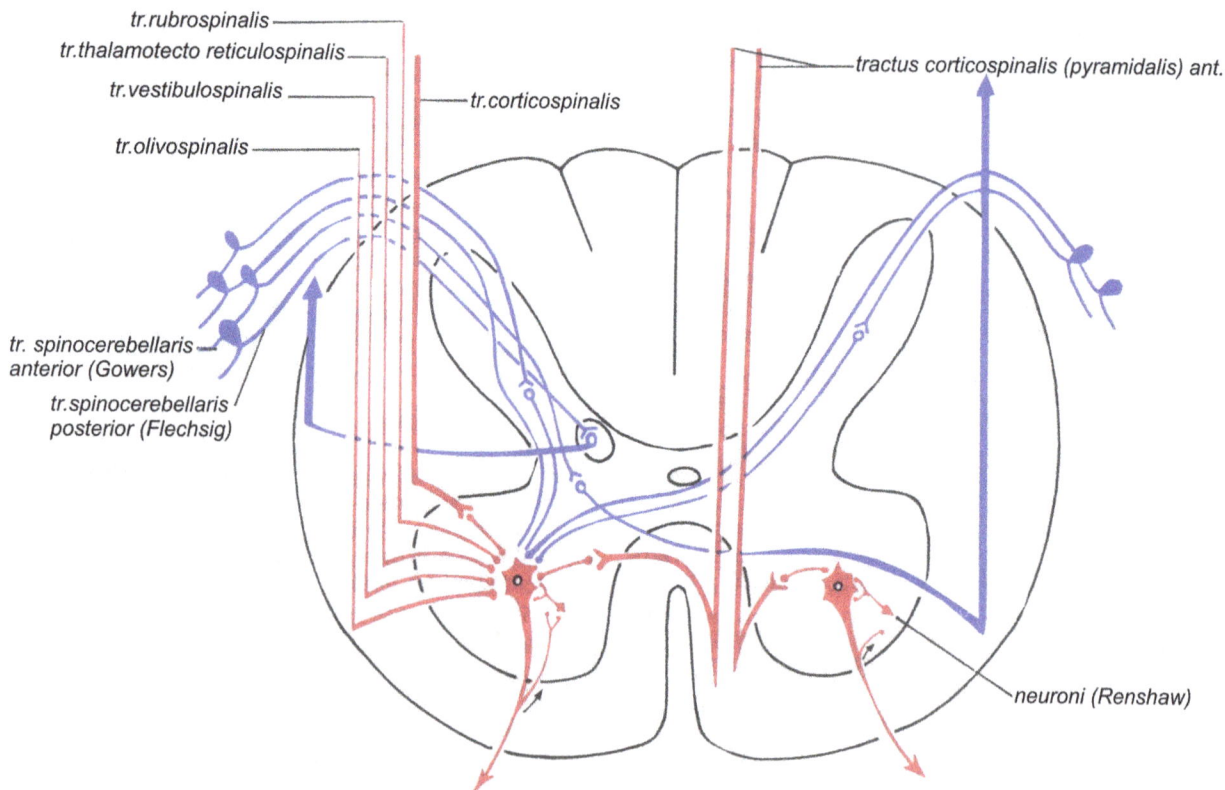

tr.rubrospinalis
tr.thalamotecto reticulospinalis
tr.vestibulospinalis
tr.olivospinalis

tr.corticospinalis

tractus corticospinalis (pyramidalis) ant.

tr. spinocerebellaris anterior (Gowers)

tr.spinocerebellaris posterior (Flechsig)

neuroni (Renshaw)

38

Fig. 28 Spinal cord – common final pathway

The intercalary or internuncial neurons, also called interneurons, represent connection nerve cells of the grey matter. They are divided into several categories:

1. *Connector cells* (Golgi cells), which are more frequent in the posterior horn and in the posterior gelatinous substance. They receive the sensory excitations of the afferent fibres and convey them to the motor cells in the anterior horn. Their dendrites end on the same side and in the same segment.

2. *Association cells*, which are more frequent in the posterior horn. They receive excitations from the peripheral sensory pathways. Their dendrites ascend several segments on the same side or descend several segments and course to the motor cells in the anterior horn.

3. *Commissural cells*, which are predominantly present in the lateral portions of the central grey matter and mediate, through their afferent processes, the passage of the influx through the anterior white commisure to the motor cells in the anterior horn on the opposite side. The commisure cells are of two kinds:

- *short commisural cells*, the processes of which course on the opposite side of the same segment;

- *long commissural cells*, the processes of which course on the opposite side, in some higher or lower situated segments.

4. *Long conduction cells*, situated in the nuclei of the posterior horn. They receive impulses through the peripheral sensory neurons. Their processes remain on the same side, ascending toward higher situated centres (for example, the posterior spinocerebellar tract).

5. *Commissural conduction cells*, situated in the nuclei of the posterior horn. They receive too, impulses from the afferent sensory neurons. Their fibres cross in the grey commissure, course toward the opposite site, ascending and ending in the brain centres (for example, the spinothalamic tract).

Nuclei

The nerve cells are not distributed uniformly in the thickness of the grey matter, but are grouped in regular systems, in the form of columns parallel to the axis of the spinal cord, called nuclei or columns. The number of motor neurons in the anterior horns is about 100,000 per column. T*heir neurites form the anterior roots*.

In the anterior horns five groups of nuclei, may be distinguished with the following innervation territories:

a. *the medioventral nucleus* supplies the ventral musculature of the trunk;

b. *the mediodorsal nucleus* supplies the dorsal musculature of the trunk;

c. *the central nucleus* supplies from the C3-C4 level the diaphragm and from the S2-S4 level the muscles of the pelvic diaphragm;

d. *the lateroventral nucleus*, situated at the level of the cervical intumescense supplies the flexor muscles of the thoracic (upper) limbs, whereas that situated at the level of the lumbar intumescence supplies the flexor muscle of the pelvic (lower) limbs;

e. *the laterodorsal nucleus* situated at the level of the cervical intumescense supplies the extensor muscles of the thoracic limb, whereas that at the level of the lumbar intumescence supplies the extensor muscles of the pelvic limb.

The motor cells of the anterior horn receive impulses through the corticospinal tract.

In the posterior horn, throughout all its length, the following nuclei are situated:

a. *the nucleus proprius of the posterior horn.* Here end the peripheral sensory neurons for the pain, thermal and sometimes tactile sensibility and from here starts the spinothalamic tract;

b. *the dorsal nucleus* or Clarke-Stilling's column is situated medially, at the boundary between the posterior column (posterior horn) and the grey commissure. Up to C7 it represents the ending of the peripheral neurons for the deep sensibility; from here starts the posterior spinocerebellar tractb (Flechsig's tract) (the Clarke-Stilling's column should not be confused with the lateral horn or with Clarke's intermediolateral tract);

c. *the intermediomedial nucleus* is situated at the base of the posterior horn throughout all the length of the spinal cord, and represents the ending of the peripheral neuron for the deep sensibility and the origin of the anterior spinocerebellar tract (Gower's tract);

d. *the intermediolateral nucleus* is contained in the anterior horn, from C8 to L3 and represents the spinal centre of the sympathetic system *(n. sympaticus)*, the origin of the efferent peripheral sympathetic fibres, as well as the ending of the afferent fibres;

e. *the sacral intermediolateral nucleus* is located at the boundary between the anterior and the posterior horn, at the S2 -S4 level. It is the spinal centre of the parasympathetic nerve (n-parasympathicus), the origin of the efferent peripheral parasympathetic fibres and the ending of the afferent fibres.

The control of all the movements is achieved directly or indirectly through the motor neurons in the anterior column of the spinal cord, which is therefore named by Sherrington the „*common final pathway*" (fig. 28). Each cell of the anterior horn supplies several muscle fibres representing a "*motor unit*".

THE RACHIDIAN (SPINAL) NERVES

Each spinal nerve consists of two *attachment roots* (anterior and posterior), a trunk and *terminal rami* (anterior and posterior). For clinical and didactic reasons, Sicard calls *spinal root* only the portion from the spinal cord to the dura mater, hence the zone *inside the dural sac* (fig. 29).

Roots of the spinal nerves

They are situated in the spinal subarachnoid space; the lower they are located, the more oblique their direction is.

The anterior or ventral root emerges from cells situated in the anterior horn of the spinal cord, representing axonic processes of these cells, which are named for this reason radicular cells. It is the motor root of the spinal nerve *(radix ventralis)*, which is usually single at the level of the spinal nerve C1.

It leaves the spinal cord at the level of the anterolateral sulcus and traverses *the pia mater, the arachnoid* and *the dura mater* under the form of filaments *(fila radicularia ventralia)*, which represent the collection of a large number of interconnected fibres that may be divided into thick middle-sized and thin myelinated fibres, accomplishing an effector function.

A. *The thick myelinated fibres* (8-18 micron-diameter fibres of alpha cells) and the middle-sized fibres (3-8 micron-diameter gamma fibres) are *somatomotor fibres*. They make up the axons of the motor neurons in the anterior horns of the spinal cord. The first supply the striated muscle fibre and the others provide the innervation of the neuromuscular spindles.

B. *The thin myelinated fibres* (under 3 microns in diameter) form the axons of the vegetative cells in the intermediolateral column of the spinal column and are classified in *sympathetic fibres*, arising from the cervicothoracic-lumbar spinal cord, and p*arasympathetic fibres*, originating in the sacral cord. They are *preganglionic fibres*, the sympathetic fibres ending in the laterovertebral ganglia and the parasympathetic fibres in the prevertebral ganglia. Functionally, they are *viscero-effector fibres*, conveying toward the vegetative ganglia stimuli with a vasomotor, lysomotor and secretory effect.

The posterior sensory root (thicker and larger) *(radix dorsalis)*, accomplishing a receptor function, enters the spinal cord at the level of the posterolateral sulcus. It is bulkier than the anterior root, with the exception of the first cervical nerve, where it is present only in 10% of cases.

Along its course *the spinal ganglion* is situated, which contains *the sensory protoneuron*. It consists of primitively bipolar cells.

The two processes separate from each other at a short distance away from the cell body; the cellulipetal or peripheral process comes from the periphery and constitutes the afferent fibre of the spinal nerve, while the celulifugal or central process penetrates into the spinal cord through the posterolateral sulcus of the cord. The spinal ganglia are situated in the intervertebral foramen, with the exception of the sacral ganglia, which lie in the vertebral canal.

The posterior root is formed of *myelinated* and *unmyelinated* fibres which, in dependence on their calibre and on the conduction speed of the nerve influx, are of four types:

A. *The thick myelinated fibres* (type I) measuring 12-20 microns in diameter characterized by a rapid conduction, transmit the unconscious proprioceptive sensibility.

B. *The middle-sized myelinated fibres* (type II), 5-12 microns in diameter, with a less rapid speed of conduction, convey the proprioceptive and the tactile sensibility.

C. *The thin myelinated fibres* (type III), 2-5 microns in diameter, with a slower speed of conduction, convey the somatic pain and the thermal sensibility.

D. *The unmyelinated fibres* (type IV), 0.3-1.3 microns in diameter, convey the visceral pain sensibility.

Both roots are included in a dural muff, up to beyond the spinal ganglion.

The trunk of the spinal nerve

It traverses the intervertebral foramen *(foramen intervertebralis)* and lies between the sites of junction of the two roots of origin and the level of division of the terminal rami.

Between the spinal nerves and the laterocervical sympathetic ganglionic chain are situated *communicating branches*: from the nerve to the ganglion extends *the white communicating branch*, containing preganglionic vegetative fibres, and from the ganglion to the nerve lies *the grey communicating branch*, containing postganglionic vegetative fibres, which assure the vasomotor, lysomotor and secretory nerve supply of the somatic area.

Thus, it results that the spinal nerve is a *mixed nerve*, containing sensory, motor and vegetative fibres.

Terminal rami

Immediately after the exit from intervertebral foramen, the root gives off a recurrent ramus which penetrates into the canal - *Luschka's spinovertebral ramus*. This ramus divides into two other rami: anterior and posterior, both mixed.

I. *The anterior rami of the spinal nerves* supply the anterior and lateral regions of the trunk and the limbs. In the thoracic region (T2-T12) they traverse the space between the pleura and the thoracic fascia and course along the inferior border of the ribs, supplying the intercostal muscles and the skin. In the cervical and lumbar region they form plexuses (cervical C1-C4; brachial C5-T1; lumbar L1-L4; sacral L4-S2; pudendal (Fig. 32,33).

II. *The dorsal rami* pass backwards and, after a short course, each dorsal ramus gives off a medial branch, which innervates *the multifidus, longissimus, semispinalis* and *trapezius muscles*, then runs along the spinous processes and supplies the skin; and a lateral branch, which crosses *the longissimus muscle* and supplies the dorsal part of the intercostal muscles and the adjacent skin. Exception to this arrangement make the branch of the C1 nerve, consisting only of motor fibres (unique in 90% of cases), and that of the C2 nerve, which is a sensory branch (Arnold's nerve).

Medullary segments

There are 31 pairs of spinal nerves, corresponding to the 31 segments of the spinal cord, which are variable in size: some of them are larger, others are smaller. The segments with the corresponding nerves are distributed as follows:

41

- *The cervical part (pars cervicalis)* consists of eight segments (C1-C8), with eight pairs of cervical nerves. The first cervical nerve leaves the vertebral canas between the atlas and the occipital zone;
- *The thoracic or dorsal part (pars thoracica)*, consisting of 12 segments (T1-T12), with 12 pairs of thoracic nerves *(nn. thoracici)*;
- *The lumbar part (pars lumbalis)*, consisting of five segments (L1-L5), with five pairs of lumbar nerves *(nn. lumbales)*.
- *The sacral part (pars sacralis)*, consisting of five segments (S1-S5), with five pairs of sacral nerves *(nn. sacrales)*;
- *The coccygeal part (pars coccygea)*, consisting of three segments, with one pair of coccygeal nerve (Fig. 30,31) *(nn.ano-coccygei)*.

The nerve of each segment passes through the corresponding intervertebral foramen. Thus, for example, the first thoracic nerve *(n. thoracicus)* passes through the intervertebral foramen between the first and the second thoracic vertebra. Up to the third month of intra-uterine life, the neural tube and the spinal cord develop concomitantly and the spinal nerves, in a segmental arrangement, emerge through the intervertebral foramen at the level of the site of origin. After this period, the lower portion of the embryo, including the vertebral column, develops much faster, the spinal cord lagging behing the growth of the column. As the cephalic part of the nervous system is situated in the cranial cavity and as the spinal cord develops much slower than the column, the cord seems drawn upwards into the vertebral canal, so that the nerves descend obliquely up to the intervertebral foramen at the level of which they have been emitted. The move downwards the spinal nerves run, the more marked is their obliquity. In a term foetus the spinal cord ends at the L3 vertebra and in the adult, at the level of L1-L2 *vertebrae*. The sacral and coccygeal nerves descend beside each other and form with *the filum terminale, the cauda equina*.

The external stimuli received by the organism are conveyed to the brain by the spinal nerves and by some cranial nerves, through the peripheral branches which come from the sensory organs and course to the peripheral neurons in the spinal cord or in the brain.

As the distribution of the nerve roots is known, it is easy to determine the level of a possible medullary lesion:

We mention below some of the more significant locations:
- the clavicular region, supplied by the sensory root C5;
- the deltoid region, supplied by C5 and C6;
- the mammillary region, supplied by T4;
- the umbilical region, supplied by T10;
- the inguinal region, supplied by T12;
- the lateral region of the arm, supplied by C3, C6 and C7;
- the medial region of the arm, supplied by C8 and C1;
- the antero-internal region of the thigh, supplied by L1, L2, L3 and L4;
- the postero-external region of the thigh, supplied by L5, S1 and S2;
- the perineal region, supplied by S2, S3, S4 and S5;
- the forearm and the hand supplied from the radial to the ulnar border by C6, C7 and C8;
- the thumb, supplied by C7; the index and the third (middle) finger, supplied by C7; the ring finger and the little (fifth) finger, supplied by C8;
- the leg and the soles supplied from the lateral to the medial border by S1, L5 and L4 (Fig.30, 31).

It ensures from the above enumerated regions that there is no clear-cut overlapping between the topography of dermatomes (Head) and the cutaneous surface supplied by the peripheral nerves.

Moreover, the dermatomes for the tactile sensibility have a wider extension than those for the painful sensibility. Likewise, between dermatome and sclerotome (including the myotome) there is no overlap since the sclerotomes are supplied by several roots.

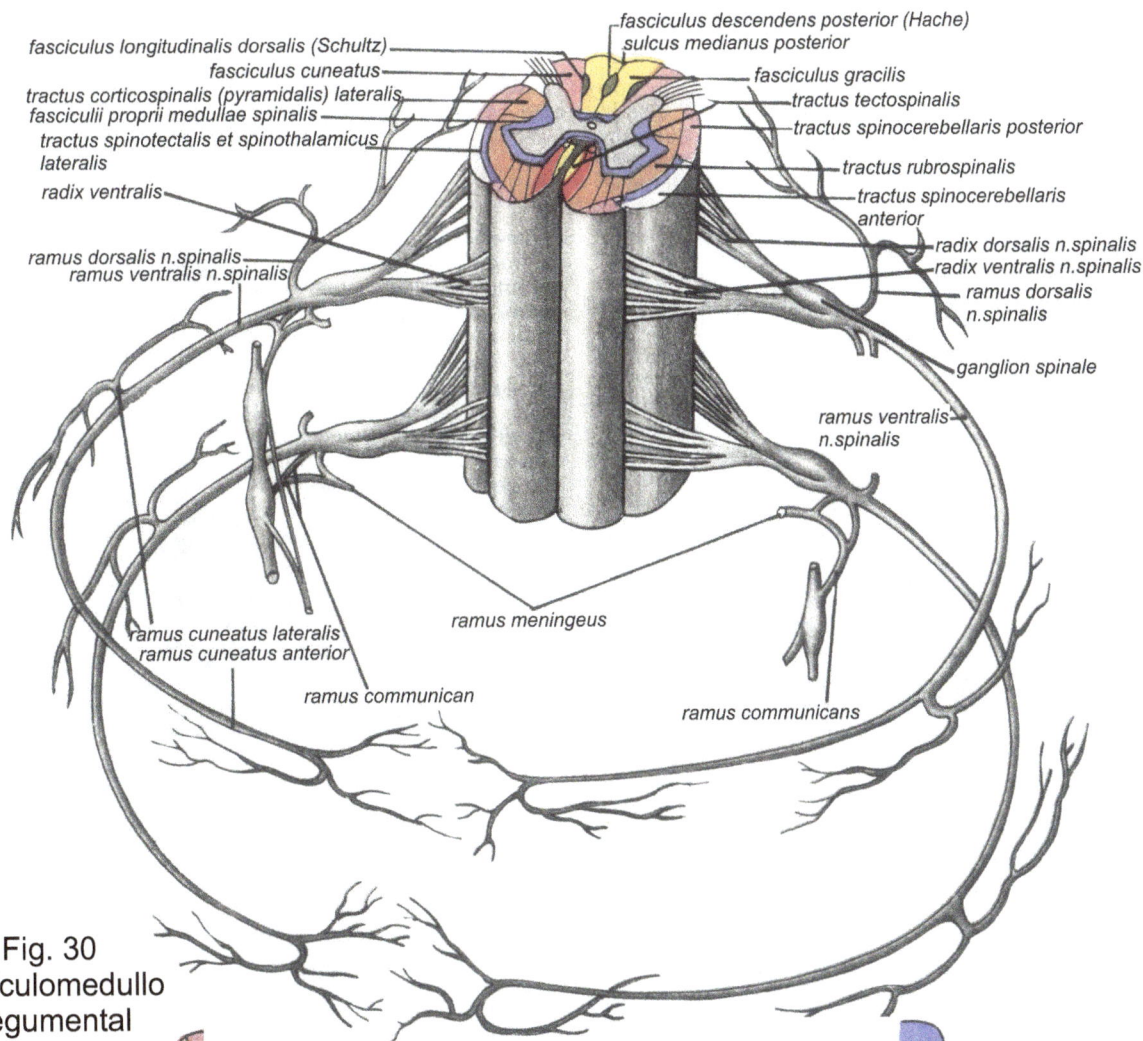

fasciculus longitudinalis dorsalis (Schultz)
fasciculus cuneatus
tractus corticospinalis (pyramidalis) lateralis
fasciculii proprii medullae spinalis
tractus spinotectalis et spinothalamicus lateralis
radix ventralis
ramus dorsalis n.spinalis
ramus ventralis n.spinalis

fasciculus descendens posterior (Hache)
sulcus medianus posterior
fasciculus gracilis
tractus tectospinalis
tractus spinocerebellaris posterior
tractus rubrospinalis
tractus spinocerebellaris anterior
radix dorsalis n.spinalis
radix ventralis n.spinalis
ramus dorsalis n.spinalis
ganglion spinale
ramus ventralis n.spinalis

ramus cuneatus lateralis
ramus cuneatus anterior
ramus meningeus
ramus communican
ramus communicans

Fig. 29 Spinal nerve.

Fig. 30 Radiculomedullo-tegumental topography (anterior view)

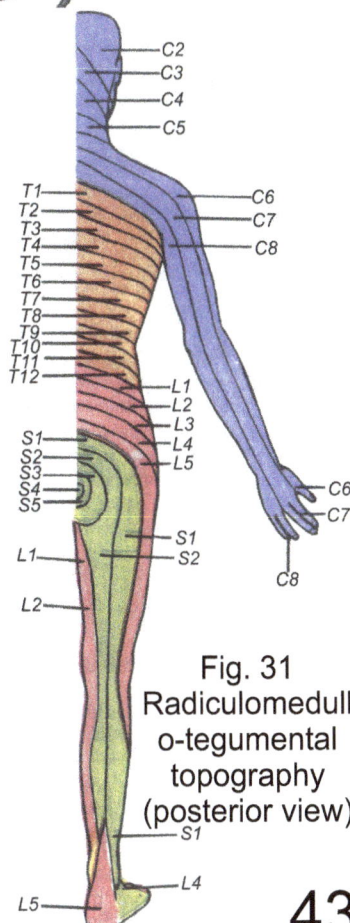

Fig. 31 Radiculomedullo-tegumental topography (posterior view)

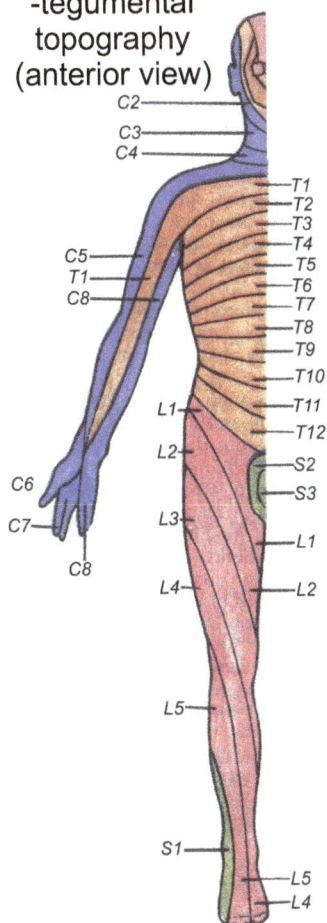

The reflex arc

Before studying the reflex arc, we mention that the large motor and sensory pathways are situated in the white matter. These pathways will be studied afterwards.

The spinal cord is not only *an organ of transmission*, but also *a nerve centre*, with an own function, that can be exerted also independently by the encephalic nerve centres. Thus, an excitation at the body surface is converted into a nerve influx and arrives through the sensory fibres (the dendrites of the neuron in the spinal ganglion), at the spinal ganglia, after which it travels through the posterior root of the spinal nerve (the axons of the same neuron), to other cells(intercalary neurons) of the anterior or posterior horn.

43

Fig. 32 Cervical plexus and brachial plexus (constitution).

Fig. 33 Lumbar plexus and sacral plexus (constitution)

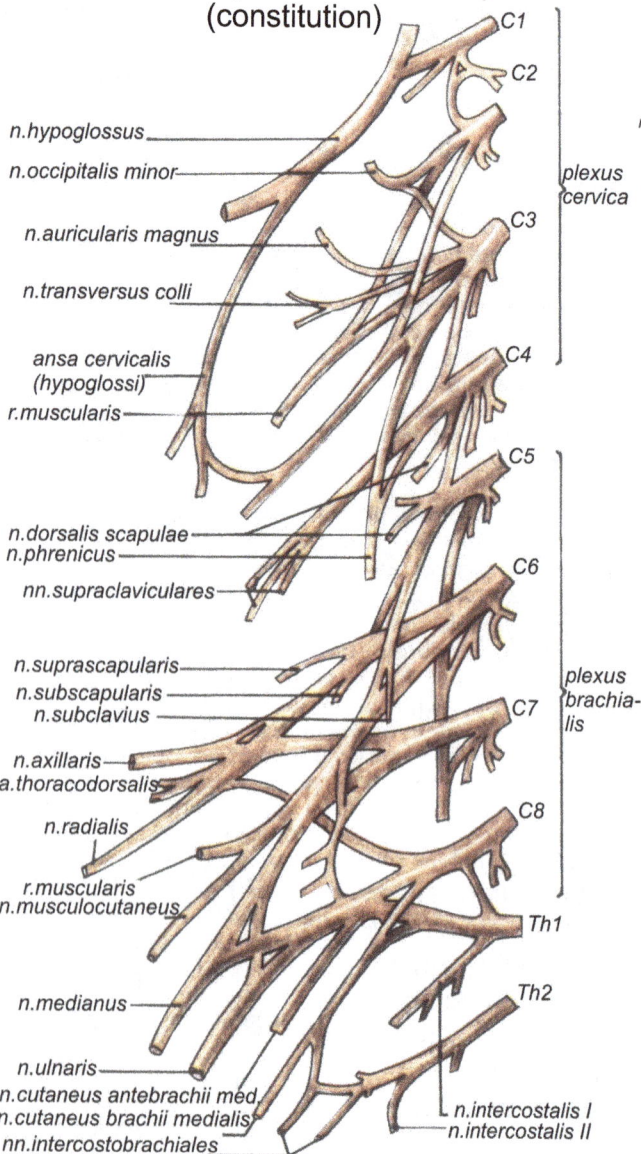

Fig. 32 labels:

n.subcostalis — Th12
n.iliohypogastricus — L1
n.genitofemoralis — L2
n.ilioinguinalis
r.genitalis n.genitofem — L3
n.cutan.femoris lateralis
r.femoralis n.genitofem
n.furcalis — L4
n.femoralis
n.obturatorius — plexus lumbalis
n.glutaeus superior — plexus lumbo-sacralis
n.ischiadicus — S1
n.peronaeus (fibularis) communis — plexus sacralis
n.glutaeus inferior — S2
n.tibialis — S3
r.muscularis — S4
n.cuneatus femoris post. — S5 — plexus coccyge
n.clunium inferior — Co — n.pudendus
nn.rectales inferiores
rr.musculares
nn.anococcygei

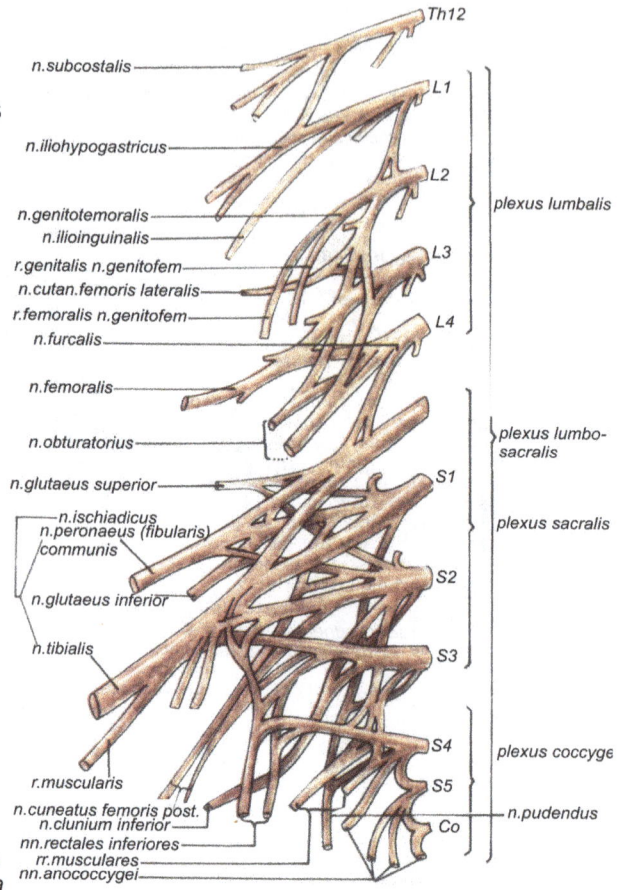

Fig. 33 labels:

n.hypoglossus — C1
— C2
n.occipitalis minor
n.auricularis magnus — C3
n.transversus colli
ansa cervicalis (hypoglossi) — C4
r.muscularis
— C5 — plexus cervica
n.dorsalis scapulae
n.phrenicus
nn.supraclaviculares — C6
n.suprascapularis
n.subscapularis
n.subclavius — C7 — plexus brachia-lis
n.axillaris
a.thoracodorsalis
n.radialis — C8
r.muscularis
n.musculocutaneus
— Th1
n.medianus
— Th2
n.ulnaris
n.cutaneus antebrachii méd.
n.cutaneus brachii medialis — n.intercostalis I
nn.intercostobrachiales — n.intercostalis II

At this level, through the reflex motor collaterals, it produces a motor reaction which arrives at the motor cells (motor neurons) of the anterior horn. Through the anterior roots of the spinal nerves, along the axon of the radicular motor neuron, it causes a muscle contraction. This whole physiological act makes up *the reflex nerve arc*.

In the formation of the reflex arc are involved at least two neurons: a sensory and a motor neuron. Some reflex arcs are more complex (between the two neurons are intercalated also other neurons).

According to the site of the centre are described the following reflexes:

- *spinal reflexes*, which end in the spinal cord, as it was shown previously;

- *reflexes with the centre in one of the nuclei of the brainstem*; the first sensory neuron is situated in one of the ganglia of the cranial nerves, from which the influx arrives at the nuclei of origin of these sensory nerves, and then passes to the motor neuron situated in the nuclei of origin of the motor cranial nerves in the brainstem.

44

The reaction (motor, secretory or vascular) produced in response to a stimulus is called reflex act.

Pavlov has divided the reflexes into two great categories:

- *unconditioned, inborn reflexes*, the natural substrate of which is the reflex arc, with the centre in the lower segments of the central nervous system. In the clinic, the first place among these reflexes is held by the myotactic reflexes;

- *conditioned reflexes, acquired in the course of the life*; they have a temporary character and in their formation is involved the cerebral cortex.

According to the involved segments, *direct* (proprioceptive) and *indirect* (exteroceptive) *reflexes* are distinguished.

In the case of *the direct or proprioceptive reflexes*, the connection is achieved at the level of the same segment; they are in general monosynaptic. An excitation with the origin in a muscle or tendon arrives, under the form of an influx, at the cells of the anterior horn and produces in response a short muscle contraction - for example, the patellar reflex. In the case of an interruption of the reflex arc, the influx can no more be propagated and hence the muscle contraction does not occur. The proprioceptive impulses produce the contraction of the extensor muscles and the relaxation of the muscles (for example, the Achilles reflex, the patellar reflex).

The exteroceptive or indirect reflexes are formed by the involvement of several segments, on the same side or on the opposite side.

The cells of the anterior horn, on the same side, receive the excitations under the form of the nerve influx, through the collaterals of the sensory afferent fibres, through the connector (Golgi) cells and the intercalary neurons. The cells on the opposite side receive the nerve influxes, following the excitations, through the short and long commisural cells. The movements produced by the stimulation of the skin or of the mucosae are usually of defence (for example, the skin reflexes - cremasteric, abdominal). The exteroceptive stimuli influence predominantly the group of flexor muscles.

The membranous coverings of the spinal cord

The spinal cord is delimited from the vertebral bony canal through its coverings: the meninges, the adipose tissue and the intrarachidian venous plexus, which occupies the space between *the dura mater* and the walls of the vertebral canal.

The meninges envelop the cerebrospinal nervous system and are constituted of three concentrically disposed membranes, which are the following from outside inwards: the spinal dura mater *(dura mater spinales)* or *the pachymeninx, the arachnoid* and the spinal pia mater *(pia mater spinalis)*. The arachnoid and the pia mater form together the *leptomeninx*.

The meningeal membranes penetrate *the intervertebral foramen* and envelop both the anterior root and the posterior root with the spinal ganglion. At the external extremity of the spinal ganglion they are continuous with the epineurium of the spinal nerves.

A. *The spinal dura mater* is glistening white and made up of dense connective tissue. It is poor in blood vessels and supplied by sensory nerves represented by the meningeal branches which emerge from the spinal nerves. To pressure and extension the dura mater reacts by intense pain. At the level of the great occipital foramen *(foramen occipitale magnum)*, the dura mater divides into two laminae, separated by a space: *lamina externa* and *lamina interna*, which will unite again at the level of the S2-S3 vertebrae. The lower extremity, acquiring the aspect of a *cul-de-sac* which surrounds the elements of *the cauda equina* and end under the form of a sheath around *the filum terminale*, is termed coccygeal ligament of the spinal cord. In operations on the rectum by sacral posterior route, the coccyx and the sacrum arc removed only up to the vertebrae S2-S3, in order to avoid the opening of the dural sac. Between *the lamina externa* and *the lamina interna* lies the epidural cavity, anteriorly almost inexistent, filled with adipose tissue and containing the internal vertebral venous plexus, that has the role of damping the shocks (protective role).

45

The epidural cavity is a capillary space situated between the dura mater and the arachnoid.

B. *The spinal arachnoid (arachnoidea spinalis)*, formed of delicate connective tissue, is a vascular and adheres throughout its length to the internal surface of the dura mater. *The subarachnoid cavity* or space is situated between the arachnoid and the pia mater. It is cylindrical in shape, forming in its upper segment the cerebello-medullary cistern *(cisterna magna)*, and descends inferiorly up to the lower extremity of the dural cul-de-sac. This space is filled with cerebrospinal fluid, secreted by the choroid plexus situated in the cerebral ventricles. By suboccipital punctures (atlanto-occipital) this fluid can be extracted for diagnostic or therapeutic purposes. In adults, the spinal cord ends at the first or second lumbar vertebra, hence it cannot be injured when the puncture is made below this level. Between the internal surface of the dura mater and the arachnoid lies the subdural space.

C. *The spinal pia mater* is a thin layer of connective tissue which adheres to the spinal cord and penetrates into *the anterior median sulcus*. It contains the nutrient vessels of the spinal cord. Between the pia mater and the dura mater is situated the denticulate ligament *(ligamentum denticulatum)* which connects them. It consists of 21 fibrous dents processes disposed in a frontal plane.

Blood vessels of the spinal cord

The vertebral canal contains a rich arterial and venous network, which assures the extra- and intramedullary blood supply.

The arterial system

The spinal cord is supplied by a double embryological arterial system of various origins: a system of longitudinal vessels arising from the spinal artery, respectively the anterior and posterior spinal arteries, parallel to the axis of the spinal cord, individualized especially at the level of the cervical cord, and a system of paired, symmetrical transverse vessels, named radicular arteries. There are numerous anastomoses between the two systems, which form a perimedullary network in the pia mater, from which arise the arteries which supply directly the spinal cord, forming the intramedullary arterial system (Fig. 37).

The longitudinal arterial system. The vertebral arteries *(aa-vertebrales)*, penetrated in the cavity, before their union in the basilar trunk, give off the following branches inside the vertebral canal:

A. The anterior spinal arteries *(aa. spinales anterior)* arise from the vertebral artery near the basilar trunk, where they detach under the form of two arteries which run - below the pia mater - downwards and medially. After a short course in the superior cervical region (C5-C6), they unite on the median line, forming the anterior spinal artery, which descends vertically into the median sulcus of the spinal cord. In the inferior cervical and dorsolumbar regions they unite with ascending and descending branches arisen from the radicular arteries (lateral spinal braches).

B. The posterior spinal arteries *(aa. spinales posterior)* arise either from the vertebral artery or from the inferior and posterior cerebellar arteries. After a short descending course on the posterior surface of the medulla oblongata, they divide into an anterior branch, situated in front of the posterior roots of the spinal nerves, and a larger posterior branch, behind the root of the spinal nerve.

The transverse arterial system is formed of the segmental radicular arteries *(aa. spinales)* arising from: the vertebral artery, the ascending cervical artery, the suboccipital artery of the cervical region, the intercostal arteries of the thoracic region and the lateral sacral arteries of the sacral region.

46

As regards to **the intermediate arterial system**, which arises from the anastomotic network formed in the pia mater, we mention that the arterial supply is much richer in the grey matter, especially around the neurons in the anterior horns and in the gelatinous substance in the posterior horns, in comparison with the white matter. They grey matter contains twice as many capillaries as the white matter and in the pyramidal tract the number of capillaries is double in comparison with Goll's and Burdach's tract. The capillary density seems to be directly proportional to the number of synapses, to the density of the neuronal mitochondria and to the abundance of the content of cytochromoxidase.

The venous system

The sinuous form of the intramedullary vessels represents a functional adaptation to the eventual elogations and twistings of the spinal cord (Fig. 36, 38).

The veins of the spinal cord *(venae spinales)* lie in the pia mater, forming at this level a plexus made up of longitudinal veins: two median, two anterolateral and two posterolateral veins. They communicate with the internal vertebral venous plexuses *(plexus venosi vertebralis interni)*, situated between the dura mater and the vertebrae, and with the intervertebral veins *(vv. intervertebrales)*. At the base of the skull, the veins of the spinal cord unite, forming several truncks which communicate with the vertebral veins and end in the inferior cerebellar veins *(vv. cerebelli-inferiores)* or in the inferior petrosal sinus *(sinus petrosi inferiores)*.

Lymph. At the level of the nerve centres there are no lymph vessels, the lymph circulating in the perivascular sheaths which open into the subarachnoid space.

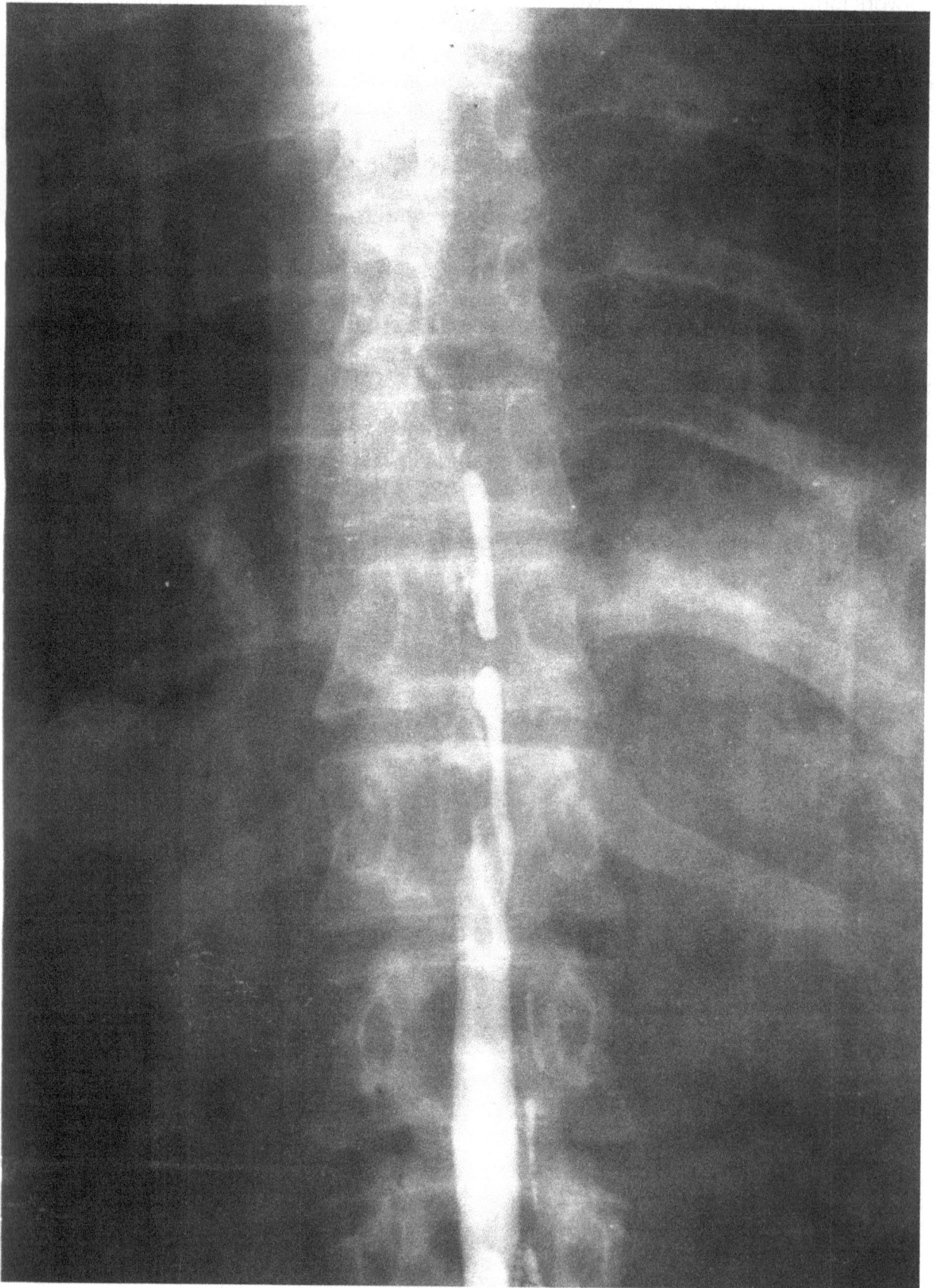

Fig. 34 Myelography – anterior view

48

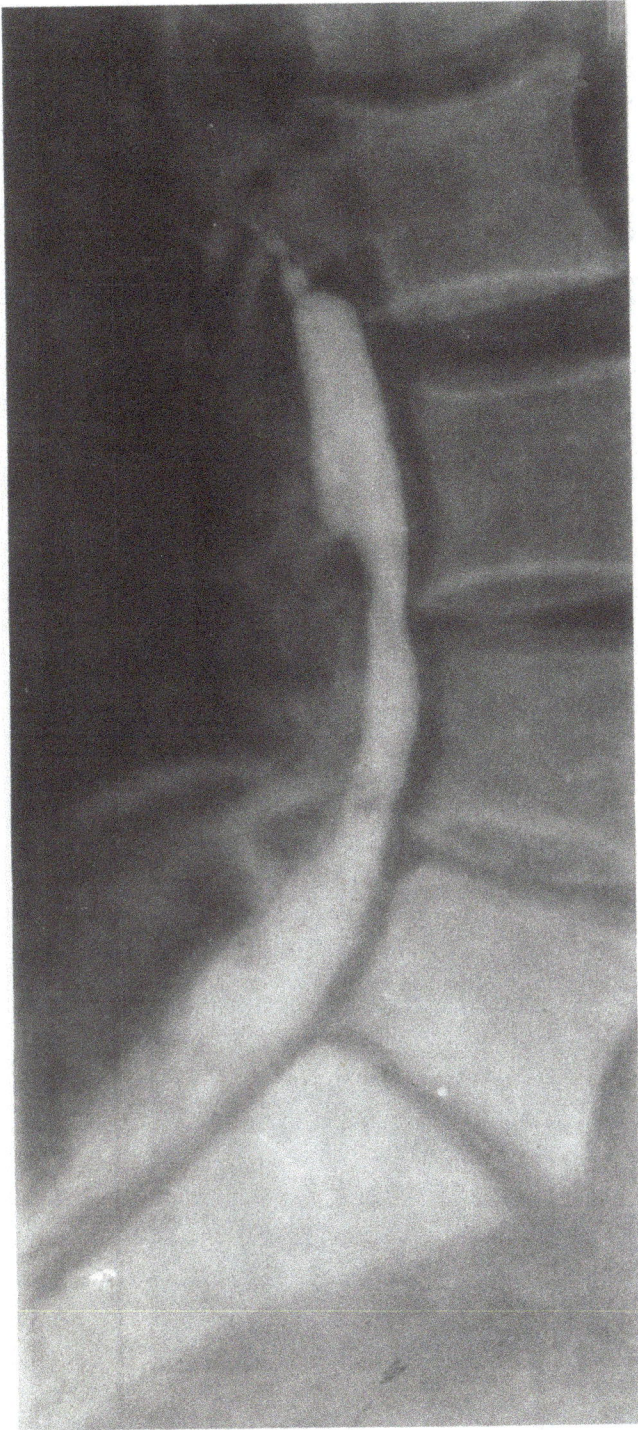

Fig. 35 Myelography – mediolateral view

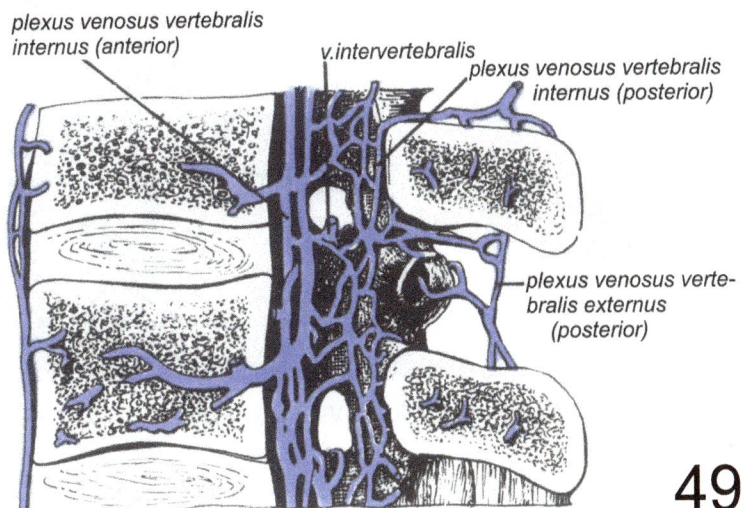

Fig. 36 Venous plexuses of
the spinal cord

plexus venosus vertebralis
internus (anterior)

v.intervertebralis

plexus venosus vertebralis
internus (posterior)

plexus venosus verte-
bralis externus
(posterior)

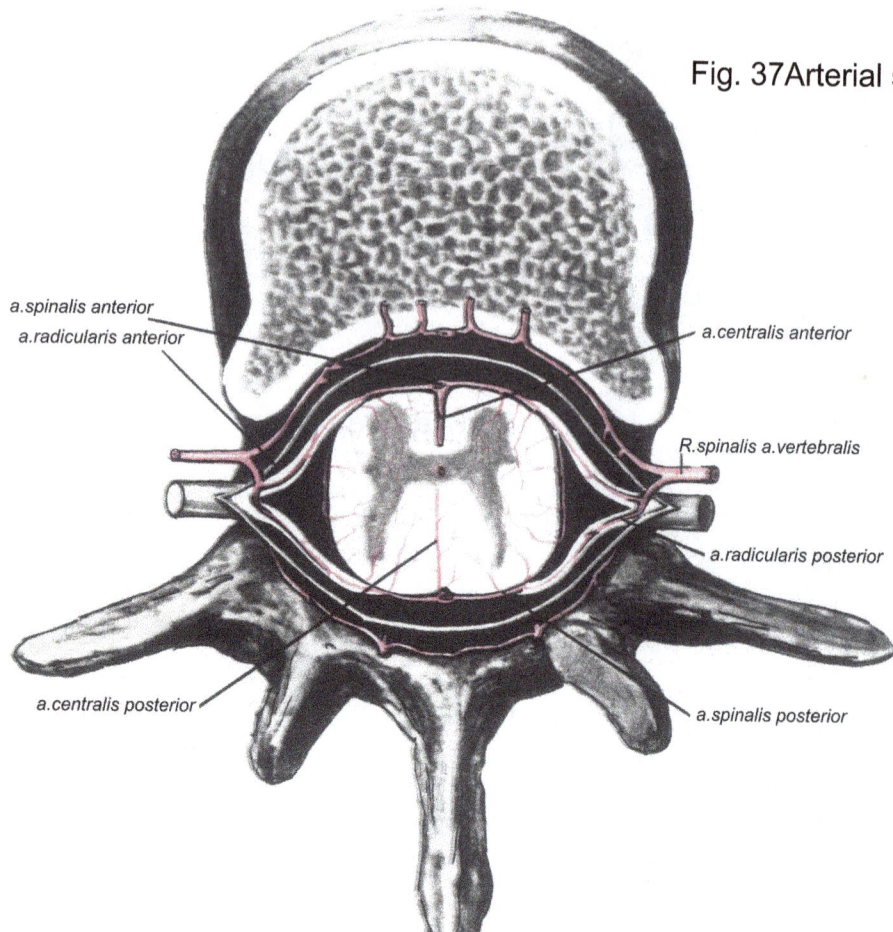

Fig. 37 Arterial supply of the spinal cord

a.spinalis anterior

a.radicularis anterior

a.centralis anterior

R.spinalis a.vertebralis

a.radicularis posterior

a.centralis posterior

a.spinalis posterior

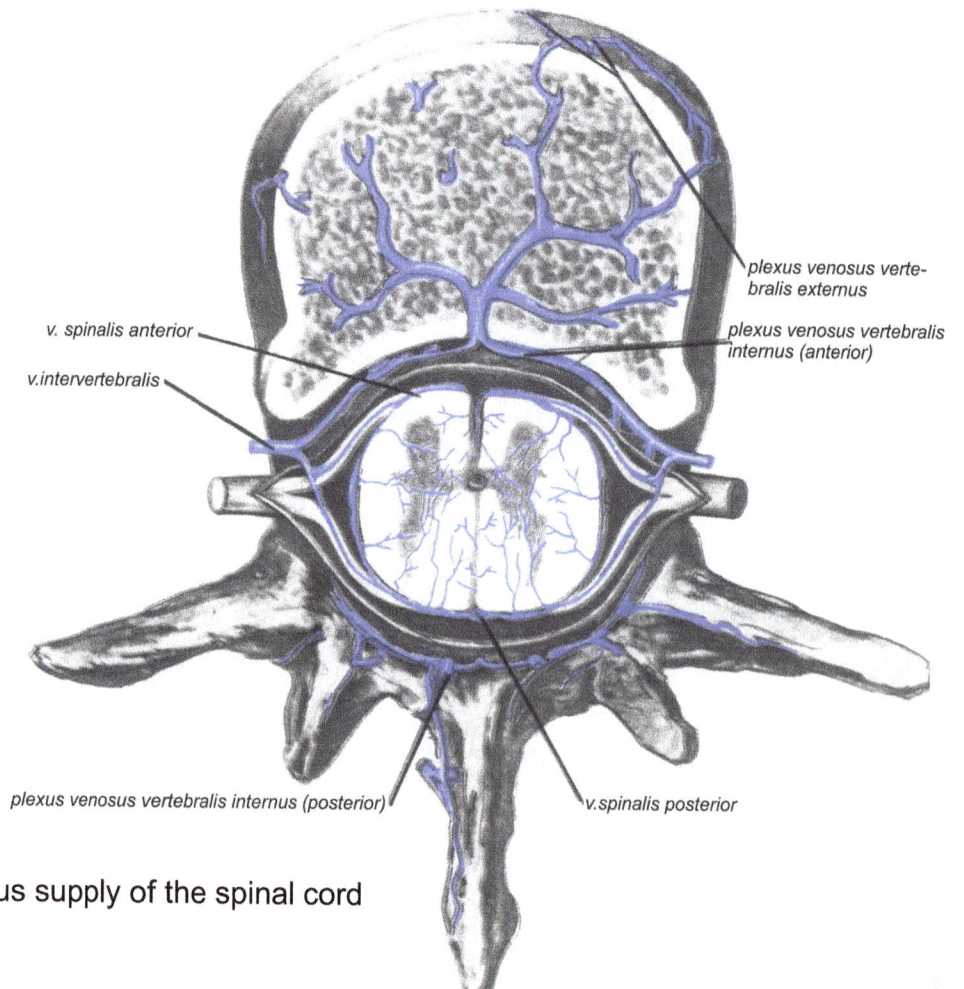

v. spinalis anterior

v.intervertebralis

plexus venosus verte-
bralis externus

plexus venosus vertebralis
internus (anterior)

plexus venosus vertebralis internus (posterior)

v.spinalis posterior

Fig. 38 Venous supply of the spinal cord

50

THE BRAINSTEM
(Fig. 39-67)

It is a portion of the central nervous system which connects the spinal cord with the diencephalon. Its posterior surface is covered by the cerebellum and the cerebellar peduncles, while the anterior and lateral surfaces are visible. It consists of longitudinal cords of white matter which fragment the grey matter into nuclei. In the median area lies a bundle of transverse white fibres, below which the longitudinal cords of white matter pass like under a bridge; this zone is called pons. The region of the brainstem situated below the pons is the medulla oblongata and the portion situated above it, the mesencephalon or midbrain. The three portions of the brainstem are successively studied below.

The medulla oblongata

The medulla oblongata is that part of the encephalon which is continuous above, without a clear-cut delimitation, the spinal cord. It extends form the C1 vertebra upwards and forwards, up to the pons with which it is continuous behind up to the medullary striae in the floor of the rhomboid fossa, and contributes to the formation of the fourth ventricle. The medulla oblongata begins at the level of the atlas and extends up to the middle of the clivus.

External configuration

The medulla oblongata has the shape of a frustum of cone, directed with the base upwards and the truncated vertex downwards, continuous with the spinal cord. It measures 28-30 mm in length and 20 mm in breadth superiorly, respectively 14 mm inferiorly.

It has an anterior surface, two lateral surfaces, a posterior surface and two extremities.

The anterior surface is marked, on the median line, by a groove, *the anterior median fissure*, which is continuous with the anterior median fissure of the spinal cord; it is interrupted, at the level of the C1 vertebra, by *the decussation of the pyramids* (the intercrossing of the bundles of the pyramidal tracts - *the lateral corticospinal tract*). *The anterior median fissure* ends at the lower border of the pons by a small depression, termed *the foramen caecum*. Above the decussation of the pyramids and laterally to *the anterior median fissure* are situated the pyramids of the medulla oblongata *(pyramis medullae oblongatae)*, which make up the two white bundles of the pyramidal tract. They are bounded, outwards, by the anterior collateral fissure, which continues, at the level of the medulla oblongata, that of the spinal cord. Through this fissure emerge 10-12 root filaments of hypoglossal nerve (XII) (Fig. 39).

The lateral surfaces continue above the lateral columns of the spinal cord and are bounded in front by *the anterior collateral sulcus* and behind by *the posterior collateral sulcus*.

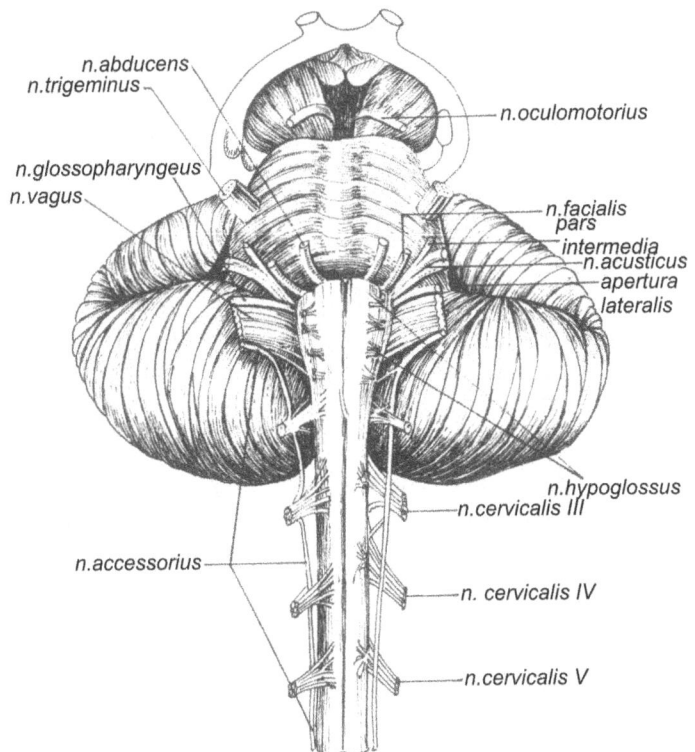

Fig. 39 Brainstem – anterior surface

In the latter lies the apparent origin of the spinal (accessory), the vagus (pneumogastric) and the glossopharyngeal nerves. At mid-way of the surface are *two sulci*, *the preolivary* and *retro olivary sulci*, which surround an elongated oval elevation, *the oliva of the medulla oblongata*, that contains *the olivary nucleus* (the termination of the cerebello-olivary tract and, at the same time, the origin of the olivospinal tract) (Fig.40).

At the extremities of this nucleus lie groups of cells constituting *the accessory olivary nuclei*; *dorsal* and *medial*. In the mass of white fibres inside the medulla oblongata, situated between the two olivary nuclei, lies *the decussation of the lemniscs*, representing the medial part of the crossing of the bulbothalamic tracts *(lemniscus medialis)*.

In the lower part of the medulla oblongata, the lateral surfaces and the anterior surfaces present loop-shaped, more or less visible fibres, named external arcuate fibres.

The posterior surface of the medulla oblongata exhibits in its lower half, *the posterior median sulcus* and *the posterior lateral sulcus*. Between them lies the posterior intermediate sulcus, which separates Goll's tract *(fasciculus gracilis)*, medially, from Burdach's tract *(fasciculus cuneatus)*, laterally (fig. 41,42).

Goll's tract and Burdach's tract form together the posterior funiculus. In the upper part of the medulla oblongata the two fasciculi diverge laterally from the median line.

In this region the Goll's tract presents a swelling, named *clava* or tubercle of *the nucleus gracilis*, which contains *the nucleus gracilis*. In the nucleus ends the sensory protoneuron of the spinobulbar tract, the medial portion, and lies also the origin of fibres of the bulbothalamic and bulbocerebellar tracts. Likewise, the Burdach's tract presents also a thickened portion, named tubercle of *the cuneate nucleus*, limited by *the posterior intermediate sulcus* and by *the posterior lateral sulcus*. It its depth lies *the nucleus cuneatus*, in which ends the peripheral sensory protoneuron of the spinobulbar tract, the lateral portion and here originate other fibres of the bulbothalamic and of the bulbocerebellar tracts.

The tuber cinereum is a prominence situated superolaterally to *the tuberculum cuneatum*. At this level may be observed, through the transparency of a thin layer of white matter, the grey bluish masses of nuclei derived from the posterior horn of the spinal cord.

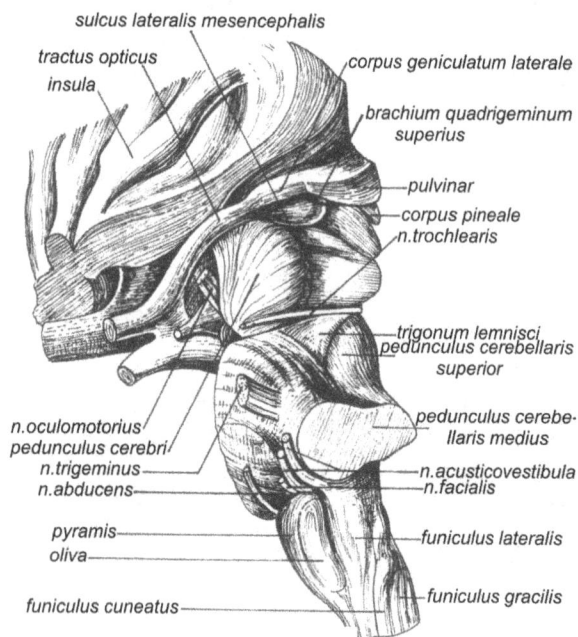

Fig. 40 Brainstem – lateral view

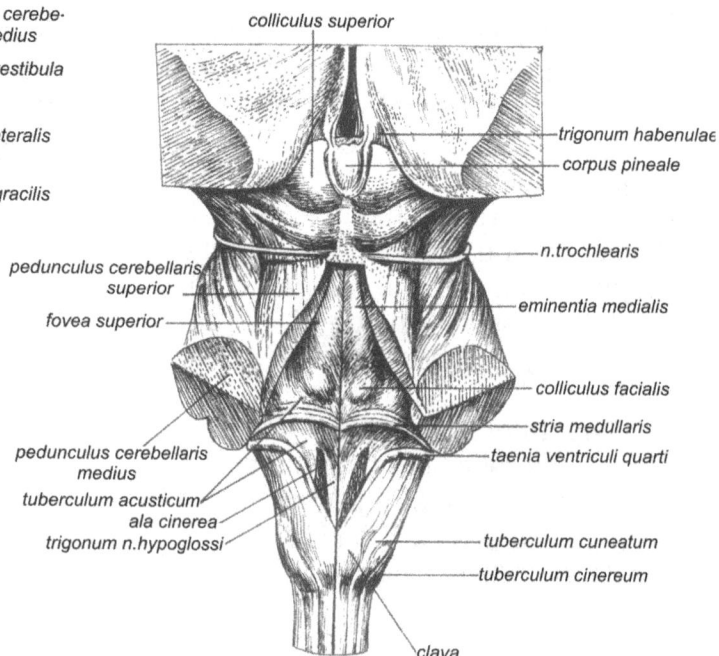

Fig. 41 Brainstem – posterior view

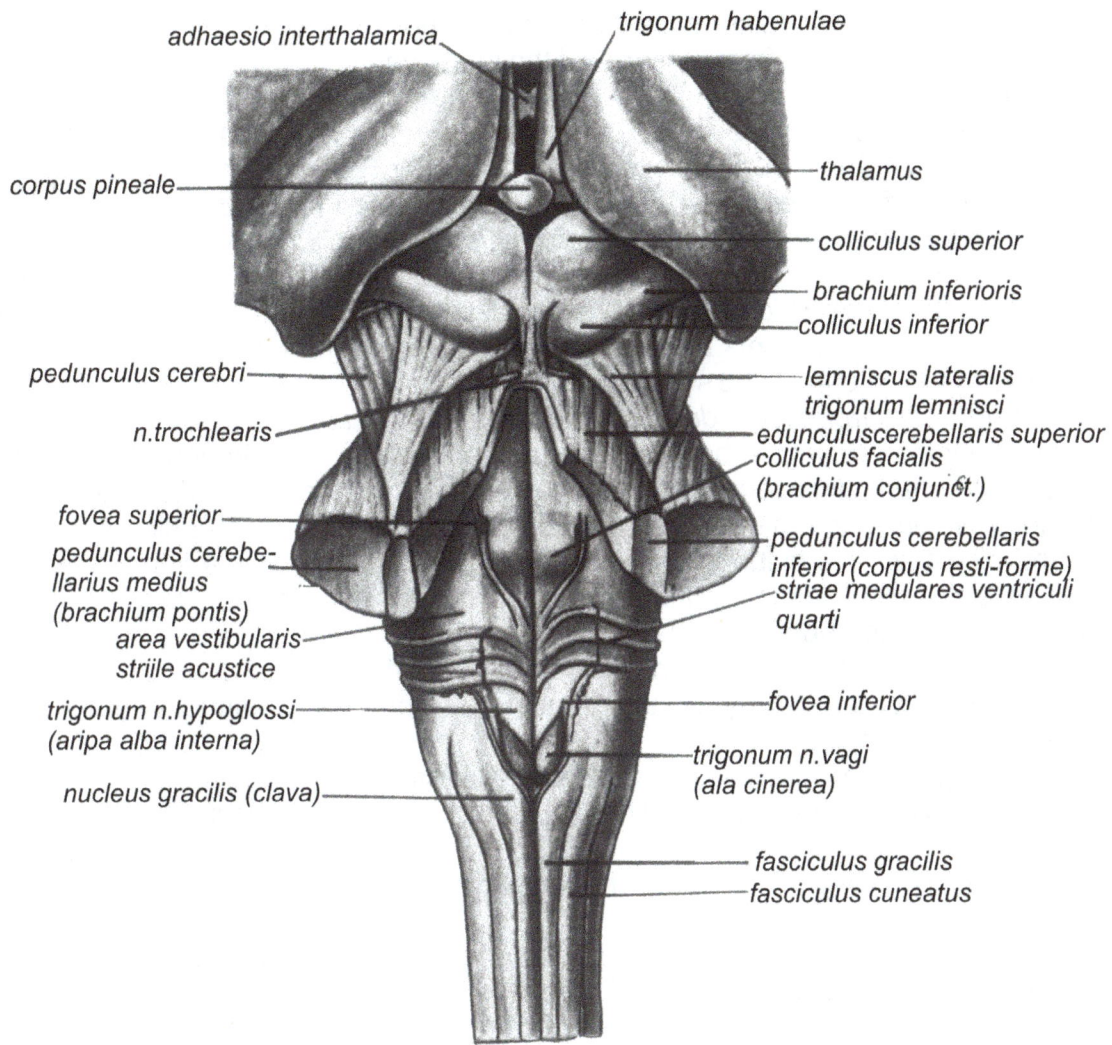

Fig. 42 Brainstem – posterior surface and thalamus

Labels for Fig. 42:
- adhaesio interthalamica
- trigonum habenulae
- corpus pineale
- thalamus
- colliculus superior
- brachium inferioris
- colliculus inferior
- pedunculus cerebri
- lemniscus lateralis
- trigonum lemnisci
- n.trochlearis
- edunculuscerebellaris superior
- colliculus facialis (brachium conjunct.)
- fovea superior
- pedunculus cerebellarius medius (brachium pontis)
- pedunculus cerebellaris inferior(corpus resti-forme)
- striae medulares ventriculi quarti
- area vestibularis striile acustice
- trigonum n.hypoglossi (aripa alba interna)
- fovea inferior
- trigonum n.vagi (ala cinerea)
- nucleus gracilis (clava)
- fasciculus gracilis
- fasciculus cuneatus

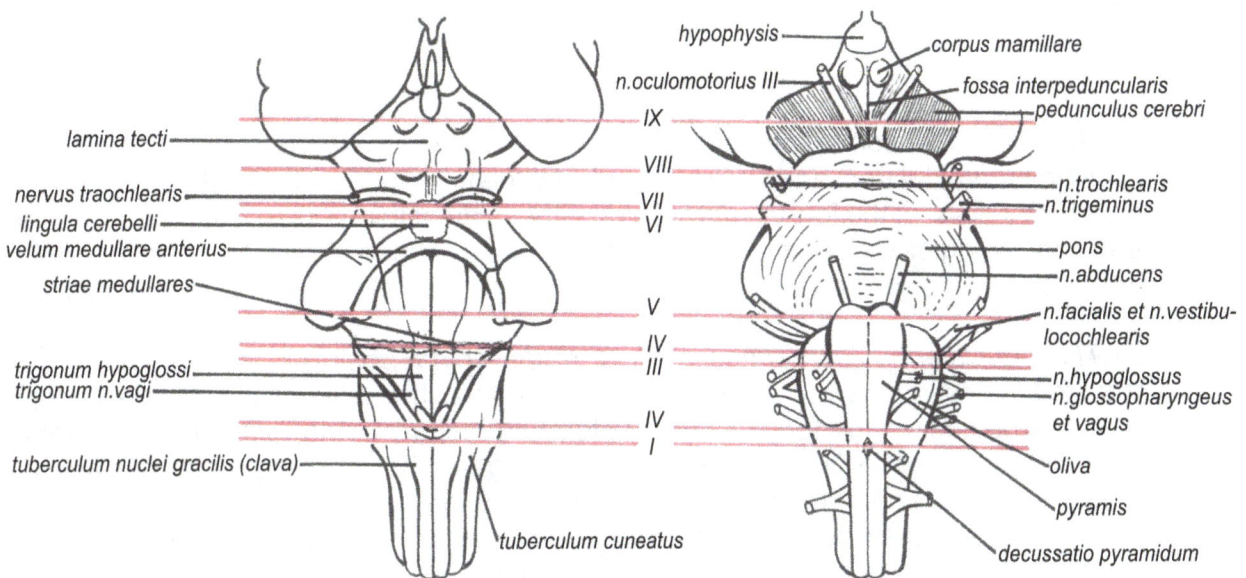

Fig. 43 Plans of transverse sections through the brainstem (after prof. R. Robacki)

Labels for Fig. 43:
- hypophysis
- corpus mamillare
- n.oculomotorius III
- fossa interpeduncularis
- pedunculus cerebri
- lamina tecti
- nervus traochlearis
- n.trochlearis
- n.trigeminus
- lingula cerebelli
- pons
- velum medullare anterius
- n.abducens
- striae medullares
- n.facialis et n.vestibu-locochlearis
- trigonum hypoglossi
- trigonum n.vagi
- n.hypoglossus
- n.glossopharyngeus et vagus
- tuberculum nuclei gracilis (clava)
- oliva
- pyramis
- tuberculum cuneatus
- decussatio pyramidum

IX, VIII, VII, VI, V, IV, III, IV, I

53

The posterior funiculi (continued from the spinal cord), which diverge from each other in the medulla oblongata, become the inferior cerebellar peduncles *(pedunculus cerebellaris inferior)*, enclosing a triangular space with the base upwards; between the peduncles lies a very thin membrane, which forms the roof of the fourth ventricle, in its portion belonging to *the medulla oblongata*.

It is the choroid lamina of the fourth ventricle *(lamina choroidea ventriculi quarti)*, covered by the cerebellum. If the cerebellum is removed, this membrane tears and its insertion on the inferior cerebellar peduncles may be seen under the form of a thin band, called *tenia choroidea ventriculi quarti* or *ponticulus*.

The choroid lamina, covered by the deep surface of the pia mater, forms with it the inferior choroid velum, which in the lower angle of the fourth ventricle, between the two cerebellar peduncles, presents *the Magendie's foramen*. Through this aperture, the fourth ventricle communicates with the subarachnoid space.

The inferior cerebellar peduncle continues *the tuberculum cuneatum* and is made up of the white matter extending on both sides to the cerebellum. It is also termed restiform body *(corpus restiforme)*. It is formed of the posterior spinocerebellar (Flechsig's) tract, the vestibulocerebellar tract, the cerebellovestibular tract fibres, the cerebello-olivary tract the olivocerebellar tract, the bulbocerebellar tract and, outside, of white matter fibres, visible to the naked eye, called *dorsal arcuate fibres* (uncrossed parts of the fibres of the neurons situated in the bulbocerebellar tract).

The medulla oblongata has an inferior and a superior extremity.

The inferior extremity or the truncated summit is continuous, without a clear delineation, with the spinal cord.

The superior extremity or the base is separated from the pons by the bulbopontine sulcus, which on the anterior medium line, presents the foramen caecum.

Internal configuration

The internal structure of *the medulla oblongata* differs from that at the level of the spinal cord, because; *the posterior funiculi* disappear gradually; the crossing of the motor and sensory bundles occurs; new elements of white and grey matter appear; the fourth ventricle is formed by widening of the ependymal canal.

As a result appear the following elements: the grey substance of the medulla oblongata is formed of two kinds of nuclei: equivalent nuclei of the spinal cord and nuclei proper (Fig. 44-47, 49-51).

The equivalent nuclei

1) *The equivalent somatomotor nuclei of the spinal anterior horns*, fragmented by the crossing of the pyramidal tract and shifted laterally by the formation of the fourth ventricle, are situated under the floor of the fourth ventricle, on either side of the median line. These are the nucleus of the hypoglossal nerve (XII) and the ambiguus nucleus, which contains motor nuclei, respectively the true origin of the glossopharyngeal (IX), vagus (X) and spinal (XI) nerves.

2) *The equivalent viscerosensory nuclei of the spinal posterior horns*: those which correspond to the base of the posterior horn are the nuclei of the solitary viscerosensory tract situated under the floor of the fourth ventricle, forming the grey wing *(ala cinerea)*, level at which the vegetative sensory fibres of the glossopharyngeal (IX), Wrisberg's intermediate (VII bis) and vagus (X) terminate, those which corespond to the head of the posterior horn are somatosensory and form the nucleus of the descending root of the trigeminal nerve.

3) *The vegetative nuclei* are the dorsal nucleus of the vagus and the inferior salivatory nucleus.

54

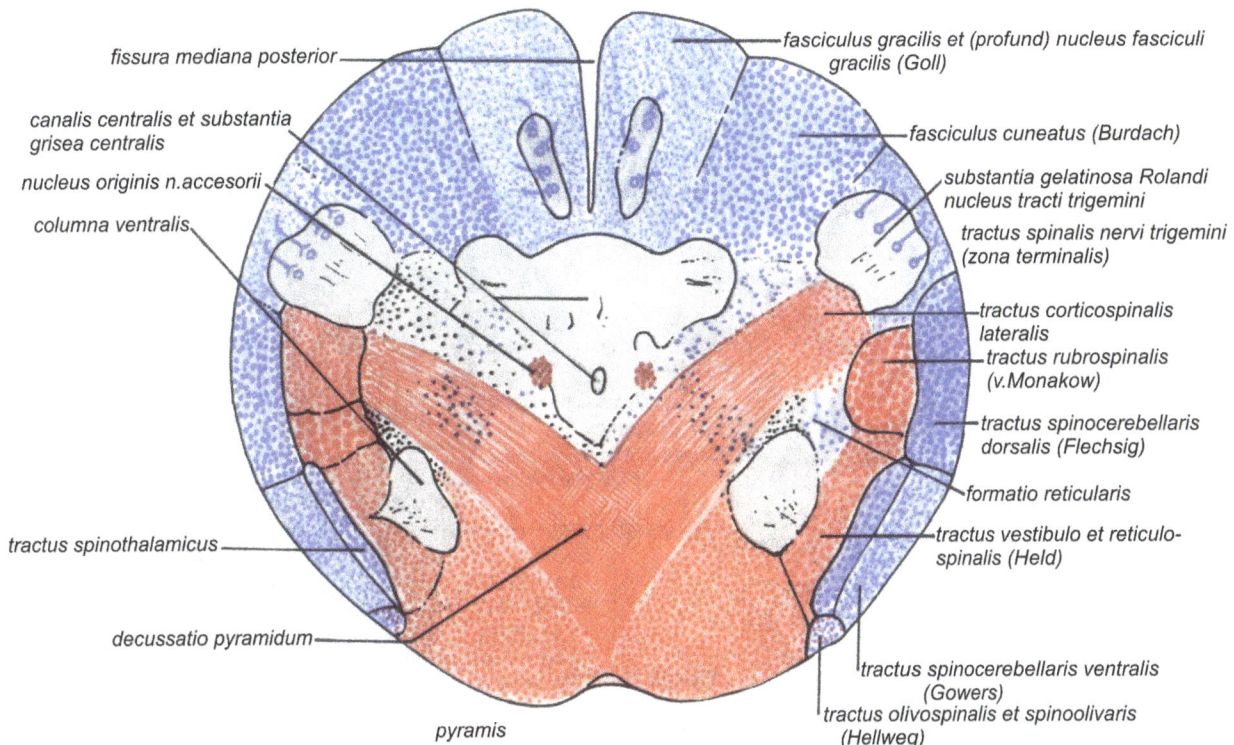

Fig. 44 Section I at the level of the pyramidal decussation (after prof. R. Robacki)

Labels in figure:
- fissura mediana posterior
- canalis centralis et substantia grisea centralis
- nucleus originis n.accesorii
- columna ventralis
- tractus spinothalamicus
- decussatio pyramidum
- pyramis
- fasciculus gracilis et (profund) nucleus fasciculi gracilis (Goll)
- fasciculus cuneatus (Burdach)
- substantia gelatinosa Rolandi nucleus tracti trigemini
- tractus spinalis nervi trigemini (zona terminalis)
- tractus corticospinalis lateralis
- tractus rubrospinalis (v.Monakow)
- tractus spinocerebellaris dorsalis (Flechsig)
- formatio reticularis
- tractus vestibulo et reticulo-spinalis (Held)
- tractus spinocerebellaris ventralis (Gowers)
- tractus olivospinalis et spinoolivaris (Hellweg)

The nuclei proper

1) *The nuclei gracilis (Goll's nucleus) and cuneatus (Burdach's nucleus)*, situated in the lower part of *the medulla oblongata*, are constituted of the second neuron, on the conscious proprioceptive pathway.

Their axons form, after crossing, Reil's median band. Laterally to Goll's and Burdach's nuclei lies the Monakow's nucleus (*accessory cuneate nucleus*, which is also proprioceptive and receives proprieceptive fibres from the head and the neck).

2) *The bulbar oliva* constituted of small polygonal cells, of the aspect of a pouch with a medial opening, is surrounded by a capsule of fibres presenting dorsally a lamina of cells forming the dorsolateral paraoliva and ventrally a similar lamina forming the ventromedial paraoliva. Phylogenetically, the paraolivae and the medial zone of the oliva are more ancient, as they are connected with the paleocerebellum, while the lateral zone of the oliva is more recent, as it is in relationship with the neocerebellum.

The connections of the oliva are *afferent*, represented by the central fasciculus of the calvaria, the tecto-olivary, dento-olivary, cerebello-olivary and medullo-olivary (from the cervical cord) fibres, and *efferent*, represented by the olivo-cerebellar fibres, which arise from each oliva and run to the cerebellar hemisphere or to the dentate nucleus, on the opposite side, where they arrive through the inferior cerebellar peduncle or through the external and internal arcuate fibres.

The main tracts of white fibres which traverse the medulla oblongata are descending or ascending.

The descending tracts are: the pyramidal rubrospinal, vestibulospinal, reticulospinal, tectospinal tracts, the central tract of calvaria and the posterior longitudinal fasciculus (which in the posterior part contains fibres that unite the nuclei of the hypothalamus with the dorsal nucleus of the vagus nerve, forming Schütz' bundle).

The ascending tracts are represented by the median lemniscus -Reil's median band - which contains the axons of the neurons at the level of the Goll's, Burdach's and Monakow's nuclei, fibres of the nucleus of the solitary tract and of the sensory nucleus of the trigeminus (V) after crossing, as well as, situated posterior, the anterior spinothalamic tract, the posterior spinothalamic tract and the posterior longitudinal tract.

55

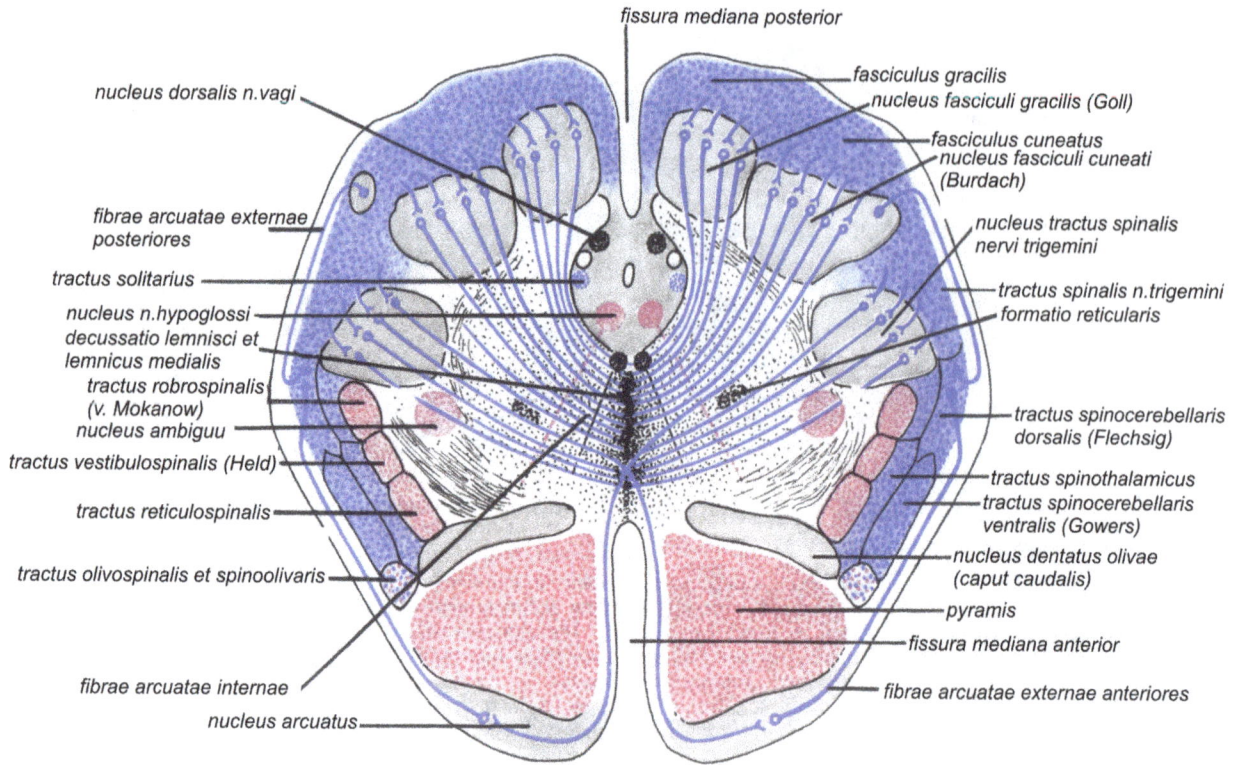

Fig. 45 Section II, immediately below the caudal extremity of the olivae (after prof. R. Robacki)

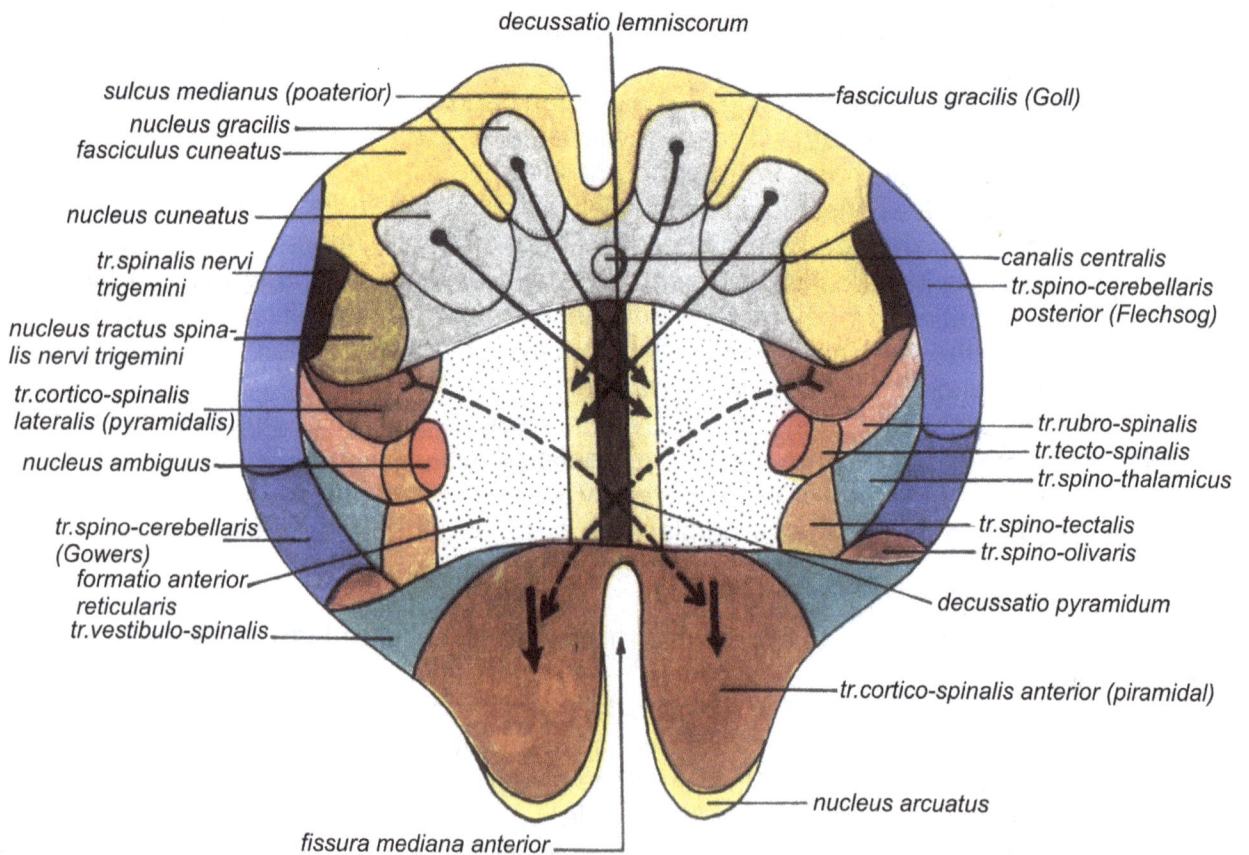

Fig. 46 Medulla oblongata – section at the level of the inferior part, at the boundary with the spinal cord

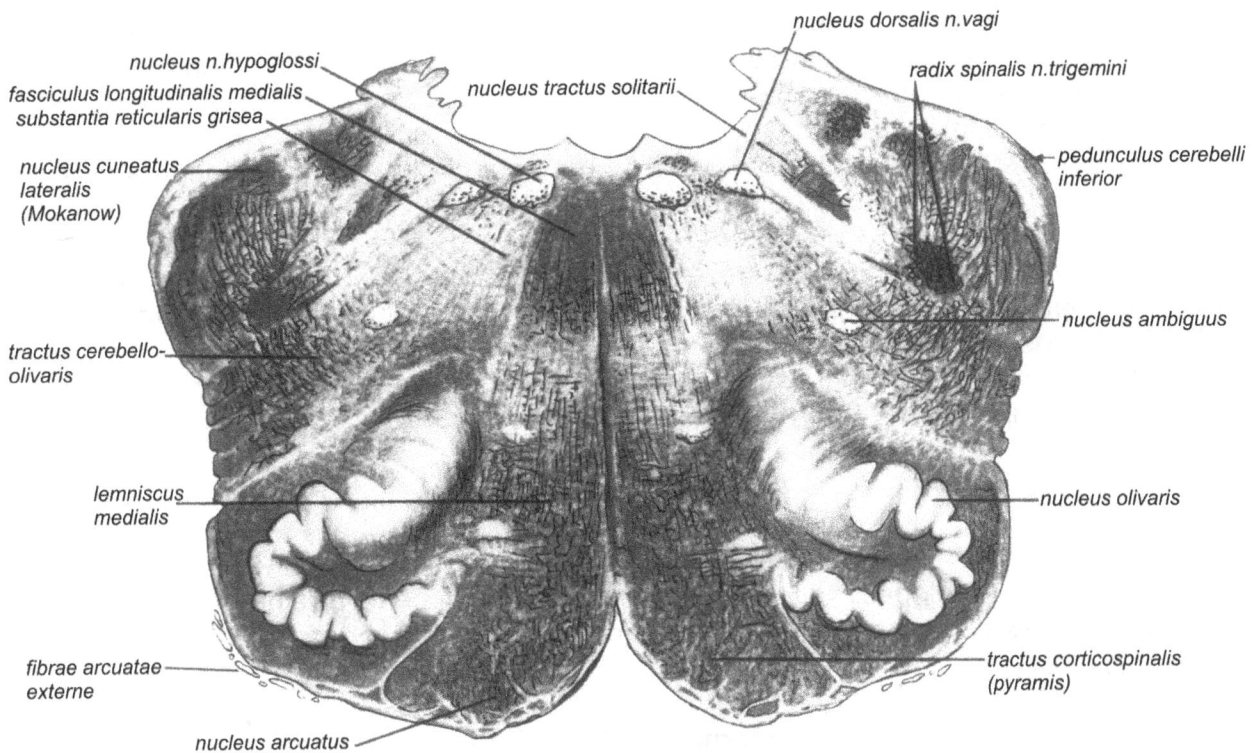

Fig. 47 Medulla oblongata – section at the level of the olivae

We mention that in the medulla oblongata are situated a number of reflex centres of an utmost importance, some of which are described below:

1. *The respiratory (inspiratory and expiratory) centres.* The action on these centres is mainly achieved through the Hering-Breuer reflex, starting from the chemical composition of blood, which excites the inspiratory centre, producing inspiration, after which the dilated pulmonary alveoli elicit nerve impulses through the vagus nerve, that inhibit the inspiratory centre, producing expiration (usually passive); only when the excitations started from the pulmonary alveoli are particularly intense, the inspiratory centre is inhibited and at the same time the expiratory centre is excited. This mechanism operates repeatedly, as the inspiratory and expiratory centres are subordinate to the law of reciprocal innervation. In the regulation of respiration play also a role the excitations related to blood pressure variations, which arrive from the carotid sinus, the carotid body *(glomus caroticum)*, the aortic arch and the aortic body *(glomus aorticum)*, through Hering's nerve, and through Ludwig's depressor nerve, as well as the excitations started from the proprioceptors of the respiratory muscles and of other muscles and joints and, finally, painful and sensory excitations.

2) *The cardioinhibitory and cardioaccelerator centres.* The cardioinhibitory centre is situated at the level of a group of cells of the dorsal nucleus of the vagus, its centripetal and centrifugal fibres coursing mainly via the vagus nerve. It is influenced by the blood pressure receptors arising from the carotid sinus and the aortic arch, which send their nerve impulses via Hering's and Ludwig's nerves. The cardioaccelerator centre, with antagonistic effects, is not yet well delimited topographically; it is especially influenced by the chemical composition of blood and its centripetal and centrifugal fibres are represented by the middle and inferior cervical cardia nerves.

3. *The vagoconstrictive and vasodilative centres*, which maintain and regulate the tone of the smooth muscles in the walls of arteries through the vasomotor nerves. These centres are in relation with the cardioinhibitory and cardioaccelerator centres.

4. *The inferior salivary centre*, from which fibres start through the glossopharingeal nerve and arrive through the small superficial petrosal nerve at the otic ganglion where they form a synapse; from here, the postganglionic fibres run via the auriculotemporal nerve to the parotid gland.

57

5. *The centre of vomiting*, which has afferences from the labyrinth and the abdominal organs and efferences through the vagus nerve, which produces the opening of the cardia and the contraction of the stomach, on the one hand, and through the sympathetic nerve, which brings about the closure of the pyloric sphincter, on the other hand.

6. *The centre of deglutition* situated on the floor of the fourth ventricle, above the respiratory centre. The afferent-efferent pathways are represented by the glossopharyngeal and the vagus nerve.

7. *The centre of cough*. The afferent pathway is represented by the vagus nerve and the efferent pathway, by the nerves of the respiratory muscles and of the constrictor muscles of the vocal folds.

All these reflex centres are situated at the level of the reticulate substance, which still involves numeorus non-elucidated aspects.

The pons
(Pons Varolii)

The pons is developed from the metencephalon and appears as a prominence situated above *the medulla oblongata*, below the cerebral peduncle *(mesencephalon)* and in front of *the cerebellum*.

External configuration

The pons Varolii presents four surfaces: an anterior surface, two lateral surfaces and a posterior surface (fig. 48).

The anterior surface *(pars basilaris pontis)* is the largest portion. It is transversally and vertically convex and is marked by a median groove, in which lies the basilar artery; it presents transverse striations determined by transverse fibres. It is in relationship with the superior part of *the clivus*. It extends up to the posterior surface of *the sella turcica* and measures 3 cm in height and 4 cm in width. On the median line is situated a groove, t*he basilar sulcus*, in which is lodged the basilar artery and laterally are two prominences traversed by *the pyramidal fasciculi (pyramis pontis)*. It is separated from the medulla oblongata through *the bulbopontine sulcus.*

The posterior surface *(pars dorsalis pontis)* presents on its lateral sides, the posterior surface of the superior *(cerebellar peduncles* which, getting gradually closer to each other, unite at the upper extremity of the pons. It is flattened and form the superior portion of the rhomboid fossa. This portion will be described in the chapter dealing with the rhomboid fossa.

The lateral surfaces represent the emergence site of the middle cerebellar peduncles *(crusa pontocerebellares)*, through which the pontocerebellar tracts pass to the cortex of the cerebellar hemispheres.

Internal configuration

The grey matter may be systematized into equivalent nuclei and proper nuclei (fig. 52-54, 56-59).

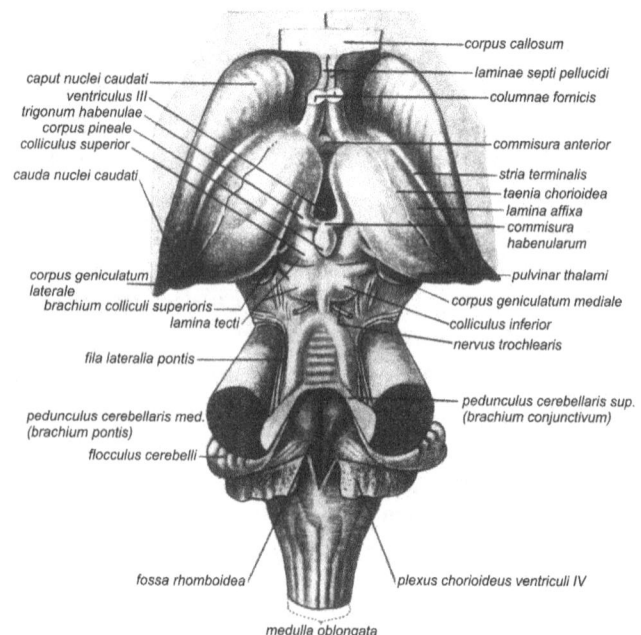

Fig. 48 Brainstem, diencephalon, and caudate nucleus – posterior view

caput nuclei caudati
ventriculus III
trigonum habenulae
corpus pineale
colliculus superior
cauda nuclei caudati
corpus geniculatum laterale
brachium colliculi superioris
lamina tecti
fila lateralia pontis
pedunculus cerebellaris med. (brachium pontis)
flocculus cerebelli
fossa rhomboidea
medulla oblongata

corpus callosum
laminae septi pellucidi
columnae fornicis
commisura anterior
stria terminalis
taenia chorioidea
lamina affixa
commisura habenularum
pulvinar thalami
corpus geniculatum mediale
colliculus inferior
nervus trochlearis
pedunculus cerebellaris sup. (brachium conjunctivum)
plexus chorioideus ventriculi IV

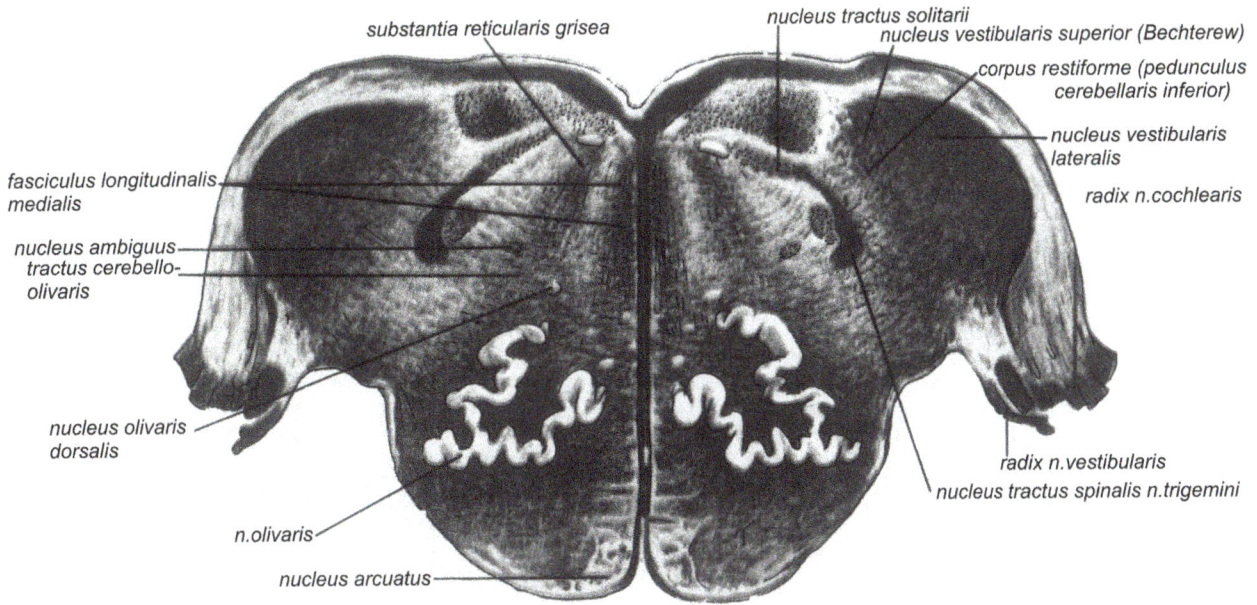

Fig. 49 Medulla oblongata – section at the level of the
vestibular nuclei

substantia reticularis grisea

nucleus tractus solitarii
nucleus vestibularis superior (Bechterew)

corpus restiforme (pedunculus
cerebellaris inferior)

nucleus vestibularis
lateralis

radix n.cochlearis

fasciculus longitudinalis
medialis

nucleus ambiguus
tractus cerebello-
olivaris

nucleus olivaris
dorsalis

radix n.vestibularis
nucleus tractus spinalis n.trigemini

n.olivaris

nucleus arcuatus

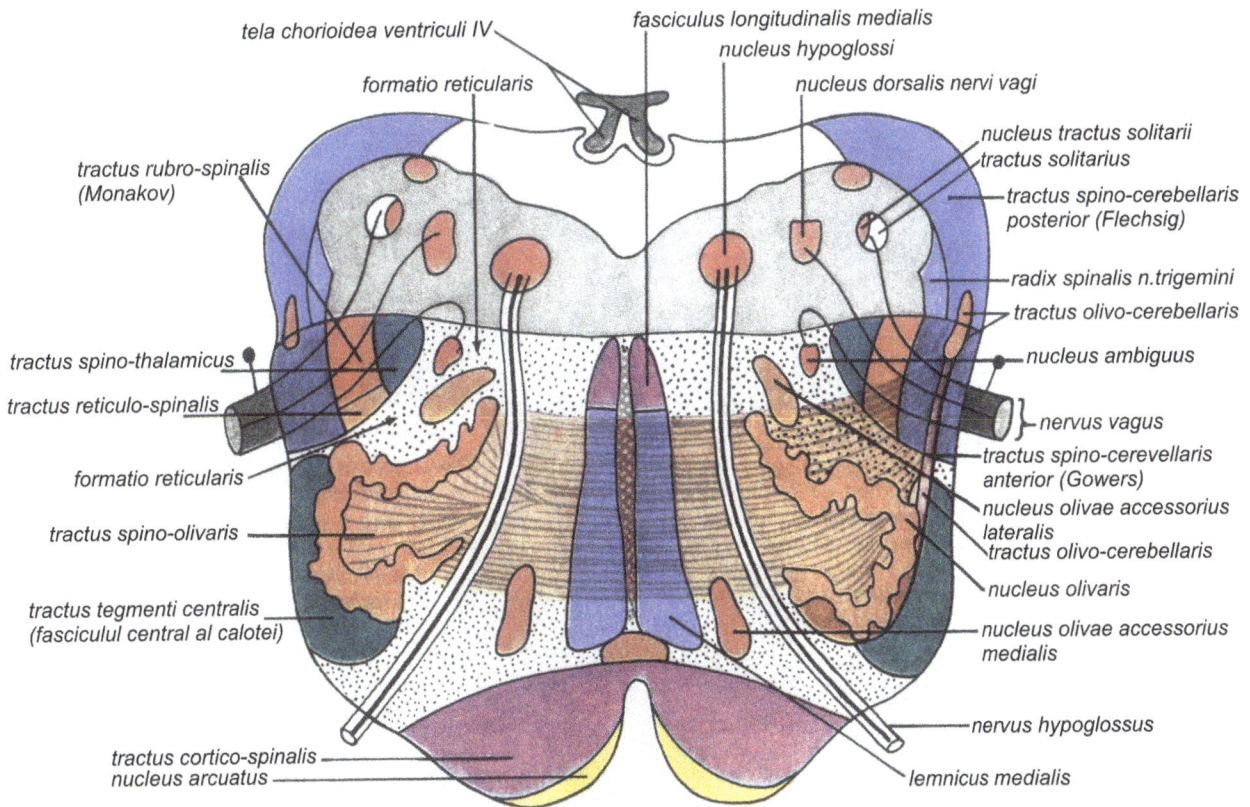

Fig. 50 Medulla oblongata –section at the level of
the olivae

tela chorioidea ventriculi IV

fasciculus longitudinalis medialis
nucleus hypoglossi

formatio reticularis

nucleus dorsalis nervi vagi

nucleus tractus solitarii
tractus solitarius

tractus rubro-spinalis
(Monakov)

tractus spino-cerebellaris
posterior (Flechsig)

radix spinalis n.trigemini

tractus olivo-cerebellaris

tractus spino-thalamicus

nucleus ambiguus

tractus reticulo-spinalis

nervus vagus

formatio reticularis

tractus spino-cerevellaris
anterior (Gowers)

tractus spino-olivaris

nucleus olivae accessorius
lateralis
tractus olivo-cerebellaris

tractus tegmenti centralis
(fasciculul central al calotei)

nucleus olivaris

nucleus olivae accessorius
medialis

nervus hypoglossus

tractus cortico-spinalis
nucleus arcuatus

lemnicus medialis

59

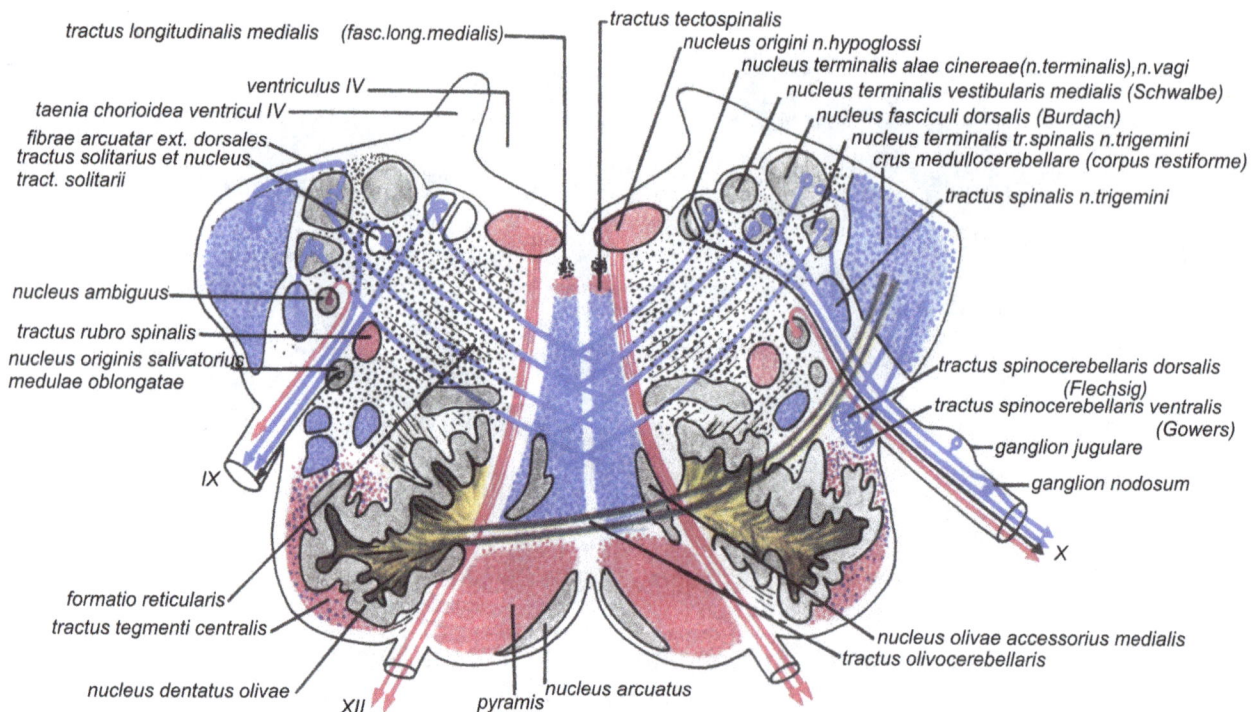

Fig. 51 Section III through the brainstem, at the level of the bulbar oliva (after prof. R. Robacki). On the right side were used the data of several adjacent and serial sections (normally, the glossopharyngeal nerve is not sited at the same level as the vagus nerve).

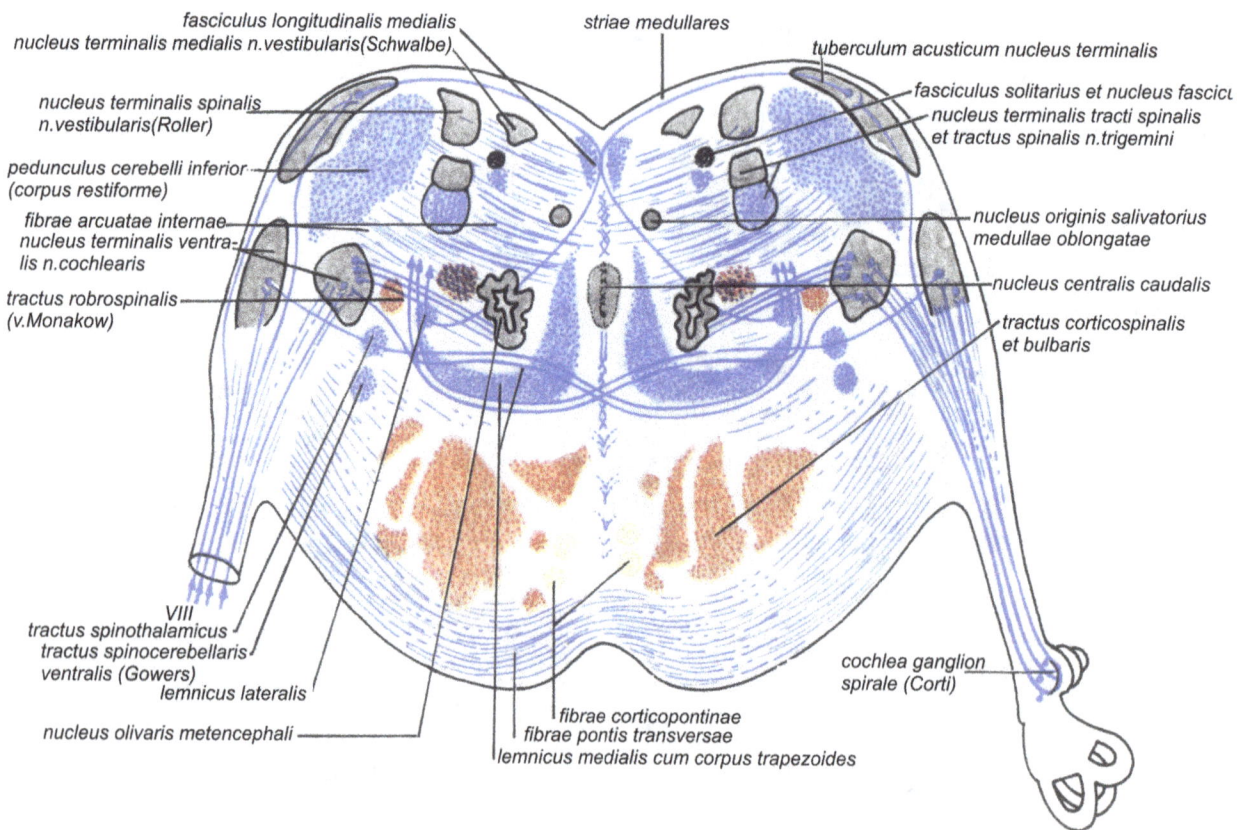

Fig. 52 Section IV through the ponce, at the level of the emergence of the cochlear nerve (VIII) – scheme (after prof. R. Robacki)

Equivalent nuclei

a) *motor nuclei*, represented by:

- the nucleus of the abducent nerve (VI), situated beneath *the eminentia teres*;

- the nucleus of the facial nerve (VII), situated anteriorly and laterally to the nucleus of the abducent nerve;

- the masticatory nucleus of the trigeminal nerve (VI), situated at the level of the superior fovea;

b) *sensitive nuclei*, represented by the sensory nucleus of the trigeminal nerve;

c) *vegetative nuclei*: the superior salivatory and the lacrimal nuclei, situated in the vicinity of the nucleus of the facial nerve.

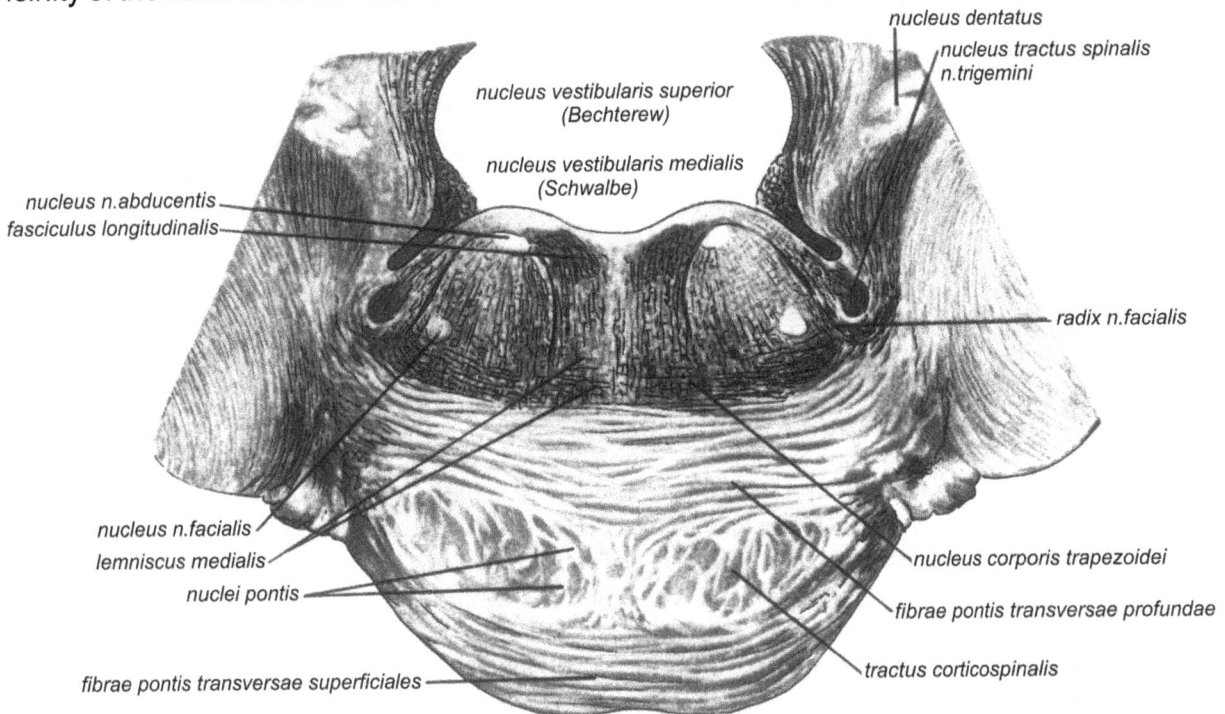

Fig. 53. Pons – section at the level of facial and abducens nerves nucleus

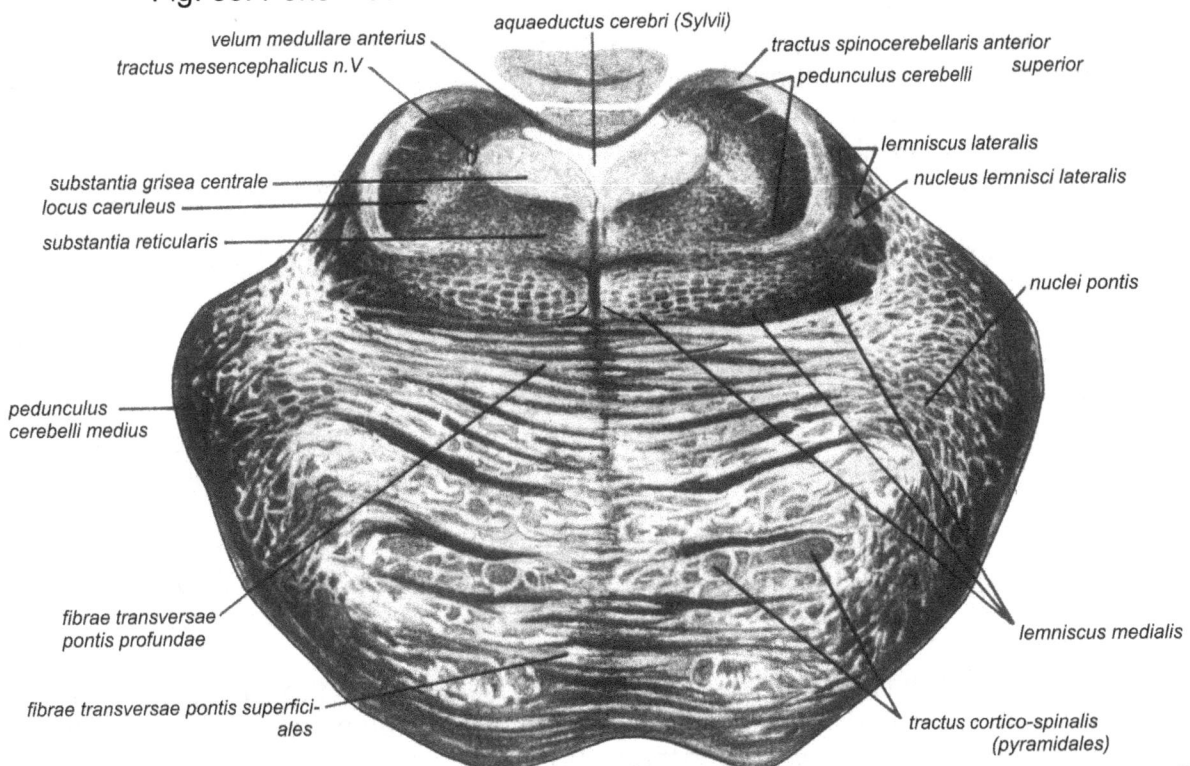

Fig. 54 Pons – section at the level of the superior medullary velum

61

Nuclei proper. In the pons, at the ending of the corticopontine tracts, formed of the frontopontine tract and of the occipitotemporopontine tract, lie the pontine nuclei. The pontocerebellar tract has also its origin in the pontine nuclei. In the pons lies also *the olivary nucleus of the metencephalon*, united with that on the opposite side by a number of transverse fibres, mixed with grey cell masses designated under the name of *trapezoid nuclei*, these formations make up in totality *the trapezoid body*. The region of the pons situated posterior to the trapezoid body, which enters into the structure of the rhomboid fossa, will be again dealt with in the chapter regarding the latter.

The white matter is grouped in descending and ascending tracts.

The ascending tracts are the following:

- the medial lemniscus (medial Reil's band)

- the spinothalamic tracts

- the crossed spinocerebellar tract (Gowers)

- the pontocerebellar tracts.

- the lateral lemniscus (Reil's lateral band), starting from the pontine oliva and possessing fibres which, some of which run towards the interval geniculate body and others to the interior quadrigeminal colliculus.

The descending tracts are:
- the pyramidal tract,
- the corticonuclear tract,
- the cortico-oculocephalo-gyral fibres,
- the corticopontine tract,
- the rubrospinal tract,
- the central tract of the tegmentum,
- the posterior longitudinal tract,
- the tectospinal tract,
- the hypothalamovagal tract (Schütz)

We mention also the more important reflex centres and their main pathways; these centres are situated at the level of the pontine reticular substance. They are the following:

1. *Lacrimal nucleus.* The lacrimal reflex has the afferent pathway represented by the trigeminal nerve, which brings the excitations to the lacrimal nucleus, while the efferent pathway, through Wrisberg's intermediate nerve (VII), brings them to the geniculate ganglion, from which through the greater superficial petrosal nerve, the pathway reaches the sphenopalatine ganglion, where it forms a synapse and then, through the lacrimal nerve, it courses to the lacrimal gland.

2. *The superior salivatoriy nucleus*, from which the excitations are transmitted through the chorda tympani, then through the lingual nerve, to the submaxillary and sublingual ganglia, where the synapse is formed, after which the postganglionic fibres reach the submaxillary and sublingual glands.

3. *The centre of the corneal reflex.* The afferent pathway is represented by the trigeminal reflex and the efferent pathway, by the facial nerve.

4. *The centre of the masseter reflex.* The afferent pathway is the sensory trigeminal nerve and the efferent pathway, the motor trigeminal nerve.

5. *The pneumotoxic centre* is excited by the inspiratory bulbar centre and, on its turn, the pneumotoxic centre excites the expiratory bulbar centre, being thus involved, in the case of hyperpnea, especially in the respiratory automatism.

We mention that the acousticovestibular nuclei, situated in the pontobulbar part of the floor of the fourth ventricle will be studied in the chapter regarding the acousticovestibular pathways.

The mesencephalon
(Mesencephalon)

The mesenchephalon or midbrain develops subsequently to the changes occurred in the evolution of the intermediate cerebral vesicle and is made up of the cerebral peduncles, the quadrigeminal tubercles *(colliculi)* and their *anexae*.

It is traversed by the aqueduct of Sylvius or cerebral aqueduct, which represents the ependymal cavity of the intermediate primary cerebral vesicle and connects the fourth with the third ventricle. It is separated from the pons by the pontopeduncular groove visible on the inferior surface of the neuraxis, and extends above up to the optic tracts (the boundary from the diencephalon and the inferior border of the mamillary bodies); posteriorly, it extends from the inferior border of the quadrigeminal lamina up to the root of the pineal body.

External configuration

It is continuous with the medulla oblongata and the pons, and oriented upwards and forwards; it has an anterior, a posterior and two lateral surfaces.

The anterior surface presents, on each side, two bulky bundles of white matter, called *cerebral peduncles.*

The cerebral peduncles (pedunculi cerebri) are cords of the thickness of the thumb, 1.5 cm in length and width, formed of white matter. They have their origin at the upper border of the pons, then diverge nearly at a right angle, coursing upwards, above the optic tracts, which cross their inferior surface. The triangular depression, with the base upwards, which lies between the two peduncles, is termed interpeduncular space - *fossa interpeduncularis* -, which extends anteriorly up to the mamillary bodies. The bottom of the depression is crossed by the vessels that penetrate into it and are designated under the name of posterior perforated substance. In the cerebral peduncles lie the tracts which come from the cortex (*pyramidal tracts and corticopontine tracts*, formed of the frontopontine and occipitotemporopontine tracts).

The boundary between the peduncles and the posterior perforated space is marked by two grooves: the lateral sulcus of the mesencephalon *(sulcus lateralis mesencephali)* and the medial sulcus *(medialis sulcus)* called also sulcus of the oculomotor nerve. The boundary between the tegmentum and the lamina of the mesencephalic tectum *(lamina tecti)* is represented by a conventional line passing through the cerebral aqueduct.

On **the posterior surface** are situated the quadrigeminal tubercles *(colliculi laminae tecti quadrigeminae)*. The posterior surface extends between the superior cerebellar peduncles and *the thalamus*, from which it is separated by a groove. The quadrigeminal tubercles are hemispherical eminences, four in number, two on each side, one superior and the other inferior. Between the superior colliculi lies the pineal body *(epiphysis)*. The inferior colliculi are covered by the cerebellum.

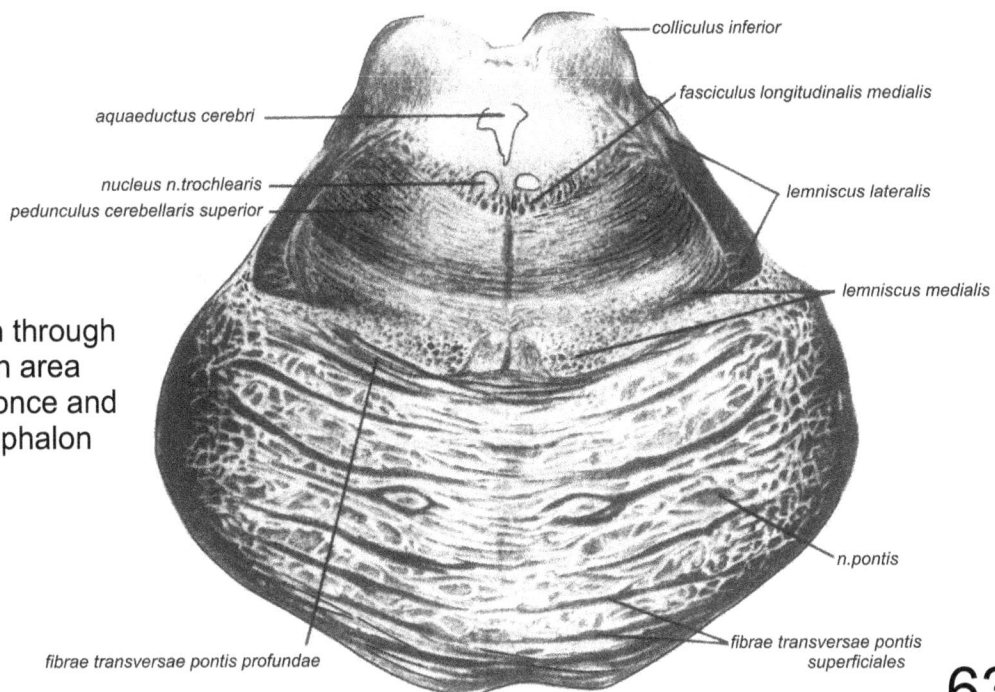

Fig. 55 Section through the transition area between the ponce and the mesencephalon

- colliculus inferior
- fasciculus longitudinalis medialis
- aquaeductus cerebri
- nucleus n.trochlearis
- pedunculus cerebellaris superior
- lemniscus lateralis
- lemniscus medialis
- n.pontis
- fibrae transversae pontis profundae
- fibrae transversae pontis superficiales

63

Each colliculus contains grey matter. In the superior colliculi the connection of the reflex optic pathway occurs. They are made up of seven layers of grey matter alternating with white matter and are an archaic integration centre. In the nucleus of the inferior colliculi occurs the connection of the reflex auditory pathway. From both colliculi (superior and inferior) start the tectospinal tract. From the superior colliculus starts a ridge, brachium of the superior colliculus *(brachium colliculi superioris)*, which passes upwards and laterally to the lateral geniculate body. From the inferior colliculus arises the brachium of the inferior colliculus *(brachium colliculi inferioris)*, which courses upwards and laterally, ending in the medial geniculate body.

Each of **the lateral surfaces** presents an oblique groove, oriented forward and outward, termed lateral sulcus, which continues the interpeduncular groove and which, at the level of the pons, separates the middle from the inferior cerebellar peduncles.

Internal configuration

On a transverse section at the level of the mesencephalon, three parts may be observed (Fig. 55, 59, 63):

- a posterior-tectal-part, formed of *the quadrigeminal colliculi*;
- a middle part - termed *mesencephalic tegmentum* and
- an anterior part - represented by *the cerebral peduncles*.

A. **The posterior-tectal-part** has been described at the chapter „External configuration".

B. **The middle part-tegmentum** - has a very complex structure, being made up of a mass of grey matter which surrounds the cerebral aqueduct.

Fig. 56 Section V, through the ponce, at the level of the origin nuclei of the vestibular nerve and the eminentiatores (medial eminence, nucleus of the abducent nerve) – scheme (after R. Robacki)

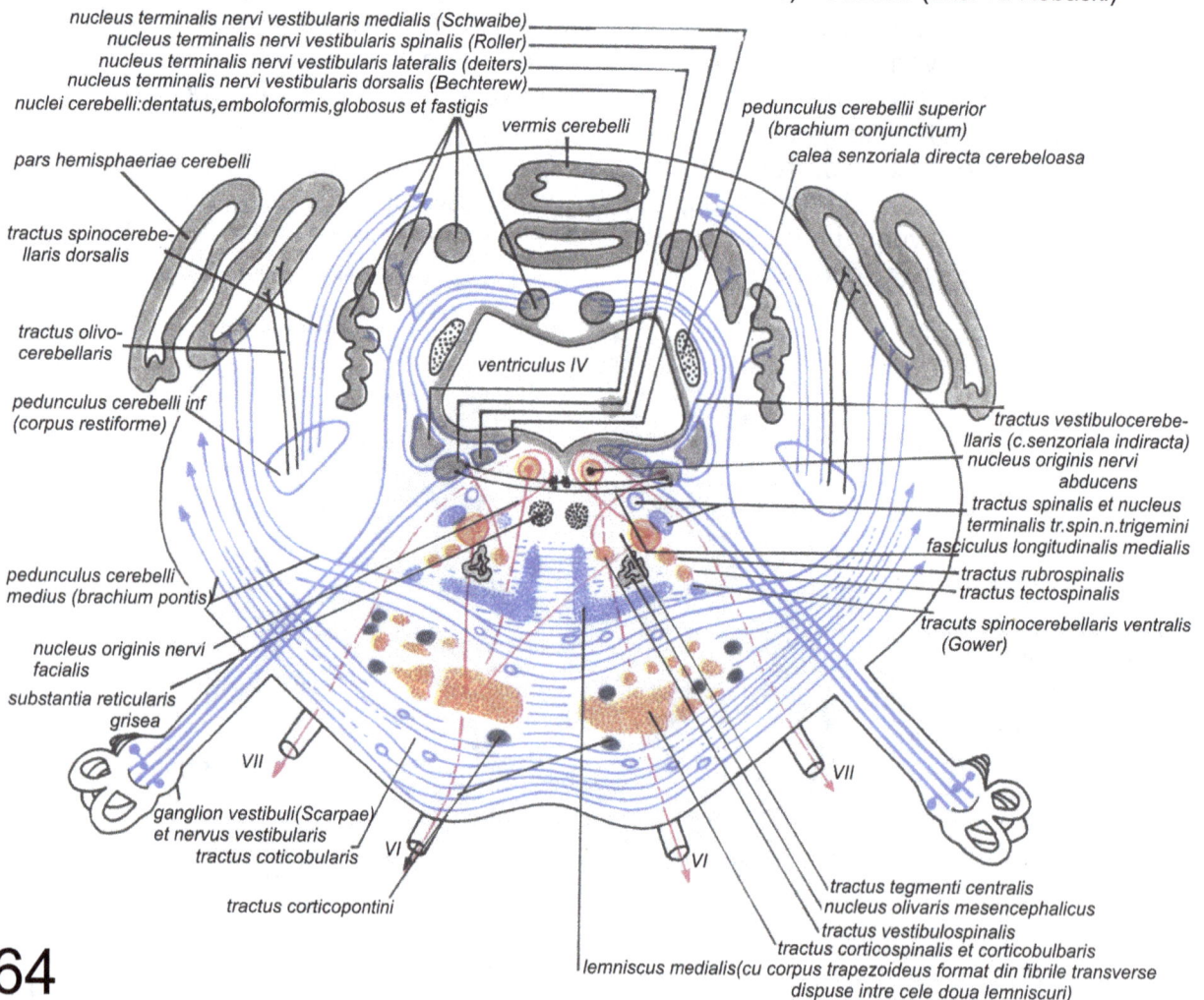

nucleus terminalis nervi vestibularis medialis (Schwaibe)
nucleus terminalis nervi vestibularis spinalis (Roller)
nucleus terminalis nervi vestibularis lateralis (deiters)
nucleus terminalis nervi vestibularis dorsalis (Bechterew)
nuclei cerebelli:dentatus,emboloformis,globosus et fastigis
vermis cerebelli
pedunculus cerebellii superior (brachium conjunctivum)
calea senzoriala directa cerebeloasa
pars hemisphaeriae cerebelli
tractus spinocerebe-llaris dorsalis
tractus olivo-cerebellaris
ventriculus IV
pedunculus cerebelli inf (corpus restiforme)
tractus vestibulocerebe-llaris (c.senzoriala indiracta)
nucleus originis nervi abducens
tractus spinalis et nucleus terminalis tr.spin.n.trigemini
fasciculus longitudinalis medialis
pedunculus cerebelli medius (brachium pontis)
tractus rubrospinalis
tractus tectospinalis
tracuts spinocerebellaris ventralis (Gower)
nucleus originis nervi facialis
substantia reticularis grisea
VII
VII
ganglion vestibuli(Scarpae) et nervus vestibularis
tractus coticobularis
VI
VI
tractus corticopontini
tractus tegmenti centralis
nucleus olivaris mesencephalicus
tractus vestibulospinalis
tractus corticospinalis et corticobulbaris
lemniscus medialis(cu corpus trapezoideus format din fibrile transverse dispuse intre cele doua lemniscuri)

64

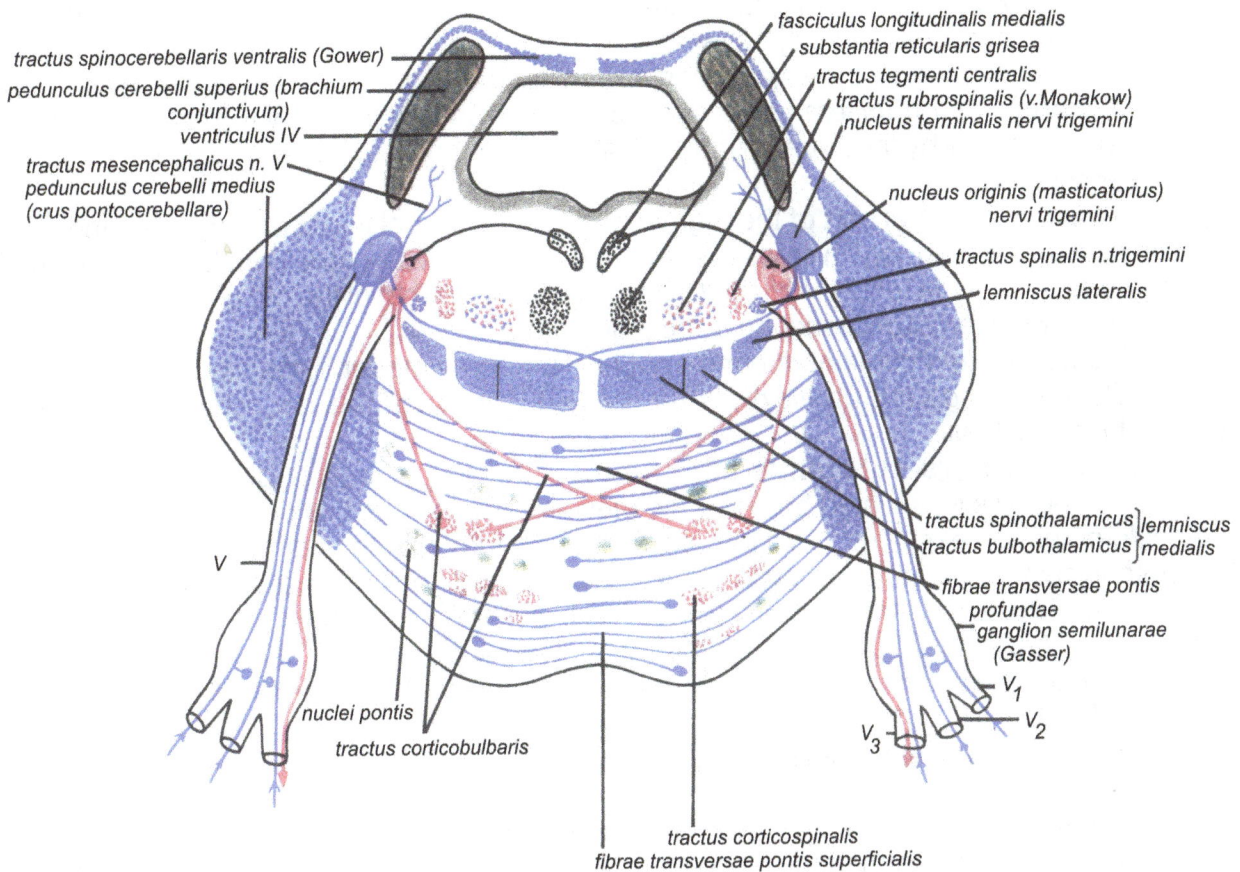

Fig. 57 Transverse section VI, through the ponce, at the level of the emergence of the trigeminal nerve – scheme (after prof. R. Robacki)

Labels (Fig. 57):
tractus spinocerebellaris ventralis (Gower)
pedunculus cerebelli superius (brachium conjunctivum)
ventriculus IV
tractus mesencephalicus n. V
pedunculus cerebelli medius (crus pontocerebellare)
V
nuclei pontis
tractus corticobulbaris
fasciculus longitudinalis medialis
substantia reticularis grisea
tractus tegmenti centralis
tractus rubrospinalis (v.Monakow)
nucleus terminalis nervi trigemini
nucleus originis (masticatorius) nervi trigemini
tractus spinalis n.trigemini
lemniscus lateralis
tractus spinothalamicus }lemniscus
tractus bulbothalamicus }medialis
fibrae transversae pontis profundae
ganglion semilunarae (Gasser)
V1
V2
V3
tractus corticospinalis
fibrae transversae pontis superficialis

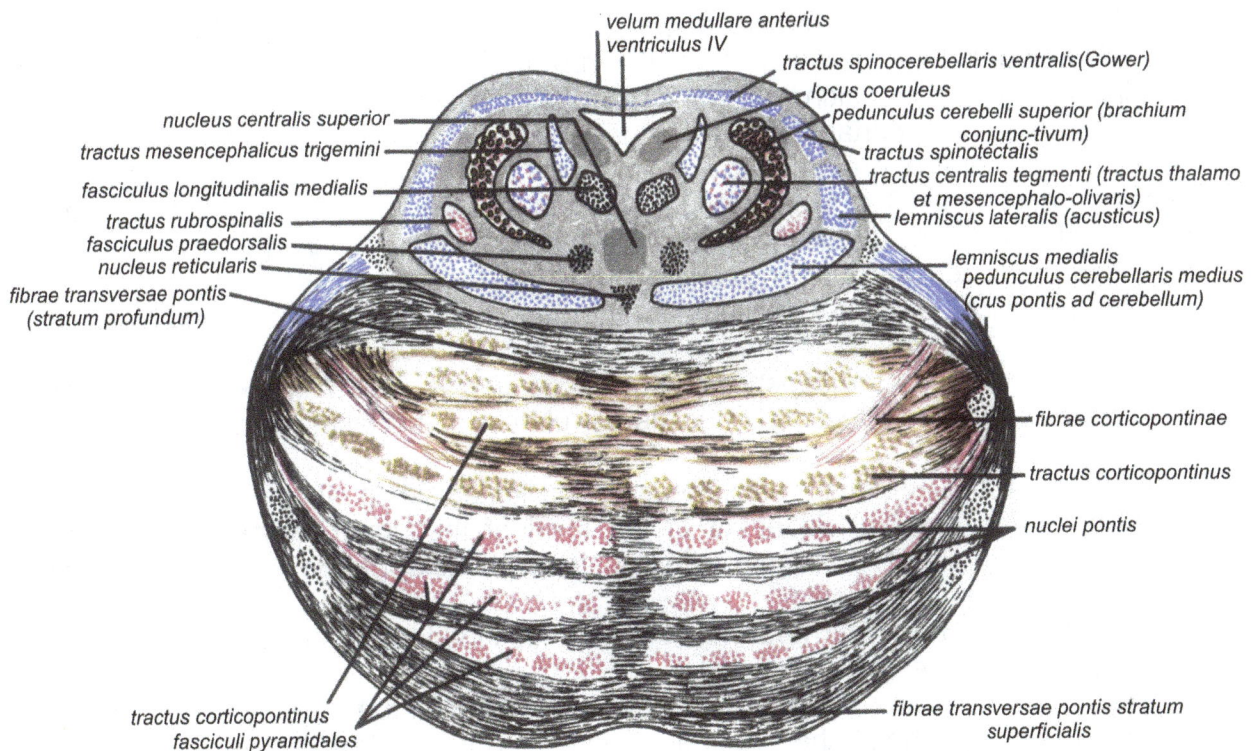

Fig. 58 Section VII, through the ponce, in front of the emergence of the trigeminal nerve (after prof. R. Robacki)

Labels (Fig. 58):
velum medullare anterius
ventriculus IV
nucleus centralis superior
tractus mesencephalicus trigemini
fasciculus longitudinalis medialis
tractus rubrospinalis
fasciculus praedorsalis
nucleus reticularis
fibrae transversae pontis (stratum profundum)
tractus spinocerebellaris ventralis (Gower)
locus coeruleus
pedunculus cerebelli superior (brachium conjunc-tivum)
tractus spinotectalis
tractus centralis tegmenti (tractus thalamo et mesencephalo-olivaris)
lemniscus lateralis (acusticus)
lemniscus medialis
pedunculus cerebellaris medius (crus pontis ad cerebellum)
fibrae corticopontinae
tractus corticopontinus
nuclei pontis
tractus corticopontinus
fasciculi pyramidales
fibrae transversae pontis stratum superficialis

65

Through the tegmentum ascend towards the cortex the sensitive pathways: the bulbothalamic tract, the nucleothalamic tract *(lemniscus medialis)*, the spinothalamic tract and the lateral lemniscus (auditory pathway).

These tracts lie on the lateral part of *the tegmentum*, close to the surface, in the lemniscal trigone *(trigonum lemnisci)*. This trigone is bounded above by the brachium of the inferior collicle, below by the superior cerebellar peduncle and laterally by the lateral sulcus of the mesencephalon.

In the anterolateral part of the tegmentum lies the lateral lemniscus (Reil's band), the rubroreticulospinal tract, which crosses at this level, forming the ventral Forel's decussation of the tegmentum, and the tectospinal tract, which also crosses that of the opposite side, forming the posterior (Meynert's) decussation of the tegmentum.

The tegmentum contains a number of nuclei of grey matter locus niger or *substantia nigra* or *Soemmerring's substantia nigra*, red nucleus, interstitial nucleus, oculomotor nucleus and trochlear nucleus.

Soemmerring's substantia nigra *(locus niger)* is a mass of pigment (melanin) - containing nerve cells, situated between the mesencephalic lateral sulcus and the medial (of the oculomotor nerve) sulcus. On a transverse section through the peduncles it appears as a macroscopically visible band owing to its dark colour. Functionally, this nucleus is a significant constituent of the extrapyramidal system. As it is known, the substantia nigra has significant connections through its fibres with the striate body, but the direction of these fibres is not precisely defined. In addition, in this nucleus end probably also the fibres coming from the spinal cord.

Locus niger receives afferent fibres also from the cerebral cortex of the precentral area (Bechterew).

The efferent fibres either follow the direction of the pyramidal tract towards protuberantial or bulbar formations, or course towards the red nucleus (Foix and Niculescu).

The red nucleus of the tegmentum *(nucleus ruber tegmenti)* appears macroscopically on fresh preparations, owing to its reddish colour. It has a rounded shape and is situated posterior to *the substantia nigra* and anterior to the nucleus of the oculomotor nerve. The red nucleus is made up of two portions: *the paleorubrum*, respectively the magnocellular portion, from which starts the rubrospinal tract, and *the neorubrum*, the parvocellular portion, from which the central tegmental tract starts to the bulbar oliva. The red nucleus represents an important structure of the extrapyramidal system. Here end the striato-rubral, cerebelo-rubral and mamilo-tegmental tracts and the rubro-reticulo-spinal and the rubro-olivar tracts begin. The rubro-reticulo-spinal tract goes down the opposite side of the spinal cord, after it cris-crosses at the level of Forel's decussation.

Likewise, the red nucleus sends efferences to the thalamus.

The afferences of the red nucleus are the dentorubral, corticorubral and nigrorubral tracts.

The intersection of the superior cerebellar peduncles - the Wernekick decussation- which consists of the intersection of the dento-rubrical paths takes place under the red nucleai.

The interstitial nucleus is also a precise and limited site of the reticular substance. It lies in front of the cerebral aqueduct at the level of the upper border of the inferior quadrigeminal tubercle. Here begins the posterior bundle of t*he medial longitudinal fasciculus*.

The nucleus of the trochlear nerve will be described in the chapter dealing with the nuclei of the cranial nerves.

The nucleus of the oculomotor nerve. The nuclei of the oculomotor nerves (third pair of cranial nerves) lie below the cerebral aqueduct, on either side of the median line. These nuclei are made up of nerve formations which innervate the extrinsic muscles of the eyeball (all the muscles of the eyeball, save the superior oblique and the external rectus muscle), as well as the intrinsic muscles of the eyeball (the Edinger-Westphal nucleus) and the Perlia
nucleus which assure the function of eyeball convergence.

C. **The peduncular -anterior -part** contains, from the medial to the lateral area, the following tracts: the corticonuclear tract mixed with fibres of the frontopontine tract, the pyramidal tract and the temporo-pontine (Türk-Meynert's) tract; there are also cortico-oculocephalogyral fibres.

Among the reflexes of a particular significance which have the centre in the mesencephalon we mention the light reflex and the convergence reflex.

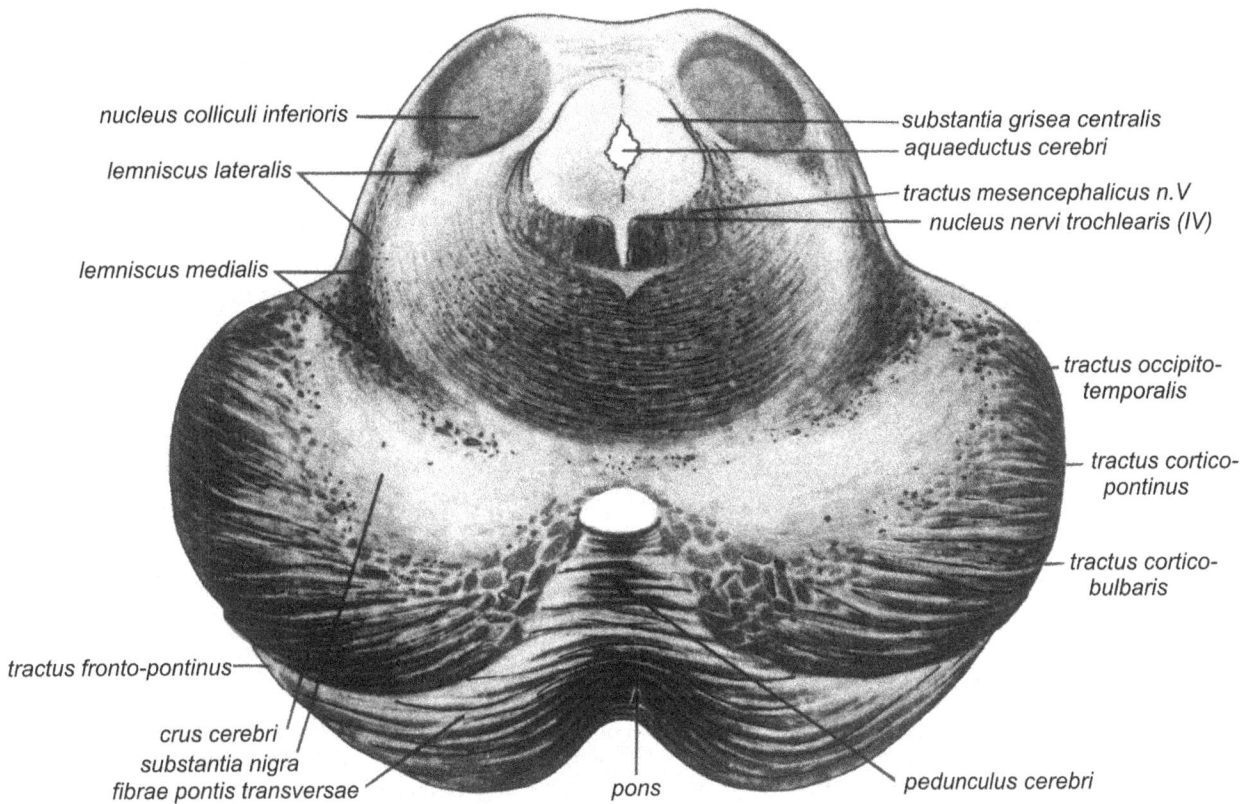

Fig. 59 Mesencephalon – section at the level of the nucleus of the inferior colliculus

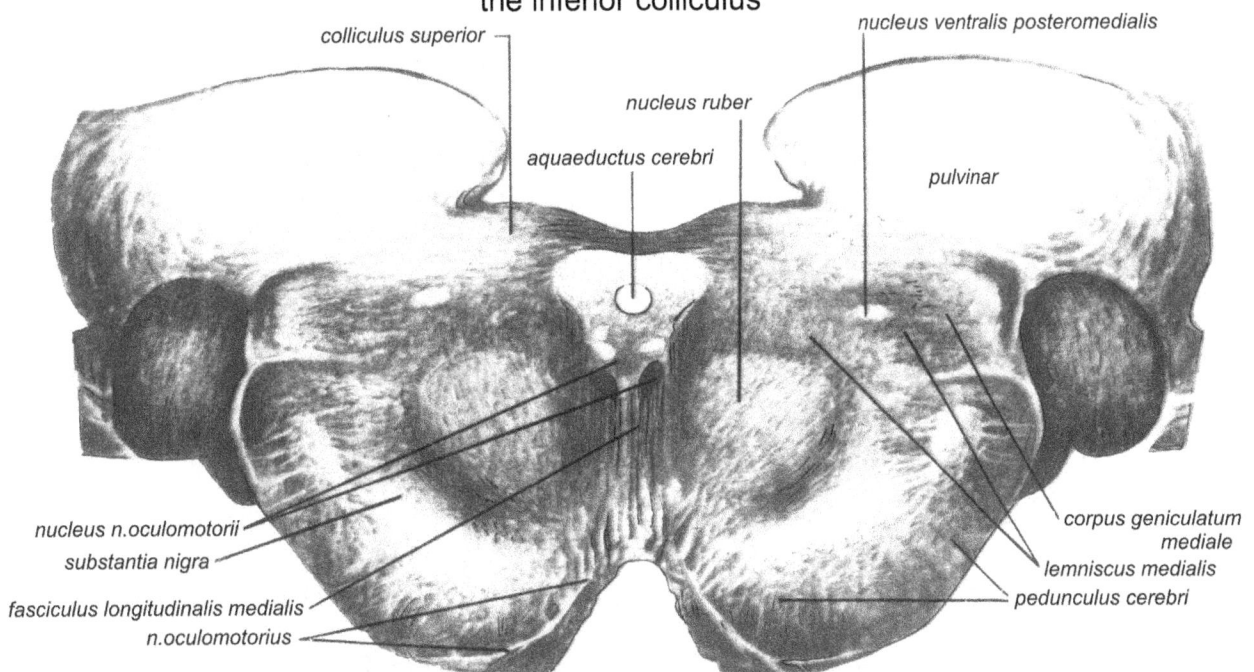

Fig. 60 Mesencephalon – section at the level of the superior colliculus

67

Light reflex. The afferent pathways are represented by the fibres of the optic nerves, by the optic tracts and by fibres reaching the transition (pretectal) area between the thalamus and the mesencephalon, where a first synapse is formed, from which the fibres course to the Edinger-Westphal nucleus, where a second synapse is formed; from here the fibres run via the oculomotor nerve to the ciliary ganglion and then to the iris, which they contract.

The lesion of these fibres leads to the appearance of the Argyll Robertson sign, respectively to the loss of the light reflex, with conservation of the convergence reflex.

The convergence reflex consists in a reflex diminishing of the pupils for viewing near objects. The excitations for this reflex reach the Perlia's nucleus, then the Edinger-Westphal nucleus and the ocular internal rectus muscles. In this way occurs the diminishing of the pupils and the convergent contraction of the internal rectus muscles. The lesion of Perlia's nucleus causes the loss of this reflex.

The brainstem - connections
The equivalent nuclei, the reticular substance

Clinically, the medulla oblongata, the pons and *the mesencephalon* form a morphofunctional unit, which connects the spinal cord with the cerebral hemispheres and which is termed brainstem, since it bears the cerebral hemispheres like the trunk of a tree which bears the crown of the latter and at the same time the cerebellum through the three pairs of cerebellar peduncles *(superior, middle and inferior cerebellar peduncles)*.

For the purpose of *didactic systematization* we mention the (ascending and descending) connections, the nuclei (sensory and motor), the own structure and those of the brainstem.

Connections

Ascending connections
-*The medial lemniscus* (Reil's band) starts from *the nucleus gracilis* and from *the cuneate nucleus*, on the posterior surface of the medulla oblongata and after its fibres cross in the medulla oblongata, it passes into the pons, where it joins the lateral lemniscus (the auditory pathway). The medial lemniscus ends in the thalamus and the lateral lemniscus, in the medial geniculate body.

- *The spinothalamic tracts* (previously described).
- *The spinocerebellar tracts* (previously described).

Descending connections
- *The pyramidal tract.* At the level of the cerebellar peduncles are sited - in the medial fifth the corticofrontopontine (Arnold) tract *(tractus corticofrontopontinus)*, in the medial portion is situated the pyramidal tract, the fibres of which cross in 80% of cases in the medulla oblongata; in the external fourth courses the corticoparieto-temporopontine Türk-Meynert's tract *(tractus temporopontinus)*, which ends in the nuclei of the pons, from which start the pontocerebellar tracts;

- *The rubrospinal tract* crosses on the midline in Forel's decussation (decussatio tegmentalis ventralis), descends and ends in the spinal cord (synapse with the alpha-motoneurons);

- *The rubro-olivary, tectospinal, reticulospinal tracts* and *Russel's hooked bundle (tractus cerebellovestibularis)* (previously described).

Association connections
- *The posterior longitudinal fasciculus (fasciculus longitudinalis medialis)* connects the vestibular nuclei, the motor nuclei of the eyebulb and those lying in the anterior horns of the cervical cord.

- *The fasciculus of Schütz (fasciculus longitudinalis dorsalis)* connects the hypothalamic nuclei of the dorsal nucleus of the vagus in the medulla oblongata.

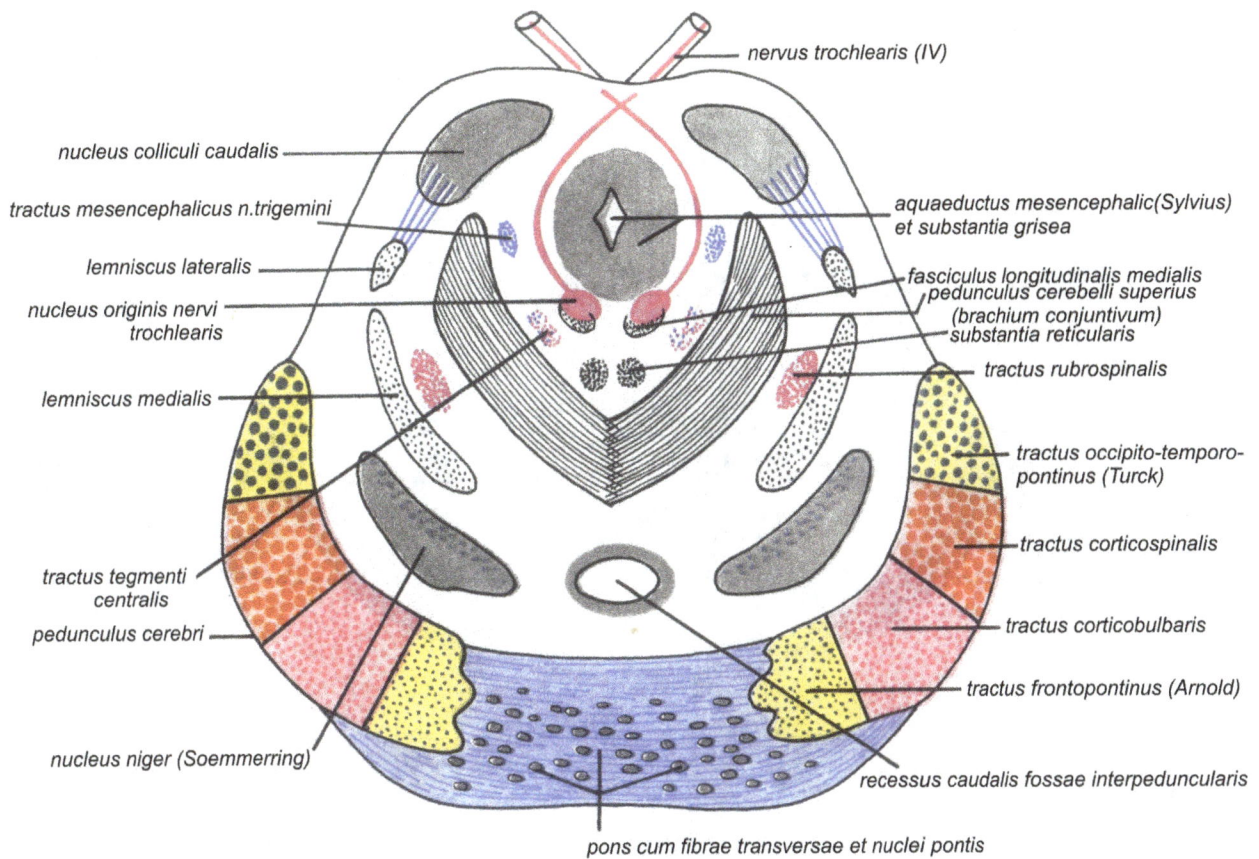

Fig. 61 Section VIII, through the quadrigeminal (inferior) colliculus, achieved through the summation of several adjacent sections (after prof. R. Robacki). The trochlear nerve appears after its decussation on a caudal section – scheme.

nervus trochlearis (IV)

nucleus colliculi caudalis

tractus mesencephalicus n.trigemini

lemniscus lateralis

nucleus originis nervi trochlearis

lemniscus medialis

tractus tegmenti centralis

pedunculus cerebri

nucleus niger (Soemmerring)

aquaeductus mesencephalic(Sylvius) et substantia grisea

fasciculus longitudinalis medialis

pedunculus cerebelli superius (brachium conjuntivum)

substantia reticularis

tractus rubrospinalis

tractus occipito-temporo-pontinus (Turck)

tractus corticospinalis

tractus corticobulbaris

tractus frontopontinus (Arnold)

recessus caudalis fossae interpeduncularis

pons cum fibrae transversae et nuclei pontis

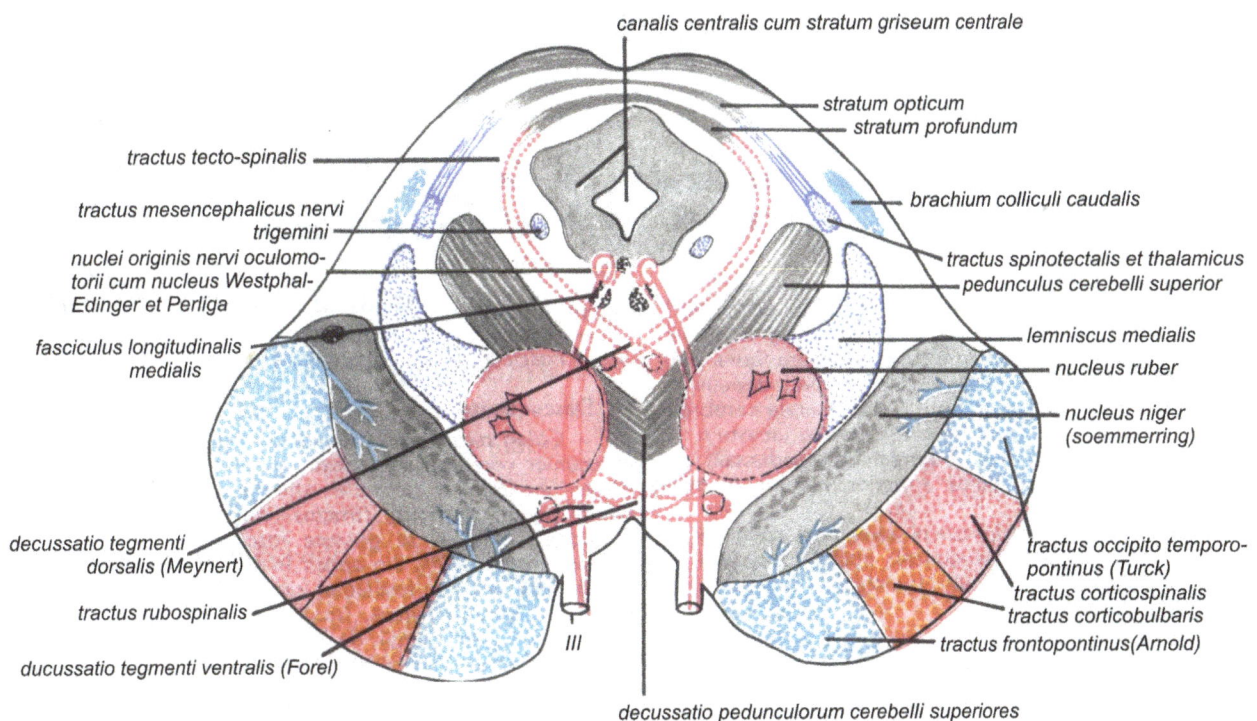

Fig. 62 Section IX, through the mesencephalon, at the level of the superior colliculi (after prof. R. Robacki)

canalis centralis cum stratum griseum centrale

stratum opticum

stratum profundum

tractus tecto-spinalis

tractus mesencephalicus nervi trigemini

nuclei originis nervi oculomotorii cum nucleus Westphal-Edinger et Perliga

fasciculus longitudinalis medialis

decussatio tegmenti dorsalis (Meynert)

tractus rubospinalis

ducussatio tegmenti ventralis (Forel)

brachium colliculi caudalis

tractus spinotectalis et thalamicus

pedunculus cerebelli superior

lemniscus medialis

nucleus ruber

nucleus niger (soemmerring)

tractus occipito temporo-pontinus (Turck)

tractus corticospinalis

tractus corticobulbaris

tractus frontopontinus(Arnold)

III

decussatio pedunculorum cerebelli superiores

69

Fig. 63 Mesencephalon – section at the level of the red nucleus - scheme

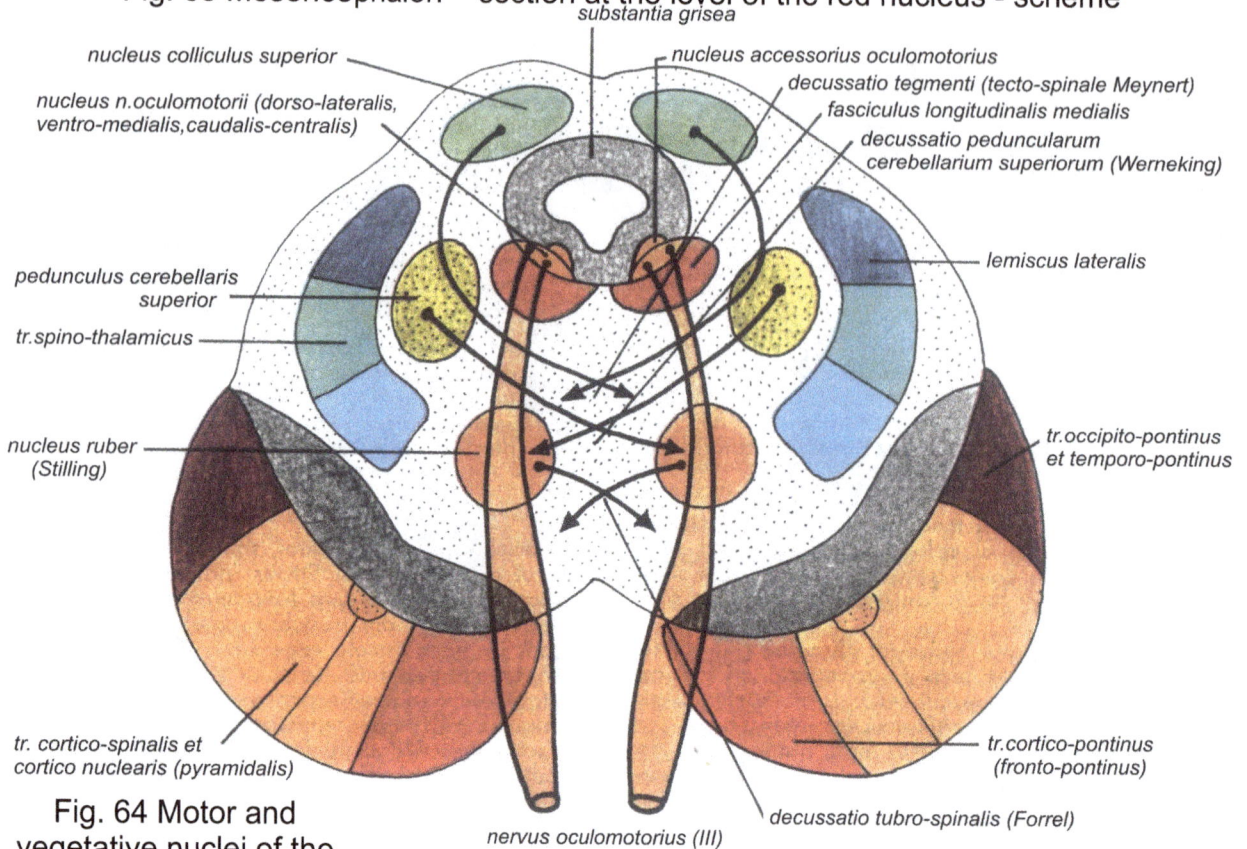

substantia grisea

nucleus colliculus superior

nucleus n.oculomotorii (dorso-lateralis, ventro-medialis, caudalis-centralis)

nucleus accessorius oculomotorius

decussatio tegmenti (tecto-spinale Meynert)

fasciculus longitudinalis medialis

decussatio peduncularum cerebellarium superiorum (Werneking)

pedunculus cerebellaris superior

tr.spino-thalamicus

lemiscus lateralis

nucleus ruber (Stilling)

tr.occipito-pontinus et temporo-pontinus

tr. cortico-spinalis et cortico nuclearis (pyramidalis)

tr.cortico-pontinus (fronto-pontinus)

decussatio tubro-spinalis (Forrel)

nervus oculomotorius (III)

Fig. 64 Motor and vegetative nuclei of the brainstem - scheme

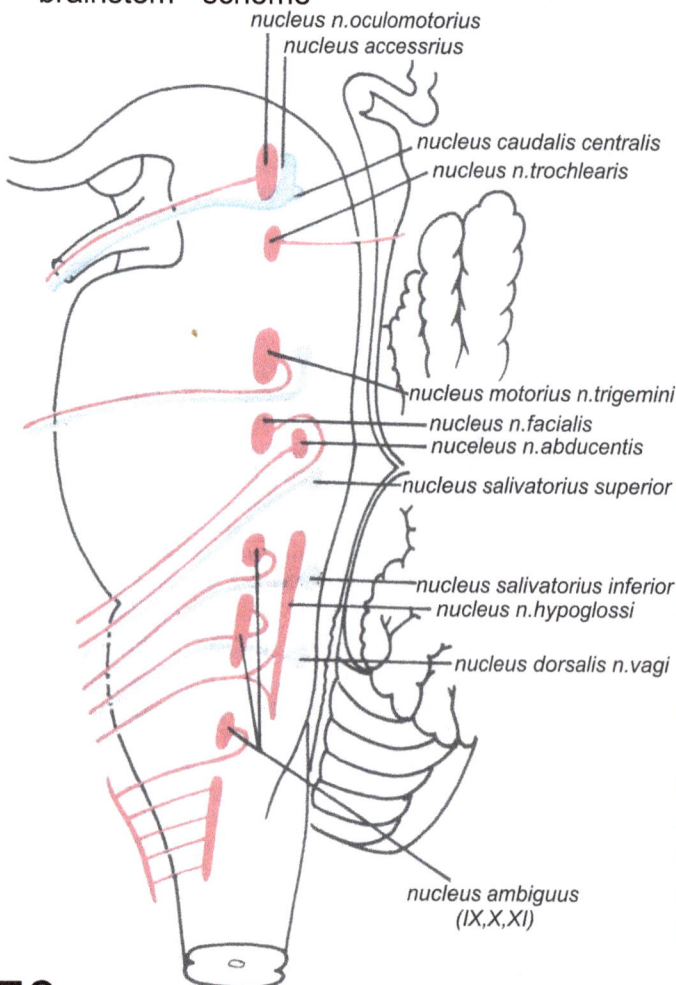

nucleus n.oculomotorius
nucleus accessrius

nucleus caudalis centralis
nucleus n.trochlearis

nucleus motorius n.trigemini
nucleus n.facialis
nuceleus n.abducentis
nucleus salivatorius superior

nucleus salivatorius inferior
nucleus n.hypoglossi

nucleus dorsalis n.vagi

nucleus ambiguus (IX,X,XI)

Nuclei

Sensitive nuclei

- In *the medulla oblongata*: the nucleus of the solitary tract, in which end the vegetative fibres of the seventh bis (Wrisberg's intermediate nerve) the ninth and the tenth pairs of cranial nerves.

- In *the medulla oblongata and the pons*: the vestibular and cochlear nuclei, Rolando's gelatinous nucleus (*nucleus tractus spinalis nervi trigemini fifth pair*).

Motor nuclei

- In *the medulla oblongata*: the nucleus ambiguus ninth, tenth and eleventh pairs and the nucleus of the hypoglossal nerve located in the hypoglossal trigone located in the floor of the fourth ventricle for the twelfth pair.

- In *the pons*: the motor nucleus of the trigeminal nerve (fifth pair) at the level of the locus ceruleus, the nucleus of the facial nerve (seventh pair) and the nucleus of the abducens nerve (sixth pair).

- In *the mesencephalon*: the nuclei of the oculomotor nerve (third pair) and the nucleus of the trochlear nerve (fourth pair).

Vegetative (parasympathetic) nuclei

- In *the medulla oblongata*: the dorsal nucleus of the vagus (tenth pair) and the superior salivary nucleus, which innervates the parotid gland.

- In *the pons*: the lacrimonasal nucleus, which innervates the lacrimal gland and the nasal mucosa, and the inferior salivary nucleus, which innervates the submandibular gland and the sublingual gland.

- In *the mesencephalon*: the Edinger-Westphal nucleus, which produces miosis (constriction of the pupil), and the Perlia nucleus, which produces the convergence of eyeballs.

The reticular substance
(Substantia reticularis)

The reticular substance ows its name to its characteristic structure of reticulum, respectively of a dense network of longitudinally and transversally oriented fibres, between which lie groups of neurons, owing to which a very great number of synapses is present; topographically it extends from the spinal cord up to the diencephalon.

This nervous structure does not play any role in the transmission of special sensory, motor or vegetative messages but, as it receives numerous informations, it associates them in a general, diffuse information, which assures to the central nervous system a background facilitating an optimum activity.

The following nuclei of reticular substance may be distinguished.

The central reticular nuclei occupy the central part *(tegmentum)* of the medulla oblongata, of the pons and of the mesencephalon and are traversed by the central tegmental tract, their mass being fragmented into a series of secondary nuclei that may be systematized according to their topography into a lateral group and a median group.

- *The lateral group*, termed sensory (Brodla) is provided with numerous transverse dendrites which connect it with the spino-thalamic tracts, the medial lemniscus, the nuclei of the trigeminal nerve, of the solitary tract, as well as with the coehlear and vestibular nuclei.

- *The median group*, with larger cells, is termed motor and is connected with the lateral sensory group; it is the site of origin of some ascending and descending nerve influxes.

The ascending influxes travel to the upper part of the mesencephalon, to the reticular formation and to the centromedian nucleus in the thalamus and, through these formations, to the hypothalamus and the cerebral cortex. The influx courses either on a longitudinal polysynaptic pathway or through the central tegmental fasciculus. It is not a specific sensory influx, but an unspecific, diffuse influx, able either to induce reflexes in the brainstem in relation with the cranial nerves, or to act upon the superior thalamic, hypothalamic and cortical centres, putting them in a state of alert (Magoun).

The descending influxes are transmitted through two tracts:

- from the central bulbal nuclei starts the lateral reticulospinal tract, which has the role to inhibit the nuclei situated in the anterior horns of the spinal cord;

- from the pontine and mesencephalic nuclei descends the medial reticulospinal tract which, on the contrary, facilitates the action of the nuclei lying in the anterior horns of the spinal cord.

These tracts, respectively the nuclei of origin, are under the influence of the cerebral cortex (the motor and cingulate areas) and of the striated bodies, with which they are connected through the corties and strioreticular tracts.

The lateral nuclei and the paramedian nucleus, which surround in the medulla oblongata the central nuclei, receive sensory afferents like the central nuclei and assure the links with the cerebellum, achieving reticulo cerebello-reticular circuits.

71

The median reticular nuclei, situated medially in the medulla oblongata, in the pons, around the cerebral aqueduct, receive afferents from the hypothalamus and the encephalon through the basal fasciculus and are connected with the vegetative nuclei of the brainstem their role in the vegetative mechanisms of the brainstem being significant.

During the last period, researches have demonstrated that, in addition to its role in the activation of the cerebral cortex (vigil state), the reticular substance acts also in the induction of sleep. Thus, the median pontine nucleus inhibits the waking mechanism, the diencephalon and the cortex producing the superficial sleep with the lowering of the electrical activity of the cortex, while the median bulbar nucleus induces the paradoxical sleep with the increase of the electrical activity of the cortex through the action which it exerts also on the motor nuclei of the eyeball.

Consequently, the activity of the reticular substance is characterized by a high complexity: it receives from a multitude of structures (cerebellum, striated bodies, red nucleus, quadrigeminal colliculi, vestibular nuclei, cerebral cortex) informations which are synthesized into a general, single information, directed towards the reception and execution areas of the nervous system.

The cranial nerves

The nuclei of the cranial nerves lie in the floor of the rhomboid fossa and of the cerebral aqueduct. There are *motor nuclei, sensitive nuclei* and *vegetative nuclei*, which are made up of well delimited cell groups (Fig. 64 - 67).

The motor nuclei are *nuclei of origin*. They are formed of the multipolar cells of the motor neuron, which are comparable to the motor cells lying in the anterior horns of the spinal cord, and are arranged in two rows: a lateral row and a medial row. The lateral motor row is made up of the following nuclei: the trigeminal nucleus, the facial nucleus, the motor part of the glossopharyngeal nerve and of the vagus nerve (these represent the nerves of the embryonic branchial archs). The medial motor row is constituted of the nuclei of the oculomotor, trochlear, abducent and hypoglossal nerves.

Their axons form the motor nerves. The motor part of the spinal nerves leaves the spinal cord always on the same side, without crossing. The cranial nerves which do not cross with each other are: the abducent nerve, the motor part of the trigeminal nerve, the facial nerve,. the motor part of the glossopharyngeal nerve, of the vagus, spinal nerve and the hypoglossal nerve. The trochlear nerve is the only cranial nerve which crosses completely.

The motor nuclei of the cranial nerves receive impulses from the cortex through the pyramidal tract *(tractus corticobulbaris)*. In the motor nuclei of the anterior horn of the spinal cord terminate only fibres of the pyramidal tract on the opposite side nerve. In contrast in the motor nuclei of some cranial nerves terminate fibres from both cerebral hemispheres (for example, the superior part of the nucleus of the facial nerve). The unilateral interruption of the central tract for the facial nerve, does not induce impairments in the musculature supplied by the superior nucleus of the facial nerve (the frontal muscle and the orbicular muscle of the eyeball), while the mimetic musculature, innervated by the inferior nucleus of the facial, is paralyzed. This is due to the fact that the inferior nucleus of the facial nerve receives impulses only from the opposite side of the cerebral cortex, while at the superior facial nucleus arrive impulses both from the same side and from the opposite side.

The sensitive nuclei behave differently from the motor nuclei. They are *terminal nuclei (nuclei terminationis)*. The protoneurons are situated in the ganglia of the cranial nerves, so as they are found also in the spinal ganglia of the spinal nerves, under the form of bipolar (pseudounipolar) cells of the peripheral sensory neuron. One of the projections comes from the periphery (skin of the face, mucosae) and the other ends in the terminal nucleus. The terminal nuclei contain the cell body of the second neuron, which courses to the central organs.

72

The gustatory, auditory and equilibrium excitations follow the same pathways as the pressure, tactile, pain and thermal excitations. In contrast, the sensory visual and olfactory pathways behave differently. In the rhomboid fossa, the sensory nuclei are situated laterally *(derived from the alar lamina)*.

Vegetative nuclei. In addition to the sensory nuclei, in the rhomboid fossa lie also parasympathetic nuclei of origin, the efferent fibres of which are associated with certain cranial nerves (the mesencephalic and the rhombencephalic part of the parasympathetic nervous system).

The nuclei of the cranial nerves are enumerated below.

Pair I and **pair II**: the nuclei of these pairs of "nerves" *(olfactory and optic)* will be described at the chapter of the respective sense organs; actually they are not nerves, but central white substance, tracts.

Pair III - nucleus of the oculomotor nerve *(nucleus nervi oculomotorii)* is situated in the mesencephalon, below the interstitial nucleus, at the level of the superior colliculus of the lamina tecti. Between the two oculomotor nuclei (from either side) lies the parasympathetic part of the oculomotor nerve, which consists of a median part (Perlia 's nucleus) and two paired lateral parts (Westphal -Edinger's nucleus).

Pair IV - *nucleus of the trochlear nerve (nucleus n. trochlearis)* is situated below the cerebral aqueduct, at the level of the inferior colliculus.

Pair V - the motor nucleus of *the trigeminal nerve (nucleus motorius n. trigemini)* is situated in the region of the superior fovea, in the upper part of the rhomboid fossa; it represents the nucleus of origin of the minor portion of the trigeminal nerve.

The superior sensory nucleus *(nucleus sensorius superior)* of the trigeminal nerve is external to the motor nucleus and is much larger than the latter. It represents the terminal nucleus of the: major portion (sensory). From it starts upwards the nucleus of the mesencephalic tract of the trigeminal nerve, which ascends up to the region of the lamina of the mesencephalic tectum. Downwards runs the nucleus of the spinal tract of the trigeminal nerve, which is continuous at the C2 level of the spinal cord with the gelatinous substance of the posterior horn. The nucleus of the spinal tract lies at the level of *the tuber cinereum* of the medulla oblongata near the surface, so that it may be observed through the transparency of the layer of white substance as a darker spot.

Pair VI - *the nucleus of the abducent nerve* lies beneath the facial colliculus, anterior and lateral to the nucleus of the facial nerve.

Pair VII - *the nucleus of the facial nerve (nucleus n-facialis)*, is situated beneath the facial colliculus, posterior and median to the nucleus of the abducent nerve.

The superior salivatory nucleus lies in the inferior part of the facial nucleus and extends up to the medullary striae of the rhomboid fossa.

The inferior salivatory nucleus is continuous with the superior salivatory nucleus and lies beneath the medullary striae. It is the parasympathetic secretory nucleus of the glossopharyngeal nerve, which sends fibres to the parotid gland, and of the vagus nerve.

Pair VII bis - *nucleus intermedius*, the sensory branch of the facial nerve, receives parasympathetic fibres from the superior salivatory nucleus. Its terminal nucleus is the nucleus of the solitary tract *(nucleus tractus solitarii)*.

Pair VIII - the stato-acustici nuclei, called also vestibulocochlear nuclei are two groups of nuclei, one for the vestibular part (static nerve) and another for the cochlear part *(acoustic nerve)*.

The vestibular part *(pars vestibularis)* contains four terminal nuclei, situated in the vestibular area of the rhomboid fossa: the lateral vestibular nucleus (Deiters' n.), the superior vestibular nucleus (Bechterew's n.), the spinal nucleus (Roller's n.) and the medial vestibular nucleus (Schwalbe's n.). In addition to these four nuclei, an inferior vestibular nucleus is also present, which penetrates into the superior portion of the spinal cord. The most important in the lateral vestibular nucleus.

The cochlear part *(pars cochlearis)* contains two terminal nuclei , located external to the ventral and vestibular nuclei, near the inferior cerebellar peduncle. They are termed the ventral cochlear nucleus *(nucleus cochlearis ventralis)* (connection of the auditory pathway) and the dorsal cochlear nucleus *(nucleus cochlearis dorsalis)* (connection of the reflex acoustic pathway).

Pair IX - nucleus of the glossopharyngeal nerve *(nucleus n. glossopharyngei)* has nuclei of origin and terminal nuclei.

The ambiguous nucleus *(nucleus ambiguus)* reaches the lower portion of the inferior salivatory nucleus. It is shared by the vagus nerve and the glossopharyngeal nerve.

The dorsal nucleus of the vagus nerve, located medially and posterior to the nucleus ambiguus, represents the origin of the efferent vagal fibres (visceral motoricity of the smooth musculature)

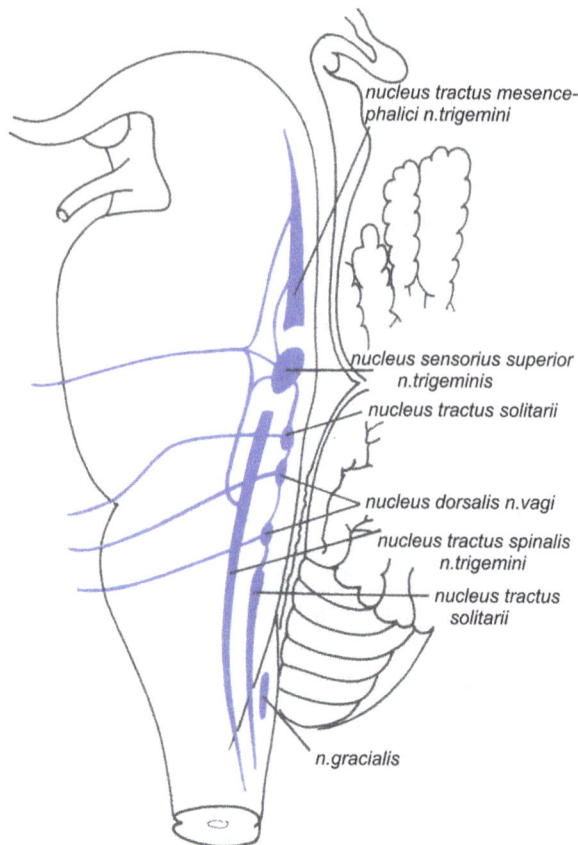

Fig. 65 Sensory nuclei of the brainstem - scheme

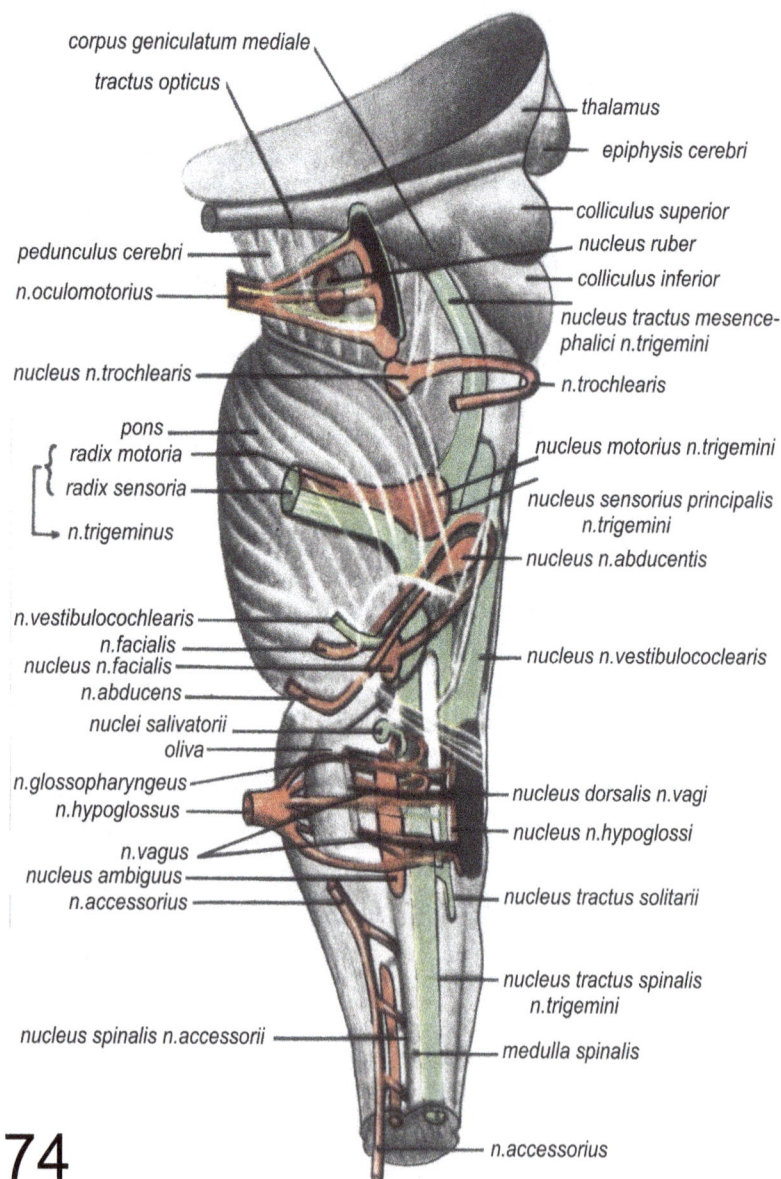

Fig. 66 Nuclei of the brainstem – lateral view

74

The inferior salivatory nucleus (the secretory nucleus which sends its fibres to the parotid gland) has been mentioned above.

The terminal nucleus of the ala cinerea (nucleus terminationis alae cinereae) is the terminal nucleus of the vagus nerve (termination of the peripheral sensory neurons of both nerves - glossopharyngeal and vagus).

The solitary tract and the nucleus of the solitary tract represent the terminations of the terminations of the gustatory sensory neurons of the glossopharygean, vagus and intermediary (Wrisberg's) nerves.

Pair X - nucleus of the vagus nerve - has its origin and termination similar to those of the glossopharyngeal nerve:
- *nucleus ambiguus* - automatic motoricity;
- *dorsal nucleus* - visceral motoricity;
- *inferior salivatory nucleus* -secretion;
- *terminal nucleus of the ala cinerea* - represents the termination of the peripheral sensory neurons;
- *solitary tractus and nucleus of the solitary tract.*

The solitary tract is formed of the sensory fibres of the intermediary, glossopharyngeal and vagus nerves, which run downwards. This bundle of fibres is accompanied by the terminal nucleus of these pathways, the nucleus of the solitary tract. The mass of fibres and the nucleus begin at the level of the trigone of the vagus nerve, *(ala cinerea)*, lateral to it, and course downwards up to the C1 level (gustatory nucleus).

Pair XI - *accessory nucleus (nucleus of the accessory nerve)* - has two nuclei of origin: the cranial nucleus and the spinal nucleus. The cranial nucleus of the superior root of the accessory (spinal) nerve is located below the nucleus ambiguus (this portion of the accessory nerve contains the fibres which are associated with the vagus nerve and supply the motor innervation of the internal muscles of the larynx). The spinal nucleus of the spinal root of the accessory nerve continues from the cranial nucleus downwards in the anterior horn of the spinal cord, up to C5-C7.

Pair XII - *hypoglossal nucleus (nucleus nervi hypoglossi)* - it forms the trigone of the hypoglossal nerve (at the level of the bulbar trigone of the rhomboid fossa).

Below is presented the functional organization of the cranial nerves according to Abraham, Kaplan, Gerhardt von Bonin and Frank Netter; the mentioned authors divide the nuclei of the cranial nerves into two groups: effector and sensory nuclei.

Effector nuclei:

1. *Edinger - Westphal nucleus* - its fibres, which travel to the oculomotor nerve, then to the ciliary ganglion, supply the intrinsic muscles of the eye.

2. *Oculomotor nucleus* - it is made up of a median cell group and of a lateral cell group; its fibres supply all the extrinsic eye muscles except the lateral rectus and the superior oblique.

3. *Trochlear nucleus* - after a course downwards, its fibres decussate with contralateral fibres and supply the superior oblique muscle.

4. *Motor muscles of the trigeminal nerve* - its fibres run laterally to the mandibular nerve and innervate the muscles of mastication.

5. *Abducens nucleus* - its fibres run forwards to innervate the lateral rectus muscle.

6. *Motor nucleus of the facial nerve* - its fibres travel posteromedially, turn round the nucleus of the abducens nerve, then run anterolaterally towards the inferior portion of the pons, leave the brainstem through the bulbopontine sulcus and innervate the facial muscles.

7. *Superior salivatory nucleus* - it sends fibres, through the cord of tympanum, to the submandibular ganglion and innervates the submandibular and the sublingual glands.

8. *Inferior salivatory nucleus* - it sends fibres to the otic ganglion and innervates the parotid gland.

9. *Hypoglossal nucleus* - it sends fibres downwards and laterally, leaving the brainstem through the sulcus between the oliva and the pyramid, and innervates the tongue.

10. *Dorsal motor nucleus of the vagus* - it sends through the medulla oblongata to the vagal nerve fibres which end in the sympathetic vagal plexuses, in the thorax and the abdomen.

11. *Ambiguous nucleus (nucleus ambiguus)* - it sends fibres through the glossopharyngeal, vagus and spinal accessory nerves and innervates the pharynx and the larynx.

12. *Spinal nucleus of the accessory nerve* - its fibres form roots which unite with the bulbar roots of the spinal accessory nerve and innervate the trapezius muscle and the sternocleidomastoid muscle.

Sensory nuclei

1. *Mesencephalic nucleus of the trigeminal nerve* - fibres of the extrinsic muscles of the eyeball and of the masticatory muscles end in this nucleus, from which arise the mesencephalic roots of these nerves.

2. *Principal sensory nucleus of the trigeminal nerve* -fibres trasmitting impulses for sensations of pain, temperature, touch from the head and the face reach the pons and divide into short ascending fibres, which end in this nucleus, and long descending fibres, end in the spinal nucleus of the trigeminal nerve.

3. *Spinal nucleus of the trigeminal nerve* - its fibres transmit exclusively painful and thermal excitations and are continuous with Rolando's gelatinous substance.

4. *Vestibular nucleus* - fibres of the vestibular bipolar ganglionic cells end in this nucleus; its short ascending fibres end in the lateral nucleus (Deiters' nucleus), while its long descending fibres end in a principal nucleus, in the spinal descending nucleus, in the superior nucleus (Bechterew's nucleus) and in the cerebellum.

5. *Cochlear nucleus* - the central fibres arising from the bipolar cells of the spinal ganglion of the cochlea end in this nucleus. From its anterior portion the fibres cross the lateral lemniscus and travel towards the inferior colliculus. From its posterior portion the fibres pass through the medial geniculate body to the inferior colliculus and end in the auditory area of the cerebral cortex.

6. *Nucleus of the solitary tract* - it receives afferent visceral fibres from the facial glosso-pharyngeal and vagus nerves.

7. *Commissural nucleus* - it unites with the nuclei of the solitary tract.

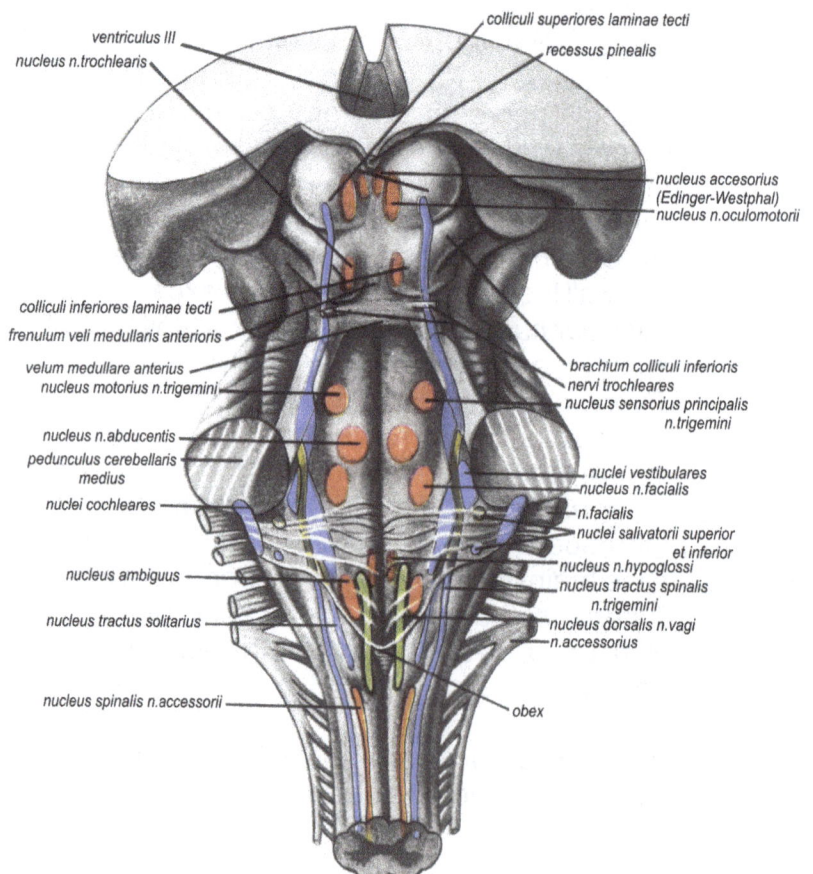

Fig. 67 Nuclei of the brainstem – posterior view

The apparent origin of the cranial nerves

- *The oculomotor nerve (pair III)* has its apparent origin in the medial sulcus of the cerebral peduncle.
- *The trochlear nerve (pair IV),* as the single posterior cranial nerve, has its apparent origin beneath the inferior colliculus.
- *The trigeminal nerve (pair V)* appears on the lateral side of the pons of Varolius, in proximity to the middle cerebellar peduncle.
- *The abducens nerve* appears between the pons and the pyramid of the medulla oblongata, above the pyramid, in the bulbopontine groove.
- *The facial nerve (pair VII), with a thinner sensory root, Wrisberg's intermediate nerve (pair VII bis) and the stato-acoustic nerve (pair VIII)* have their apparent origin between the medulla oblongata and the pons, above and lateral to the oliva (in the pontocerebellar angle). Among these nerves, the innermost is the facial nerve, the most lateral the stato-acoustic nerve and between them lies the intermediate nerve.
- *The glossopharyngeal nerve (pair XI), the vagus (pair X) and the accessory nerve (pair XI)* with its superior root emerge, one below the other, from the posterior lateral sulcus of the medulla oblongata. *The spinal root of the accessory nerve (pair XI)* emerges between the posterior roots of the cervical nerves and the denticulate ligament, from where it runs upwards.
- *The hypoglossal nerve (pair XII)* emerges between the pyramid and the oliva in the preolivary sulcus.

The cerebellum
(Cerebellum)
(Fig. 68-76)

From the covering plate and the roots of the alar lamina of the neural tube develops the cerebellum, a significant regulation centre of fine movements, of involuntary automatic movements and of equilibrium.

It controls the automatism and coordinates the tone of the antagonistic and synergistic muscles. The cerebellum informs the cortex permanently of the position of the body, owing to the afferent pathways of the proprioceptive unconscious sensibility and receives from the cortex motor response impulses which are processed in the cerebellum and then, through the efferent pathways and the extrapyramidal system, arrive at the periphery. The sudden exclusion of the cerebellar activity gives to severe impairments of the movements, without inducing a paralysis or other sensibility changes.

The cerebellum is situated in the cerebellar occipital fossae in the posterior area of the base of the skull, beneath the *tentorium cerebelli*: The cerebellum has a weight of 120-150g.

External configuration
Constituant parts (Fig. 68-73)

The cerebellum is made up of a median, narrow, elongated part, called vermis (because it resembles a silk worm), and of two lateral parts: the cerebellar hemispheres. Between them are located two notches situated above and below, termed the anterior cerebellar notch and the posterior cerebellar notch. The posterior notch *(posterior incisura cerebelli)* continues on the infero-anterior part of the cerebellum and forms a depression, termed vallecula cerebelli, which overlaps the dorsal part of the medulla oblongata.

The surface of the cerebellum is crossed by a great number of transverse furrows (fissurae cerebelli), which delineate thin lobes, termed cerebellar lobes, and in continuation lobules, laminae and lamellae (folia). One of these fissures, which is also the deepest,

Parts of the vermis	Parts of the hemispheres	They control
Superior surface		
Lingula	Vinculum lingulae cerebelli	Movements of the eye
Lobulus centralis	Ala lobuli centralis	Expression of the face
Culmen	Quadrangular lobule	Tongue, mastication
Declive	Lobulus simplex	Deglutition, larynx
Folium	Superior semilunar lobule	movements of the neck
Horizontal fissure		
Inferior surface		
Tuber	Inferior semilunar lobule	Movements of the upper limb
Pyramid	Biventer lobule	Movements of the trunk
Uvula	Tonsil	Movements of the lower limb
Flocculus	Peduncle of flocculus Flocculus	Equilibrum.

the horizontal fissure *(fissura horizontalis cerebelli)*, divides the surface of the cerebellum into a superior surface *(facies superior hemispherii cerebelli)* and an inferior surface *(facies inferior hemispherii cerebeli)*.

The portion of the cerebellum which develops phylogenetically the first is the vermis *(archicerebellum)*. The hemispheres appear phylogenetically later *(neocerebellum)*. Embryologically, both formations develop from two symmetrical buds, united on the median line. To each portion of the vermis corresponds a part of the cerebellar hemispheres. Moreover, it is belived that a precise location exists for the coordination of movements in certain territories of the body. Below is presented a scheme in this respect (according to Alverdes).

In the case of isolated , sudden diseases of the vermis, appear disturbances of equilibrium, respectively forwards and backwards oscillations, whereas in lesions at the level of hemispheres the oscillations are lateral.

Surfaces

The cerebellum has three surfaces: superior, inferior and anterior.
- On **the superior surface**, on its median line, is situated the superior vermis and on each side lie the almost flat cerebellar hemispheres, slightly sloping laterally and downwards. This surface is bounded by a circumferential margin (Vicq d'Azyr), which separates it from the two other surfaces. It presents an anterior median notch and a posterior median notch (deep and narrow). The latter deepens on the anterior surface and is named *valeculla cerebelli*; in it project the posterior and inferior parts of the vermis.
- The **inferior surface** is crossed in the middle by the posterior notch, the great median notch of the cerebellum, in the bottom of which -as shown above -is prominent the inferior vermis. Laterally are the convex surfaces of the cerebellar hemispheres, separated from the inferior vermis by two very deep furrows.
- The **anterior surface** is occupied by a projection of the fourth ventricle, which is bounded by anterior extremity of the superior vermis or the *lingula* and a projection of it - Vieussens' valve (upwards) - , by the anterior extremity of the inferior vermis or the nodulus (downwards and medially), by Tarin's valvulae or membranes (downwards and laterally) and by the cerebellar peduncles (laterally).

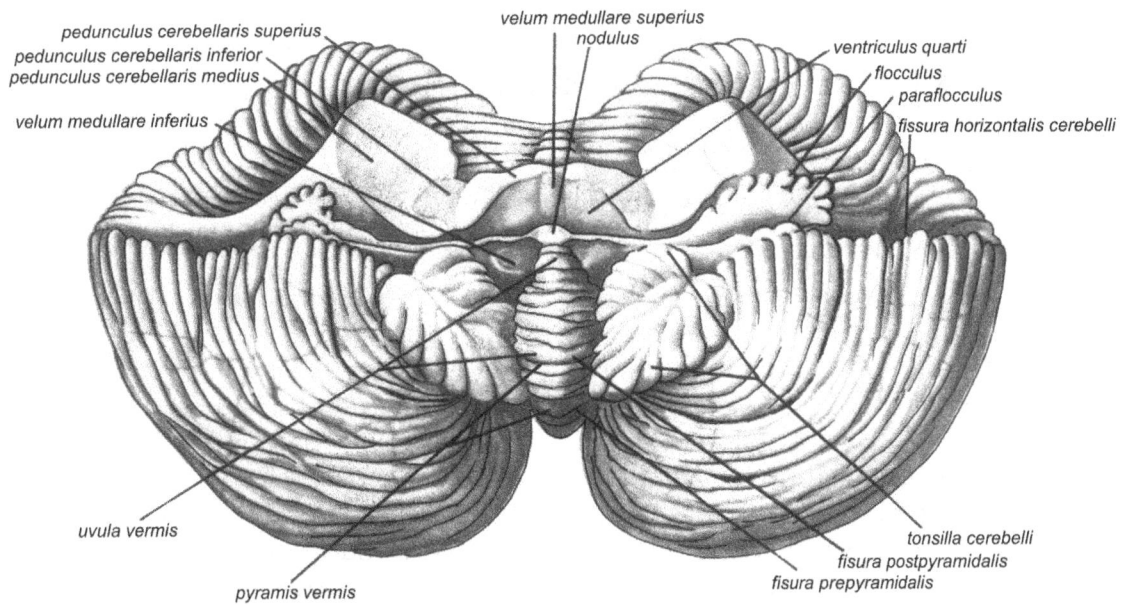

Fig. 68 Cerebellum – antero-inferior view, with maintenance of the tonsils

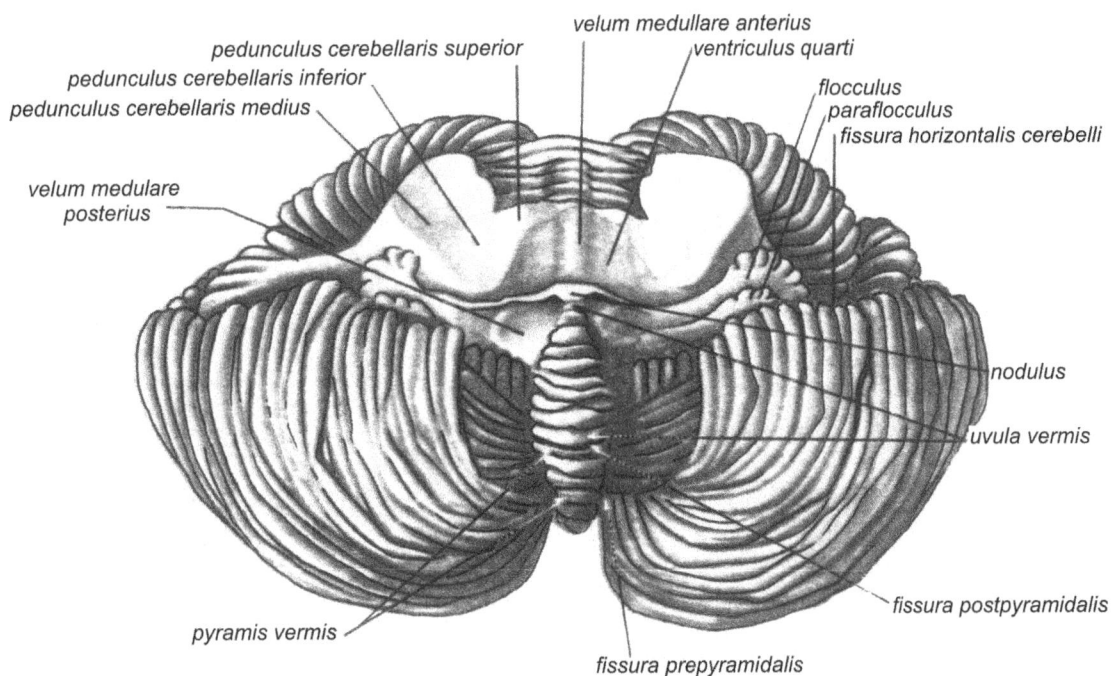

Fig. 69 Cerebellum – antero-inferior view; the cerebellar tonsil has been removed
and the whole medullary velum may be seen

The cerebellar peduncles
(Pedunculi cerebelli)

They represent the connections of the cerebellum with the neighbouring regions and contain the afferent and efferent tracts.

The superior cerebellar peduncle (*brachium conjunctivum cerebelli*) starts from the upper part of the cerebellum. Both peduncles converge upwards and course towards the lamina tecti, below which they penetrate. They contain the anterior spinocerebellar tract (Gower's tract), an afferent pathway, and the cerebellocerebral tract, an efferent pathway, termed also cerebellorubral tract.

79

The middle cerebellar peduncle *(brachium pontis)* connects the lateral parts of the cerebellum with the pons. Its place of origin is represented by the apparent origin of the trigeminal nerve. It contains the pontocerebellar tract, an afferent pathway.

The inferior cerebellar peduncle *(corpus restiforme)* is a flattened cord which connects the cerebellum with the medulla oblongata. It contains the following afferent pathways: the posterior spinocerebellar tract (Flechsig's tract), the bulbocerebellar tract and the vestibulocerebellar tract, as well as two efferent pathways: the cerebellovestibular tract and the cerebello-olivary tract.

The medullary velum

The anterior (superior) medullary velum *(velum medullare anterius)* is a thin lamina of white matter which stretches between the two superior cerebellar peduncles. It forms the upper part of the roof of the fourth ventricle and is covered with the lingula of the cerebellum.

The posterior (inferior) medullary velum is a narrow strip of white matter, continuous medially with the nodule and laterally with the peduncle of the flocculus. Like the anterior medullary velum, it participates in the formation of the fourth ventricle, respectively of its inferior portion. The tectorial lamina, a component of the tela choroidea of the fourth ventricle, is attached to the posterior medullary velum.

Internal configuration

Macroscopic structure *(Fig. 74-76)*

The grey matter of the cerebellum is present in two regions of the cerebellum: one part lies on the surface of the cerebellum, forming its cortex, while the other part lies inside the cerebellum, forming the grey nuclei of the cerebellum.

There are four pairs of cerebellar nuclei:

1. *The dentate nucleus (nucleus dentatus)* is the largest and the most lateral of the cerebellar nuclei; it consists of a lamina of grey matter in the shape of a bag with many folds, the opening of which is directed medially. The fibres arising from the cortex of the vermis end at the cells of this nucleus and here have also their origin the cerebellocerebral or cerebellorubral tract and the cerebello-olivary tract.

2. *The nucleus of the roof, called the fastigial nucleus (nucleus fastigii)*, is situated in the ceiling of the fourth ventricle, on each side of the midline. Here occurs the connection of a part of the vestibulicerebellar tract with a part of the cerebellovestibular tract.

3. *The emboliform nucleus (nucleus emboliformis)* lies on the medial side of the dentate nucleus. Its function is not known.

4. *The globose nucleus (nucleus globosus)* consists of several spheroid nuclei and is situated between the fastigial nucleus (medially) and the emboliform nucleus (laterally). Neither its function is known.

The white matter, poorly developed in the vermis, is very well developed in the hemispheres. On a mediosagittal section through the vermis, the aspect of the white matter is similar to the branching system and the arrangement of leaves on the conifer Thuja occidentalis (tree of life) and therefore this treelike disposition of the white matter was termed by the ancient anatomists arbor vitae.

The cerebellum receives stimuli from the peripheral receptors through the spinal cord and vestibular stimuli from the internal ear and from the cerebral cortex, so that it is informed both from the periphery of the body and from the cortex, representing a "servomotor" in the regulation of the motor acts of the body.

Microscopic structure

The white matter of the cerebellum consists of myelinated nerve fibres, without neurilemma, and of glial cells.

80

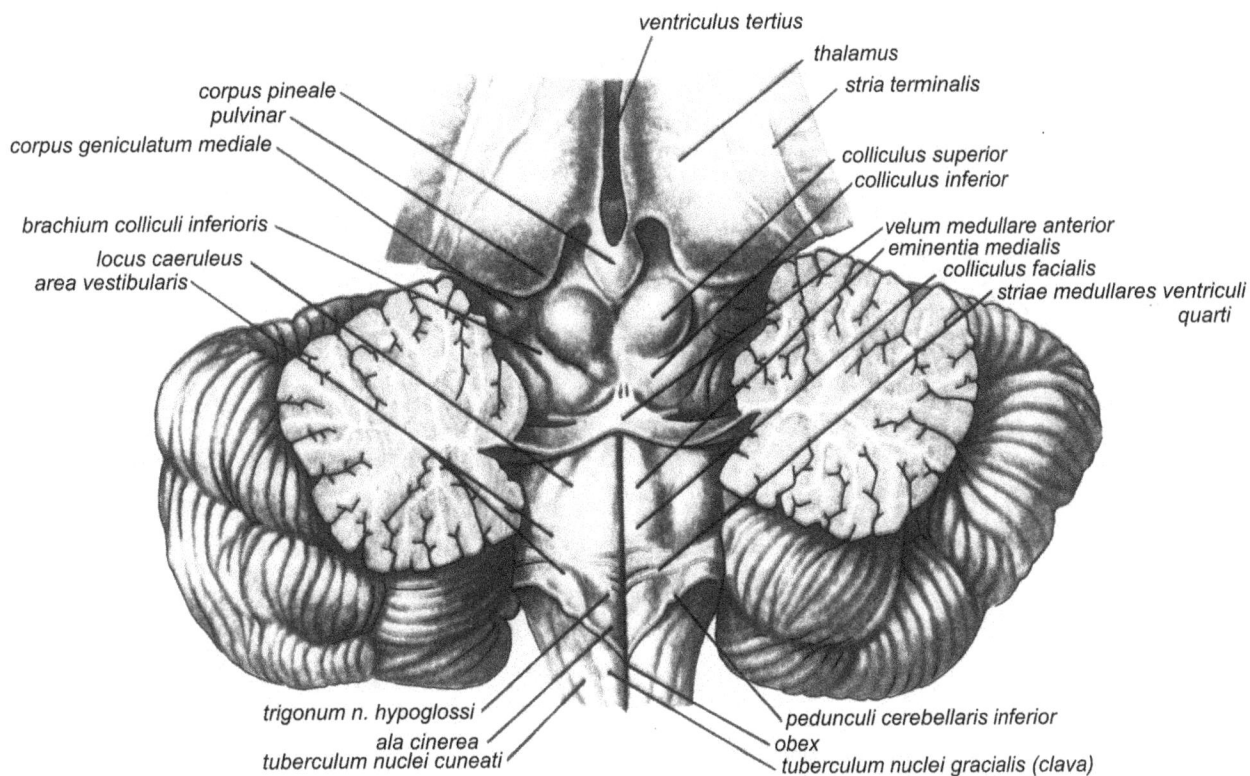

ventriculus tertius
thalamus
stria terminalis
corpus pineale
pulvinar
corpus geniculatum mediale
colliculus superior
colliculus inferior
brachium colliculi inferioris
velum medullare anterior
eminentia medialis
locus caeruleus
colliculus facialis
area vestibularis
striae medullares ventriculi quarti
trigonum n. hypoglossi
pedunculi cerebellaris inferior
ala cinerea
obex
tuberculum nuclei cuneati
tuberculum nuclei gracialis (clava)

Fig. 70 Cerebellum and diencephalon – posterior view

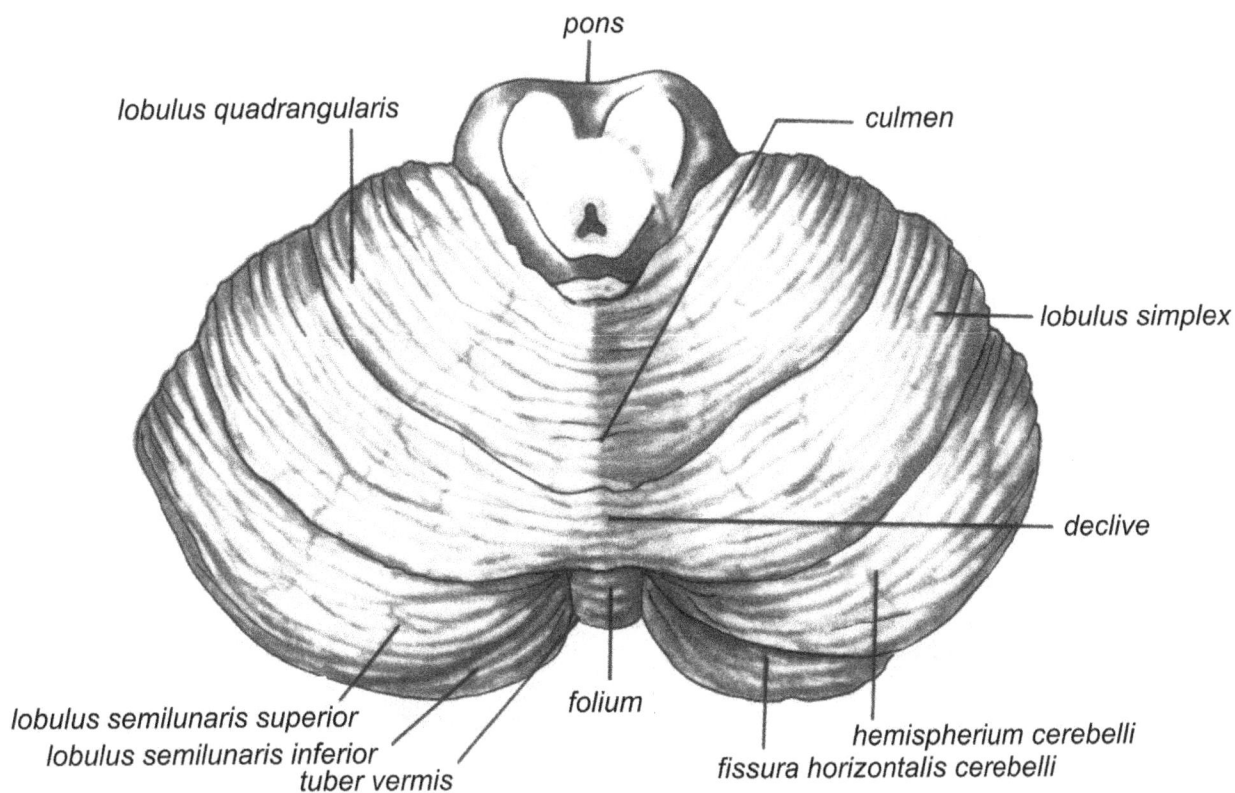

pons
lobulus quadrangularis
culmen
lobulus simplex
declive
lobulus semilunaris superior
folium
lobulus semilunaris inferior
tuber vermis
hemispherium cerebelli
fissura horizontalis cerebelli

Fig. 71 Cerebellum, its external surface, superior view

81

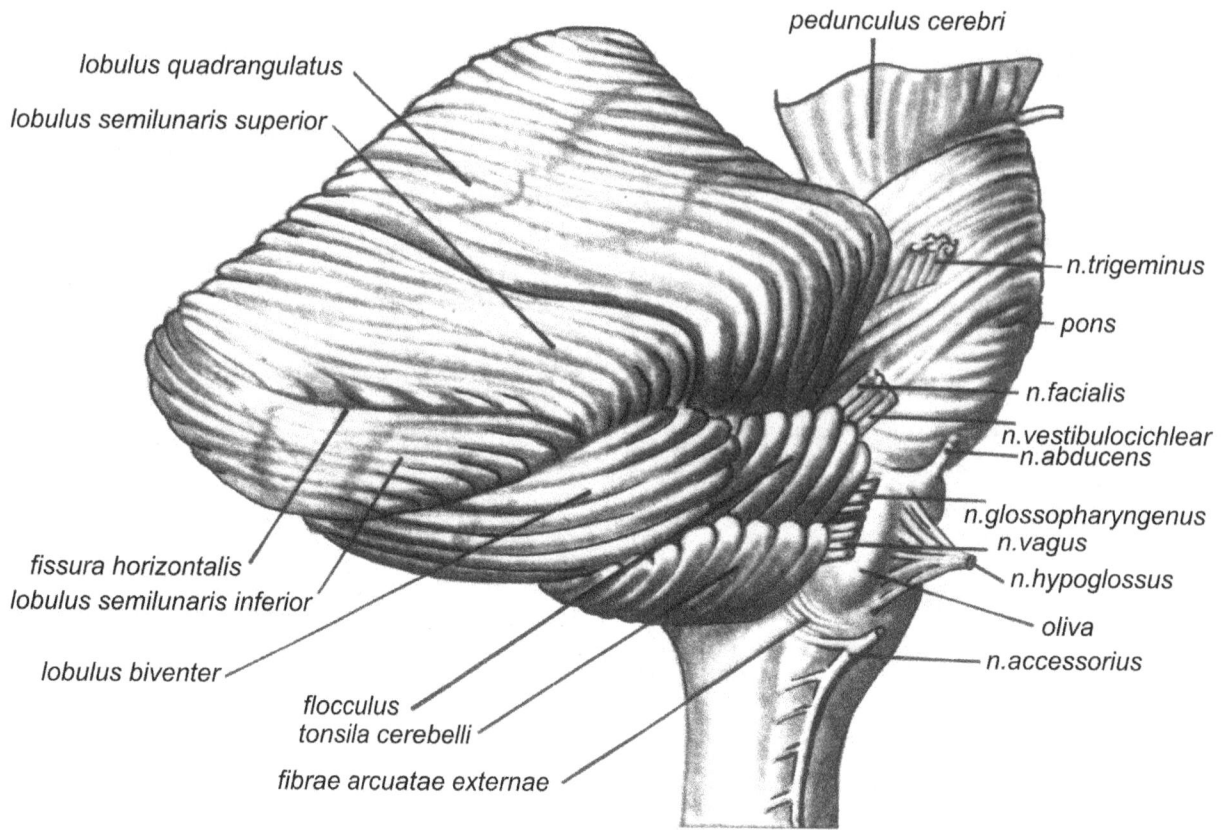

lobulus quadrangulatus
lobulus semilunaris superior

pedunculus cerebri

n.trigeminus
pons
n.facialis
n.vestibulocichlear
n.abducens
n.glossopharyngenus
n.vagus
n.hypoglossus
oliva
n.accessorius

fissura horizontalis
lobulus semilunaris inferior

lobulus biventer

flocculus
tonsila cerebelli

fibrae arcuatae externae

Fig. 72 Cerebellum and brainstem – lateral view

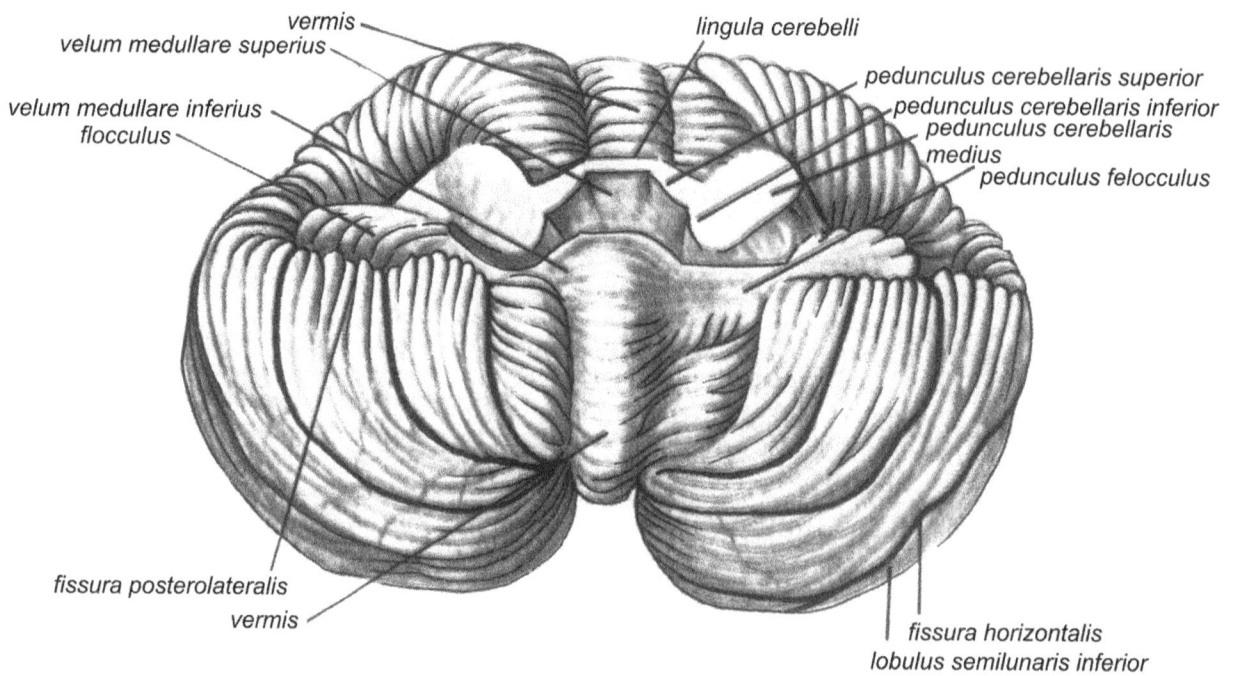

vermis
velum medullare superius
velum medullare inferius
flocculus

lingula cerebelli

pedunculus cerebellaris superior
pedunculus cerebellaris inferior
pedunculus cerebellaris
medius
pedunculus felocculus

fissura posterolateralis
vermis

fissura horizontalis
lobulus semilunaris inferior

Fig. 73 Cerebellum – anterior and inferior view

82

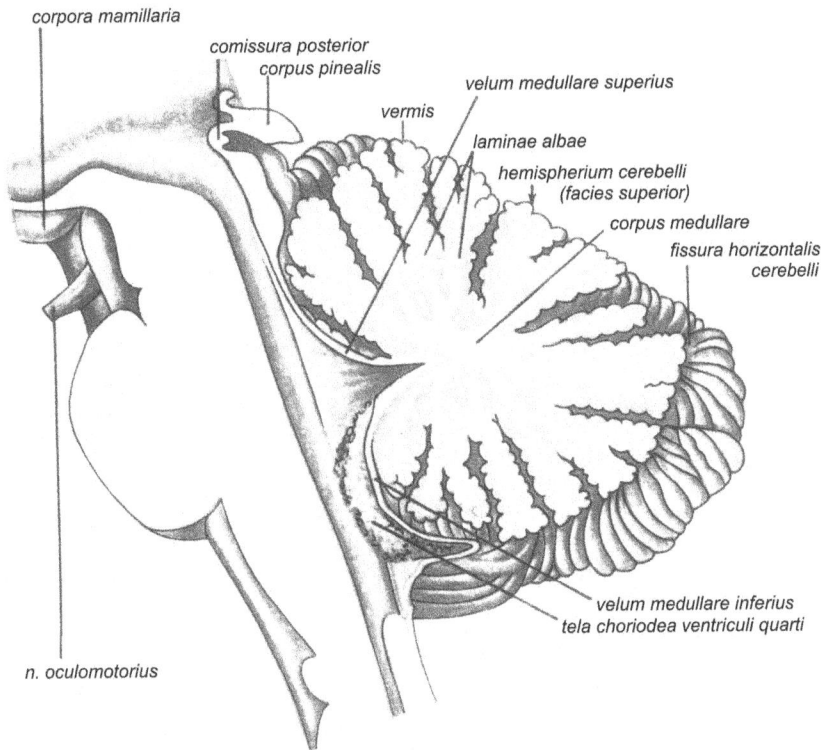

corpora mamillaria

comissura posterior
corpus pinealis

vermis

velum medullare superius

laminae albae

hemispherium cerebelli
(facies superior)

corpus medullare

fissura horizontalis
cerebelli

Fig. 74 Cerebellum and
brainstem – mediosagittal
section

velum medullare inferius
tela choriodea ventriculi quarti

n. oculomotorius

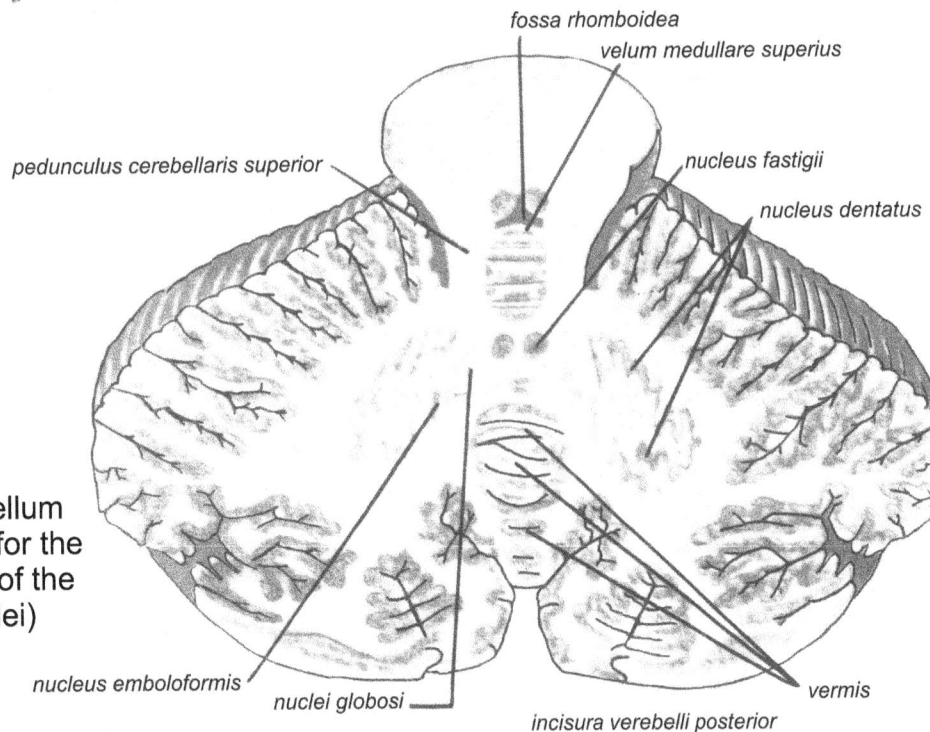

fossa rhomboidea
velum medullare superius

pedunculus cerebellaris superior

nucleus fastigii

nucleus dentatus

Fig. 75 Cerebellum
(frontal section for the
demonstration of the
intrinsic nuclei)

nucleus emboloformis

nuclei globosi

incisura verebelli posterior

vermis

The cerebellar cortex is made up of three cell layers:
- the molecular layer *(stratum moleculare)* at the surface, followed by
- the ganglionic layer (Purkinje cell layer) *(stratum gangliosum)* and
- in the depth, the granular layer *(stratum granulorum)*.
The molecular layer is mostly formed of glial tissue, in which unmyelinated dendrites from the ganglia layer penetrate.
The ganglionic layer consists of Purkinje cell bodies arranged in a single row - The cells. The cells are provided with numerous dendrites, richly arborized in a single plane, perpendicular on the fissures of the cerebellar cortex.

The dendrites spread up to the superficial border of the molecular layer. From the side opposite to the emergence site of dendrites arises a thin projection, which ends in the nuclei of the cerebellum and represents the axon (the efferent cell process). The whole morphophysiology of the cerebellum is centred around the Purkinje cell, the basic neuron of the cerebellum structure. It is believed that these cells process the impulses arrived in the cerebellum and then transmit them to the cerebellar nuclei. Thus, it seems that the influxes are transferred from the cerebellar pathway to the extrapyramidal system.

The granular layer contains two nerve cell types:

- the small granule cell, which establishes the contact with the dendrites of the Purkinje cells;

- the Golgi cell.

It should not be forgotten that the appearance and functional complexity of the various cerebellar formations are in a direct relation with the appearance of the cerebellar formations on the phylogenetic scale.

Thus, the oldest part of the cerebellum, the archicerebellum *(archaeocerebellum) (archaic cerebellum)* represents, on the animal scale, the region to which arrive the impulses from the vestibular system. In mammals it is represented by the lingula and the flocconodular lobe.

The paleocerebellum (paleocerebellum) (old cerebellum) is represented by a large portion of the vermis (the central lobule, the culmen, the ala of the central lobule, the quadrangular lobule, the pyramid and the uvula). The paleocerebellum cotains also the fastigial, globose and emboliform nuclei. Here arrive the ecitations from the mesencephalon and the hypothalamus.

The neocerebellum (the most recent portion) appears in mammals and is represented by the cerebellar hemispheres and the largest part of the dentate nucleus. These formations have significant connections with the cerebral hemispheres and the bulbar olivae (corticopontocerebellar and cerebellorubro-thalamocortical fibres).

The fourth ventricle *(ventriculus quartus cerebri)*

The fourth ventricle results from the dilation of the ependymal canal and is situated between the medulla oblongata, the pons and the cerebellum *(rhombocephalon)*. It is a flattened space, filled with cerebrospinal fluid, rhomboidal in shape and has an anterior wall (the floor), made up of the rhomboid fossa, a posterior wall (the roof) and four boundaries: two inferior and two superior boundaries.

The rhomboid fossa *(fossa rhomboidea)*

The rhomboid fossa represents the floor of the fourth ventricle and owes its name to its rhomboidal shape. In its formation participate both the medulla oblongata (in the inferior portion of the fossa) and the pons (in the superior portion of the fossa); both portions have the shape of isoscele triangles united through their bases. The lateral angles of the rhomboid fossa extend on the inferior cerebellar peduncles, forming lateral recesses *(recessus lateralis)*, which open by way Luschka's foramena into the subarachnoid space. The superior angle is acute and represents the site where the fourth ventricle is continuous with the cerebral aqueduct *(aquaeductus cerebri, aqueductus Sylvii)*. The inferior angle corresponds to the pointed end calamus scriptorius and is continuous downwards with the central or ependymal canal of the spinal cord, at the site where between the gracile tubercles is stretched a band of the posterior commissure of the spinal cord, called obex.

The median sulcus of the rhomboid fossa *(sulcus medianus fossae rhomboidae)* divides it into a right half and a left half. In proximity to the lateral recesses, between the recesses and the median sulcus lie, visible transversally, bundles of white fibres, the medullary striae *(striae medullares)*, which penetrate into the depth of the median sulcus.

84

Together with this sulcus, the *medullary striae* are shaped like a pen and therefore this lower part of the floor of the fourth ventricle has been termed *calamus scriptorius* (writer's pen). They form the reflex auditory pathway. The areas above and beneath these fibres, situated in proximity to the lateral recesses, are termed vestibular areas *(areae vestibulares)* (they contain the terminal nuclei of the vestibular nerves). *The medullary striae* form at the same time the boundary between *the mesencephalon* and *the telencephalon* (the pontine tringle and the bulbar triangle).

The pontine triangle. On each side of the median sulcus, an elongated formation, called medial eminence *(eminentia medialis)*, is putting out, bounded laterally by *the sulcus limitans fossae rhomboideae*, which in its middle portion presents a small hemispherical prominence, *the facial colliculus (colliculus facialis)*, which corresponds to the nucleus of origin of the abducent nerve, surrounded by the internal genu of the facial nerve. Lateral to *the facial colliculus* lies a small depression, the superior fovea or t*he fovea of the trigeminus*, continuous upwards with *the locus coeruleus* (of a bluish colour). Beneath *the locus coeruleus* lie the masticatory nucleus of the trigeminal nerve, the inferior part of the accessory masticatory nucleus and the autonomic nuclei attached to the nerves VII, VII bis and VIII.

The bulbar triangle. On each part of the median line, in the inferior part of *the rhomboid fossa*, is situated the hypoglossal trigone *(trigonum hypoglossi)* (the internal white ala), below which lies the nucleus of origin of the hypoglossal nerve. Lateral to it is a small elongated depression, ala cinerea, of a grey colour, in the depth of which lie important vegetative nuclei (vagal, glossopharyngeal). The vagal trigone is bounded laterally and below by a thin white strip, *the funiculus separans*.

The abundance and significance of the nerve centres lying underneath the floor of the fourth ventricle explain the severity of the lesions which could involve this level.

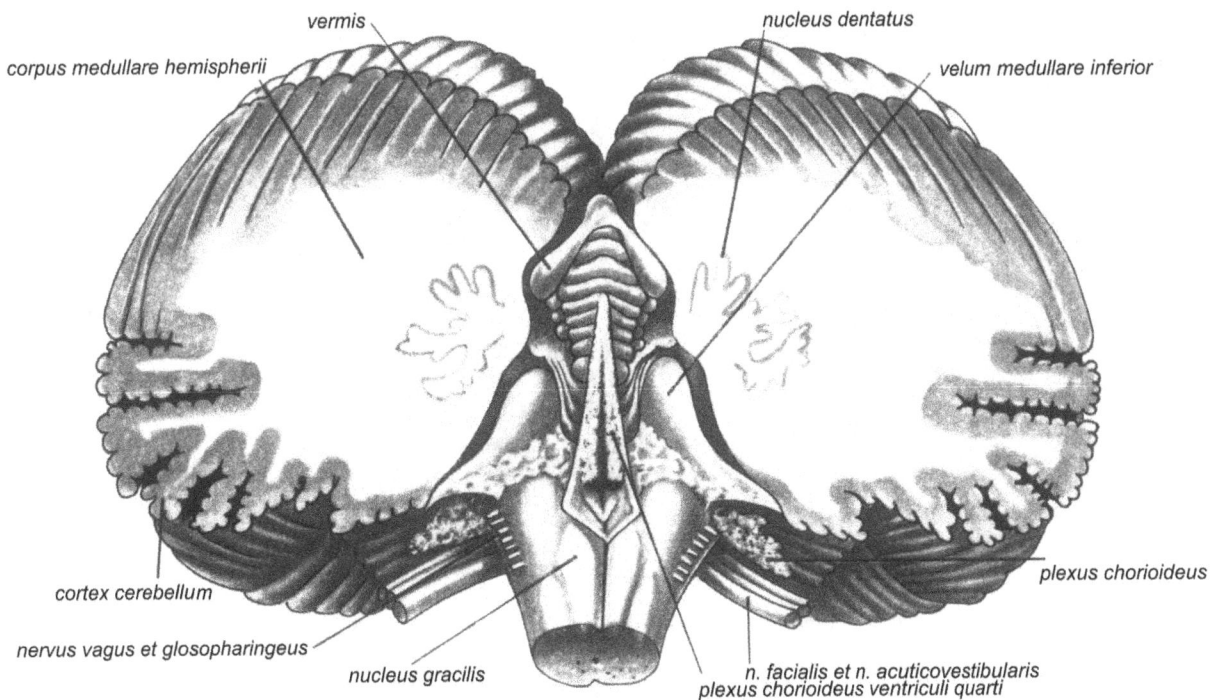

Fig. 76 Cerebellum – posterior view, after the removal of the tonsils, and horizontal section, for the demonstration of the dentate nucleus and of the choroid plexuses of the fourth ventricle

The roof *(posterior wall)*

It is formed of three parts.

The median (or cerebellar) part is made up of the middle portion of the cerebellum, comprised between *the lingula* (superiorly), the nodule and the Tarin's valve or *inferior medullary velum* (inferiorly) and the cerebellar peduncles (externally).

The superior part is formed by the Vieussen's valve, a lamina of white matter which stretches from one superior cerebellar peduncle to the other, covering the pontine triangle. In the upper portion may be observed *the frenulum of the valve*, which connects its superior extremity to the groove between *the inferior quadrigeminal colliculi*. The trochlear nerves emerge lateral to *the frenulum of the valve*.

The inferior part is represented by *the lamina choroidea*, to which *the choroid tela* of the fourth ventricle *(tela choroidea ventriculi quarti)* adheres deeply.

The borders of the fourth ventricle

They are four in number: two inferior and two superior. The inferior borders correspond to the insertion line of *the taenia choroidea* on the inferior cerebellar peduncles. The superior boders are formed by the internal edges of the superior cerebellar peduncles.

The tela choroidea of the fourth ventricle is a triangular lamina with the apex directed downwards. Towards the ventricle it is covered by the epithelial lamina and contains the choroid plexus of the fourth ventricle *(plexus choroideus ventriculi quarti)*. *The tela choroidea* is attached to the nodule, the peduncle of the flocculus and the posterior medullary velum, then stretches on the inferior cerebellar peduncles (restiform body) and ends on the obex. This line of insertion forms the taenia of the fourth ventricle.

The tela choroidea has three apertures: one unpaired, on the median line, at the level of the obex (Magendie's foramen), and two paired at the level of the lateral recesses; the lateral apertures of the fourth ventricle (Luschka's foramen).

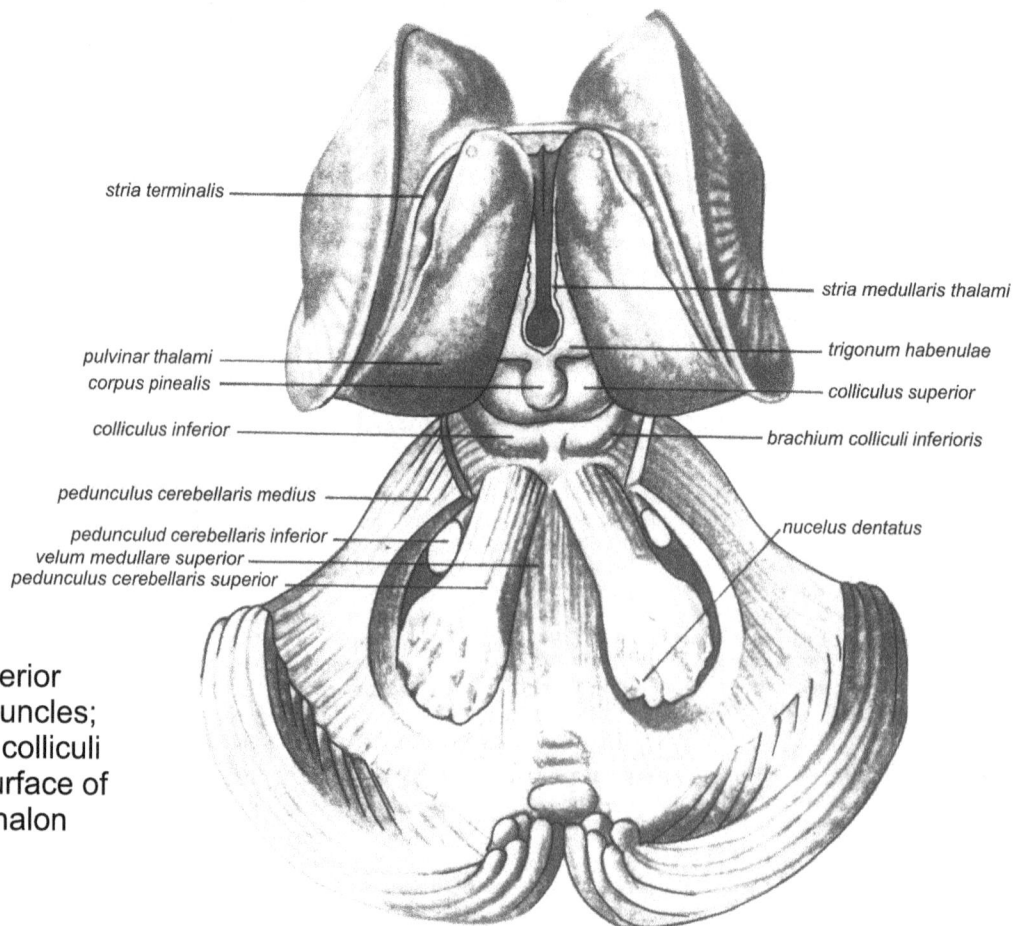

Fig. 78 Superior cerebellar peduncles; quadrigeminal colliculi and superior surface of the diencephalon

stria terminalis

stria medullaris thalami

pulvinar thalami

trigonum habenulae

corpus pinealis

colliculus superior

colliculus inferior

brachium colliculi inferioris

pedunculus cerebellaris medius

nucelus dentatus

pedunculud cerebellaris inferior
velum medullare superior
pedunculus cerebellaris superior

Through these openings the ventricular system communicates with the subarachnoid space. If these apertures are absent (owing to a deficient embryological development, or to their later closure), the dilation of the ventricles (internal hydrocephalus) occurs. Through the lateral apertures, the tela choroidea with the choroid plexus emerges in the form of bunches from the fourth ventricle and penetrate into the subarachnoid space (Bochdalek's flower basket).

THE DIENCEPHALON
(Diencephalon) (Fig. 77-80)

In the course of the embryological development, this region is covered by *the telencephalon*, so that only a small part remains visible at the base of the brain.

The diencephalon is represented by several structures grouped around the third ventricle: the epithalamus (corresponding to the roof of the third ventricle) the thalamus and the metathalamus (corresponding to the anterior, lateral and posterior of the third ventricle walls), as well as the hypothalamus (corresponding to the floor of the third ventricle).These structures are basic, phylogenetically old parts. The part which underwent the most significant transformations, related to the increasing development of the telencephalon, is the thalamus, which became an important relay of the latter.

The diencephalon contains:
- the nervous formations making up the walls of the median ventricle;
- the cavity of the median ventricle or the third ventricle.

The nervous formations which make up the walls of the median ventricle are five in number and constitute the walls of the third ventricle, as follows:
- the lateral walls are formed by *the thalamus*, t*he hypothalamus* and *the subthalamus*;
- the superior wall or the roof of the third ventricle is formed by the pineal gland and the very thin epithelial lamina of *the tela choroidea* of the third ventricle;
- the postero-inferior wall or the floor is made up of the posterior perforated plate, the mamillary tubercles, *the tuber cinereum* with the hypophysial stalk, the hypophysis and the optic chiasma;
- the anterior wall is constituted of the anterior pillars of the trigone, t*he lamina terminalis*, Monro's foramina and the anterior white commissure.

We shall not follow "*ad litteram*" this description, but we shall continue to refer to the connections of these structures - in relationship with the anatomic structures -, in order to facilitate the understanding of the physiological aspects and clinical syndromes specific to this area, of a particular pathophysiological significance.

The thalamus
(Thalamus)

The thalamus develops from the alar plates and is formed of an anterior and a posterior portion. The cavity of the third ventricle separates the left from the right thalamus. Each thalamus is ovoid-shaped, with the large diameter directed from before backwards, measuring about 4 cm; the most voluminous extremity is oriented backwards. Portions of the thalamus may be explored only after the removal of the fornix and the opening of the third ventricle (in this way its superior and medial parts may be observed).

The thalamus has four convex surfaces and two ends. Its medial surface is in relationship with the lumen of the third ventricle, forming its lateral wall; the lateral surface is separated from the lenticular nucleus by the posterior limb of the internal capsule: the superior surface forms with the caudate nucleus the floor of the lateral ventricle (the choroid plexuses, which secrete the cerebrovascular fluid, lie on it); the inferior surface is in relationship with the infundibulotubal and suboptic areas. The anterior end (pole) is situated ventrally and medially, while the posterior end is directed dorsally and laterally.

Superiorly and laterally, the thalamus is in relationship, upwards, with *the caudate nucleus* and backwards with a mass of white fibres of the internal capsule.

The thalamus is made up of grey matter (cells), which forms its nuclei. The free surfaces of the thalamus is covered by a thin layer of white matter, called *stratum zonale* or *zonular layer*. In the limiting groove between *the caudate nucleus* and the thalamus lies a thin layer of white fibres, termed *stria terminalis* (terminal stria), through the transparence of which may be seen a vein, the thalamostriate vein.

The ependyma of the lateral ventricle continues from *the caudate nucleus* over the *terminal stria* on the thalamus, to the surface of which it adheres firmly and is therefore called *lamina affixa*. To its medial border, which is formed by *the choroid taenia*, is attached the choroid plexus of the lateral ventricle.

The anterior extremity (pole) of the thalamus is slightly vaulted and contains the anterior tubercle *(tuberculum anterius thalami)*.

The dorsal part (pole) of the thalamus has the appearance of a marked blunt prominence (the pulvinar); it is the place where a number of fibres of the optic tract terminate (not all the authors agree with this opinion) and where the optic radiation begins.

Near the pulvinar a round prominence may be observed, the lateral geniculate body *(corpus geniculatum laterale)*, where 4/5 of the fibres of the optic tract end and where the optic radiation begins. The medial geniculate body *(corpus geniculatum mediale)* is situated on the anterior and posterior border of the pulvinar. It represents the relay centre of the auditory pathway. The two geniculate bodies form the metathalamus.

From the superomedial surface of the thalamus, at the superior border of the interventricular (Monro's) foramen, starts a band of white matter, called medullary stria of thalamus *(stria medullaris thalami)*, which terminates in *the habenulae* of the pineal body (which represent parts of the olfactory reflex pathway). On the medullary stria and on the habenulae is attached *the lamina tectoria*, which represents the epithelial roof of *the tela choroidea* and of the choroid plexus of the third ventricle. These structures contribute to the formation of the roof of the third ventricle. The free margin of the medullary stria forms *the taenia thalami*.

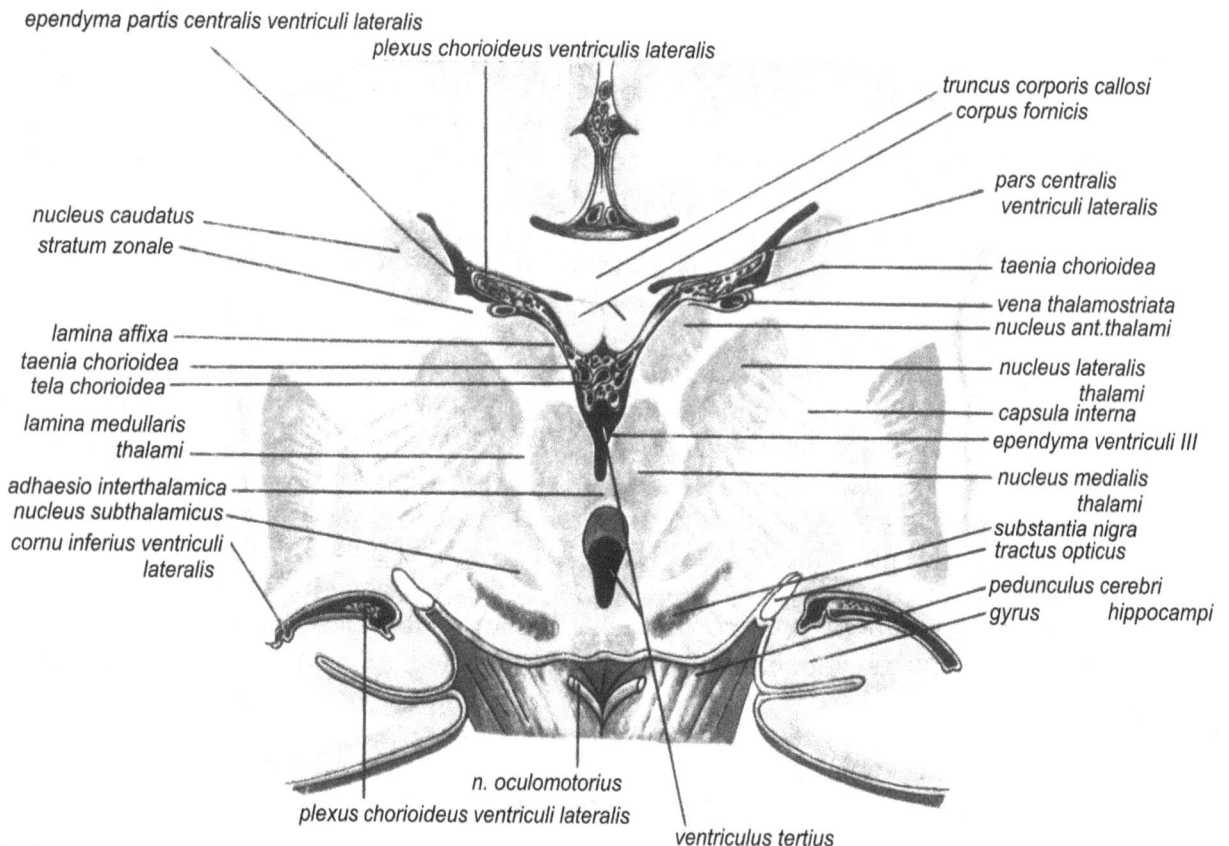

Fig. 79 Tela choroidea – frontal section

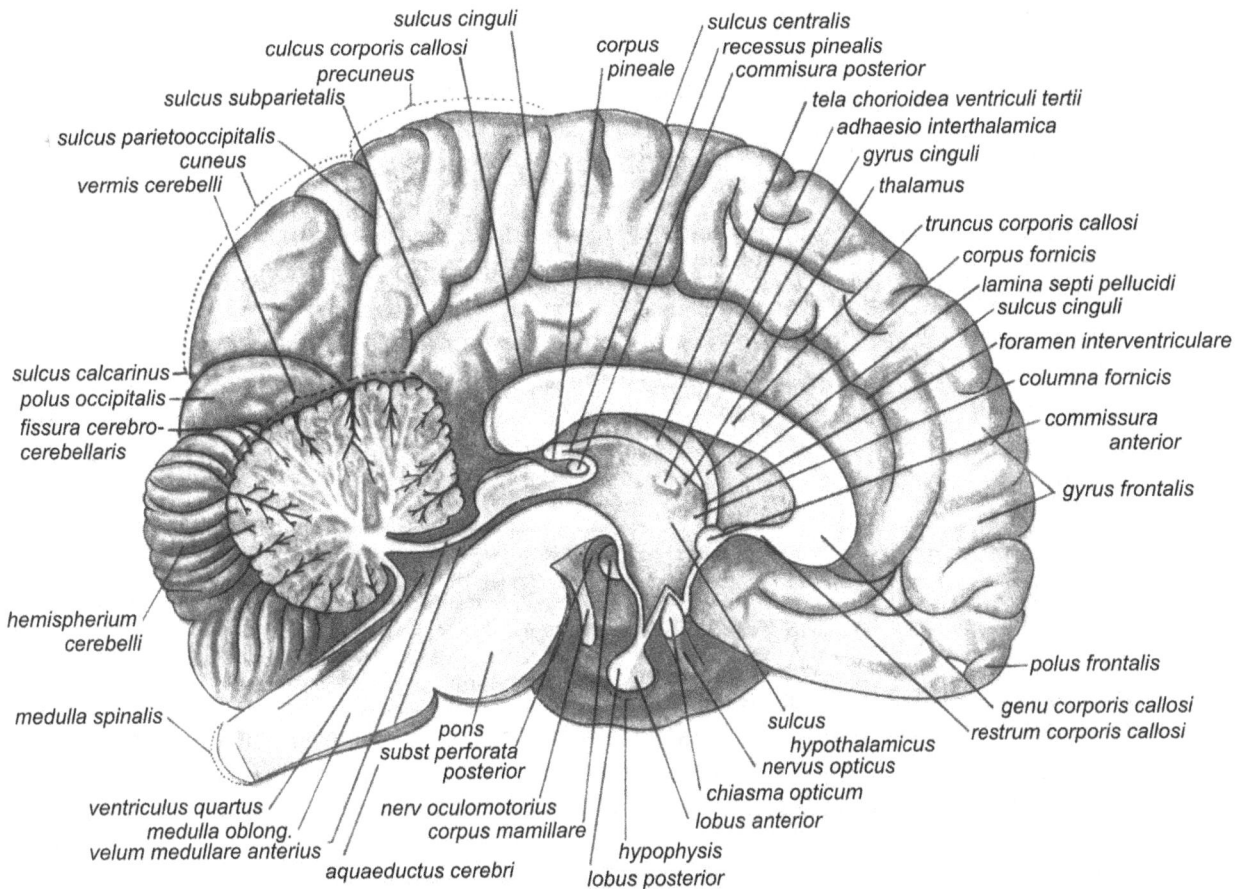

Fig. 80 Sagittal section through the encephalon, with demonstration of the thalamus, the pituitary gland (hypophysis) and the interthalamic adhesion

The right thalamus is sometimes connected with the left one by a band of grey matter, called interthalamic adhesion, which bulges transversally into the third ventricle and unites in this way the two thalamic masses.

The thalamus plays also a determining role in the wakefulness-sleep cycle and represents, according to Villablanca (1974), the "ganglion of sleep".

Nuclei and connections of the thalamus

The thalamus, which is a significant relay centre, is constituted of phylogenetically different areas, as follows:

- *the paleothalamus*, an old sensory area situated in the prolongation of the brainstem, from which it receives the most afferences. It has connections especially to the striated bodies.

- *the neothalamus*, on area of more recent development, connected to the cerebral cortex through afferent and efferent pathways;

- *the archithalamus*; *the paleo* - and *the neothalamus* are surrounded and traversed by diffuse or reticular grey formations, considered by some authors as remainders of the primitive wall of the diencephalon of fishes and reptiles, which form *the archithalamus*.

These structures seem to be specialized. Thus, the archithalamus is a diffuse activator of the cortex, the paleothalamus is a relay of the sensory pathways and the neothalamus, connected to the cerebral cortex, is a selection and regrouping centre of the messages arrived from the underlying regions.

The nuclei of the thalamus are separated from each other by the white matter - the medullary (internal and external) lamina of the thalamus; they are connected to various portions of the cortex by fibres termed radiate crown (corona radiata), which contribute to the formation of the internal capsule.

89

The thalamus is constituted of approximately 60 nuclei, which may be grouped as follows:

I. The anterior group, which receives mainly ascending fibres of the mamillothalamic tract stemming from the mamillary tubercle, considered nowadays as having connections with the rhinencephalon. It sends efferents to the limbic system *(gyrus cinguli)* and plays a role in emotiveness.

II. The lateral group, comprised between the internal medullary lamina and the thalamic reticular formation. This group, on turn, is made up of the ventral paleothalamic nuclei and the dorsal neothalamic nuclei.

A. The ventral nuclei, considered from behind forwards, are the following:

1) the lateroventral nucleus, situated behind, the afferences of which define it as the great sensory centre of thalamus. Among the ascending sensory tracts which have here a relay we mention: the medial lemniscus, the spinothalamic tract, the fibres of the nucleus of the solitary tract (gustatory fibres) and the sensory fibres of the nucleus of the trigeminal nerve. It sends efferences into the sensory cortex at the level of the ascending parietal gyrus;

2) the intermediate lateroventral nucleus, which receives afferents from the cerebellum, respectively paleo- and especially neocerebellar afferents, and sends efferents into the precentral cortex assuring the control of movements on the basis of informations from the cerebellum;

3) the lateroventral nucleus, a paleoencephalic nucleus, situated in front, united with the striate bodies through the lenticular fasciculus; it sends efferents towards the frontal cortex and is involved in the muscle maintenance and regulation mechanisms.

B. The dorsal nuclei, situated above the ventral nuclei, are associative neothalamic, namely:

1) the proper laterodorsal nucleus, connected to the parietal cortex; it plays a role in the control of the extrapyramidal system;

2) the pulvinar nucleus, which receives afferents from the posterior lateroventral nucleus and the geniculate bodies and sends efferents to the parietotemporo-occipital cortex.

III. The medial nuclei are situated between the internal medullary lamina and the subependymal nuclei of the third ventricle. They are considered as reticular structures, since they are diencephalic reticular centres. They are the following:

1) the medioventral (centromedial) nucleus, whose great development in man suggests that it plays a significant part in the mechanisms not only of thalamic, but also of cortical integration and activation. It has connexions to the striate bodies and to the reticular substance;

2) the mediodorsal nucleus, situated on the internal surface of the thalamus is connected to the centromedial nucleus, to the autonomic centres in the hypothalamus and to the prefrontal cortex, which achieves affective integration circuits.

IV. The reticular thalamic nuclei proper, situated on the inferior surface of the thalamus, form the zona incerta, They line also the external and internal surfaces (the paraventricular group) of the thalamus.

Inside the thalamus are the intrathalamic nuclei at the level of the internal medullary lamina. They have the role to assure the link between all the thalamic centres and have also an activating role.

V. The inferior nuclei form the metathalamus, respectively the lateral and medial geniculate bodies, the first of which correspond to the optic tract and the second to the auditory pathway.

Connections. Between the thalamic nuclei and the other encephalic portions are established mainly the following connections:

I. Subcortical connections with the thalamus, the hypothalamus and the extrapyramidal nuclei through:

1) the midline nuclei, which have connections with the hypothalamus and the pretectal region; they give off fibres to the intralaminar nuclei;

2) the intralaminar nuclei, especially the centromedian nucleus of Luys, send fibres to the extrapyramidal system, mainly to t*he globus pallidus*;

3) the anterior lateroventral nucleus situated in front, has also connections with the globus pallidus.

II. Cortical connections, achieved through the following nuclei:

1) the anterior nuclei receive fibres through the mamillothalamic fasciculus or Vicq d'Azyr's bundle and send fibres to the areas 23 and 24 of the cerebral cortex at the level of the *gyrus cinguli*, as well as - according to some autors - to Brodmann's area 32;

2) the lateroventral nuclei receive fibres of the superficial and deep sensibility tracts, of the cerebrorubrothalamic tract and of t*he medial lemniscus* (Reil's ribbon). From here start thalamocortical fibres which terminate in Brodmann's areas 3,1,2,5 and 7 of the parietal cortex. They represent the cortical projection of the sensory analyzer. They send also thalamocortical fibres to Brodmann's areas 4 and 6 at the level of the frontal lobe. They represent the cortical projection of the motor analyzer;

3) the lateral and medial geniculate bodies are relays of the optic and auditory tracts.

- The lateral geniculate bodies receive fibres from the optic tract, which is composed of homolateral temporal fibres and heterolateral nasal fibres. From here start Gratiolet's optic radiations to Brodmann's area 17 in the occipital lobe; fibres of the superior half of the retina project on the superior margin of the calcarine fissure, and fibres of the inferior half of the retina project on the inferior margin of the same scissure. The lateral geniculate bodies, through *the anterior brachium conjunctivum*, establish the connection with the anterior quadrigeminal tubercles, thus achieving pathways of optic reflexes.

- The medial geniculate bodies receive fibres through the lateral lemniscus (Reil's lateral band) and send fibres to the areas 41 and 42 in the temporal lobe.

Through the posterior conjunctival brachium, the medial geniculate bodies send fibres towards the posterior quadrigeminal tubercles, thus achieving the pathways cephalogyric of auditory reflexes.

III. Intrathalamic and cortical connections are made through the following nuclei:

1) the dorsomedial nucleus, which receives fibres from the hypothalamus and sends fibres to Brodmann's area 14, represented by the orbital cortex of the frontal lobe;

2) the lateral thalamic nuclei, which establish connections with the posterior central nucleus and the dorsomedial nucleus, seding fibres to Brodmann's areas 5, 7 and 6.

3) the pulvinar, which seems to have connections with the visual and the auditory cortex.

It is assumed that a sleep centre is sited in the floor of the third ventricle, in proximity to the entrance into the cerebral aqueduct, owing to the concentration of so many nuclei, with different functions, in a rather small space, the thalamus is an extremely significant centre. It represents a site of convergence of the sensitive and of the sensory pathways. These types of stimuli are processed at this level and only after this are directed towards the cortex. Thus, it depends on the thalamus whether and how the influxes of the excitations from the environment penetrate into the conscience. Through the autonomic nuclei and their afferent pathways the thalamus receives stimuli also during the performance of movements, so that here is the place of confluence of the whole sensibility from the external world and from the entire organism.

The hypothalamus
(Hypothalamus)

The hypothalamus has a particular clinical significance. Researches of the last period have shown that it contains integrative mechanisms which contribute to the regulation of the vital functions of the organism and that it should be considered as an important link in the circuits of the various encephalic regions, being in correlation with the hypophysis, too.

The hypothalamus is bounded: anteriorly, by *the lamina terminalis* with the anterior commissure (above) and the optic chiasma (below), posteriorly, by *the interpeduncular fossa*; dorsally, by the hypothalamic sulcus, which marks the connection with *the thalamus*; ventrally, by *the tuber cinereum* (a prominence which narrows towards the infundibulum, acquiring the appearance of a funnel); laterally it is in relationship with the interval capsule - the subthalamic nucleus and the cerebral peduncles (Fig.81). The boundary between the thalamus and the hypothalamus is indicated by the hypothalamic sulcus *(sulcus hypothalamicus)*, which is sited on the lateral wall of the third ventricle. It extends from the inferior border of the interventricular (Monro's) foramen to the cerebral or sylvian aqueduct *(aqueductus cerebri)*. The hypothalamus is structured of several portions which be seen on the basal surface of the brain, forming the floor of the third ventricle.

The anterior boundary is represented by the optic chiasma, which surrounds the cerebral peduncles (crura cerebri) and ends in the lateral geniculate body, from which the optic radiation starts towards the striate area. Fibres traval also to the pulvinar and to the superior colliculus of the tectal lamina of the mesencephalon *(lamina tecti mesencephali)* (this part of the optic pathway forms the boundary between the mesencephalon and the diencephalon).

Nuclei and connections of the hypothalamus

We successively describe the nuclei of the hypothalamus and then the connections with the regions associated to the hypothalamus.

The nuclei of the hypothalamus. Inside the hypothalamus lie aggregations of nuclei, clearly distinguishable especially in animals and human embryos. In adults they are more diffuse, with the exception of the supraoptic and paraventricular nuclei and of the mamillary complex. The diffuse appearance is due to the fact that many of these nuclei contain various, histologically different cell types.

Sections through the hypothalamus make possible a better demonstration of the aggregations of nuclei, as well as of the afferent and efferent pathways.

These nuclei are described in a slightly different manner by the researchers who have studied them.

I. We consider that the most systematic and scientifically best substantiated topographic study of the hypothalamus is that performed by Gerhard von Bönin and Frank Netter, who have demonstrated these formations by three sections at variable levels.

1) *The supraoptic nucleus (nucleus supraopticus)* is situated in the optic recess and is involved in the water and mineral salt metabolism, as well as in the sensation of thirst.

2) *The paraventricular nucleus (nucleus paraventricularis)*, situated below the hypothalamic sulcus, near the posterior margin of the interventricular foramen, coordinates the vasomotoricity and the sweat and urine secretion.

3) *The subthalamic nucleus (nucleus subthalamicus, Luys' body)*, situated below the hypothalamic sulcus and above the mamillary bodies, is involved in the carbohydrate metabolism, the fat metabolism, the genital functions and the production of hunger sensation.

4) *The tuberal nucleus (nucleus tuber cinereum)*, situated between the infundibulum and the mamillary bodies, is the thermal centre; it is involved in the production of the vomiting sensation.

The tridimensional schematic reconstruction offers even more data in this respect, permitting to differentiate the following groups of nuclei:

a) The anterior group, formed of the supraoptic and the paraventricular nuclei.

b) The median group, made up of the ventromedial *(n. ventromedialis)* and dorsomedial *(n. dorsomedialis)* nuclei.

c) The lateral group contains the lateral hypothalamic nucleus *(nucleus subthalamicus, Luys' body)* and the lateral nuclei *(nuclei tuber cinereum)*.

d) The posterior group is represented by the posterior hypothalamis nucleus and the mamillary body.

92

II. Other authors consider that we can distinguish, at the level of the hypothalamus, three successive layers arranged around the cavity of the third ventricle: a paraventricular, a medial and a lateral layer.

1) **The periventricular layer**, improperly called the grey commissure, forms, below the Monro's prominence, the ependymal organs (the secretory diencephalic apparatus):

2) **The medial layer** makes up the autonomic hypothalamus proper and consists of:

a) the anterior hypothalamus with three nuclei: preoptic, above the anterior commissure; supraoptic, around the optic chiasma; paraventricular, which surrounds the anterior column of the fornix, at the site where it penetrates the ventricular wall. The anterior hypothalamus is trophotropic (Hess), stimulating the respiration and the pulse and exerting parasympathetic effects;

b) the posterior hypothalamus, which contains the following nuclei: dorsal, ventral and a third nucleus, situated behind the first two nuclei; this zone is considered as ergotropic (Hess), stimulating the orthosympathetic component of the autonomic nervous system;

c) the mamillary bodies, formed of several nuclei.

3) **The lateral layer** forms the tuberal nuclei, which border the optic tract, lie laterally along the column of the trigone and the mamillo-thalamic tract and continue in the subthalamic and sublenticular areas. It is assumed that this layer, as it is in connection with the rhinencephalon controls some phenomena related to the nutrition function (appetite, hiccup, nausea etc.). If we take into account that these structures are uninterruptedly continuous with a layer of grey matter accomplishing an anatomic role - which begins at the terminal cone and lines the ependymal canal up to the opening of the cerebral aqueduct into the third ventricle - we may assimilate these formations with a true autonomic encephalon, in a functional relation with the visceral cortex, by means of which one becomes aware of same emotion-related autonomic phenomena.

Connections with the areas associated with the hypothalamus

Although it contains very significant integrating mechanisms, necessary to the vital activities, and although it "condenses" these neuron circuits in a unitary manner, the hypothalamus has, nevertheless, connections also with other encephalic structures through afferent and efferent pathway.

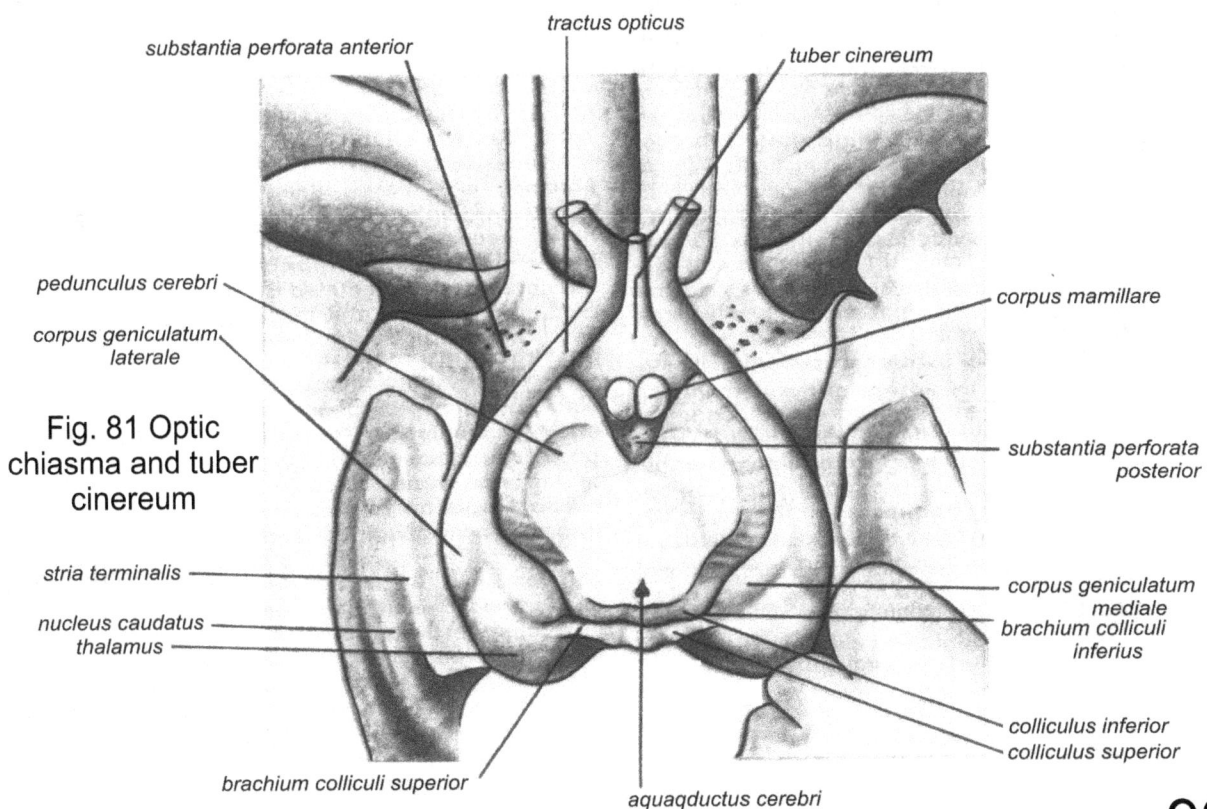

Fig. 81 Optic chiasma and tuber cinereum

substantia perforata anterior

tractus opticus

tuber cinereum

pedunculus cerebri

corpus geniculatum laterale

corpus mamillare

substantia perforata posterior

stria terminalis

nucleus caudatus

thalamus

corpus geniculatum mediale

brachium colliculi inferius

colliculus inferior

colliculus superior

brachium colliculi superior

aquaqductus cerebri

I. Afferent pathways

1) Thalamohypothalamic fibres which are arising from the medial nuclei of the thalamus and terminating in the anterior and medial hypothalamus; they play a role in the general neuro-endocrine reflexes.

2) Corticothalamo-hypothalamic fibres arise from the cortex of the frontal lobe passing through the dorsomedial nucleus of the thalamus and the median centre of Luys and terminating in the hypothalamus.

3) Connections through the fibres starting from Ammon's horn *(cornu Ammonis)* and from the amygdaloid nucleus or almond nucleus (amygdaloid complex) and terminating in the mamillary bodies.

4) Vegetative connections, through the dorsal nucleus of the vagus, the fasciculus of Schütz and the supraoptic and paraventricular nuclei.

5) Optic connections, arising from the optic tract and terminating in the supraoptic nucleus.

6) Fibres arising from various rhinencephalic cortical centres, which explains the primary importance of the diencephalon, dominated by the olfactory sphere, in the functioning of the vegetative activity and of instincts (Laruelle). The olfactory stimuli arrive at the hypothalamus, either by direct pathways (the basal fasciculus) at the level of the preoptic nucleus of the laterobasal nucleus of the tuber or indirectly by the nervous connections between the hippocampal cortical centres, the tuberal nuclei and the mamillary bodies, situated at the level of the anterior pillars of the trigone, or by the interpeduncular nucleus, respectively through Meynert's retroreflex fasciculus.

7) Proprioceptive afferences, which arrive at the hypothalamus through the collateral fibres of the cerebellorubro-thalamic pathways.

8) Extrapyramidal afferences starting from the striate (striohypothalamic) bodies and arriving at the lateral hypothalamus through the pallidohypothalamic tract.

II. Efferent pathways

1) The fasciculus of Schütz starting from the anterior hypothalamus and terminating in the dorsal nucleus of the vagus in the medulla oblongata.

2) Supraopticoretinal fibres which arise from the supraoptic nucleus of the anterior hypothalamus and travel up to the retina via the optic tracts; they play a trophic and excitofunctional role through the retinohypothalamo-retinal reflexes (Roussy and Mosinger).

3) Vicq d'Azyr's mamillothalamic tract, starting from the mamillary body and ariving in the anterior nucleus of the thalamus.

4) Hypothalamocortical fibres, which may travel along two pathways: some arise from the supraoptic nucleus and arrive at the cortex of the frontal lobe (O. Sager), while others arise in the paraventricular region, pass through the dorsomedian nucleus of the thalamus and terminate also in the frontal lobe.

5) Mamillomesencephalic tract, which arise in the mamillary body and terminate in the mesencephalon, influencing the motor nuclei of the brainstem.

6) Hypothalamohypophysial tract, arising from the supraoptic and paraventricular nuclei and terminating in the posterior hypophysis thus is formed the hypophysial or pituitary stalk, which Roussy has assimilated with a true peripheral nerve and which he termed peduncular nerve of hypophysis, with a neurosecretory significance.

7) Connections to the thalamus and to various extrapyramidal centres.

The above described connections point out the important role of the hypothalamus in the control of the various functions of the organism.

94

The hypophysis
(Pituitary gland)
(Hypophysis cerebri)

Taking into account the very close connections of the hypophysis with the hypothalamus, we describe this gland below, while the problems of neuroendocrinology will be resumed in the chapter dealing with the autonomic nervous system.

The hypophysis cerebri is suspended on the pituitary stalk, continuous with the infundibulum, under the tentorium of hypophysis, between the two cavernous sinuses. It is a gland of internal secretion or endocrine gland, dominating all the other glands and influencing the physical and psychic constitution of the organism.

It is ovoid -shaped, measures 1.5 cm transversally and 5-7 mm anteroposteriorly and weighs 0.6-0.7 g may be divided into the adenohypophysis and the neurohypophysis, It is situated in the Turkish saddle *(sella turcica)* of the sphenoid bone and is separated from the inferior surface of the encephalon by a dural diaphragm.

It is made up of three parts or lobes, of a different origin:

- the anterior lobe, layer, crescent-like shaped, in the concavity of which are situated the intermediate and the posterior lobes;

- the intermediate lobe which, with the anterior lobe, forms the adenohypophysis;

- the posterior or cerebral lobe, which is rounded and connected through the pituitary stalk with the diencephalon; they form together the neurohypophysis.

The hypophysis is connected with the hypothalamus through the infundibulum, the prolongation of which is termed infundibular recess. On the posterior side of the infundibulum lies a small prominence which contains the anatomic centres.

Behind the tuber cinereum are sited the mamillary bodies *(corpora mamillaria)*, two white hemispheric prominences, containing the nuclei termed nuclei of the mamillary bodies (the connection of the olfactory afferent pathway with the reflex olfactory efferent pathway), which pass over the fornix into the mamillothalamic tract. The two mamillary bodies border the rostrum of the interpeduncular fossa, whose base is formed by the posterior perforated substance.

Histologically, the adenohypophysis is formed of cell cords, consisting of chromophobe, acidophilic and basophilic cells separated by vascular sinusoids, while the neurohypophysis consists of pituicytes, neuroglial cells; basophilic cells migrated from the adenohypophysis and hypothalamohypophisal nerve fibres. In the neurohypophysis are also sited Herring's hyaline bodies and tubular structures derived from the buccal epithelium, the significance of which is not completely elucidated.

Fr. I. Rainer has discovered the portohypophysial system by dissections performed on an abundant necroptic material. Gr. T. Popa and Unna Fielding have demonstrated by histological studies the existence of the portohypophysial system.

The posterior lobe of the hypophysis, to which the intermediate lobe is attached, has an own blood circulation, differing from that of the anterior lobe, but connected with it through anastomoses:

- the arteries of the posterior lobe are the inferior hypophysial arteries, which arise from the internal carotid artery and penetrate in the lobe on its posterior side, travelling on the inferior surface of the hypophysis; they form:

- a capillary system in the posterior lobe;

- vascular (venous) trunks in the hypophysial stalk;

- a second capillary system in the tuber cinereum and in the walls of the third ventricle. Hence, the circulation has a hypophysothalamic direction.

The blood supply of the anterior lobe is assured by the superior hypophysial arteries, which have their origin in Willis' polygon, termed also the arterial circle of Willis, enter the anterior lobe through its anterior border and through the stalk and anastomose with the inferior hypophysial arteries, but form also a capillary network with the veins of the adenohypophysis, through a venous trunck, which connects the capillaries of the tuber cinereus of the diencephalon with the sinusoids of the adenohypophysis. In this case, the blood flow is cerebrofugal. The veins run into the venous system of the base of the skull *(meningeal veins - cavernous sinus)*.

According to the connections which they establish, the hypothalamic neurons were grouped into two systems: the neurohypophysial magnocellular system and the adenohypophysial parvocelullar system.

The hypothalamic neurohypophysial magnocellular system

This system is represented by the supraoptic and the paraventricular nuclei. As it is shown also by their name the neurons of these nuclei are characterized, by a large size, by an intense captation capacity of marked methionine and by a selective slaining with chrome-alum hematoxylin (Gomori's methods).

The progressive accumulation of observations (Cajal, 1894; Scharrer and Gaup, 1933; Bargmann and Scharrer, 1951) led to the elucidation of the aspects related to the structure and function of this system which has become classical; the neurosecretory material produced by the magnocellular nuclei migrates along the axons of the supraoptic - hypophysial tract into the neurohypophysis, where it is stored, after which it enters into the systemic circulation under the form of active hormones of the posterior hypophysis.

The theory „a cell - a hormone", according to which the two chief hormones of the hypothalamic, magnocellular system, vasopressin (ADH = antidiuretic hormone) and oxytocin (OXT), are secreted by separate cell groups (Olivecrona, 1957; Bisset et al., 1913, Sokol and Valtin, 1967), was invalidated by the studies of Leclerc and Pelletier (1974), who have demonstrated immunohistochemically that the supraoptic and the paraventricular nuclei are to the same extent involved in the synthesis of vasopressin, while the oxytocin is synthesized only by the supraoptic nucleus (Bisset et al. 1973).

In addition these two hormones the hypothalamic magnocellular nuclei secrete also some carrier proteins, called neurophysins, marked with A and B or I and II. It is not know whether these neurophysins are specific for vasopression and respectively oxytocin. Likewise, the site of their synthesis is also not elucidated, although most of the more recent data plead for the conclusion that both the supraoptic nucleus and the paraventricular nucleus are involved in the secretion of the two neurophysins.

The hypothalamic adenohypophysial parvocellular system.

Taking into account the peptidic nature of the hypothalamic hormones, the property of the neurons of secreting peptides is conventionally called neurosecretion.

The number and types of the existent hypophysiotropic hormones is not exactly known yet. According to a more recent hypothesis, to each hypophysial hormone correspond two regulating hormones: one which stimulates its release and another which inhibits it. Rather recently, three hypothalamic hormones playing a regulating role were isolated and characterized chemically: the thyrotropin-releasing hormone (TRH), the luteinizing hormone-releasing hormone (LH-RH) and the somatostatin (or somatotropin release - inhibiting hormone). Afterwards, other hypothalamic factors, which influence the anterior hypophysis, such as a substance, neurotensin, opioid peptides etc, were also identified.

Moreover, peptides discovered first in the hypophysis, such as ACTH (adrenocorticotropic hormone) and a-MSH (melanocyte -stimulating hormone), were detected in the hypothalamus.

Immunohistochemical methods were used to localize morphologically in the parvocellular regions of the hypothalamus the isolated peptides.

In 1975 endogenous peptides with narcotizing properties, known under the name of endorphins, were discovered. A pair of pentapeptides: leucin -and methionine - enkephalin, are the first endorphins identified.

96

Endorphins act not only as analgesics or as behaviour modulators (Bloom et al. 1976; Jacquert and Marks, 1976), but also as modulators of the neuroendocrine functions. The beta-endorphin is distributed in small amounts in the brain and in the hypophysis (Bloom et al., 1978).

Nerve fibres and endings containing enkephalin are sited in most hypothalamic nuclei, supporting the idea that these peptides are acting as neurotransmitters. Especially the basal ganglia contain large amounts of enkephalins. In the caudate nucleus, the locus ceruleus and the nuclei of the medulla oblongata, the enkephalins appear in the dendrites.

Outside of the central nervous system, enkephalins are present in the myenteric plexus, in some sympathetic ganglia, in the endocrine cells of the digestive tract and in many cells of the medulla of the adrenal gland.

In the median eminence, in addition to the group of tuberoinfundibular nuclei which have their origin in the hypophysiotropic zone of the hypothalamus, a group of nerve endings secreting biogenic amines projects, too.

This aminergic neuronal system has its cell bodies sited in the nuclei of the brain, as serotonergic neurons in the raphe nuclei of the mesencephalon, as dopaminergic neurons in the substantia nigra and as noradrenergic neurons in the locus ceruleus. Moreover, there is a main dopaminergic pathway from the arcuate periventricular region.

The epithalamus

The olfactory pathway arrives at the temporal cerebral cortex, where it ends, without a thalamic relay. However, the olfactory cortical centres are united with a primitive brain, the rhinencephalon, that has projection pathways some of which reach the thalamus (the anterior nucleus), others the hypothalamus and others the epithalamus.

Strictly speaking, the epithalamus is understood as a small nervous dorsomedial area of the thalamus, a distribution centre of the reflex pathways of the rhinencephalon, corresponding to the habenula and to the associated structures (*medullary stria, pineal body and commissure of the habenula*). The habenula is the peduncle of the pineal body and is made up of the adjacent group of nerve cells with which the pineal body is associated; it receives afferences from the medullary stria. Via the retroflex (habenulo-interpeduncular) fasciculus, connections are established with the interpeduncular nucleus and with the group of paramedian cells of the mesencephalic tegmentum.

The medullary stria of the thalamus is a narrow strip of nerve fibres attached to the roof of the third ventricle, stretched from the thalamus on each side and terminating behind in the habenular nucleus. The fibres arise from the septal area, in the anterior perforated plate, in the preoptic nucleus, and in the globus pallidus.

The commisure of the habenula is the connection between the habenular nuclei on each side; the decussatio of the fibres of the two medullary striae form the dorsal portion of the peduncle of the pineal body.

The epiphysis (*pineal body*)
(Corpus pineale seu Epiphysis cerebri)

The epiphysis, or pineal body, or pineal gland, is a small pine cone-shaped, unpaired structure, of 2-8 mm in length. Through the two habenulae, the epiphysis is connected with the two thalamic medullary striae.

It is situated between the superior (or anterior) quadrigeminal colliculi of the lamina tecti. At the place where the habenula passes on the pineal body, a small triangle, the habenular trigone (trigonum habenulae), is formed, within which lies the habenular nucleus, in which the connection between the reflex olfactory pathways occurs. The two nuclei are connected through the habenular commissure. The choroid tela of the third ventricle continues on the internal surface of the habenula and forms with the anterior border of the latter the pineal recess *(recessus pinealis)*. The pineal body and the adnexa described form the epithalamus.

The posterior commissure is sited beneath the habenula and the pineal body. It is a white, round band, which joins the two habenular nuclei (left and right). It is crossed by the extrapyramidal tracts from the striate body to the red nucleus on the opposite side.

Histologically, the pineal body consists of pinealocytes and neuroglial cells. The pinealocytes are perivascular polyhedral endocrine cells, with an eccentric nucleus, and contain pigments, vitamin C, lipids and numerous mitochondria. In addition to these cells in the epiphysis are also present fibrocytes, mast cells, plasmocytes, lymphocytes, macrophages and pigment cells.

The epiphysis represents a portion of the brain with multiple implications in the economy of the organism. Beside the gonadal function, the epiphysis is correlated with the light (which inhibits the epiphyseal activity) - darkness (which stimulates the epiphyseal activity) succession. A significant role is ascribed to the autonomic nerve supply of the ephiphysis (Kappers, 1965) and especially to the retinohypothalamic projection, a recently described neural tract (Hendrickson et al., 1972; Reiter 1973; Moore, 1978). Taking into account the retino-ophthalmic connections, Gusek (1981) suggested the inclusion of the pineal body into an ophthalmopineal system. Moreover, the epiphysis contains numerous substances of an utmost biochemical and endocrine importance.

The epiphysis is structured in the form of lobules delineated by connective septa (Şt. M. Milcu, 1957; Şt. M. Milcu and Vrejoiu, 1959), containing vessels and unmyelinated nerve fibres, sending off projections to the inside of the lobules. The pineal cells are grouped into dense nests and *trabeculae*, sometimes around a small lumen. At the periphery of the lobules, the projections of the pinealocytes form a marginal plexus, establishing close contacts with the capillary network. The unmyelinated nerve fibres situated between the pineal cells possess granulations of 450-1,200Ĺ, containing noradrenalin and serotonin. The exact site of the synaptic contact between the autonomic nerve fibres and the pineal cell is not elucidated yet. An interesting finding is the presence of synaptic structures between the adjacent pineal cells, which have been designated by Hopsu and Arstila (1965) under the name of somatosomatic synapses. These parenchymatous anatomic interrelation give the endocrine morphological character to the epiphysis, which assures photosensorial and maybe also thermal regulation by a complex mechanism. The term „*neuroendocrine transducer*" (Wurtman and Anton-Tay, 1969) expresses better the functional relationship. Like other neuroendocrine transducers, the pineal body converts a neural impulse, i.e. neurotransmitter released at a synapse, into an endocrine response: the secretion of melatonin. Beside other transducing cells, such as the chromaffine cells of the adrenal medulla, the releasing factor secreting cells of the median eminence, the vasopressin and oxitocyn producing cells and the beta-pancreatic cells, the pinealocytes respond also both the humoral and nervous impulses, translating the language of nerves into the „language of glands" (Wurtman, 1979).

A structural characteristic of the human epiphysis is the presence of calcareous concretions, termed corpora arenacea *(acervulus cerebri)*, sometimes visible also macroscopically under the form of yellowish granulations. These concretions appear at any age (Şt. M. Milcu and Vrejoiu, 1959), often even in the newborn. They are made up of a mixture of proteins and polysaccharide complexes, with a rich indole content, displaying a secondary calcification process. The calcification begins usually in the cytoplasm and the nucleus, sometimes in the extracellular space.

The pinealocytes contain a great number of tubular mitochondria, ribosomes, lysosomes, smooth and granular endoplasmic reticulum, a well developed Golgi apparatus, microtubuli and liposomes, which represent probably the support of secretion products.

The ultrastructural aspects, which attest the endocrine character of the epiphysis, are based on the presence of an acinous architectonics, on the structure of capillaries delimited by a pericapillary space provided with a basal membrane, as well as on the release of cell granulations (lysosomes) in the perivascular space.

The metathalamus

Two nuclei situated beneath the pulvinar constitute the metathalamus. These nuclei, termed the medial and lateral geniculate bodies, make up diencephalic relays of the auditory and visual pathways.

The medial geniculate body. The deutoneuron of the auditory pathway has its site in the anterior and posterior cochlear bulbar nuclei. Its axon runs along the lateral lemniscus and the trapezoid body and forms in the medial geniculate body a synapse with the third thalamo-cortical neuron; via the acoustic radiation, it arrives in the temporal cortex. We mention that the medial geniculate body has a certain topography, so that the nerve impulses generated by high-pitched sounds emerge through the medial part and those induced by low-pitched sounds, through the lateral part. The medial geniculate body has connections with the posterior quadrigeminal tubercle.

The lateral geniculate body. We mention from the onset that the optic nerve, the chiasma and the optic tract represent, embryologically, expansions of the diencephalon, therefore the optic tract should be described at this level. Nevertheless, for the unity of presentation, we shall describe it concomitantly with the optic analyzer.

The lateral geniculate body makes up the relay where the connection between the retinodiencephalic neuron and the diencephalocortical neuron occurs.

Some authors consider that there are also fibres of the retinodiencephalic neuron which run directly into the pulvinar or into the anterior quadrigeminal tubercles. The pupillary fibres cross the lateral geniculate body without stopping and terminate in the anterior quadrigeminal tubercles. The lateral geniculate body has nerve connections with the anterior quadrigeminal tubercles. Some authors consider that the pupillary fibres end in the pretectal formation.

The subthalamus

The subthalamus lies in the subthalamic (suboptic) region, it is continuous with *the peduncular tegmentum* and projects, at the level of opening of the third ventricle beneath Monro's hypothalamic sulcus being embryologically derived from the basal lamina of the primitive neural tube.

The subthalamus integrates in the paleoencephalic motor circuits and is under the dependence of the striate nuclei *(putamen and pallidum)*. We include at this level the formations already studied when we dealt with the mesencephalon (the superior reticular nuclei, the red nucleus, *locus niger*) and the formations called Luys' body (subthalamic nucleus) and *zona incerta*.

Luys' body is a biconvex mass of grey matter, situated between *the pallidum* and *the locus niger*; it seems to be the centre of the swinging movements in locomotion.

The zona incerta, situated immediately beneath the thalamus, is continuous with the thalamic reticular substance, respectively with the intralaminar and centromedian nuclei, and the same functions are ascribed to it.

The diencephalic secretory apparatus

In addition to the fact that they are the place where the neurovegetative centres and pathways are grouped, the walls of the third ventricle represent also that region of the neuraxis where important secretory phenomena playing a significant role in the function of the nervous system and in the maintenance of the homeostasis of the organism. The secretion products are neurotrophic, neuroregulating and psychoregulating.

The secretory phenomena at the level of the diecenphalon are the neurosecretion, the neurocrinia and the ependymocrinia.

Neurosecretion. It has been proved that the magnocellular groups of the supraoptic and paraventricular nuclei exert an endocrine activity. These neurons elaborate a neurosecretion which descend along the axons up to the posterior hypophysis (Scharrer). The neurosecretion is accumulated in the hypophysis and then released in the blood; it seems that the site of elaboration of oxytocin and vasopressin does not lie in the hypophysis but in the hypothalamus.

The hypothalamic nuclei secrete hormones which pass into the tuberal region, stimulating and regulating the activity of the adenohypophysis. Likewise the existence of a hypothalamic neurosecretion running along the nerve fibres disposed in the region of the epiphysis, of the mesencephalon and of *the amygdalian nucleus*, is also assumed.

Neurocrinia. The term neurocrinia defines the phenomenon of excretion, inside the nervous system, of the products elaborated by the glandular or secretory organs, such as the hypophysis, the epiphysis (pineal gland) or the ependyma (the opposite of the term hemocrinia, which means the discharge of the glandular secretion products into the blood). It is the reverse of the phenomenon of neurosecretion (See also the chapter „The diencephalo-hypophysial system).

Ependymocrinia. The ependyma of the third ventricle is actively involved in the secretory phenomena of the diencephalon. In this respect, the ependymal organs (Legait) or the neurocrine glands of the encephalon (Roussy and Mosinger), three in number: the subcommissural organ, the subfornix organ and the paraventricular organ.

- *The subcommissural organ* is formed by the ependymal epithelium of the inferior (ventricular) surface of the posterior commissure.

- *The subfornix organ* is situated between the two Monro's foramina, above the optic chiasma, hence at the level of the anterior wall of the median region of the telencephalon.

- *The paraventricular organ* corresponds to that portion of the ependyma which covers the paraventricular nucleus

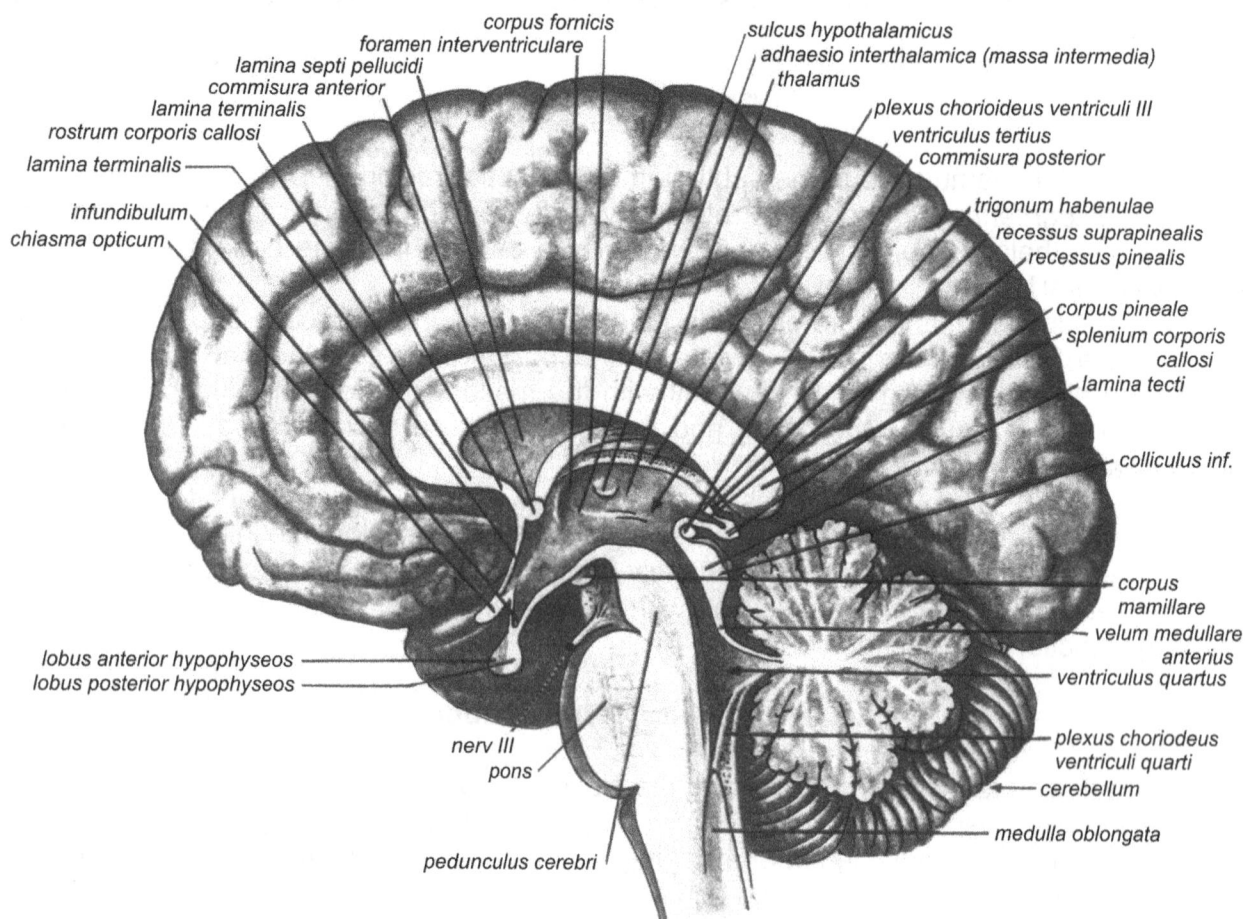

Fig. 82 Mediosagittal section through the encephalon, with demonstration of the fourth ventricle

100

The ependymal cells of these organs situated at the entrance (Monro's foramen) and at the exit (opening of the cerebral aqueduct) of the third ventricle, elaborate colloidal products.

Some authors (ascribe a secretory role also to the ependyma which covers the fourth ventricle).

In the context should be also mentioned Reissner's fibres, attached above on the subcomissural organ and passing through the cerebral aqueduct, the fourth ventricle and the central canal of the spinal cord, up to its terminal part. Maybe they play a part in the achievement of a unity of the organs which have a role in ependymocrinia. Moreover, there are numerous connection nerve fibres between these ependymal glandular organs, as well as between these organs and the vegetative formations of the diencephalon and of other parts of the nervous system, which explains their great pathophysiologic significance.

Olson's recent researches prove that the subcommissural organ, by its secretions demonstrated by histochemical methods, plays a role in the enzymatic activities and in the detoxification processes of the nerve cell.

In addition, the idea that the ependyma of the third ventricle has not only a secretory role, but also a function of sensitive receptor surface, was recently acknowledged; it seems that this is the explanation of the vegetative clinical phenomena (somnolence, showing down of the pulse rate, vomiting) in ventricular hypertension syndromes.

The third ventricle
(Ventriculus tertius)

The third ventricle is an elongated, vertical, median cavity, with the largest diameter anteroposterior. In it opens posteriorly the cerebral aqueduct (aqueductus cerebri); through the interventricular or Monro's foramina (foramina interventriculares), this ventricle communicates with the lateral ventricles. Each interventricular foramen is bounded (Fig. 82,83):
- anteriorly by the free part of the column of the fornix (pars libera columnae fornicis);
- posteriorly, by the thalamus;

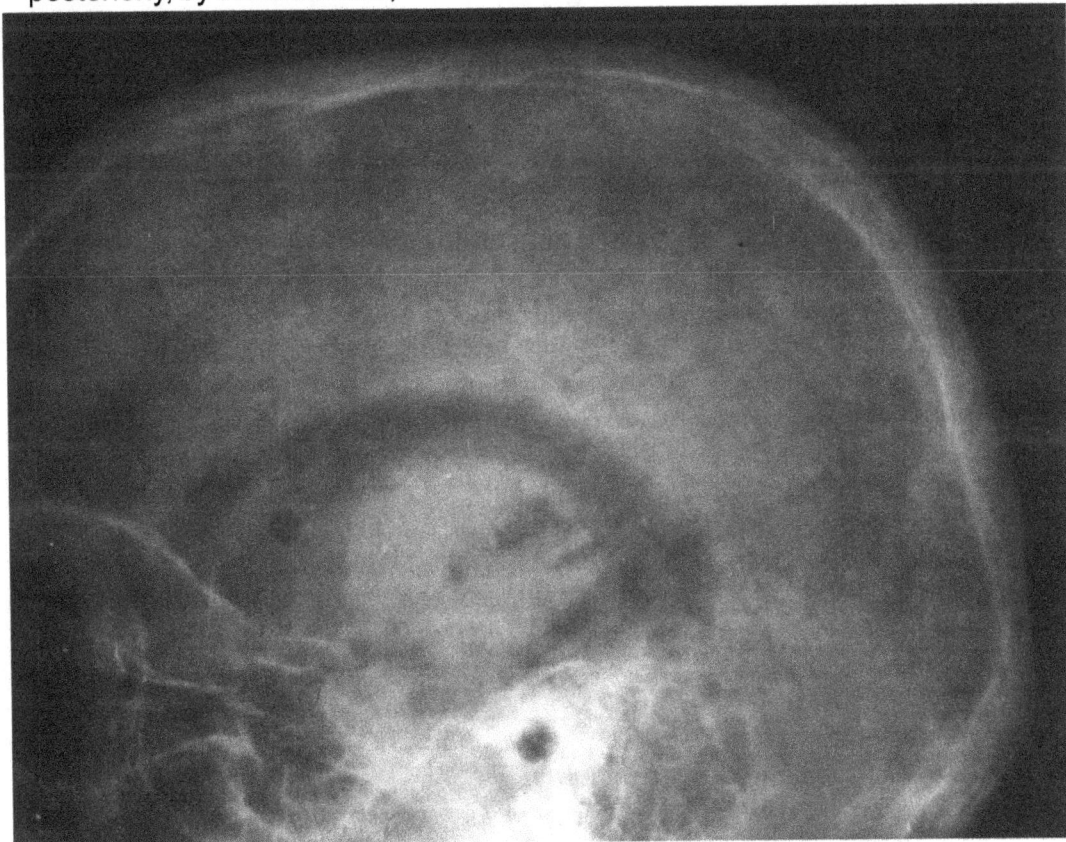

Fig. 83 Ventriculography – lateral view

- inferiorly, by the hypothalamic *(sulcus hypothalamicus)*;
- superiorly by the lamina of the tectum of the mesencephalon with *the tela choroidea* of the third ventricle.

The third ventricle has the following walls:
- the anterior (rostral) vertical wall consists of: the free part of the column of the fornix, the anterior commissure, the terminal lamina (a thin layer of grey matter which during the embryonic life makes up the most cranial part of the encephalon and the site where the anterior neuropore is assumed to close) and the optic recess *(ricessus opticus)* which represents a prolongation of the third ventricle between the chiasma and *the terminal lamina*;
- the inferior (caudal) wall represents the floor of the third ventricle and has connections with, the optic chiasma, *the infundibulum*, the infundibular recess, *the tuber cinereum*, the mamillary bodies, the posterior perforated substance and the cerebral peduncles;
- two lateral walls, through which the third ventricle has connections with the thalamus, the interventricular foramina and the hypothalamic sulcus;
- the superior (cranial wall) represents the roof of the third ventricle and has connections with *the tela choroidea ventriculi III* (the free edge of *the thalamic taenia*);
- the posterior (dorsal wall), oriented obliquely downwards and forwards, consists of the pineal recess, the posterior commissure, the entrance into cerebral aqueduct and the tegmental wall of the mesencephalon.

THE TELENCEPHALON
(Telencephalon seu cerebrum)

The telencephalon is the anterior part of the prosencephalon, which gives rise, by its development to the cerebral hemispheres *(cerebrum)*.

The higher we ascend the phylogenetic scale, the more developed in the telencephalon, the highest degree being attained in man. Its mean weight (1,300 g, togetter with the cerebellum and the brainstem) is proportional to the size of the body; however, exceptions are very frequent.

The telencephalon is formed of the two cerebral hemispheres, separated by the very deep longitudinal cleft of the brain, the longitudinal fissure *(fissura longitudinalis cerebri)*, in which *the falx cerebri*, a derivative of *the dura mater*, penetrates. The left hemisphere is more developed in right-handed subjects.

Across the median plane, the two cerebral hemispheres are connected by *the corpus callosum* and the white commissure *(commisura anterior)* which forms the strongest link between them. Below the hemispheres and continuous with them is the brainstem. The brain is separated from the cerebellum, posteriorly, by a horizontal indentation, called the transverse incisure or notch of the cerebrum.

By their base *(basis cerebri)*, the cerebral hemispheres correspond to the anterior and middle region of the base of the skull, as well as to *the tentorium cerebelli* formed by the dura mater. Each hemisphere has three surfaces (lateral, medial and inferior) and two poles (anterior and posterior).

The lateral surface (facies convexa seu superolateralis cerebri) is in relation with the skull cap (the cranial vault and the lateral parts of the skull); the medial surface *(facies medialis)* is situated sagitally and is in relationship with that of the opposite side. At the level of the longitudinal fissure, the lateral surface and the medial surface are continuous with each other by a rounded border.

The inferior surface (facies inferior cerebri), in connection with the base of the skull, is divided into two parts by the lateral sulcus.

Each hemisphere has an anterior pole *(polus frontalis)* and a posterior pole *(polus occipitalis)* and a temporal pole *(polus temporalis)*.

The scissures which furrow the surface of the cerebral hemisphere delineate on the surface of the brain four crebral lobes: the frontal *(lobus frontalis)*, the parietal *(lobus parietalis)*, the temporal *(lobus temporalis)* and the occipital *(lobus occipitalis)* lobes.

The brain, like the whole cerebrospinal system is made up of grey matter (40%) and white matter (60%). Most of the grey matter lies in the cortex (33% on the surface of the hemispheres) and a smaller amount (7%) in the depth of the base of the brain under the form of grey nuclei (striate body) or basal ganglia.

In each hemisphere is sited a lateral ventricle. The constituent parts of the telencephalon are: the cortex *(cortex cerebri)*, the grey nuclei *(corpus striatum)*, the white matter *(centrum ovali)* and the first and second ventricles *(ventriculus lateralis)*.

The cerebral cortex
(Cortex cerebri, Pallium)

Phylogenetically, the cerebral cortex has developed by a double process: of telencephalization and of corticalization.

The telencephalization consists in the migration, into the endbrain, of the sensory and motor zones of the subcortical centres, which remain subordinated to the cortex.

The corticalization represents the morphofunctional differentiation process of the cortical zones (Fig. 84 - 89).

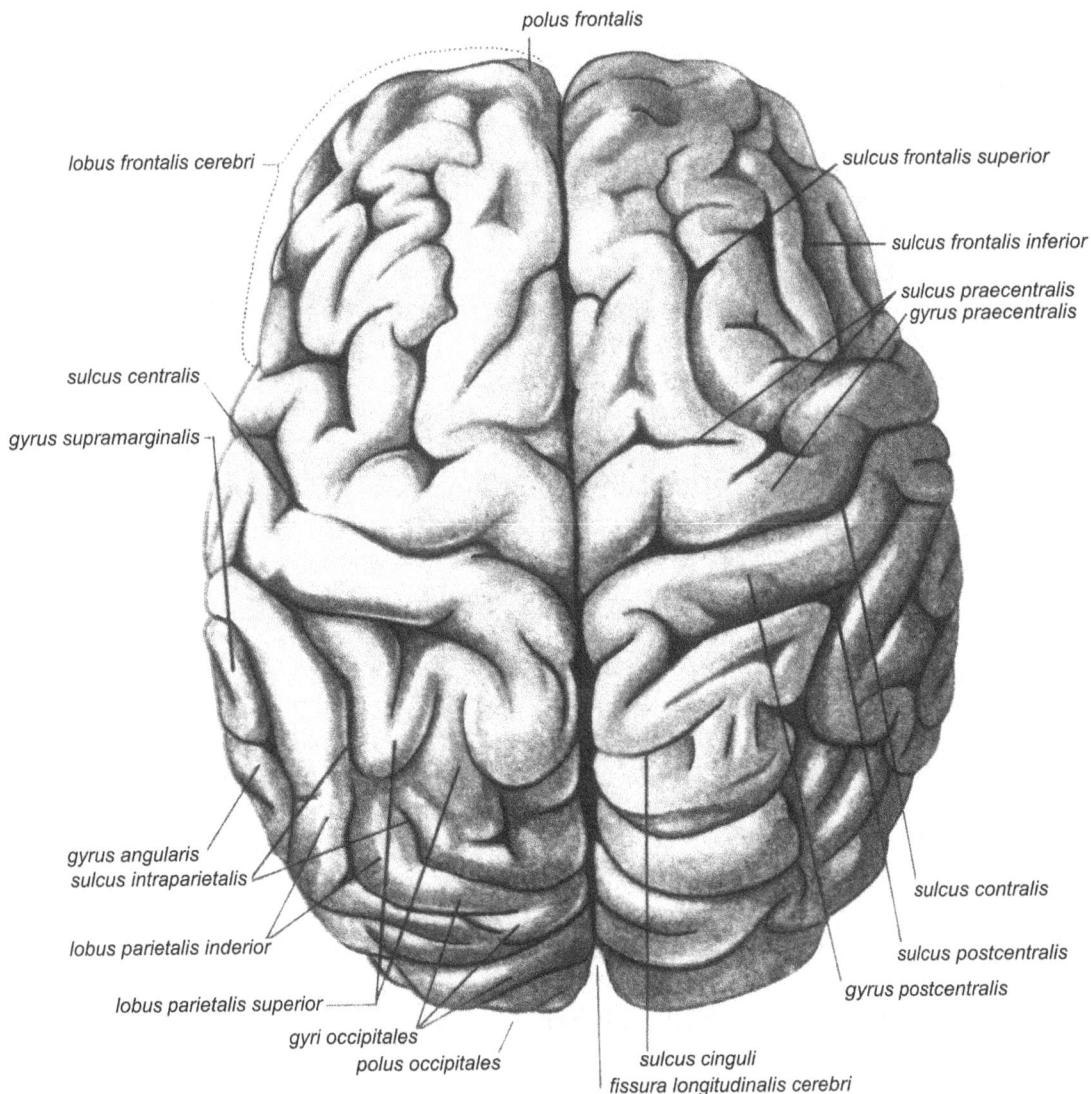

Fig. 84 Superior surface of the cerebral hemispheres

External configuration

The cerebral cortex consists of eminences named *gyri (gyri cerebri)* and of irregular furrows *(sulci cerebri)*, which separate them. In some mammals, the encephalon is smooth, called lisencephalon (rabbit), while in others, especially in man, as a result of the unequal growth of some territories of the cortex, as well as owing to the disproportionality between the development of the cerebral functions and the capacity of the neurocranium, convolutions are developing and the folding phenomenon, called gyrencephaly, occurs, the opposite of the smooth encephalon. *The gyrencephaly* has as effect the increase of the cortical surface and of the amount of grey cortical matter.

The gyri and *sulci* show variations between individuals, but certain *sulci* and convolutions are constant and play a significant role in the delineation of the cerebral lobes and in the location of the cortical fields. The result, as it was shown, is the formation of four lobes: frontal, parietal, temporal, and occipital, separated by constant sulci.

The lateral cerebral sulcus or fissure, called also sylvian fissure *(sulcus cerebri lateralis, Sylvii)* is the deepest. It separates the frontal lobe from the temporal lobe. It begins latero-inferiorly, then extends obliquely backwards and upwards up to above the middle of the insula. The lateral cerebral sulcus divides into three rami:

- the anterior ramus, which runs forwards into the frontal lobe;
- the ascending ramus, which courses upwards towards the inferior frontal gyrus and;
- the posterior ramus, which is the main sulcus and runs upwards and backwards, ending in the parietal lobe, in the supramarginal gyrus.

Rolando's central sulcus *(sulcus centralis Rolandi)* begins at the junction of the anterior third with the middle third of the longitudinal fissure and runs downwards and forwards towards the middle of the lateral cerebral sulcus, but without having a connection with the latter. The central sulcus forms the posterior boundary of the frontal lobe.

The parieto-occipital sulcus *(sulcus parieto-occipitalis)*, 1-5 cm in length, starts from the junction of the middle third with the posterior third of the superior border of the cerebral hemisphere and lies at about 5 cm before the occipital lobe.

It has a short course visible on the lateral surface of the hemisphere, after which it extends on the medial surface of the hemisphere, running forwards and downwards up to its junction with the calcarine sulcus *(sulcus calcarinus)*. It separates the parietal from the occipital lobe.

The transverse occipital sulcus *(sulcus occipitalis transversus)* runs parallel to the occipital portion of the superior border of the cortex and lies at a fingerbreadth from the latter. It is often medially continuous with the intraparietal sulcus, but it does not extend laterally up to the free border of the hemispheres. It borders imprecisely the occipital , parietal and temporal lobes.

The external surface of the brain

The following gyri and sulci are sited on the surface of each lobe:

I. **The frontal lobe** *(lobus frontalis)* has the following boundaries: laterally the lateral cerebral sulcus and posteriorly the central sulcus.

On the frontal lobe lie three sulci which separate from each other gyri (convolutions, according to the old nomenclature):

1) The superior frontal sulcus *(sulcus frontalis superior)*, made up of two parts: a superior (posterior) part and an inferior (orbital) part. This furrow is parallel to the longitudinal fissure.

2) The inferior frontal sulcus *(sulcus frontalis inferior)*, which has a course similar to the former and lies laterally to it.

104

Betwen the two sulci are situated three frontal gyri (frontal convolutions);

- the superior frontal gyrus *(gyrus frontalis superior)* (first or superior frontal convolution);

- the middle frontal gyrus *(gyrus frontalis medius)* (second or middle frontal convolution);

- the inferior frontal gyrus *(gyrus frontalis inferior)* (third or inferior frontal convolution, Broca's convolution), which is subdivided by the anterior, ascending and posterior rami of the lateral cerebral fissure into: the orbital part *(pars orbitalis)* (the anterior part), the triangular part *(pars triangularis)*(the middle part) and the opercular part *(pars opercularis)* (the posterior part), the latter forming Broca's speech centre.

3) The prerolandic vertical sulcus or precentral sulcus *(sulcus praecentralis)* has the same direction as the central sulcus and is anterior to it; it participates in the formation of the fourth gyrus.

- The precentral gyrus *(gyrus praecentralis* (ascending frontal convolution)) is sited between the precentral sulcus and the central sulcus.

The lesions of the frontal lobe manifest themselves clinically as motility disturbances, speach impairments, mental and vegetative disorders. The motor impairments manifest themselves either under the form of cortical hemiplegia (destruction of the areas 4 and 6), or as epileptic Jacksonian scizures (excitation of Brodman's area 4). Mental disorders appear as a consequence of tumours or of vascular or traumatic lesions of the frontal lobe and are located in areas 9, 10, 11 and 12.

They assume either the aspect of affectivity impairments - Gastrowitz type „moria" characterized by euphoria, infantilism etc . - or that of intellectual activity disorders (lack of interest) etc.

II. **The parietal lobe** *(lobus parietalis)* has the following boundaries: anteriorly the central (Rolando's) sulcus, the pars marginalis of the cingulate sulcus (on the medial surface of the hemisphere); posteriorly, the boundaries are less clear-cut (the parieto-occipital and the transverse occipital sulcus; inferiorly - the lateral cerebral (Sylvian) sulcus. On the surface of the parietal lobe may be distinguished the following sulci and gyri:

1) The postcentral sulcus *(sulcus postcentralis)*, situated behind the central sulcus, parallel to it; together with the latter, it delineates the postcentral gyrus *(gyrus postcentralis)* - the ascending parietal convolution.

- The ascending parietal gyrus unites with the ascending frontal convolution - turning round the inferior extremity of Rolando's scissure - and they form together the frontal operculum *(operculum frontalae)* or Rolando's operculum. As it has been shown, in the depth of the Sylvian scissure, beneath the Rolando's operculum and the first temporal gyrus lies the lobe of the insula (Reil's insula).

2) The intraparietal sulcus *(sulcus intraparietalis)*, T - shaped, parallel to the longitudinal fissure, unites anteriorly with the postcentral sulcus and posteriorly usually, with the transverse occipital sulcus. Above it lies the superior parietal lobe and beneath it, the inferior parietal lobe.

- The supramarginal convolution *(gyrus supramarginalis)* or the parietotemporal (Broca's) fold surrounds the posterior ramus of the lateral cerebral sulcus; it is formed by the union of the superior temporal gyrus with the anterior portion of the second parietal gyrus.

- The curved fold called angular gyrus *(gyrus angularis)* turns round the superior temporal sulcus; formed by the union of the posterior portion of the first temporal gyrus with that of the second temporal gyrus.

The lesions at the level of the parietal lobe are characterized by impairments of the superficial and deep sensibility (equally distributed, contralateral hemihypesthesia), speech disorders (a lesion due to damage caused to Wernicke's area in the left parietal cortex) and, especially, abnormalities in the body scheme representation (wrong appreciations of the body size), often induced by the presence of a tumoral process in the right parietal lobe.

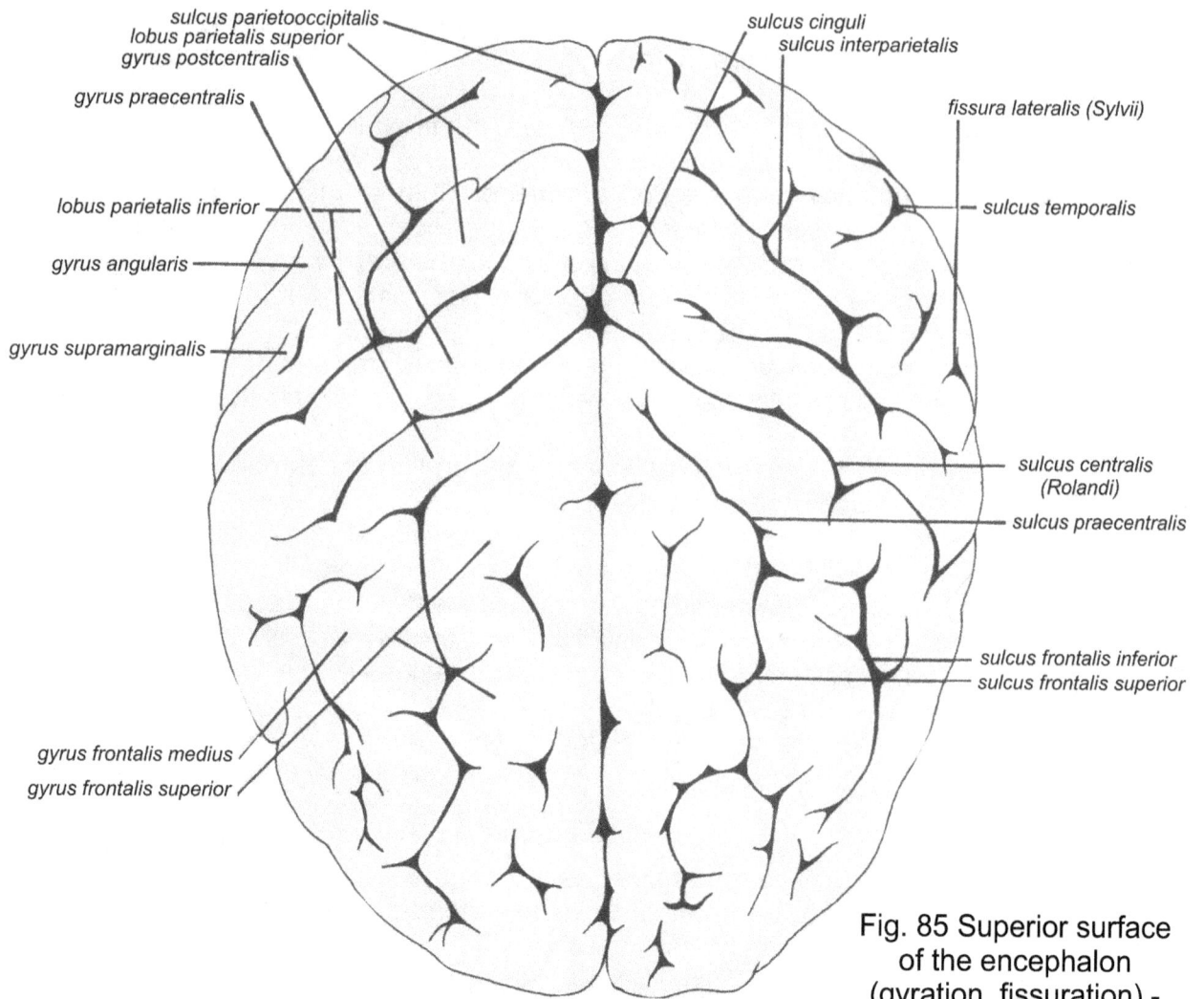

sulcus parietooccipitalis
lobus parietalis superior
gyrus postcentralis
gyrus praecentralis

lobus parietalis inferior

gyrus angularis

gyrus supramarginalis

sulcus cinguli
sulcus interparietalis

fissura lateralis (Sylvii)

sulcus temporalis

sulcus centralis (Rolandi)
sulcus praecentralis

sulcus frontalis inferior
sulcus frontalis superior

gyrus frontalis medius
gyrus frontalis superior

Fig. 85 Superior surface
of the encephalon
(gyration, fissuration) -
scheme

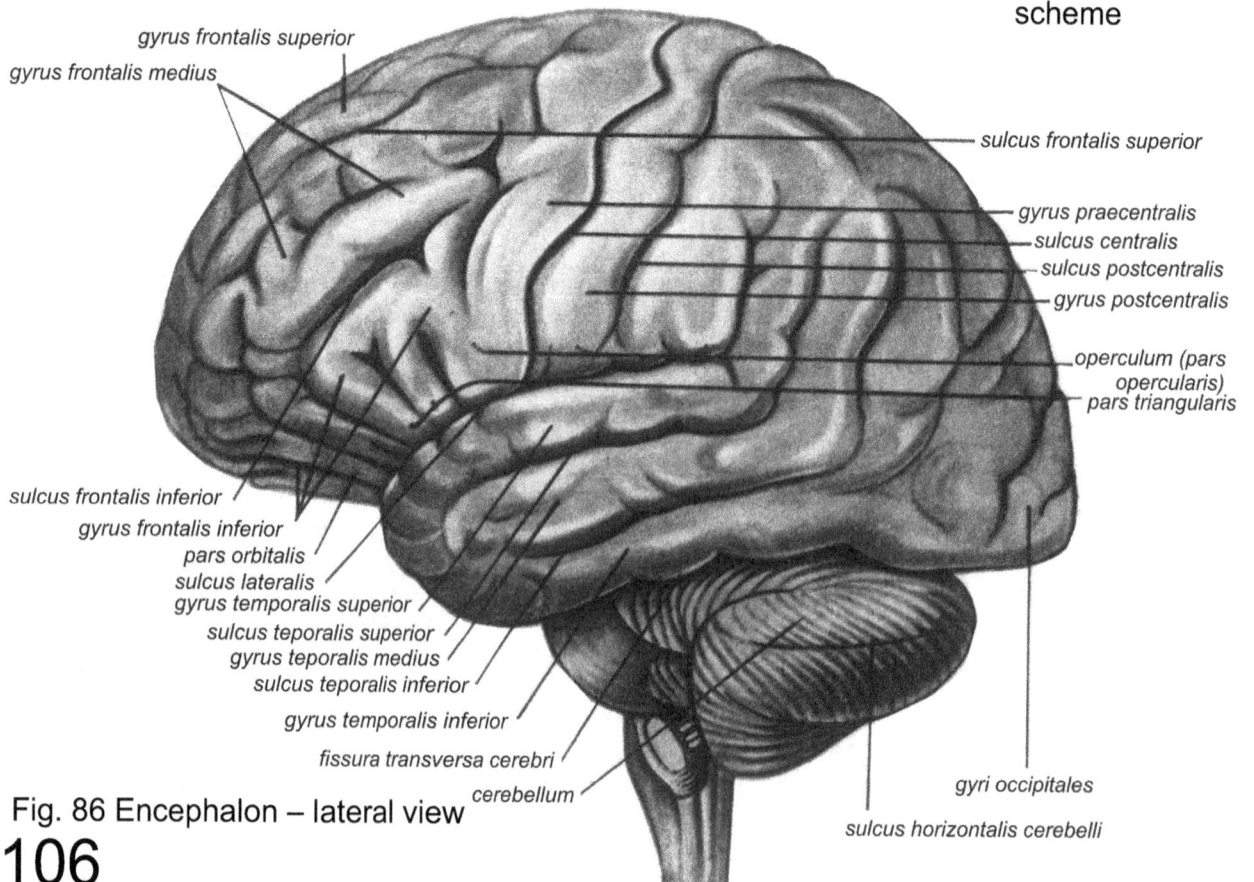

gyrus frontalis superior
gyrus frontalis medius

sulcus frontalis superior

gyrus praecentralis
sulcus centralis
sulcus postcentralis
gyrus postcentralis

operculum (pars opercularis)
pars triangularis

sulcus frontalis inferior
gyrus frontalis inferior
pars orbitalis
sulcus lateralis
gyrus temporalis superior
sulcus teporalis superior
gyrus teporalis medius
sulcus teporalis inferior
gyrus temporalis inferior
fissura transversa cerebri
cerebellum

gyri occipitales
sulcus horizontalis cerebelli

Fig. 86 Encephalon – lateral view

106

Fig. 87 Lateral surface of the cerebral hemispheres (gyration, fissuration) - scheme

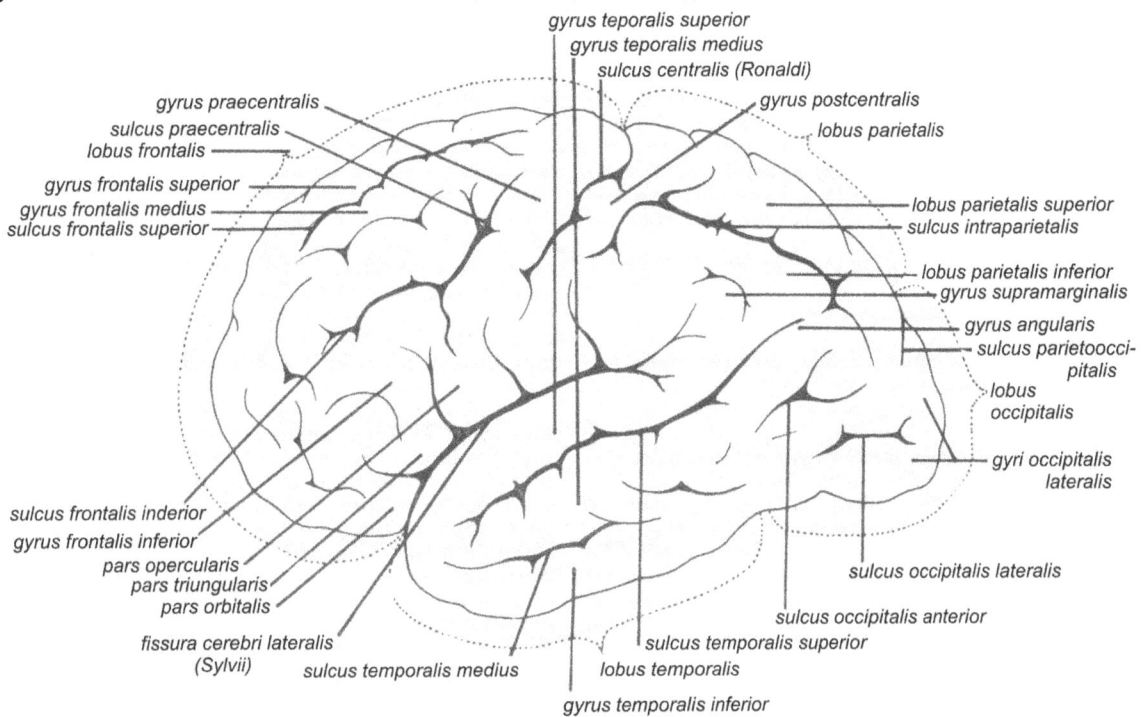

gyrus teporalis superior
gyrus teporalis medius
sulcus centralis (Ronaldi)
gyrus postcentralis
lobus parietalis
gyrus praecentralis
sulcus praecentralis
lobus frontalis
gyrus frontalis superior
gyrus frontalis medius
sulcus frontalis superior
lobus parietalis superior
sulcus intraparietalis
lobus parietalis inferior
gyrus supramarginalis
gyrus angularis
sulcus parietoocci-
pitalis
lobus
occipitalis
gyri occipitalis
lateralis
sulcus frontalis inderior
gyrus frontalis inferior
pars opercularis
pars triungularis
pars orbitalis
fissura cerebri lateralis
(Sylvii)
sulcus temporalis medius
sulcus occipitalis lateralis
sulcus occipitalis anterior
sulcus temporalis superior
lobus temporalis
gyrus temporalis inferior

Fig. 88 Lateral surface of the cerebral hemispheres (gyration, fissuration)

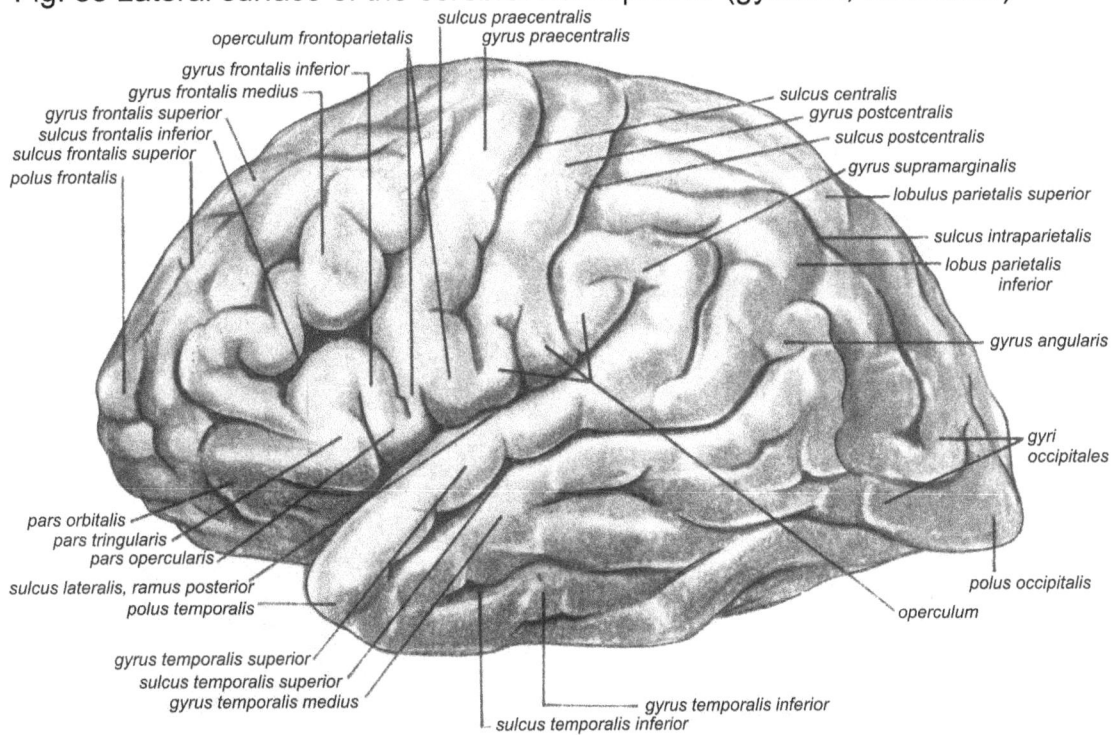

sulcus praecentralis
gyrus praecentralis
operculum frontoparietalis
gyrus frontalis inferior
gyrus frontalis medius
gyrus frontalis superior
sulcus frontalis inferior
sulcus frontalis superior
polus frontalis
sulcus centralis
gyrus postcentralis
sulcus postcentralis
gyrus supramarginalis
lobulus parietalis superior
sulcus intraparietalis
lobus parietalis
inferior
gyrus angularis
gyri
occipitales
pars orbitalis
pars tringularis
pars opercularis
sulcus lateralis, ramus posterior
polus temporalis
polus occipitalis
operculum
gyrus temporalis superior
sulcus temporalis superior
gyrus temporalis medius
gyrus temporalis inferior
sulcus temporalis inferior

III. **The temporal lobe** *(lobus temporalis)* is situated below the sylvian fissure and has the following boundaries: above - the lateral cerebral sulcus; behind - the transverse occipital sulcus; below - the hippocampal sulcus (on the base of the brain). On its surface lie the following sulci and gyri.

1) The superior temporal sulcus *(sulcus temporalis superior)*, sited beneath the lateral cerebral sulcus and parallel to it.

2) The middle temporal sulcus *(sulcus temporalis medius)*, parallel to the superior temporal sulcus.

3) The inferior temporal sulcus *(sulcus temporalis inferior)*, situated on the base of the brain.

107

- The superior temporal gyrus is bounded by the lateral cerebral fissure and the superior temporal sulcus. On its anteromedial surface, oriented towards the insula, lie 3-4 small transverse temporal gyri, named Heschl's transverse gyri, separated from each other by transverse grooves *(sulci transversi)*.

- The middle temporal gyrus (the second temporal convolution) *(gyrus temporalis medius)* is bounded by the superior and the middle temporal sulci.

- The inferior temporal gyrus (the third temporal convolution) *(gyrus temporalis inferior)* is bounded by the middle and the inferior temporal sulci.

The temporal lobe contains distinct areas of cerebral neocortex, the lesions of which induce auditory impairments under the form of tinnitus (noises, crackling, whizzing in the ears), speech disorders (Wernicke type aphasia) and hallucinations (auditory, visual, olfactory, gustatory). Characteristic are the appearance of a dream state, the revived memory of events experienced in the past, the transient loss of the present etc.

IV. **The occipital lobe** *(lobus occipitalis)* is bounded anteriorly by the transverse occipital sulcus and by the parieto-occipital sulcus.

On its surface, two transverse sulci lie, named lateral occipital sulci *(sulci occipitales laterales)*: the superior occipital sulcus, which is a prolongation of the intraperietal sulcus, and the inferior occipital sulcus. They delineate irregularly three occipital gyri (I, II and III) *(gyri occipitales laterales superior, medius et inferior)*. (Fig. 86, 87).

The lesions of the occipital lobe, both of the occipital cortex and of the geniculocalcarine fibres, are characterized by impairments of the visual field, which may acquire various forms: cortical blindness (loss of sight), visual agnosia (loss of the ability to recognize objects by sight), visual hallucinations (bright or coloured spots) or oculomotor impairments.

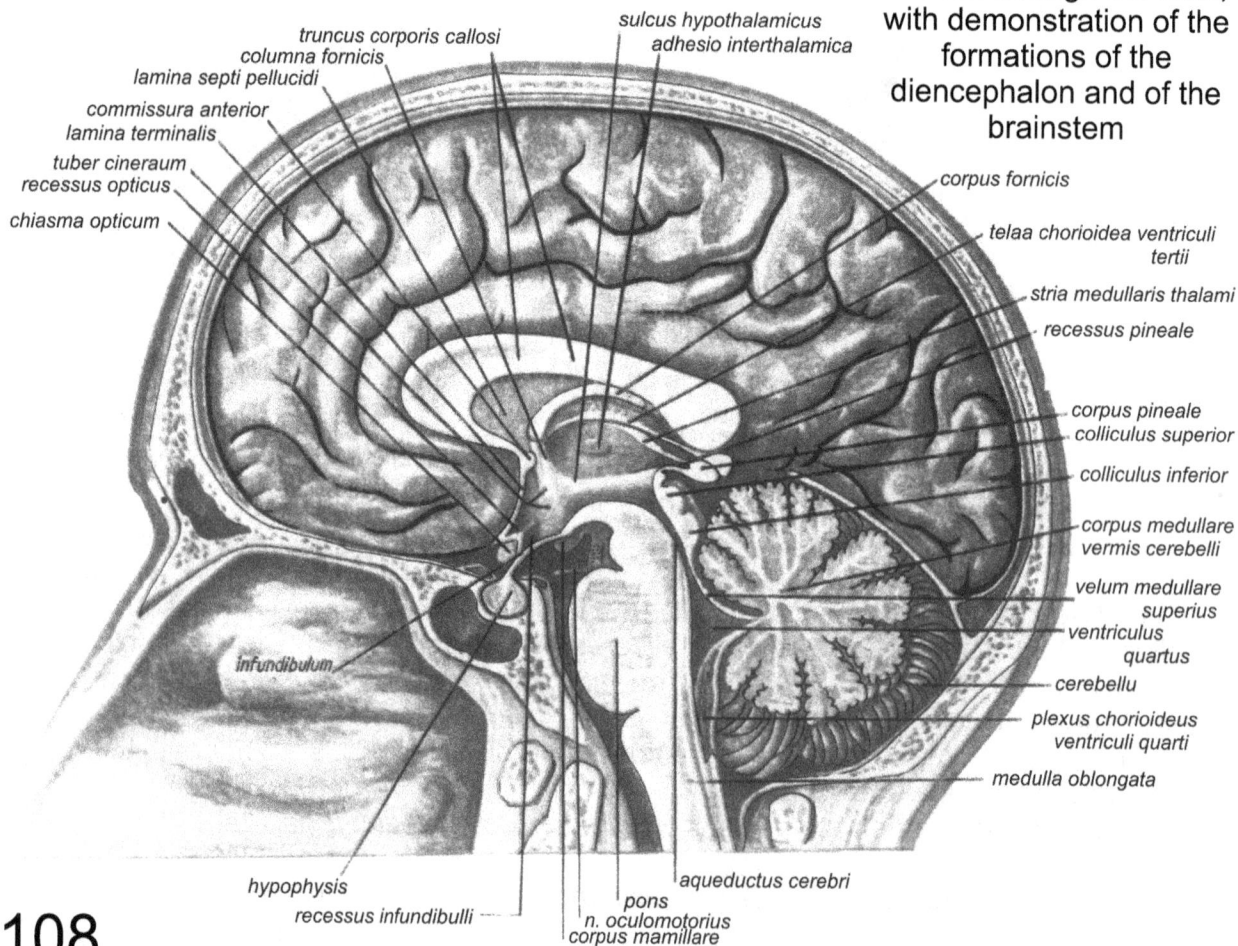

Fig. 89 Mediosagittal section through the skull, with demonstration of the formations of the diencephalon and of the brainstem

truncus corporis callosi
columna fornicis
lamina septi pellucidi
commissura anterior
lamina terminalis
tuber cineraum
recessus opticus
chiasma opticum
sulcus hypothalamicus
adhesio interthalamica
corpus fornicis
telaa chorioidea ventriculi tertii
stria medullaris thalami
recessus pineale
corpus pineale
colliculus superior
colliculus inferior
corpus medullare vermis cerebelli
velum medullare superius
ventriculus quartus
cerebellu
plexus chorioideus ventriculi quarti
medulla oblongata
infundibulum
hypophysis
recessus infundibulli
pons
n. oculomotorius
corpus mamillare
aqueductus cerebri

Fig. 90 Medial surface of the cerebral hemispheres (gyration, fissuration) - scheme

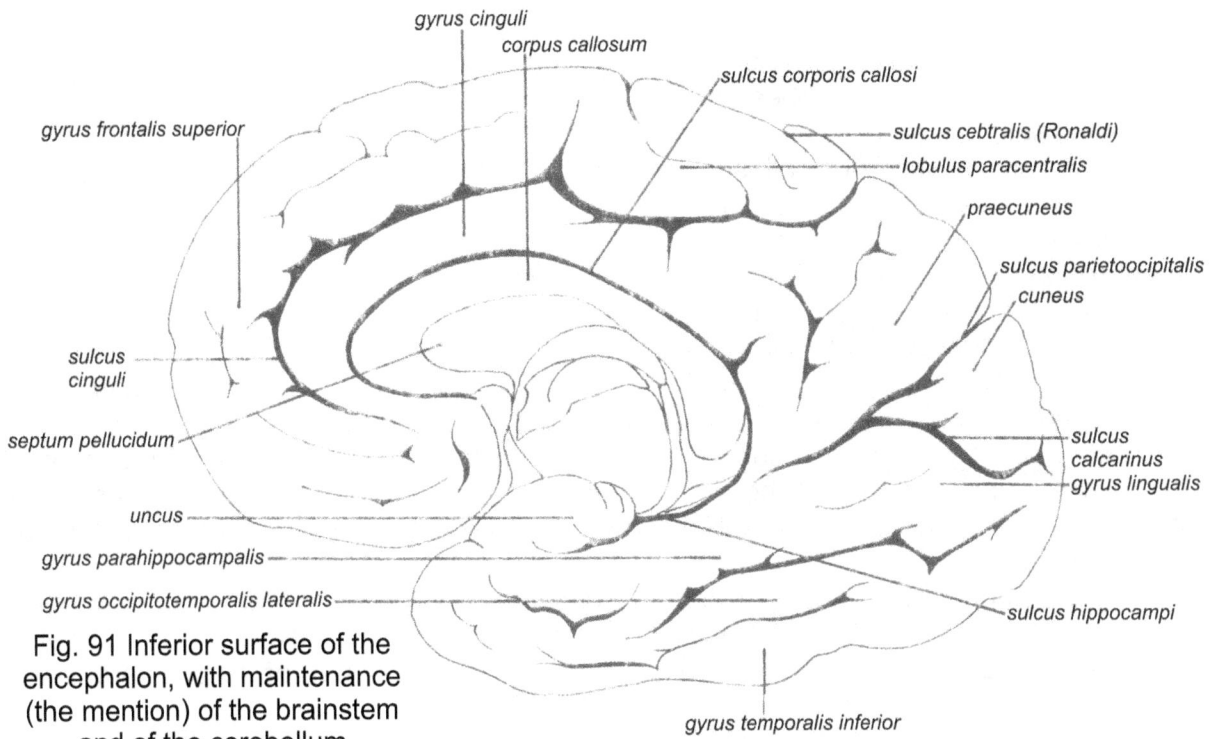

gyrus cinguli

corpus callosum

sulcus corporis callosi

gyrus frontalis superior

sulcus cebtralis (Ronaldi)

lobulus paracentralis

praecuneus

sulcus parietoocipitalis

cuneus

sulcus cinguli

septum pellucidum

sulcus calcarinus

gyrus lingualis

uncus

gyrus parahippocampalis

gyrus occipitotemporalis lateralis

sulcus hippocampi

gyrus temporalis inferior

Fig. 91 Inferior surface of the encephalon, with maintenance (the mention) of the brainstem and of the cerebellum

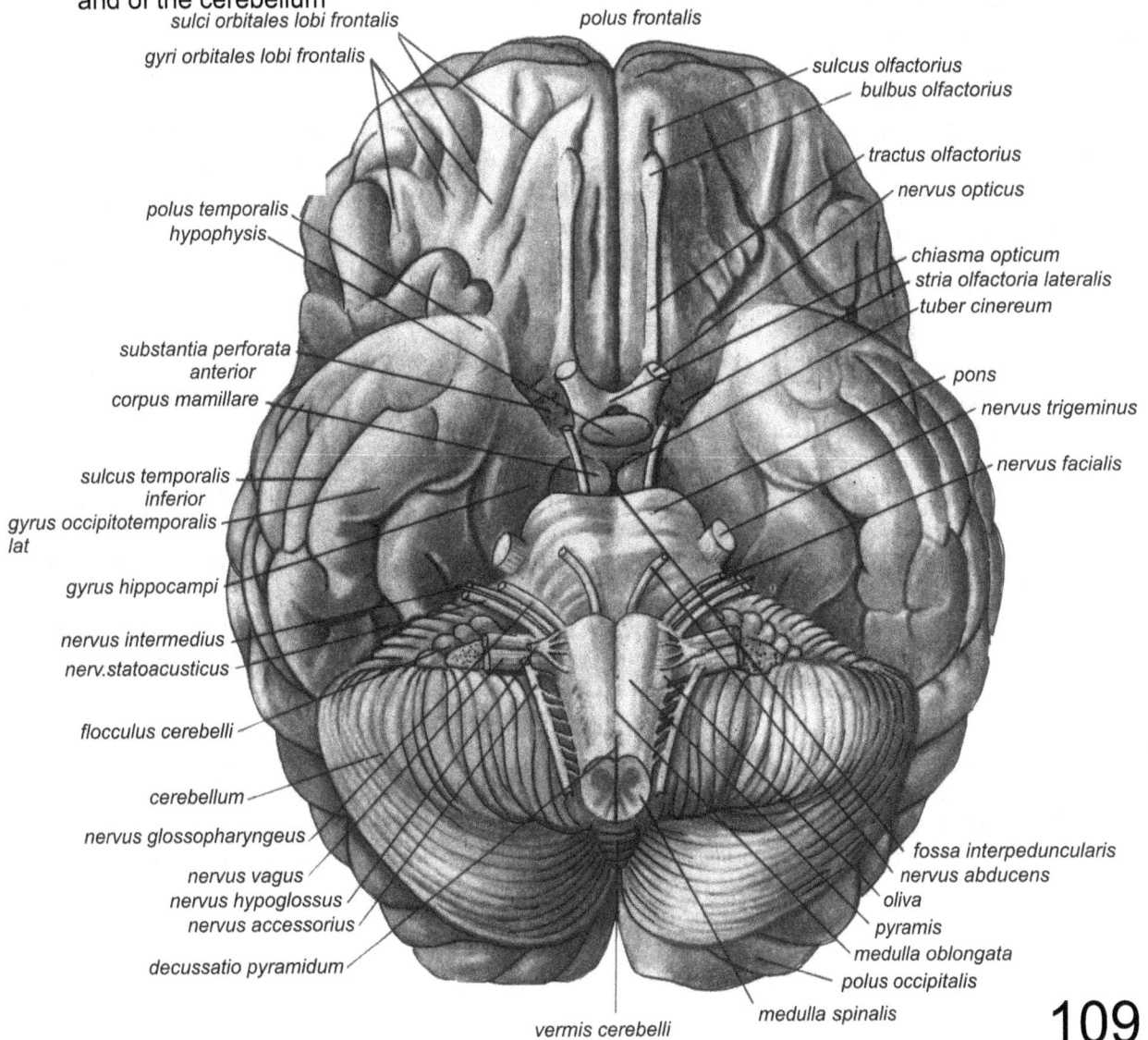

sulci orbitales lobi frontalis

gyri orbitales lobi frontalis

polus frontalis

sulcus olfactorius

bulbus olfactorius

tractus olfactorius

nervus opticus

polus temporalis

hypophysis

chiasma opticum

stria olfactoria lateralis

tuber cinereum

substantia perforata anterior

corpus mamillare

pons

nervus trigeminus

nervus facialis

sulcus temporalis inferior

gyrus occipitotemporalis lat

gyrus hippocampi

nervus intermedius

nerv.statoacusticus

flocculus cerebelli

cerebellum

nervus glossopharyngeus

fossa interpeduncularis

nervus abducens

oliva

pyramis

medulla oblongata

polus occipitalis

medulla spinalis

nervus vagus

nervus hypoglossus

nervus accessorius

decussatio pyramidum

vermis cerebelli

109

The medial surface of brain

On the medial surface, the delineation of the lobes is less clear-cut. Some gyry pass beyond the boundaries of the lobes and are independent unitary elements, under the form of the following sulci and gyri:

1) The sulcus of corpus callosum or callosal sulcus *(sulcus corporis callosi)* is situated above *the corpus callosum*, parallel to it.

2) The hippocampal sulcus *(sulcus hippocampi)* continues *the sulcus of corpus callosum* on the basal surface of the cerebral hemisphere;

3) The cingulate sulcus *(sulcus cinguli)* is arched in the anterior part and lies between the sulcus of the corpus callosum and the superior margin of the hemisphere. It begins beneath the anterior part of the genu of the corpus callosum *(genu corporis callosi)*, by the subfrontal portion, while above the middle of the corpus callosum it sends upwards the paracentral sulcus *(sulcus paracentralis)*. It ends on the superior margin of the cerebral hemisphere, somewhat behind the central sulcus, by a concavity oriented forwards. The cingulate sulcus presents a small projection on the lateral surface of the brain, called pars marginalis sulcus cinguli.

4) The subparietal sulcus *(sulcus subparietalis)* continues on the temporal lobe the cingulate sulcus, with some interruptions, then it is continuous with the calcarine sulcus *(sulcus calcarinus)* and forms, in its terminal portion, the anterior part of the collateral sulcus *(sulcus collateralis)* on the basal surface of the brain.

- The paraterminal gyrus *(gyrus paraterminalis)* is small and lies beneath the rostrum of the corpus callosum.

- The subcallosal area *(area subcallosa)* is situated anteriorly to the paraterminal gyrus, both structures belonging to the rhinencephalon.

- The fornicate gyrus *(gyrus fornicatus)* extends from the level of the frontal lobe to the temporal lobe, between the sulcus of the corpus callosum and the hipocampal sulcus, on the one hand, and the cingulate, subparietal and calcarine sulci and the anterior part of the collateral sulcus, on the other hand.

This gyrus is made up of two portions: the first portion, the cingulate gyrus, described above, and a second portion, the isthmus of the cingulate gyrus *(isthmus gyri cinguli)*, which is narrow and situated between the calcarine sulcus and the sulcus of the corpus callosum.

- The parahippocampal gyrus *(gyrus parahippocampalis)* is situated in the temporal lobe, between the hippocampal sulcus and the collateral sulcus. Its hook-shaped terminal portion is named uncus gyri hippocampalis.

5) The calcarine sulcus *(sulcus calcarinus)* begins at the occipital pole and, after a curved course, arrives behind and below the splenium of the corpus callosum and unites with the parieto-occipital sulcus. The calcarine sulcus is so deep, that it forms, on the medial wall of the posterior horn of the lateral ventricle, an elevation resembling the leg of a bird, called calcar avis.

- The paracentral lobule *(lobulus paracentralis)* is situated in the frontal lobe, between the cingulate sulcus *(pars marginalis)* and the paracentral sulcus (non-homologated in the [N.A.])

- The precuneus *(praecuneus)* is a quadrilateral area with irregular borders, situated in the parietal lobe, between the pars marginalis of the cingulate sulcus, the subparietal sulcus, the parieto-occipital sulcus and the superior margin of the cerebral hemisphere.

The cuneus is a triangular area, situated behind the precuneus in the occipital lobe, between the parieto-occipital sulcus, the calcarine sulcus and the superior margin of the cerebral hemisphere (of course, bilaterally, in both hemispheres).

The basal surface of the brain

On this surface, the following sulci and gyri may be distinguished :

1) The orbital sulci *(sulci orbitalis)* are short, irregular furrows, delineating the orbital gyri *(gyri orbitales)*, which are also small and variable in shape and number.

2) The olfactory sulcus *(sulcus olfactorius)* is paralled to the longitudinal cerebral fissure and is covered by the olfactory tract and olfactory bulb, which are lodged inside it.

- The straight gyrus *(gyrus rectus)* is bounded by the cerebral longitudinal fissure and by the olfactory sulcus.

3) The hippocampal sulcus *(sulcus hippocampi)* is a groove, situated on the medial border of the temporal lobe. It continues the callosal sulcus (sulcus of the corpus callosum) and brings about the appearance of an elevation on the internal surface of the inferior horn of the lateral ventricle, termed foot of hippocampus *(pes hippocampi)*.

4) The collateral sulcus *(sulcus collateralis)*, situated between the hippocampal sulcus and the inferior temporal sulcus, begins at the occipital pole and runs towards the anterior margin of the temporal lobe, without reaching the anterior margin of the temporal lobe. Its depth produces the appearance of the collateral trigone *(trigonum collaterale)* and of the collateral eminence *(eminentia collateralis)* on the internal surface of the lateral ventricle.

Fig. 92 Inferior surface of the encephalon after the section of the brainstem at the level of the mesencephalon and the removal of the brainstem and of the cerebellum

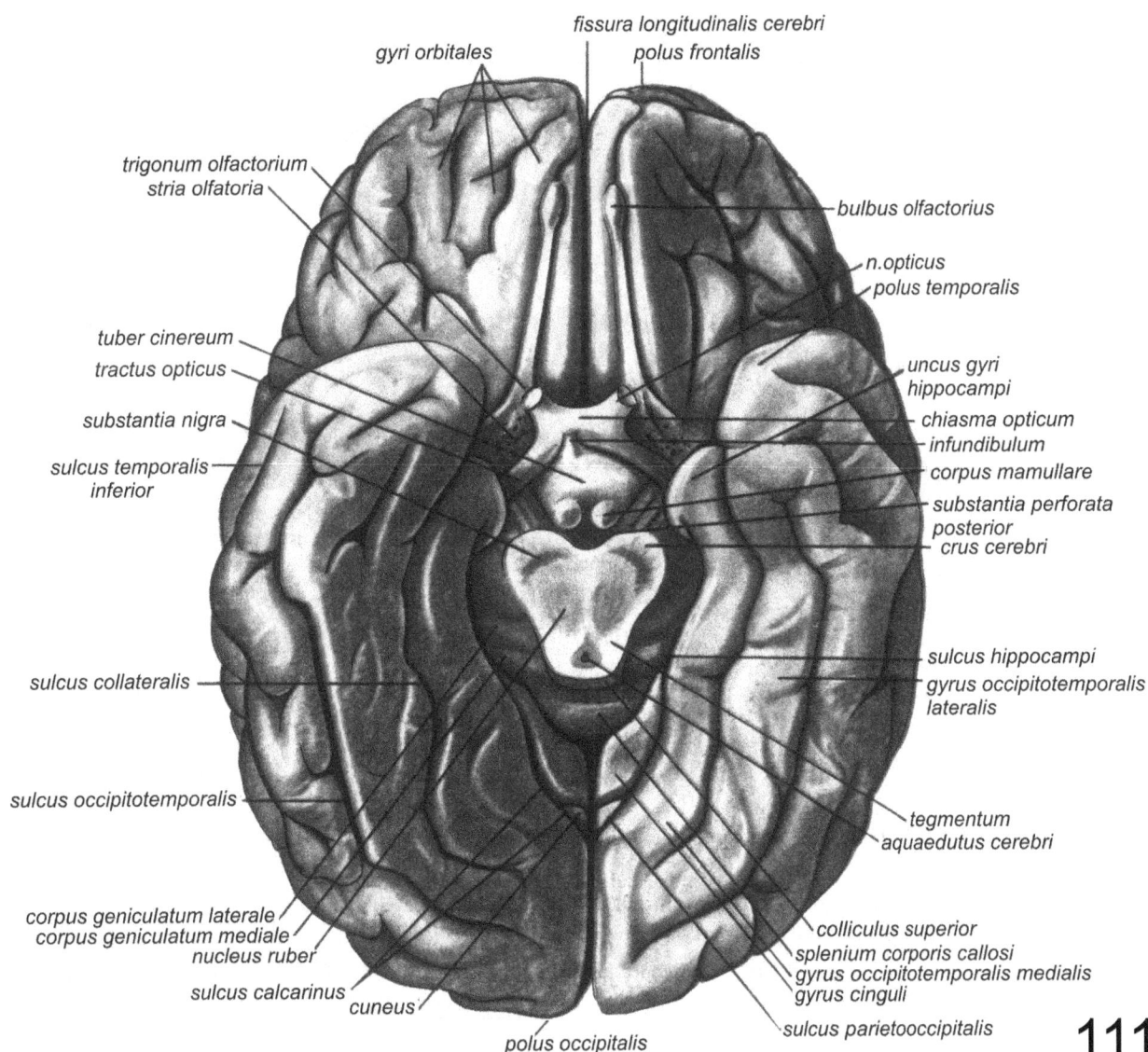

111

- The parahippocampal gyrus *(gyrus parahippocampalis)* represents the terminal portion of the fornicate gyrus and is bounded by the hippocampal sulcus and the anterior part of the collateral sulcus. Its free extremity is hook-shaped and termed uncus gyri hippocampalis.

- The medial occipitotemporal gyrus *(gyrus occipitotemporalis medialis seu gyrus lingualis)* is situated between the calcarine sulcus and the collateral sulcus. It is continuous anteriorly with the parahippocampal gyrus.

- The lateral occipitotemporal gyrus *(gyrus occipitotemporalis lateralis seu fusiformis)* is situated between the collateral sulcus and the inferior temporal sulcus.

- The indusium griseum *(supracallosal gyrus)* is a thin lamina of grey substance , situated on the corpus callosum, displaying medially and laterally *the longitudinal striae*. It is a rudimentary structure, a component of the hippocampus, and continues caudally around the splenium of the corpus callosum with the fasciolar gyrus, which is in relationship with the dentate gyrus of the hippocampus. *Indusium griseum* curves around the genu and the rostrum of the corpus callosum and extends ventrally towards the olfactory trigone, as *the taenia tecta (rudimentum hippocampi)* situated in the depth of the posterior paraolfactory sulcus, which marks the anterior border of the subcallosal gyrus.

The insula

It is a part of the cerebral cortex situated in the depth of the sylvian fissure, hidden in the course of the embryonal evolution by the projections of the neighbouring lobes, called opercula; the frontal operculum, the frontoparietal operculum and the temporal operculum. It derives from the floor of the sylvian fossa *(lateral cerebral fossa)* and is known also under the name of Reil's insula or Reil's island.

The insula has an oval elongated shape and its free surface at the junction with the other structures is separated from these by Reil's circular sulcus *(sulcus circularis)*, which lacks in the anterior portion of the insula. At this level, the insula is connected with the anterior perforated substance through the falciform fold *(limen insulae)*.

The insula is composed, on average, of five short gyri directed vertically and, at its inferior margin, a long gyrus *(gyrus longus insulae)*. The gyri are separated by the corresponding sulci of the insula *(sulci insulae)*.

The microscopic structure of the cerebral cortex

The cerebral cortex *(cortex cerebri)* is the grey layer which covers the whole surface of the brain and only at the level of the choroid tela remains a one-layered epithelial lamina.

Owing to the presence of sulci and gyri, the surface of the brain is considerably increased, attaining on the average 1,800 - 2,200 cm^2, a size which corresponds to a square with the side of 45 cm.

The thickness of the cortex is variable in the different regions: the maximum thickness is observed in the precentral gyrus (4.5 mm) and the minimum, in the frontal and the occipital lobes (1.5mm), the number of nerve cells attaining 14-18 billion.

According to the arrangement of the nerve cells and fibres, a number of 47 or even more cortical areas may be distinguished, which differ more or less from each other (Brodmann's and Vogt's architectonics) and have various functions, as it was shown previously.

The basic type of the cortex is made up of six layers) which differ in the various regions.

The six layers are the following (see Fig. 322 - 332):

1) *the molecular layer*, in connection with the pia mater, is formed of small, fusiform, horizontal (Cajal) cells, with short dendrites and with an axon dividing into two branches which depart from each other like the letter T and of parallel nerve fibres, arising both from the cells

112

Fig. 93 Lobe of the insula demonstrated after removal of a part of the lateral surface of the cerebral hemispheres

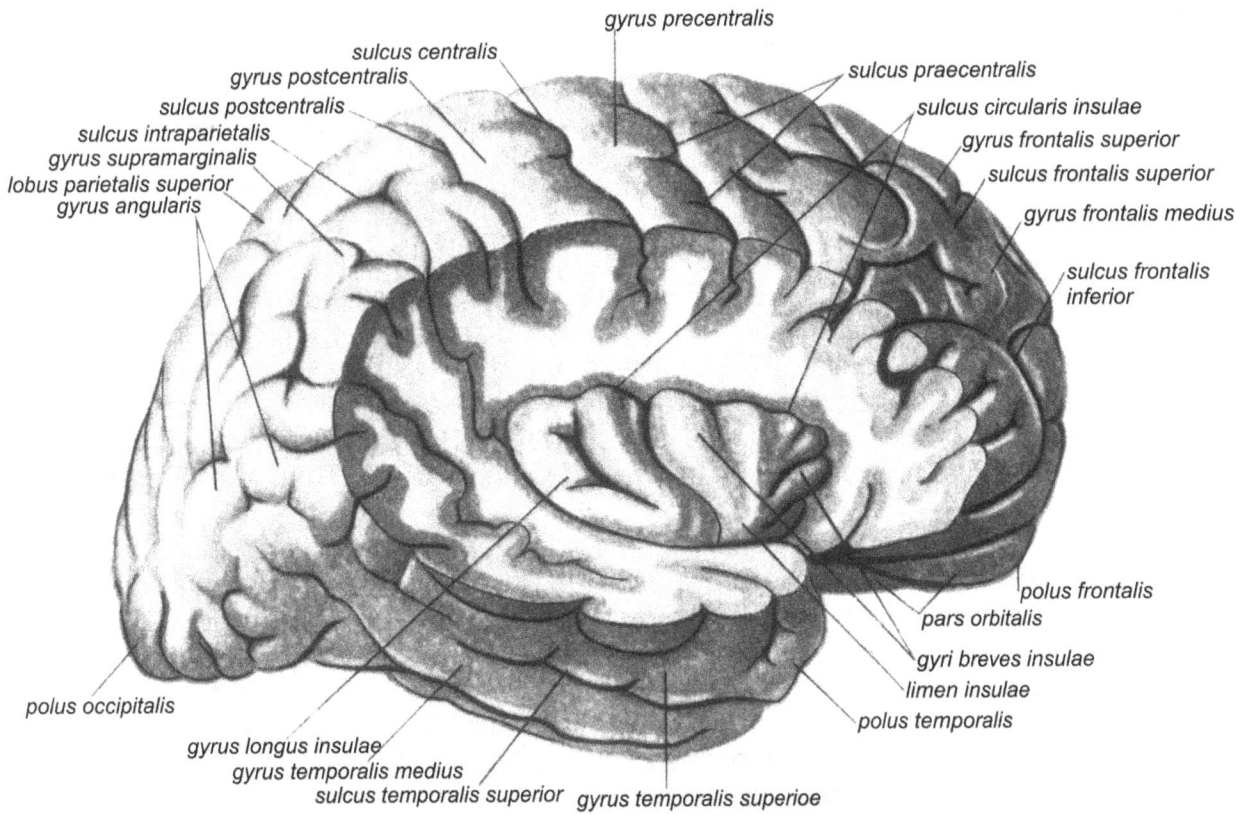

gyrus precentralis
sulcus centralis
gyrus postcentralis
sulcus postcentralis
sulcus intraparietalis
gyrus supramarginalis
lobus parietalis superior
gyrus angularis
sulcus praecentralis
sulcus circularis insulae
gyrus frontalis superior
sulcus frontalis superior
gyrus frontalis medius
sulcus frontalis inferior
polus frontalis
pars orbitalis
gyri breves insulae
limen insulae
polus temporalis
polus occipitalis
gyrus longus insulae
gyrus temporalis medius
sulcus temporalis superior
gyrus temporalis superioe

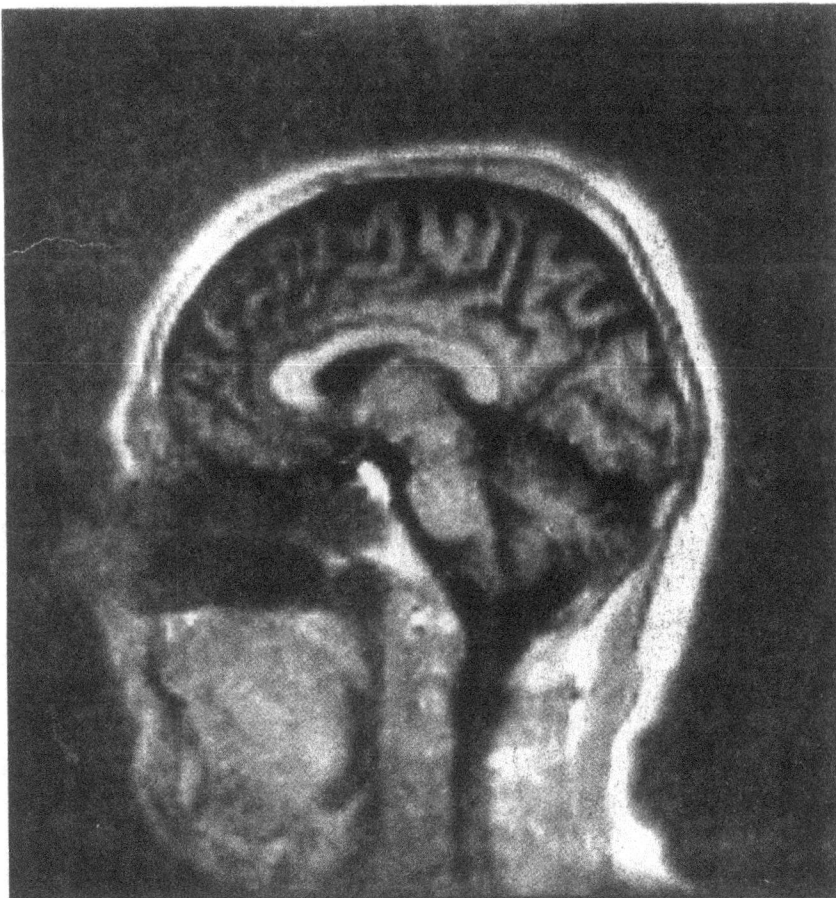

Fig. 94 Tomography of the encephalon in mediosagittal position (nuclear magnetic resonance) (Siemens collection)

113

present in a small number in this layer and from the cells of the deep layers. In this way, complex connections between the various layers are formed;

2) *the external granular layer* is constituted of small, round, polygonal or pyramidal cells, with a large nucleus, belonging to the association neuronal type. Owing to the fact that it receives fibres from the thalamic nuclei, it represents one of the sites of sensibility;

3) *the external pyramidal layer* is formed of middle-sized pyramidal cells which give off dendrites that reach the molecular layer, while the axons pass into the white matter. It is one of the sites of motoricity;

4) *the internal granular layer* is made up of polymorphous nerve cells; their transverse myelinated projections form the Baillarger's or Gennari's external layer (the external band or the external stria), taking into account the staining of the myelin sheath.

5) *the internal pyramidal layer* is formed of large pyramidal Betz cells, lying in the precentral gyrus. The myelinated horizontal collaterals of the large pyramidal cells form - after staining - Baillarger's inner layer (the inner band or the inner stria). The axon of these cells forms the projection and association pathways in the white matter.

6) *the multiform (polymorphous) layer* is made up of polymorphous nerve cells. Some cells send their axons into the molecular layer (Martinotti cells). They form the boundary towards the white matter.

At the level of the calcarine sulcus, this arrangement is different since three other layers appear, inserted between the above described layers 4 and 5. They are the following: a layer of small pyramidal cells, another one formed of medium-sized pyramidal cells and the third made up of polymorphous nerve cells. Hence, on the whole there are nine layers at the level of the calcarine sulcus.

By the interposition of tangentially orientated white fibres, the visual cortex presents, on the transverse section, a striate apparance (*striate area* - calcarine type). In the depth of the calcarine sulcus, the cortex is very thin (1.5 mm).

The following structural cortical types are described according to the arangement of the described layers.

The allocortex (archicortex) is formed of only three layers: a granular external layer (large-sized neurons with a receptor role), an internal pyramidal layer (large - and medium - sized neurons , with a receptor and association role) and the stratum oriens; the allocortex represents approximately the twelfth part of the cerebral cortex.

The paleocortex resembles the allocortex, but the layers cannot be clearly distinguished, because of the interpenetration, of the intermingling of the constituent cells. The allocortex and the paleocortex form that zone of the cerebral cortex which is called archipallium and which represents the oldest zone (the archicortex seems to be older than the paleocortex).

The isocortex represents a more recent structure of the cerebral cortex, as it enters in the constitution of the neopallium (neocortex). This structure contains all the six layers and, according to the predominence of certain cell types, it is divided into two categories:

- *The heterotypic isocortex*, characterized by the absence of some layers, may be either granular (at the level of the sensitive-sensors regions), where small granular cells predominate, or agranular (at the level of the motor regions, where the small granular cells are lacking).

- *The homotypic cortex*, situated at the level of the frontal and parietal cortex and at the two poles (anterior and posterior) of the central hemispheres, in which the six layers are proportionally disposed, this cortical type is present in the association areas.

The problem arises whether the study of the brain after the death of an individual permits to draw reliable conclusions regarding his intellectual capacity. Of course, this is not possible, since the intellect (depends on a lot of factors, most of which are unknown). However, the frontal cortex plays a decisive role. Moreover, the weight of the brain can not make up a conclusive index, although it is known that oligophrenics have a small brain.

114

Furthermore, the thickness of the cortex and the abundance of the cerebral microcirculation should be also taken into account. In some old persons, the sclerosis of the cerebral arteries leads to an impairment of memory, a change of the character (senile dementia). The number of nerve cells in the cortex does not represent the most important factor. Of course, the intimate cell constitution could have a theoretical value, but this assumption has not been histologically confirmed yet.

The cortical areas

The cerebral cortex is the centre of consciousness, of memory of educational behaviour and of emotional states. Although there is a relationship between all the portions of the cortex and although the portions of the cortex act jointly at the manifestations of the organism, there are aslo regions more or less clear-cut delineable, in which precise functions are located.

These regions are termed cortical areas and they were outlined by Brodmann by numbering with figures and by Von Economu by marking with letters. The exclusion or loss of an area leads to the loss of the respective function. Between the cortical areas lie cortical surfaces whose role is not elucidated yet and whose lesions do not bring about obvious functional impairments, which explains why they are designated as "silent areas".

These areas will be described below taking into account the cortical structure types, namely we shall begin with the areas of the archipallium, represented by the archi-and the paleocortex, and continue with the areas of the neopallium, represented by the homo - and the heterotypic isocortex.

Areas of the archipallium
(Fig. 94 - 97)

The term allocortex or achipallium cortex has been for a long time, anatomically considered synonymous with olfactory cortex, by analogy with the situation existent in lower macrosmatic mammals, in which the rhinencephalon is developed. Recent researches have proved that, in higher vertebrates, in the archipallium may be distinguished paleocortical formations whose development is related to the sense of small and archicortical formations which are not related to this sense.

Hence, the term rhinencephalon does not refer to the whole allocortex, but it should be reserved strictly to the olfaction related areas. The archicortex, situated in the neighbourhood of the olfactory paleocortex, seems to be a fundamental part of the brain of vertebrates, which influences their general behaviour and their instinctual life. In man, this arhicortex receives various impulses, which acquire at this level, an agreeable or disagreeable character; the manifestations of the archicortex have a somatic and visceral character, as it may be observed in the sexual activity or in the various aggressive reactions of the individual. The projection pathways of the archicortex are connected to the hypothalamus to the brainstem and to the neighbouring cortical neopallium. There is a permanent interrelation between the archipallium and the neopallium; although the archipallium is under the control of the neopallium, it may inhibit or potentiate the activity of the neopallium and is responsible for some sudden and unforeseeable actions.

By systematyzing the above mentioned data, we consider that two territories lie at the level of the archipallium:

a) the territory of the paleocortex belonging to the olfactory system, with its paleocortical pathways and centres;

b) the territory of the archicortex or of the behavioural brain, whose hippocampal structures represent its main parts.

The territory of the paleocortex (olfactory territory). The olfactory formations lie on the orbital surface of the frontal lobe, making up the olfactory tracts, whose anterior parts are termed olfactory bulbs. The tracts end behind at the level of the anterior perforated space by the olfactory trigone, which divides into three olfactory striae, each coursing to a well delimited cortical territory, as follows:

- the medial stria runs medially and upwards, arriving to the septal area formed of the parolfactory (Broca's) area and the subcallosal gyrus;

- the lateral stria courses laterally, reaching the fifth lateral convolution *(uncus)* and terminates forming the entorhinal area;

- the intermediate stria reaches the anterior perforated space, forming an elevation called olfactory tubercle.

Inside these striae lie the olfactory tracts. We mention that, according to the opinion of most researchers, they have no diencephalic relay before reaching the cortex.

The protoneuron is represented by the bipolar (Schultze's) cells situated in the olfactory mucosa of the nasal cavity. Their axons, equivalent to a posterior root, form the olfactory nerves which pass through the cribriform plate of the ethmoid bone and form a synapse with the second neuron, the mitral cell of the olfactory bulb. The axons of the mitral cells form the olfactory tracts, which reach the various cortical zones. The cortical areas are divided into sensory areas (the entorhinal area on the temporal lobe), association areas and reflex centres (the septal area and the anterior perforated plate on the frontal lobe). The septal area has connections with the archicortex of the hippocampus and the amygdaloid nucleus, which have no olfactory value, but are association and projection centres towards the overlying centres. Through these connections the olfactory tracts can elicit a reflex pathway, general especially vegetative mechanisms. Lesions of the olfactory territory may bring about anosmia (complete loss of the sense of smell), hyposmia (diminished sense of smell), hyperosmia (abnormally increased sense of smell), parosmia (confusion of odours), cacosmia (all odours seem disagreeable), olfactory hallucination (subjective perception of nonexistent odours).

The hippocampal formations or the archicortex. The hippocampal formations make up a poorly visible ring-shaped structure formed of grey matter and fibres, which surround each hemisphere. The hippocampal scissure separates them from the neocortex and the paleocortex. A posterior and an anterior hippocampus are distinguished.

The posterior hippocampus has the appearance of a strip, arises at the level of the subcallosal convolution, covers the corpus callosum and is continuous posteriorly with the fasciola cinerea, a grey band which may be followed up on the dentate gyrus of the anterior hippocampus.

The anterior hippocampus is much more developed – its shape in the cavity of the lateral ventricle represents a wrapping up of the archicortex in the hippocampus – and runs along the dentate gyrus.

The posterior and the anterior hippocampus are united anteriorly through a tract called Giacomini's band and in the temporal part through a slender band named Broca's band, which borders the anterior perforated plate.

The posterior hippocampus, getting in contact with the subcallosal convolution, enters in relationship with the medullary stria of the olfactory tract, while the anterior hippocampus is in relationship with the olfactory paleocortex at the level of a zone of the fifth temporal convolution *(subiculum).*

The hippocampus receives afferens from multiple sources, respectively from the adjacent entorhinal area, from the neopallium, from the limbic gyrus – brought close to it through the fifth temporal convolution -, from the hypothalamus.

The hippocampus has a single efferent pathway, the fimbria, which is continuous with the trigone – that penetrates into the hypothalamic wall – and terminates in the mamillary bodies, a significant connection centre.

cavum septi pellucidi
lamina septi pellucidi

corpus callosum

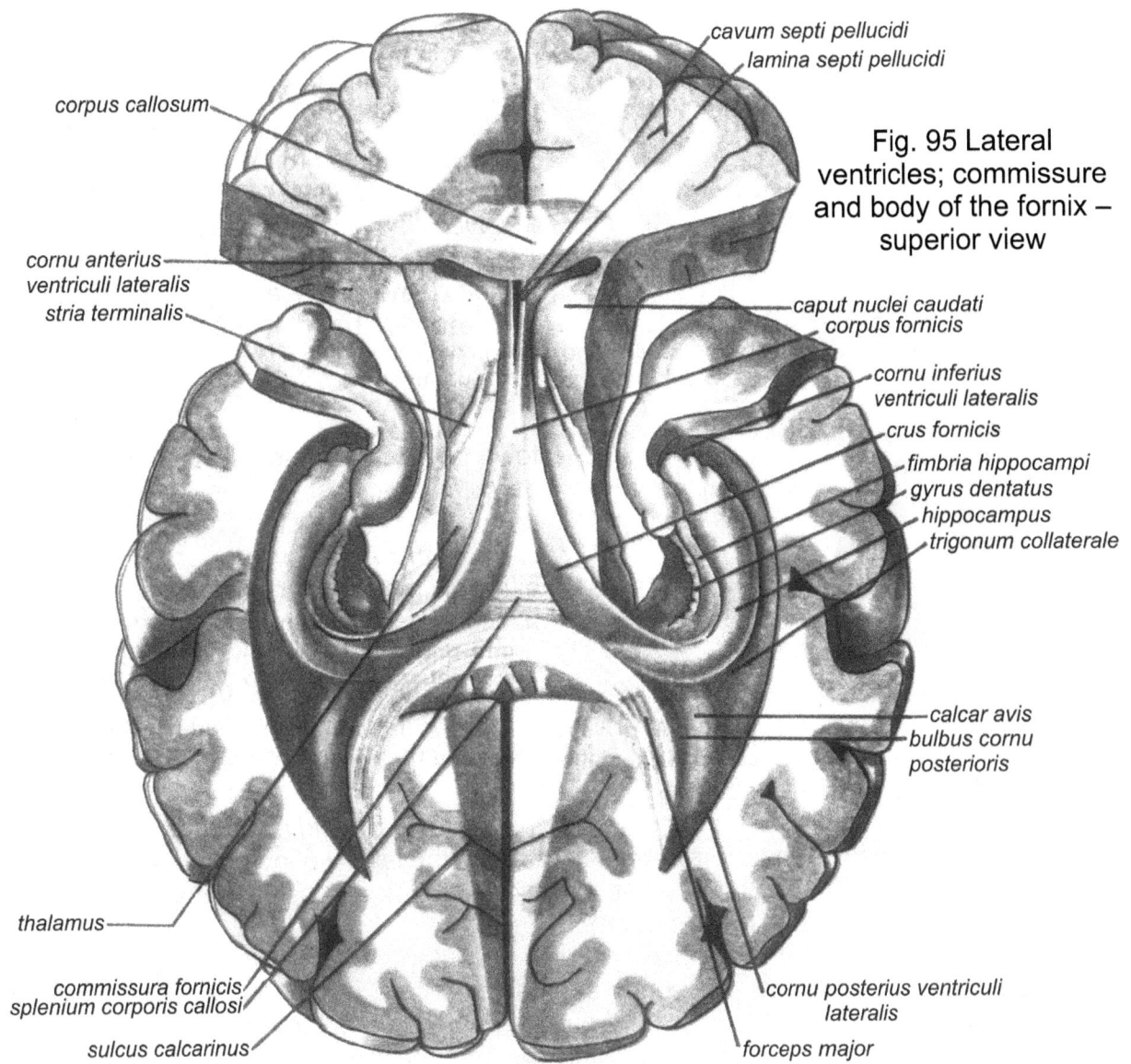

Fig. 95 Lateral
ventricles; commissure
and body of the fornix –
superior view

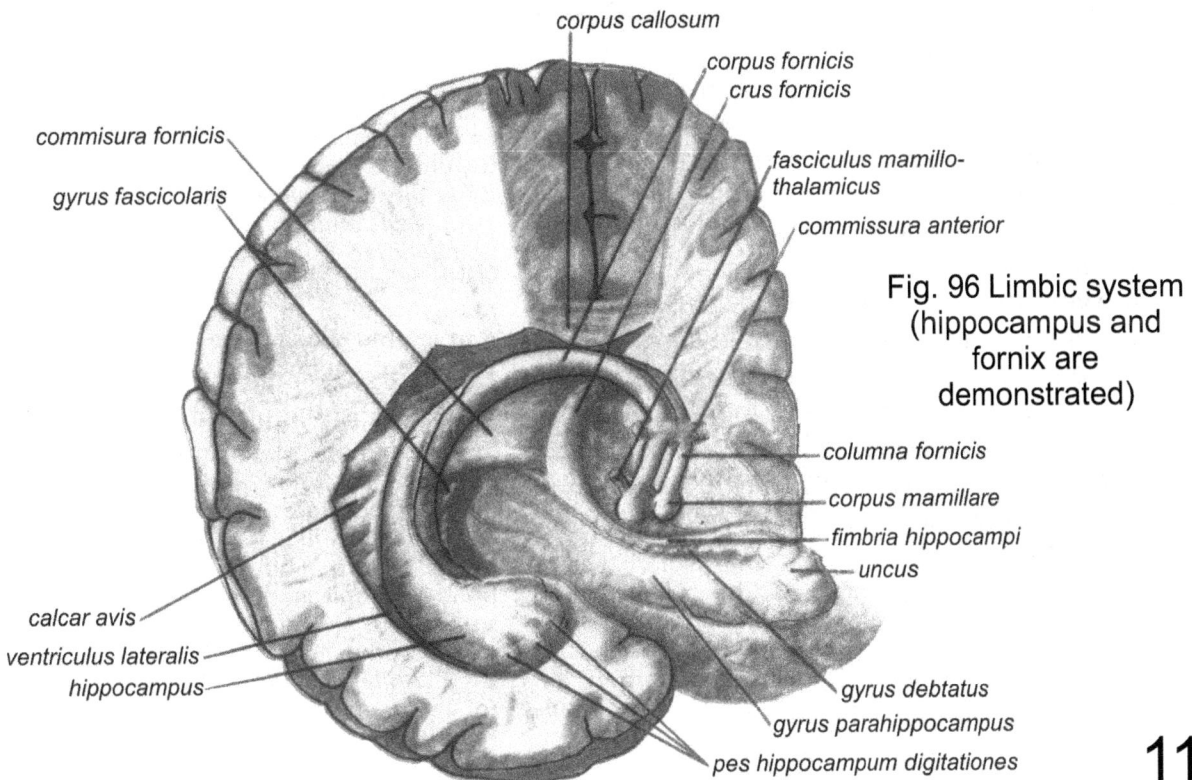

cornu anterius
ventriculi lateralis
stria terminalis

caput nuclei caudati
corpus fornicis

cornu inferius
ventriculi lateralis

crus fornicis

fimbria hippocampi
gyrus dentatus
hippocampus
trigonum collaterale

calcar avis
bulbus cornu
posterioris

thalamus

commissura fornicis
splenium corporis callosi

sulcus calcarinus

cornu posterius ventriculi
lateralis

forceps major

corpus callosum

corpus fornicis
crus fornicis

commisura fornicis

gyrus fascicolaris

fasciculus mamillo-
thalamicus

commissura anterior

Fig. 96 Limbic system
(hippocampus and
fornix are
demonstrated)

columna fornicis

corpus mamillare

fimbria hippocampi

uncus

calcar avis

ventriculus lateralis

hippocampus

gyrus debtatus

gyrus parahippocampus

pes hippocampum digitationes

117

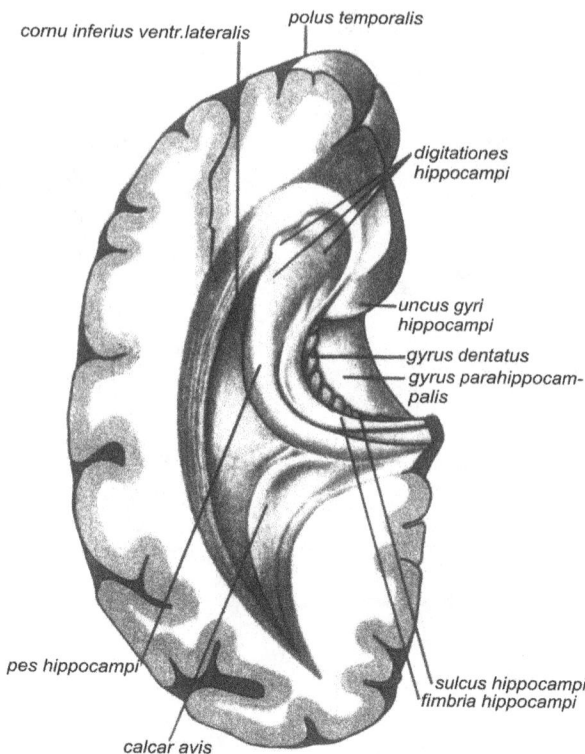

Fig. 97 Hippocampus

From this level start:

- *an ascending mamillothalamic* (Vicq d'Azyr) *fasciculus*, which reaches the anterior nuclei of the thalamus, where it synapses with a neuron that projects in the limbic or cyngulate convolution; it seems to be involved in the mechanisms of emotion. Starting from the limbic area, the nerve influx travels to the frontal and to the parietal lobes or returns to the superior temporal convolution, controlling the hippocampal centres;

- *a descending fasciculus* (Gudden) towards the brainstem, which through the reticular nucleus of the mesencephalon terminates in the nuclei of the cranial nerves;

- *the fibres of the fornix* run directly either in the nuclei of the septum or in the hypothalamic nuclei, eliciting the vegetative phenomena which accompany the emotions.

The amygdaloid nucleus. The nuclei of the septum. Taking into account the existent connections, we mention also the amygdaloid nucleus and the nuclei of the septum.

The paleocortex lines, in front of the uncus, *the amygdaloid nucleus* (a component of *the archistriatum*) which, on turn, receives fibres from the lateral olfactory stria and gives off the semicircular band called terminal stria, which courses alongside the thalamocaudate sulcus and reaches the nuclei of the septum.

The septal nuclei are situated in front of to the anterior commissure, below the genu of the corpus callosum, beneath the septal area. They receive numerous afferents: from the olfactory bulb (through the medial stria), from the hippocampus through the fornix, from the amygdaloid nucleus through the terminal stria and from the prefrontal cortex. They represent a very significant association centre between the rhinencephalon, the visceral cortex, the hypothalamus and the epithalamus.

They send their efferents to the hypothalamus (through a bundle of poorly myelinated fibres, which courses between the mediolateral nuclei of the hypothalamus and is distributed to the mamillary bodies up to the nuclei of the tegmentum of the brainstem) and to the epithalamus (through the thalamic stria – the habenula -, which terminates posteriorly in the ganglion of the habenula, the site of the synapse with the cells from which Meynert's retroreflex fasciculus starts; the latter reaches the interpeduncular ganglion, from which fibres start to the vegetative centres through the brainstem).

The importance of the archipaleal system results from the above-mentioned data. In man, the paleocortical olfactory areas are reduced, but through their relationships with the archicortex, the behavioural brain, the amygdaloid nucleus and the septal nuclei they can elicit complex motor and vegetative mechanisms.

The archicortex plays a role in the behavioural phenomena, often emotive, sometimes aggressive, sometimes of obedience, evident especially in the course of conflict situations and sexual activity.

118

The areas of the neopallium

The pyramidal areas. The region of voluntary motoricity is made up by the precentral gyrus and extends also on the adjacent gyri, in proximity to the lateral cerebral sulcus, on the opercular part of the frontal lobe and on the parietal part. Moreover, it extends also beyond the superior border of the hemisphere and ends in the posterior part of the paracentral lobe. From all this region arises the pyramidal pathway (the corticobular tract and the corticospinal tract).

The cortical extremity of the centre of motoricity is situated predominantly in areas 4 and 6 (motor frontal cortex), as well as at the periphery of this region, respectively in the areas 8, 9, 10, 11, 12, 14, 45, 46 and 47 (the cortex of the prefrontal lobe); area 4 comprises Betz's pyramidal cells, while areas 3 and 5 contain the cortical basket cells, which play an inhibitory role. Area 6 is devoid of Betz's cells.

The projection of the various portions of the body on the cortex (homunculus motor) occurs as follows: the portions of a half of the body project on the cortex of the contralateral hemisphere, in a reverse order, so that the order of the areas is the following: the head downwards to the sylvian fissure, then the upper limbs with the hand, the trunk, while the lower limbs are situated the closest to the convexity of the hemisphere. This fact, although it seems paradoxical, may be explained on the basis of the embryological development. The portion from which the cerebrum will arise is initially flattened in the form of a neural plate. The motor centres which have the shortest course to the musculature of the head are situated medially; lateral to them lie the centres from which start the motor pathways towards the neck, the thoracic limb, the trunk etc.

As a result of the closure of the neural tube, the medial regions remain in the inferior portion of the future hemispheres, while the lateral regions are, logically, forced to shift into this superior portion, together with the borders of the neural groove. As a consequence, the regions of motoricity are situated in the following order, from above downwards:

- in the paracentral lobe: the musculature of the sphincters of the urinary bladder (and of the rectum; the muscles of the pelvis;

- in the precentral gyrus: the muscles of the foot, of the pelvic (lower) limb, of the trunk, of the thoracic (superior) limb, of the hand, the neck, the mimetic muscles;

- in the opercular part of the frontal and parietal lobes: the muscles of the tongue, the larynx, the muscles of mastication.

The surfaces of the motor regions are not represented proportionally to the size of the muscles, but only the functional value finds its expression in the spatial extension of these zones. Thus, for example, in the sphere of motoricity for the hand, the thumb occupies a much larger area than the fourth finger, owing to its much greater movement possibilities than those of the fourth finger, which performs a considerably reduced activity. The greater the movement possibilities, the larger the respective area. These regions do not represent centres for the muscles taken as individual units, but for muscle groups acting on the various articulations. The exclusion of certain parts of the anterior central gyrus (for example, in the cerebral stroke) leads to a spastic paralysis of the musculature of the respective region on the opposite side.

With the motor disturbances occured at the level of the respective half of body is often associated an impairment of movements on the contralateral side, since the corresponding regions, on either side of the two hemispheres, are connected through the commissural fibres of the corpus callosum. The impulses at the level of area 4 produce localized convulsions under the form of Jacksonian epileptic scizures.

The cortical extrapyramidal areas

Beside the effector pyramidal bineuronal system, with two fasciculi, corticospinal and corticonuclear, there is also, as we have shown, the extrapyramidal system, whose control formations are situated on the entire extent of the neuraxis, in the subcortical segments:

the striated nuclei, the subthalamic nuclei, the quadrigeminal colliculi, the red nucleus, the substantia nigra, the vestibular nuclei, the olivae, the cerebellum – by excellence a regulating organ, which receives the vestibular and proprioceptive influxes. However, these centres function under the influence of the cortex and, in turn, they influence the cortical activity, the extrapyramidal system being a servomechanism which regulates the motor influx at the level of the common final pathway. In this complex participates actively also the reticular substance, which sends from the thalamus, fibres ascending up to the level of the cortex,while others descend in the brainstem and the spinal cord. The cortical neurons which regulate the extrapyramidal system and the reticular neurons occupy a large part of the cortical surface, not only of the motor cortex of the ascending frontal convolution, but also of the parietal, occipital and temporal cortex. Some of their fibres (the parapyramidal fibres) accompany the pyramidal tract, whereas others reach the reticular pontine (facilitory) or bulbar (inhibitory) nuclei. Thus the cortex, held in a vigil state by the reticular system, influences, in turn, the reticular facilitory and inhibitory centres of the brainstem, modifying through them the activity of the spinal cord. The surface occupied by the cortical area of the extrapyramidal system (extrapyramidal areas) represents 85% of the motor cortex, as follows:

1. *Inhibitory areas, called so because they may inhibit the function of the main motor area.* They are situated on either side of the precentral area (areas 4 and 2 in the frontal lobe, 8 in the occipital lobe around the areas 19 and 24, in *the cyngulate gyrus* on the internal surface of the hemisphere). The influx from these areas reaches the caudate nucleus, then *the globus pallidum*, from where it is sent back into the precentral cortex (areas 4 and 6); at the same time, the nerve influx from *the globus pallidum* is sent to the ventral anterolateral nucleus of the thalamus and then, through the thalamocortical fibres, to the cortex. This represive system is corticostriocortical, the thalamus playing the role of a filter.

The cortex of the areas 4 and 6, influenced by the striate nuclei acts upon the spinal cord, the nerve inflow being transmitted through the extrapyramidal (Bucy's) fibres, which accompany the corticospinal fibres at the level of the anterior horns, where they terminate on the common final neuron.

2. *Integration areas of the neocerebellar circuits, playing a motor role.* The neocerebellum achieves the second regulation mechanism of the cortical motor influx. It involves two large afferent tracts from the cortex to the cerebellum: frontopontine and parietotemporopontine tracts.

The frontopontine tract arises from the cortical areas 6 a and 6 b, the parietotemporopontine tract from the postcentral and superior parietal gyrus (areas 2, 1 and 5) and from the first temporal gyrus (area 22). The fibres of these tracts situated in the internal capsule and the cerebral peduncles reach the cerebellar cortex, the dentate nucleus, after which they course towards the thalamus, arrive again at the level of the motor cortex and achieve in this way a corticocortical circuit, the cerebellum playing thus also the role of a servomotor.

We mention that the parietotemporopontine tract, which at the level of its cortical area has connections with the zones from which sensitive, sensory, auditory and visual inflows arise, is involved in the body representation in the space, indispensable for directing the movements.

Moreover, the prevision, the projection in the time of the movement, are facilitated by the action of the frontopontine tract, which in its cortical area is in connection with the prefrontal areas concerned with prevision (Bonin).

3. *Integration areas of corticosubcortical circuits accomplishing a motor role.* The extrapyramidal cortex controls the striate nuclei and the subthalamic paleoencephalic centres by direct or collateral pathways arising from the descending tracts.

The first category of direct fibres is represented by the corticostriate, corticorubral, corticonigral and corticomesencephalic reticular tracts, arising especially from areas 4, 5 and 6; likewise, the descending tracts, at the level of the internal capsule, give off collaterals to the same formations.

120

4. *Cortical integration areas of oculocephalogyric movements.* We have shown that the synergism of movements of the two eyeballs is assured through the posterior longitudinal fasciculus which unites the oculomotor nucleus. The voluntary movements of the eyes require a unique command, started from the cortex. This cortico oculo-cephalogyric area lies at the level of the second frontal convolution, an area which has proved to emit commands through direct nerve tracts to the heterolateral *nucleus abducens* of the sixth cranial nerve pair; the latter, in turn, acts homolaterally, through the posterior longitudinal tract, on the centre of the internal *rectus abdominis* muscle of the abdomen and on the centre of the homolateral eleventh cranial nerve pair. This explains why a hemiplegic patient, in deep coma, regards towards his lesion situated in the right or left cerebral hemisphere (conjugated deviation of the head and of the eyes).

In addition, there is also an area concerned with the command of eye movements, subserving directly the action of the visual centres at *the calcarine sulcus*, situated at the level of field 19 and united with the frontal motor area through the fronto-occipital fasciculus.

Systematizing, it may be considered that the brainstem, through the posterior longitudinal fasciculus regulates the automatic movements of the eyeballs receiving commands from the frontal area for the voluntary movement in a certain direction and from the occipital area for the regulation of the voluntary movements related to a visual gnosia, for example, reading, following up of an object etc).

Of course, this regulation of eye movements is much more complex and researches aimed at the discovery of anatomical pathways of mimetic, sensorireflex, optokinetic, postural movements etc. are under way.

Thus, from the above-mentioned data ensues the complexity of the motor regulation. In order to be precisely performed, a motor act requires a succession of muscle contractions regulated by the extrapyramidal and cerebral centres, upon the cortex devolving, through the areas studied, the role of controlling and unifying all the nerve inflows, for the purpose of obtaining the harmony of the desired gesture.

The sensibility area. The sensibility area is sited in the post central gyrus of the parietal lobe. This cortical field extends on the lateral surface of the brain posteriorly and on the anterior portions of the superior parietal gyrus, reaching the medial surface of the brain up to the precuneus. Similarly to the motor region and for the same phylogenetic and embryologic reasons, the sensory projection surfaces for the head are situated at the lowest level, followed by those of the upper limb, while the projection surfaces for the pelvis and the lower limbs are located at the highest level, on the retrocentral gyrus, i.e. at the convexity of the cerebral hemispheres *(sensory homunculus)*.

The sensibility sphere makes up the end-point of the whole sensibility of the organism with the exception of the proprioceptive unconscious sensibility (of muscles and tendons). Thus, the tactile, pressure, vibration, pain and temperature stimuli are here converted into *"conscious perceptions"*, into sensations. Here are also located the perception memories of the respective stimulus.

The stronger sensibility, such as pain and temperature sensibility, is sited deeper, on the posterior portion of the central sulcus, whereas the other sensibilities are disposed closer to the surface, in the postcentral gyrus.

The cortical end of the sensory analyzer lies in the Brodmann's areas 3, 1 and 2 and at the periphery of areas 5 and 7.

The researches carried out by Petit-Dutaillès show that the somatosensory area may be, in turn, divided into a general sensibility area of the body (supplementary general sensibility – Penfield's area), on the internal surface of the hemisphere, a segmental sensibility area, situated on the superior part of the ascending parietal gyrus, and an expression sensibility area in the inferior part of the ascending parietal gyrus.

Beside the somatosensory area, which receives the primary sensations, there is the somatopsychic area, situated in the posterior half of the ascending parietal gyrus, in which the interpretation of these primary sensations and the perception of the element which has generated them, take place, and finally a gnosic area of the tactile sensibility, situated at the base of the parietal superior and inferior gyri, where a superior stage of sensory impulse integration is achieved, respectively the recognition and identification of the element that has generated them.

The gustatory area. The gustatory area has been, for a long period, considered as solidary with the olfactory area, although it could not be demonstrated how the gustatory nerve inflows, which course along the seventh, the ninth and the tenth cranial nerves, arrive at the allocortex.

It is nowadays known that the cortical projection taste area lies in the neocortex, at the level of the base of the ascending parietal convolution. A gustatory lingual area is described also above the sylvian fissure, associated with the lingual sensibility area, where an interference of the general sensibility of the tongue is achieved, the nerve inflow of which travels via the trigeminal nerve, with the gustatory sensibility transmitted via the glossopharyngeal (ninth), the vagus (tenth) and Wrisberg's intermediate (seventh bis) nerves.

The acoustic analyzer area. The acoustic centre of the analyzer is represented by a very small field, measuring a few mm^2, situated on the upper part of the inferior temporal gyrus, at the level of the Heschl's transverse gyri (areas 41 and 42).

An exact projection of Corti's organ is described at this level: high-pitched sounds in the deepest part and low-pitched sounds in the external part (for this reason, the area is, also, called cortical cochlea).

This primary area makes the perception of rough sounds possible, and by analogy with the other cortical functions, it is considered that here, too, there are superordinate perception and gnosia areas, which permit the interpretation and recognition of sounds.

Wernicke's speech centre (centre of heard), in which the memories of heard sounds are localized, is situated in the centre on the posterior part of the superior temporal gyrus (areas 39 and 40). A bilateral lesion to this centre or a lesion only to the centre on the left side, leads to the occurrence of Wernicke type sensory aphasia, which manifests itself under two forms: word agnosia or amnestic aphasia.

In the case of word agnosia or verbal deafness, the patient does no more perceive the words which he has heard because he lacks the of words. Such a patient can speak, but he does not utter the words in the logical order. He behaves like an individual who does not understand the language spoken by those with whom he gets in touch. The amnestic aphasia consists in the inability to call by their name cernain objects or notions.

The vestibular area. The vestibular field of cortical projection, too, is located at the level of the first temporal gyrus, which is also the site of origin of the temporopontine tract. Moreover, some authors consider that a cortical vestibular centre exists also at the level of the ascending parietal gyrus, which receives proprioceptive afferents and is also the site of origin of extrapyramidal fibres.

The area of the optic analyzer. It is situated on the lips of *the calcarine sulcus* (area 17) and in the adjacent regions (18, 19). On the upper lip of the calcarine sulcus lies the striate field for the superior half of the retina (hence for the inferior part of the visual field) and on the lower lip, that for the inferior part of the retina. The region of the macula lutea is situated on the posterior extremity of the calcarine scissure and on the occipital pole.

It may be considered that this calcarine area of visual protection (termed also "striate area" owing to the presence of Vicq d'Azyr's stria) is characterized by a topographic systematization similar to that of the retina, as it is actually a cortical retina.

Around this sensorimotor projection area, where all the optic radiations terminate, lies, like in the case of the other analyzers, visuopsychic association area, situated mainly on the external surface of the occipital lobe, where the synthesis of the primary sensations recorded in the sensorivisual zone occur, permitting first to perceive and then to recognize the objects. The destruction of the association area, brings about psychic blindness. We mention the recent researches which suggest that the visuopsychic area has a role in the building up of the notion of space with consequences in orientation, in the postural image integration etc.

122

The region for the visual memory is mostly situated in *the cuneus* and extends also, on the convex surface of the occipital lobe. The lesions in the region of the calcarine sulcus are followed by the more or less severe loss of vision, in dependence on the extent of their location. Lesions in the cuneus cause gaps in the visual memory: the patient sees the object, without being aware of the representation in reality of the respective object.

This syndrome is termed *visual agnosia* – the inability to recognize objects by sight, although sensitivosensory lesions are not present. *Visual agnosia* may be *partial* (wandering of the patient in unknown places, due to lesions in area 19 of the left hemisphere), *for objects* (inability to identify the objects) and visual *for the written word*.

The centre of optic memory for writing (visual agnosia for the written word) is sited in the angular gyrus (areas 19 and 39) – "optic centre for reading and writing". It is assumed that the centre on the left side is dominating. The lesions in these regions induce a complete inability to write (*letter alexia* – patient is unable to recognize the word).

The cortical area of the second signalling system (language). The problems related to the comprehension of the language are very complex, as they suppose the associated intervention of many of the above-mentioned cortical areas. For the purpose of a systematization, the clinicians classify the aphasia – the main impairment in verbal expression – in anarthria and agraphia, for the motor expressive aphasias, and in verbal blindness and verbal deafness, for sensory or comprehension aphasias; the occurrence of aphasia is due to the lesion of cortical area which we shall describe after the presentation of the new viewpoint in this problem.

Thus, the idea that the lesion of these areas leads to apraxia and agnosia phenomena which do not belong to the field of strictly speaking aphasia gets accredited in recent years. They represent a much more complex process than the impairments of the automatism in the expression or comprehension of the language, as the language is the means of communication with other persons, especially resembling each other, and with the own person. Under these conditions, the language involves the entry into action of emission and reception mechanisms.

Emission mechanisms are movements (mimicry being the most suggestive form), sounds and the language itself, while reception mechanisms are the vision (interpretation of forms, images, writing), the hearing (interpretation of noises, of the voice etc.), the touch. Under the term aphasia should be understood only impairments in the emission or comprehension of the spoken or written language but there are also other forms of language impairment, such as, for example, the tactile agnosia, which, for a blind person, accustomed to the Braille system represents a true verbal blindness (Delmas).

However, limiting ourselves to aphasia, in its classical unanimously admitted meaning, the areas of verbal expression are the following:

- the crus of the second frontal gyrus (anterior to the motor centres of fingers) for agraphia;

- the crus of the third frontal gyrus (anterior to the motor pneumolinguo-pharyngolaryngian centres) for anarthria;

- the middle part of the first temporal gyrus (anterior to the projection area of cochlear fibres), for verbal deafness;

- the angular area situated around the end of the first temporal gyrus, anterior to the occipital lobe, for verbal blindness.

Between the four areas lie association fibres, the lesion of which is involved in aphasia processes.

Some authors consider that there is an ideation area which assemblates the activity of these four areas (Delmas). Moreover, the fact that these four areas are present in both hemispheres, but that only the left hemisphere in right-handed persons and the right hemisphere in left-handed persons are involved in the speech, associated with the finding that the entire phonatory musculature, with motor representations on both hemispheres, participates synchronously, raises a series of problems which for the time being are not elucidated.

The cortical integration areas of vegetative manifestations. We emphasize the significance of the cyngulate area in the production of vegetative phenomena related to emotions, hence in the external expression of internal states (Yakovlev).

Moreover, the orbital gyrus of the frontal lobe, the uncus of the temporal gyrus, the lobe of the insula, the lateral surface of the frontal lobe are involved in the control of vegetative phenomena. The visceral frontal cortical areas are in relationship with the anterior hypothalamus (trophotropic), while the orbital areas are in relationship with the posterior hypothalamus (ergotropic). All these areas make up the so-called visceral brain. The visceral brain is in relationship with the rhinencephalon, either directly or through the amygdaloid nucleus.

1. *Representation of memory on the cortex.* Starting from the fact that memory is a very complex phenomenon, conditioned by sensory elements (images, sounds, touch, pain etc.), motor elements, as well as by elements of the affective and emotive sphere and by merely intellectual elements (ideas, thoughts etc.), it is very difficult to establish a cortical representation. However, taking into account that the excitation phenomena may be elicited by the stimulation of the lateral and medial surfaces of the temporal lobe (Penfield) and that the regions are connected with the prefrontal areas through the uncinate fasciculus, with the acoustic and visual areas of the body schema, recent researches have the tendency to situate the area of memory at this level (Delmas). Moreover, as regards the fixation in the memory of recent facts and the possibility of evoking them, it is considered that the hippocampus plays an essential role.

2. *Integrative representation of the whole proper body.* The human individual becomes aware of his own body in various ways: by palpation, which permits us to feel the form and consistence of the body, by proprioceptive influxes which inform us about the attitude and the situation in the space, by visual and auditory influxes, which permit us to recognize the voices etc. It has been proved that the areas of sensibility (parietal), the acoustic (temporal) areas and the visual (occipital) areas are connected at the level of the body projection area, situated in the convolutions surrounding the posterior extremity of the sylvian fissure and of the first temporal fissure.

In this region projects also the pulvinar, which, in turn, receives inflows from the lateral and medial geniculate bodies and from the posterior lateroventral nucleus, hence diencephalic relays of the auditory, visual and tactile pathways.

3. *Cortical areas related to the prevision capacity.* Action requires prevision. Penfield's researches confirm the links of the prefrontal cortex with the medial nucleus of the thalamus, with the motor, the sensory, the occipital and the temporal cortex. These anatomic elements, associated with the clinical facts, suggest the existence of prevision areas at the level of the anterior portion of the prefrontal cortex and it is proved that the involvement of the prefrontal region is achieved in relation with mental processes able to lead to foreseeable actions.

4. *Cortical emotiveness areas.* The latest electrophysiological and clinical researches permit to elucidate some cortical areas of emotiveness.

The affectivity and the emotions are most frequently produced on the basis of sight, hearing, smell etc. The emotions elicit motor and vegetative manifestations. Two regions of the cortex are directly related to the regulating centres of motoricity and to the hypothalamic nuclei, respectively to the prefrontal area and to the cyngulate (limbic) area, a finding which on the basis of electrophysiological and clinical facts led to the conclusion that these zones are area of emotiveness representation.

The prefrontal cortex is in a double-way connection with the precentral cortex and the posterior nuclei of the hypothalamus (ergotropic-Hess), the stimulation of which elicits hypertension, tachycardia and stress reactions.

The cortex of the cyngulate area (Boca's limbic lobe), through its anterior part, inhibits the motor activities by its action on the caudate nucleus and on the reticular substance of the brainstem.

124

Fig. 98 Caudate nucleus, corpus callosum, hippocampus and inferior horn of the lateral ventricle – superior view (preparation achieved by horizontal section and removal of a part of the temporal and occipital lobes)

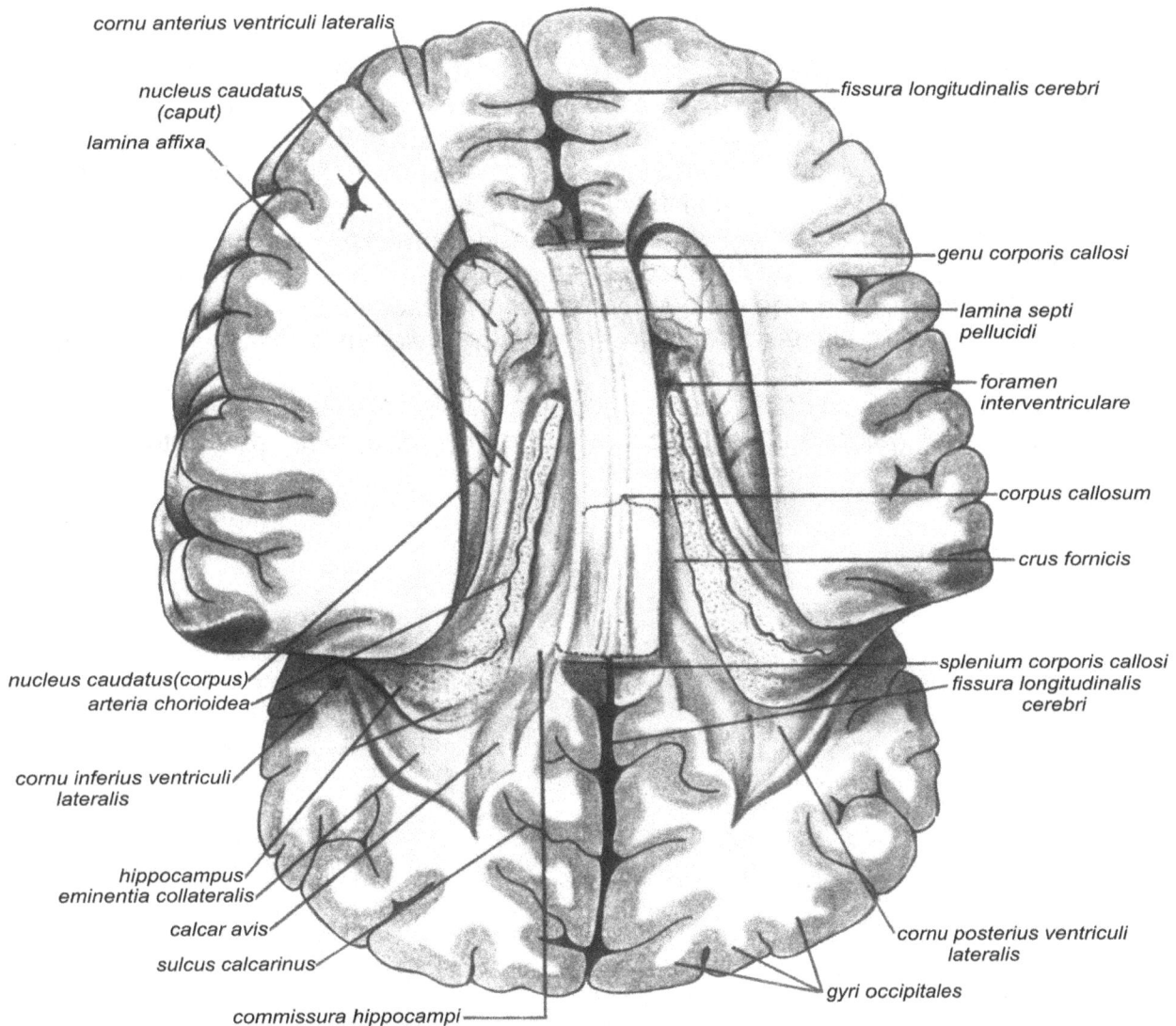

cornu anterius ventriculi lateralis

nucleus caudatus (caput)

lamina affixa

fissura longitudinalis cerebri

genu corporis callosi

lamina septi pellucidi

foramen interventriculare

corpus callosum

crus fornicis

nucleus caudatus(corpus)

arteria chorioidea

splenium corporis callosi

fissura longitudinalis cerebri

cornu inferius ventriculi lateralis

hippocampus

eminentia collateralis

calcar avis

sulcus calcarinus

cornu posterius ventriculi lateralis

gyri occipitales

commissura hippocampi

Moreover, the cyngulate area receives afferents from all the suppressor areas (frontal, parietal, preoccipital) of the cortex and from the hippocampus, through the anterior nucleus of the thalamus. This last afferent indicates that the rhinencephalon may elicit vegetative effects through its projection fibres on the anterior hypothalamic (trophotropic – Hess) nuclei, and as we have mentioned above, may also exert a direct action – through the cortex of the cyngulate area, on the subcortical motor centres.

The cyngulate area, activated in this way by the neocortex and the archicortex, sends, in turn, inflows to the inhibitory centres; this area represents the site where all the mechanisms, achieving the state called emotion (Papez) take place.

No region of the cerebral cortex functions isolatedly, but each of them is based upon the fiability principle and is in a close relationship with the other cortical areas. Thus, for an example, in the thinking-speaking, reading and writing processes, which are proper only to humans, a great number of cortical areas is involved, among which functional relationships are established.

The grey nuclei of the telencephalon
The striate body – corpus striatum (Fig. 98 – 117)

The grey nuclei of the telencephalon are represented by groups of cells situated in the depth of the encephalon (subcortical region) and included in the white matter. These nuclei, called also *basal ganglia of the brain*, are paired. They are the following: *the caudate nucleus (nucleus caudatus)*, the lentiform or *lenticular nucleus* (these two nuclei form together t*he striate body*, called so owing to its striate appearance on section, caused by the myelinated fibres which cross it), *the claustrum* and *the amygdaloid body*.

The caudate nucleus *(nucleus caudatus)* has the shape of an arch with the convexity caudal. It begins rostrally by a thickened portion, termed the head *(caput nuclei caudati)*, which is continuous dorsally with the body *(corpus nuclei caudati)* and the tail *(cauda nuclei caudati)*. The tail of the caudate nucleus is fused with the thalamus at the level of the terminal stria and ends in the amygdaloid body.

The superior surface of the caudate nucleus, situated in the lateral ventricle, forms a portion of its lateral wall and is continuous with the anterior terminal part of the lenticular nucleus.

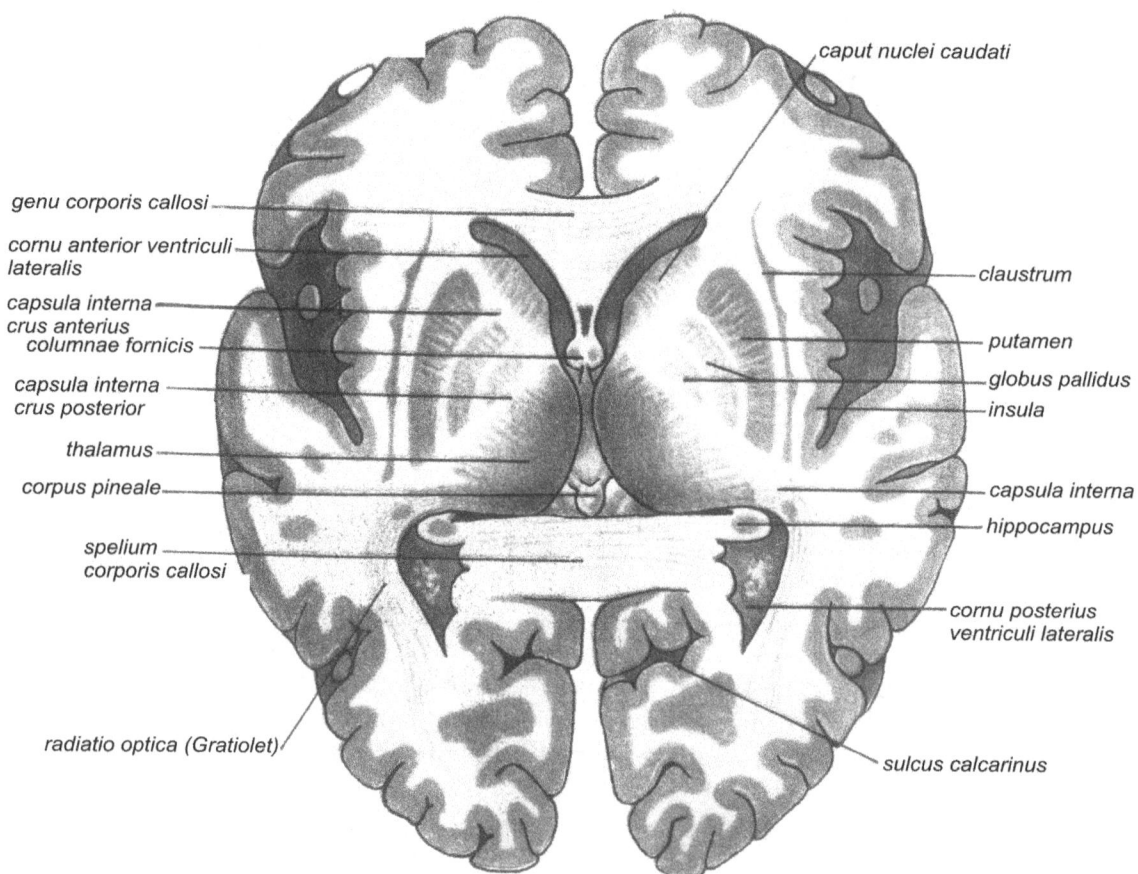

Fig. 99 Horizontal section through the encephalon at the level of the interventricular foramen – superior view; the pineal gland, the third ventricle, the pellucid septum and the lateral ventricles may be seen

126

Fig. 100 Horizontal section through the encephalon at the level of the commissure of the fornix

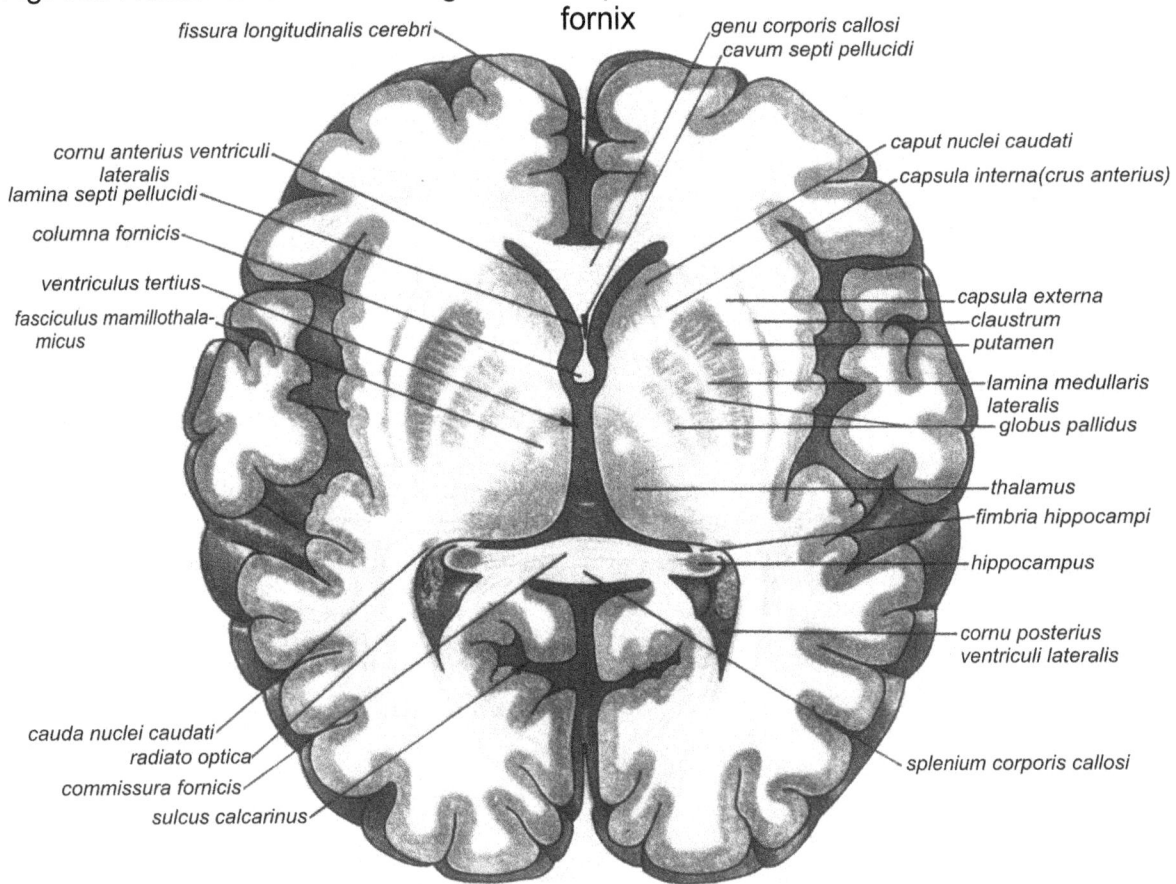

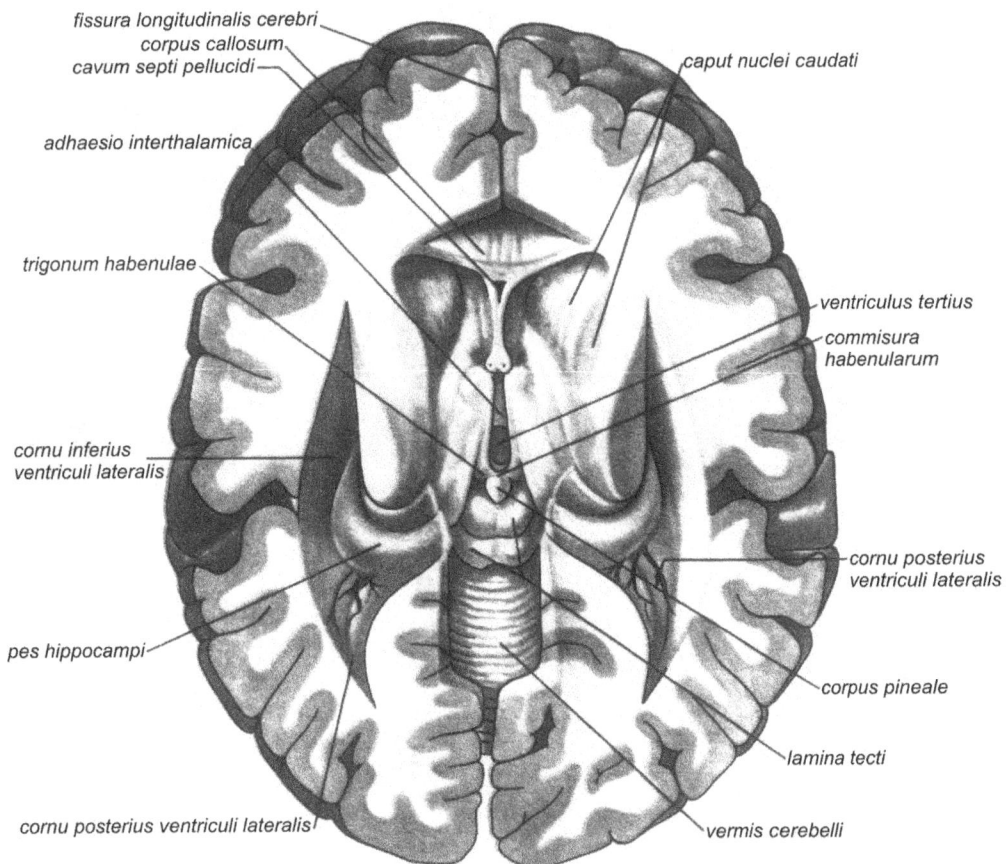

fissura longitudinalis cerebri

genu corporis callosi
cavum septi pellucidi

cornu anterius ventriculi lateralis
lamina septi pellucidi
columna fornicis
ventriculus tertius
fasciculus mamillothalamicus

caput nuclei caudati
capsula interna(crus anterius)

capsula externa
claustrum
putamen
lamina medullaris lateralis
globus pallidus

thalamus
fimbria hippocampi
hippocampus

cornu posterius ventriculi lateralis

cauda nuclei caudati
radiato optica
commissura fornicis
sulcus calcarinus

splenium corporis callosi

Fig. 101 Horizontal section through the cerebral hemispheres, with the removal of the splenium of the corpus callosum, the column of the fornix and the tela choroidea of the fourth ventricle (the third ventricle and the lateral ventricles may be seen)

fissura longitudinalis cerebri
corpus callosum
cavum septi pellucidi

caput nuclei caudati

adhaesio interthalamica

trigonum habenulae

ventriculus tertius
commisura habenularum

cornu inferius ventriculi lateralis

cornu posterius ventriculi lateralis

pes hippocampi

corpus pineale

lamina tecti

cornu posterius ventriculi lateralis

vermis cerebelli

127

The lenticular or lentiform nucleus (nucleus lentiformis) is the largest portion of the striate body and has on the transverse section, the shape of a triangle with the basis orientated laterally and the summit medially. Both nuclei are crossed here and there by white fibres and are separated from each other by a bulky mass of white fibres, the internal capsule. The lentiform nucleus is formed of two parts: laterally, the putamen, of a dark colour, and medially the globus pallidus, of a lighter colour.

The claustrum is a thin layer of grey matter, with a sinuous course, corresponding to the gyri and sulci of the insula; it lies sagitally between the cortex of the insula and the putamen. The white matter between the cortex of the insula and the claustrum forms the extreme capsule; the white matter between the claustrum and the putamen makes up the external capsule.

The amygdaloid body or amygdaloid nucleus (corpus amygdaloideum) is situated in proximity to the of the temporal lobe, anterior to the ending of the inferior horn of the lateral ventricle.

It unites with the claustrum, with the anterior perforated substance and has connections with the cortex of the hippocampal gyrus and with the tail of the caudate nucleus. The amygdaloid body and the claustrum are in a close relationship with the olfactory formations.

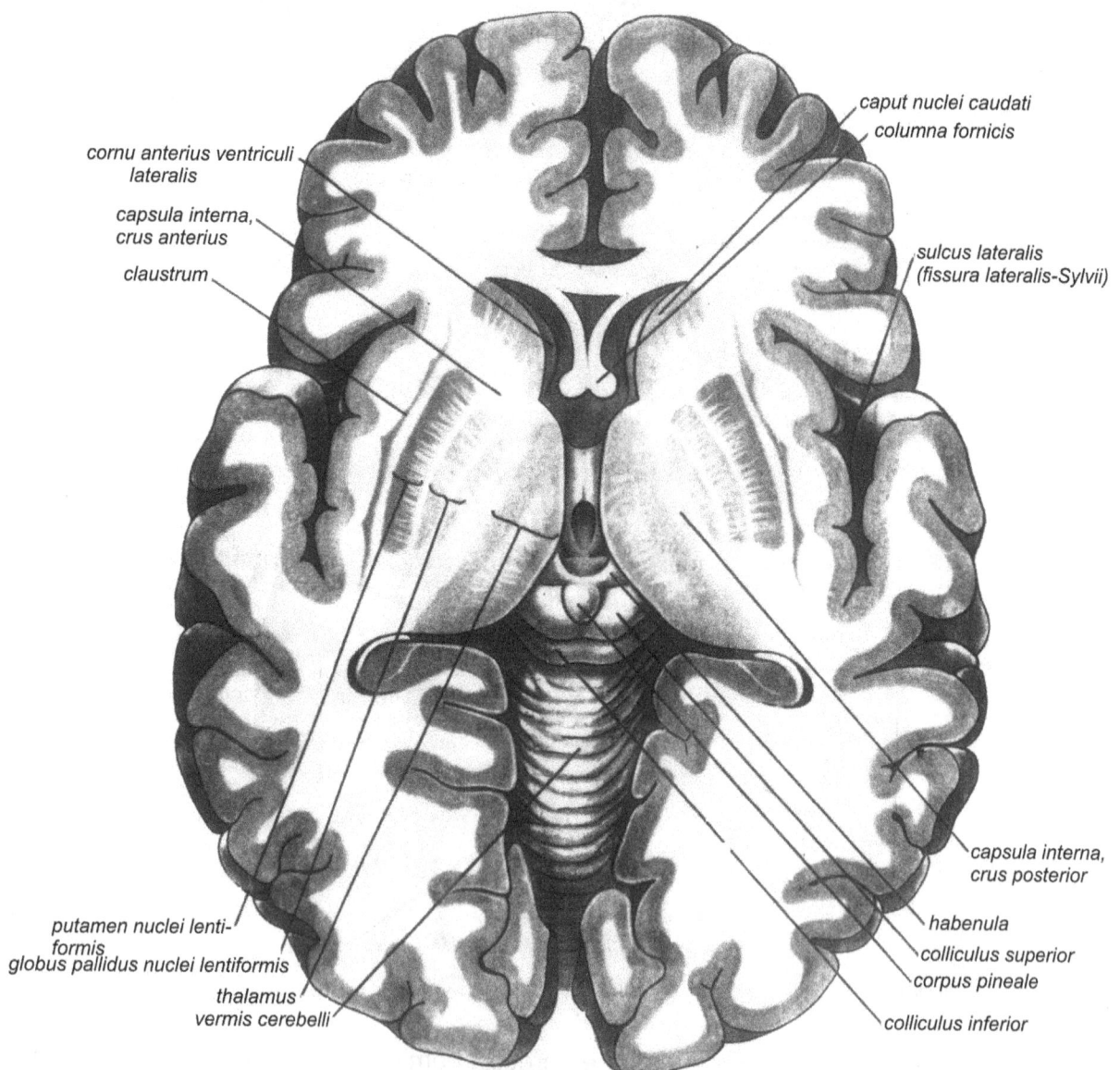

Fig. 102 Horizontal section through the encephalon at the level of the striate body and of the thalamus

128

We mention that, on the basis of the anatomic connections, of the function and of the embryonic origin, *the globus pallidus* is more ancient on the phylogenetic scale, representing *the paleostriatum*, whereas *the caudate nucleus*, a more recent formation, represents *the neostriatum*.

The striate body is an important centre of the extrapyramidal system and participates in all the automatic movements. It receives motor impulses from the precentral gyrus (areas 4, 6 and 8 of the frontal lobe, which are processed and individualized at this level and then conducted towards the red nucleus. The striate body governs the motor automatism, the muscle coordination and tone, as well as the *"play"* of all the elements of the locomotor system and of the mimetic muscles, making up the *"personal note"* of the individual. For example, if two individuals perform the same movement, this movement is, nevertheless, not identical, since one of them carries it out faster and the other one slower, one is more skilful, the other less skilful; some elements may become identical by means of training, but the *"personal tempo"* remains the same.

The striate body acts as a whole, in a unitary way, but its components accomplish antagonistic functions. *The caudate nucleus* and *the putamen* represent somehow a break to the movements. *The globus pallidus* and the medial nucleus of the thalamus, which functionally belongs to it (they appear concomitantly on the phylogenetical scale), accelerate the movements. Both systems impact the *"personal character"* by the different processing of the impulses that pass through them (the phlegmatic type acts rapidly in the case of danger, like the dynamic type).

The afferents and efferents of the basal nuclei are systematized below.

The afferent pathways derive mainly from the thalamus and from the cerebral cortex. The thalamic afferents are achieved through the thalamostriate fibres which arise from the centromedian and intralaminar nuclei, old parts of the thalamus, and which course into *the caudate nucleus* and *the putamen*, traversing *the internal capsule*. There are sometimes putaminopallidal associations which assure the relationship between these nuclei, regardless the level of termination of the thalamostriate fibres. The cortical afferents have their origin in the sensory and motor cortex assuring the unity between the voluntary and the automatic movements.

The efferent pathways. The fibres from the caudate nucleus and the putamen reach *the globus pallidus*, which is the single efferent centre (it ensues that the neostructure of *the caudate nucleus* and of *the putamen* achieves the control of the paleostriatum, respectively of *the globus pallidus*, which is the effector). The efferents to *the globus pallidus* are following an anterior fasciculus, the lenticular loop *(ansa lenticularis)* and a posterior fasciculus, the lenticular fasciculus, which join, surround anteriorly the internal capsule and enter into the subthalamic region (fasciculus H2), from which they travel to: the vegetative nuclei of the hypothalamus (the pallidohypothalamic fasciculus), the anterior ventrolateral nucleus of the thalamus (the thalamic fasciculus H4) the subthalamic nuclei (*zona incerta*, the red nucleus – fasciculus H); *locus niger* – fibres forming Forel's subthalamic field).

In the clinic, other terms than the anatomic ones are used for the basal ganglia, which we mention owing to the practical necessity: *the caudate nucleus* and *the putamen* are called *the striatum*; *the globus pallidus* is termed *the pallidum*.

In the lesions of the pallidum and of the striate appear hyperkynesias assuming various forms in dependence on the site of the pathological process:

- purposeless head and trunk rotation, facial deformation, disordered movements of the hands, like in *chorea minor* (acute form – Sydenham's chorea, or chronic form – Huntington's chorea);

- permanent rhythmic contractions of certain muscle groups, especially at the level of fingers (like in Parkinson's disease), accompanied by the impairment of the automatism of movements, explained by the blocking of the cortical inhibition at the level of the reticular substance in the mesencephalic tegmentum. A significant role in the occurrence of such a tremor devolves also upon impairments in the dopaminergic mechanism;

Fig. 103 Horizontal section through the telencephalon at the level of the mesencephalic tectum and of the subthalamic nucleus

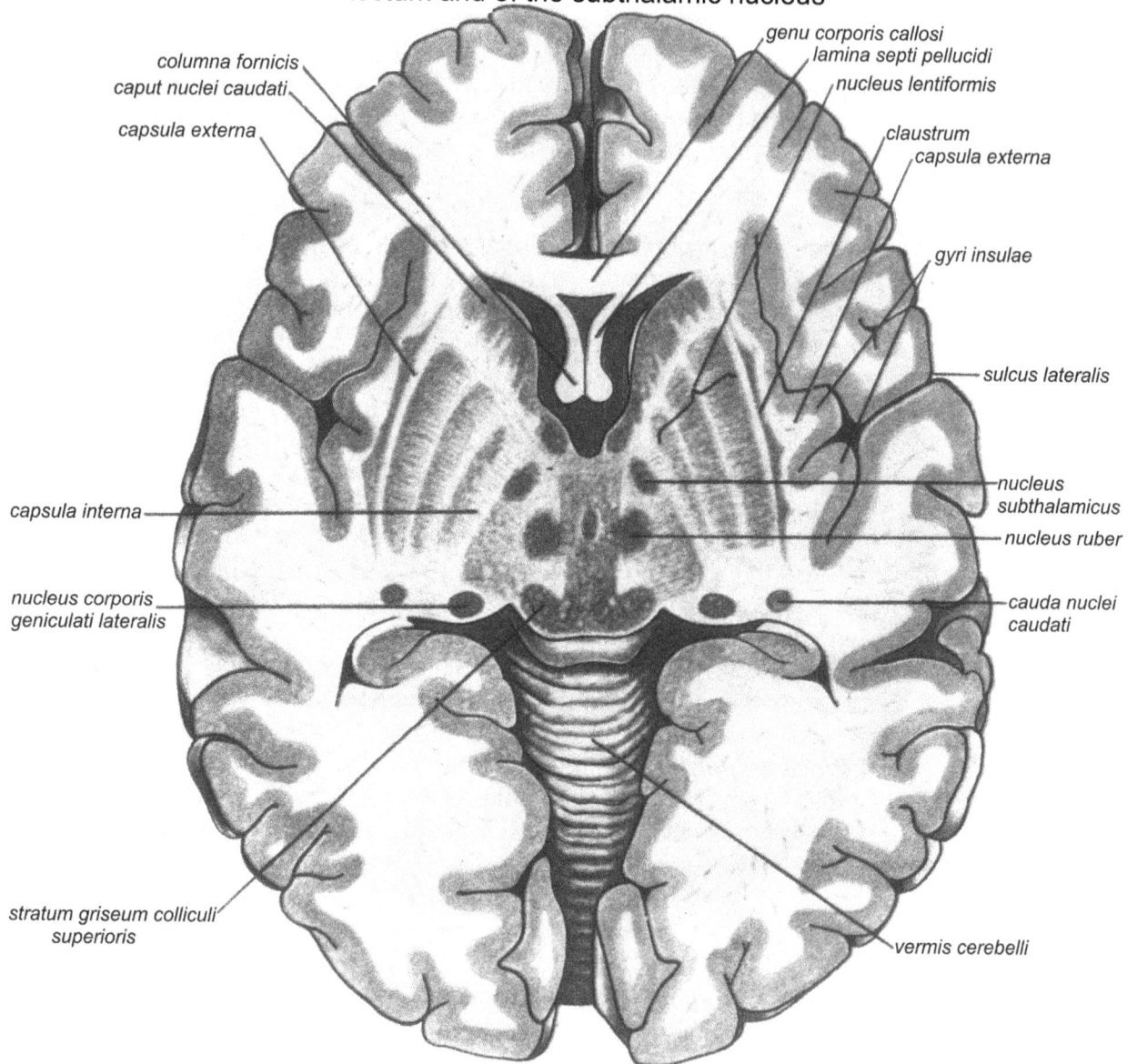

columna fornicis
caput nuclei caudati
capsula externa

genu corporis callosi
lamina septi pellucidi
nucleus lentiformis

claustrum
capsula externa

gyri insulae

sulcus lateralis

nucleus subthalamicus
nucleus ruber

capsula interna

nucleus corporis geniculati lateralis

cauda nuclei caudati

stratum griseum colliculi superioris

vermis cerebelli

- hypokinesias which appear often after a lethargic encephalitis. The patient knows how to perform all the movements, but he carries them out very slowly, as the automatism of basal ganglia is lacking. In the facial expression, too, the imobility of the mimetic muscles may be observed, the patient acquiring the characteristic appearance of *"mask-like face"*.

These lesions are rarely situated only at the level of the striate body or only on *the globus pallidus*. Usually both nuclei are affected, making possible the appearance of a large range of complex hypertonic-hyperkinetic symptoms.

By correlating the manifestation forms of the disease with the level of the lesion, it ensues that the lesion of *the globus pallidus*, which is an older centre regulating the muscle tone and the coordination of elementary automatic movements, leads to muscle rigidity, akinesia and defective coordination of movements, whereas the lesion of *the caudate nucleus* and of *the putamen* – structures more recently appeared on the phylogenetical scale, which have the role of controlling and inhibiting the activity of *the pallidum* – causes the lowering of the muscle tone, and the appearance of incoordinated choreo-athetotic movements, as a consequence of the escape of the pallidum from this control.

Taking into account the close functional and topographical relationship, we describe here, also, the sublenticular and subthalamic areas.

130

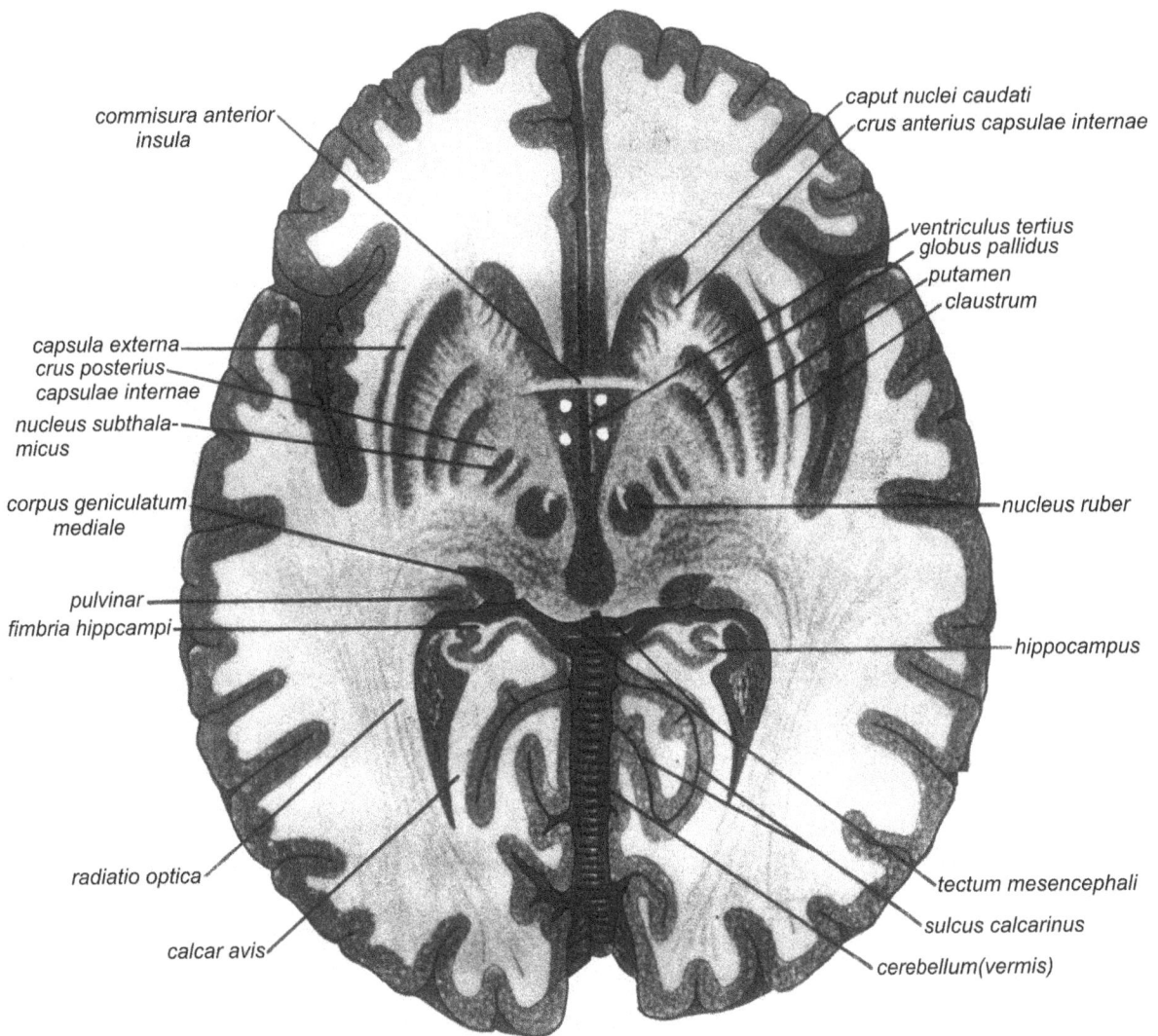

commisura anterior
insula

caput nuclei caudati
crus anterius capsulae internae

ventriculus tertius
globus pallidus
putamen
claustrum

capsula externa
crus posterius
capsulae internae

nucleus subthala-
micus

corpus geniculatum
mediale

nucleus ruber

pulvinar
fimbria hippcampi

hippocampus

radiatio optica

tectum mesencephali

sulcus calcarinus

calcar avis

cerebellum(vermis)

Fig. 104 Horizontal section through the telencephalon
at the level of the anterior commissure and of the
medial geniculate body

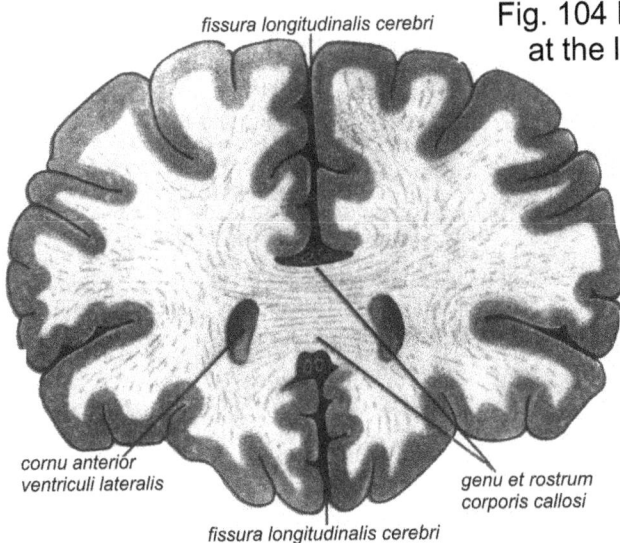

fissura longitudinalis cerebri

cornu anterior
ventriculi lateralis

genu et rostrum
corporis callosi

fissura longitudinalis cerebri

Fig. 105 Frontal section through the
genu of the corpus callosum at the level
of the anterior horn of the lateral
ventricle

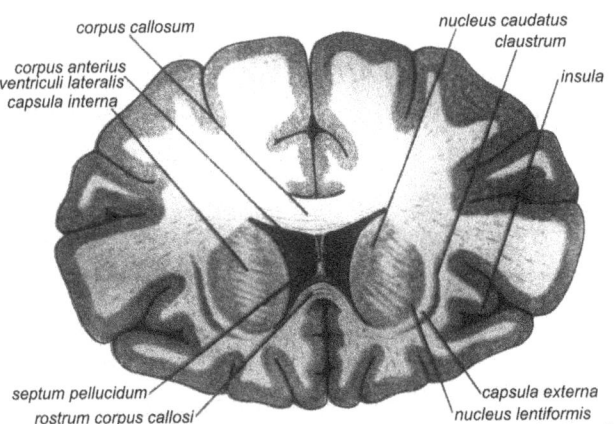

corpus callosum

nucleus caudatus
claustrum

corpus anterius
ventriculi lateralis
capsula interna

insula

Fig. 106 Frontal section through
the pellucid septum at the
claustrum

septum pellucidum
rostrum corpus callosi

capsula externa
nucleus lentiformis

131

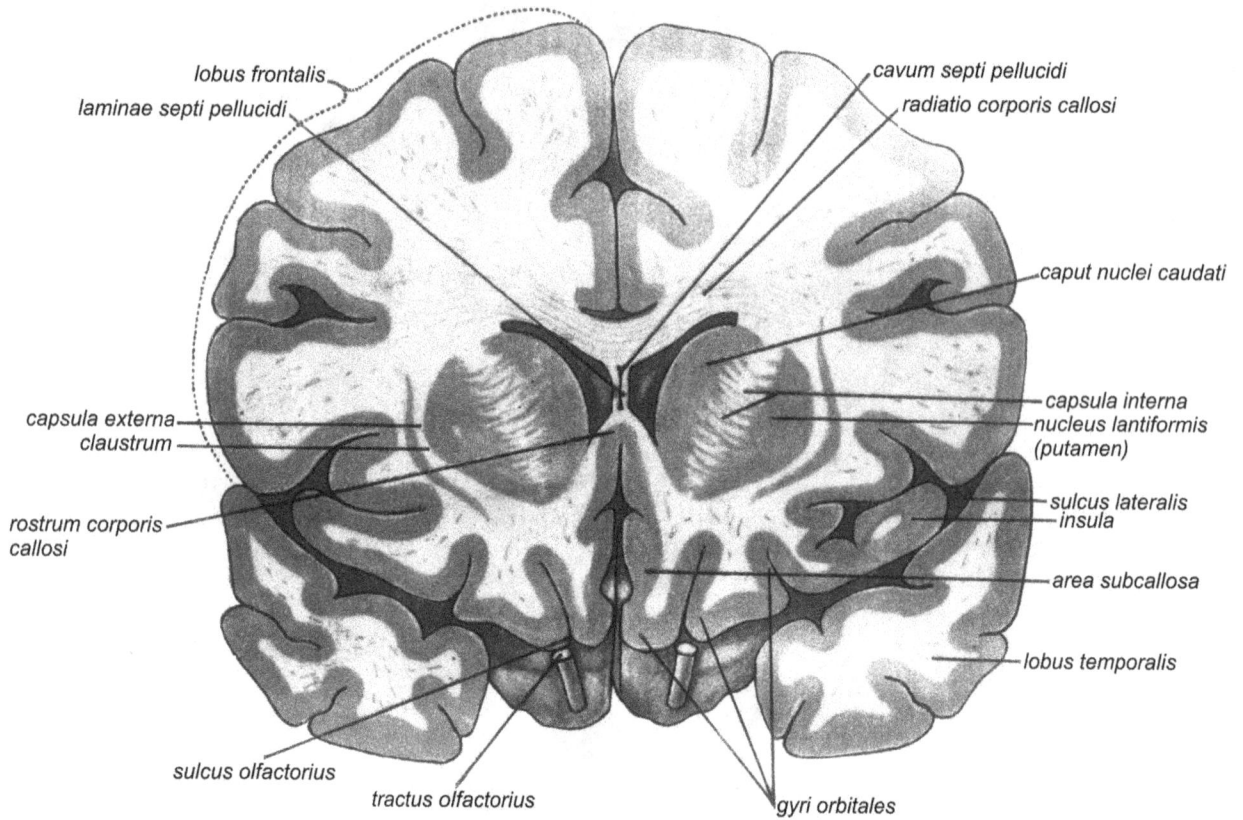

Fig. 107 Telencephalon – frontal section in front of the thalamus, at the level of the anterior part of the pellucid septum

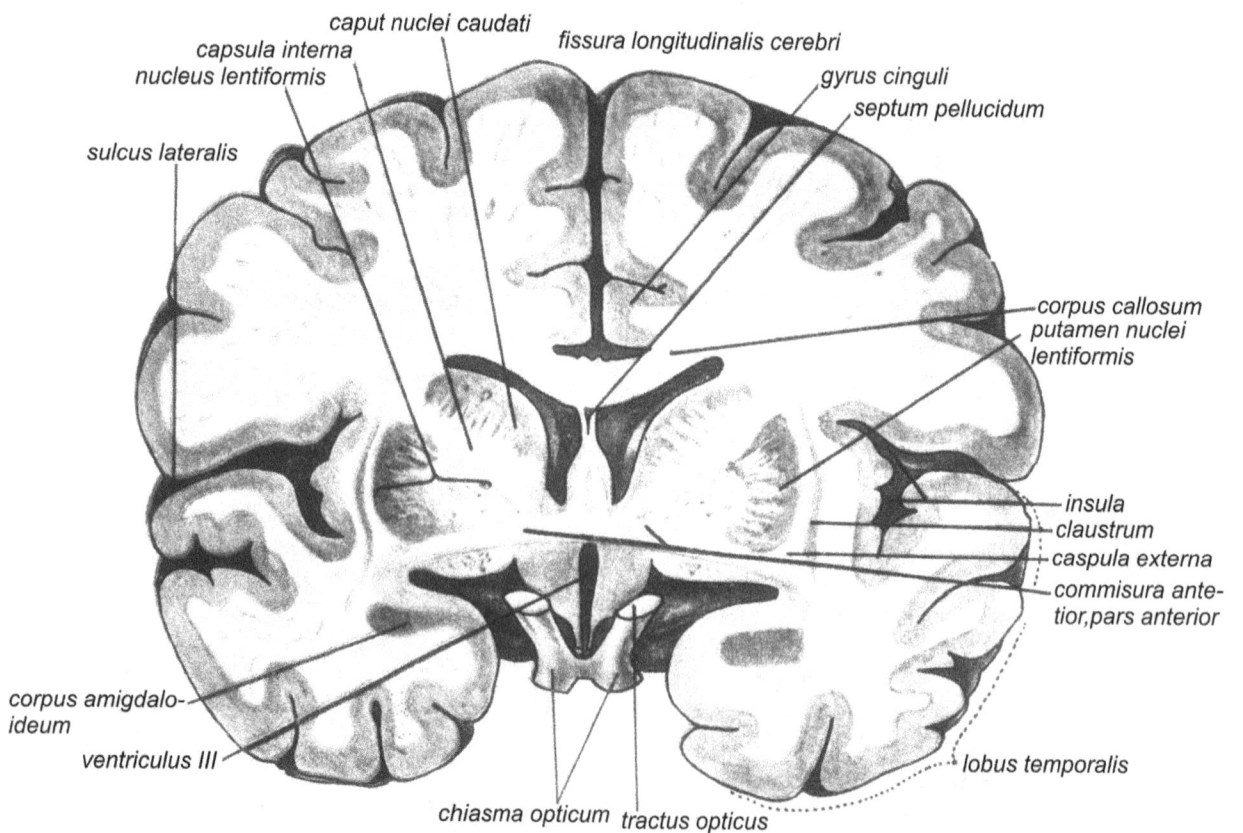

Fig. 108 Telencephalon – frontal section at the level of the anterior commissure

132

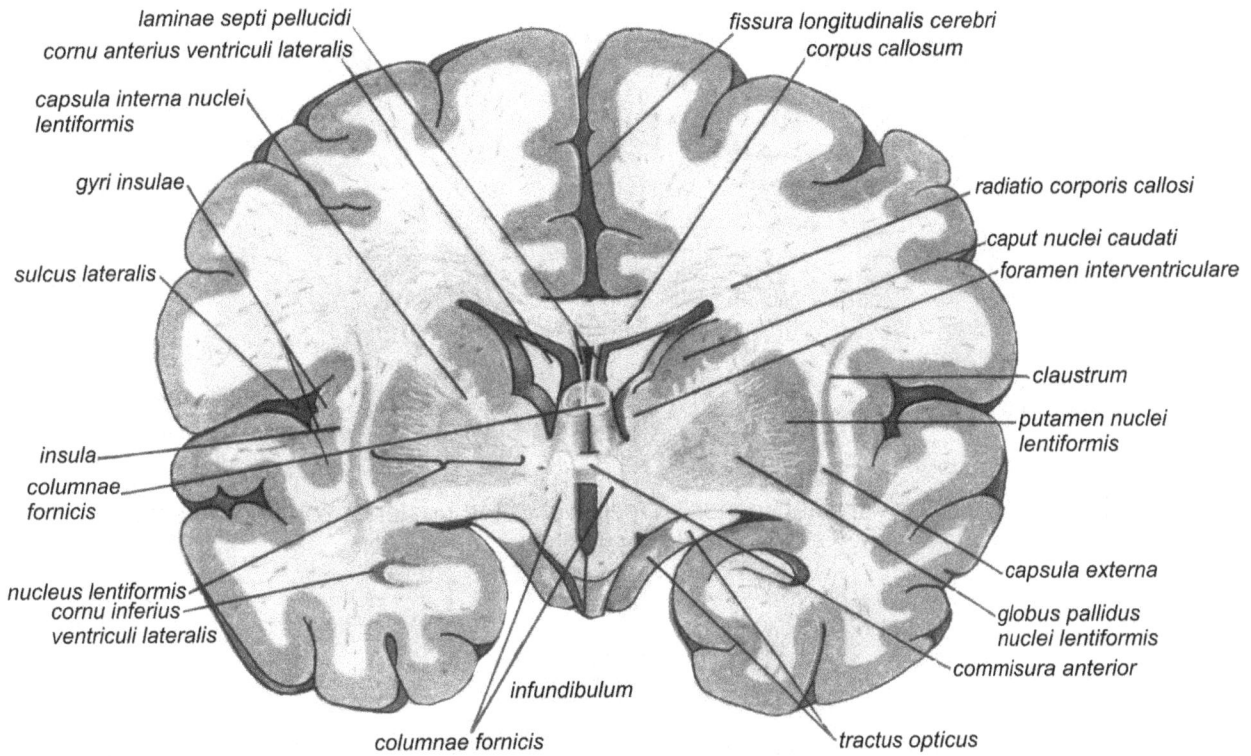

Fig. 109 Telencephalon – frontal section at the level of the columns of the fornix

laminae septi pellucidi
cornu anterius ventriculi lateralis
capsula interna nuclei lentiformis
gyri insulae
sulcus lateralis
insula
columnae fornicis
nucleus lentiformis
cornu inferius ventriculi lateralis
columnae fornicis
infundibulum

fissura longitudinalis cerebri
corpus callosum
radiatio corporis callosi
caput nuclei caudati
foramen interventriculare
claustrum
putamen nuclei lentiformis
capsula externa
globus pallidus nuclei lentiformis
commisura anterior
tractus opticus

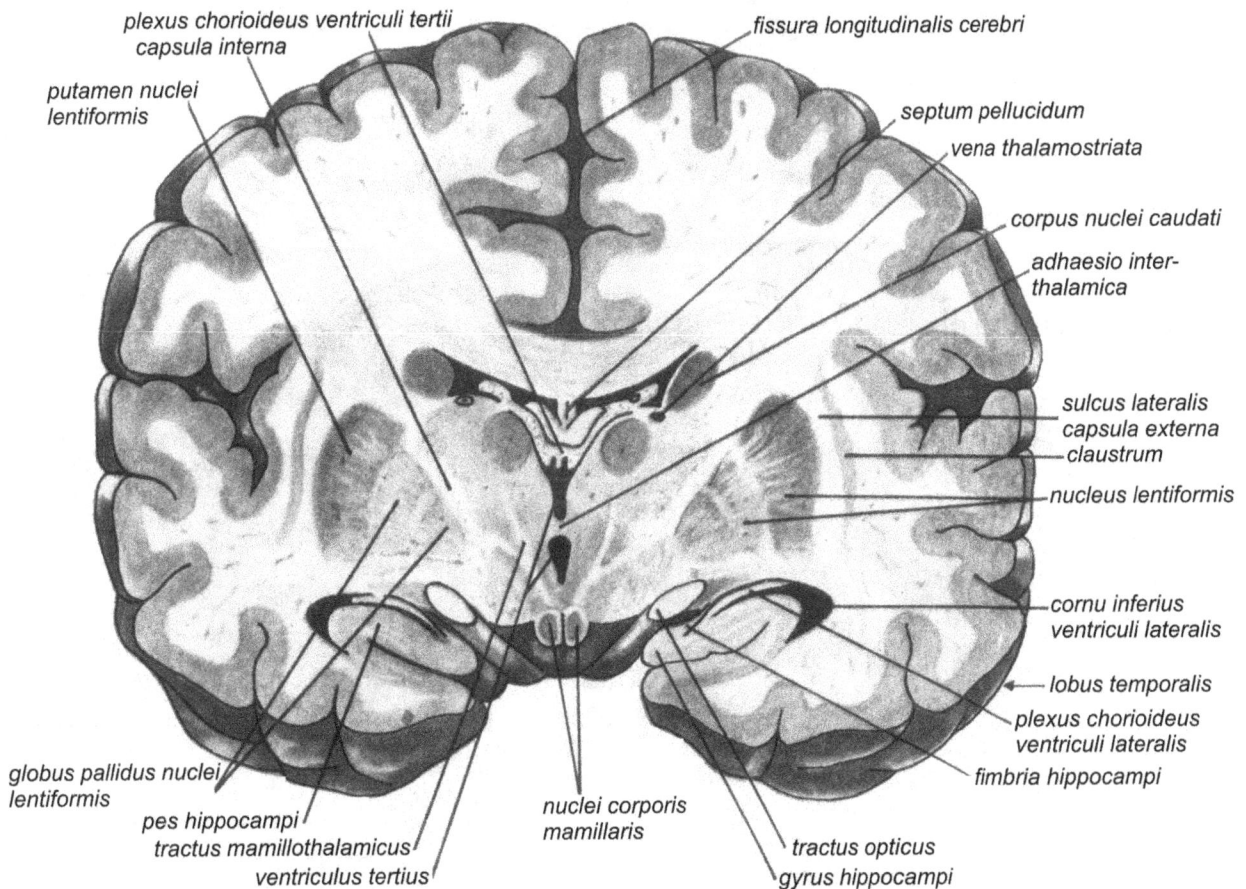

Fig. 111 Telencephalon – frontal section at the level of the interthalamic adhesion and of the mamillary bodies

plexus chorioideus ventriculi tertii
capsula interna
putamen nuclei lentiformis
globus pallidus nuclei lentiformis
pes hippocampi
tractus mamillothalamicus
ventriculus tertius
nuclei corporis mamillaris
tractus opticus
gyrus hippocampi

fissura longitudinalis cerebri
septum pellucidum
vena thalamostriata
corpus nuclei caudati
adhaesio inter-thalamica
sulcus lateralis
capsula externa
claustrum
nucleus lentiformis
cornu inferius ventriculi lateralis
lobus temporalis
plexus chorioideus ventriculi lateralis
fimbria hippocampi

133

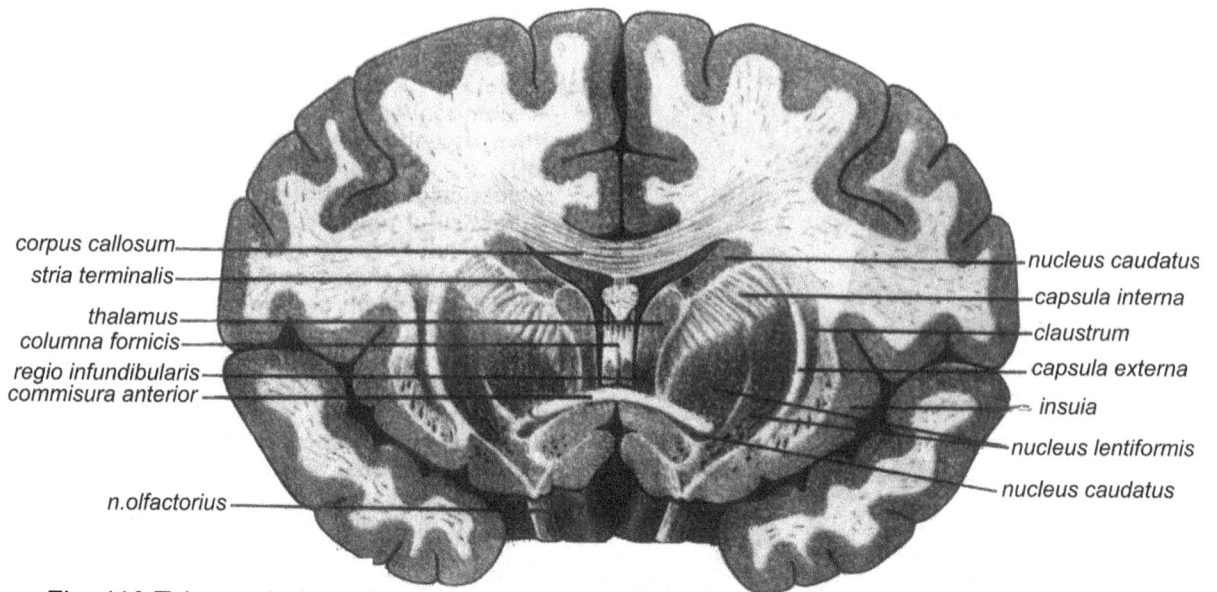

corpus callosum
stria terminalis
thalamus
columna fornicis
regio infundibularis
commisura anterior

n.olfactorius

nucleus caudatus
capsula interna
claustrum
capsula externa
insuia
nucleus lentiformis
nucleus caudatus

Fig. 110 Telencephalon – frontal section at the level of the posterior zone of the anterior commissure

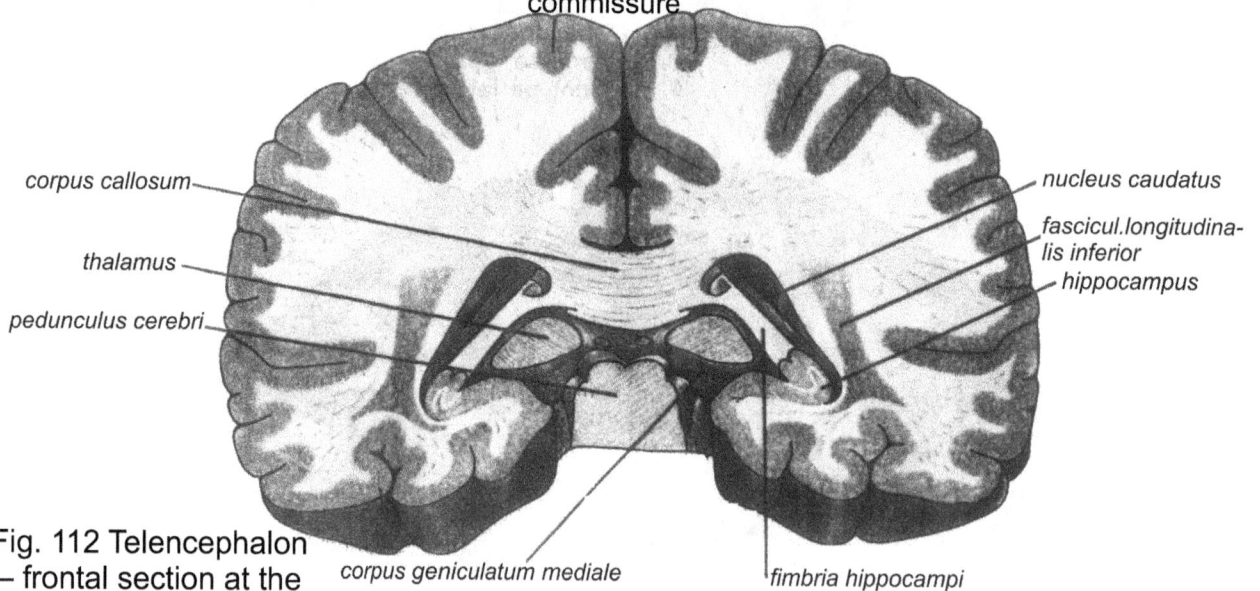

corpus callosum

thalamus

pedunculus cerebri

nucleus caudatus
fascicul.longitudinalis inferior
hippocampus

corpus geniculatum mediale

fimbria hippocampi

Fig. 112 Telencephalon – frontal section at the level of the hippocampus and of the metathalamus

Fig. 113 Telencephalon – frontal section at the level of the corpus callosum, of the dentate gyrus and of the posterior horn of the lateral ventricle

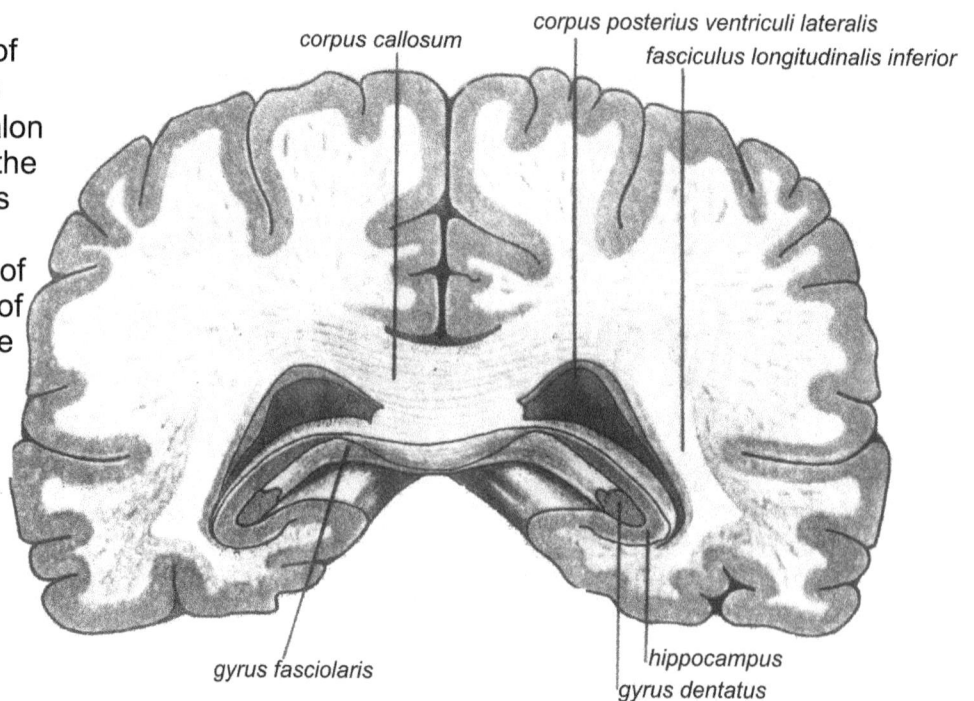

corpus callosum

corpus posterius ventriculi lateralis
fasciculus longitudinalis inferior

gyrus fasciolaris

hippocampus
gyrus dentatus

134

The sublenticular and subthalamic areas

A frontal section at the penetration site of the two cerebral peduncles in the two hemispheres, shows that the white matter of the peduncles is disposed in three areas: a first area, situated below the thalamus, another one at the level of the inferior surface of the lenticular nucleus and the last between the thalamus and the lenticular nucleus, forming the internal capsule. The subthalamic and the sublenticular areas are very important, since they represent a crossroad at the level of which intersect pathways which are oriented in all directions and which for the time being are stil incompletely elucidated. These pathways may be systematized (Delmas) as follows:

1. *Thalamopetal pathways* coursing towards the anterior nucleus of the thalamus:
- the medial lemniscus and the spinothalamic tracts (ascending sensory pathways coming from the spinal cord and the brain stem);
- the dentatothalamic tract;
- the pallidothalamic tract and
- the mamillothalamic tract.

2. *Thalamofugal pathways* represented by fibres arising from the posterior nuclei of the thalamus (neothalamus) and coursing to the cerebral cortex.

3. *Striopetal pathways* arising from the cerebral cortex either directly or through the corticonuclear or corticospinal tracts.

4. *Striofugal* pathways represented by:
- caudopallidal and putaminopallidal neurons;
- the lenticular ansa, formed of fibres arising from the medullary lamina of the lenticular nucleus;

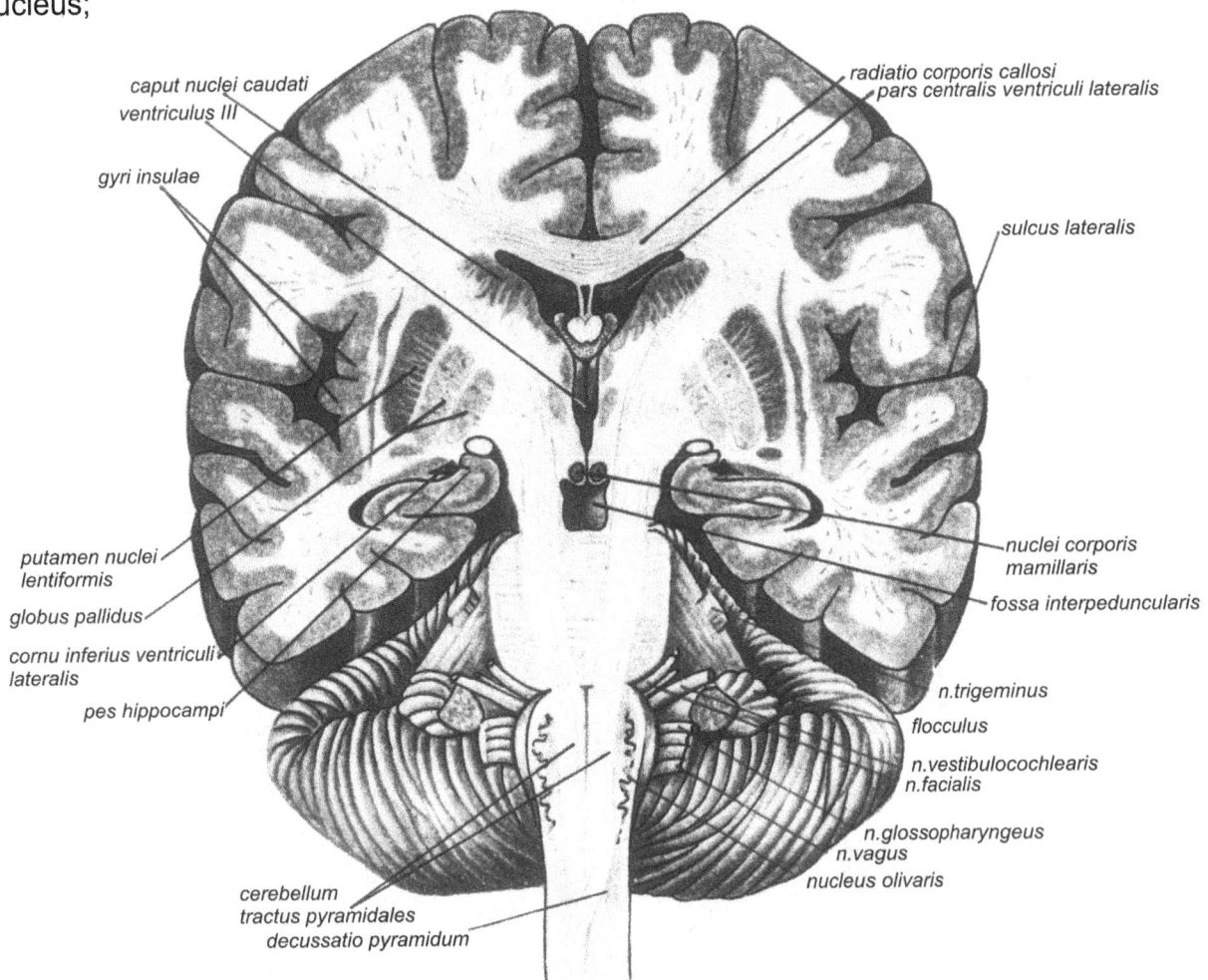

Fig. 115 Section through the telencephalon, parallel to the cerebral peduncles and passing at the level of the nuclei of the mamillary bodies and of the bulbar olivae. The internal capsule and the pyramidal tracts may be seen

135

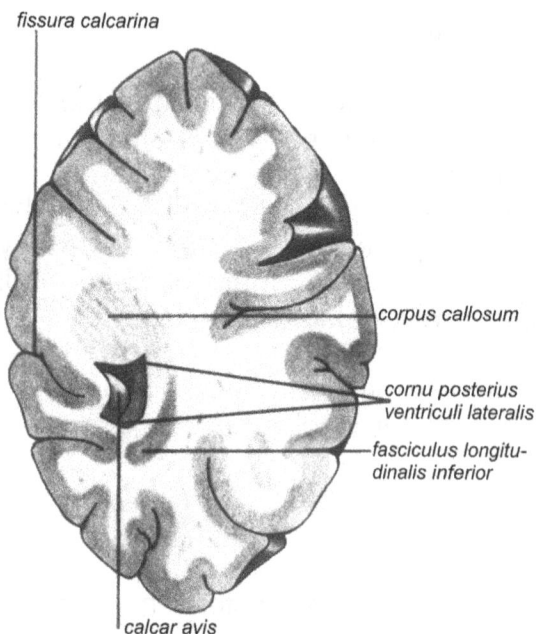

fissura calcarina

corpus callosum

cornu posterius
ventriculi lateralis

fasciculus longitu-
dinalis inferior

calcar avis

Fig. 114 Frontal section of the
occipital pole of the telencephalon

Fig. 116 Section through the
telencephalon, parallel to the
cerebral peduncles, at the level of
the head of the caudate nucleus,
of the substantia nigra and of the
bulbar olivae

- the lenticular fasciculus arising from the posterior surface of *the pallidum*, then oriented towards the subthalamic region, where it is distributed to the anterior contralateral nucleus of the thalamus (thalamic tract), to *the zona incerta*, to the red nucleus, to the subthalamic nuclei, *the locus niger* and the hypothalamic vegetative nuclei (the pallidohypothalamic tract).

Topographically, at this level, in the sublenticular region, lies also the anterior commissure, which forms a groove on the inferior surface of the pallidum.

Moreover, taking into account the neighbourhood relationships, we mention, also, that in the interpeduncular space, at the level of the small interpeduncular ganglion is sited a centre which regulates the cortical activities and which performs a filtration of the nerve inflows and a cortical inhibition. Its lesions produces mental confusion, dementia praecox (according to some authors) and hallucinations. Moreover, the electrophysiological researches of Magoun and Moruzzi have demonstrated that a significant zone of reticular substance, playing a similar role in the regulation of the cortical activities, lies around the interpeduncular ganglion.

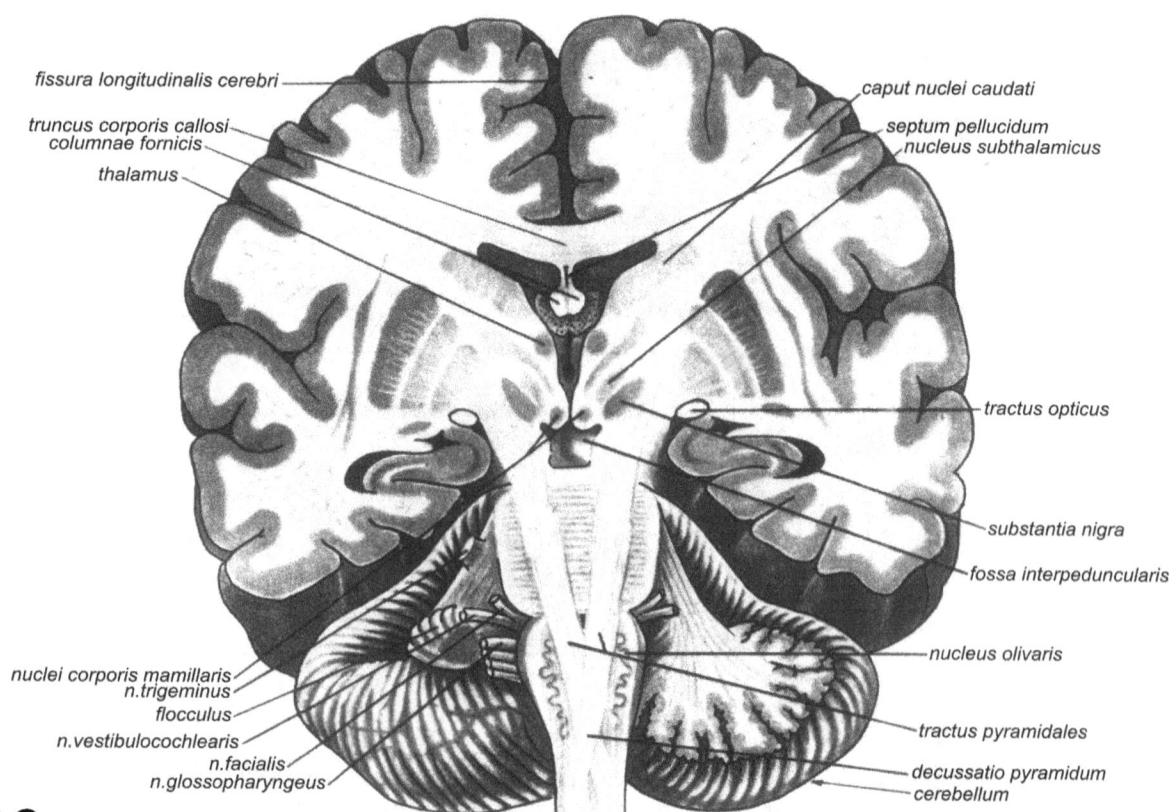

fissura longitudinalis cerebri

truncus corporis callosi
columnae fornicis

thalamus

caput nuclei caudati

septum pellucidum
nucleus subthalamicus

tractus opticus

substantia nigra

fossa interpeduncularis

nucleus olivaris

nuclei corporis mamillaris
n.trigeminus
flocculus
n.vestibulocochlearis
n.facialis
n.glossopharyngeus

tractus pyramidales

decussatio pyramidum
cerebellum

According to topographical criteria, the commissures existent in the striate bodies and of the thalamus are described below:

1. *Meynert's commissure*, which crosses the median line at the level of the anterior wall of *the tuber cinereum* and unites the two sublenticular regions.

2. *The posterior subthalamic commissure*, which unites the two subthalamic regions and crosses the median line posteriorly to the mamillary tubercles.

3. *Gudden's commissure*, situated between the medial geniculate bodies (and considered by some authors as extending also between the posterior quadrigeminal tubercles), topographically in relationship with the posterior surface of the optic tract and of the optic chiasma, which are involved in the acoustic system.

All these commissures are sited at the level of the floor of the third ventricle.

4. *The posterior commissure*, which passes through the mediosagittal plane above the opening of the sylvian aqueduct into the third ventricle. Its structure is complex, comprising fibres which join the two pulvinar nuclei, fibres which join the pulvinar to the anterior quadrigeminal tubercles, the lateral geniculate bodies and the nuclei of the oculomotor nerves, as well as fibres uniting the red nuclei with *the substantia nigra* of one side with those on the opposite side.

5. *The interhabenular commissure* connects the two ganglia of the habenula and crosses the midline above the pineal recess.

The white matter of the telencephalon
(Centrum semiovale)

The white matter of the telencephalon in horizontal section, somewhat above the callosal bodies, has at the level of each cerebral hemisphere a semioval shape (to which it owes its name of *centrum semiovale*) and is surrounded by grey matter. It is also called Vieussens' semioval centre.

Vieussens' oval centre is formed by the junction – through the white matter of *the corpus callosum* – of the two semioval centres. On a preserved brain, the white matter fibres can be prepared and even dissociated under the magnifying-glass by means of a blunt instrument.

The white matter has a regular architectonics, within which may be differentiated: a commissural system, formed of transverse fibres, a projection system, the fibres of which are oriented vertically and an associational system, formed of fibres with a more or less anteroposterior disposition.

The commissural system

The commissural system is constituted of transverse fibres connecting the two cerebral hemispheres and includes the following structures (Fig. 118 – 120):

The corpus callosum or *the great commissure* of the brain represents the strongest connection of the two cerebral hemisphere and is sited in the floor of the median or sagittal fissure, between the two cerebral hemispheres. It is covered by the occipital, parietal and frontal lobes (Fig. 119).

The corpus callosum has the shape of an arch with a hook at the anterior extremity. It begins at the anterior commissure, by a thin plate, the rostral lamina *(lamina rostralis)*, which is continuous with the rostrum of the corpus callosum *(rostrum corporis callosi)*. This is followed by the part with the convexity anterior, termed the genu, then by the trunk *(truncus)* and finally by the posterior, thickened end, the splenium of the corpus callosum *(splenium corporis callosi)*.

It measures 8 cm in length, 1 cm in breadth, anteriorly, 2 cm in breadth posteriorly and 1 cm in thickness. It has two surfaces – superior and inferior – and two extremities – anterior and posterior.

137

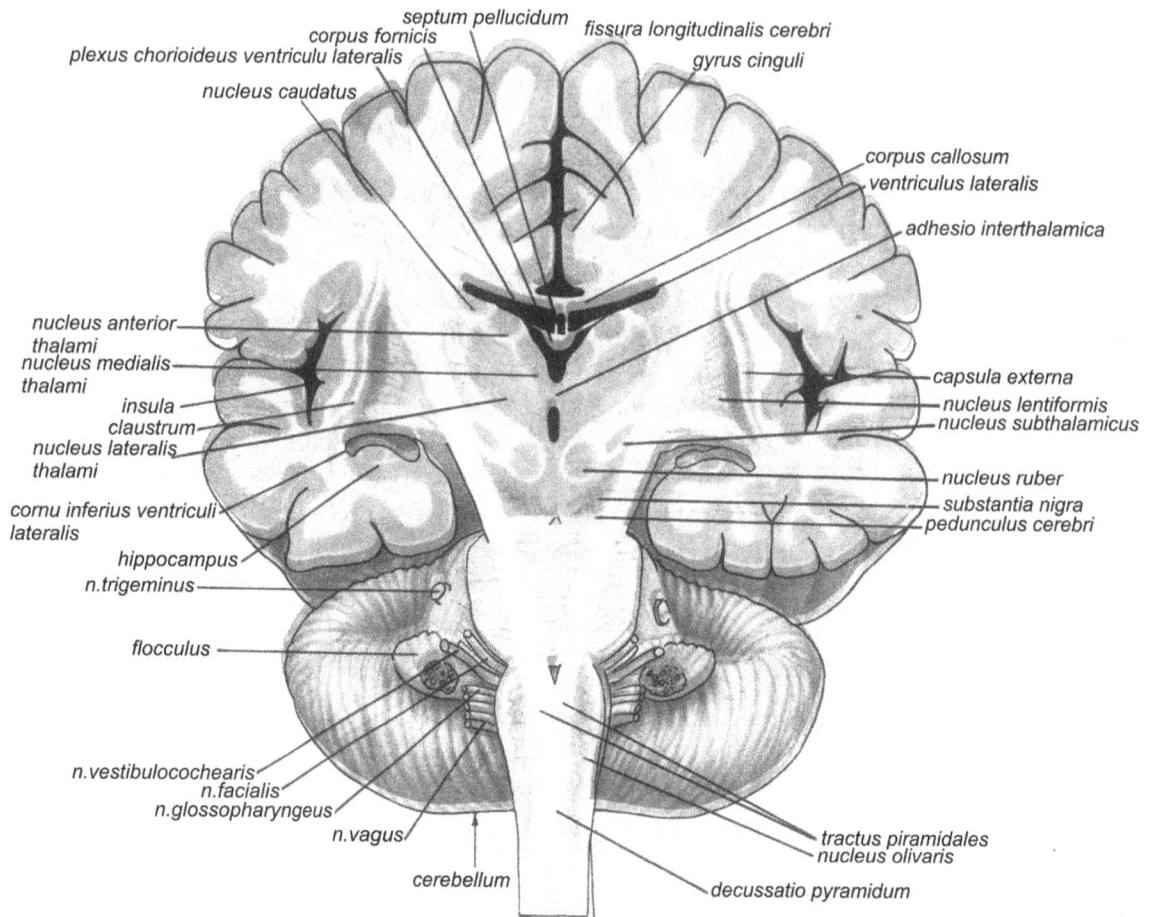

Fig. 117 Encephalon – section at the level of the internal capsule, the pyramidal tracts and the red nucleus

-The superior surfaces is convex anteroposteriorly and slightly concave transversally, corresponding on the midline to the interhemispheric fissure and laterally to *Lancisi's longitudinal striae* and *the indusium griseum*;

-The inferior surface is concave anteroposteriorly and slightly convex transversally and is connected behind with the posterior border of the fornix. On the midline it is connected with *the septum pellucidum* and forms the vault of the lateral ventricle;

-The anterior extremity is situated at about 3 cm behind the frontal pole of the brain;

-The posterior extremity lies at about 6 cm away from the occipital pole and is separated from the quadrigeminal tubercles by Bichat's fissure.

On the surface of the corpus callosum lie rudimentary parts of grey matter: indusium griseum and Lancisi's longitudinal medial and lateral striae. The corpus callosum is delineated from the cingulate gyrus by the sulcus of the corpus callosum (sulcus corporis callosi).

The radiation of the corpus callosum starts from the corpus callosum and courses towards the cortex, forming the frontal part *(forceps minor)*, the parietal, the temporal and the occipital parts *(forceps major)*. The fibres of the temporal and the occipital part form the tapetum, the superior and lateral boundaries of the inferior and posterior horns of the lateral ventricle. The commissural fibres form connections between the cortical areas of both sides of the cerebral hemispheres.

Significant in this respect is the connection between the precentral gyri, the left one being more developed in the right-handed subjects. In lesions of *the corpus callosum*, situated in the region of the precentral gyri (e.g., the cortical fields for the hand) may occur dispraxias of the left hand. These impairments manifest themselves by the inability to perform both complex and simple movements, such as those of making a sign with the hand (salutation), or a gesture of threatning; however, the ability of performing uncomplicated movements (lifting, kissing) is preserved.

138

The anterior commissure *(commissura anterior cerebri)* consists of a band of white fibres arranged in the form of a transversally directed cylindrical bundle, which rest upwards upon *the rostral lamina* of the corpus callosum and downwards upon *the lamina terminalis cinerea* of the third ventricle. The anterior commissure joins, by its anterior part, the anterior perforated substance and the olfactory bulbs and, by its posterior fibres, *the uncus* and the amygdaloid nuclei of the two sides of the cerebral hemispheres and it belongs to the olfactory formation.

The hippocampal commissure *(commissura hippocampi)*, called also the interammonian commissure, is represented by a lamina of white matter which joins the two hippocampal formations, at the level of Ammon's horns (Fig. 118).

The posterior commissure is a part of the diencephalon and contains the crossed parts of the striatorubral tract.

The fornix or **cerebral trigone** contains, mainly, the olfactory pathways. It makes up a paired horseshoe-shaped tract of white fibres (two nervous tracts), with the opening inferior (Fig. 118).

In their middle portion, the two tracts get close to each other, forming the body of the fornix, after which they diverge, both anteriorly and posteriorly giving rise to the anterior pillars or columns *(columnae fornicis)* of the fornix, respectively to the posterior pillars or *crura* of the fornix *(crura fornicis)*. Owing to its shape, the fornix is also called the *four-pillar-vault*.

Sited beneath *the corpus callosum*, it has relationships with this structure, to which it is connected through its posterior border. It has two surfaces and three borders.

-The superior surface is adherent on the midline, to the posterior border of the septum pellucidum, contributing to the formation of the inferior wall of the frontal projection of the lateral ventricle.

-The inferior surface is in relationship with the summit of *thela choroidea* of the lateral ventricle.

-The posterior border or the base of the fornix is intimately adherent to the corpus callosum.

-The thin lateral borders are laterally continuous with the optic layer.

The fornix is formed of *columns, body* and *crura*.

The column of the fornix begins on each side in the mamillary body. The initial portion is discernible on the mediosagittal section, but is not possible through the hypothalamus *(pars tecta fornicis)*. Then it emerges on the lateral surface of the third ventricle and bears, in this portion, the name of pars libera, which borders the interventricular (Monro's) foramen on the rostral part.

The body of the fornix (corpus fornicis) is formed by the junction of the two sides and lies beneath the corpus callosum as shown above.

The crura of the fornix, two in number, arise from the posterior portion of the body of the fornix and diverge under the form of an arch laterally and downwards. Each crus of the fornix penetrates, after the pulvinar of the thalamus, into the inferior horn of the lateral ventricle and is continuous as a thin band *(fibria hippocampi)* with *the pes hippocampi* (foot of hippocampus), which is the anterior thickened extremity of the hippocampus.

The tela choroidea of the lateral ventricle is attached to the lateral, pointed edge of the body of the fornix and to the pillars of the fornix; its free border forms t*he taenia fimbriae hippocampi*.

A tract of special fibres of the fornix Vicq d'Azyr's mamillothalamic tract emerges from the mamillary body and ends in the anterior nucleus of the thalamus.

The septum pellucidum is a thin plate of white matter (lamina of *the septum pellucidum*), stretched sagittally from *the corpus callosum* to ther fornix. It is attached to *the rostral lamina*, to the genu and the trunk of the corpus callosum, as well as to the superior part of the column and to the body of the fornix. The septum borders medially the anterior horn of the lateral ventricle. Initially it is unpaired. During the postfoetal life, the substance inside the septum may be resorbed, forming a slit-like space, the cavity of the septum pellucidum *(cavum septi pellucidi)*, considered as a rudimentary cerebral ventricle.

columnae fornicis

commissura fornicis
crus fornicis

Fig. 118 Fornix

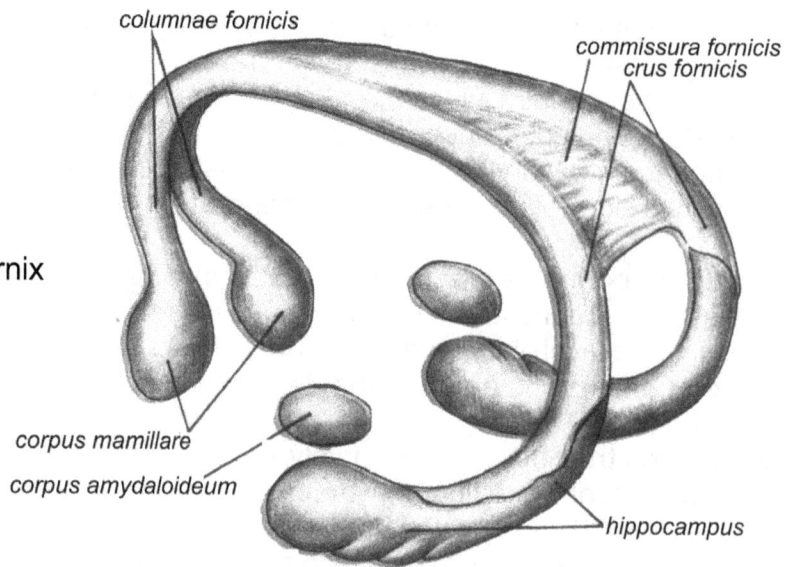

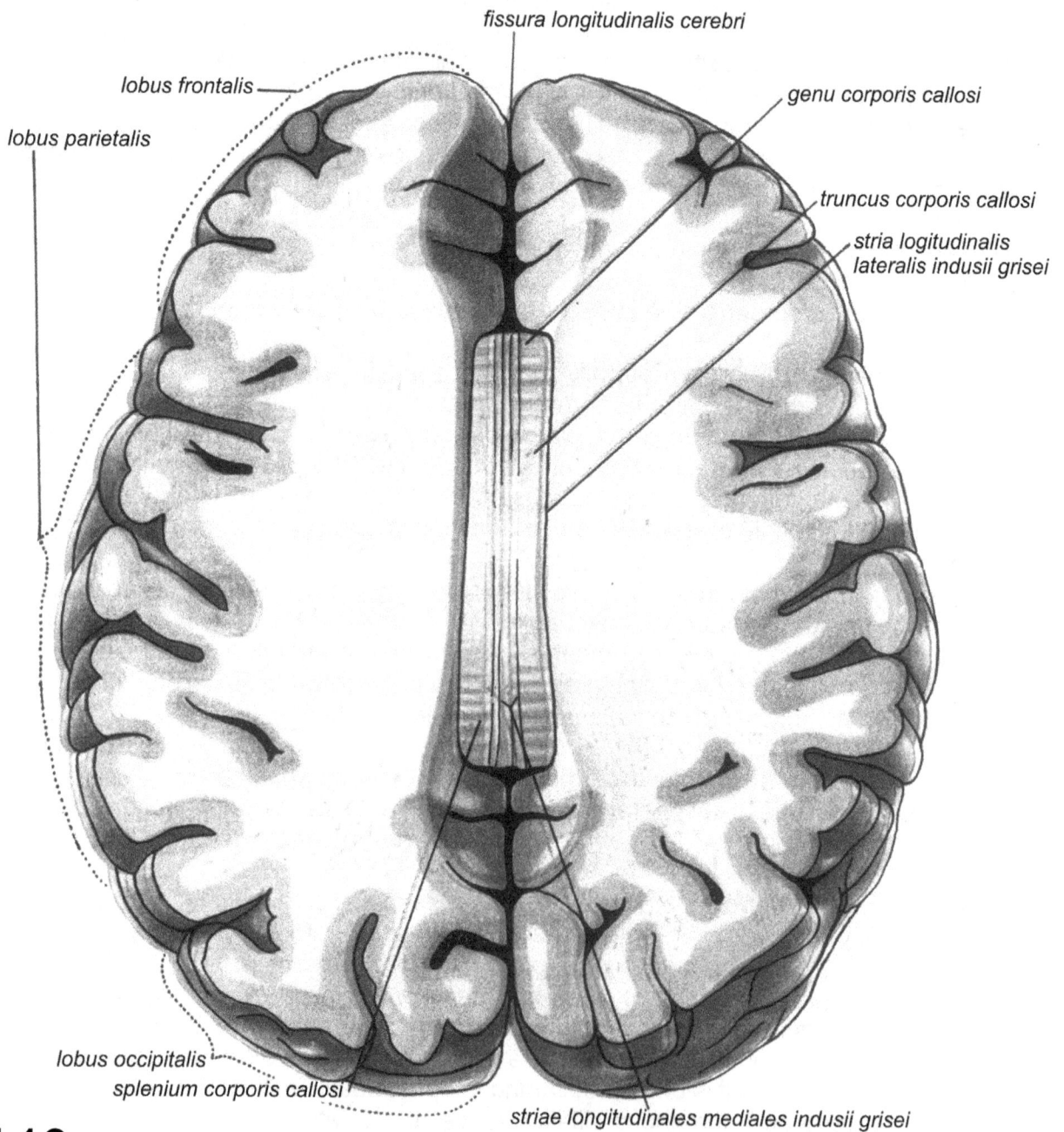

corpus mamillare

corpus amydaloideum

hippocampus

fissura longitudinalis cerebri

lobus frontalis

lobus parietalis

genu corporis callosi

truncus corporis callosi

stria logitudinalis
lateralis indusii grisei

lobus occipitalis

splenium corporis callosi

striae longitudinales mediales indusii grisei

140

Fig. 119 Corpus callosum – view from above

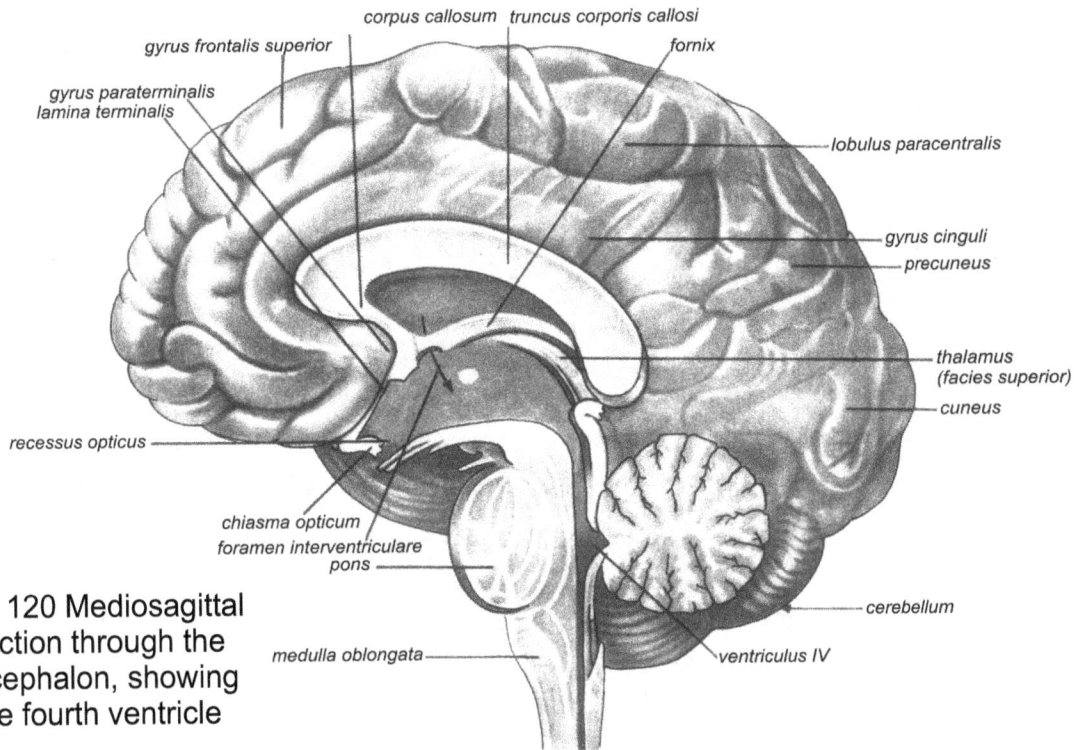

Fig. 120 Mediosagittal section through the encephalon, showing the fourth ventricle

Labels on the figure:
corpus callosum truncus corporis callosi
gyrus frontalis superior
fornix
gyrus parateminalis
lamina terminalis
lobulus paracentralis
gyrus cinguli
precuneus
thalamus (facies superior)
cuneus
recessus opticus
chiasma opticum
foramen interventriculare
pons
medulla oblongata
cerebellum
ventriculus IV

It is bordered by the right and left lamina of the pellucid (lamina septi pellucidi dexter et sinister). The cavity of the septum pellucidum does never communicate with the lateral ventricles. The pellucid septum contains fibres of the olfactory pathway.

At the level of the inferior part of the septum lies a nucleus of grey matter, termed the ganglion of the pellucid septum.

As regards the appearance of the phylogenetical scale, we mention that the fornix (trigone) and the anterior commissure are ancient, archipallial commissures, whereas the corpus callosum is a more recent, neopallial.

Projection systems

The projection fibres are long, more or less vertical tracts, either arising from the cerebral cortex (for example, the pyramidal tract) and from the inferior regions of the neuraxis (thalamus, striate bodies, cerebellum, brainstem) or ascending towards the cerebral neuraxis (for example, the axons of the third neuron of the sensitive pathways, the thalamocortical neuron) (Fig. 125).

The fibre masses in the cerebral peduncles form, in the region of the basal ganglia, the internal capsule, a lamina of white matter of 5-10 mm in thickness, with the following borders: medially, the caudate nucleus and the thalamus and laterally the lentiform nucleus. It is continuous inferiorly with the vertebral peduncle and superiorly and posteriorly with the semioval centre, its fibres radiating widely under the shape of a fan, called Reil's corona radiata (radiate crown).

On a horizontal section through the splenium of the corpus callosum the internal capsule appears in the form of an obtuse angle, opening laterally. The apex of the angle or the genu of the internal capsule (genu capsular internae) divides this structure into two limbs: the anterior or lenticulostriate limb (crus anterius), situated between the head of the caudate nucleus and the anterior part of the striate nucleus, and the posterior limb (crus posterius), situated between the thalamus and the lenticular nucleus. It is continuous with the retrolenticular segment, composed by Gratiolet's optic radiations. The sublenticular region contains, also, a layer of grey matter, called Reichert's substance, crossed by fibres of the anterior white commissure. In the posterior part of this region course the corticopontine tracts.

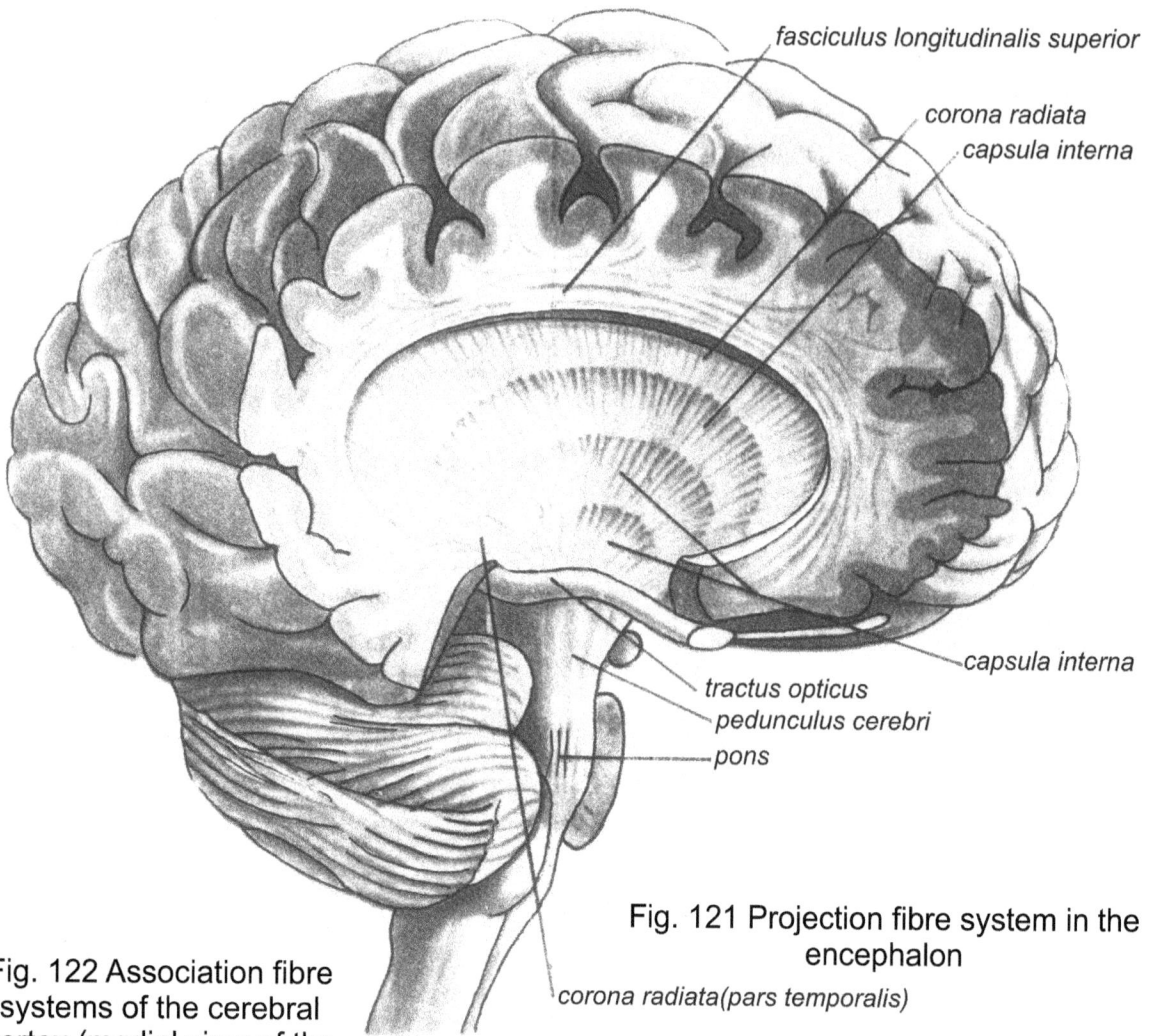

Fig. 121 Projection fibre system in the encephalon

- fasciculus longitudinalis superior
- corona radiata
- capsula interna
- capsula interna
- tractus opticus
- pedunculus cerebri
- pons
- corona radiata(pars temporalis)

Fig. 122 Association fibre systems of the cerebral cortex (medial view of the right cerebral hemisphere)

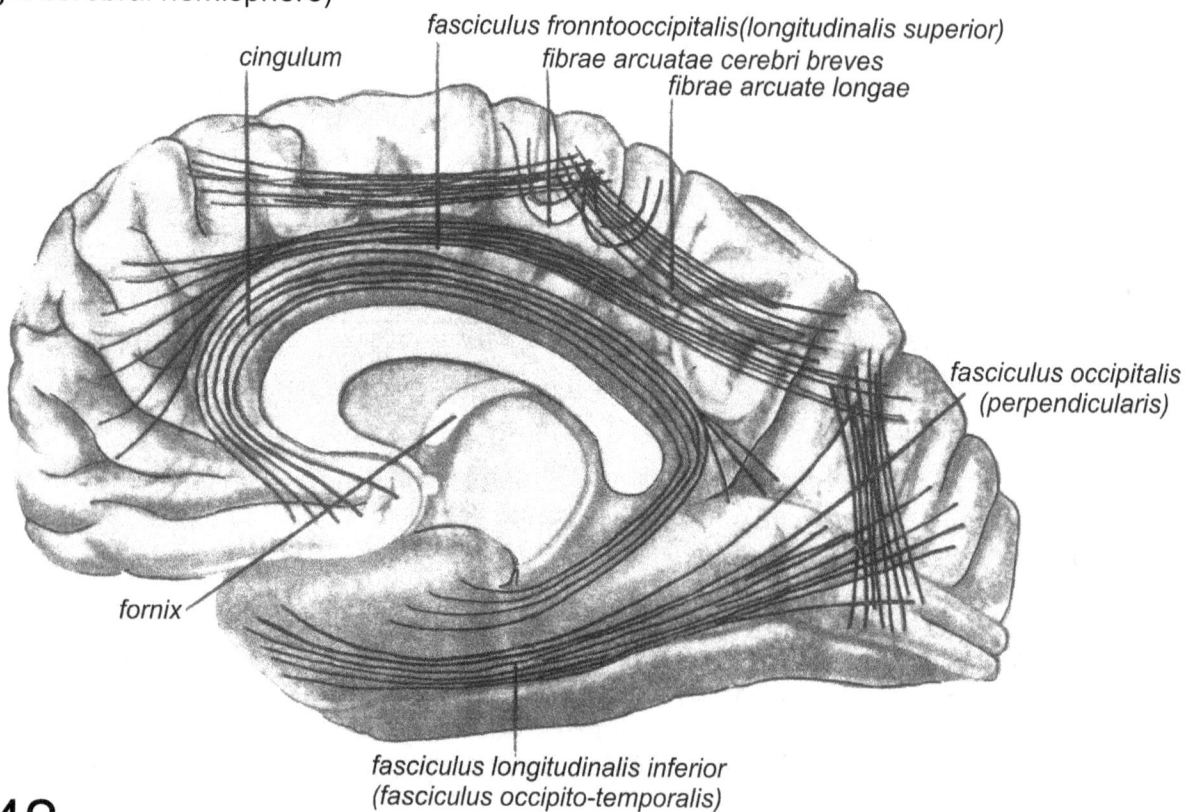

- cingulum
- fasciculus fronntooccipitalis(longitudinalis superior)
- fibrae arcuatae cerebri breves
- fibrae arcuate longae
- fasciculus occipitalis (perpendicularis)
- fornix
- fasciculus longitudinalis inferior (fasciculus occipito-temporalis)

142

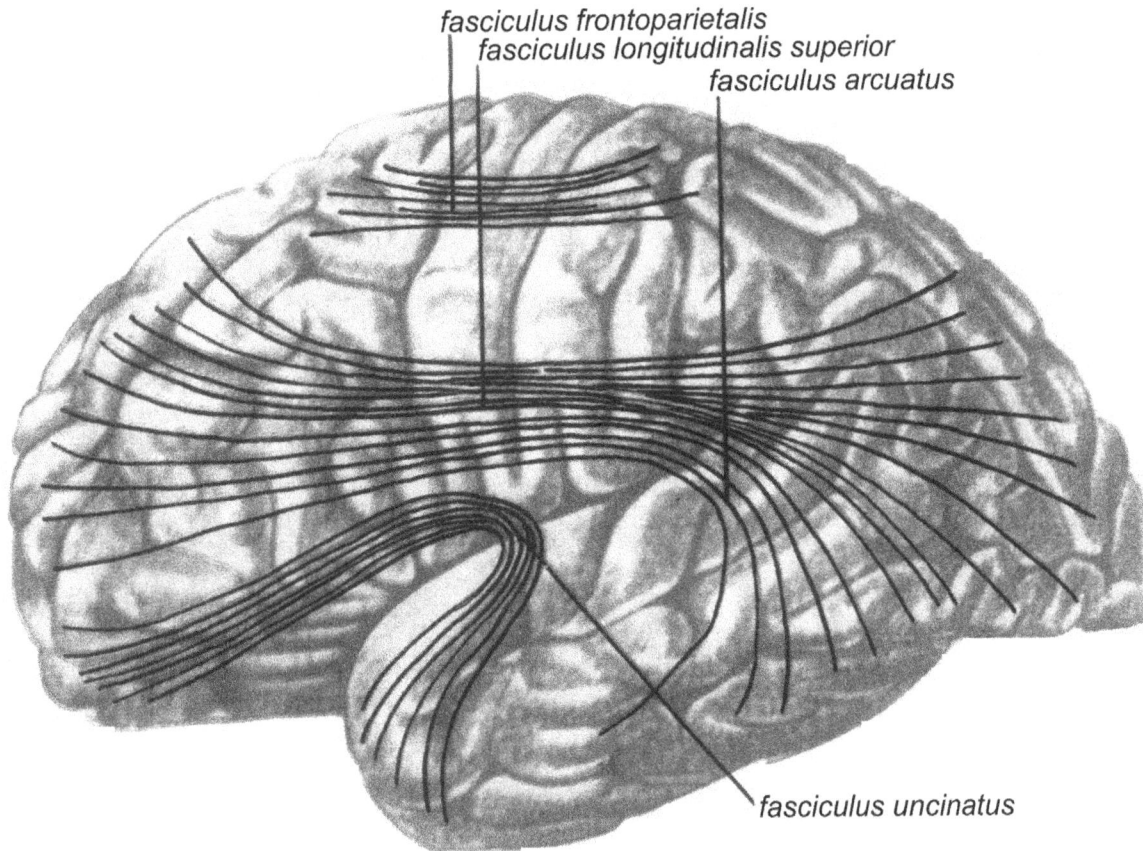

fasciculus frontoparietalis
fasciculus longitudinalis superior
fasciculus arcuatus

fasciculus uncinatus

Fig. 123 Association fibre systems of the cerebral cortex (medial view of the left cerebral hemisphere)

The pathways forming the internal capsule are directed from before backwards in the following order:

-in the anterior limb, the frontothalamic, the frontorubral and the frontopontine tracts;

-in the genu. The corticobulbar (seventh and twelfth cranial nerves) or corticonuclear tract;

-in the posterior limb, the corticospinal tract (the pyramidal pathway) directed from before backwards, for the upper and lower extremities and for the trunk, the thalamocortical tract (sensory pathway), the occipitotemporo-pontine tract, the auditory pathway and the optic tract.

The arrangement of the fibres in the external capsule (between *the putamen* and *the claustrum*), as well as for those in the extreme capsule (between *the claustrum* and the cortex of *the insula*) will be studied in the chapter dealing with the description of the conduction pathways in the nervous system.

The tracts joining the lentiform nucleus and the thalamus form the lenticular loop *(ansa lenticularis)*.

In addition to the above-described fibres, the formations of the internal capsule include, also, fibres coming from or starting towards the basal ganglia and the thalamus, forming Reil's radiate crown *(corona radiata thalami)*. The corona radiata consists of a frontal, a parietal and an occipital part.

It ensues from the above described structures, that at this level, in a very small space, course the most important afferent and efferent pathways, so that in the case of a possible lesion of this region, especially in the case of hemorrhages (from the lenticulostriate artery), extensive paralyses occur, often associated, also, with other disorders.

143

Intrahemispherical association systems

The connections between the cortical portions of the same hemisphere are achieved through short association fibres, usually directed anteroposteriorly: short arcuate fibres of the cerebrum *(fibriae arcuatae cerebri brevis)*, which connect adjacent gyri to one another and long arcuate fibres of the cerebrum *(fibrae arcuatae cerebri longae)*, which connect the convolutes of some lobes of each hemisphere to one another and which have a long course through the white matter. The most important fibre bundles of the last group are (Fig. 122, 123):

- *the superior longitudinal fasciculus (fasciculus longitudinalis superior)* connecting the frontal, occipital and temporal lobes;
- *the fronto-occipital fasciculus (fasciculus fronto-occipitalis)* extending from the frontal to the occipital lobe;
- *the uncinate fasciculus (fasciculus uncinatus)*, coursing in the form of an arch from the inferior frontal gyrus towards the uncus of the parahippocampal gyrus;
- *the inferior longitudinal fasciculus (fasciculus longitudinalis inferior)* extending from the occipital to the temporal lobe;
- *the cingulum*, a bundle of white matter below the fornicate gyrus, following the course of the latter, from the paraterminal gyrus to the uncus of the parahippocampal gyrus. In its course, the cingulum gives off fibres to the adjacent gyri, from which, in turn, start fibres that penetrate in the cingulum. So that an intense exchange of neighbourhood relationships takes place.

The lateral ventricles
(ventriculi laterales)

Each hemisphere contains a lateral ventricle which is in relationship with the third ventricle through the interventricular (Monro's) foramen. The lateral ventricles corresponding to the four lobes of the brain, the lateral ventricles consist of four portions: the ventricular brain or central part *(pars centralis)* and three horns, respectively frontal or anterior *(cornu anterius)*, occipital or posterior *(cornu posterius)* and temporal or inferior *(cornu inferius)* (Fig. 124 - 130).

The central part is narrow and long, extending from the interventricular foramen to the emergence place of the anterior and posterior horns. Its walls are the following: above – the radiation of the corpus callosum; below and medially – the fornix, *the tela choroidea* with the tenia fornicis and the tenia choroidea, the lamina affixa and the terminal stria; below and laterally – the tail of the caudate nucleus.

The anterior horn penetrates deeply into the frontal lobe. It has the following walls:

-a superior concave wall (the vault), formed of the radiation of *the corpus callosum*;

-the internal (medial) wall, constituted of *the septum pellucidum*, situated between *the corpus callosum* upwards and the trigone downwards;

-the inferior much more complex wall, formed of two segments (anterior and posterior) and Monro's foramen, at the junction of the interval with the inferior wall. At the level of the anterior segment lies the caudate nucleus, its head *(caput nuclei caudati)* bulging into the lateral ventricle. In continuation run the fibres of the genu of the corpus callosum and those of the reflected (inferior) lamina of the genu of the corpus callosum. The posterior segment is represented succesively by the body of the caudate nucleus, the optostriate sulcus, the lateral choroid plexus (which bulges into the ventricular cavity) and the lateral half of the fornix.

The posterior (occipital) horn is short and terminates by tapering in the occipital lobe. It has two walls: a superolateral, concave wall, corresponding to the radiation of the corpus callosum *(tapetum)* and to the optic radiations, and an inferomedial wall with two prominences: the bulb of the posterior horn and the calcar avis, an elevation formed by the calcarine sulcus.

144

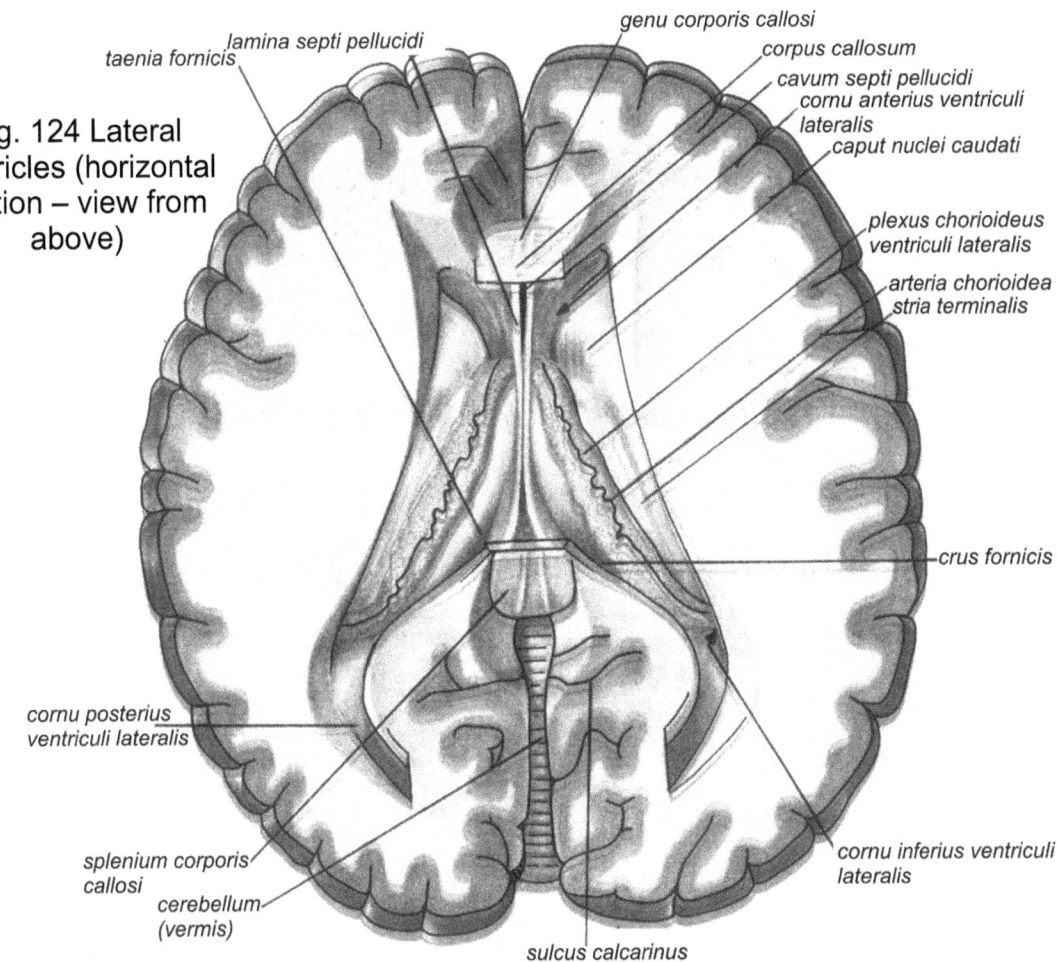

Fig. 124 Lateral ventricles (horizontal section – view from above)

taenia fornicis
lamina septi pellucidi
genu corporis callosi
corpus callosum
cavum septi pellucidi
cornu anterius ventriculi lateralis
caput nuclei caudati
plexus chorioideus ventriculi lateralis
arteria chorioidea
stria terminalis
crus fornicis
cornu posterius ventriculi lateralis
splenium corporis callosi
cerebellum (vermis)
sulcus calcarinus
cornu inferius ventriculi lateralis

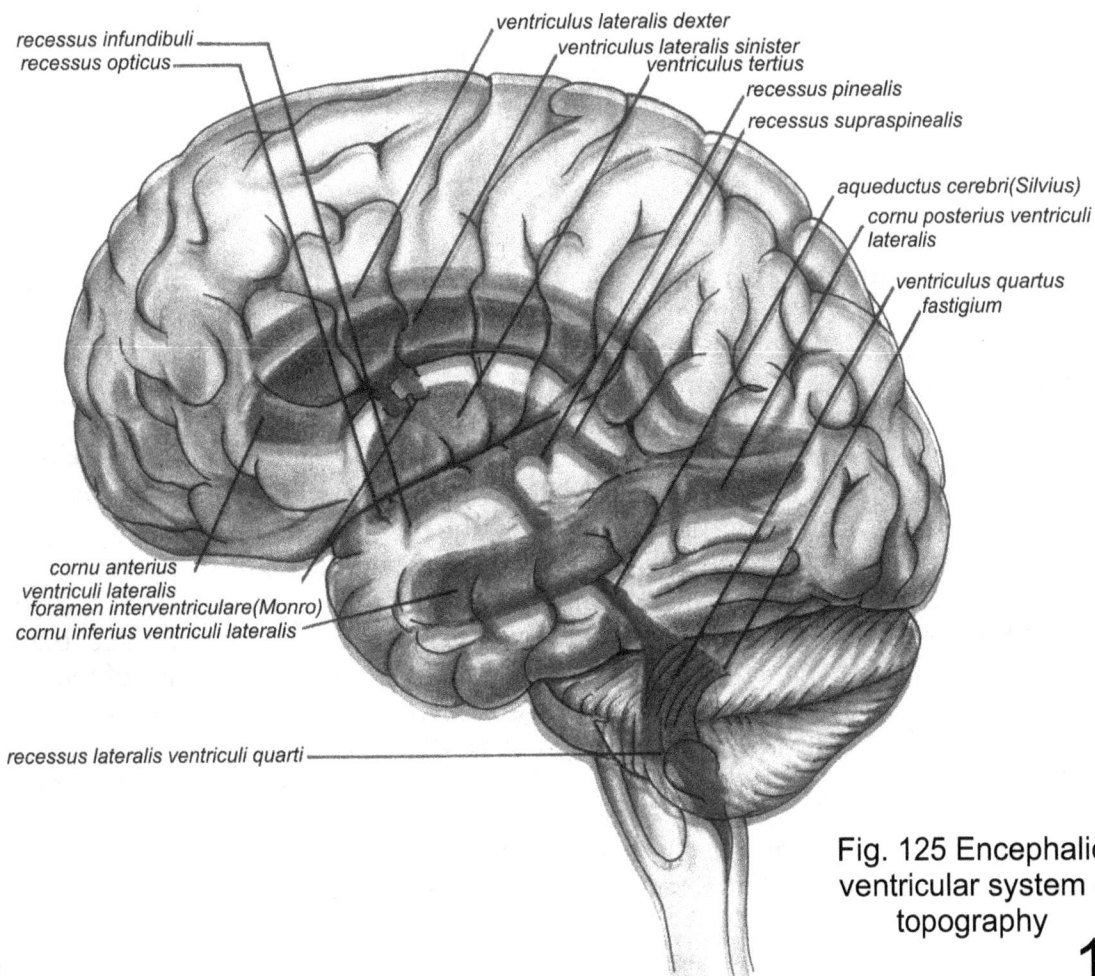

recessus infundibuli
recessus opticus
ventriculus lateralis dexter
ventriculus lateralis sinister
ventriculus tertius
recessus pinealis
recessus supraspinealis
aqueductus cerebri(Silvius)
cornu posterius ventriculi lateralis
ventriculus quartus
fastigium
cornu anterius ventriculi lateralis
foramen interventriculare(Monro)
cornu inferius ventriculi lateralis
recessus lateralis ventriculi quarti

Fig. 125 Encephalic ventricular system - topography

145

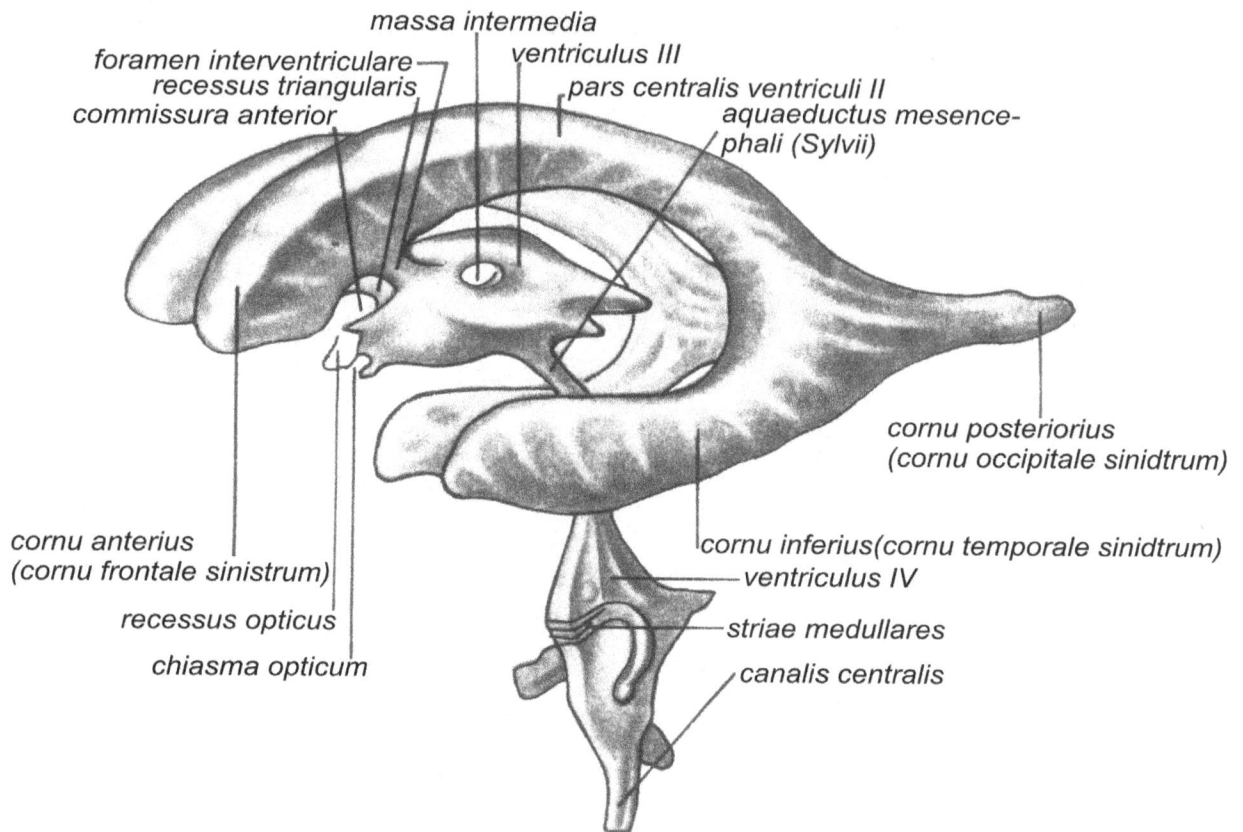

massa intermedia
foramen interventriculare
recessus triangularis
commissura anterior
ventriculus III
pars centralis ventriculi II
aquaeductus mesence-
phali (Sylvii)
cornu posteriorius
(cornu occipitale sinidtrum)
cornu anterius
(cornu frontale sinistrum)
recessus opticus
chiasma opticum
cornu inferius(cornu temporale sinidtrum)
ventriculus IV
striae medullares
canalis centralis

Fig. 126 Cerebral ventricles

The inferior (temporal) horn extends in the temporal lobe nearly up to *the uncus* of the hippocampal gyrus, along the lateral side of Bichat's fissure. It terminates at about 2 cm behind the anterior extremity of the temporal lobe and is crescent-shaped with the concavity inferomedial. It has two walls: a supero-external and an infero-internal wall.

-The supero-lateral wall is formed of a lamina of nervous substance, containing the radiation of *the corpus callosum (tapetum)* and the small semicircular band which separates the inferior horn of the lateral ventricle from the inferior surface of *the lenticular nucleus* and which Déjérine includes in the internal capsule under the name of sublenticular segment.

-The convex infero-medial wall is made up of the collateral trigone *(trigonum collaterale)*, of the collateral eminence *(eminentia collateralis)*, often included in a groove, called the collateral sulcus, of *the pes hippocampi*, provided at its thickened rostral extremity with small prominences *(digitationes)*, of *the hippocampal fimbria* and, finally, of *the choroid tela* of the third ventricle with *the taenia choroidea* and *taenia fimbriae hippocampi*.

The choroid tela of the lateral ventricle is formed of the epithelial lamina tectoria (which is continuous with the ependyma of the ventricle), of *the choroid plexus* and of loose connective tissue, adjacent to the pia matter of the encephalon.

The tela choroidea of the lateral ventricle exhibits, at the boundary between the central part and the inferior horn, an enlargement termed choroid glomus *(glomus choroideum)*.

146

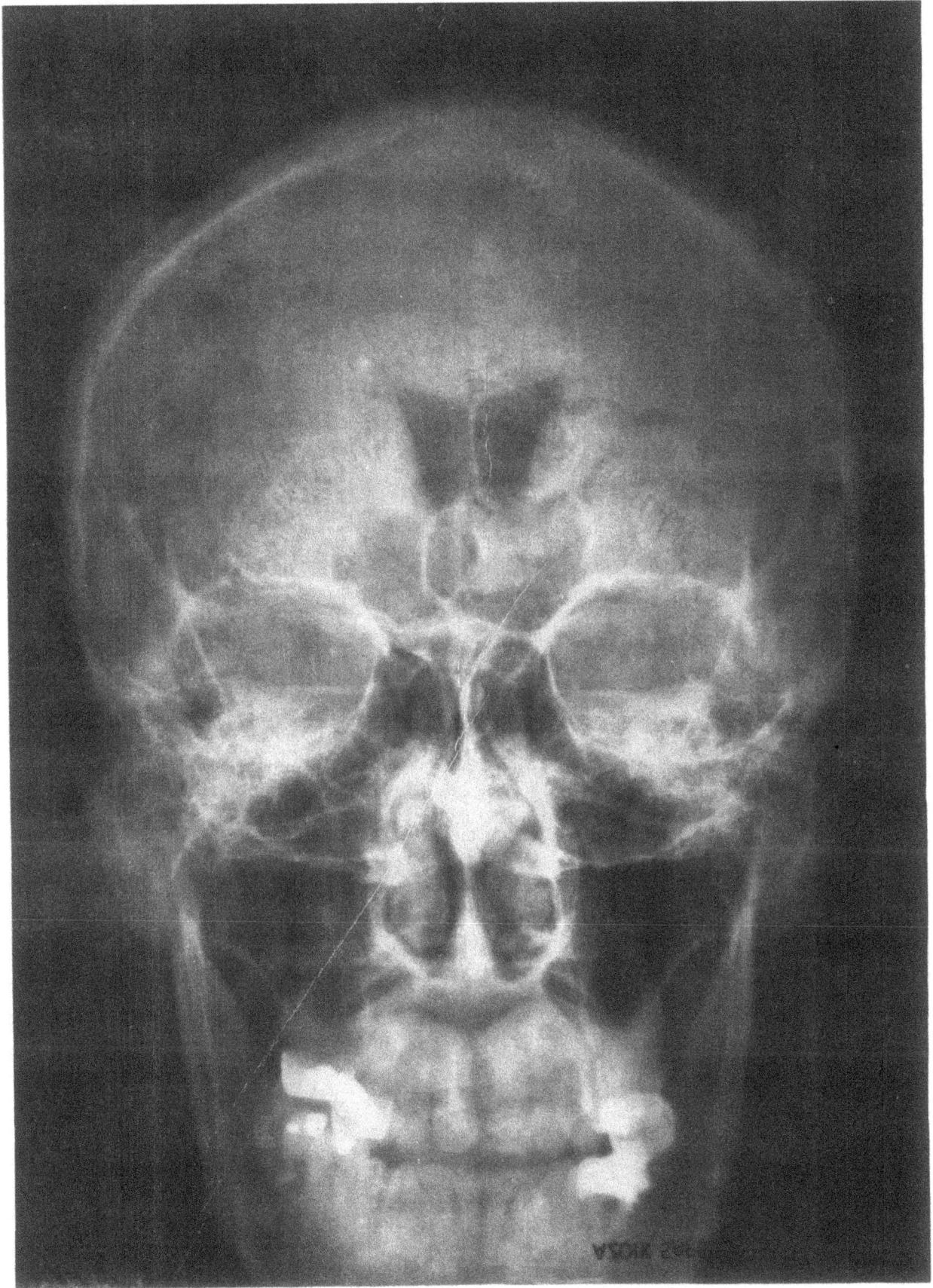

Fig. 127 Ventriculography – anterior view

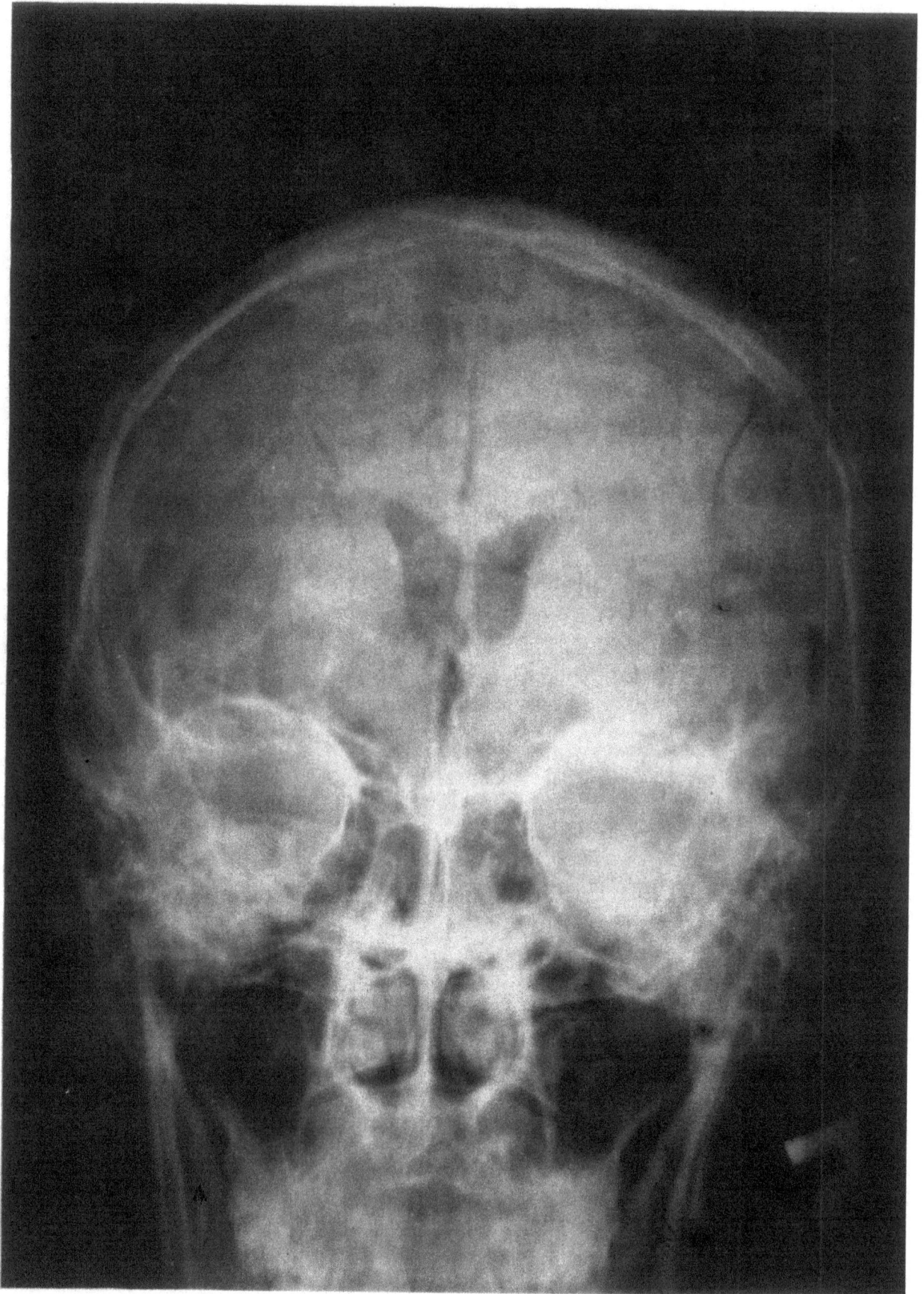

Fig. 128 Ventriculography – anterior view

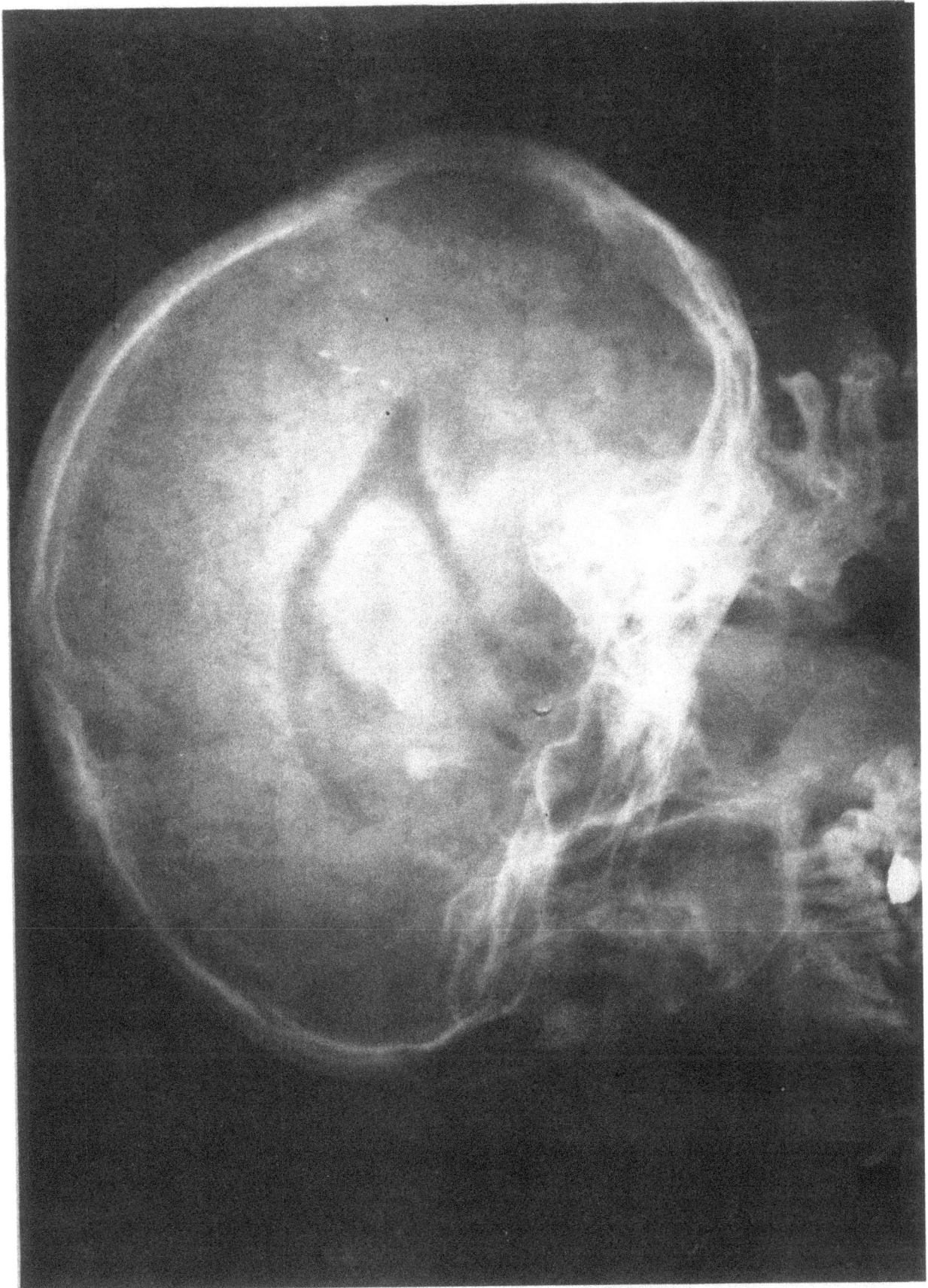

Fig. 129 Ventriculography – lateral view

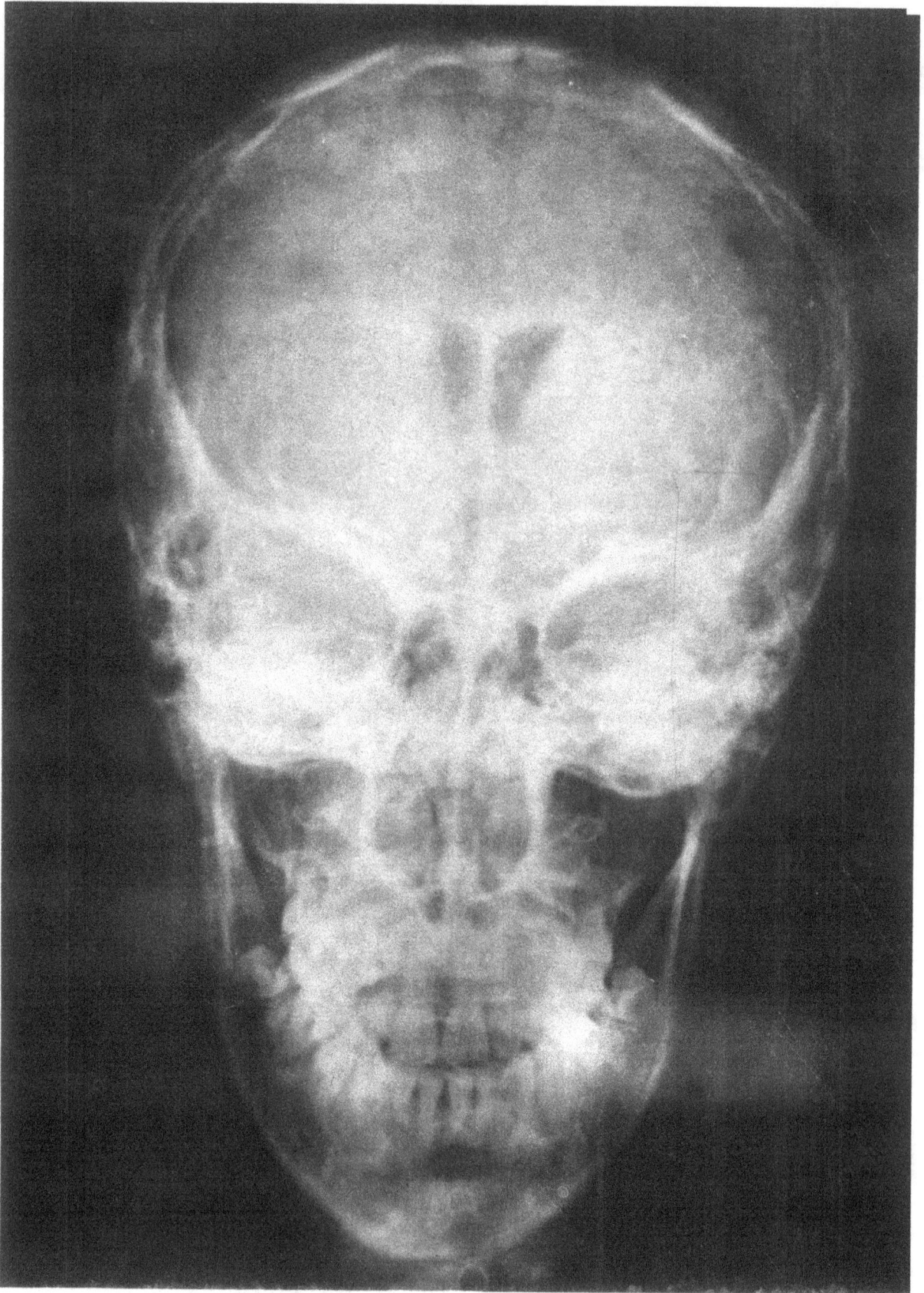

Fig. 130 Ventriculography – anterior view

Computed tomography of the encephalon

In 1979, the Nobel Prize for Medicine and Physiology was awarded to the American Physicist Cormack Allan MacLeod and to the British electronics engineer Sir Hounsfield Godfrey Newbold for the invention of the system and device of computed transverse axial tomography EMI- Scanner, used in X-ray investigations of the organism and their recording on the computer.

Revolutionizing the classical X-ray systems, applied till now, the computed tomographies – name under which they were introduced in medicare – are used with remarkable results in the detection and localization of the concealed tumours, fractures, hematomas and haemorrhages.

We reproduce some computed tomographies performed at the level of the encephalon (Fig. 131 - 147).

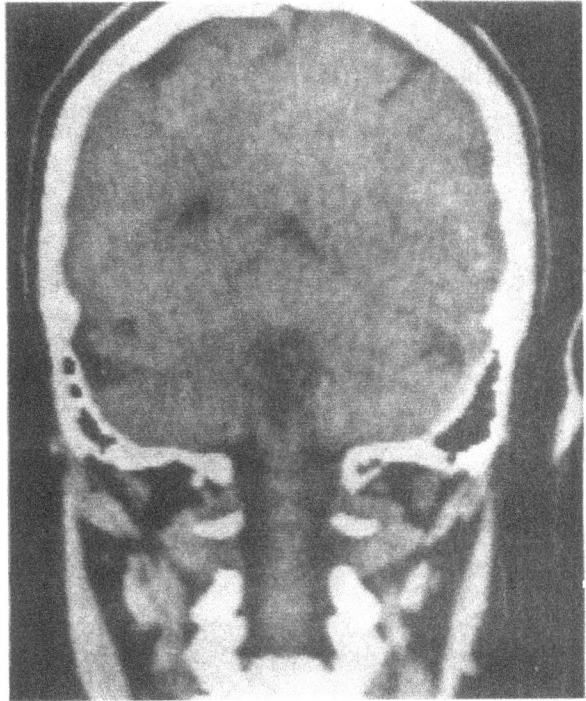

Fig. 131 Computed tomography at the level of a frontal section through the posterior third of the encephalon (Collection Dr. Şerban Georgescu – "Fundeni" Hospital)

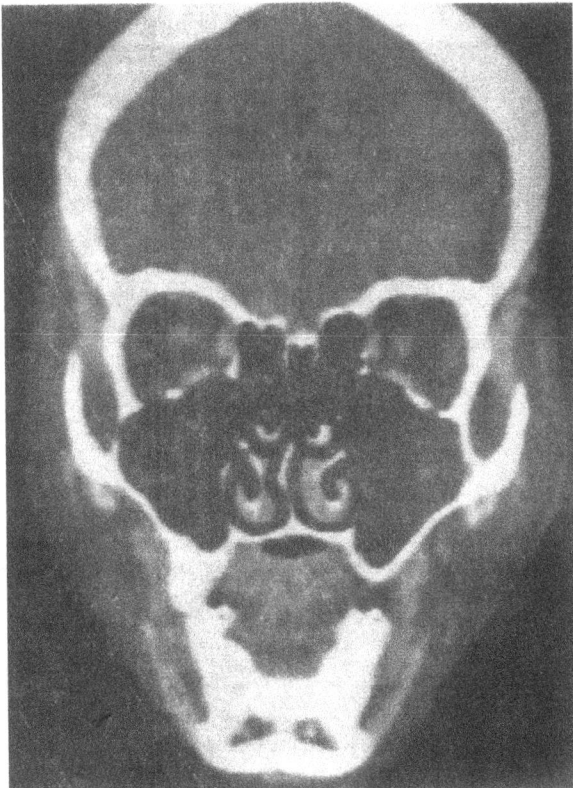

Fig. 132 Computed tomography at the level of a frontal section through the middle of the encephalon (Collection Dr. Şerban Georgescu – "Fundeni" Hospital)

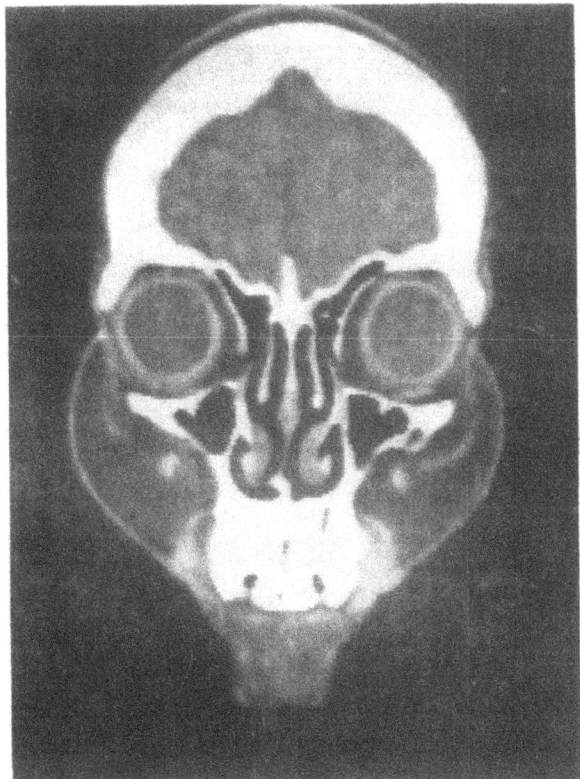

Fig. 133 Computed tomography at the level of a frontal section through the anterior third of the encephalon (Collection Dr. Şerban Georgescu – "Fundeni" Hospital) 151

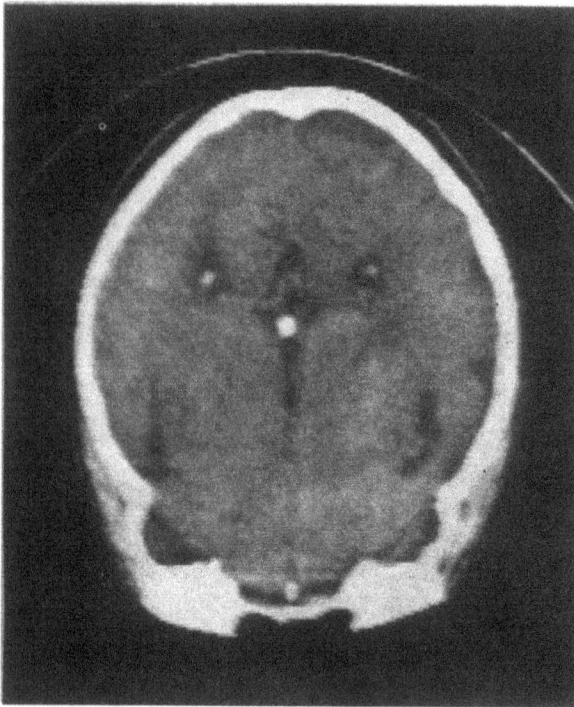

Fig. 134 Computed tomography at the level of a horizontal section through the frontal eminences (Collection Dr. Şerban Georgescu – "Fundeni" Hospital)

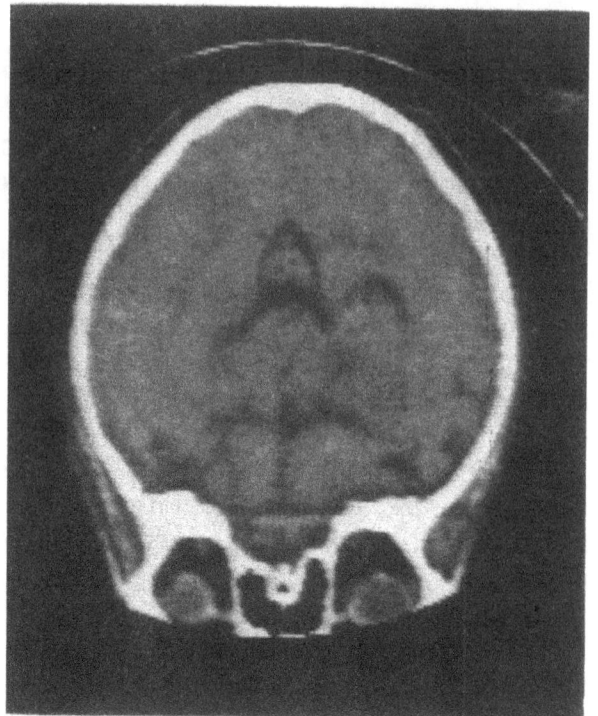

Fig. 135 Computed tomography at the level of a horizontal section through the superior zone of the orbit (Collection Dr. Şerban Georgescu – "Fundeni" Hospital)

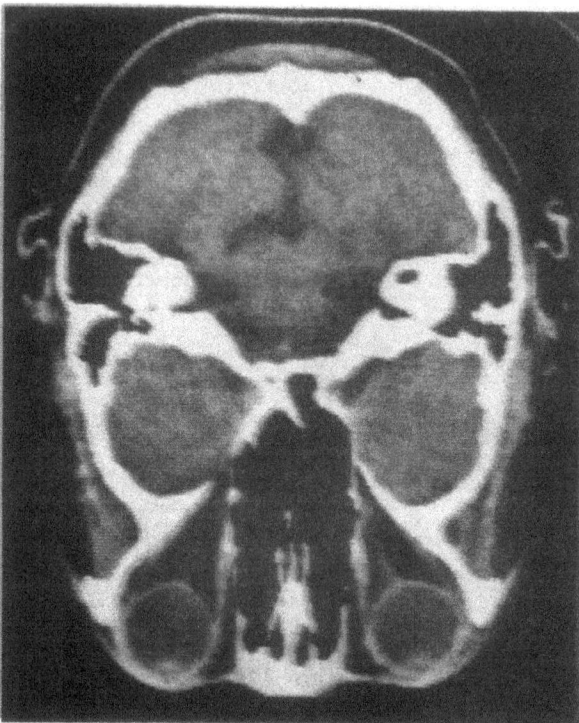

Fig. 136 Computed tomography at the level of a horizontal section through the junction zone of the superior third with the middle zone of the orbit (Collection Dr. Şerban Georgescu – "Fundeni" Hospital)

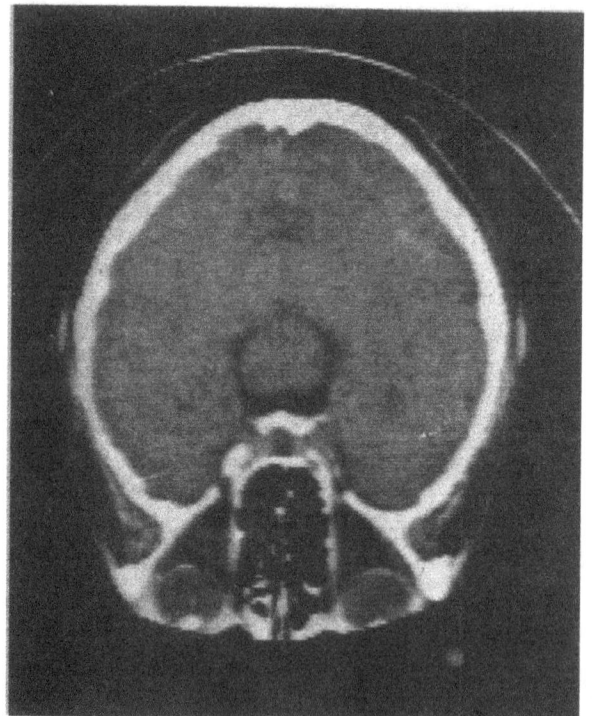

Fig. 137 Computed tomography at the level of a horizontal section through the lower part of the middle third of the orbit (Collection Dr. Şerban Georgescu – "Fundeni" Hospital)

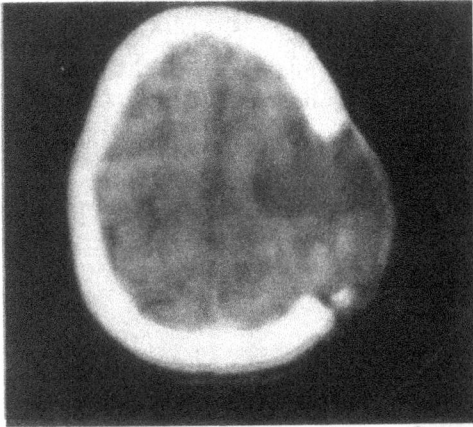

Fig. 138 Right frontal meningioma. Computed tomography at the level of a horizontal section through the squamofrontal suture (Collection Dr. Şerban Georgescu – "Fundeni" Hospital)

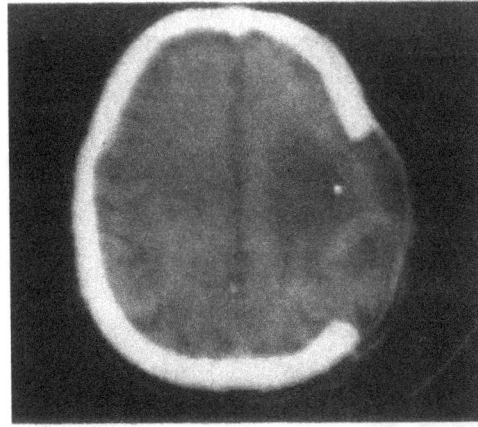

Fig. 139 Right frontal meningioma. Computed tomography at the level of a horizontal section below the squamofrontal suture (Collection Dr. Şerban Georgescu – "Fundeni" Hospital)

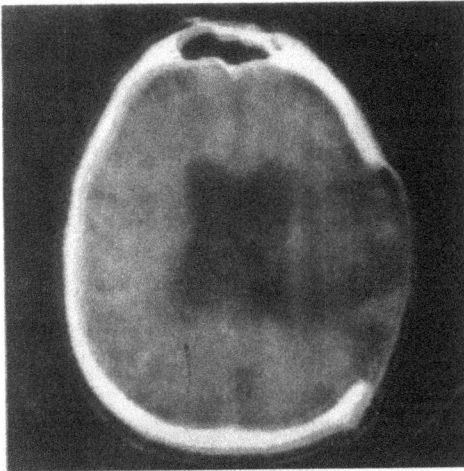

Fig. 140 Right frontal meningioma. Computed tomography at the level of a horizontal section through the encephalon (Collection Dr. Şerban Georgescu – "Fundeni" Hospital)

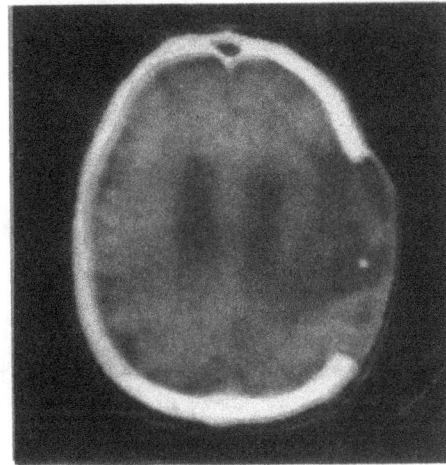

Fig. 141 Right frontal meningioma. Computed tomography at the level of a horizontal section through the encephalon (Collection Dr. Şerban Georgescu – "Fundeni" Hospital)

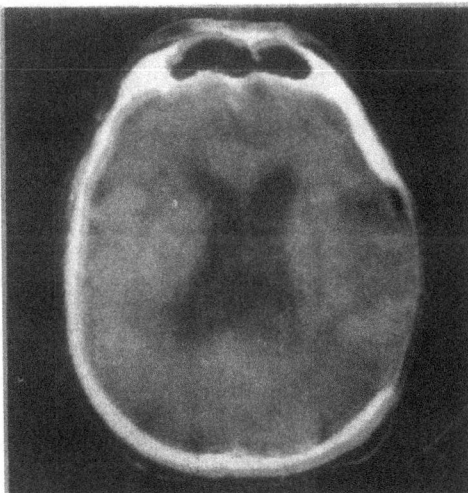

Fig. 142 Right frontal meningioma. Computed tomography at the level of a horizontal section through the encephalon. The lateral ventricles may be seen (plan 5) (Collection Dr. Şerban Georgescu – "Fundeni" Hospital)

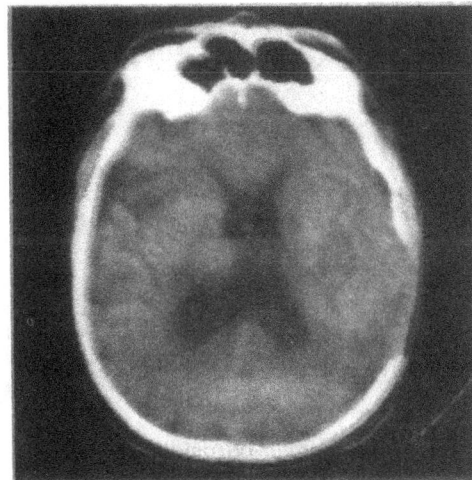

Fig. 143 Right frontal meningioma. Computed tomography at the level of a horizontal section through the encephalon. The lateral ventricles may be seen (plan 6) (Collection Dr. Şerban Georgescu – "Fundeni" Hospital)

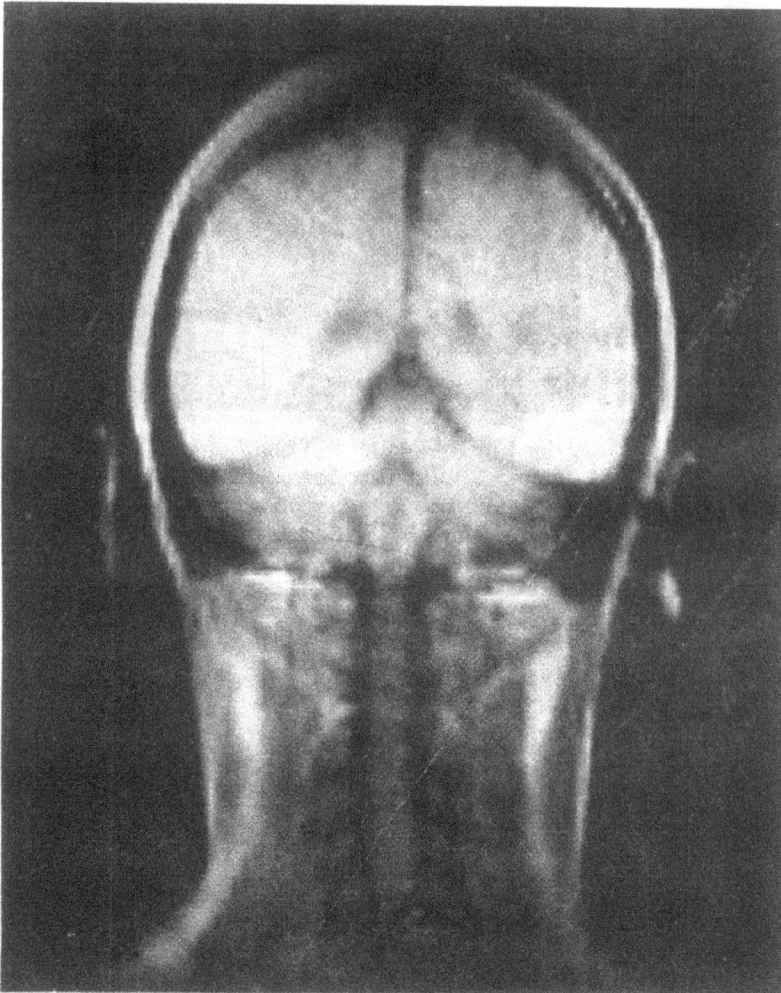

Fig. 144 Mediofrontal tomography of the encephalon (nuclear magnetic resonance) (Siemens Collection)

Fig. 145 Horizontal tomography of the encephalon at the level of the lateral ventricles (nuclear magnetic resonance) (Siemens Collection)

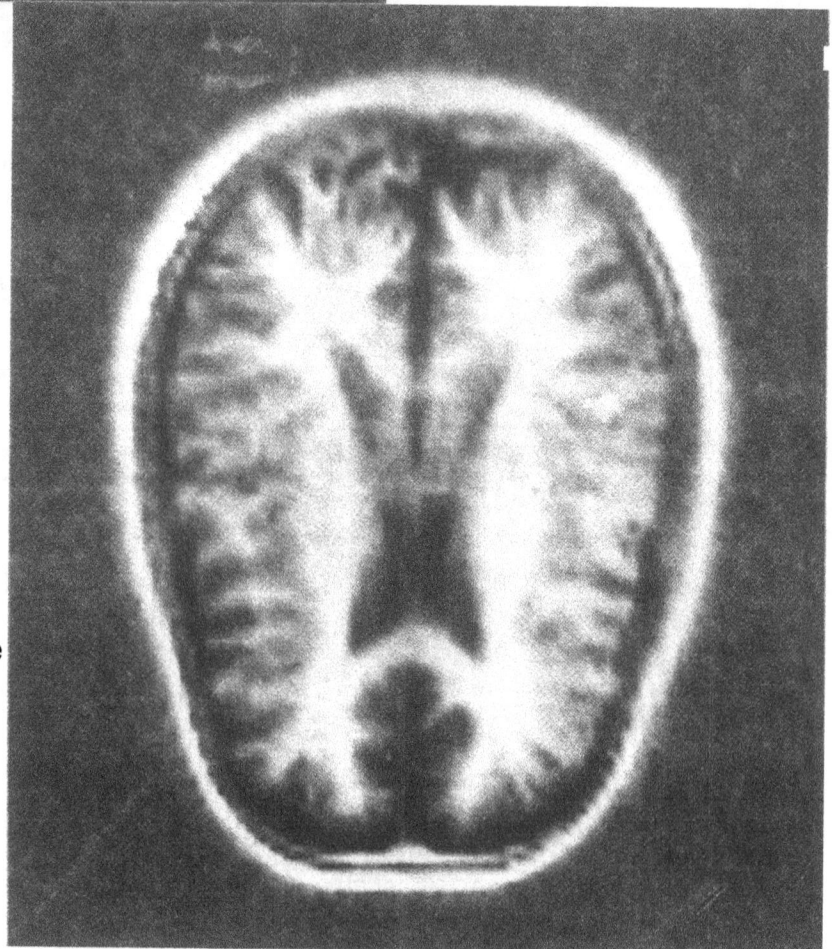

154

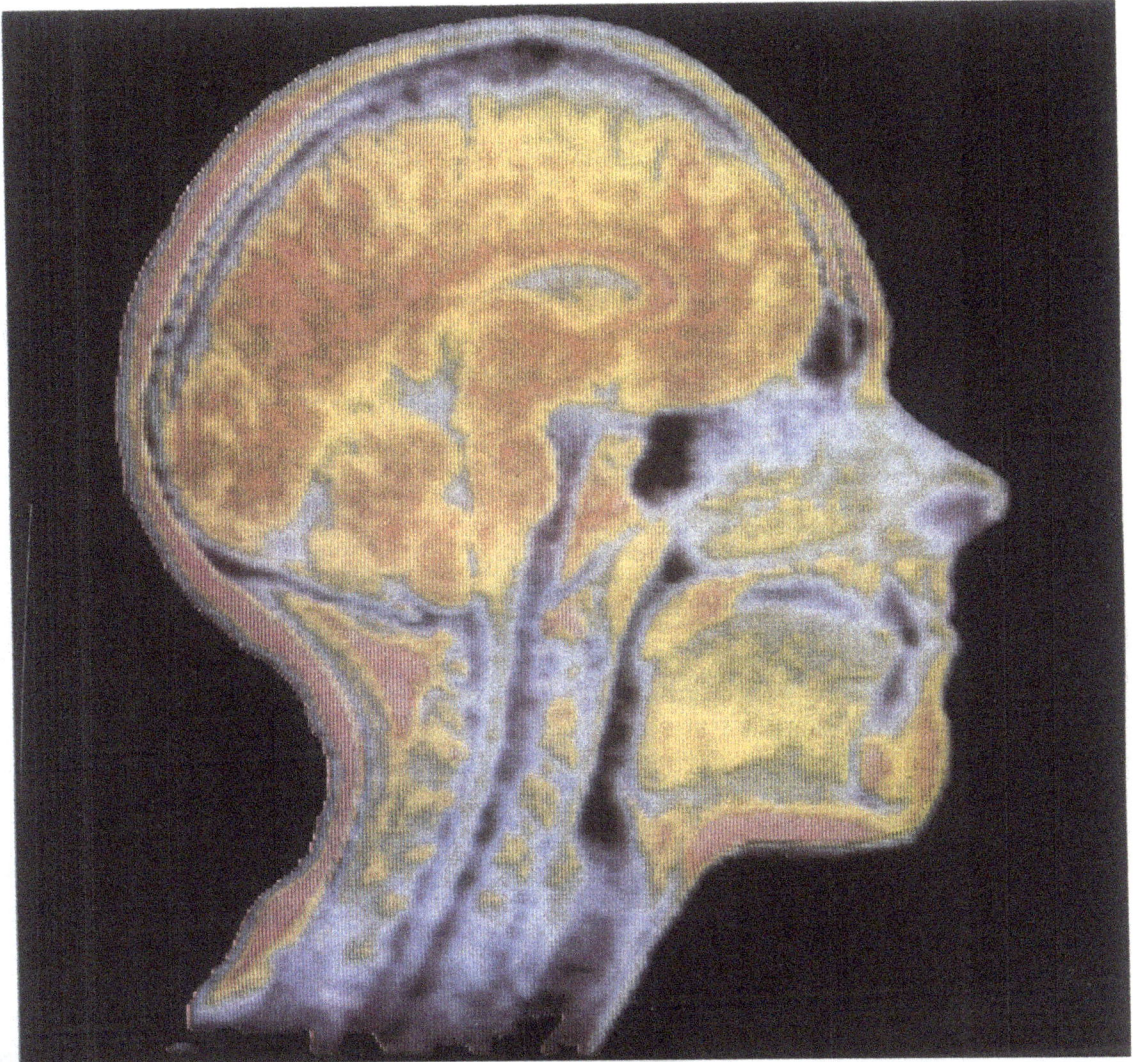

Fig. 146 Tomography of the encephalon in mediosagittal position (nuclear magnetic resonance) (Siemens Collection, color)

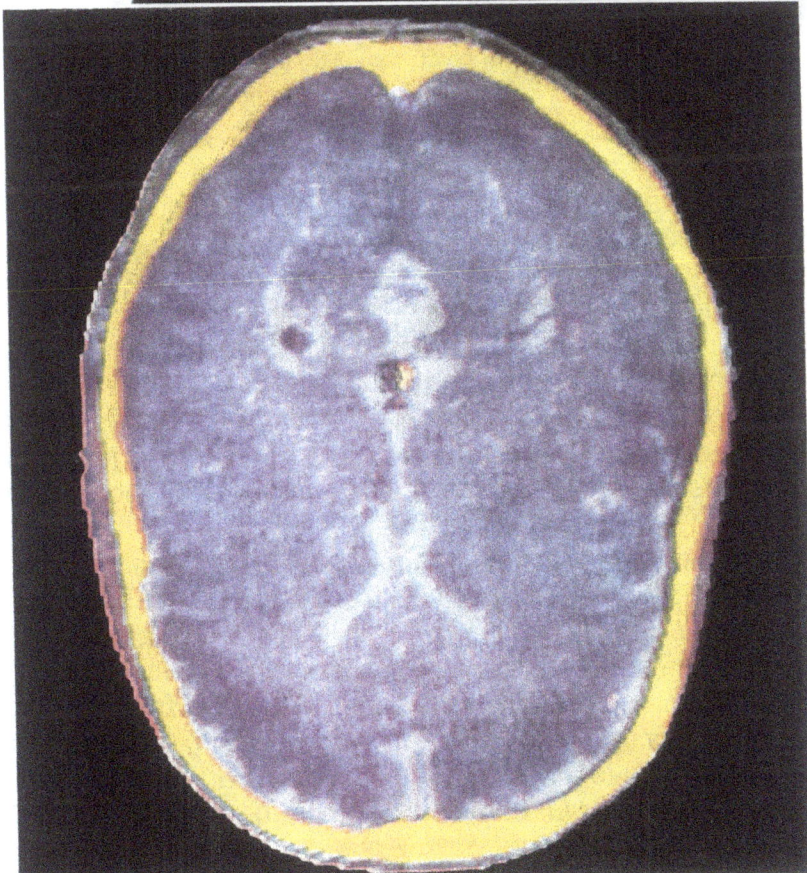

Fig. 147 Horizontal tomography of the encephalon made at the level of the lateral ventricles (nuclear magnetic resonance) (Siemens Collection, color)

155

Vessel supply of the encephalon

The vessel supply of the encephalon consists of an arterial circulation *(aa. cerebri)* and a venous circulation *(vv. cerebri)*. Lymph vessels do not exist in the brain.

The arterial circulation

The arterial circulation *(aa. cerebri)* arises from the anastomosis of four arteries: the carotid system, formed of two internal carotid arteries *(aa. carotides internae)* and the vertebrobasilar system made up of the two vertebral arteries *(aa. vertebrales)* (Fig. 148-156).

The carotid system

The carotid system is represented by the two internal carotid arteries *(arteriae carotides internae)*, which arise from the common carotid arteries.

At the level of the upper border of the thyroid cartilage (C4 vertebral level), the common carotid artery bifurcates into the external carotid artery and into the internal carotid artery. At this level lies the carotid glomus *(glomus caroticum)*, a very significant reflexogenic area, in relation with the blood chemism changes.

On the course of the internal carotid artery, immediately above the bifurcation, lies the carotid sinus *(sinus caroticus)*, which is a reflexogenic area, too, sensitive to blood pressure changes.

The internal carotid artery penetrates into the skull through the carotid canal situated in the petrous part of the temporal bone, then it passes at the level of the anterior extremity of the cavernous sinus, medially to the anterior clinoid process, and has relationships with the oculomotor nerves (third, fourth and sixth pairs) and with the ophthalmic branch of the trigeminus.

The sinocavernous segment of the artery, known under the name of "carotid siphon", has the shape of the letter S and plays a significant role in lowering the blood pressure (by about 20 mmHg) and in damping the puse wave.

Alongside its course, the internal carotid artery gives off:

-**collateral branches**:

-osteoperiosteal and caroticotympanic arteries, at the level of the petrous part of the temporal bone;

-the arteries to *the cavernous sinus*, the oculomotor nerves, the gasserian ganglion, the hypophysis and the meninges, in the cavernous area;

-the ophthalmic arteries, in the extrasinusal portion *(cavernous sinus)*, which enter in the orbits through *the optic foramina*;

-**terminal branches**, four in number, which are given off immediately after having passed beyond *the optic chiasma*; they are the following:

-the anterior cerebral artery *(arteria cerebri anterior)*, which runs forwards and medially, passes above the optic nerve and, anteriorly to the chiasma, is joined to the artery on the opposite side, through a short transverse anastomosis, the anterior communicating artery *(a. communicans anterior)*;

-the posterior communicating artery *(a. communicans posterior)* runs backwards, crosses the inferior surface of the optic tract and anastomoses with the posterior cerebral artery *(a. cerebri posterior)*, a branch of the basilar trunk.

-the anterior choroidal artery *(a. choroidea anterior)* courses backwards and laterally, penetrates into the anterior part of Bichat's fissure and passes into the lateral choroidal plexuses nearly up to the Monro's foramen, giving up branches to the villosities of choroidal plexuses and to the walls of the lateral ventricle;

-the middle or sylvian cerebral artery *(a. cerebri media)* will be described in the cerebral hemispheres vessel supply chapter.

The vertebrobasilar system

The vertebrobasilar system is made up of the two vertebral arteries *(aa. vertebrales)*, which arise from the subclavian artery, penetrate at the level of the transverse process of the sixth cervical vertebra into the transverse canal, then into the cranium, through the occipital foramen, after traversing the atlanto-occipital membrane. At the level of the bulbopontine sulcus, the two arteries join on the midline, forming the sinusoidal basilar trunk *(a. basilaris)*, which also plays a significant role in the blood pressure regulation.

Along their course the basilar trunk gives off:

-**collateral branches**, which assure the arterial supply to:

-the brainstem through the paramedian arteries, the short and the long circumferential arteries;

-the cerebellum through: the superior cerebellar arteries *(aa. cerebelli superiores)*, the middle cerebellar arteries *(aa. cerebelli inferiores et anteriores)* and the inferior cerebellar arteries *(aa. cerebelli inferiores et posteriores)*, which will be described in the chapter dealing with the arterial supply of the cerebellum;

-**terminal branches**, which are its bifurcation branches, formed at the level of the superior border of the pons, namely the two posterior cerebral arteries *(aa. cerebri posteriores)*. The terminal branches of the arteries and the internal carotid arteries, together with the anastomoses which unite them, form at the base of the skull and around the Turkish saddle *(sella turcica)*, a hexagonal arterial polygon, termed the polygon of Willis or the arterial circle of cerebrum.

The polygon of Willis
(circulus arteriosus cerebri)

The polygon of Willis is made up of the following arteries:

-**the anterior communicating artery** *(a. communicans anterior)*, which joins to each other the two anterior cerebral arteries;

-**the anterior cerebral arteries** *(aa. cerebri anteriores)*, running on both sides and directed anteromedially;

-**the posterior communicating arteries** *(aa. comunicantes posteriores)*, which join the carotid system to the vertebrobasilar system;

-**the posterior cerebral arteries** *(aa. cerebri posteriores)*, branches of the basilar trunk.

The cerebral blood supply is assured by this rich arterial circulation.

The arterial distribution at the level of the encephalon
The arteries of the medulla oblongata

The blood vessel supply of the medulla oblongata is assured by the anterior and the posterior spinal arteries, which arise from the vertebral arteries. They pass on the surface of the medulla oblongata and give off branches, which penetrate into the medulla oblongata, where they divide into four groups:

-the anterior middle arteries or nuclear arteries, which penetrate into the medulla oblongata through the anterior median sulcus and terminate in the grey substance of the ventricular floor;

-the posterior middle arteries, which penetrate through the posterior median sulcus and supply the subventricular part of the medulla oblongata;

-the radicular arteries, which follow the course of the nerve roots;

-the accessory arteries, which penetrate into the medulla oblongata through various areas.

Part of the arterial supply of the medulla oblongata is derived from the postero-inferior cerebellar artery.

The arteries of the pons

The pons is supplied by arteries which arise either directly from the basilar trunk *(a. basilaris)* or from the inferior and anterior cerebellar arteries *(aa. cerebelli inferiores et anteriores)* and the superior cerebellar arteries *(aa. cerebelli superiores)*.

At the level of the pons they divide into three groups:

-the median arteries, branches of the basilar trunk, which reach the floor of the fourth ventricle;

-the radicular arteries, which arise both from the basilar trunk and the cerebellar arteries, that follow the course of nerve roots;

-the accessory arteries, which terminate at the periphery of the pons.

The arteries of the cerebellum

As it was shown, the blood supply of the cerebellum is assured by the three cerebellar arteries:

-the superior cerebellar artery *(a. cerebelli superior)*, arises from the superior extremity of the basilar trunk, near the bifurcation, winds round the lateral surfaces of the cerebral peduncles and ramifies on the superior surface of the brain;

-the middle cerebellar artery*(a. anterior et inferior)*, starts from the median part of the basilar trunk and ramifies into the anterior and inferior part of the cerebellum. It gives off the labyrinthic artery *(a. labyrinthi)*, which may, sometimes, arise directly from the basilar trunk;

-the inferior cerebellar artery *(a. cerebelli posterior)* has its origin in the vertebral artery and after winding round the lateral surfaces of *the medulla oblongata* supplies the posterior part of the inferior surface of the cerebellum.

These arteries anastomose with each other in the depth of the pia mater, at the surface of the cerebellum, where it breaks up into numerous arterioles which penetrate perpendicularly into the cerebellum.

The arteries of the cerebellar peduncles derive from the basilar trunk and from the posterior cerebellar arteries. At the level of the peduncles they give off the following branches:

-the median arteries, which penetrate into the peduncles through the openings of the posterior perforated substance;

-the radicular arteries, which follow the course of the roots of the oculomotor nerve and of the trochlear nerve;

-the accessory nerves, which penetrate into the grey substance surrounding the cerebral aqueduct.

The arteries of the cvadrigeminal colliculi are the following:

-the anterior and middle quadrigeminal arteries derive from the posterior cerebral arteries, wind around the cerebral peduncles and are distributed to the anterior quadrigeminal tubercles and to the anterior half of the posterior tubercles;

-the posterior quadrigeminal arteries arise from the superior cerebral arteries and supply the posterior half of the posterior quadrigeminal tubercles, the valve of Vieussens and the superior cerebellar peduncles.

The arteries of the cerebral hemispheres

According to the territory of distribution, the cerebral arteries constitute the following arterial systems: the superficial arterial system (cortical branches) *(rami corticales)*; the deep system (central branches)*(rami centrales)*; the choroidal system (the anterior and posterior choroidal artery) *(a. choroidea anterior et posterior)* and the posterior communicating system *(a. communicans posterior)*.

158

The superficial arterial system (cortical branches)
(rami corticales)

It is made up of *the three cerebral arteries (anterior, middle and posterior)*, which ramify at the level of the cerebral hemispheres, both on the surface of the gyri (convolutions) and in the depth of *the scissures* and *sulci*.

This vascular network lies either in the subarachnoid tissue (capillaries exceeding 1 mm in diameter – Charpy) and has a perpendicular direction on the cortical surface.

It gives off two kind of branches *(terminal rami)*:

-the cortical-short-arteries, which form in the grey substance a fine and rich vascular network;

-the subcortical-long-arteries, which course up to the white substance, without supplying the grey central nuclei and form a poorer vascular network.

The anterior cerebral artery *(a. cerebri anterior)*, a terminal branch of the internal carotid artery, runs forwards and medially towards the interhemispheric scissure, above the optic nerve, and unites through the communicating artery, with that on the opposite side. Then it winds round the genu of *the corpus callosum* and passes on the internal surface of the respective hemisphere, giving off the following branches:

-the pericallosal artery runs round the corpus callosum and terminates through the posterior pericallosal artery;

-the arteries supplying the corpus callosum;

-the corticosubcortical arterial system constituted of: the inferior frontal arteries, the internal anterior frontal arteries, the internal middle frontal arteries, the posterior internal frontal arteries and the internal parietal arteries (the paracentral, precuneal and parieto-occipital arteries).

These branches supply:

-the internal surface of the cerebral hemisphere, from the anterior extremity of the frontal lobe up to the parieto-occipital scissure;

- the superior portion of the external surface of the hemispheres, an area occupied by the frontal lobe, in which are situated the first two frontal gyri and a third of the ascending frontal gyrus;

- the orbital portion on the inferior surface of the frontal lobe;

The middle or sylvian cerebral artery *(a. cerebri media)*, a terminal artery of the internal carotid artery, directed laterally, crosses the anterior perforated substance and penetrates into the sylvian scissure.

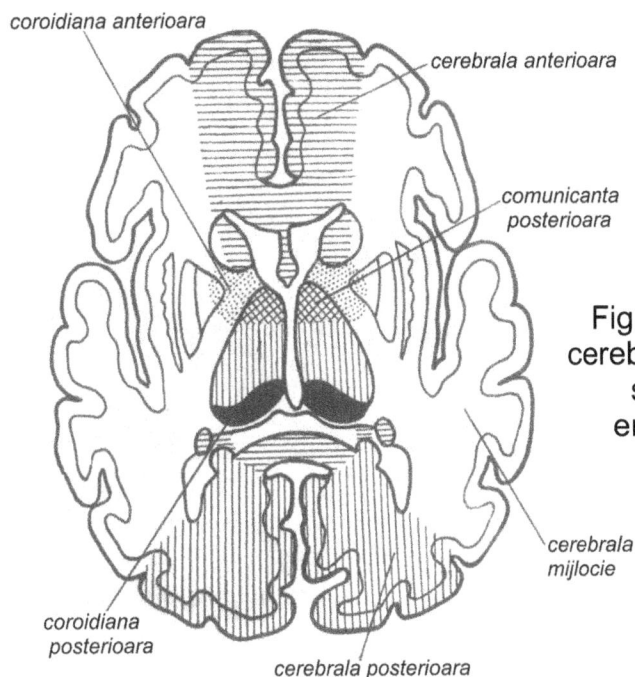

Fig. 148 Territories of the cerebral arteries – horizontal section through the encephalon (scheme)

coroidiana anterioara

cerebrala anterioara

comunicanta posterioara

cerebrala mijlocie

coroidiana posterioara

cerebrala posterioara

159

a.cerebri anterior
a.cerebralis media
a.carotis interna
a.comunicans posterior
a.cerebri posterior
a.cerebelli superior
a.basilaris
a.cerebelli inferior anterior
a.cerebelli inferior posterior
a.vertebralis
a.spinalis ventralis

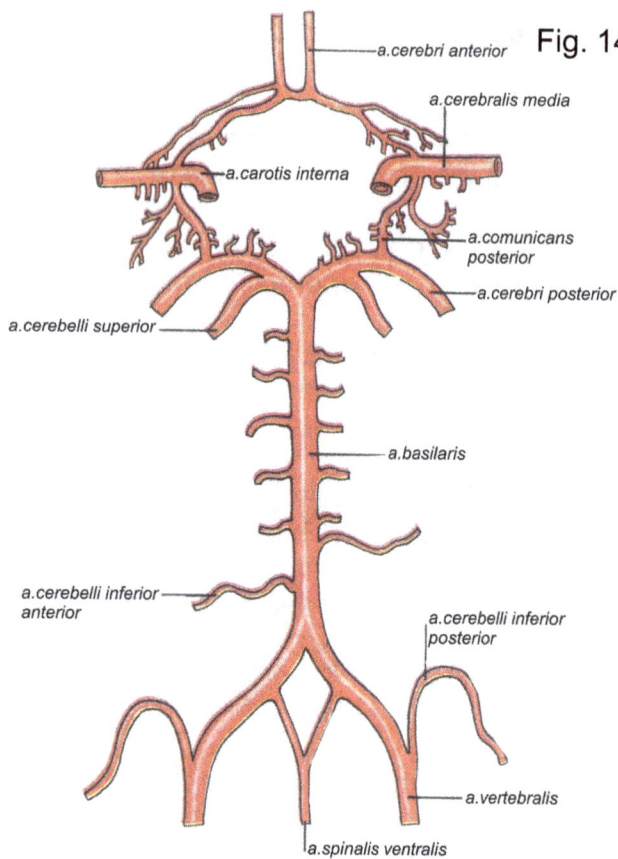

Fig. 149 Scheme of the main arterial sources of the encephalon

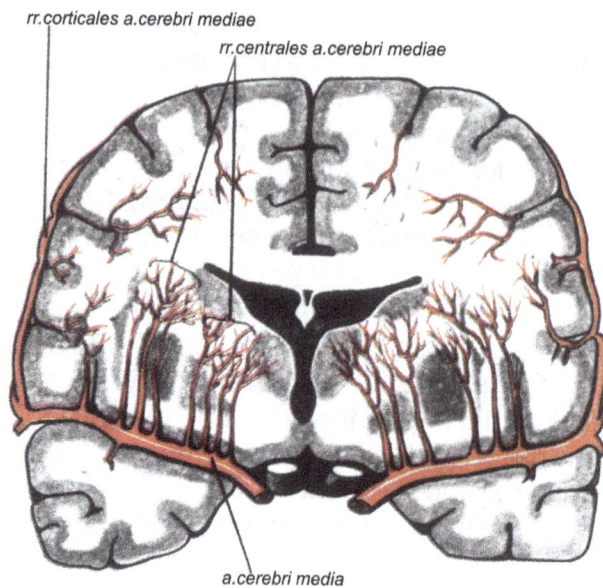

rr.corticales a.cerebri mediae
rr.centrales a.cerebri mediae
a.cerebri media

Fig. 150 Middle cerebral artery - territory

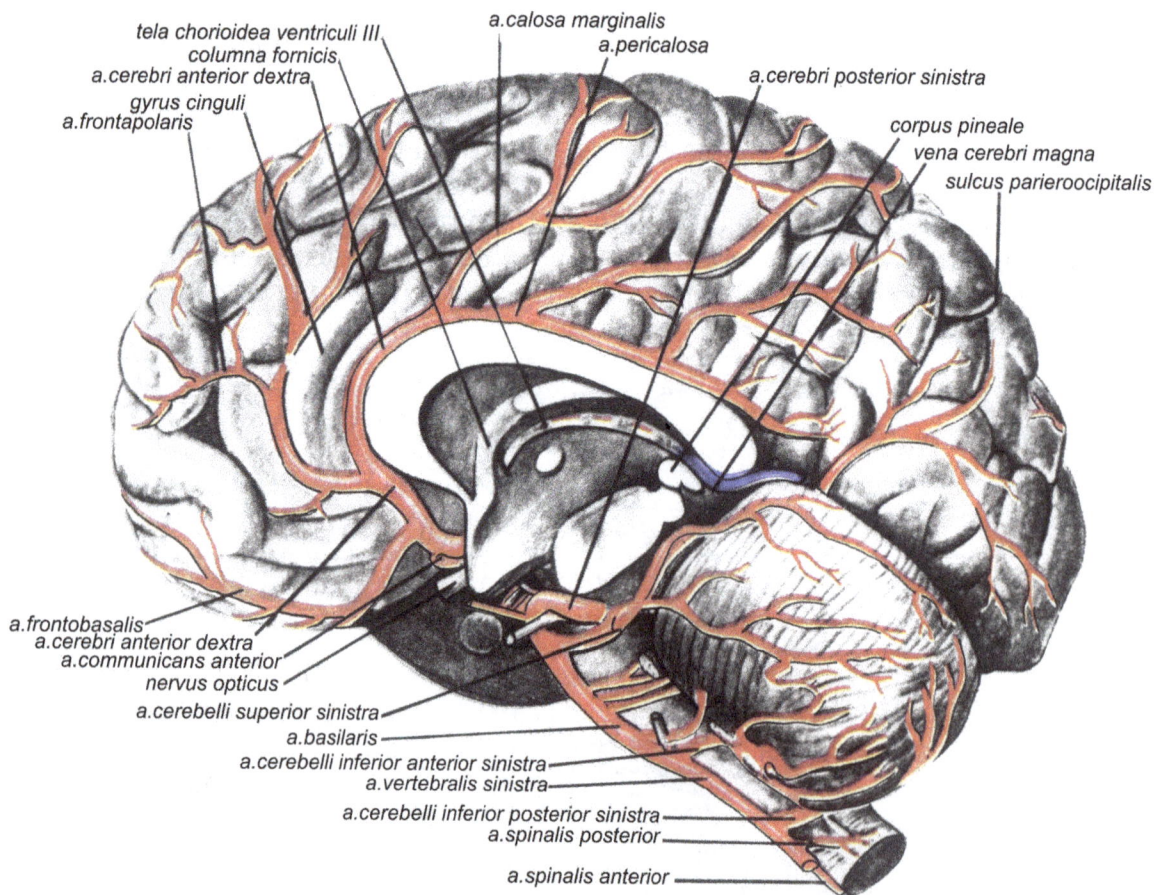

tela chorioidea ventriculi III
columna fornicis
a.cerebri anterior dextra
gyrus cinguli
a.frontapolaris
a.calosa marginalis
a.pericalosa
a.cerebri posterior sinistra
corpus pineale
vena cerebri magna
sulcus parieroocipitalis

a.frontobasalis
a.cerebri anterior dextra
a.communicans anterior
nervus opticus
a.cerebelli superior sinistra
a.basilaris
a.cerebelli inferior anterior sinistra
a.vertebralis sinistra
a.cerebelli inferior posterior sinistra
a.spinalis posterior
a.spinalis anterior

Fig. 151 Arterial supply of the encephalon (medial surface of the cerebral hemisphere, with the maintenance of the brainsystem and of the cerebellum on the preparation)

160

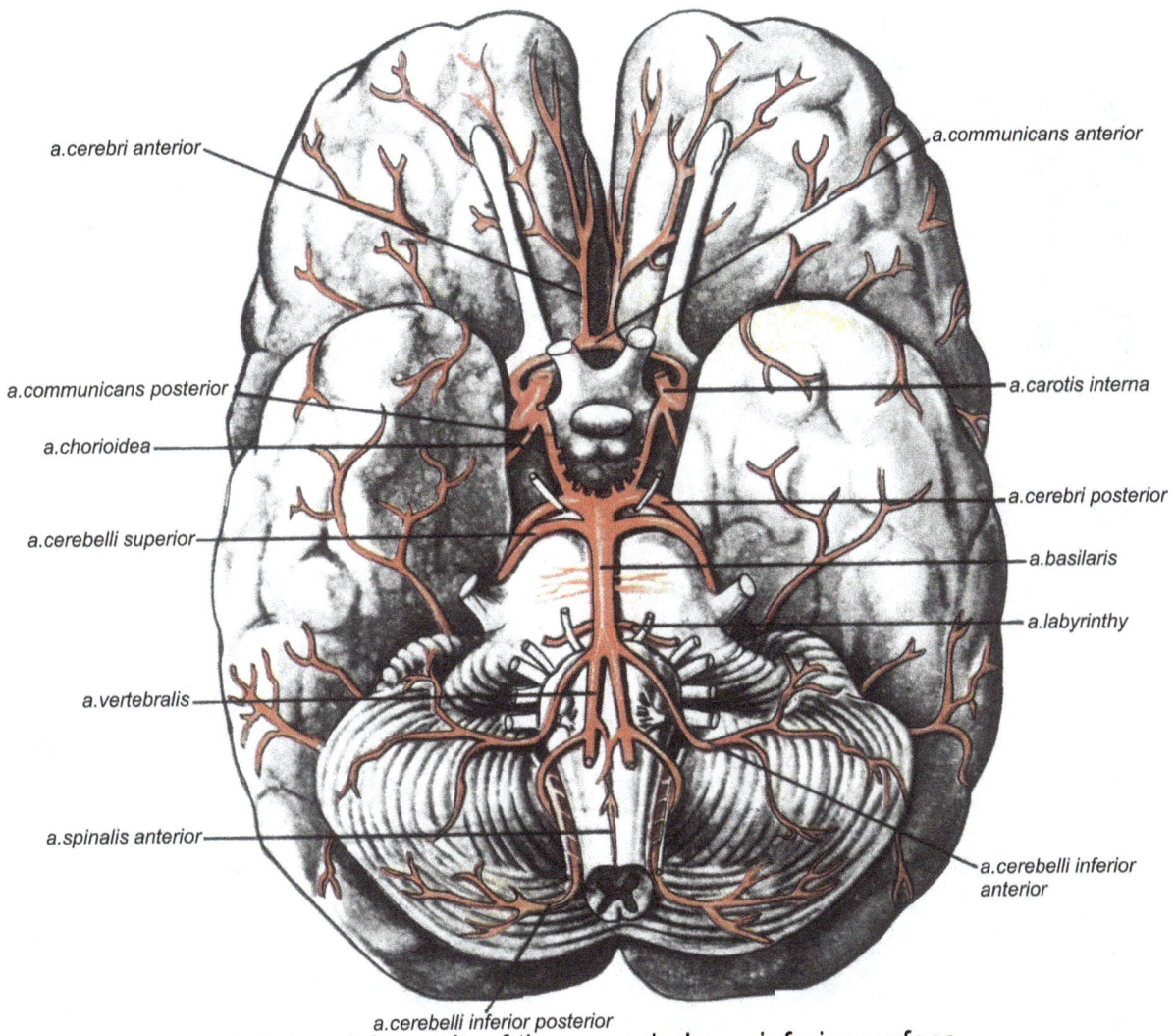

a.cerebri anterior
a.communicans anterior
a.communicans posterior
a.carotis interna
a.chorioidea
a.cerebri posterior
a.cerebelli superior
a.basilaris
a.labyrinthy
a.vertebralis
a.spinalis anterior
a.cerebelli inferior anterior
a.cerebelli inferior posterior

Fig. 152 Arterial supply of the encephalon – inferior surface

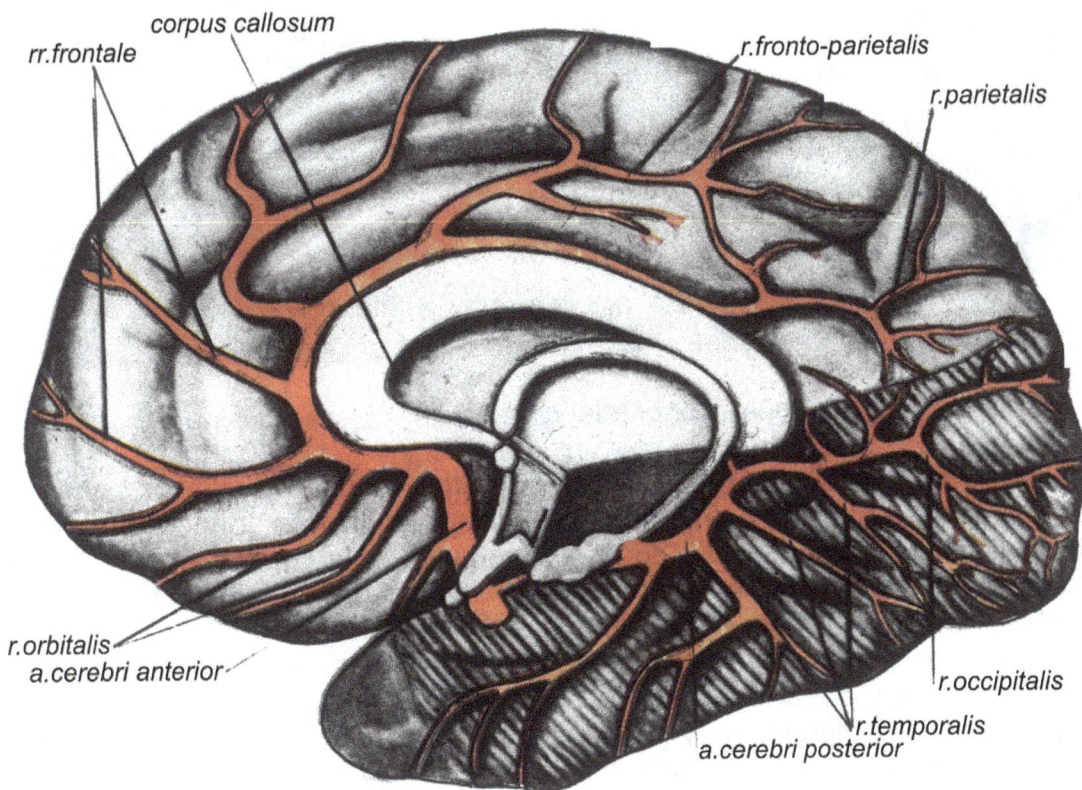

corpus callosum
rr.frontale
r.fronto-parietalis
r.parietalis
r.orbitalis
a.cerebri anterior
r.occipitalis
r.temporalis
a.cerebri posterior

Fig. 153 Arterial supply of the encephalon – internal surface

161

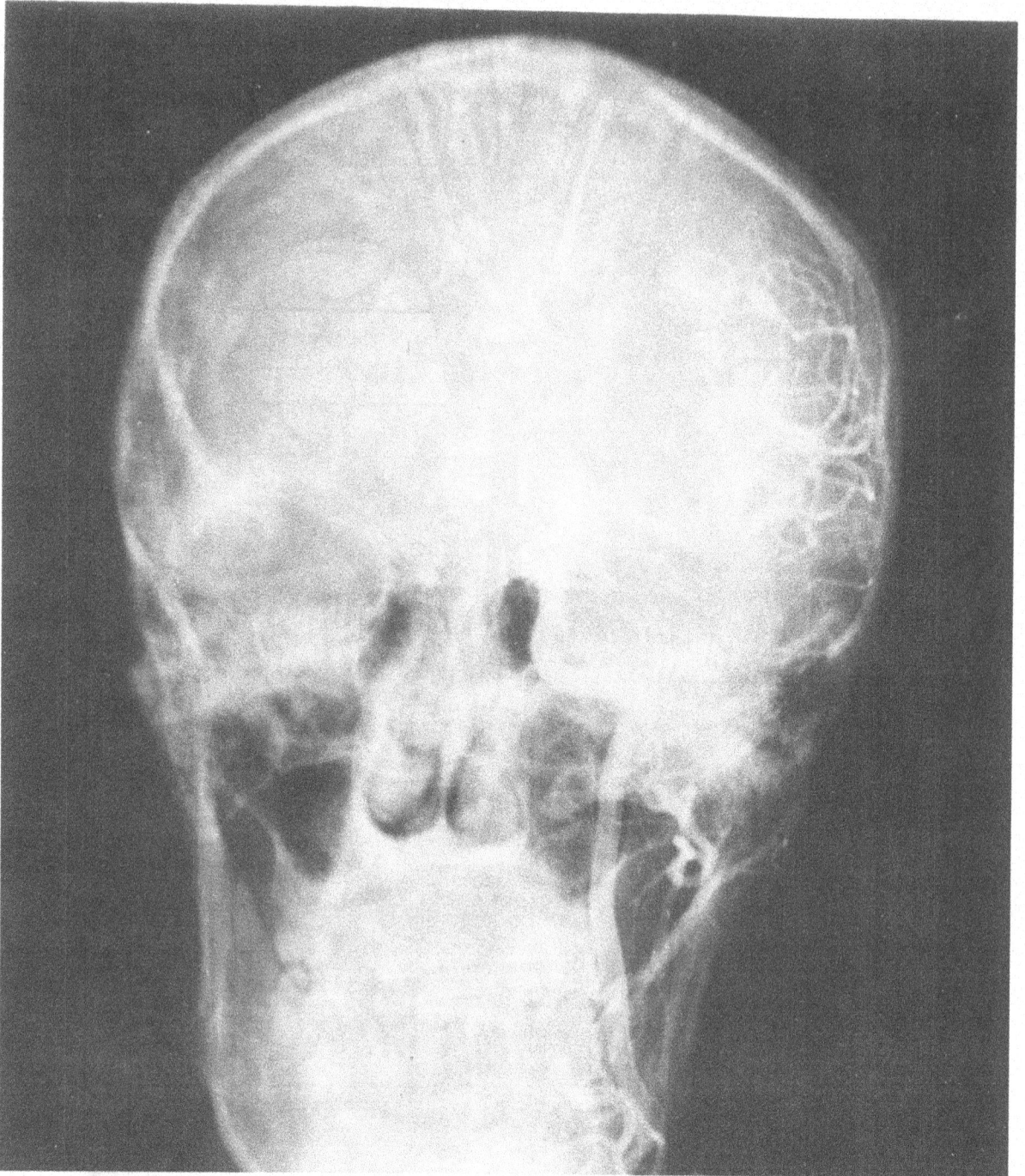

Fig. 154 Cerebral arteriography – anterior view

This artery gives off:

- the insular artery;

- the orbitofrontal or orbital inferior artery;

- the anterior temporal artery, which runs to the T1 and T2 temporal gyri;

- the common trunk of the ascending arteries, formed of four branches, which sometimes may have a separate emergence: the ascending frontal artery to the gyrus of the third frontal convolution; the rolandic artery, which is the artery of the rolandic fissure; the ascending parietal artery or of the retrorolandic fissure, distributed to the anterior part of the parietal lobe; and the artery of the prerolandic fissure;

- the posterior parietal artery, distributed to the posterior part of the parietal lobe;

- the middle temporal artery;

- the posterior temporal artery (the main artery which through its obstruction courses Wernicke's aphasia, that runs towards the posterior part of the T1 and T2 gyri;

- the terminal artery.

162

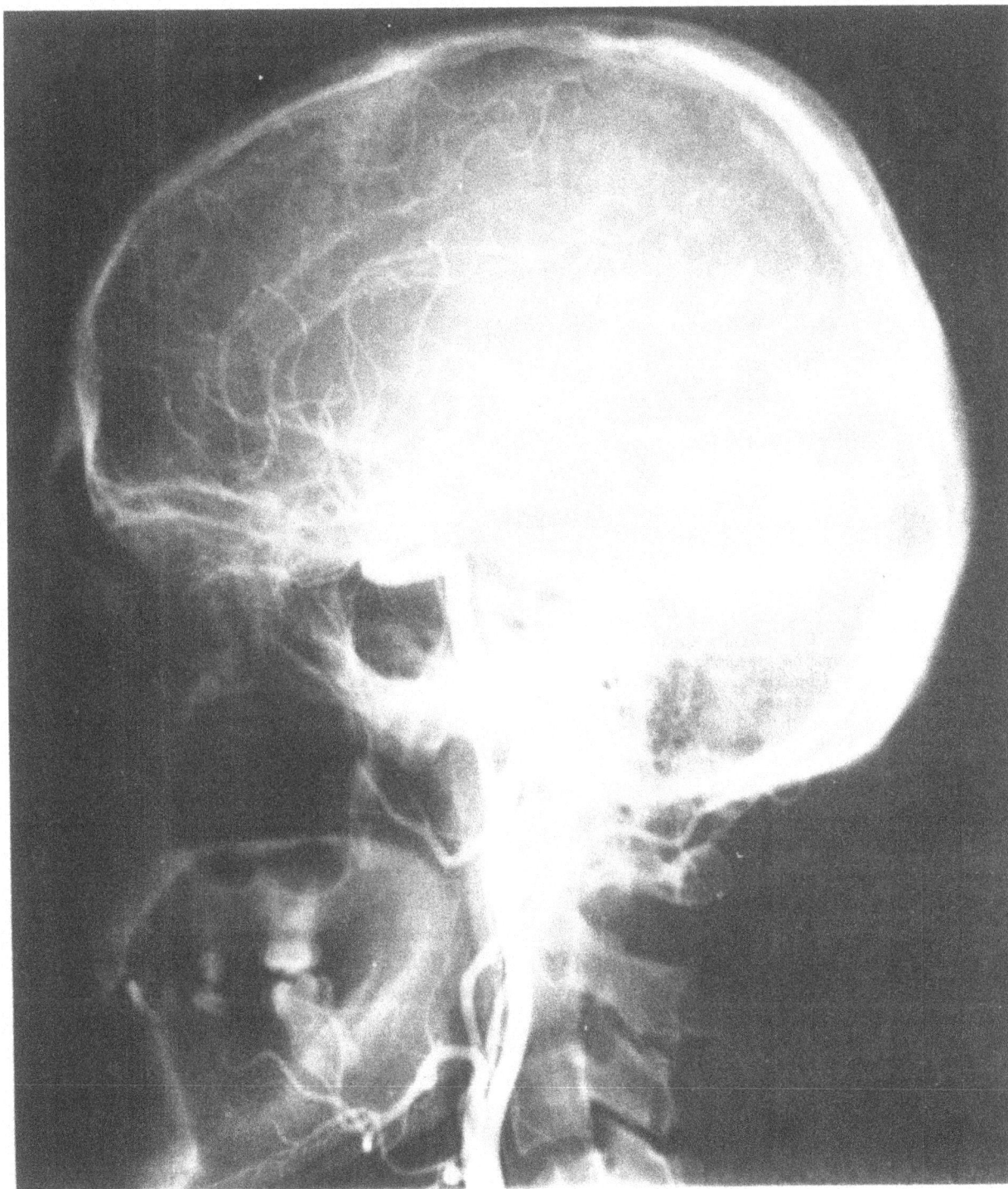

Fig. 155 Cerebral arteriography – lateral view

Hence, the sylvian artery assures the vessel supply of:

- the external surface of the cerebral trigone hemispheres, with the exception of its superior portion and of the occipital pole (the external surface of the terminal lobe, the anterior part of the occipital lobe and the orbital position of the frontal lobe);

- the inferior surface of the frontal lobes (the third frontal gyrus, two inferior thirds or three inferior fourths of the ascending frontal gyrus), of the temporal and of the occipital lobes;

The posterior cerebral artery *(a. cerebri posterior)* is a bifurcation branch of the brainstem, surrounds the inferior surface of the temporal lobe, where, at the level of its inferior surface, it gives off two branches: the anterior temporal artery and the middle temporal artery, to terminate through the temporo-occipital and the calcarine artery, at the level of the inferior surface of the occipital lobe.

Thus, visuosensory centres and the peripheral visuognosic zone are completely supplied.

163

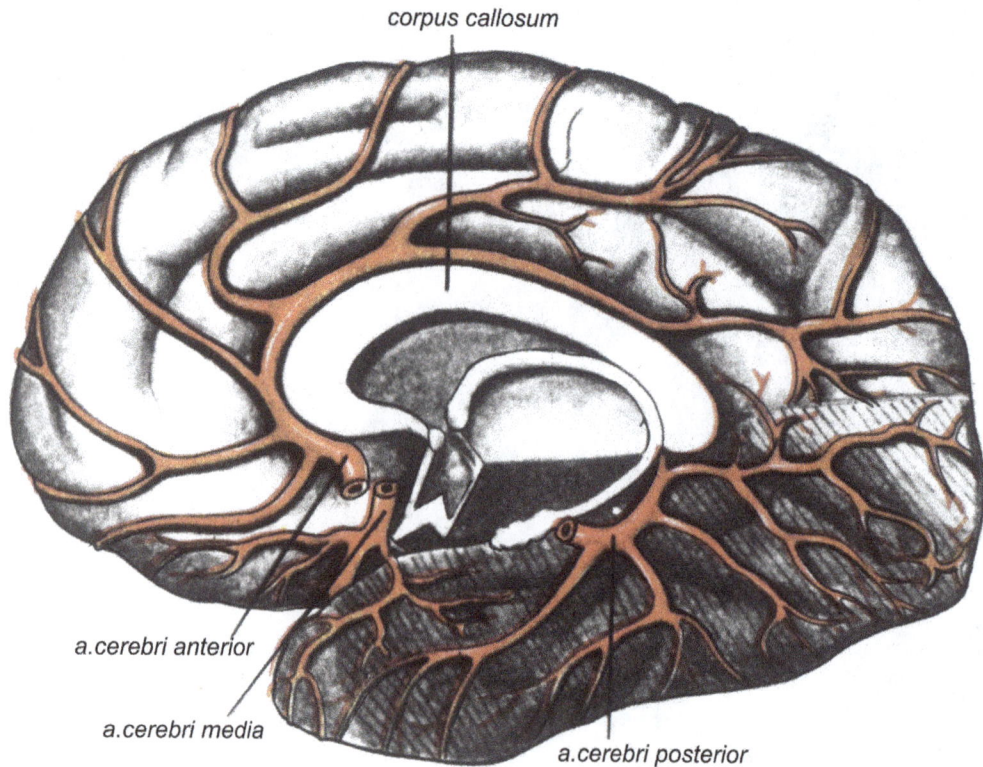

corpus callosum

a.cerebri anterior

a.cerebri media

a.cerebri posterior

Fig. 156 Arterial supply of the encephalon – internal surface, with demonstration of the middle cerebral artery

The deep arterial system
(Rami centrales)

It supplies the grey nuclei and the inferior wall of the middle ventricle, arises from the cerebral arteries and respectively from the posterior communicating artery, as follows:

The anterior cerebral artery gives off, just at its origin, the anterior communicating artery and then Heubner's long central artery *(a. recurrens)*, which penetrates into the cerebral substance through the perforated substance, supplying the anterior limb of the internal capsule, *the putamen* and the portion adjacent to the head of the caudate nucleus.

The middle cerebral artery achieves the most extensive supply to the grey nuclei through its two branches:

- the internal striate arteries (external pallidal arteries), which ramify in the external segment of *the globus pallidus* in *the lentiform nucleus*, while its internal segment is supplied by branches of the anterior choroid artery;

- the external striate arteries (putaminocapsulocaudate arteries, which run towards *the putamen* or the lateral segment of *the lentiform nucleus*, then pass through the anterior segment of the internal capsule and terminate in *the caudate nucleus* (lenticulostriate arteries). Of the group of these arteries, a particular attention attracts an artery which ascends the external border of the lentiform nucleus and which Charcot has called the artery of cerebral haemorrhage as it is often the site of haemorrhages in this region. Another group of arteries, the lenticulo-optic arteries, cross the posterior segment of the internal capsule and ramify in the thalamus (O. Sager). Recently, Percheion's studies contested the supply of the thalamus by the middle central artery.

Thus, the middle cerebral artery is involved in the supply of the putamen, the globus pallidus, the caudate nucleus and the internal capsule.

The posterior cerebral artery gives off the internal optic arteries, for the postero-internal part of the thalamus, and the external optic arteries, for the inferior part of the thalamus and the geniculate bodies (Foix and I.T. Niculescu).

164

It ensues that, through its deep, interpeduncular, quadrigeminal and thalamic branches, the posterior cerebral artery supplies the posterior hypothalamus, the postero-inferior thalamus, the epiphyseal region, the Luys' body, *the locus niger* and the red nucleus.

The posterior communicating artery, through the middle central arteries or the arteries of the inferior wall of the middle ventricle supplies the various parts of the ventricular floor (chiasma, *tuber cinereum*, mamillary tubercles), the optic tracts and the infero-internal part of the anterior third of the thalamus.

The choroidal system
(Aa. choroidea anterior et posterior)

It is formed of three arteries on each side:

- The anterior choroidal arteries *(aa. choroidea anteriores)*, which are terminal branches of the internal carotid artery and end in Bichat's fissure, crossing the lateral choroidal plexus nearly up to Monro's foramen. They give off branches to the villosities of the choroidal plexuses and to the walls of the lateral ventricle.

- The posterior and lateral choroidal arteries *(aa. choroidea posteriores et laterales)* derive from the posterior cerebral artery and run up to Monro's foramen.

- The posterior and median choroidal arteries *(aa. choroidea posteriores et mediales)* arise from the superior cerebellar artery, surround the pineal gland and distribute to the internal choroidal plexuses.

The posterior communicating system
(A. communicans posterior)

It plays a part in the formation of the anastomosis between the carotid and the vertebrobasilar systems. It supplies the hypothalamus, the thalamus and the optic chiasma.

The venous circulation
(Fig.157-160)

The venous circulation *(vv. cerebri)* is very complex, forming networks which afterwards converge the venous sinuses, either directly or through Trolard's venous polygon.

The veins on the surface of *the medulla oblongata* form a network which runs: upwards to the venous network of the pons, downwards to that of the spinal cord and laterally to the condylar veins.

The veins on the surface of *the pons* course, through the pontine venous network, towards the posterior communicating vein, then towards the cerebellar veins and terminate in the petrosal sinus and in the occipital transverse sinus.

At the level of *the cerebellum* run the following veins:
- the superior median vein, which ends in Galen's vein;
- the inferior median vein, which empties into the lateral sinus and sometimes into the occipital sinus;
- the lateral veins empty into the lateral sinus and the superior petrosal sinus.

The veins at the level of *the encephalon* are characterized by:
- a course independent of the arterial course;
- they empty completely in the venous sinuses of the skull;
- they anastomose with each other;
- they are thin-walled;
- they are avalvular.

From the viewpoint of the topographical diposition may be distinguished: superficial veins, deep veins, Trolard's polygon, anastomotic veins and cerebral sinuses.

1. The superficial cerebral veins *(vv. cerebri superficiales)* (the veins of the gyri) are distributed on the external, internal or inferior surface for each cerebral hemisphere, as follows:

a) *The veins of the external surface* converge towards the superior longitudinal venous sinus, termed also superior sagittal sinus *(sinus sagittalis superior)*, the lateral sinus and the superior petrosal sinus.

- The afferents of the superior longitudinal sinus are: the frontal, the rolandic, the parieto-occipital and the occipital veins.

- The afferents of the lateral sinus are the descending occipital veins.

b) *The veins of the internal surface* are the following:

- the ascending veins, which drain into the superior longitudinal sinus, namely the anterior veins at a right or acute angle opened forwards and the posterior ones, at an acute angle opened backwards;

- the descending veins are collected by the inferior longitudinal sinus and by the Galen's great cerebral vein *(vena cerebralis magna)*;

- the anterior cerebral vein unites through the anterior communicating vein with that on the opposite side and each is collected by the basilar vein;

- the sylvian vein collects the blood from the level of the sylvian fissure and from the deep striate veins and empties into the cavernous sinus and the basal vein;

- the posterior cuneolimbic vein empties into Galen's ampulla.

c) *The veins of the inferior surface* run: anteriorly towards the superior longitudinal sinus, behind towards the superior and internal petrosal sinus, the basilar veins and the Galen's veins.

2. The deep cerebral veins *(vv. cerebri profundae)*, under the form of two venous trunks - Galen's veins - collect the blood from the level of the central grey nuclei *(septum pellucidum*, striate body) and from the ventricular walls (choroidal plexuses). On their anteroposterior course they receive veins from the thalamus, the fornix and Ammon's horn. The two veins fuse, forming the Galen's ampulla, which is continuous with the straight sinus.

3. The Trolard's polygon is similar to the arterial hexagon of Willis (cerebral arterial circle of Willis) and is formed of:

- the basilary veins, in number of two, situated along the Bichat's cerebral fissure. Each of them is formed by the junction - in front of the anterior perforated substance - of the anterior cerebral vein with the insular vein, collecting in this way the venous blood from the inferior wall of the middle ventricle, the inferior part of the central grey nuclei, the posterior extremity of the optic layer and of the geniculate bodies and from the inferior and internal parts of the temporal lobe. They are united by the posterior communicating vein;

- the anterior cerebral veins are united anteriorly by the anterior communicating vein.

4. The anastomotic veins assure the connection between the two venous systems of the two cerebral hemispheres, as it is known that each hemisphere has two main venous areas: a superior area, which drains into the superior longitudinal sinus, and an inferior area, which empties into the other sinuses and into the veins of Trolard's polygon.

The connection is achieved by the following anastomotic veins:

- the superior anastomotic vein or Trolard's vein *(v. anastomotica superior)* assures the communication between the superior longitudinal sinus and the cavernous sinus, either directly or through the spheropalatine sinus;

- Labbé's anastomotic vein or inferior anastomotic vein *(v. anastomotica inferior)* may be inconstant; it relates the superior longitudinal sinus to the lateral sinus;

- the above described communicating veins of Trolard's polygon;

- the anastomotic veins between the superficial and the deep system.

5. The cerebral sinuses. At the level of the dura mater lie the following venous sinuses: the superior longitudinal sinus, the inferior longitudinal sinus, the straight sinus, the lateral sinus, the superior petrosal sinus, the cavernous sinus, the sphenoparietal sinus (Brachet's sinus) etc.

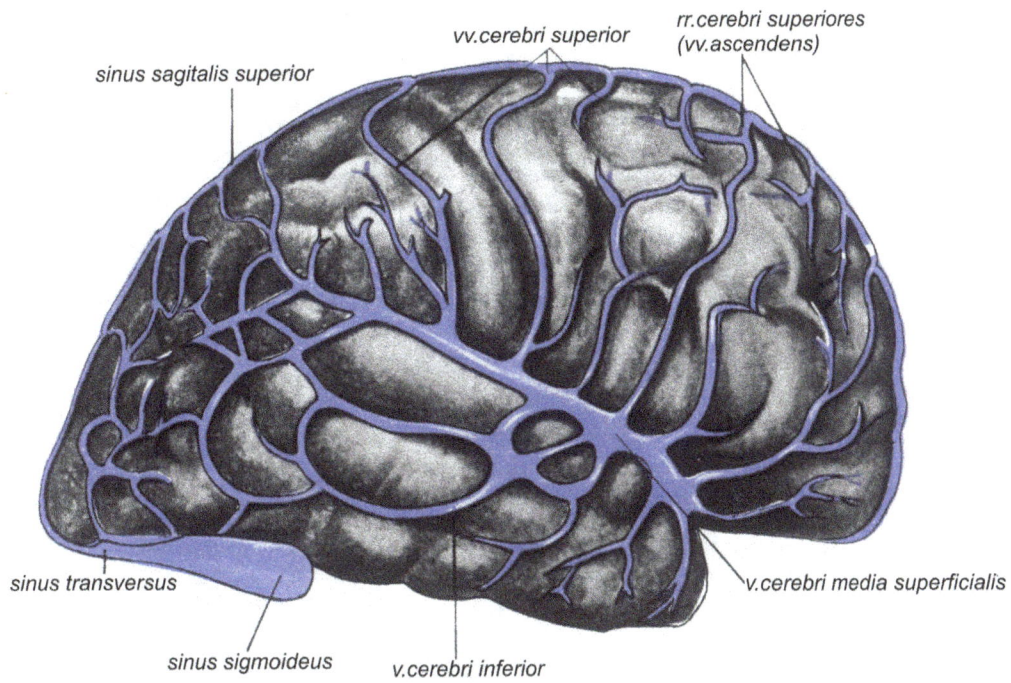

Fig. 157 Venous drainage of the encephalon – lateral view

sinus sagitalis superior

vv.cerebri superior

rr.cerebri superiores (vv.ascendens)

sinus transversus

sinus sigmoideus

v.cerebri inferior

v.cerebri media superficialis

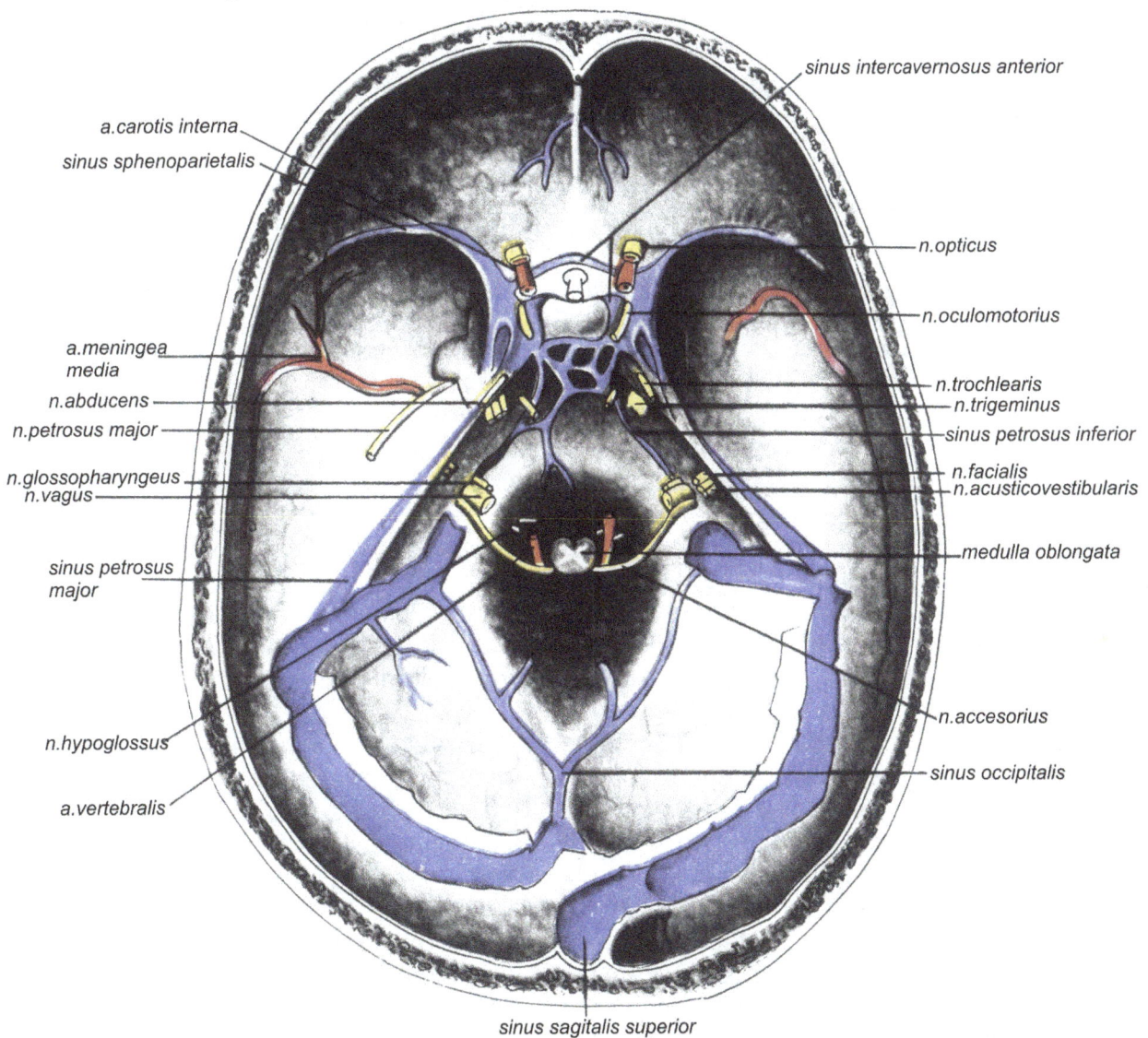

sinus intercavernosus anterior

a.carotis interna

sinus sphenoparietalis

n.opticus

n.oculomotorius

a.meningea media

n.abducens

n.petrosus major

n.glossopharyngeus

n.vagus

sinus petrosus major

n.trochlearis

n.trigeminus

sinus petrosus inferior

n.facialis

n.acusticovestibularis

medulla oblongata

n.accesorius

sinus occipitalis

n.hypoglossus

a.vertebralis

sinus sagitalis superior

Fig. 158 Sinuses of the dura mater – inferior view after removal of the encephalon

167

Fig. 159 Sinuses of the dura mater – sagittal section of the cranium

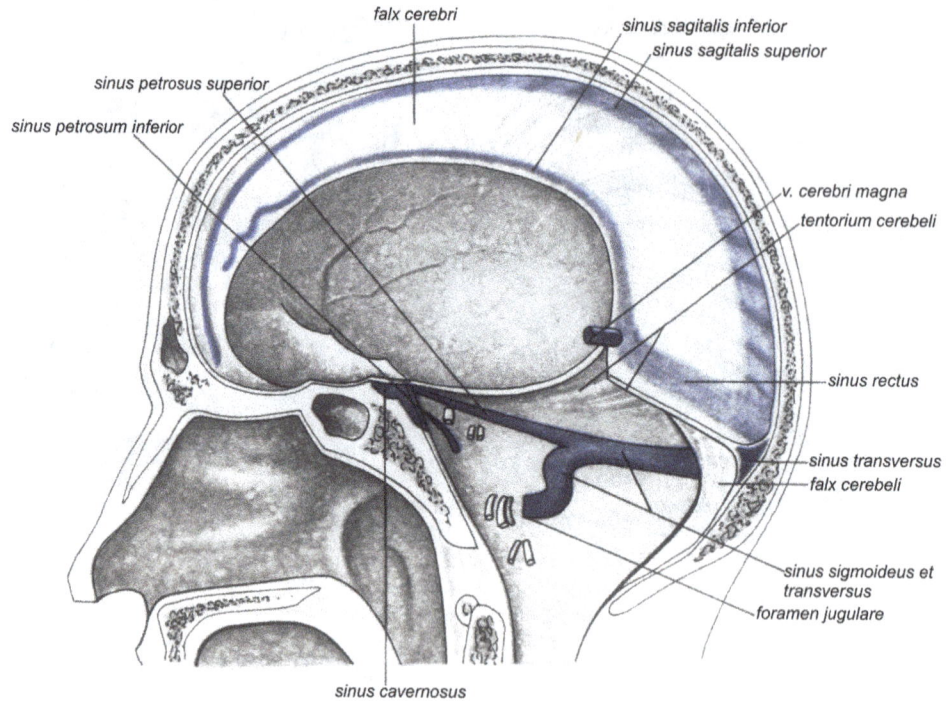

The venous blood which comes from these sinuses and from the cerebral veins is collected in the jugular bulb and, through the posterior lacerated foramen *(foramen lacerum)*, leaves the neurocranium (brainbox) through the internal jugular vein.

The capillary system

The latest researches with the electron microscope have shown that the endothelium of these vessels is of a continuous type, the junctions between the epithelial cells being of the *"zonula occludens"* type (tight junctions) and not *"macula occludens"* type (gap junctions), as the capillaries of the encephalon are surrounded by processes of the astrocytes. Between the astrocytes and the capillary is interposed the basement membrane. This arrangement explains the blood-brain barrier function of the capillary endothelium.

The envelops of the brain *(the meninges)*
(Meninges encephali)

Like the spinal cord, the brain, too, has three envelops: the dura mater *(pachymeninx)*, the arachnoid *(arachnoidea)* and the pia mater *(leptomeninx)*.

The dura mater encephali is a dense, white, glistening membrane, formed of connective tissue which adheres very tightly to the outline of the great foramen *(foramen magnum)*, as well as to the bones of the neurocranium, whose internal periosteum *(endocranium)* it makes up.

In the middle cranial fossa, at the level of the middle meningeal artery and of its branches, the dura presents a detachable, less adherent area (Marchant's zone), which may be frequently damaged in the case of injuries, making up the source of subdural and extradural haemorrhages; after an asymptomatic period, concomitantly with blood accumulation, appear cerebral compression phenomena requiring surgery, which consists in the removal of the hematoma and the ligation of the bleeding vessel.

From *the dura mater* detach and extend inside the cranial cavity several fibrous structures which, on the one hand, separate the various component parts of the encephalon and, on the other hand, offer them the possibility to maintain a stable position even in the case of sudden and impetuous movements.

168

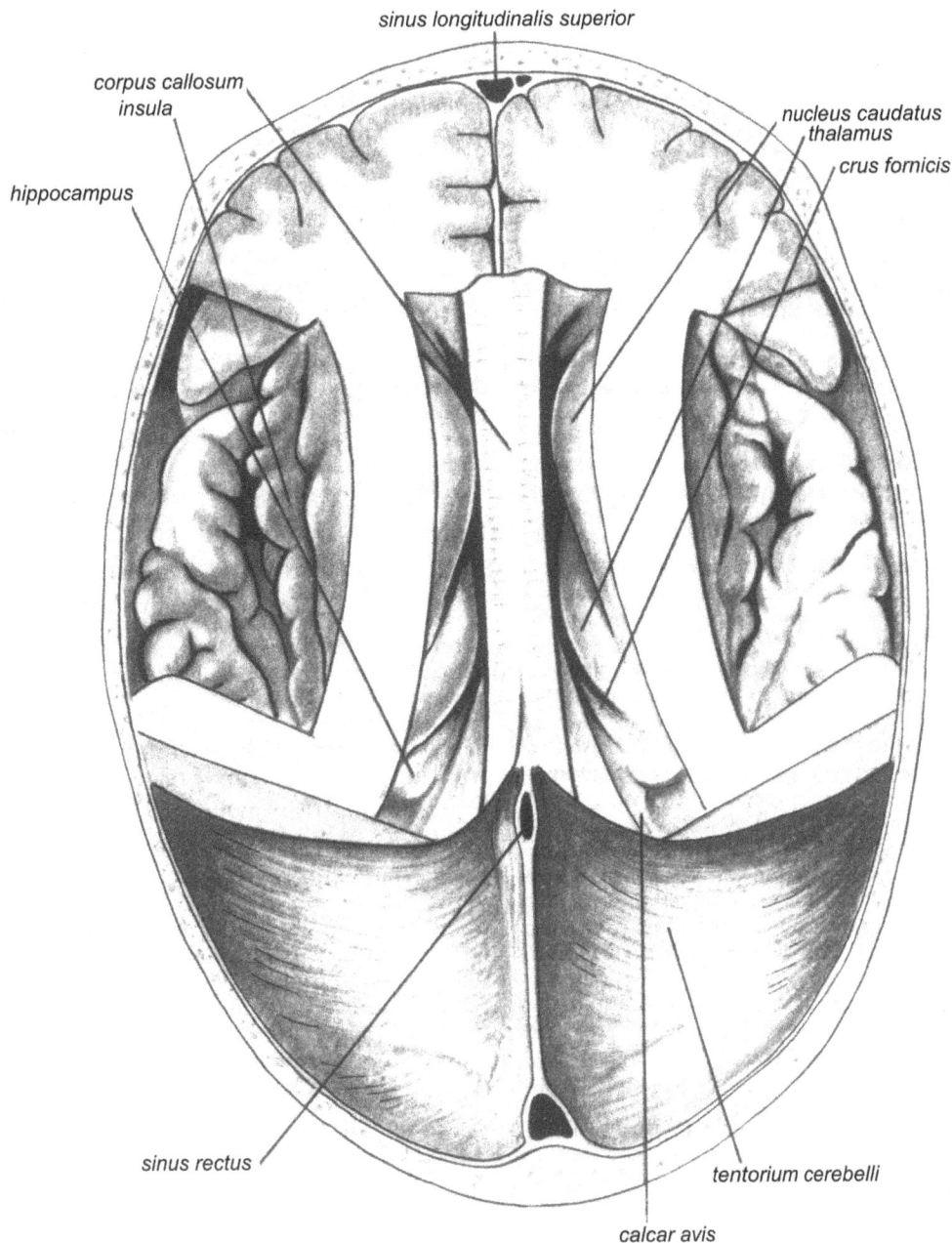

Fig. 160 Corpus callosum, lobe of the insula, cerebral
cortex – superior view (anatomic preparation)

These structures are *the falx cerebri*, *the falx cerebelli*, *the tentorium cerebelli*, the diaphragm of *the sella turcica (diaphragma sellae)* and the trigeminal or Meckel's cavity *(cavum trigeminale Meckeli)*.

The falx cerebri is a projection of the dura mater, 3 cm in width, situated sagittaly, in the median plane, inside the skull, between the two hemisheres. It is attached to the process of crista galli, to the frontal crest *(crista frontalis)* of the frontal bone, to the lateral borders of the sagittal sulcus formed by the two parietal bones, to the internal occipital protuberance and to the margin of *the tentorium cerebelli*.

The lower concave margin is free and penetrates into the longitudinal fissure of the brain, in proximity of *the corpus calossum*. At the insertion site, on the calvaria, lies the superior sagittal sinus *(sinus sagittalis superior)*, while near the face border is situated the inferior sagittal sinus. *The falx cerebri* delineates the site of the hemisphere and presents their displacements.

169

The falx cerebelli is a prolongation of *the dura mater,* with the concavity anterior, which begins at the internal occipital crest and extends up to the great occipital foramen *(foramen occipitale magnum)*. At this level, a part of it penetrates into the posterior cerebellar notch or incisure.

The tentorium cerebelli is a well stretched fibrous plate, situated in the transverse fissure of the brain *(fissura transversa cerebri)*. It represents a kind of floor for the occipital lobe and constitutes the roof which maintains *the cerebellum* in *the cerebellar fossae* of the posterior region of the base of the skull. *The tentorium* has its origin on the anterior and the posterior clinoid processes and is attached also to the superior border of the petrous part of the temporal bone, to the margins of the transverse sulcus of the occipital bone and of the parietal bones, as well as to the internal occipital protuberance, its edge lying on *the falx cerebri*. Its free anterior border has the anterior concavity crescent-shaped, termed tentorial incisure, extending nearly up to *the splenium* of *the corpus callosum*. At the junction with the *falx cerebri* lies the straight sinus *(sinus rectus)*.

The diaphragm of the turkish saddle *(diaphragma sellae turcicae)* or *the tentorium* of hypophysis is an expansion of the dura mater, extending between the anterior and posterior clinoid processes *(processus clinoidei anteriores and posteriores)*, beyond the hypophysial fossa *(fossa hypophyseos)*; it has an opening, hiatus diapragmatis, which gives passage to the infundibulum *(infundibular stem)*.

The cranial nerves, at their passage through the dura mater, are surrounded by a dural sheath. The semilunar (gasserian) ganglion of the trigeminal nerve is situated in a pouch formed by the dura mater on the trigeminal impression *(impressio trigeminalis)* on the petrous part of the temporal bone and is named trigeminal cave or *cavum trigeminale Meckeli*.

Fr. I. Rainer considered that the functional structure of the dura mater may be explained by the action of the two groups of factors: internal and external.

In the group of internal factors he has included the pulsations of the brain, the weight of the brain and the action of respiration (increase of the volume of brain during inspiration and decrease of it in expiration); among the external factors he ranged the action of the muscles of mastication and of the thoracocranial muscles with insertion on the skull. Thus, it ensues that the inner table of the skull *(lamina interna cranii)* is submitted to the action of the brain, while on the outer table *(lamina externa cranii)* the action of musculature is exerted. Each of the tables, on turn, acts through forces on the opposite table.

At the level of the great foramen *(foramen magnum)* the encephalic dura mater is continuous with the spinal dura mater.

The dura mater is abundantly supplied by branches arising from the arteries mentioned below.

The middle meningeal artery, a branch of the internal maxillary artery, is the most important and the most jeopardized in the fractures of the calvaria. It enters into the braincase through *the foramen spinosum* of the sphenoid bone. Sometimes, an accessory meningeal artery is present, which derives from the middle meningeal artery, but - in this situation - it enters in the skull through *the foramen ovale*, ramifying to the dura mater and to the gasserian ganglion. In addition, the anterior part of the dura mater is supplied by the anterior meningeal artery, which derives from the anterior ethmoidal artery. The posterior part of the dura mater is supplied by the posterior meningeal artery *(lesser)* which arises from the ascending pharyngeal artery. All the arteries are accompanied by the homonymous veins.

The above mentioned arteries supply also the bony tissue which they cross and their terminal branches anastomose with the contralateral branches.

The return veins anastomose with the diploic veins, which drain into sinuses. The dural veins communicate with the basal cerebral veins and the emissary veins.

The venous sinuses of the dura mater (sinus durae matris) are situated between the layers of the dura and are lined by endothelium; in cross section they exhibit a form closely resembling that of a triangle. The haemorrhage from these formations is very severe, as the value of the local venous pressure is rather high.

170

The venous sinuses can be distributed into two groups: the superoposterior and the antero-inferior (Kaplan's) groups.

1. The superoposterior group includes the following sinuses:

a) *The superior sagittal sinus (sinus sagittalis superior)* has its origin in *the foramen caecum*, lies on the superior margin of the *falx cerebri* and ends near to the internal occipital protuberance, where it unites with the transverse sinus *(sinus transversus)*. At its origin, at the level of *the crista galli*, it receives venous blood from the nasal cavities through the emissary veins and then tributary sources from the superior cerebral veins, the result being its progressive anteroposterior increase in size. Sometimes, the sinus may be double presenting lateral prominences named lacunae.

b) *The inferior sagittal sinus (sinus sagittalis inferior)* is situated at the level of the inferior margin of *the falx cerebri* and also increases in size owing to tributary sources. It empties posteriorly in the straight sinus.

c) *The straight sinus (sinus rectus)* passes through a dural fold - at the junction of *the falx cerebri* with *the tentorium cerebelli* - to the occipital protuberance and empties into the transverse sinus, after receiving the great cerebral vein. The existence of a confluence between the sagittal and the straight sinus may be often observed.

d) *The transverse sinus (sinus transversus)* is double; it is the largest in size and lies on the external margin of *the tentorium cerebelli*, along the transverse sulcus of the occipital bone. It represents the continuation of the superior sagittal sinus and of the straight sinus. At the level of the sigmoid sulcus it describes a curve forwards and downwards and is continuous with the sigmoid sinus. It receives blood from the superior petrosal sinuses, the mastoid and condylar emissary veins, the cerebellar and the diploic veins.

e) *The occipital sinus* begins at the confluence at the level of the great foramen, of the small veins present in this area; it increases gradually in size, runs on the inferior margin of *the falx cerebelli* towards the occipital bone and empties into the confluence of the sinuses.

2. The antero-inferior group comprises the following sinuses:

a) *The cavernous sinus (sinus cavernosus)* is double; it is situated on both sides of the body of the sphenoid bone, on the sides of the turkish saddle. It presents a reticular structure and has its origin at the level of the superior orbital fissure, through the tributary vessels of the superior ophthalmic veins and some small cerebral veins. Between the sinuses pass transversally the anterior and the posterior intracavernous sinuses, forming together a ring - the circular sinus *(sinus circularis)*, in which, in beside the above-mentioned ophthalmic veins, empty also the sphenoparietal, the inferior petrosal and the superior petrosal sinuses (the latter unite the cavernous with the transverse sinus). Medially to the cavernous sinus passes the internal carotid artery, surrounded by the carotid plexus and by numerous veins. Laterally on the artery are situated the oculomotor, trochlear, ophthalmic and maxillary nerves. The lesion of the carotid artery at this level may bring about the formation of an arteriovenous aneurysm. The presence of an infection of *the ala nasi*, of the upper lip or of the internal angle of the eye is very dangerous, as it may cause an extremely severe thrombophlebitis of the cavernous sinus (which was lethal before the antibiotic era).

b) *The anterior and posterior intracavernous sinuses* form the above-mentioned circular sinus.

c) *The superior petrosal sinus (sinus petrosus superior)* lies on the margin of *the tentorium cerebelli* and joins the cavernous sinus to the transverse sinus.

d) *The inferior petrosal sinus (sinus petrosus inferior)* begins at the level of the cavernous sinus, passes through the jugular foramen and ends in the internal jugular vein.

e) *The basilar plexus* is situated at the level of the basilar region of the occipital bone and is formed of the veins which communicate with the two inferior petrosal sinuses. It drains the venous blood into the anterior vertebral plexus.

The sensory nerve supply is assured by the meningeal branches of the ophthalmic, maxillary and mandibular nerves. Consequently the dura mater is very sensible to pain, especially near the vessels.

171

The subdural cavity is a capillary space situated between the arachnoid of the encephalon *(arachnoidea encephali)* and *the dura mater*.

The arachnoid *(arachnoidea cerebri)* is situated between *the dura mater* and *the pia mater* and delineates, with the latter, a space named subarachnoid cavity or subarachnoid space *(cavum subarachnoideum)*, which is filled with cerebrospinal fluid *(liquor cerebrospinalis)*(CSF). Histologically, it is formed of fine plate of avascular connective tissue; it is thicker and more opaque towards the base of the skull. It covers the whole surface of the cerebral hemispheres, without penetrating into the sulci of the cortex. On the basal surface of the brain, too, it does not follow the surface relief of the latter. Thus are formed spaces which are called subarachnoid cisternae *(cisternae subarachnoidales)*, filled with cerebrospinal fluid.

Special formations of the arachnoid are the arachnoid granulations *(granulationes arachnoidales)*, termed also Pacchionian granulations, which are millet-seed-or pepper-berry-sized projections of the arachnoid. They are formed of connective tissue bundles crossing with each other. They may often contain a small cavity, called subarachnoid space or cavity *(cavum subarachnoideum)* and are devoid of blood vessel supply. by their bulging they narrow the dura mater up to a thin layer and they penetrate as small prominences into the dural venous sinus or into the diploic veins. At this level, the cranial bones present small pits, called granular pits *(foveolae granulares)*, which are also dependent on the sinus. Usually these granulations are sited along the superior sagittal sinus and in its vicinity. At this level, the bone may get thinner to such a degree, that only its external lamina is preserved. Such areas are especially jeopardized in traumatic lesions of the calvaria. The arachnoid granulations are more numerous and larger in old people, while in children up to 5 years of age they are completely lacking.

The cerebral pia mater *(pia mater cerebri)* intimately covers the surface of the brain, following exactly all the irregularities of the cortex and penetrating even in its sulci. It is formed of thin collagenous tissue and contains the vessels of the brain.

The cerebrospinal fluid; anatomical spaces and circulation

The cerebrospinal fluid (CSF) *(liquor cerebrospinalis)* is contained in the anatomical spaces resulted from the ontogenetic evolution of the primitive neural tube (the ventricular system of the encephalon and the spinal ependymal canal), as well as in the subarachnoid space. These areas communicate with each other at the level of the fourth ventricle, thus making up a unique system which assures the CSF circulation from the level of the cerebral ventricles towards the pericerebral and perimedullary subarachnoid space.

The CSF is secreted by the choroid plexuses of the cerebral ventricles and is resorbed by the nervous system through *the arachnoid villi* and the Pacchionian granulations. In this way is achieved the circulation of the fluid, which passes through the lateral ventricles, Monro's foramen, the third ventricle, the cerebral (sylvian) aqueduct, the fourth ventricle and the ependymal canal. The connection with the subarachnoid spaces occurs at the level of the fourth ventricle through the two lateral Luschka's foramina and the medial Magendie's foramen.

In adults, the total amount of CSF is of 100-150 ml, the fluid being permanently secreted and 3-4 times/day resorbed (about 400 ml / 24 hours).

This plasma dialysate, in addition to its role of brain and spinal cord protection, is also involved in the maintenance of a constant intracranial pressure and in the elaboration of a constant perineural molecular biochemical environment, indispensable to the activity of the central nervous system. Some authors ascribe to the CSF the quality of a tissue, with a specific function and activity. Moreover, it achieves the transfer of nutritional substances to the brain, beside the elimination of some metabolites. The above-mentioned functions are possible to the existence of permanent relations between the perivascular Virchow-Robin's spaces.

The blood-brain barrier (BBB), constituted of the vascular endothelium, the reticuloendothelial system and the neuroglia, assures a selective permeability for the various blood components.

As it was shown before, the anatomical spaces containing normally CSF are divided into two groups:

- spaces derived from the primitive neural tube and
- the subarachnoid space.

The spaces derived from the primitive neural tube form, at the level of the encephalon a series of more dilated zones, called ventricles, which continue in the spinal cord, through the ependymal canal.

Inside the tissue mass of the encephalon, four ventricles are described: the lateral or telencephalic ventricles, the third or diencephalic ventricle and the fourth or rhombencephalic ventricle.

The lateral ventricles *(pars lateralis ventriculi)* are present under diverse anatomical variants, according to the form, the size and the volume of the encephalon. Usually, each lateral ventricle may be compared with a horseshoe, which is made up of the following three elongated portions, situated in proximity to the poles of the cerebral hemispheres: the frontal or anterior horn *(pars frontalis seucornu frontale)*, the occipital horn *(pars occipitalis seu cornu occipitale)* and the temporal or sphenoid horn *(pars temporalis seu cornu temporale)*. All these extensions join in an area called central part *(pars parietalis)*, known also under the term ”carrefour” used by the French authors or Schwalbe's ventricular trigone.

Most frequently, the lateral ventricles are not perfectly symmetrical, neither identical in shape and volume. This is due to the total or partial coalescence of the walls of a horn, most often being involved the occipital horn; more recently appeared phylogenetically. When only the middle portion of the horn is involved in the coalescence process a ventricular island is formed, called ependymal vesicle.

The lateral ventricles make up about 80-90% of the total ventricular volume; they represent the bulkiest mass of the endoneural area or of the fluid-filled spaces. On the average, in the young adult, each ventricle has a volume of approximately 20-30 cm^3, which doubles after 60-70 years of age.

Owing to the presence of the choroid plexuses in the central part and in the anterior horn, these parts are the main site of CSF secretion.

The third ventricle *(ventriculus tertius)* is situated on the median, interhemispheric line, under the form of a high, sagittaly oriented cleft, and separates the anatomical structures of the diencephalon. Some authors consider that its rostral part *(pars telencephalica ventriculi tertii)* belongs to the telencephalon.

It communicates with the lateral ventricles through Monro's (interventricular) foramina. These openings, located in the rostral zone of the lateral walls, are true channels of 3-4 mm in diameter and 4-5 mm in length.

On the ventricular roof *(lamina tectoria ventriculi tertii)* lie also choroid plexuses.

The fourth ventricle *(ventriculus quartum)* is situated in the posterior part of the brainstem, at the level of the rhomboid fossa, and consists of a paramedian longitudinal part and a transverse part.

It is connected with the third ventricle through the cerebral (sylvian) aqueduct, a remnant of the cerebral cavity of the mesencephalon.

The aqueduct is of about 2 cm in length and 1.5 mm in width; it represents one of the ”sensible” areas of the ventricular system.

The fourth ventricle has a great significance in the CSF circulation, since the connection between the endoneural and the subarachnoid areas is achieved at this level through three foramina:

- Magendie's foramen as the medial aperture of the fourth ventricle, situated in the median zone of the roof of the fourth ventricle, represents the main way of commnication between the ventricular system and the subarachnoid space;

173

- Lushka's foramina or the lateral apertures of the fourth ventricle are situated at the distal extremity of the lateral recesses of the fourth ventricle and achieve the communication between this ventricle and the basal or interpeduncular cistern *(cisterna pontomedullaris)*. A part of the choroid plexus of the fourth ventricle penetrates into the subarachnoid space through Lushka's foramina and constitutes an anatomical structure called "horn of plenty" or "horn of abundance" (Bochdalek), through which a certain amount of CSF is secreted directly into the subarachnoid space. As in about 20% of the individuals these foramina are lacking, it is appreciated that their functional significance is rather low and therefore some authors consider them as "pseudoforamina".

Choroid plexuses are present also in the inferior area of the roof.

In the case of absence or later fusion of these orifices, appears *the internal hydrocephaly* with distention of the ventricles. A permanent hyperproduction in children leads to the development of *the external hydrocephaly (hydrocephalus externus)*, with a considerable enlargement of the skull cap calvaria, associated also with an internal hydrocephaly.

The ependymal canal, described in the chapter regarding the spinal cord, presents in the adult, both above (Arantius ventricle) and below (Krause's terminal ventricle), more dilated portions, the remainder being obliterated by the ependymal proliferations.

The whole surface of the endoneuronal fluid-filled space - with the exception of the choroid plexuses - is covered with an endothelium called ependyma, formed of cubic or prismatic ependymal cells *(ependymal glia)*.

The subarachnoid space is bounded by the two membranes of *the leptomeninx*: pia mater and arachnoid. The arachnoid lines the internal surface of the dura mater along his whole extent, while the pia mater adheres to the external surface of the anatomical formations, which make up the central nervous system, penetrating in all the fissures, sulci and in their depth, with the exception of some sulci of the cerebral hemispheres. It results that the whole surface of the brain and of the spinal cord, with the exception of an area at the level of the convexity of the cerebral hemispheres, where the two membranes form - by coalescence - a unique pia-arachnoid membrane, is bathed by CSF.

Between the two leptomeningeal membranes, the arachnoid achieves an areolar system of intercommunicating *lacunae* through which CSF is circulating.

The more dilated zones, whose volume varies between 1 and 10 mm^3, were called cisterns, lakes or confluences, in accordance with the great diversity of their shape and size. This system of intercommunicating cisterns is connected, on turn, with the spinal subarachnoid space. The following dilations are described at the level of the intracranial subarachnoid space.

- **The cerebellomedullary cistern** or *cisterna magna*, unpaired is situated between the inferoposterior surface of the cerebellum and the medulla oblongata, at the level where the arachnoid of the brain is continuous with the spinal arachnoid. It is the largest cistern, and at the same time the most important for the medical practice. By means of suboccipital taps, cerebrospinal fluid may be extracted or the cerebral ventricles may be filled with air for an encephalography;

- **The pontomedullary cistern** *(cisterna pontomedullaris)*, situated at the level of the bulbopontine sulcus, presents a basilar confluent *(cisterna basalis sive Waterbed-Hilton's cisterna pontis)* and a transverse confluent (bulbopontine confluent). Some authors divide this cistern into two portions: the bulbopontine cistern, containing the origin of the basilar trunk, and the prepontine or the basilar cistern, situated between the basilar process of the occipital bone (in front) and the pons (behind);

- **The basal cistern** *(cisterna basalis)* lies between the internal surface of the temporal lobes the optic chiasm and the bulbopontine sulcus; it is crossed by the pituitary stalk. Two subdivisions of this cistern are described: *the interpeduncular cistern (cisterna interpeduncularis)* and *the optochiasmatic cistern (cisterna chiasmatis)*, a zone situated between the optic tracts and considered by some authors as a distinct morphological entity.

174

- **The ambient cistern** or cistern of the great cerebral vein *(cisterna ambiens)* or the superior cerebellar lake is bounded by the cerebellum, *the corpus callosum* and the quadrigeminal tubercles;

- **The interhemispherical cistern** *(cisterna interhemispherica)*, sited between *the corpus callosum the falx cerebri* and the cerebral hemispheres, is made up of the callosal lake *(cisterna laminae terminalis)*, lying between the optic chiasm and the anterior space of the interhemispherical sulcus, and the pericallosal cistern *(cisterna corporis callosi)*, placed between *the falx cerebri, the corpus callosum* and *the cyngulate gyrus*;

- **The sylvian lake** or cistern of the lateral fossa of the brain *(cisterna fossae lateralis cerebri sive cisterna fossae Sylvii)* is situated at the level of the sylvian scissure (fossa of Sylvius).

The subarachnoid space extends inferiorly, acquiring the form of a sac which begins at the level of the occipital foramen and ends in the dural cul-de-sac, situated between the L2 and S2 vertebrae; it is called the lumbar cistern or the inferior medullary lake, an anatomical site of choice for lumbar puncture, as it lies in the inferior lumbar region.

The denticulate ligament *(ligamentum denticulatum)* divides the subarachnoid space into two compartments (anterior and posterior), which communicate with each other.

Expansions of the subarachnoid space occur both along the nerve trunks and in the intimacy of the nervous tissue.

The expansions of the intracranial subarachnoid space accompany extracranially some nerves. Thus they accompany the olphactory filaments which cross the cribriform plate of the ethmoid bone up to the neighbourhood of the nasal mucosa, the optic nerve in the orbital cavity up to the posterior pole of the eyeball, as well as along the sheath shared by the vestibulocochlear facial and intermediate (Wrisberg's) nerves up to the fundus of *the internal acoustic meatus*, dug in the petrous portion of the temporal bone.

These expansions make possible the communication of the subarachnoid space with the lymphatic networks of *the nasal fossae*, of the orbit and of the internal ear and, at the same time, they explain the possible pathological correlation in the occurrence of *meningitides* starting at this level.

Furthermore, expansions are also described around the roots of the spinal nerves, stopping at the level of *the conjugate foramina*, communications being in this way created between the subarachnoid space and the lymphatic network of the nerve sheaths or of the epidural space, as well as with the vast venous plexus, situated in the vicinity of the emergence of the spinal nerve roots, through the structures, resembling to *the arachnoid villi*.

As regards the expansions in the intimacy of the nervous tissue, they appear in the form of the perivascular Virchow-Robin's spaces which, extend up to the level of arterioles and venules.

The anatomical elements assuring every 6-12 hours the CSF renewal are the choroid plexuses and the arachnoid villi or pacchionian arachnoid granulations.

The choroid plexuses are anatomical vasculo-epithelial structures sited in the first and second lateral ventricles, in the third and in the fourth ventricles, resulted from the coalescence of the pia mater with the ependyma. They are formed in the regions where the choroid webs (the superior *tela choroidea* or *tela choroidea prosencephali* and the inferior *tela choroidea* or *tela choroidea rhombencephali*) come into contact with t*he laminae tecti* of the third and fourth ventricles or with the choroid area at the level of the medial wall of the first and second ventricles.

They have the aspect of thin reddish balls, bulging into the ventricles; their colour being due to their rich vessel supply. On their surface, numerous fringes and villous processes may be seen, which increase up to 200 cm^2 the exchange area, exceeding four times the surface of the whole ventricular system.

The morphofunctional unit of the choroid plexus is the choroid villus, made up of a connective-vascular axis resulted from the evagination of the pia mater and covered with a cell layer originating in the ependyma.

The vessel supply has its origin in the anterior, middle and posterior, choroidal arteries, as well as in some branches of the inferior cerebellar artery, for the choroid plexus of the fourth ventricle. The vascular capillaries form a very rich network.

The cerebral circulation is particularly intense. Thus, although the brain represents about 2.5% of the body weight, 15% of the cardiac output is directed towards this level.

This explains the main role of choroid plexuses in the CSF elaboration and in the removal of noxious substances from the fluid-filled spaces..

The arachnoid villi or arachnoid (Pacchionian) granulations are digitiform or clubbed evaginations of the arachnoid which, after crossing the dura mater, open into the endocranial venous sinuses, predominantly on both sides of the superior longitudinal sinus or into the blood lakes of the diplo.

They were described also at the level of the venous plexuses associated to the roots of the spinal nerves. They have the role to drain the CSF into the venous system.

The CSF elaborated in this way is colorless, as clear as spring water and represents one part of the interstitial fluid, 1.5% of the total body water and 7% of the extracellular fluid of the young adult.

The CSF is separated from the circulating fluid (blood plasma) by a morphofunctional system represented by the choroid epithelium, the external meningoblastic layer of the arachnoid, the epithelium of the arachnoid villi and the endothelium of the vessels that cross the subarachnoid space (D. Chimion).

It prevents the penetration of molecular substances into the fluid-filled spaces, thus making up an obstacle to the diffusion of smaller-sized molecule substances, but permitting the free diffusion of water. In this way, the fluid-filled spaces are largely open towards the extracellular spaces of the central nervous system, which would represent about 17% of the nervous tissue mass.

The fluid-filled spaces and the extracellular space of the CNS form a unique system, as it is demonstrated by the similar composition of the CSF and of the extracellular fluid of the CNS. These two fluids are separated from the intravascular circulating space by the blood-cerebrospinal fluid and blood-brain barriers.

Two movement types are described in the CSF dynamics: a longitudinal movement, from the choroid plexuses to the arachnoid villi and a transverse, transependymal and transspinal movement, by which a continuous exchange between the fluid filled spaces and the extracellular space of the CNS takes place (D. Chimion).

It may be investigated by examining the specimens obtained by lumbar puncture, either by suboccipital or by ventricular puncture, performed only in neurosurgery departments.

Following the puncture, the colour and the pressure are appreciated, and various laboratory (cytological, chemical and bacteriological) examinations of the extracted CSF are carried out.

Its color may be *red* (subarachnoid haemorrhage or haemorrhagic meningitis), *yellow* (medullary compression or ancient subarachnoid haemorrhages), *opalescent* (viral meningitides) or even *turbid* (bacterial meningitides).

The CSF pressure is measured with the Claude manometer through lumbar puncture and is of 30-45 cm (specimens obtained puncture).

An increase in the CSF pressure is recorded in cerebral tumors, subarachnoid haemorrhages, meningitides etc. Moreover, the laboratory CSF investigations may indicate the presence of neoplastic cells, of a positive Bordet-Wassermann seroreaction (diagnosis of syphilis).

Normally, no cells or only very few cells (1-3 lymphocytes/mm^3) are present in the CSF.

Conduction pathways in the central nervous system

The phylogenetic development of the nervous system, of its formerly studied organization levels, led to the appearance of connection routes between the specialized receptors and these structures, on the one hand, and between the nervous structures themselves, called conduction pathways on the other hand.

They are formed on the axonal or dendritic processes of the cells situated in the various regions of the nervous system. We consider that the topographical, integrative description of the conduction pathways, in a separate chapter, after the study of the nervous structures, on the occasion of which also some analytical notions regarding these pathways are given, responds better to the didactic requirements and to the clinical interpretations, although it may be questionable.

We shall describe hereafter the great integrative pathways of the nervous system, with the exception of the connections between the various segments of the encephalon, which we have extensively studied in the respective chapters (the afferents and efferents of the diencephalic nuclei, of the basal nuclei etc), and of the sensory pathways, which will be discussed together with each sense organ in order to make possible a better understanding of each analyzer.

The medullary pathways
Ascending pathways *(Fig.161-169)*

These pathways convey the nerve inflows towards the superior areas of the central nervous system and they are grouped in three categories, according to the nature of the conducted sensibility: proprioceptive, exteroceptive and interoceptive pathways.

Proprioceptive pathways

The receptors are proprioceptors sited in muscles, tendons and joints and they are represented by Golgi's corpuscles, neuromuscular spindles, Pacini's corpuscles and free nerve endings.

It is considered that there are two categories of proprioceptive sensibility: conscious and unconscious.

The unconscious proprioceptive sensibility has its first neuron in the spinal ganglion, from which the fibres penetrate into the spinal cord where they synapse with the second neuron, sited in the posterior horn of the spinal cord *(the Clarke-Stilling's dorsal nucleus for the direct spinocerebellar tract and the intermediomedial nucleus for the indirect spinocerebellar tract)*. From this level, some axons of the cells of the posterior horn course in the lateral fasciculus of the spinal cord, on the same side, forming the direct or posterior spinocerebellar tract *(tractus spinocerebellaris dorsalis)*, called also *Flechsig's tract* which reaches the cerebellum through the inferior cerebellar peduncles, whereas other axons pass into the lateral funiculus on the opposite side, forming the indirect or anterior spinocerebellar or *Gower's tract (tractus spinocerebellaris ventralis)*, which reaches the cerebellum through the superior cerebellar peduncles. Both tracts arrive at the level of *the paleocerebellum*.

The conscious proprioceptive sensibility has the first neuron at the level of the spinal ganglion. The axons of the cells of the spinal ganglion penetrate into the posterior funiculi of the spinal cord (without forming a medullary synapse), making up most of the fibers at this level (we mention that the medial fibers form *the fasciculus gracilis* while the lateral fibers constitute *the fasciculus cuneatus*) and reach the second neuron, situated in the medulla oblongata, at the level of the Goll's and Burdach's nuclei *(gracile and cuneate nuclei)* after which, via the median Reil's ribbon *(lemniscus medialis)*, they reach the third neuron, sited in the thalamus, from which they project into the parietal cortex.

We mention that the medial part of the spinobulbar tract (fasciculus gracilis, Goll's tract) conveys the sensibility of the lower half of the body. It begins in the sacral cord. The first afferent neuron is sited in the spinal ganglion, while the second afferent neuron lies in the clava or tubercle of the gracile nucleus.

The lateral part of the spinobulbar tract (fasciculus cuneatus or Burdach's tract) begins at the third thoracic vertebra and courses upwards, lateral to the medial part (Goll), from which it is separated through the posterior intermediate septum. Into it run the sensitive fibres of the superior half of the body. The first neuron ends in *the cuneate nucleus* of the tubercle of *the cuneate nucleus* in *the medulla oblongata.*

The degenerative and inflammatory processes of *the posterior funiculi* (for example, tabes) cause the nearly complete loss of the sensibility of the lower extremities. The patient's gait becomes unsteady and the ocular control should compensate for the deep sensibility, which is lacking.

With closed eyes or in the dark, during walking or in static position, patients lose their equilibrium. This indicates *a positive Romberg's sign.*

Exteroceptive pathways

The anterior spinothalamic tract *(tractus spinothalamicus ventralis)* also called the spinal lemniscus, which conveys *the protopathic tactile sensibility,* has the first neuron in the spinal ganglion, from which the axons run and synapse with the second neuron, situated in the head of the posterior horn of the spinal cord; from here, the axons of the second neuron pass into the lateral funiculus of the spinal cord, on the same side, and then ascend up to the third neuron, situated in the thalamus, from which they project into the parietal cortex.

The epicritic tactile sensibility follows the route of Goll's and Burdach's tracts *(gracilis and cuneatus),* their relays and topography being the same.

The lateral spinothalamic tract *(tractus spinothalamicus lateralis),* which conveys *the thermal and painful sensibility,* has the first neuron also in the spinal ganglion. The second neuron is sited in the posterior horn of the spinal cord on the same side and the fibers of this neuron pass into the lateral funiculus of the spinal cord, on the opposite side, behind the anterior spinothalamic tract and in front of the crossed pyramidal tract. These fibers reach the third neuron, situated in the thalamus, from which start the cortical projection fibers in the parietal cortex.

Interoceptive pathways

It is admitted that the periependymal grey matter, which has a vegetative role, and in which the interoceptive protoneurons synapse, is constituted of a multitude of sympathetic neurons, whose intrinsic fibers may have an ascending course along several successive segments thus imparting to the vegetative grey matter a conduction function (Laruelle). In this way, the vegetative interoceptive nerve inflows reach the thalamus.

As regards *the visceral pain* it is considered that it follows the route of *the lateral spinothalamic tract,* the inflow transfer from the level of the interoceptive neurons the exteroceptive neurons stopping either at the level of the spinal ganglion or at that of the posterior horn.

In the case of transfer to the spinal ganglion, the Dogiel's cells (basket cells) assume the connection between the intero- and exteroceptive T cells, whereas in the case of transfer into the posterior horn, the connections achieved through the collateral branches of the interoceptive fibers which travel to the head of the posterior horn; this explains also why the visceral pain projects on the integument as follows: heart pain at the C8-T6, dermatomes, gastralgias at T6-T9, intestinal pain at T7-T10, pain in the genital glands and the uterus at T10-T12, renal pain at T11 – L and proctalgias at S2-S4.

Descending pathways
Motor pathways - *(Fig. 165 - 169)*

The pyramidal or corticospinal tract *(tractus corticospinalis)* which conveys the voluntary mobility, presents the first neuron in the frontal cortex, on the Betz's pyramidal cells of the precentral gyrus, from which it descends and enters in the constitution of the internal capsule), then of the brainstem, at the level of the medulla oblongata, about 80% of its constituent fibers intercross in the pyramidal decussation, forming the lateral corticospinal tracts (intercrossed pyramidal bundles). Then these fibres travel in the lateral funiculi of the spinal cord and synapse with the second neuron (neuron) situated in the anterior horn of the spinal cord, at various levels. A proportion of 19% of the fibers do not intercross at this level, but descend through the anterior funiculus of the spinal cord, crossing at the level of each medullary segment, where they synapse with the second neuron of the anterior medullary horn (cells), thus forming the direct pyramidal tracts (the anterior corticospinal tracts).

A small rate, of about 1%, of the pyramidal tract remains uncrossed and is sited in the anterior funiculus of the spinal cord, forming the homolateral direct pyramidal tract.

We mention that a group of fibers of the pyramidal tract courses to the motor nuclei of the cranial nerves, where it synapses with the second neuron; this fibre group forms the geniculate (corticonuclear) tract.

Monakow's or rubrospinal tract *(tractus rubrospinalis)*. It is the most significant pathway of the extrapyramidal system. It contains conduction pathways which assure the automation and coordination of movements. The exclusion of these pathways results in a poverty of movement. The fibres of this pathway arise from the magnocellular part of the red nucleus *(nucleus ruber)* in the cerebral peduncle and decussate on the midline, achieving the Forel's *"anterior tegmental decussation" (decussatio tegmentalis ventralis)*.

The rubroreticulospinal tract (tractus rubrocorticospinalis) passes through *the pontine peduncular tegmentum*, reaches the medulla oblongata, then enters into the lateral funiculus of the spinal cord and terminates in the alpha cells of the anterior horn of the cord.

The vestibulospinal tract *(tractus vestibulospinalis)* derives from the lateral vestibular nucleus *(Deiter's nucleus)* and descends into the anterior funiculus, between the anterior lateral sulcus and the anterior median fissure. It ends in the alpha moto cells in the anterior horn.

The tectospinal tracts or Löwenthal's tract *(tractus tectospinalis)* arises in *the superior colliculus* (reflex optic center) and in *the inferior colliculus* (reflex acoustic center) of *the lamina tecti*. Its fibers intercross completely, forming Meynert's or tegmental decussation *(decussatio tegmentalis dorsalis)*. In the anterior funiculus, the tract courses directly along the borders of the anterior median fissure and end in the motor cells of the anterior horn.

Helweg's olivospinal tract *(tractus olivospinalis)* is a pathway which coordinates the zone of cervical muscles and the movements of the head. This tract has its origin in the olivary nucleus of the medulla oblongata. It is an uncrossed tract, which descends along the external side of the anterior funiculus. This pathway ends in the alpha motor cells of the anterior horn.

The reticulospinal tract *(tractus reticulospinalis)* arises in the reticular substance and contains crossed and uncrossed bundles; it is situated in the anterior funiculus between the anterior lateral sulcus and the anterior median fissure. It ends in the alpha motor cells of the anterior horn. Along this pathways lie many funicular cells which are widely connected with the cells of the anterior horn. In this way are established connections between the various segments of both sides of the spinal cord.

Vegetative pathways

The descending tracts which convey the vegetative inflow from the hypothalamus into the vegetative centers in the mesencephalon, the occipital protuberance and the medulla oblongata continue in the spinal cord under the form of fibers distributed in the ground bundle

Fig. 161 Conscious proprioceptive pathways
(scheme)

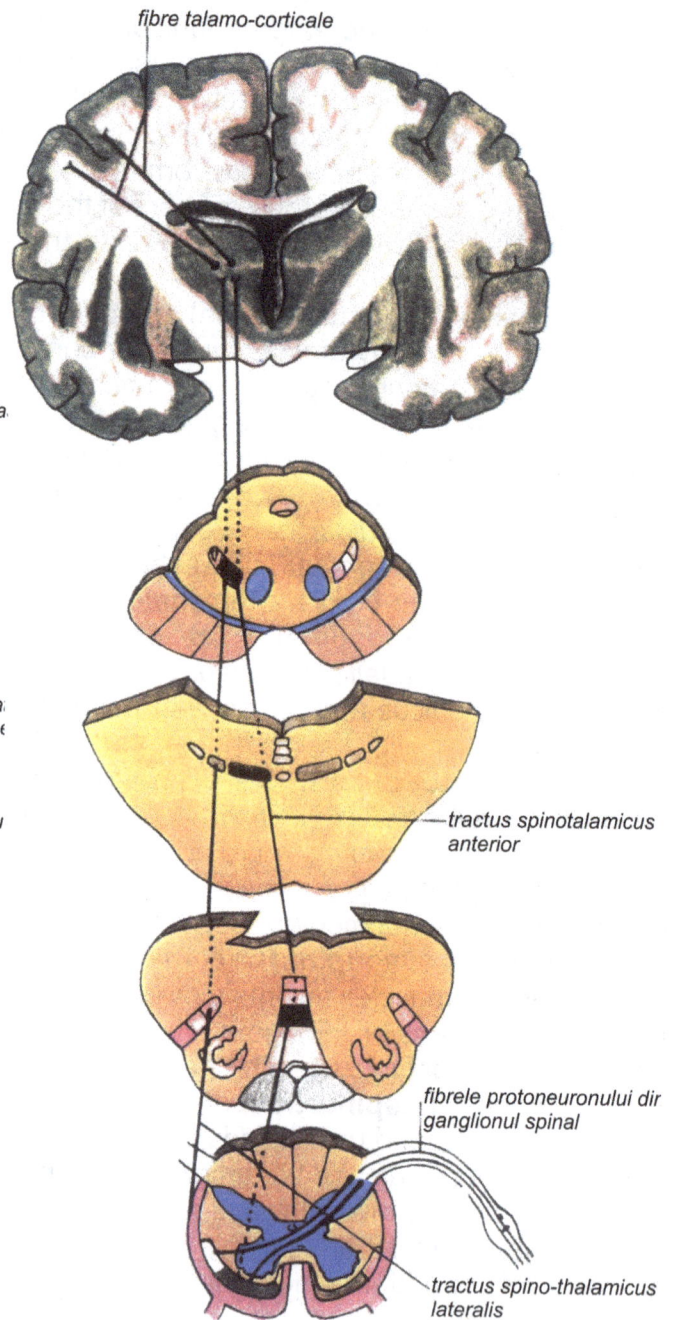

fibre talamo-corticale ale
neuronului terminal

fibre talamo-corticale

lemniscus medialis
(panglica lui Reil sau fa.
bulbo-talamic)

nucleus gracilis et cunea.
(Goll si Burdach, deuto-ne

decusatio lemnisci

protoneuron pseudo-u
din ganglionii spinali

tractus spinotalamicus
anterior

fibrele protoneuronului dir
ganglionul spinal

tractus spino-thalamicus
lateralis

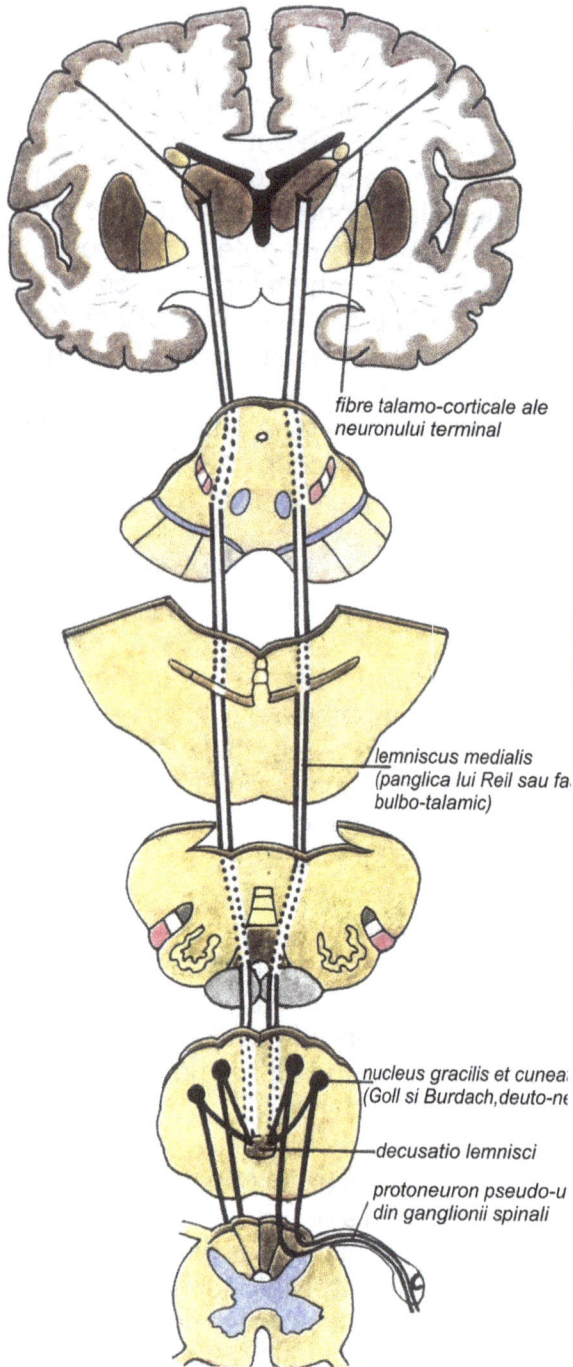

Fig. 162 Thermal, tactile and
painful sensibility pathways
(spinothalamic tracts)

180

lobus parietalis

thalamus
lobus parietalis

mesencephalon

thalamus

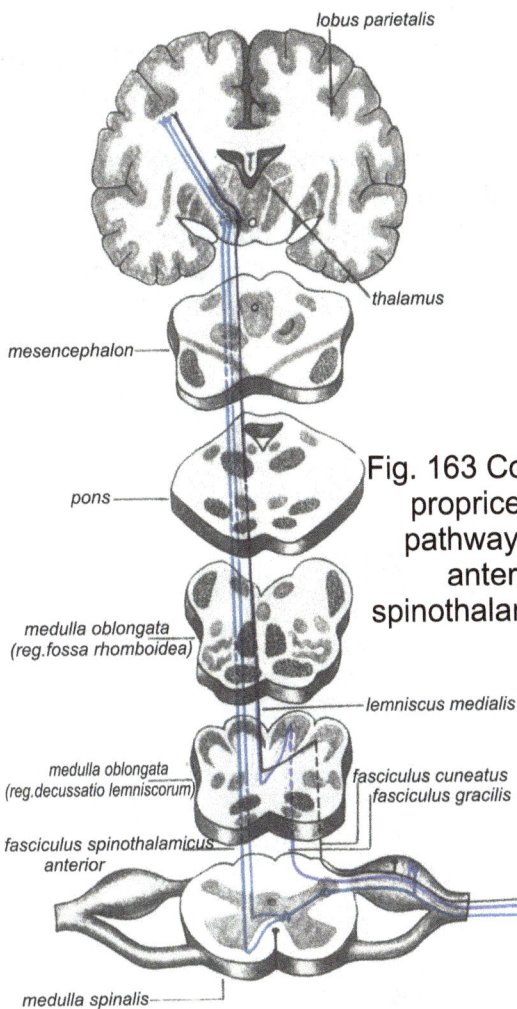

Fig. 163 Conscious propriceptive pathways and anterior spinothalamic tract

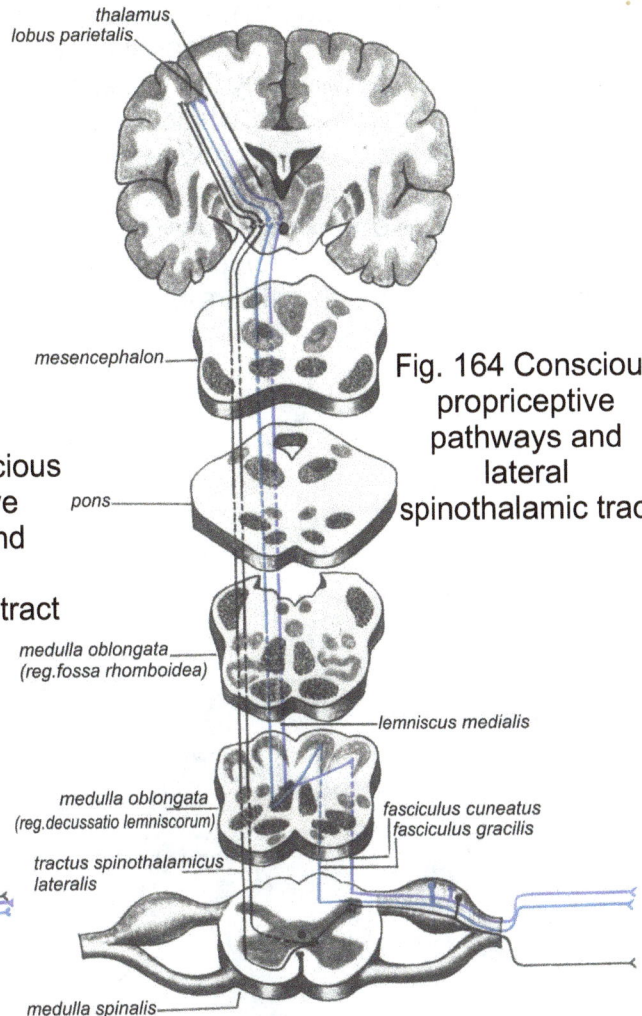

pons

medulla oblongata (reg.fossa rhomboidea)

lemniscus medialis

medulla oblongata (reg.decussatio lemniscorum)

fasciculus cuneatus
fasciculus gracilis

fasciculus spinothalamicus anterior

medulla spinalis

mesencephalon

Fig. 164 Conscious propriceptive pathways and lateral spinothalamic tract

pons

medulla oblongata (reg.fossa rhomboidea)

lemniscus medialis

medulla oblongata (reg.decussatio lemniscorum)

tractus spinothalamicus lateralis

fasciculus cuneatus
fasciculus gracilis

medulla spinalis

and it ends in the lateral horn, thus assuring the coordination of the medullary vegetative centers by the overlying centers.

Brainstem pathways

Situated in continuation of the spinal cord, the brainstem is crossed by all the corticospinal and spinocortical pathways. These pathways, if they are ascending, receive during their passage through the brainstem a group of fibers from the sensitive nuclei of the cranial nerves of the brainstem; if they are descending, various extrapyramidal pathways are associated with them. In addition, association pahtways are also present in the brainstem.

Ascending pathways

1. The pathway of the conscious proprioceptive and epicritic tactile exteroceptive sensibility is represented by *the medial lemniscus* (medial Reil's ribbon), which originates from the gracile and cuneate nuclei, a synapsis site between the proto – and the deutoneuron, and reaches the thalamus. When it passes through the brainstem, it receives fibres from the deutoneurons of the sensory nuclei of cranial nerve (Wrisberg's intermediate, IX, X, and V).

2. The pathways of the exteroceptive thermal painful and protopathic tactile sensibility is represented by *the lateral spinothalamic tracts*, for pain and temperature, and by *the anterior (ventral) spinothalamic pathway* for the crude tactile sensibility.

The anterior (ventral) spinothalamic tract is included in the medial lemniscus, whereas the lateral spinothalamic tract remains in a lateral position and courses with the crossed spinocerebellar and with the rubrospinal tract.

181

3. *The cochlear pathways* is sited in *the lateral lemniscus*; the protoneuron of the pathway ends in the anterior cochlear nucleus, after which the fibers of the deutoneuron form an anterior bundle, respectively the trapezoid body, and then, passing on the opposite side, they form the lateral lemniscus, which reaches the medial geniculate body.

4. *The pathway of the unconscious proprioceptive sensibility* is represented by *the direct (posterior) spinocerebellar or Flechsig's tract*, which crosses only the medulla oblongata and then enters in the constitution of the inferior cerebellar penduncle, and by *the anterior (ventral) crossed spinocerebellar or Gower's tract*, which crossed the medulla oblongata and the pons and enters in the constitution of the superior cerebellar peduncle.

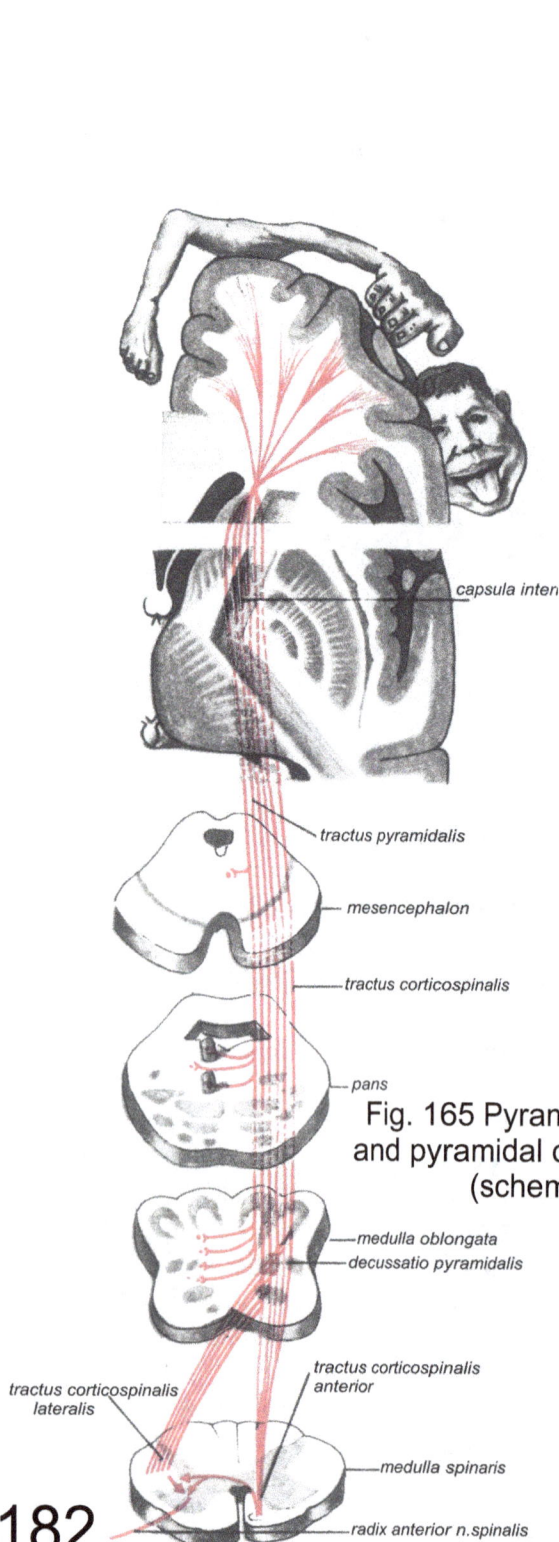

capsula inten

tractus pyramidalis

mesencephalon

tractus corticospinalis

pans

Fig. 165 Pyramidal tracts and pyramidal decussation (scheme)

medulla oblongata
decussatio pyramidalis

tractus corticospinalis lateralis

tractus corticospinalis anterior

medulla spinaris

radix anterior n.spinalis

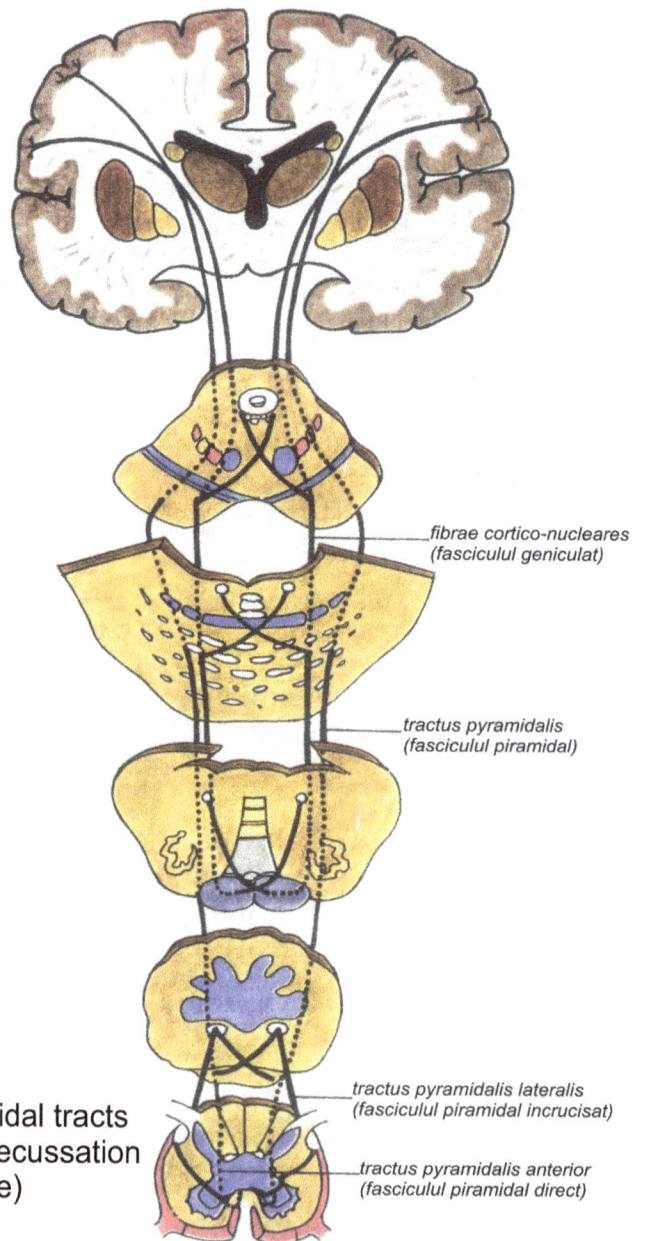

fibrae cortico-nucleares (fasciculul geniculat)

tractus pyramidalis (fasciculul piramidal)

tractus pyramidalis lateralis (fasciculul piramidal incrucisat)

tractus pyramidalis anterior (fasciculul piramidal direct)

Fig. 166 Pyramidal tracts and geniculate fasciculus

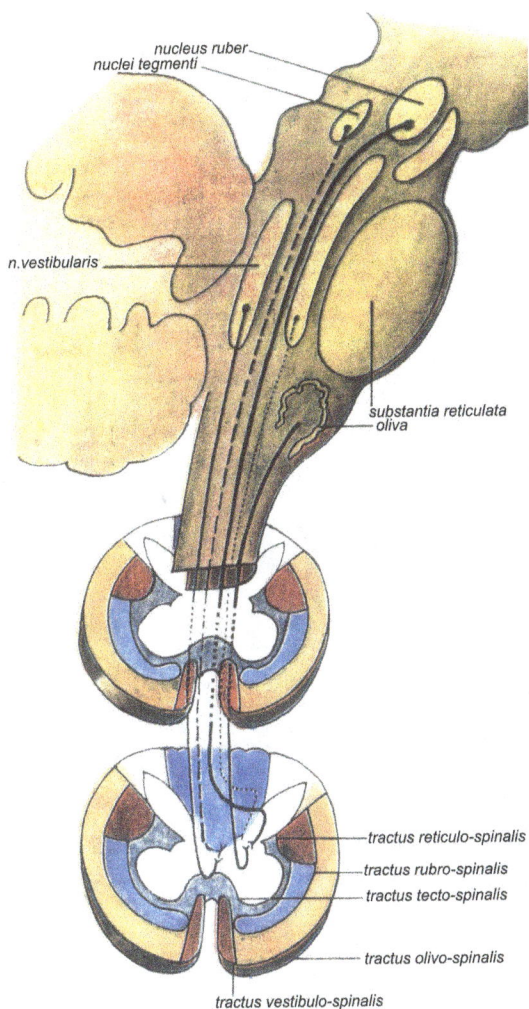

Fig. 167 Intrapyramidal tracts (scheme)

nucleus ruber
nuclei tegmenti
n.vestibularis
substantia reticulata
oliva
tractus reticulo-spinalis
tractus rubro-spinalis
tractus tecto-spinalis
tractus olivo-spinalis
tractus vestibulo-spinalis

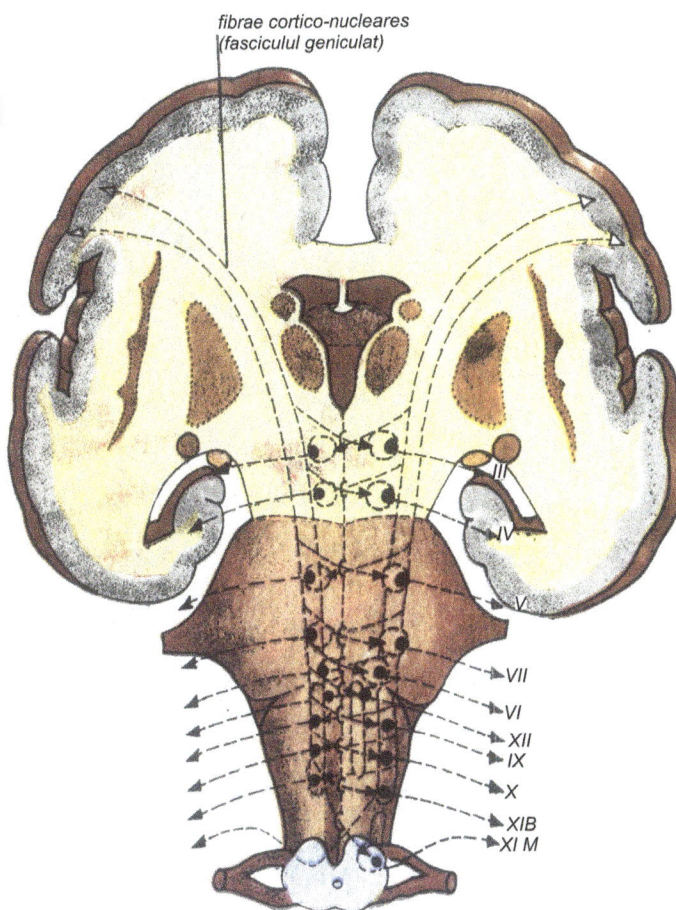

Fig. 168 Corticonuclear tract (scheme)

fibrae cortico-nucleares
(fasciculul geniculat)
III
IV
V
VII
VI
XII
IX
X
XIB
XI M

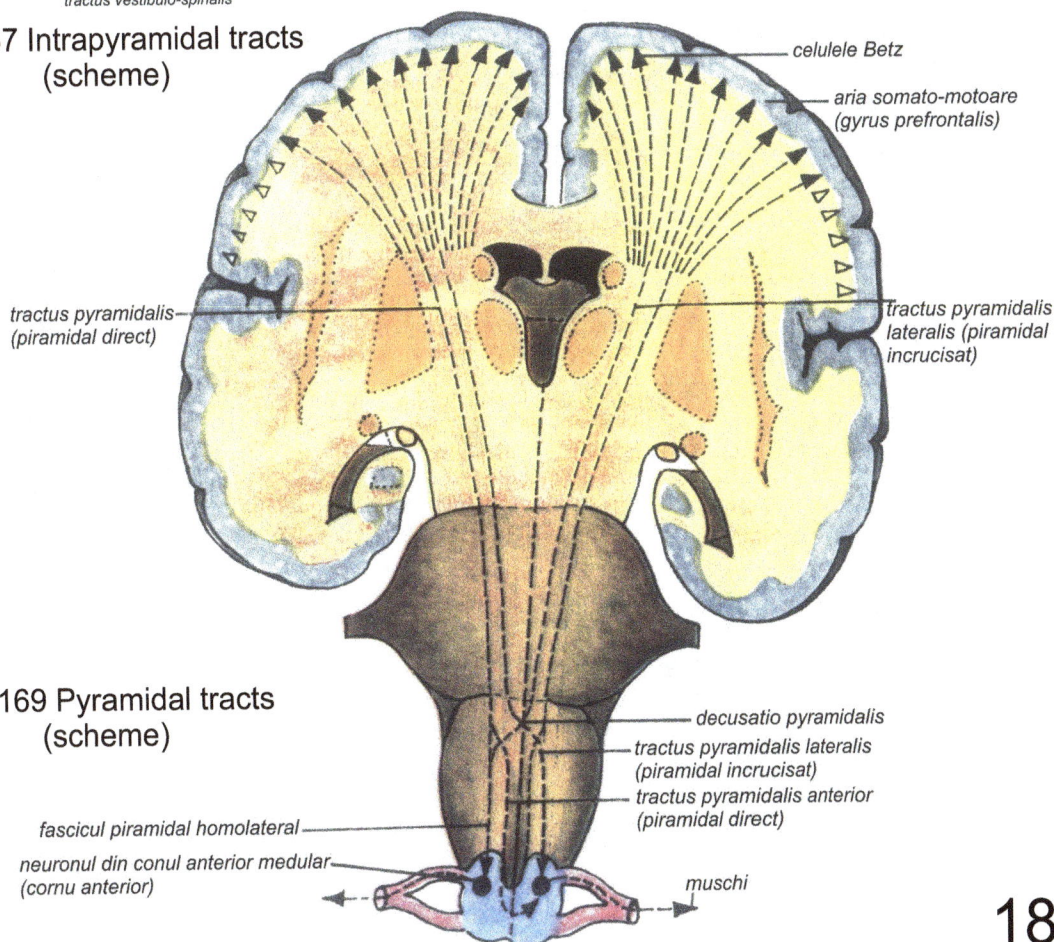

Fig. 169 Pyramidal tracts (scheme)

celulele Betz
aria somato-motoare
(gyrus prefrontalis)
tractus pyramidalis
(piramidal direct)
tractus pyramidalis
lateralis (piramidal
incrucisat)
decusatio pyramidalis
tractus pyramidalis lateralis
(piramidal incrucisat)
tractus pyramidalis anterior
(piramidal direct)
fascicul piramidal homolateral
neuronul din conul anterior medular
(cornu anterior)
muschi

183

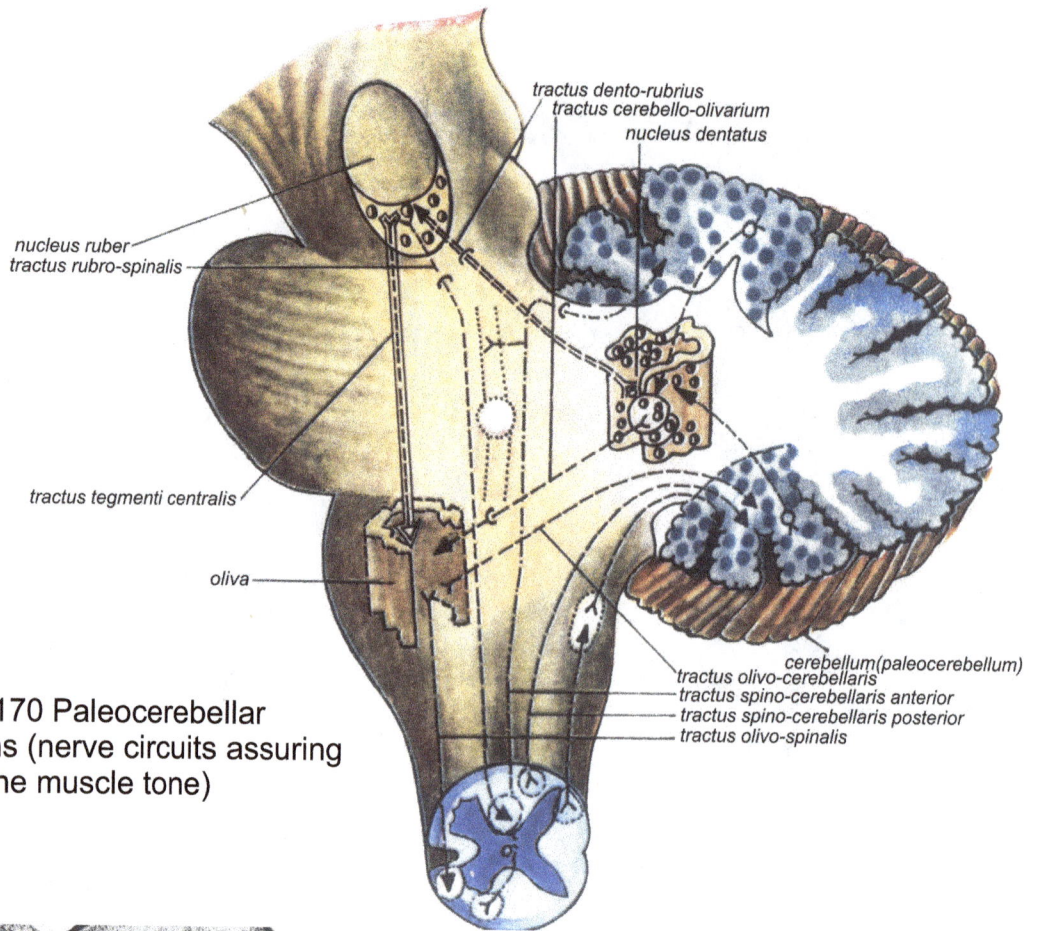

tractus dento-rubrius
tractus cerebello-olivarium
nucleus dentatus

nucleus ruber
tractus rubro-spinalis

tractus tegmenti centralis

oliva

cerebellum(paleocerebellum)
tractus olivo-cerebellaris
tractus spino-cerebellaris anterior
tractus spino-cerebellaris posterior
tractus olivo-spinalis

Fig. 170 Paleocerebellar
connections (nerve circuits assuring
the muscle tone)

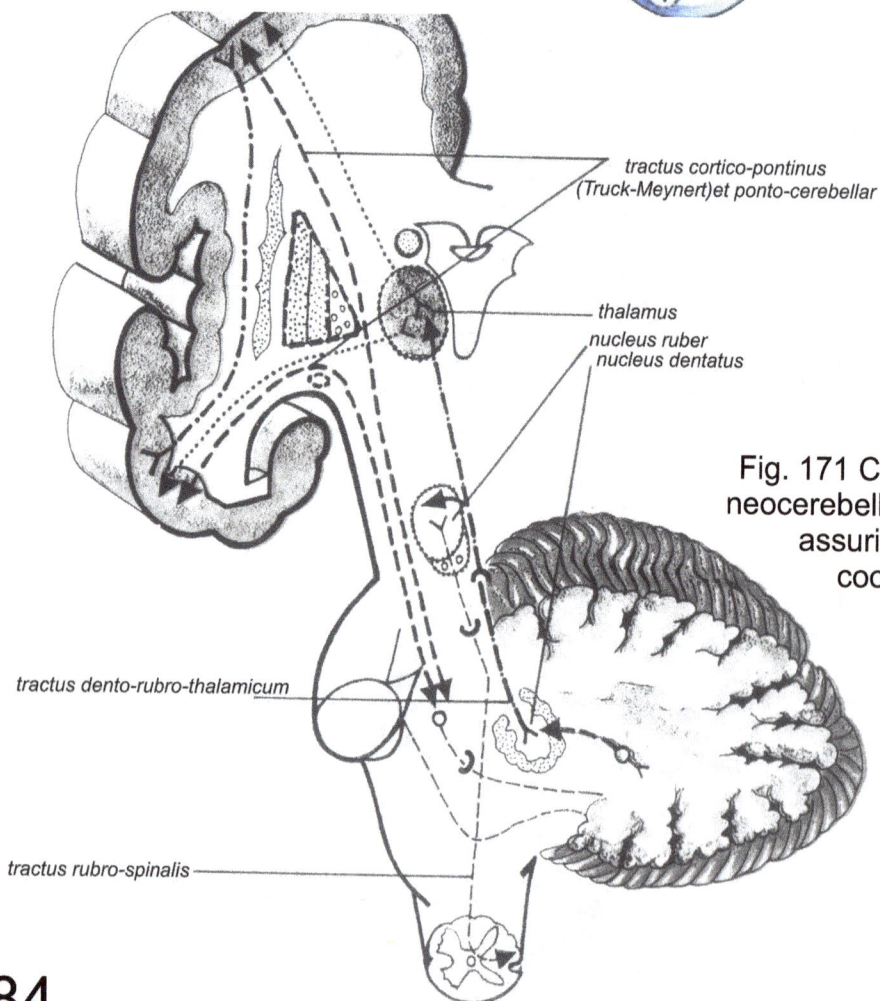

tractus cortico-pontinus
(Truck-Meynert)et ponto-cerebellar

thalamus
nucleus ruber
nucleus dentatus

Fig. 171 Connections of the
neocerebellum (nerve circuits
assuring the motor
coordination)

tractus dento-rubro-thalamicum

tractus rubro-spinalis

184

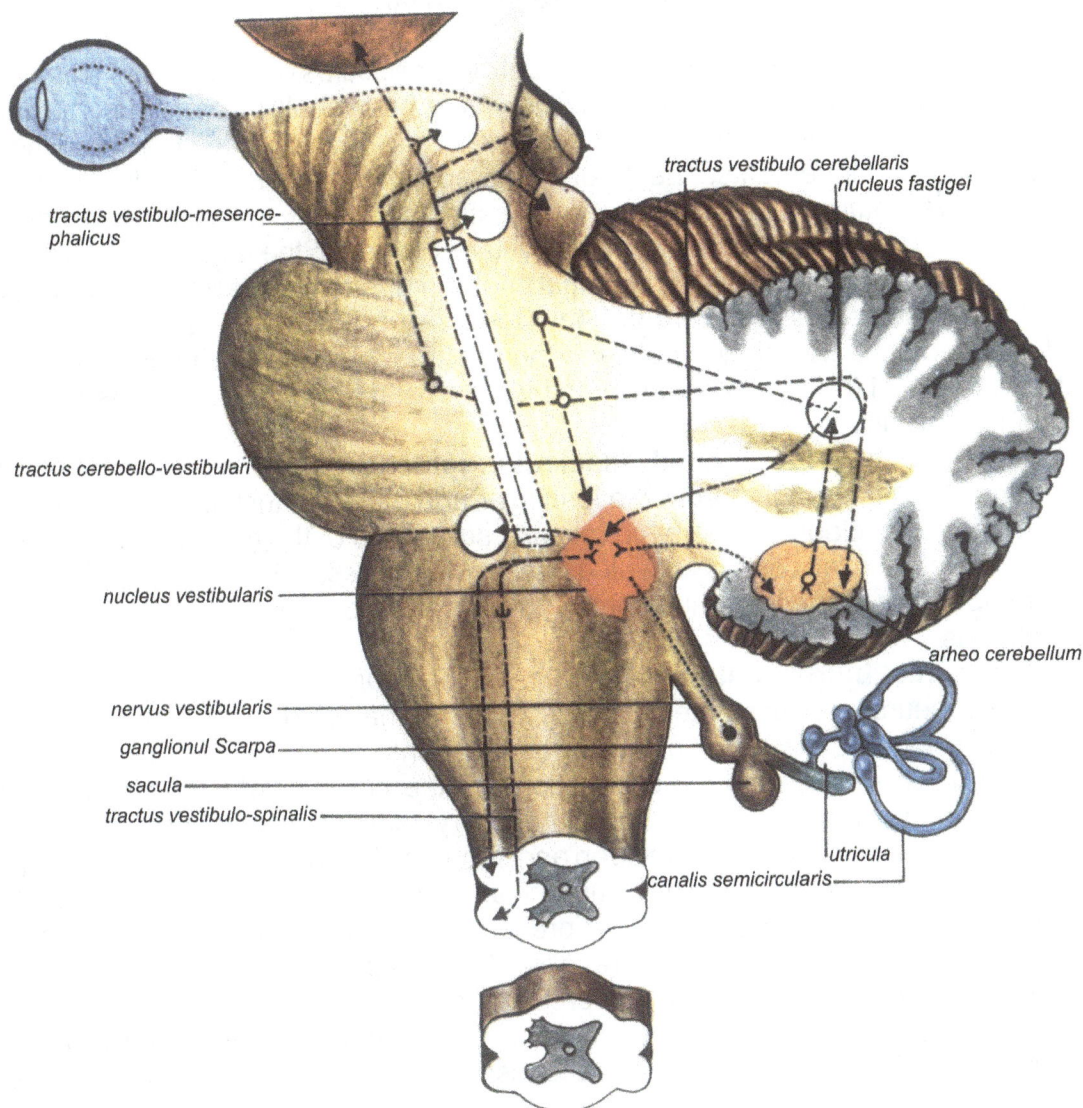

Fig. 172 Connections of the arhicerebellum (nerve circuits assuring the equilibrium)

Descending pathways

1. The pyramidal (corticospinal) tract, together with the fibers destined to the cranial nerves, form the geniculate (corticonuclear) fasciculus.

The geniculate fasciculus, which originates in the base of the ascending frontal gyrus, gives off fibers to the motor nuclei of the cranial nerves in the brainstem, after intercrossing partially, so that these nuclei receive fibers both homo- and heterolaterally, which explains why the facial paralysis is not total in the case of lesion to the motor area at the exclusive level of one cerebral hemisphere. The pyramidal tract, unlike the geniculate fasciculus which decussates at each level of cranial nerve, presents a single decussation (the pyramidal decussation in the medulla oblongata).

2. The extrapyramidal pathway contains the following tracts:
- the parapyramidal tract, topographically interpenetrated with the pyramidal tract;
- the reticulospinal tract;
- the rubrospinal tract;
- the tectospinal tract, topographically interpenetrated with the medial lemniscus;
- the olivospinal tract;
- the nigrospinal tract;
- the corticopontine tracts belonging to the neocerebellar circuits.

185

Associative pathways

1. The medial longitudinal fasciculus *(fasciculus longitudinalis medialis)* is a pathway connecting the nuclei of the motor nerves of the eyeball, the vestibular nuclei and the ambigous nucleus (eleventh cranian nerve); it is involved in the carrying out of oculocephalogyric movements. The group of extrapyramidal fibers enters into this fasciculus through the vestibular nuclei that are interconnected and it gives off also homo – and heterolateral fibers to the other nuclei of the system. However, the fasciculus has connections with the anterior and posterior quadrigeminal bodies (superior, respectively inferior colliculi).

2. The posterior longitudinal fasciculus (of Schütz) *(fasciculus longitudinalis dorsalis)* assures the connection between the posterior nuclei of the hypothalamus and the vegetative nuclei of the third, fifth, seventh bis, ninth and tenth cranial nerves and even with the hypoglossal nucleus.

3. The central tegmental fasciculus *(tractus tegmentalis centralis)* contains descending fibers proceeding from the lenticular and caudate nuclei, from the parvocellular portion of the red nucleus and from the reticular substance of the brainstem, which reach the olive and the cerebellar cortex, as well as ascending fibres, from the olive, coursing to the reticular substance of the thalamus, through which they activate the cerebral cortex and the hypothalamus.

4. The vestibulospinal tract *(tractus vestibulospinalis)*, running from the vestibular nuclei to the neurons of the anterior horns of the spinal cord, has a role in assuring the equilibrium through the regulation of muscle contractions.

In addition, there are also connections of the cerebellum with the brainstem, the cerebellum being in relation with the corticomedullary and medullocortical pathways.

Among these we mention: the deep and superficial arcuate fibers, proceeding from the gracile (Goll's) nucleus, the cuneate (Burdach's) nucleus and the sensory nuclei of the fifth pair and of the solitary tract, which cross previously vestibulocerebellar fibers which do not cross; reticulocerebellar fibers; olivocerebellar fibers; all these reach the cerebellum through the inferior cerebellar peduncles; pontocerebellar fibers, which arise from the pontine nuclei and reach the cerebellum through the middle cerebellar peduncles; the cerebellorubral fibers which cross forming the Wernekinck's decussation reach through the superior cerebellar peduncles the red nucleus.

The cerebellar pathways

They may be systematized according to the three portions of the cerebellum, respectively the archicerebellum, the paleocerebellum and the neocerebellum (Fig.170-172).

The pathways of the archicerebellum, represented by the flocconodular lobe, are involved in the equilibrium function. They are the following:

1. The vestibulocerebellar tract (sensory cerebellar pathway) connects the organ of equilibrium to the cerebellum. The cells of the direct pathways and those of the first neuron of the indirect pathway lie in the vestibular nucleus located in the fundus of the internal acoustic meatus. They are bipolar ganglionic cells, whose peripheral projection proceeds from the semicircular ducts, the saccule and the utricle. The two cerebellar pathways are the following:

a) *The direct sensory cerebellar pathway*, runs from the vestibular (Scarpa's) ganglion into the brainstem, sending only collaterals to the bulbopontine vestibular nucleus. Its crossed and uncrossed fibers reach through the inferior cerebellar peduncle, the nodule and the flocculus, without synapsing in the vestibular nuclei.

b) *The indirect sensory cerebellar pathway*. The first neuron is sited in the vestibular ganglion of the vestibular part of the statocoustic (vestibulocochlear) nerve and its fibres reach the bulbopontine vestibular nucleus. The second neuron proceeds from this level and its crossed and uncrossed fibers course through the inferior cerebellar peduncle, reaching the nuclei of the cerebellar cortex (at the level of the flocculus and of the nodule).

186

2. The cerebellovestibular tract, called also unciform fasciculus of Russell or hook bundle of Russell. The first neuron courses from the nodular and floccular cortex to the fastigial nucleus (flocculofastigial neuron). The second neuron runs (crossed and uncrossed), from the fastigial nucleus (fastigiovestibular neuron), through the inferior cerebellar peduncle, to the bulbopontine vestibular nucleus, from which the vestibulospinal fasciculi continue their route directly and crossed.

The pathways of the paleocerebellum, a more recent structure involved mainly in the muscle tone control and made up of the lingula, the central lobule, *the culmen, the uvula* and the pyramid situated on *the vermis* and on the anterior and amygdalian lobe, on the cerebel hemispheres, are the following:

- *afferent pathways*, represented by the direct and crossed spinocerebellar tracts, with which are associated fibers from the level of the gracile and cuneate nuclei, of the fifth pair of cranial nerves in the brainstem, as well as fibers from the reticular nuclei in the medulla oblongata and from the bulbar olive;

- *efferent pathways*, represented by the cerebellodentorubral tract, at the level of the red nucleus, achieving further connections, through the rubrospinal tract, with the nuclei of the cranial nerves and with the cells of the anterior horn of the spinal cord and, through the rubrothalamic fibres, with the thalamus.

The pathways of the neocerebellum, the phylogenetically most recent cerebellar structure, play a role in the fine regulation of movements and are represented by:

- *afferent pathways*, issued from the cortex, respectively the frontopontine and parieto-occipitopontine tracts and

- *efferent pathways*, cerebello (nuclei of the neocerebellar cortex) – dento (*dentate nucleus* – possibly as relay) – thalamo (latero-anterior-intermediate nucleus) – cortical (in the areas from which the afferents start towards the cerebellum).

Beside this corticocortical circuit, there is also a shorter, subcortical circuit, which courses from the thalamus to the caudate nucleus, then to the red nucleus – the magnocellular portion -, from which, through the reticular substance, it reaches the anterior horns of the spinal cord, regulating the muscular contractions and assuring their necessary precision and accuracy.

The vegetative (autonomic) nervous system

The nervous system makes up a unitary whole; its clear-cut separation into two systems, ecotropic and idiotropic, very intimately intricate, is arbitrary and dictated by didactic consideration. Actually, these two systems represent the simultaneous aspects of the apparatus which harmonizes the organism with the external environment.

Moreover, the two subdivisions of t*he vegetative nervous system, the ortosympathetic* or *sympathetic* subdivision, which is the commander and regulator of the metabolism, the ergotropic, dynamizing group. And *the parasympathetic* subdivision, which is the trophotropic, endophylactic, anabolizing – term used by W. Hess – group are in a close correlation with the vital manifestations occurring between activity (wear of the living matter) and rest (its regeneration).

The activity of the orthosympathetic system is predominant during an effort or in the invasion stage of disease, whereas the parasympathetic system intereses especially during sleep and rest as well as in the convalescence stage of a disease. This functional differention is closely connected with the rhythmicity and polarity of life, a periodicity, conditioned by the diurnal cosmic rhythm on the earth surface, which can be maintained only by a flowing equilibrium, evolving between wear and restoration.

On the functional plan, these two antagonistic systems (where the orthosympathetic systems act as an accelerator, the parasympathetic plays the role of a moderator and vice versa) function „*like the reins with which the coachman changes the speed of the animal or like the pedals of a piano by means of which the musician strengthens or muffles its sounds*".

187

The basic functions of the viscera (organs of metabolism) are also assured by the intramural apparatuses, developed in the walls of hollow organs or in parenchymatous. Only these apparatuses deserve the name *"autonomic nervous system"*, by which Langley has initially the designated the whole vegetative nervous system. The ortho – and parasympathetic innervation properties exerts only a diminishing or strengthening influence of these. This double action permanently exerted upon the organs results in an state of equilibrium, a physiological constant, the vegetative nervous tonus, which, like a swinging pendulum, oscillates between two extreme limits. In this way, the reaction capacity of the whole organism adapts itself to the variable excitations of the external environment by protective reflexes proper to the living matter.

The same tonus directs also the energy balance of the organism, which from this viewpoint is nothing else than an energy transforming machine. Therefore, involvement is in the state of disease, where affected individual has to fight against the aggression of the pathogens, the organism responds by a general effort under the control of the orthosympathetic system.

The following parts may by distinguished in the organization scheme of the vegetative nervous system.

1. The intramural apparatuses assure the autonomatism of organs, hence their basic properties, even when all their nervous connections are intercepted towards the neuroaxis. These apparatuses make up the intramural system (Daignel Lavastine's metasympathetic system) and they are scattered in all the organs; in some of them (glands) they are represented by isolated or grouped ganglionic cells, with a rich nervous network. Although it has its own activity, the intramural system, designated by the Anglo-Saxon school also under the name of *"enteric system"*, is positively or negatively influenced by the sympathetic system; this is the true autonomic system, a name especially by the Anglo-Saxon school for the whole vegetative nervous system, an improper term, introduced nevertheless by Langley into the usage, into the tradition of this school.

2. The vegetative system is the regulating element of the intramural system and consists of two subgroups: the orthosympathetic (neosympathetic) and the parasympathetic (paleosympathetic) subgroups, according to the terminology of Rousey and Mosinger.

These subdivision have a various extent along the neuroaxis, a different histotope and morphological characters which may be correlated with their physiological properties.

The sympathetic (orthosympathetic) system
(Sympathicum)

The central portion of the sympathic system lies in the intermediolateral nucleus of the spinal cord, at the (C8) T1-L3 level. This nucleus contains cells of the first efferent neuron and the endings of the afferent neuron in the viscera.

The peripheral portions of the sympathetic system contain two kinds of ganglia, respectively those which enter in the constitution of the sympathetic trunk and the prevertebral ganglia, situated on the bodies of the lumbar *vertebrae (coeliac ganglia,* superior and inferior mesenteric ganglia).

The sympathetic nervous system *(systema nervosum sympathicum)*; which is the largest subdivision of the vegetative nervous system, comprises two ganglionated sympathetic trunks *(ganglia trunci sympathici),* the branches between the ganglia *(rami interganglionares),* the plexuses *(plexus sympathici)* and their auxiliary ganglia *(ganglia plexus sympathicorum).*

Within the sympathetic system are located three groups of ganglia, which, according to their site, are classified as follows: a) latero-vertebral ganglia (which form the sympathetic chains); prevertebral or splanchnic ganglia (which lie in the various plexuses from which the nerves of organs arise); the intramural or local ganglia (and plexuses) which are situated in the walls of hollow organs.

188

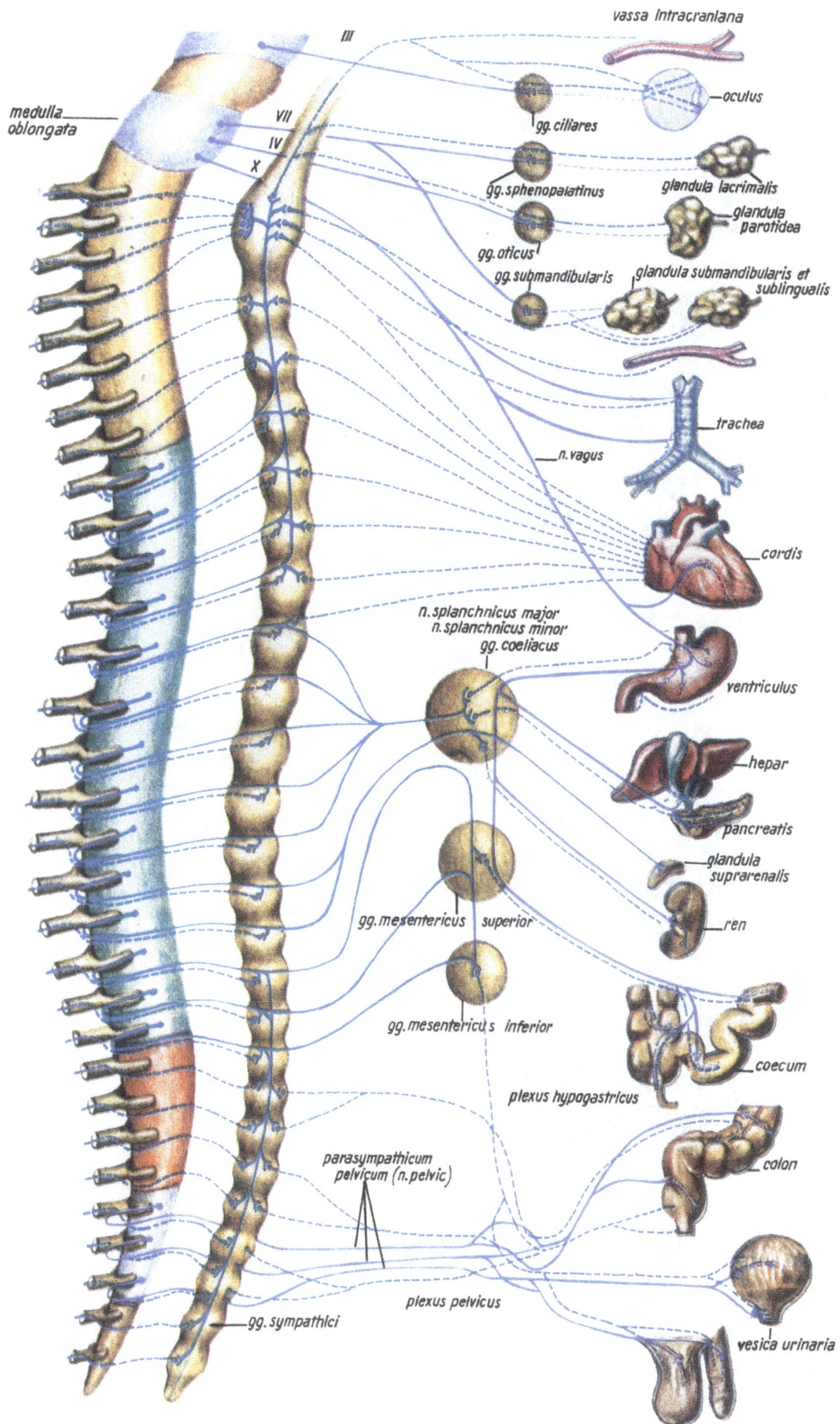

vassa intracraniana

medulla oblongata

III

VII

IV

X

oculus

gg. ciliares

glandula lacrimalis

gg. sphenopalatinus

glandula parotidea

gg. oticus

gg. submandibularis

glandula submandibularis et sublingualis

trachea

n. vagus

cordis

n. splanchnicus major
n. splanchnicus minor
gg. coeliacus

ventriculus

hepar

pancreatis

glandula suprarenalis

gg. mesentericus superior

ren

gg. mesentericus inferior

plexus hypogastricus

coecum

parasympathicum pelvicum (n. pelvic)

colon

plexus pelvicus

vesica urinaria

gg. sympathici

Fig. 173 Vegetative nervous system (overall scheme)

189

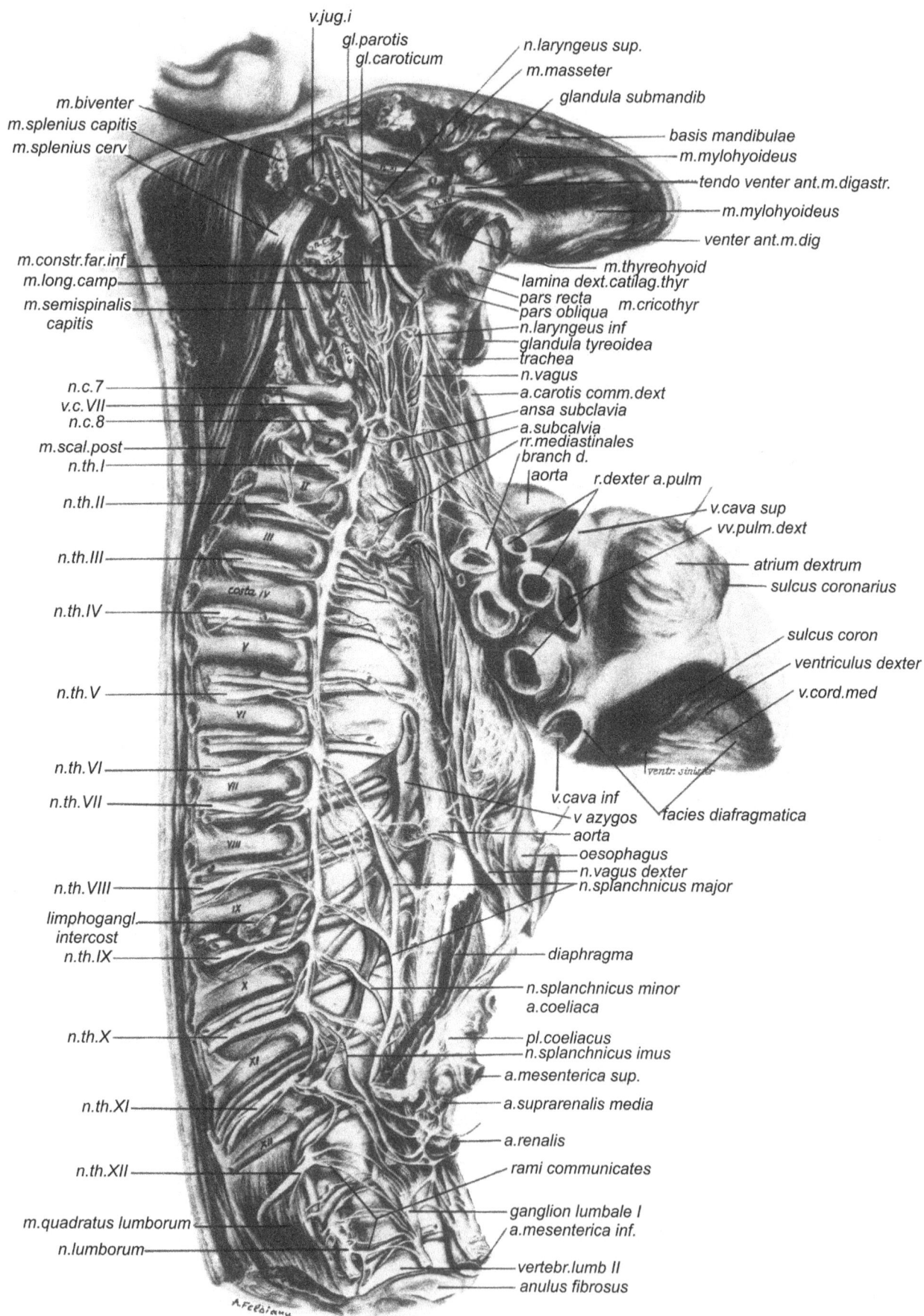

Fig. 174 Vegetative nervous system
(after Fr. I. Rainer)(overall view)

It is nowadays known that the ganglia which are connected to the cranial nerves (the ciliary ganglion, the sphenopalatine ganglion etc), although considered in the past as connected to the sympathetic system, belong to the parasympathetic system. Their connections should be the following: the ciliary ganglion to the oculomotor nerve, the sphenopalatine ganglion to the facial nerve; the submaxillary ganglion (and, when it exists, the sublingual ganglion) to Wrisberg's intermediate nerve; the otic ganglion to the glossopharyngeal nerve.

The sympathetic trunk *(truncus sympathicus)* is made up of a chain of 22-25 ganglia, united with each other by *the interganglionic rami*. It begins with the superior cervical ganglion below the base of the skull and ends with the coccygeal unpaired ganglion (impar ganglion) at the first coccygeal vertebra. One distinguishes the cephalic (cranial) cervical, thoracic, abdominal and pelvic parts of the sympathetic trunk. In the thoracic part and in the abdominal part, the sympathetic trunk is segmented in connection with the spinal nerves through the communicating branches *(rami communicantes)*.

The preganglionic fibers of the first neuron, the white communicating rami which emerge through the anterior motor roots, end in the ganglion of the sympathetic trunk. Here occurs the synapse (between the first and the second neuron of the sympathetic pathway) and it is also here that begins arising from the cells of the ganglion, the second neuron, which is associated with a spinal nerve through the postganglionic branches – grey communicating rami *(rami communicantes grisei)* -, giving off fibres for vassoconstriction, for the secretion of sudoriferous glands and for the erector muscles of hairs *(musculi erectores pilorum)*. The corresponding fibres for those parts of the body in which segmental sympathetic centers do not exist (head, neck and pelvis) have their origin in the superior or inferior sympathetic trunk and reach as *plexus sympathici*, via the blood vessels, the region, which they supply.

Inside the neuraxis may be systematized: the vasomotor, pilomotor and sudoral (T1-T12) centers, the ciliospinal (Budge's centre) or pupillodilator (C8-T3) center, the cardio-accelerator (T1-T4) centers, the intestin inhibitory (T6-L1) centers, the inhibitory zone of the vesicospinal center (L1-L3), the inhibitory zone of the anospinal centre (L2-L4) and the ejaculatory zone of the genitospinal center (L4-L5).

The cephalic part *(pars cephalica)* of the sympathetic system begins on each side as the internal carotid nerve, which is continued upwards, from the superior cervical ganglion of sympathetic trunk. This nerve ascends beside the internal carotid artery, entering the carotid canal in the temporal bone, and divides into two branches:

-the lateral branch, which forms the lateral part of the internal carotid plexus and

-the medial branch, which forms the medial part of the internal carotid plexus *(cavernous plexus)*.

The internal carotid plexus (plexus caroticus internus) surrounds the carotic artery and, sometimes, contains a carotid ganglion. The plexus has connections with the sphenopalatine ganglion, with the tympanic branch of the glossopharyngeal nerve and with the ciliary ganglion.

The communication with the sphenopalatine ganglion is affected by a branch named the deep petrosal, which unites with the superficial greater petrosal nerve to form the nerve of the pterygoid canal, that courses to the sphenopalatine (pterygopalatine) ganglion. The communication with the tympanic branch of the glossopharyngeal nerve is achieved by the superior and inferior caroticotympanic nerves.

The medial branch of the internal carotid plexus (cavernous plexus) supplies branches to the internal carotid artery and anastomoses with the oculomotor, trochlear and abducens nerves, as well as with the ciliary ganglion.

The fibers to the ciliary ganglion pass through the ciliary ganglion without being interrupted and course in the short ciliary nerves, to be distributed to the dilatory muscle of the pupillae.

The terminal fibers of the internal carotid plexus are prolonged plexuses around the anterior and middle cerebral arteries and around ophthalmic arteries.

The cervical part *(pars cervicalis)* of the sympathetic system is situated in front of the transverse process of the cervical *vertebrae*, on the long muscle of the head *(longus capitis musculus)* and on the long muscle of the neck *(longus coli musculus)*, below the prevertebral lamina of the cervical fascia and is fused with it. The cervical portion is situated in front of the vertebral artery and is separated from the deep cervical fascia through the common carotid artery and the vagus nerve. The cervical part of the sympathetic trunk is crossed either anteriorly or posteriorly by the highest point of the arch of the inferior thyroid artery. The cervical part consists of three ganglia, from which are sent grey communicating rami to the cervical nerves.

The superior cervical ganglion *(gl. Cervicalis superius)* has the shape of an elongated oval and is situated at the level of the second and third cervical vertebrae. In this ganglion occurs the connection between the first and the second neuron to the following efferent fibres:

1) C8-T1 of the ciliospinal center, to the dilatory muscle of the pupillae, the orbital muscle and the tarsal muscles of the eyelids.

2) C8, T1, T2 and T3 give off fibers to blood vessels, sudoriferous (sweat) glands *erector pilorum capitis muscles* and of the neck, as well as to the viscera of the neck.

Among the branches of the superior cervical ganglia, the most important are mentioned. From the superior extremity of the ganglion arise three nerves: the jugular nerve, the external carotid nerve and the internal carotid nerve. From the inferior extremity start the laryngopharyngeal branch and the superior cervical cardiac nerve.

The jugular nerve runs into the superior ganglion of the vagus nerve and into the inferior ganglion of the glossopharyngeal nerve. Its fibers pass through these ganglia without any connection and continue their course with the branches of the respective nerve.

The external carotid nerves form the external carotid plexus, which runs with the artery of the same name and its branches to the external surface of the head. Special branches of the external carotid plexus are:

- the sympathetic branch to the submandibular gland, it derives from the plexus of the facial artery, passes through the mandibular ganglion and reaches the submandibular and sublingual glands;

- the sympathetic branch to the otic ganglion; it derives from the plexus of the middle meningeal artery and runs, through the auriculotemporal nerve, to the parotid gland.

The laryngopharyngeal rami course downwards to the larynx and the pharyngeal plexus.

The superior cervical cardiac nerve courses on the right side of the brachiocephalic trunk and on the left side of the common carotid artery up to the cardiac plexus.

The middle cervical ganglion *(ganglion cervicale medium)* is situated on the sixth cervical vertebra, at the level of the inferior thyroid artery. It is of very small size and may be sometime absent; it rend as a more important branch the middle cervical cardiac nerve *(nervus cardiacus cervicalis medius)* to the cardiac plexus.

The inferior cervical ganglion *(ganglion cervicale inferius)* is situated on the transverse process of the seventh cervical vertebra and on the neck of the first rib; it is usually fused (75%) with the first thoracic ganglion, forming the stellate ganglion *(ganglion stellatum)*. The latter is situated between the transverse process of the seventh cervical vertebra and the head of the first rib, behind the subclavian and the vertebral arteries.

The ganglionic branch between the middle cervical ganglion and the stellate ganglion is sited on the right side posteriorly and on the left side anteriorly to the subclavian artery and is embraced like a loop by this vessel – the subclavian loop *(ansa subclavia)*. In this ganglion has its origin the inferior cervical cardiac nerve, which unites with the cardiac plexus. The cardiac nerves of both sides ramify in the shape of a network, in which cardiac branches of the vagus nerve penetrate.

Thickenings may be observed at the level of blood vessels in the coronary sulcus (coronary plexus). The coronary plexus receives afferent fibers from T3 and T4 and the efferent fibers reach T1, T2, T3, and T4.

192

The stellate (cervicothoracic) ganglion is significant connection center between the first and the second neuron. The distribution of the fibers or segments is the following:

C8-T1(T2) – at the level of the head and of the neck – assure the vasoconstriction, the secretion of the sudoriferous glands and the function of the erector muscles of the hair.

T2-T4 are distributed to the bronchi and lungs;

T3-T4 run to the heart.

The connection with the second neuron may possibly also occur in the ganglion of the sympathetic trunk. Therefore, the stellate ganglion has also a great clinical significance, as it is the crossing of the sympathetic pathways to the head, the neck and the heart.

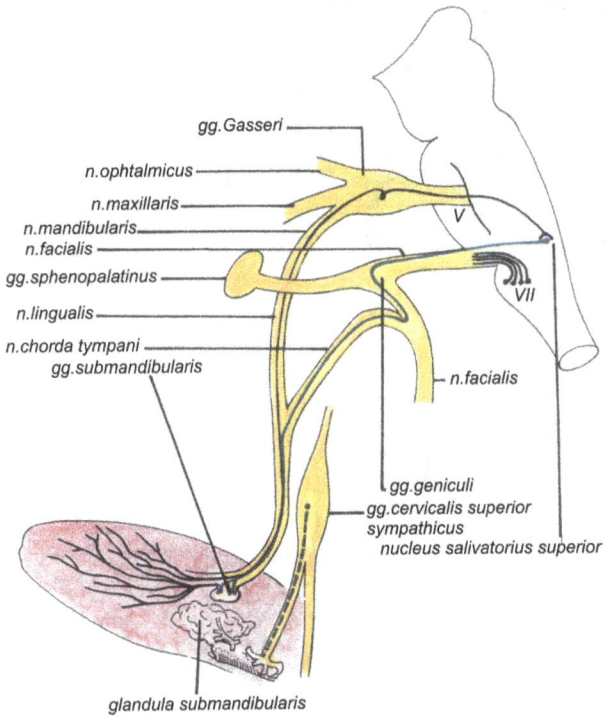

Fig. 175 Parasympathetic pathways of the submandibular gland

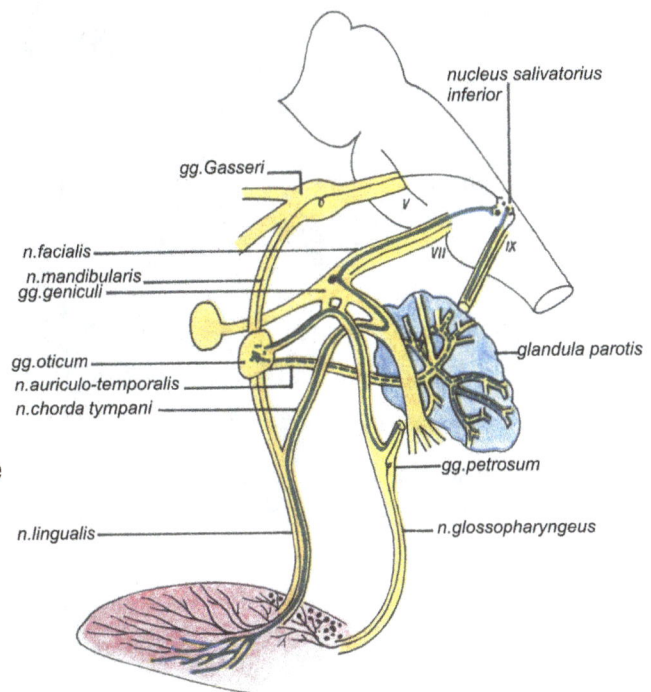

gg.Gasseri
n.ophtalmicus
n.maxillaris
n.mandibularis
n.facialis
gg.sphenopalatinus
n.lingualis
n.chorda tympani
gg.submandibularis
n.facialis
V
VII
gg.geniculi
gg.cervicalis superior sympathicus
nucleus salivatorius superior
glandula submandibularis

nucleus salivatorius inferior
gg.Gasseri
n.facialis
n.mandibularis
gg.geniculi
gg.oticum
n.auriculo-temporalis
n.chorda tympani
n.lingualis
V
VII
IX
glandula parotis
gg.petrosum
n.glossopharyngeus

Fig. 176 Parasympathetic pathways of the parotid gland

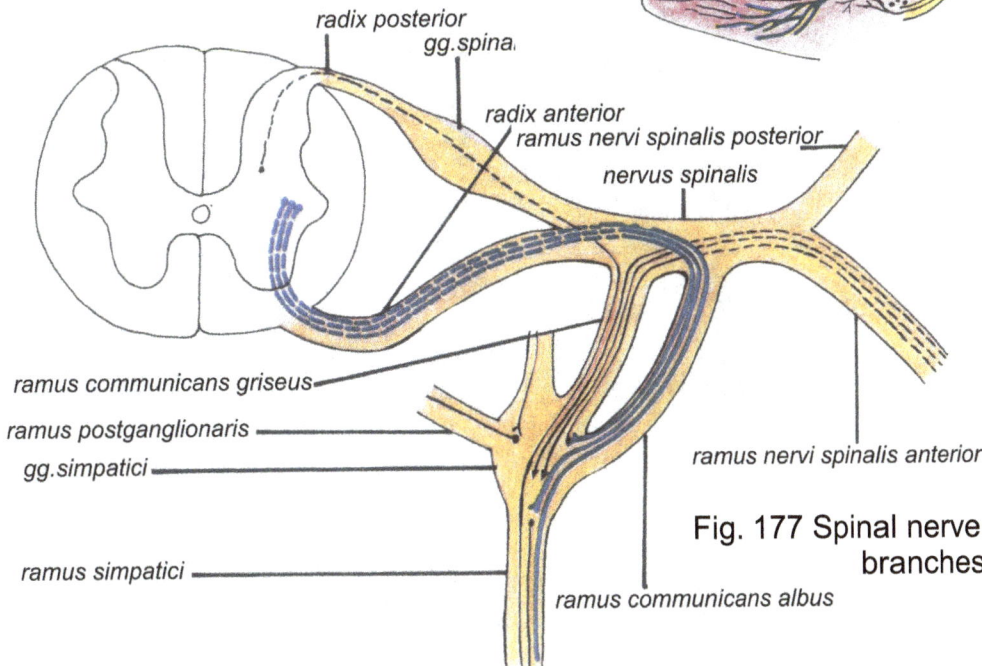

radix posterior
gg.spinal
radix anterior
ramus nervi spinalis posterior
nervus spinalis
ramus communicans griseus
ramus postganglionaris
gg.simpatici
ramus nervi spinalis anterior
ramus simpatici
ramus communicans albus

Fig. 177 Spinal nerve – vegetative branches

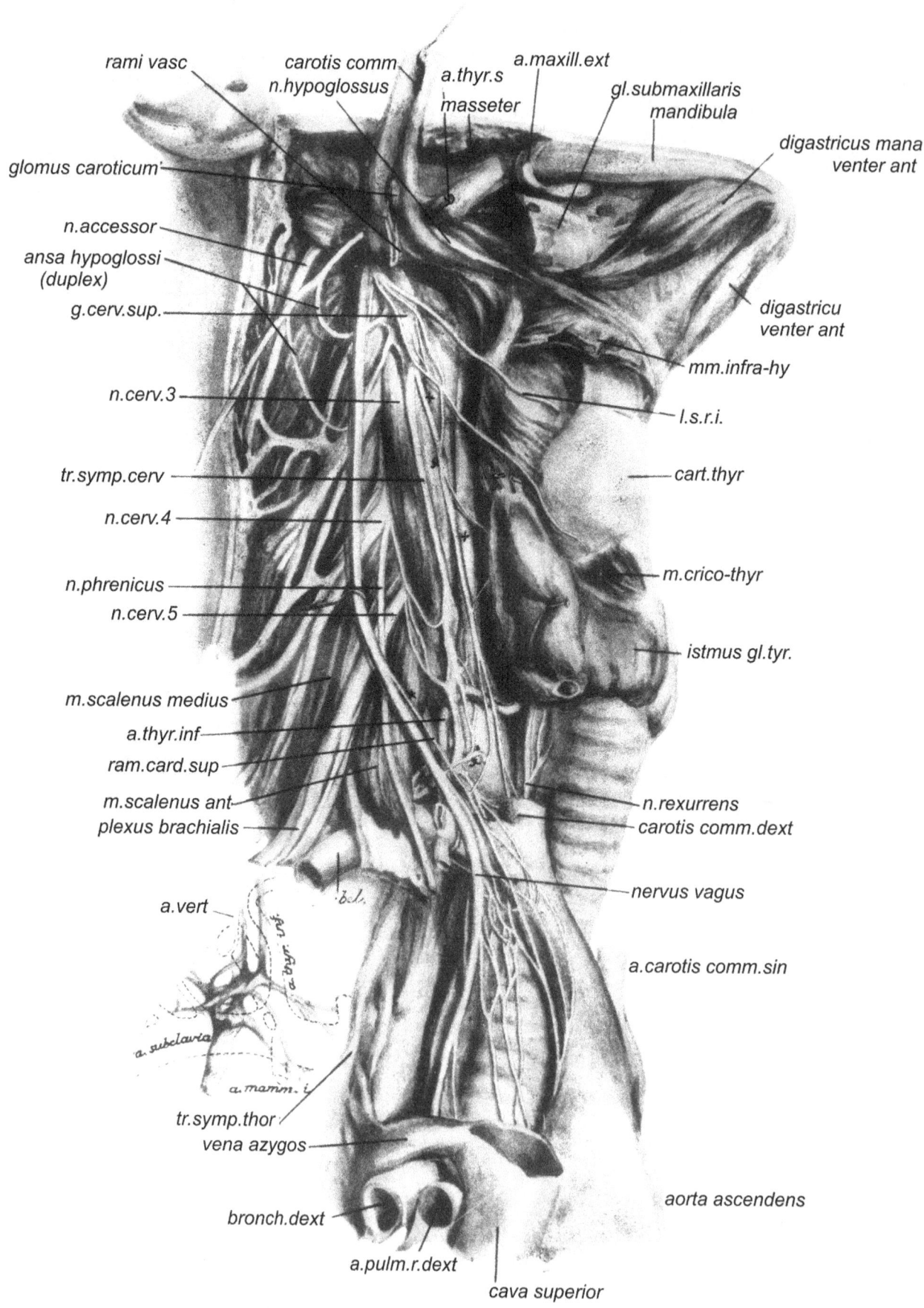

rami vasc

carotis comm
n.hypoglossus

a.thyr.s
masseter

a.maxill.ext

gl.submaxillaris
mandibula

digastricus mana
venter ant

glomus caroticum

n.accessor

ansa hypoglossi
(duplex)

g.cerv.sup.

n.cerv.3

tr.symp.cerv

n.cerv.4

n.phrenicus

n.cerv.5

m.scalenus medius

a.thyr.inf

ram.card.sup

m.scalenus ant

plexus brachialis

a.vert

tr.symp.thor

vena azygos

bronch.dext

a.pulm.r.dext

digastricu
venter ant

mm.infra-hy

l.s.r.i.

cart.thyr

m.crico-thyr

istmus gl.tyr.

n.rexurrens

carotis comm.dext

nervus vagus

a.carotis comm.sin

aorta ascendens

cava superior

a.iug.inf

a.subclavia

a.mamm. i

Fig. 178 Cervical sympathetic trunk
(after Fr. I. Rainer)

194

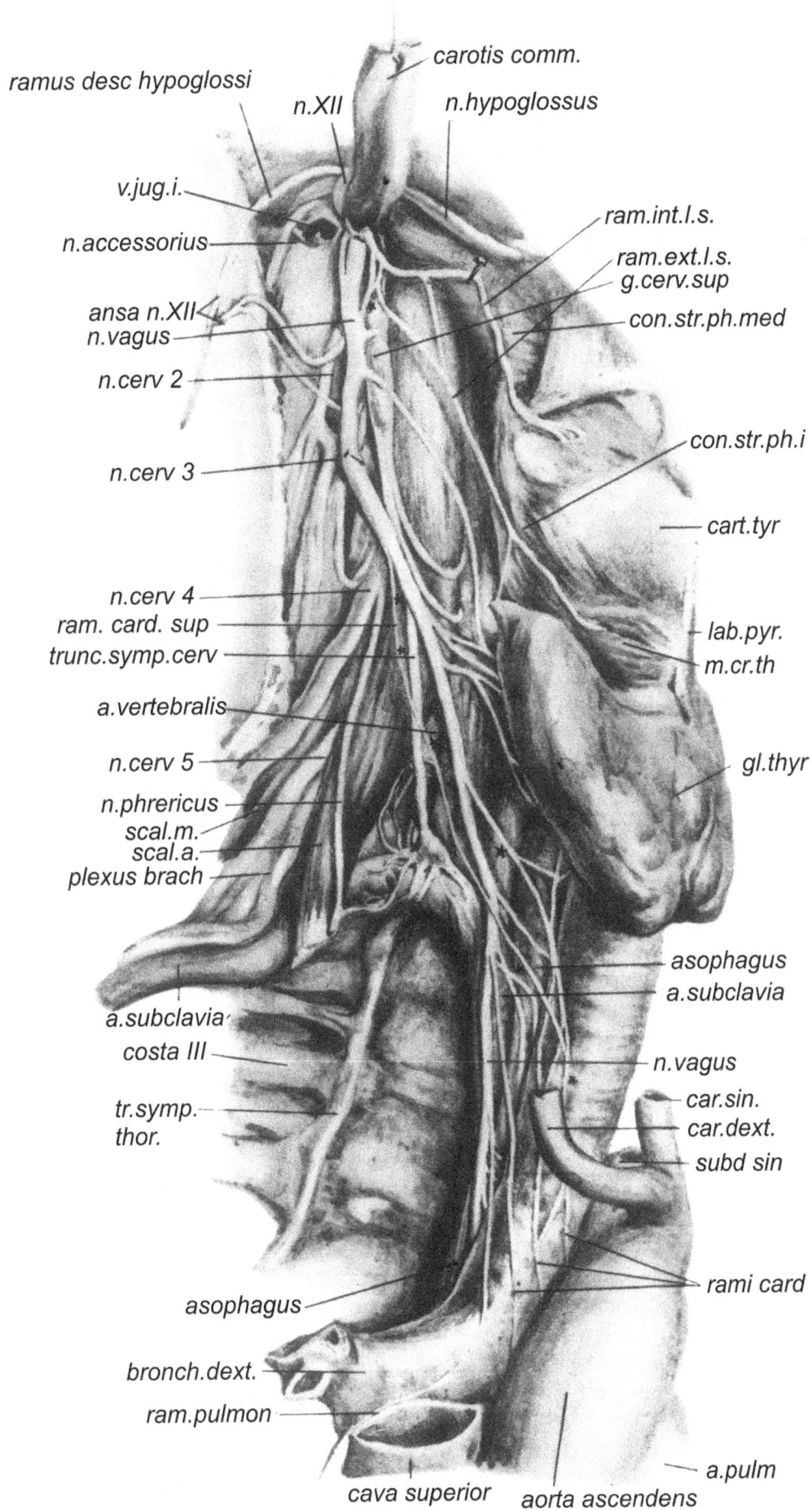

Fig. 179 Cervical sympathetic trunk and vagus nerve with the cardiac branches (after Fr. I. Rainer)

195

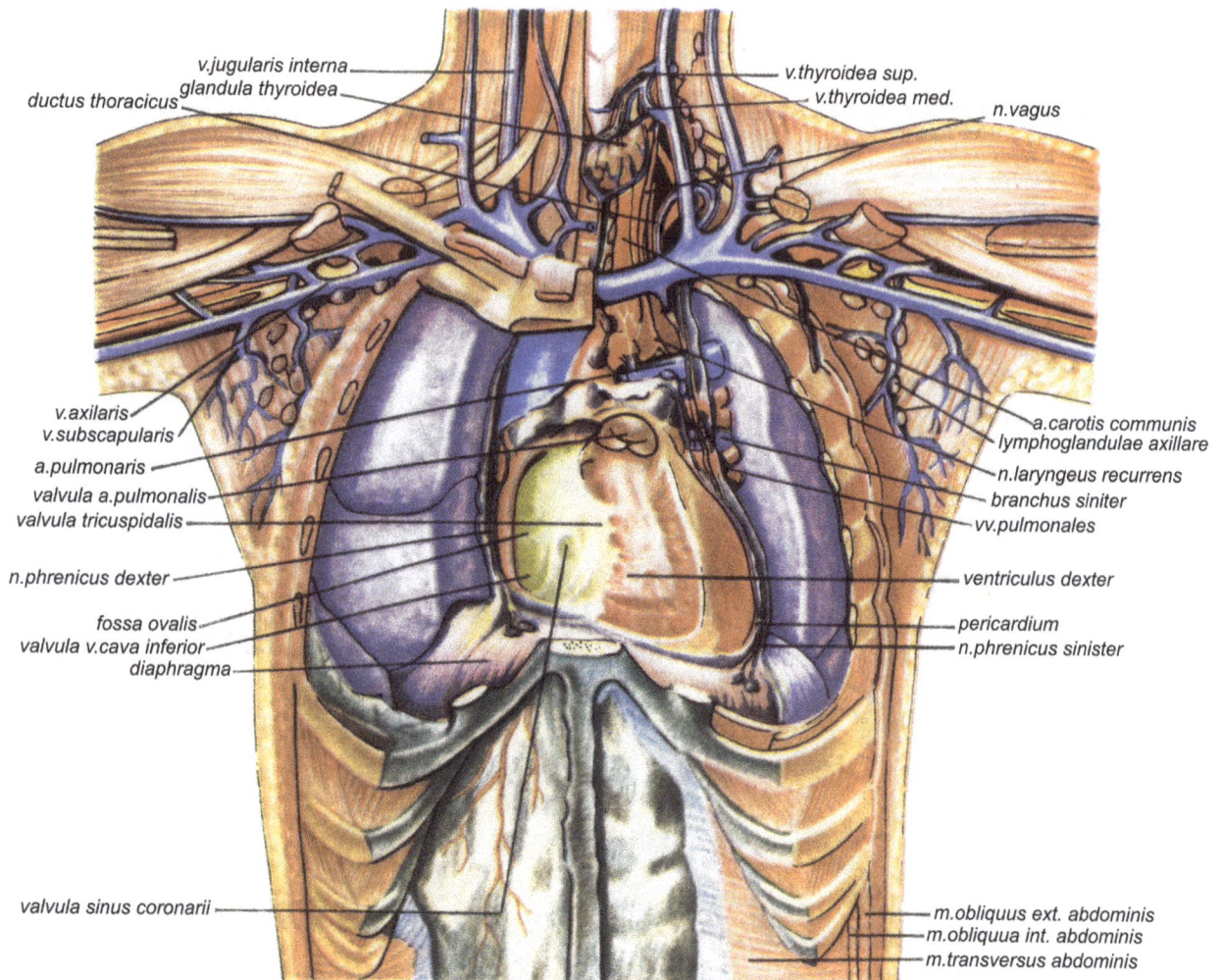

Fig. 180 Vagus nerve and inferior laryngeal recurrent branch (cervical and thoracic relations)

The thoracic part *(pars thoracica)* of the sympathetic system is constituted of 10-12 thoracic ganglia, situated in front of the heads of the ribs and immediately below the costal pleura and the endothoracic fascia, it is crossed behind by the intercostal nerves arteries and veins. The thoracic part crosses usually the diaphragm between the medial and the lateral crus and is continuous with the abdominal part.

Branches and ramification:

1. communicating branches to the nerves of the arm;
2. communicating branches to the intercostal nerves;
3. bronchial communicating branches to the bronchi and the pulmonary plexus;
4. oesophageal branches to the oesophagus;
5. the thoracic aortic plexus *(plexus aorticus thoracicus)* forms a fine plexus around the aorta and is continuous through the aortic hiatus with the abdominal aortic plexus;
6. the greater splanchnic nerve *(n. splanchnicus major)* arises with its roots from the fifth to the ninth thoracic ganglia, descend medially below the pleura and the endothoracic fascia, passing with the azygos or the hemiazygos vein across the medial crus of the diaphragm, and runs to *the coeliac ganglion*. Possibly it may course also to the superior mesenteric ganglion;
7. The lesser splanchnic nerve *(n. splanchnicus minor)* is made of the ninth-tenth thoracic ganglia, sometimes with the participation of the eleventh twelfth ganglia. It courses medially – downwards, covered with the endothoracic fascia and the pleura, through the medial crus of the diaphragm, and enters the coeliac ganglion. Some of its fibers run to the superior and inferior mesenteric ganglia. The splanchnic nerves pass uninterruptedly through the prevertebral ganglia.

196

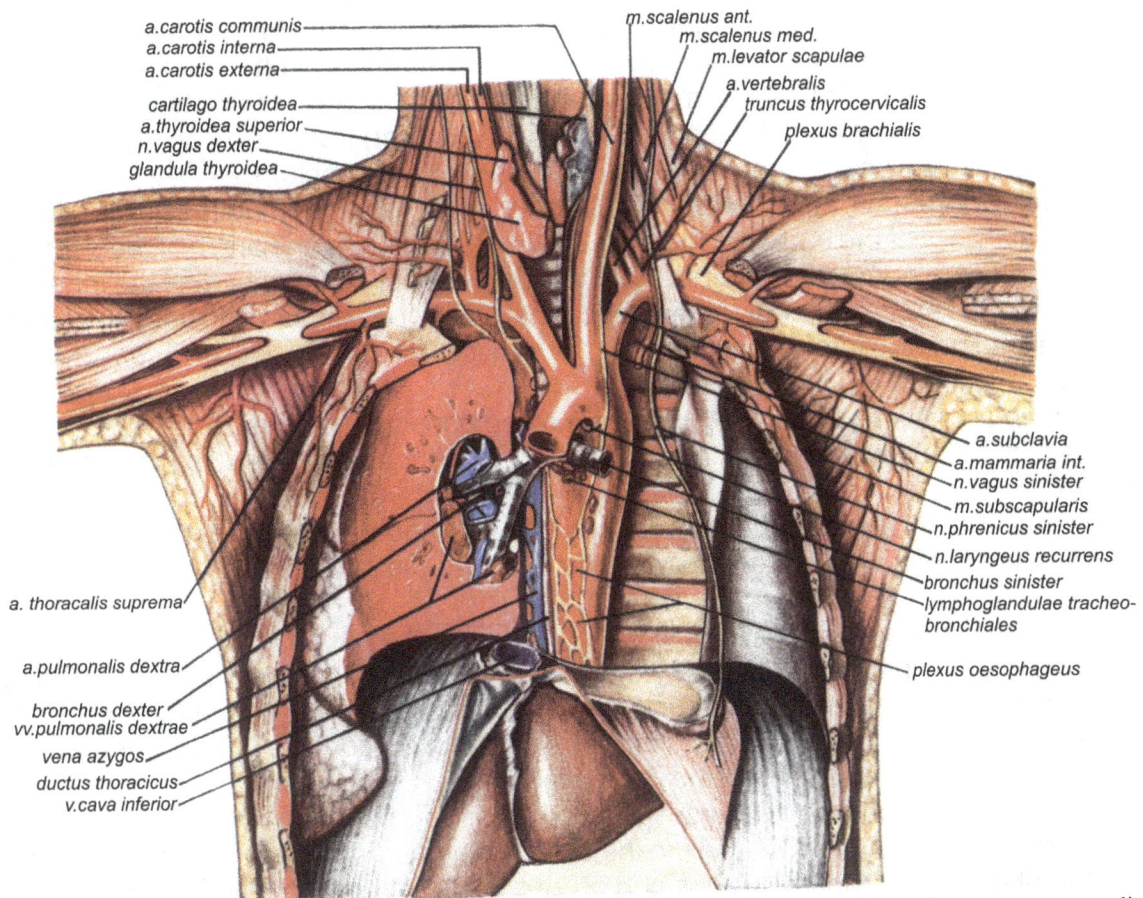

Fig. 181 Vagus nerve – oesophageal plexus – relations (the heart has been removed)

The cells of these neurons are bipolar and lie in the spinal ganglia. Their central prolongation is associated with the posterior root of the spinal nerve, enters with it in the spinal cord and intermingles, under the form of terminal arborizations, in the intermediolateral nucleus at the level of the cells of the first efferent neuron of the same segment. Thus, the sympathetic reflex arc to the viscera is closed.

The cells of the sympathetic afferent neurons in the viscera send at the level of the spinal ganglia branches to the ganglionic cells of the sensitive peripheral nerves in the same segment. In the case of occurence of impairments at the level of viscera, the excitation comes to the cells of the spinal ganglia and, from here, it is converged through the spinothalamic tract and then through the thalamocortical tract to the postcentral gyrus and, thus the sick individual becomes aware of the existence of the painful phenomenon (for example, colics induced by spasms of the intestinal smooth musculature, pain in appendicular colic, in biliary colic etc.).

By the junction of the sensitive posterior root with the motor cells of the anterior horn is achieved the intercalation in the reflex arc of the central nervous system and, consequently, the reflex contraction of the abdominal wall (for example, in appendicitis and peritonitis). Moreover, may be the excitations in the viscera of the abdominal cavity, through the afferent sympathetic fibers, integrate the peripheral prolongation of the ganglionic spinal cells and, in this ways, the patient's pain is projected into the respective cutaneous region. Thus appear hyperesthesic cutaneous zones, which contribute to the elucidation of the diagnosis in diseases of the abdominal organs (for example, Boas' pressure point in patients with gastric ulcer, to the left of the vertebral column, near the twelfth thoracic vertebra; the cholecystic point; the appendicular points; the urethral points etc., described in volume II).

The fibers which form the splanchnic nerves, after being connected with each other in the prevertebral ganglia, travel to the following organs: from T5-T8 to the stomach and the small intestine; from T9-T10, to the liver and the bile ducts; from T11-T12, to the large intestine.

197

The afferent pathways from the splanchnic nerves run to the following segment: to Th6-Th9 from the stomach; to Th8 (the left) from the pancreas; to T8- T11 (to the right) from the liver and the gall bladder; to Th10-L1 from the small intestine and the colon up to the left 1/3 of the transverse colon (Cannon-Böhm's point); to L1 (on the right) from the appendix vermiformis; to L1 and L3 from the descending colon, the sigmoid colon, the rectum.

The abdominal or lumbar part *(pars lumbalis)* of the sympathetic trunk is situated on the lateral surfaces of the bodies of lumbar *vertebrae*, medially to the origin of *the psoas major* muscle, and has a downwards course. On the right side, it is situated behind the inferior *vena cava* and on the left side, laterally to the abdominal aorta. It continues with the sacral part. The abdominal part has 4-5 lumbar ganglia.

Branches and ramifications:

1. *rami communicantes* from the origin of the psoas major muscle in T12-L1-L2-L3, to the parts of the lumbar plexus and, in continuation, on various pathways, to the nervous of the lower limb;

2. the abdominal aortic plexus is immediately continuous with the thoracic aortic plexus and has connections with *the coeliac ganglia* and with the superior and inferior mesenteric ganglia;

3. the efferent pathways, after their connection in the prevertebral ganglia – mostly by the branches of the pelvic part of the sympathetic trunk and by the plexus which is formed by the fibres arising from the prevertebral ganglia of the L1-L2-L3 segments course to the transverse colon, from the left 1/3 along the whole inferior portion, the kidney, the suprarenals, the urinary bladder, the rectum and the internal genitalia;

4. the afferent pathways course to the following segments to: T9-L2, from the ureter; to T10-L1, from the kidney; to T11-L3 from the urinary bladder; to T11-L3, from the left 1/3 of the colon together with its entire inferior portion up to the rectum; to L1-L3, from the internal genitalia.

The pelvic part *(pars pelvina seu pars sacralis)* of the sympathetic trunk is made up of four sacral ganglia with interganglionic rami between them as well as of an unpaired coccygeal ganglion. The terminal lower part of the sympathetic trunk is sited on the pelvic surface of the coccyx bone, medial to the sacral foramina *(foramina sacralis pelvina)* behind the rectum. The right and left side join at the first coccygeal vertebra, on the midline, in the coccygeal ganglion.

As the central portion of the sympathetic trunk in the spinal cord terminates at L3, the pelvic part is formed exclusively of fibres of the lumbar segments; these fibres are associated with the nerves of the pudendal plexus *(plexus pudendalis)* and with the sympathetic portion in S2, S3 and S4.

Branches and ramifications:

1. communicating branches to the sacral nerves;

2. the rectal plexus *(plexus rectalis sive haemorrhoidalis)*

3. the vesical plexus *(plexus vesicalis)*

4. in the man – the cavernous nerves of penis *(nn. cavernosi penis)*, the prostatic plexus *(plexus prostaticus)* and the diferential seminal vesicles – in the woman – the cavernous nerves of clitoris *(nervi cavernosi clitoridis)* and the uterovaginal (Frankenhäuser's) plexus.

The sympathetic system forms several plexuses, among which we mention the following:

The cardiac plexus *(plexus cardiacus)*, situated at the base of the heart, is made up of a superficial and of a deep part. The superficial part, situated below the aortic arch, is formed by the cardiac branch of the left superior cervical ganglion, the inferior cervical cardiac branch of the vagus nerve (a small Wrisberg's ganglion) and is probably part of the relay for the vagus fibers. The deep part, situated in front of the bifurcation of *the trachea*, is formed by the cardiac nerves derived from the cervical and upper thoracic ganglia and by the cardiac branches of the vagus and of the recurrent laryngeal nerves.

The left coronary plexus (*plexus coronarius cordis posterior*) formed by fibers of the deep part of the cardiac plexus, gives branches to the left atrium and ventricle.

The right coronary plexus (*plexus coronarius cordis anterior*) gives branches to the right atrium and ventricle.

The coeliac or solar plexus (*plexus coeliacus*) situated at the level of the upper part of the first lumbar vertebra, behind the stomach, between the adrenal glands, is formed by a dense network of nerve fibers which unite coeliac or semilunar ganglia. The plexus and the ganglia receive the greater and lesser splanchnic nerves from both sides and some filaments from the right vagus and they give off numerous secondary plexuses along the neighboring arteries.

The hypogastric plexus (*plexus hypogastricus*), situated in front of the left internal vein, of the last lumbar vertebra and of the promontory, is formed by the junction of the „*presacral nerves*" which descends from the aortic plexus and from the lumbar ganglia. It divides into the right and left pelvic plexus.

The pelvic or inferior hypogastric plexuses supply the viscera of the pelvic cavity and are sited on the sides of the rectum in the man and on the sides of the rectum and of the vagina in the woman.

They are formed by the continuation of the hypogastric plexus into the two first ganglia of the sacral portion of the sympathetic trunk and by the perisympathetic branches of the second and third sacral nerve.

Then, from the pelvic plexuses arise:
- the rectal plexus (*plexus haemorrhoidalis*);
- the vesical plexus (*plexus vesicalis*);
- the prostatic plexus (*plexus prostaticus*);
- the lesser cavernous nerves of the penis (*nn. cavernosi penis minores*) from which arises also the greater cavernous nerve;
- the vaginal plexus (*plexus vaginalis*)
- the uterine plexus (*plexus uterinus*).

The parasympathetic system
(*Parasympathicus*)

The centers of the parasympathetic system are located in the brainstem, in the so-called encephalic part, and in the sacral spinal cord, which is the so-called sacral part; they contain the cells of the first preganglionic efferent neuron. The emerging fibers do not form independent nerves; usually these fibers are associated to other nerves. The fibers of the first neuron is connected, in proximity to the supplied organ, in a parasympathetic or intramural ganglion.

The parasympathetic fibers lie in (1) the oculomotor, (2) the facial, (3) the glossopharyngeal, (4), the vagus, (5) the second and third sacral nerves.

(1) The parasympathetic oculomotor fibers arise from the Edinger-Westphal nucleus; they enter the ciliary ganglion, leave it as short ciliary nerves and supply the ciliary muscles and the sphincter of the papilla.

(2) The facial nerve contains efferent parasympathetic fibers of the inferior salivatory nucleus, which run through Wrisberg's intermediate nerve as the nerve VII – bis, pass into the cord of the tympanum (*chorda tympani*) and reach the lingual nerve; they continue then course in the submandibular region, where they enter then the submandibular (submaxillary) ganglion form a synapse and then run to the submandibular and sublingual glands. Moreover, the facial nerve contains the efferent parasympathetic secretomotor fibers for the lacrimal gland. They run through the pterygoid canal and synapse in the sphenopalatine ganglion.

(3) The glossopharyngeal nerve contains efferent parasympathetic fibers which are secretomotor to the parotid gland.

They have their origin in the inferior salivatory nucleus, run in the glossopharyngeal nerve, then enter the lesser petrosal nerve, reaching thus the optic ganglion where they synapse, after which they pass to the auriculotemporal nerve and reach the parotid gland.

(4) The vagus nerves is the main representative of the parasympathetic system in the organism. Its fibers run into the sympathetic plexus, the first neuron connects with the second in the small parasympathetic ganglia, near the supplied organ. It supplies the abdominal organs and the digestive tract, up to the beginning of the distal third of the transverse colon (Cannon-Böhm's point) (described in volume II).

(5) The sacral part of the parasympathetic system is situated in the sacral intermedio-lateral nucleus of segments S2, S3 and S4. The preganglionic fibers of the first neuron, united in the pelvic splanchnic nerves, form vast plexuses and are associated with the ramifications of the pudendal plexus *(plexus pudendalis)*, as well as with some sympathetic pathways. The second neuron lies in numerous parasympathetic ganglia scattered in the pelvis and its fibers to the colon and to all the viscera in the pelvis. The vegetative innervation was described at each organ separately.

The visceral parasympathetic group are located only in certain parts of the spinal cord and trunk, which may subdivide into a superior part, the cephalic parasympathetic part, and an inferior part, the sacral parasympathetic part. This group is formed of vegetative fusiform cells, too, with their large axis oriented longitudinally. The reason for the separation of the two groups of the parasympathetic system from each other at the two extremities of the neuraxis, lies in the anaboligenic action of the parasympathetic system separated in oral, respectively anal reflexes, at the two poles of the body which actually represent the entrance and exit poles of the apparatuses of the organic economy (of matter importations and discharge). These reflexes are separated in the organism and up to a certain point, even antagonistic, which is understandable as the introduced matter should have the necessary time to be digested and assimilated, before the discharge of waste matter. In the same way may be explained the fact that the cephalic parasympathetic part extends its territory of action through to vagus up to the level from which the influence of the sacral parasympathetic part, i.e. of that part that assures the discharge of feces from the digestive tract, begins to manifest itself.

Functionally, the cephalic parasympathetic part assures the myosis, the lacrimal, salivary, gastric, biliary and intestinal secretion, the peristalsis of the digestive tract, bronchoconstriction and bradycardia whereas the sacral parasympathetic part commands the erection and ejaculation, the defecation, micturition, and the parturition. It is known, physiologically, there is even a certain antagonism between the functions of the parasympathetic system. Thus, the micturition inhibits the defecation and the ejaculation exerts an inhibitory influence upon the micturition.

Hence, these anabolic reflexes were polarized out of a vital necessity which should assure the accomplishment of the restoration function of body resources. As regards the segmental vegetative, cervicothoracic-lumbar innervation, extended along the respective parts of the spinal cord, which in the past was considered as a cutaneous parasympathetic innervation, it is nowadays known that it is represented by vegetative segmental vasodilator fibres, which assure the inhibition of perspiration and pilo-erection, taking into account that the vasoconstriction, piloerection and perspiration are elicited by the orthosympathetic system.

Moreover, it is nowadays known that the place of the synapsis between the two neurons – pre and postganglionic – is variable. Thus, for the orthosympathetic system, in the ganglionic chain (sympathetic trunk) to the organs of the head, the neck and the thorax and in the prevertebral ganglia to the abdominopelvic organs; briefly, the orthosympathetic synapsis is formed as near as possible to the neuraxis and as far away as possible from the organ. For the parasympathetic group, the synapsis is formed in the peripheral or justamural ganglia, hence the farthest possible from the neuraxis and the nearest possible to the organ. The synapsis of the cutaneous vegetative segmental fibres is achieved, according to C. Elze and Ken-Kuré, in the spinal (hypersympathetic, according to Roussy and Mosinger's terminology) ganglia.

200

The diencephalohypophysial system

The diencephalohypophysial system (see also the chapter 4 - The secretory diencephalic apparatus) represents the command area of the whole vegetative life, hence constituting the vegetative, endocrine brain. It contains the hypothalamic and the tuberal nuclei described histologically, in a close anatomical and functional relationship – through neurocrinia - with the hypophysis. There are some authors who admit that the posterior part of the hypothalamus could have prominent orthosympathetic functions, whereas its anterior part could exert parasympathetic effects. We know nowadays, as a result of the researches carried out, that the hypothalamic nuclei do not act separately in the regulation of isolated function, but they are involved in a common activity, as a great functional complex. The diencephalo –hypophysial system guides the metabolism, through numerous descending pathways, towards the ortho and parasympathetic centres of the neuraxis, as well as through the neurosecretion and the hypophysial secretions.

This system orients the physicochemical reactions in the body, such as the water balance, the electrolyte, protein, carbohydrate and lipid metabolism, a series of physiological reactions such as vasomotoricity (adaptation means of the circulation to the needs of the whole organism), perspiration (aimed at thermogenesis regulation), diuresis (which promotes the discharge of toxic substances), vigil state and sleep – this intermitent function restores the vital organic capacity -, as well as somatotropic functions in morphogenesis and growth of the individual or genitotropic functions, in the service of the species perpetuation. The topography of the described hypothalamic nuclei and of those of *the tuber cinereum* does not correspond exactly to the functions studied by Hess by their stimulation with a microelectrode in the cat. Moreover, as Hess has demonstrated, in the hypothalamus there are adjacent areas of stimulation or inhibition of the functions.

The hypothalamus functions in a close correlations with the hypophysis, with which it has also anatomical relations, pointed out in two Romanian researches, respectively the hypothalamohypophysial tract, described in 1925 by I.T.Niculescu and Răileanu, and the venous portohypophysial system, discovered by Fr. Rainer and demonstrated in 1930 by Gr.T. Popa and U. Fielding. The discovery of the hypothalamohypophysial tract, as well as of secretory granules detected alongside it by Gomori's histochemical method, which have been for a long time considered as artefacts, permitted in 1949 to Bargmann and Scharrer to substantiate the neurocrinia (see also the chapter "The diencephalic secretory apparatus") intuited many years before by R. Collin. Especially the supraoptic nucleus and the paraventricular nucleus contribute by the neuritis of the perikaryons to the formation of the hypothalamohypophysial tract. The neurosecretory function also the above-mentioned chapter of the hypothalamohypophysial system appears only after birth; nowadays are known the two hypothalamic hormones: vasopressin (adiuretin) and oxytocin (OXT). This neurocrinia was demonstrated by Bargmann, Hild and Zelter, who obtained from the hypothalamic nuclei of the dog and of man an alcoholic extract containing these two hormones. As regards the portohypophysial system, complementary researches (Wilocki and King) have shown that the circulation direction of the venous blood is from the hypothalamus to the hypophysis and not from the hypophysis to the hypothalamus as shown by Popa.

This system was afterwards confirmed also in amphibians, birds and mammals. The intermediate lobe of the hypophysis secretes a melanophorotropic hormone which is involved in the mimetic adaptation of the color of animals to the environment, by migration of the melanotic pigment from the skin into the depth, vehiculated by melanophores. The anterior hypophysis, the adenohypophysis secretes endocrinotropic hormones, such a ACTH (adrenocorticotropic hormone) and TSH (thyroid – stimulating hormone), gondadotropic hormones such as FSH (follicle stimulating hormone) and LH (luteinizing hormone), and somatotropic hormone (STH); the physiologic and therapeutic action of these hormones is known.

1. Endocrinotropic hormones
- the adrenocorticotropic hormone (ACTH) stimulates the adrenal cortex;
- the thyrotropic or thyroid – stimulating hormone (TSH) induces thyrocine formation;
2. Gonadotropic hormones
- the follicle stimulating hormone (FSH) acts upon ovogenesis and spermatogenesis.
- the luteinizing hormone (LH) in association with FSH activates the maturation of the ovule the formation of estrogens (estrone, estradiol, estriol) and progesteron and the ovulation in the woman and induces testosterone formation in the interstitial Leydig's cells;

-the luteotropic hormone (LTH), alone, induces the transformation of the follicular cells, remainer after ovulation, into corpus luteum (yellow body) and activates the progesterone secretion; moreover, it acts upon the mammary gland, producing the proliferation of the secretory epithelium and stimulating catalytically the milk secretion.

3. Somatotropic hormone
STH is a growth hormone, acting upon the fat metabolism and the protein metabolism: its activity depends also on other hormones (thyroxine, corticoids and sexual hormones etc.)

It is interesting to mention the stimulating action exerted on the hypothalamus under the influence of light, which produces, on its turn, the activation of the adenohypophysis and of the above mentioned hormones of this gland. Hence, beside the retino-optic pathway, there is also a nervous retinohypothalamic pathway, which stimulates the gonadotropic ACTH, TSH and the somatotropic hormone, as well as the water-electrolyte metabolism, in the regulation of which it is involved.

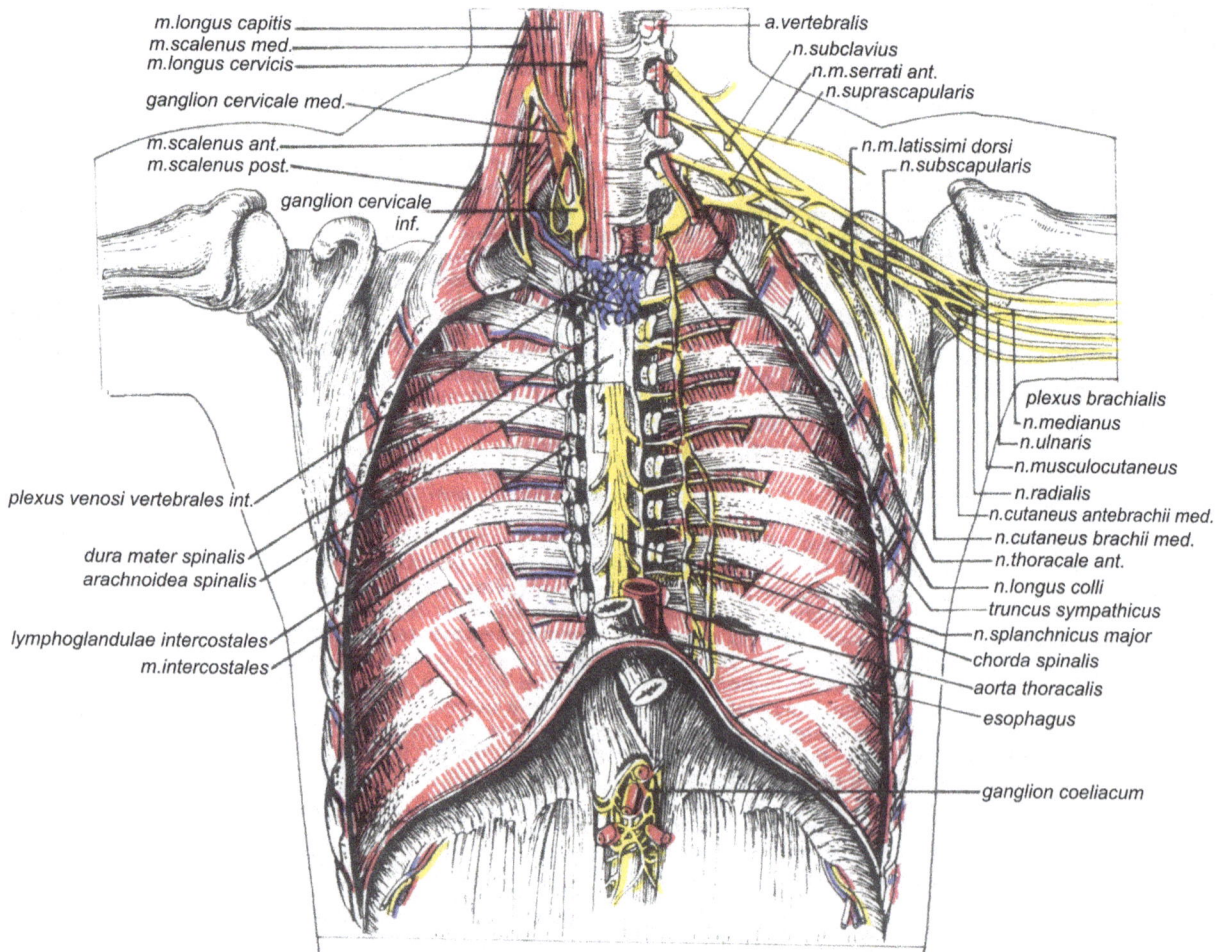

Fig. 182 Sympathetic thoracic trunk and membranes of the spinal cord

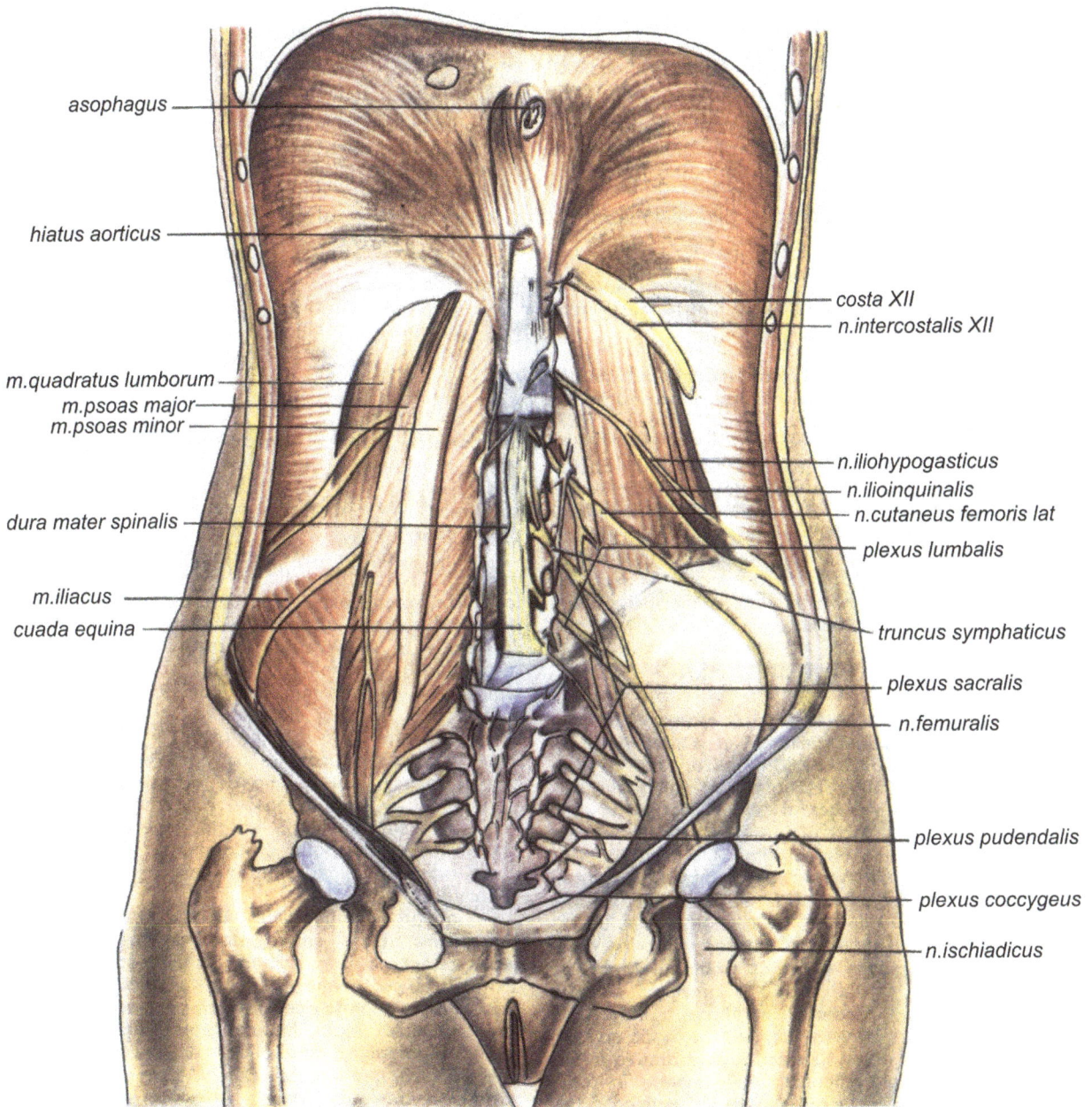

asophagus

hiatus aorticus

m.quadratus lumborum
m.psoas major
m.psoas minor

dura mater spinalis

m.iliacus
cuada equina

costa XII
n.intercostalis XII

n.iliohypogasticus
n.ilioinquinalis
n.cutaneus femoris lat
plexus lumbalis

truncus symphaticus

plexus sacralis
n.femuralis

plexus pudendalis

plexus coccygeus

n.ischiadicus

Fig. 183 Lumbosacral spinal cord, lumbosacral nerve plexuses and
lumbosacral ganglionic sympathetic chain (relations)

203

The reticular formation

Between the ecotropic segmented (metameric) and non-segmented system and the vegetative ortho- and parasympathetic territories, on the one hand and the diencephalohypophysial system, on the other hand, lies the reticular formation of the brainstem, which represents in the brainstem, longitudinally, the associative apparatus of encephalomeres (branchiomeres), while its homologous part in the spinal cord represents, longitudinally and symmetrically the associative apparatus of myelomeres (the neural part of metameres). This reticular formation is particularly important for the physiology of the organism, since on the one hand it is the organic substance of all the somatovegetative vital reflexes, as the physiologist Pfall has shown, and on the other hand it represents the ascending activating system of the central cortex, pointed out by Magoun. A somatic or vegetative reflex cannot take place without the involvement of this system, formed of a chain of reticuloreticular neurons; it is involved in all the vital manifestations, in the vigil state, in the course of activity or at rest, as well as during sleep, and it intervenes also in the orientation of the biophysicochemical mechanism of homeostasis. Furthermore, it may be influenced by physical or chemical stimuli and intoxication and can intercept temporarily some connections between the cortical centres and the hypothalamus, on the one hand, and the vegetative ortho –and parasympathetic in the spinal cord, on the other hand.

The last level of the vegetative innervation is comprised in cerebral cortex, which contains three areas with visceral functions:

a) the prefrontal area, situated in front of the somatomotor zone *(precentral gyrus)*, with a viscero-affective function;

b) the orbital area corresponding to the inferior surface of the frontal lobe, with a viscero-regulating function and

c) the cyngulate area, containing the gyrus of *the corpus callosum* or fornicate gyrus *(gyrus fornicatus)* and the hippocampal gyrus.

Homeostasis

In the frame work of his studies regarding the organization of the vegetative nervous system, Ferdinand Hoff has elaborated a very interesting scheme referring to homeostasis regulation. In its scheme, the vegetative self regulation at the level of the organic and humoral periphery, of the endocrine system, of the nervous ortho- and parasympathetic system, of the vegetative endocrine brain (diencephahypophysial system) and of the cerebral cortex is described. It should be mentioned – that a series of biophysicochemical constants are correlated with the neuro-endocrine regulation and integration mechanisms. The superordinate position of the cerebral cortex, which achieves the connection with the external environment through the somatic and sensitive-sensorimotor areas, is clear. Its result is the praxis, i.e. the practical activity by which we modify the nature and the internal environment through the above-mentioned vegetative zones and the product of this activity is the *coenesthesia*, i.e. the normal organic state which we perceive. *The organic coenesthesia* is under the integrative control of the cerebral cortex which, in the case of intero- and exteroceptive excitations, manifests its influence on the diencephalo-hypophysial system. The latter is linked to organs and to the humoral environment by two kinds of connections:

- nervous connections, through the dorsal longitudinal fasciculus (Schütz's bundle) and the Bock's paraventricular fasciculus (proved only physiologically), to the ortho and parasympathetic centers of the spinotruncal system.

- nervous and endocrine connections, through the hypothalamohypophysial tract, to the hypophysis (the endocrine brain), which influences the function of the organohumoral periphery and, in the last analysis, of the cells and tissue, through the endocrine ortho- and parasympathicomimetic glands: the parathyroid glands, the adrenal glands and the thyroid, of the first group, and respectively the insular pancreas, of the second group.

204

The balance between the ortho and the parasympathetic system achieves the vegetative zone, a state of equilibrium between the two antagonistic systems, for the exploration of which Danielopolu and Carniol have introduced the atropine and the orthostatism test.

With respect to a series of homeostatic constants, Hoff mentions the electrocytes, the blood pH, the thermoregulation, the metabolism and the glycemia. By switching the neuro-endocrine ortho- and parasympathetic influences, the mentioned constants have the following evolution between two extremes:

- in orthosympathicotonia, the mentioned constants represent the so-called A phase or Cannon's necessity (alarm) reaction, like in muscle effort and in the invasion phase of infection. This makes up the vegetative systole, characterized by a predominant orthosympathetic tonus, hypercalcemia, acidosis, myeloid reaction, hyperthermia, catabolism and hyperglycemia;

- in parasympathicotonia (vagotonia) which is a parasympathetic inbalance, the same constants represent the B phase or vegetative diastole, which characterizes restoration through rest and cure in convalescence, a phase which occure „per crisis", i.e. suddenly, or „per lysis", i.e. gradually; this phase manifests itself by: vagal tonus increase, hyperkaliemia, alkalosis, lymphoid reaction, hypothermia, anabolism and hypoglycemia. Sometimes, according to the compensation principle, this phase passes considerably below the normal limit.

Thus, the function under normal conditions of the human organism requires the maintenance within certain limits of these constants and also of others, permitting the optimum carrying out of the biochemical reactions lying at the basis of the vital functions of the organism (as an example we mention the self regulation of the cardiac muscle activity).

Endocrine receptors

We emphasize that, according to recent researches, the encephalon may be considered an endocrine gland (Claude Laroche). Thus, a series of hormones produced by the cerebral nerve cell, such as somatostatin, neurosthenin etc., united under the generic term cybernines (Lambiotte), were discovered. The equilibrium between these peptide hormones produced by the cells of the encephalon is at the basis of the behavior and reactions of the human individual. The maintenance mechanism of this equilibrium is the cybernetic feedback.

The most recent researches furnish a series of new data regarding the molecular structure of some endocrine receptors. Thus specific macromolecules named receptors, were discovered, which possesses properties of captation and concentration of specific hormones (or other substance) and which may form complexes with them, mediating in this way certain specific cell actions.

In this acception, the receptor is a biological transducer which converts the input information into a biological response. Biochemically, the receptor is a protein molecule, which binds selectively a ligand, in a saturable way and with a high affinity. Hence, the receptor is a molecular entity, formed of a receptor segment, a coupling mechanism and an effector segment.

The receptor segment in the interaction site of the hormone of the molecule belonging to the target cell, this interaction being followed by the induction of the initial stimulus; the coupling mechanism is represented by transduction, modulation and amplification processes by which the initial stimulus is transmitted to the effector segment, the effector segment is the molecular part, at the level of which an effect is generated.

The endocrine receptors are either situated in certain target tissues (endogenous and estrogenic hormone receptors) or are widely distributed in almost all the cells of the organism, the action of the receptive hormones being very extensive (insulin receptors, thyroid hormone receptors, adrenocortical somatotropic homoreceptors etc).

The receptor's concentration and affinity charges in dependence on the cell biology, the regulation of the receptors – by metabolic, genetic and endocrine factors – with respect to the differentiation state or the cell cycle being a mechanism significant for the control of the cell sensibility to the hormone.

There are a lot of receptors, among which we mention: the oestrogenic receptors, sited in the uterus, the vagina, the mammary gland and the hypothalamus; the progesteronic receptors lying generally at the same sites as the estrogenic receptors, the androgenic receptors, situated mainly at the level of the male genitalia; the water soluble receptors, the gonadotropic hormone receptors etc.

The presence of a series of hormone receptors in the brain has been recently demonstrated. The role of brain as an endocrine organ appeared clearly concomitantly with the discovery of the hypothalamic hormones and with the isolation of endomorphins and enkephalins. Moreover, it was proved that the nervous system is a target organ also for various hormones.

The brain possesses receptors for locally produced hormones, such as: serotonin, catecholamines (dopamine, norepinephrine), enkephalins, endorphins, hypothalamic hormones, P substance, somatostatin) steroid hormones (estrogens, progesteron, androgens, corticosteroids), thyroid hormones (triiodothyronine, thyroxine) and peptide hormones (ACTH).

A problem which arises with respect to the steroid, thyroid and peptide hormones refers to the way in which they pass beyond the blood brain barrier (BBB) or to their possible passage through the posthypothalami–hypophysial system, from the hypophysis to the brain, or through the cerebrospinal fluid. These hormones exert their action on the brain, producing multiple effects: a positive or negative feed-back effect on the hypothalamic and hypophysial hormone secretion, effects on the instinctual and adaptive behavior, metabolic effects etc.

As regards their topographical distribution, we mention: the target neurons for estrogens lie in the ventral part of the brainstem and of the spinal cord, in the hypothalamus and in the limbic structures; the target neurons for progesteronic hormones, as well as those for androgenic hormones have the same topography. This overlapping in the distribution of these hormones suggests that, in this case, the receptor mechanism of the nervous system depends less on the specificity of these three categories of receptors, but rather on the relative estrogen and androgen concentration.

Furthermore, the distribution of the corticosteroid and peptide hormon receptors at the level of the brain was determined. It was proved that the peptidic hormones function as neurotransmitting or neuromodulating hormones and receptors specific for each peptide are present in the corresponding regions of the brain.

The peptides could play a neuro-endocrine role, the neurons of the brain being at the same time source and target. Thus, the beta-endorphin, located around the ventricle, increases in concentration by excitation of this area and produces analgesia, while naloxone has reverse effect. Likewise, the beta-endorphin, as a central hormone, modulates the neuron excitability, like the P substance.

The skin is also an important endocrine receptor for a series of hormones: androgenic, oestrogenic, progesteronic, thyroid hormone, growth hormones etc.

Their topography has been demonstrated. For example the androgenic receptors lie at the level of the sebaceous glands and of the hair follicle. Thus, the sebaceous glands distributed predominantly at the level of the face, in the superior area of the back and of the chest are dependent on the stimulation of androgens for the sebum secretion; at puberty, the glands increase in size and the excessive sebum secretion represents a factor favoring the appearance of pubertal acne.

The anti-androgen treatment, competitive at the level of the receptor with progesterone, is very indicated both in men and in women and has no systemic secondary effects.

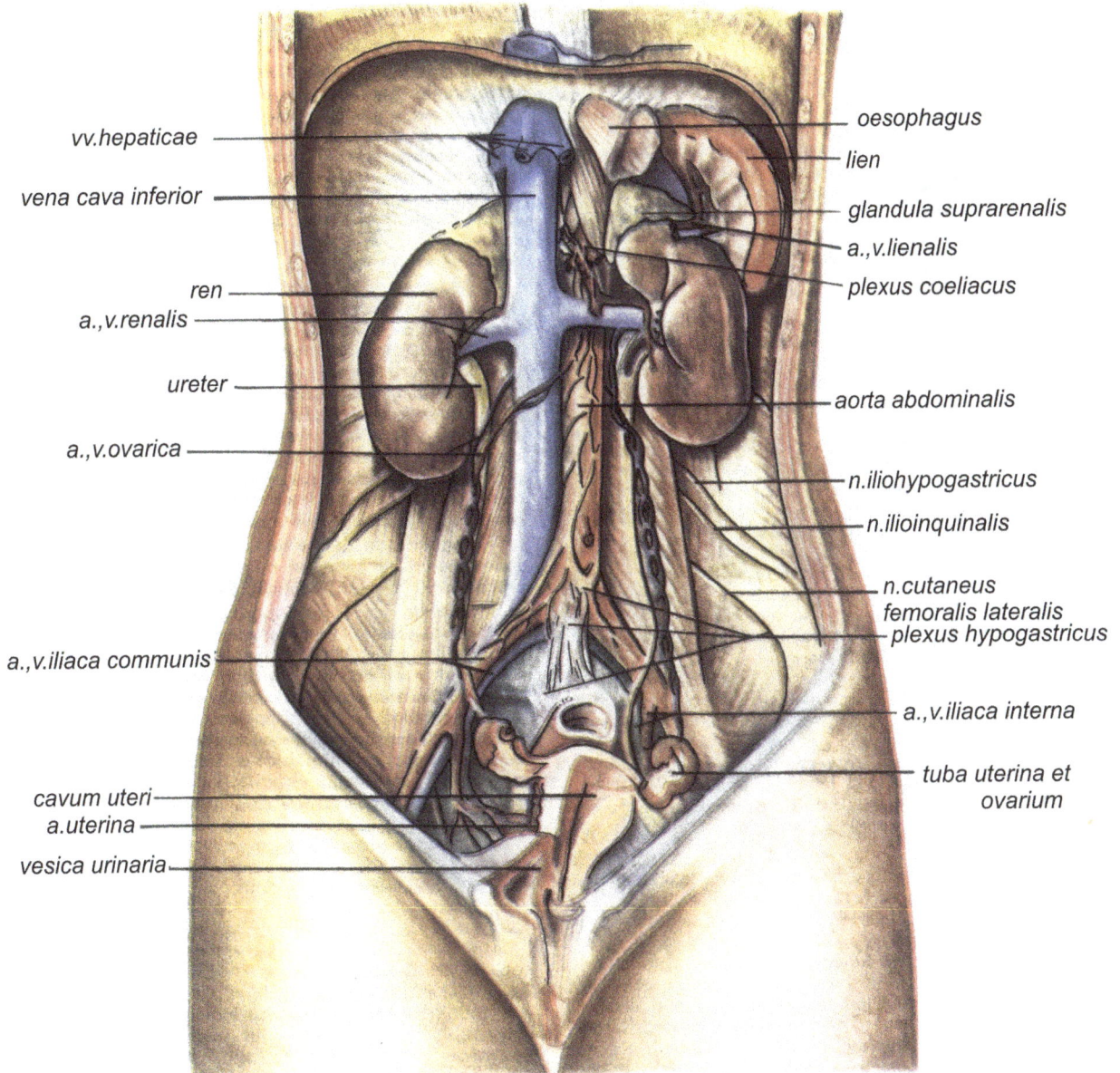

vv.hepaticae

vena cava inferior

ren

a.,v.renalis

ureter

a.,v.ovarica

a.,v.iliaca communis

cavum uteri

a.uterina

vesica urinaria

oesophagus

lien

glandula suprarenalis

a.,v.lienalis

plexus coeliacus

aorta abdominalis

n.iliohypogastricus

n.ilioinquinalis

n.cutaneus
femoralis lateralis

plexus hypogastricus

a.,v.iliaca interna

tuba uterina et
ovarium

Fig. 184 Lumbosacral sympathetic trunk (relationships)

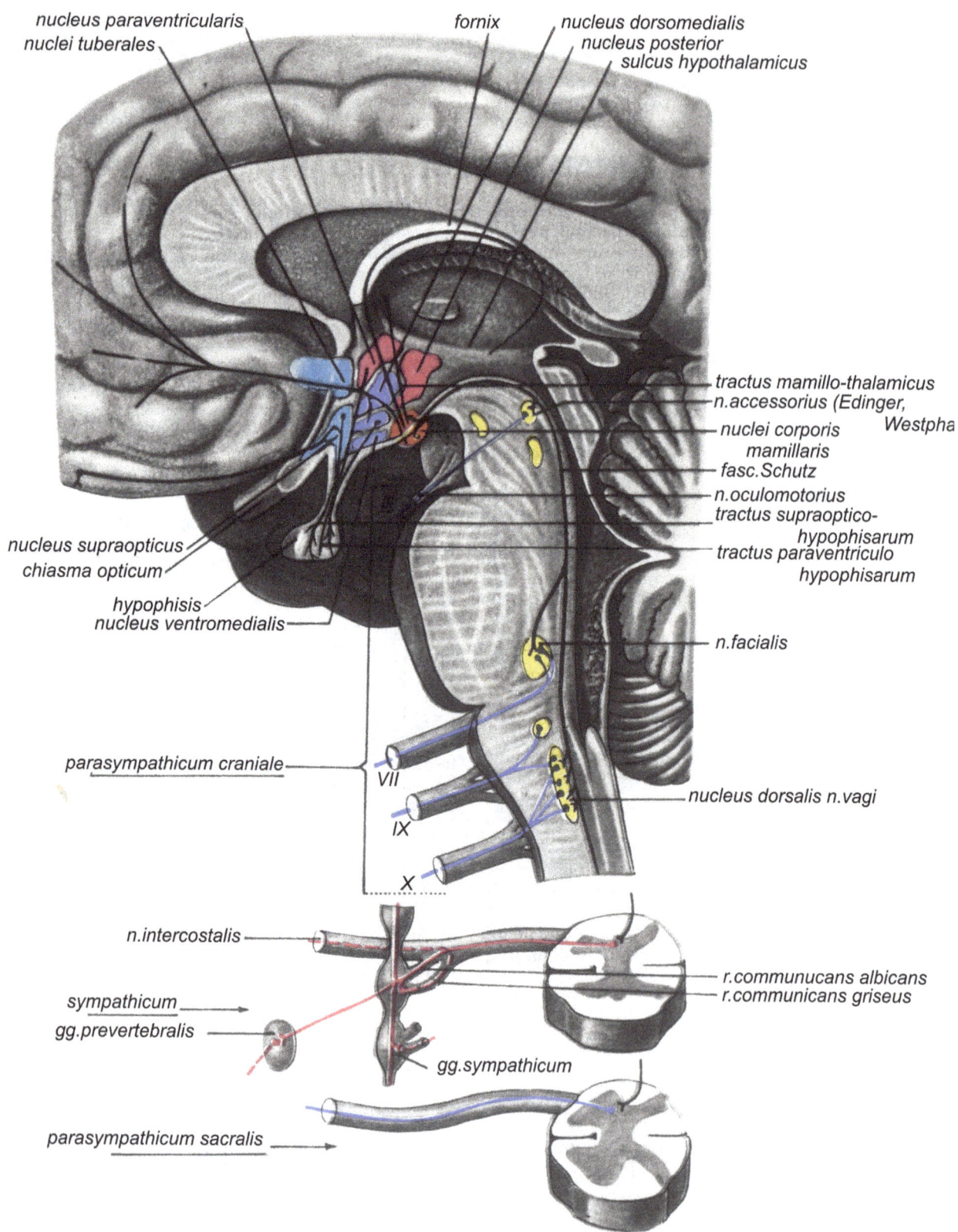

nucleus paraventricularis
nuclei tuberales
fornix
nucleus dorsomedialis
nucleus posterior
sulcus hypothalamicus

tractus mamillo-thalamicus
n.accessorius (Edinger, Westpha
nuclei corporis mamillaris
fasc.Schutz
n.oculomotorius
tractus supraoptico-hypophisarum
tractus paraventriculo hypophisarum

nucleus supraopticus
chiasma opticum

hypophisis
nucleus ventromedialis

n.facialis

parasympathicum craniale

VII

IX

X

nucleus dorsalis n.vagi

n.intercostalis

r.communucans albicans
r.communicans griseus

sympathicum
gg.prevertebralis

gg.sympathicum

parasympathicum sacralis

Fig. 185 Hypothalamohypophyseal nerve connections, parasympathetic and sympathetic systems (scheme)

208

The analyzers
(The sense organs)

THE VISUAL ANALYZER

The visual analyzer is constituted of the organ of sight (the receptor *peripherae*, segmental receptor), represented by the eyeball with all its adnexa, the conduction pathways or the intermediate segment and the cortical projection segment (central segment) (Fig. 186-203).

The eye or visual organ *(organ of sight)*
(Oculus seu organon visus)

The organ of sight is sited in the orbital region, made up of the two orbital cavities *(cavum orbitae)* (vol.I). According to the topographical distribution of the eyeball and of its adnexa the orbital region is divided into the palpebral, bulbar and retrobulbar regions.

The palpebral region

It is a superficial region which is made of the eyelids (superior and inferior) and, anexed to them, the lacrimal apparatus.

The eyelids
(Palpebrae) (Fig. 186)

They are musculo- tegumental creases which are limited by their free borders, the palpebral fissures *(rima palpebrarum)*. They defend the ocular bulb and, by their movements, wet the surface of the ocular bulb by spreading the secretion of the lacrimal grand. The opening and closing of the eyelids it is not a simple movement of rising and lowering, but a sweeping movement oriented medially and more obvious at the lower eyelid; this movement is possible owing to the structure of the eyelids. Each eyelid consists of convex ocular or tarsal part extending up to the free margin, and of an orbital part, that corresponds to the orbital fat–pad (fat body of the orbit); the two parts are separated by the superior and inferior palpebral fissures *(rima palpebrarum)*. The ends of the eyelids form, by uniting with each other, the medial palpebral commissure and the lateral palpebral commissure. The free margin presents an anterior border *(limbus palpebralis anterior)* on which lie the eyelashes, and a posterior border *(limbus palpebralis posterior)*, which delineates, with the cornea, *the rivus lacrimalis*, the pathway by which the tears reach the lacrimal lake. The anterior border is continuous with the skin. The posterior border presents 20-30 *foramina* of the meibomian or tarsal glands. In the internal quarter of the free margin of the eyelids is situated a small elevation, the lacrimal papilla, on the apex of which opens an orifice, *the punctum lacrimale*. Medially to the papilla there is a portion deprived of eyelashes, which delineates, with the opposite portion of the other eyelid, the lacrimal lake *(lacus lacrimalis)*, the bottom of which is occupied by a small prominence the lacrimal caruncle *(caruncula lacrimalis)* and by the semilunar fold of the conjunctiva *(plica semilunaris conjunctivae)*

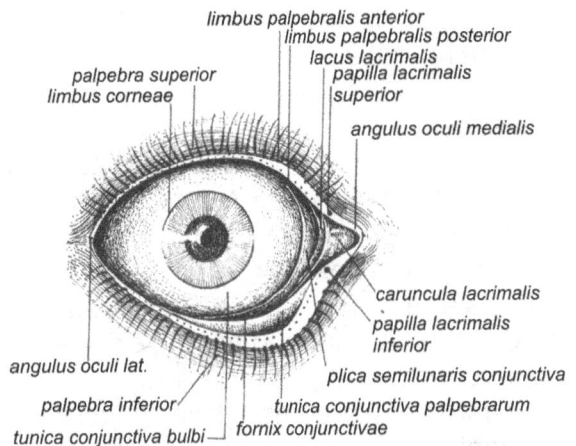

Fig. 186 Eyes – anterior view

209

(the third eyelid, analogous to the nictitating membrane of animals) (Fig. 186).

The eyelids consist of the following stratigraphic planes: skin, subcutaneous cell tissue, muscular layer, tarsus, glandular layer and conjunctiva.

The skin is thin and movable on the underlying layer.

The subcutaneous tissue is constituted of a thin layer of connective fibers and may be easily infiltrated with blood, pus, oedema; it communicates with the subcutaneous layer of the face and with the subaponeurotic layer of the calvaria.

The muscular layer is formed of the orbicular muscle of the eyelids and of the tarsal muscles.

I. The orbicular muscle of the eyelids *(m. orbicularis oculi)* is a muscle of mimicry around the palpebral fissure, and consists of:

1) a palpebral portion (from one ligament to the other), formed of four segments: preciliary marginal, retrociliary marginal, Riolan's muscle or the pretarsal portion and the preseptal portion or the tensor muscle of the lacrimal dome;

2) a peripheral orbital portion, which covers the superior and inferior border the orbit and is attached to the superior and inferior margins of the medial palpebral ligament (direct tendon);

3) a lacrimal portion comprised between the disrupted tendon of the medial palpebral ligament and the posterior orbital septum (Horner's muscle); it is inserted behind the medial palpebral ligament and surrounds the lacrimal canaliculus.

II. The superior and inferior tarsal or palpebral muscles are smooth, vertically oriented muscles. They unite the peripheral borders of the tarsal portions with the tendons of the levator muscle of the upper eyelid and of the inferior rectus muscle. The superior tarsal muscle (Muller's muscle), attached to the levator muscle of the upper eyelid, is fastened on the superior margin of the tarsus by the central and medial portions; it surrounds the palpebral region of the lacrimal gland by its lateral portion. The inferior tarsal muscle forms an expansion which arises from the inferior rectus muscle on the tarsus and reaches the neighborhood of the inferior orbital margin.

The fibro-elastic layer is formed of the tarsus and the orbital septum.

I. The tarsus of the eyelids is represented by two fibrous connective dense, resistant convex plates, which are lodged in the ocular part of the eyelids. Their ends are united with the lateral palpebral ligament, attached below the frontozygomatic suture and with the medial palpebral ligament, attached to the two the lacrimal rivus (the internal indirect tendon and the posterior reflected tendon. For the examination of the superior conjunctival cul-de-sac, termed also conjunctival fornix, it is necessary to raise the tarsus.

II. The orbital septum is a broad ligament, a fibro-connective plate, which unites the posterior fold of the orbital edge with the peripheral margin of the tarsus (outside the levator muscle of the eyelid, which the septum accompanies up to the margin).

The palpebral glands (the glandular layer) are the following:

1) the tarsal meibomian glands are developed in the region of the tarsus and upon on the free margins of the eyelids;

2) Moll's ciliary glands are modified sudoriferous glands, situated between the eyelashes;

3) the sebaceous glands of Zeiss are sebaceous glands attached to the eyelashes.

The conjunctiva (tunica conjunctiva) is a thin, glossy, transparent mucous membrane, which starts from the free margin of the eyelids, behind the openings of the tarsal glands, adheres to the posterior surface of the eyelids *(tunica conjunctiva palpebrarum)*, from which it reflects and passes on the anterior surface of the eyeball *(tunica conjunctiva bulbi)*. At the site of reflection are formed the superior and inferior conjunctival fornices *(fornices conjunctivae superior et inferior)*. The conjunctiva covers the anterior part of the sclera, without adhering to it, and that of the cornea, from which it cannot be dissociated. In the medial angle of the eye, the conjunctiva forms a vertical, falciform fold, with the lateral margin free and concave: this is the conjunctival semilunar fold *(plica semilunaris conjunctivae)*, which covers the lacrimal

210 caruncle.

The lacrimal apparatus
(Apparatus lacrimalis)

It consists of the lacrimal gland and the lacrimal pathways.

The lacrimal gland is situated in the lateral and superior angles of the orbit. Is secrets the tears, which moisten, the anterior pole of the eyeball. In the medial angle of the eye, the tears collect in the lacrimal sac and are conveyed into the inferior meatus of *the nasal fossae* through the lacrimal ducts (Fig. 187).

The lacrimal gland *(glandula lacrimalis)* consists of a main, orbital part *(pars orbitalis)* and an accessory, palpebral part *(pars palpebralis)*, separated from each other through the tendon of the levator muscle of the upper eyelid.

The orbital part, with the infero-medial surface concave, is lodged in a space bordered above and laterally by the lacrimal fossa of the frontal bone, below and medially by the tendon of the levator muscle of the eyelid, anteriorly by the orbital septum and behind by a thin mucous membrane, which separates it from the orbital fat-pad (fat body of the orbit).

The palpebral part of the lacrimal gland is flattened projection situated between the tendon of the levator muscle of the eyelid and the superior conjunctival fornix; behind it extends into the orbital region.

The excretory ducts (15-20 in number) have artero-inferiorly a slightly oblique course and open into the superior conjunctival fornix.

The nerve supply is assured via the lacrimal nerve, a branch of the ofthalmic nerve (V1), a sensitive nerve distributed to the conjunctiva and the upper eyelid. However, the course of the nerve inflow which triggers the lacrimal secretion is complex and arises from the lacrimal (lacrimomuconasal) nucleus of *the ponce*. From the lacrimal nucleus *(nucleus lacrimalis)* a pontine vegetative centre attached to Wrisberg's intermediate nerve (Wrisberg's intermediate nerve), the nerve inflow reaches the pterygopalatine (sphenopalatine) ganglion through the respective greater petrosal nerve (greater superficial petrosal nerve), a branch of the facial nerve. After the formation of a synapse in the ganglion, the inflow follows the course of the maxillary nerve and passes into the zygomatic nerve, the superior branch of which (the orbital branch) anastomoses with the lacrimal nerve, which is a branch of the ophthalmic nerve. The filaments of the anastomotic loop *(ramus communicans cum n. zygomaticus)* supply directly the gland.

The same nerve inflow involves equally *the nasal fossae* through the pterygopalatine ganglion and the nasopalatine nerve, this explains why the excitation of the pituitary mucosa induces in a reflex way the tear secretion. It elicits also a short reflex at the level of the pterigopalatine ganglion, if it is admitted that the vegetative sensibility follows in a reverse sense the course of the visceromotor fibers. The connexions between the nucleus of the facial nerve (nerve of facial expression) and the lacrimal nucleus explains the convoluted expression in states of sadness and joy.

The lacrimal ducts. The tears are collected by the *lacrimal canaliculi*. The latter begin at the level of the *lacrimal punctum* situated medially, on the free margin of the eyelids, *the inferior punctum* being slightly more medial than the superior one. They have a curved course, at an almost acute angle. The first segment, which is vertical measures 2 mm, while the following segment, which is horizontal measures 6 mm, dimensions that are important in the catheterization manoeuvre of the lacrimal ducts. The two lacrimal ducts are lodged in the lacrimal part of the orbicular muscle of the eyelids (Horner's muscle) and may empty independently from each other into a diverticulum of the lacrimal sac or they can unite in a common duct which empties into the lacrimal sac.

The lacrimal sac *(saccus lacrimalis)* is a cylindrical membranous reservoir, oriented obliquely downwards and slightly posterolaterally. It is situated in the lacrimal sulcus, between the direct and the reflected tendon of the orbicular muscle. The upper end of the sac (the fornix) lies at 1 cm away from the pulley of the superior oblique muscle. Clinically important is the relationship of the lacrimal sac with the ethmoidolacrimal cells, situated medially.

It is supplied by the angular artery, which anastomoses the facial artery – a branch of the external carotid artery – with the ophthalmic artery – a branch of the internal carotid artery.

The nasolacrimal duct *(ductus nasolacrimalis)* continues downwards the lacrimal sac, with which it forms a morphofunctional complex. It is contained in an osseus canal. It is of about 15 mm in length and is provided with inconstant mucous valves. Its inferior orifice opens at 1 cm behind the anterior extremity of the inferior meatus and is guarded, at its opening, by a fold of the mucosa, called Hasner's valve or fold *(plica lacrimalis)*. The puncture of the maxillary sinus is made at 2 cm behind this orifice, which becomes in this way a topographical landmark.

The bulbar region

It corresponds to the eyeball *(bulbus oculi)*, which is lodged in a space delineated by the peribulbar capsule (Tenon's capsule), that is continuous with the periosteum of the orbit, suspending the eyeball like in a hammock and separating the bulbar region from the retrobulbar region (Fig. 188-198).

The eyeball
(Bulbus oculi)

It has the shape of a sphere *"linked"* to the optic nerve. It is formed of three concentric tunics, with the anterior part more prominent and contains the transparent media of the eye; their tension is necessary to sight and imparts to the eyeball a firmness which can be appreciated and measured clinically. In some cases, the eyeball may be more elongated causing in this case the state of myopia (shortsightedness); if on the contrary the anteroposterior axis of the eyeball is shorter, occurs hypermetropia.

The eyeball is lodged in the anterior part of the orbit and is slightly closer to its lateral wall. The axes of the two eyes are almost parallel, while those of the orbits are clearly divergent; the axis of the eyeball does not coincide with that of the orbit.

The wall of the eyeball is formed, from outside inwards, of three tunics:
- a fibrous tunic, composed of the sclera and the cornea;
- a vascular tunic, formed of the choroid, the ciliary body and the iris, and
- a nervous tunic, composed of the optic retina, *(pars optica retinae)* and the blind retina *(pars caeca retinae)*.

Its contents are represented by the transparent media: the cornea, *the aqueous humour*, the lens and the vitreous body.

The ocular fibrous tunic (the sclera)
(Tunica fibrosa bulbi)

The sclera is the outer connective, fibrous, resistant, thick, inextensible membrane of the eyeball, it is the homologue of the dura mater, because the eyeball is considered from the viewpoint of its genesis, as a specialized diverticulum of the diencephalon. Schematically, the sclera represents the five posterior sixths of a sphere. Its external surface is white, whereas its internal surface is brown owing to a pigmented layer, called *lamina fusca sclerae*, which connects it to the vascular tunic. At the surface it is in relationship with the motor muscles of the eyeball and with Tenon's capsule *(vagina bulbi)*. It is provided with numerous orifices:

a) at the superior pole are sited the orifices of the optic nerve, a perforated plate, called *the lamina cribrosa sclerae*, through which pass the axons of the ganglionic cells that form the optic nerve. The orifices lie at 3 mm medially from the posterior pole of the eye. Owing to this position, the overlapping of the emergence of the optic nerve (the optic papilla) with the most sensible zone of the retina *(macula lutea)*, situated at the posterior end of the visual axis, is avoided. Around this chief orifice are sited the orifices of the posterior ciliary arteries and nerves (15-20).

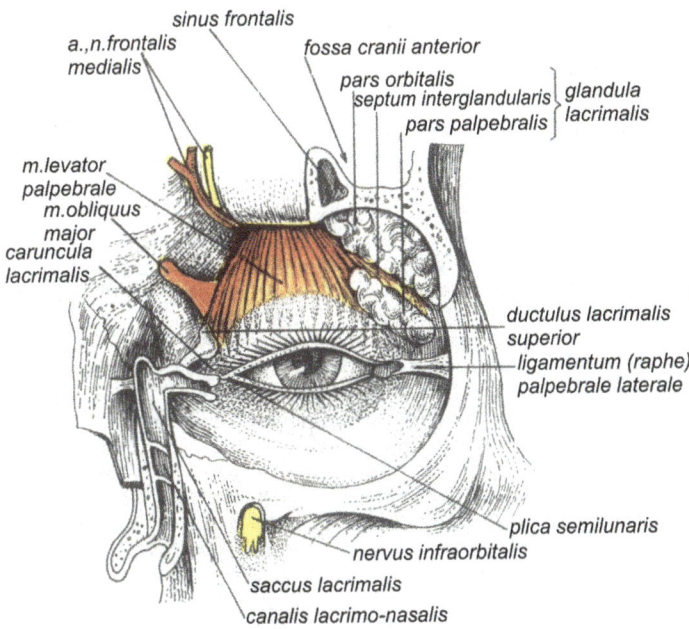

Fig. 187 Lacrimal system

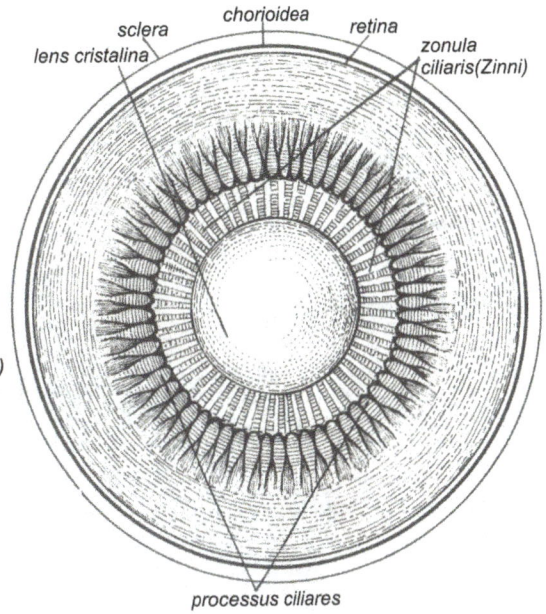

Fig. 188 Iris, ciliary zone (Zinn)(scheme)

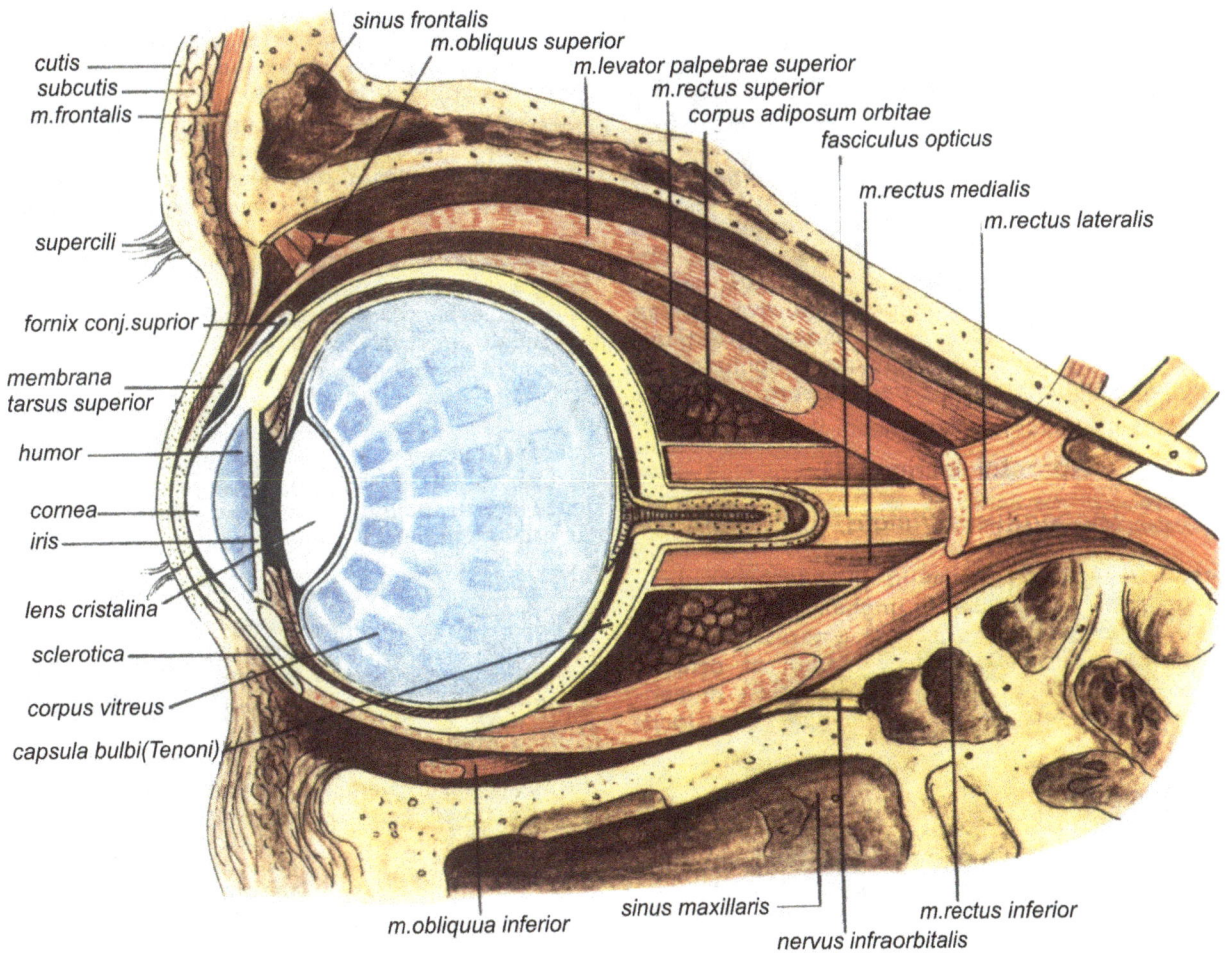

Fig. 189 Mediosagittal section through the orbit with demonstration of the eyeball and of its annexae

213

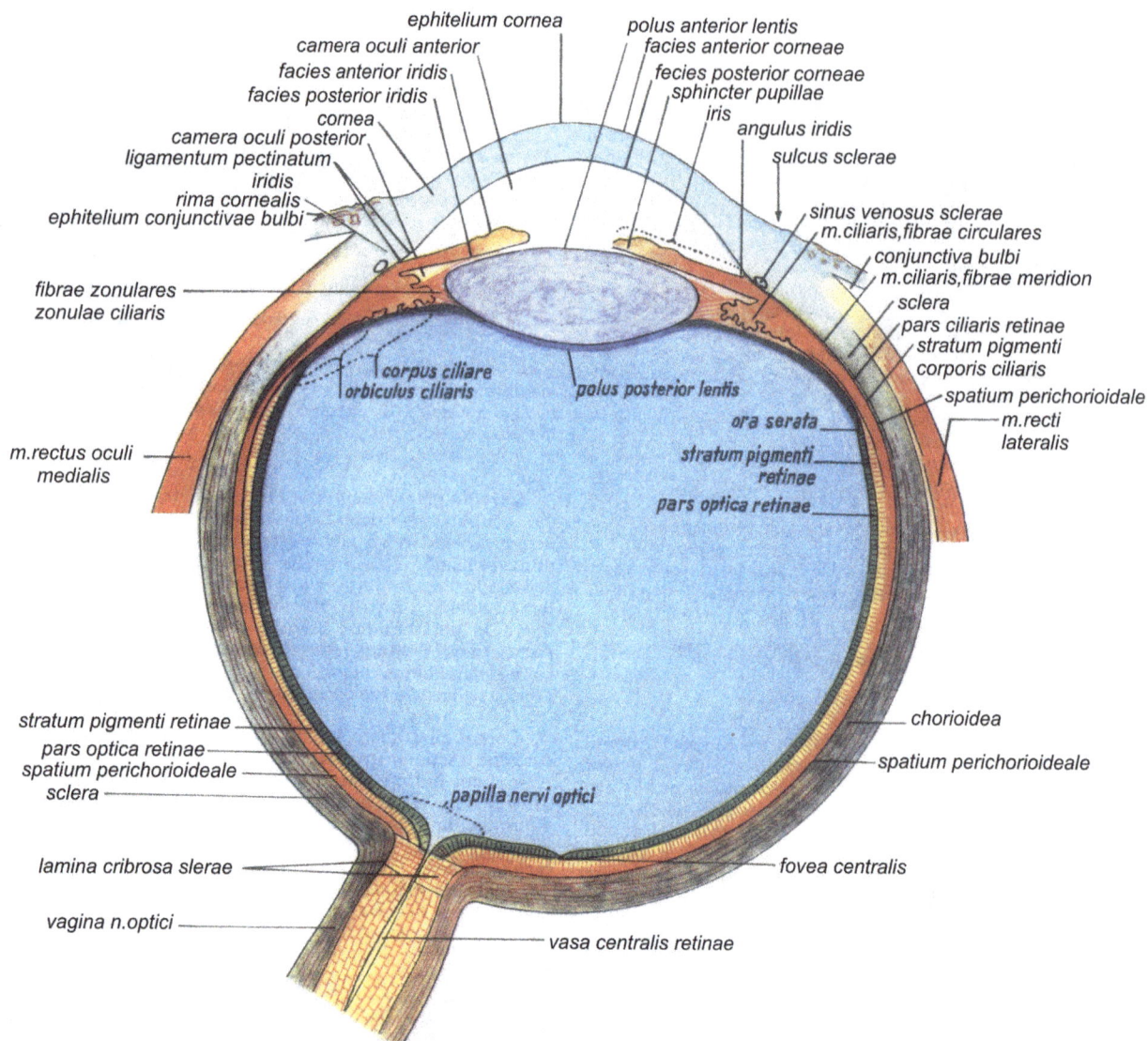

ephitelium cornea
camera oculi anterior
facies anterior iridis
facies posterior iridis
cornea
camera oculi posterior
ligamentum pectinatum
iridis
rima cornealis
ephitelium conjunctivae bulbi
fibrae zonulares
zonulae ciliaris
corpus ciliare
orbiculus ciliaris
m.rectus oculi
medialis
stratum pigmenti retinae
pars optica retinae
spatium perichorioideale
sclera
lamina cribrosa slerae
vagina n.optici
polus anterior lentis
facies anterior corneae
fecies posterior corneae
sphincter pupillae
iris
angulus iridis
sulcus sclerae
sinus venosus sclerae
m.ciliaris,fibrae circulares
conjunctiva bulbi
m.ciliaris,fibrae meridion
sclera
pars ciliaris retinae
stratum pigmenti
corporis ciliaris
spatium perichorioidale
m.recti
lateralis
polus posterior lentis
ora serata
stratum pigmenti
retinae
pars optica retinae
chorioidea
spatium perichorioideale
papilla nervi optici
fovea centralis
vasa centralis retinae

Fig. 190 Eyeball (tunics) - scheme

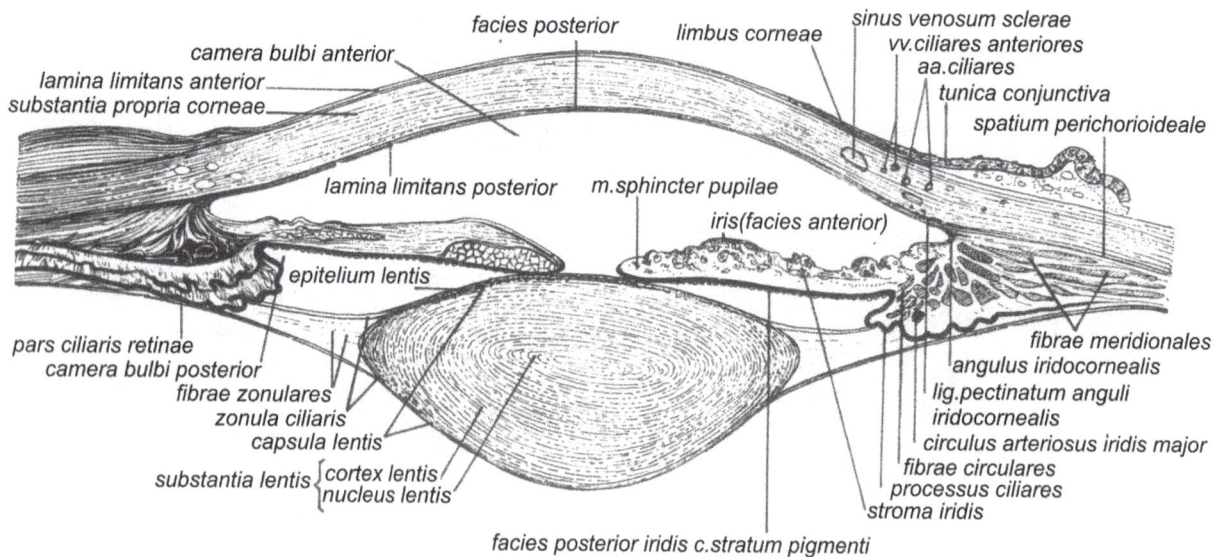

camera bulbi anterior
lamina limitans anterior
substantia propria corneae
facies posterior
limbus corneae
sinus venosum sclerae
vv.ciliares anteriores
aa.ciliares
tunica conjunctiva
spatium perichorioideale
lamina limitans posterior
m.sphincter pupilae
iris(facies anterior)
epitelium lentis
pars ciliaris retinae
camera bulbi posterior
fibrae zonulares
zonula ciliaris
capsula lentis
substantia lentis { cortex lentis
nucleus lentis
fibrae meridionales
angulus iridocornealis
lig.pectinatum anguli
iridocornealis
circulus arteriosus iridis major
fibrae circulares
processus ciliares
stroma iridis
facies posterior iridis c.stratum pigmenti

Fig. 191 Lens and ciliary body

214

a) Somewhat posteriorly to the median frontal plane, named *equator*, four apertures are present through which pass *the venae vorticosae* or *vortex veins* (veins of the choroid).

b) At the anterior pole of the eyeball lies, the corneal aperture, a large orifice in which the cornea is lodged; small orifices of the anterior ciliary arteries and veins are situated around.

The cornea. It is transparent portion included in the sclera, completing the anterior sixth of the sclera. The cornea bulges anteriorly, since the radius of its curvature is smaller (8 mm) than that of the sclera (the radii of which are of about 12 mm); the irregularities of its curvature contribute to the occurence of astigmatism.

The upper surface of the cornea bounds the anterior chamber of the eyeball, which contains *the aqueous humour*.

The cornea is devoid of blood vessels, but it is richly supplied by the ciliary nerves; during general anaesthesia, its sensibility is the last which disappears.

The so-called *limbus corneae* is the site of the sclerocorneal junction the vascular tunic adheres here to the deep part. Connective elastic fibers are organized in a pectinate ligament of the iridocorneal angle, which extends posteriorly through the ciliary muscle and the periphery of the iris.

A ring-shaped canal, devoid of an own wall encircles the limbus; this is the venous sinus of the sclera *(sinus venous sclerae)* termed also Schlemm's canal.

Structurally the cornea consists of five layers: the stratified pavement epithelium, which is continuous with the epithelium of the bulbar conjunctiva; the anterior limiting lamina of cornea *(lamina limitens anterior corneae)* or Bowman's membrane, *the substantia propria*; the posterior limiting lamina of cornea *(lamina limitans posterior corneae)*; called also Descemet's membrane and the endothelium of the anterior chamber.

The vascular tunic of the eyeball
(Tunica vasculosa bulbi)

It is a musculovascular membrane, lining the inner surface of the fibrous membrane. It adheres to the deep surface of the sclera, but remains away from the cornea. It consists behind of the choroid, the ciliary body and the iris, all comprised sometimes under the term uveal tract.

The choroid *(choroidea)*. Situated between the sclera and the retina it occupies the two posterior thirds of the eyeball. As is the homologue of the pia mater, it has essentially a vascular structure. The vessels (the posterior short ciliary arteries) pass through *the lamina fusca (suprachoroid lamina)* to reach the choroid. Its smooth, pigmented internal surface correspond to the retina, without adhering to it. Behind, the choroid is pierced by the optic nerve. Anteriorly, it is continuous with the ciliary portion – an irregular circular line, named *ora serrata*, delineates the two above-mentioned thirds of the vascular tunic.

The ciliary body *(corpus ciliare)*. It has the shape of a flattened ring, which on the transverse section is triangular and lies in front of the equator of the eyeball, between t*he ora serrata* and the iris. The external surface corresponds to the sclera; the internal surface consists of two segments: an anterior segment, formed of the ciliary processes *(processus ciliares)* and a posterior segment constituted of *the orbiculus ciliaris* (ciliary ring, *pars plana*).

The ciliary processes are vascular balls dependent on the long ciliary arteries and contained in a loose connective tissue, they are arranged in the shape of a ciliary crown (corona ciliaris).

The orbiculus ciliaris is a plicated concentric zone, following the ciliar crown. The ciliary muscle *(m. ciliaris)* occupies the external part of the ciliary body and inserts on the anterior extremity of the sclera; it is constituted of circular fibers (Rouget's or Müller's muscle) and anteroposterior meridional fibers (Brücke's muscle), which continue with radial fibers. The ciliary muscle is richly supplied with nerves and, by its action, is the muscle of accommodation to near and to far vision, acting on Zinn's zonule and, through it, on the lens.

The iris *(iris)*. It is a vertical, circular diaphragm, which controls the amount of light entering the eye. It is slightly concave behind and presents in the center an orifice named pupil *(pupilla)*.

The large circumference of the iris is continuous, at the level of the corneal limbus and of the pectinate ligament *(lig. pectinatum anguli iridocornealis)*, with the ciliary body. Its posterior surface is black and oriented towards the anterior surface of the lens. The anterior surface of the iris exhibits radial prominences; its color varies individually according to the amount of pigment contained in its cells. The iris is bathed in *the aqueous humour* and divides the space containing this fluid into an anterior and a posterior chamber. It contains muscle fibers which form a circular muscle, the sphincter of the pupil *(m. sphincter pupillae)*, and a radial dilatory muscle *(m. dilatator pupillae)*.

It is supplied by the long ciliary and anterior ciliary arteries, which form, at the periphery of the iris, a large arterial circle called the greater arterial circle of the iris *(circulus arteriosus iridis major)*. The iris is supplied by the short ciliary nerves, branches of the ciliary or ofthalmic ganglion, and by the long ciliary nerves, branches of the nasal nerve. The nature of the light influx or the distance from the object regarded induce in a reflex way the contraction or relaxation of the muscles of the iris, a reflex frequently investigated in the course of the clinical examination.

The nervous tunic (the retina)
Tunica interna bulbi

It is made up by the retina, which is functionally the chief part of the eyeball and is, at the same time, the deepest part. It extends from the optic nerve to the pupil and consists of three segments:
- the optic part of the retina *(pars optica)*, up to *the ora serrata*;
- the ciliary part *(pars ciliaris)*, up to the iris; *pars caeca retinae*;
- the iridial part *(pars iridica)*, up to the pupil

The optic retina is the reception apparatus of visual excitations. It is situated posteriorly, in relation with the choroid, but without adhering to it.

The visual retina extends from the entrance point of the optic nerve into the eyeball nearly up to the ciliary body, where it ends by a clear-cut margin, visible to the naked eye, named *ora serrata* (toothed margin).

The visual retina presents a small region, called the yellow spot *(macula lutea)*, adapted for the precise vision of objects.

The optic retina is a membrane, the thickness of which diminishes from behind forward, from 350 in the neighborhood of the pupil, the thickness decreases to 100µ near *the ora serrata*. It consists of ten superimposed layers which are from outside inwards, the following:

-The pigmented layer, situated immediately below the choroid and formed of pigment cells.

-The layer of rods and cones, called also striate layer 50-60µ in thickness;

-The outer limiting lamina, which is a very thin membrane, formed of the external projections of the neuroglial (Müller's) cells, situated in the layer of bipolar neurons;

- The external granular layer (or external nuclear layer) 60µ thick, consisting of 5-6 rows of nuclei belonging to the visual cells-rods and cones;

- The external plexiform layer, formed of the synaptic connections between the visual cells and the dendrites of bipolar neurons;

- The internal granula layer (or internal nuclear layer), less thick than the external granular (nuclear) layer; it is formed of nuclei of the bipolar nerve cells, nuclei of the association cells (horizontal cells and amacrine cells) and nuclei of the Müller's neuroglial cells;

- The internal plexiform layer, resulted like the outer plexiform layer from the synaptic connections of the bipolar and multipolar cells;

216

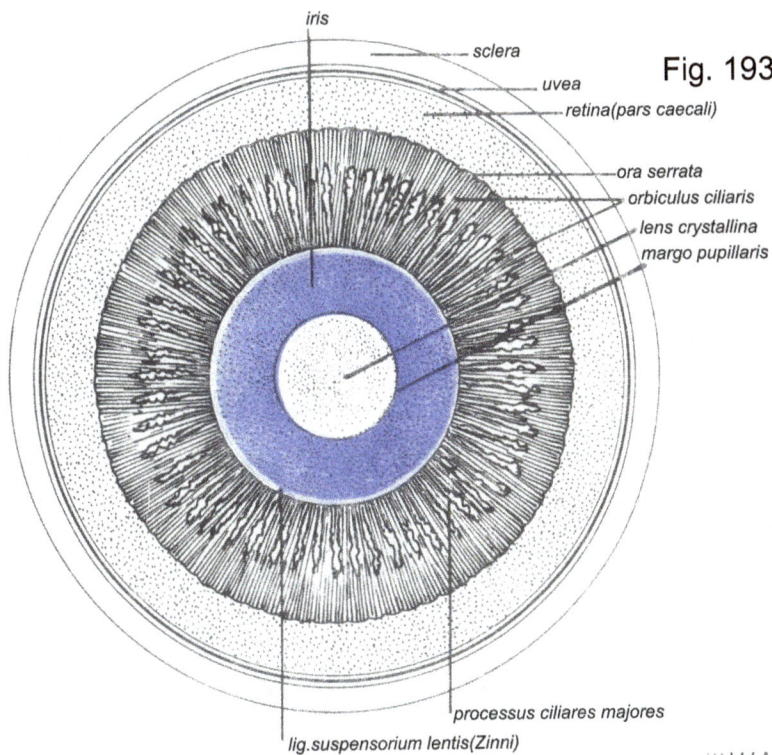

Fig. 193 Iris, ciliary processes (scheme)

iris
sclera
uvea
retina(pars caecali)
ora serrata
orbiculus ciliaris
lens crystallina
margo pupillaris
processus ciliares majores
lig.suspensorium lentis(Zinni)

Fig. 192 Orbit and eyeball

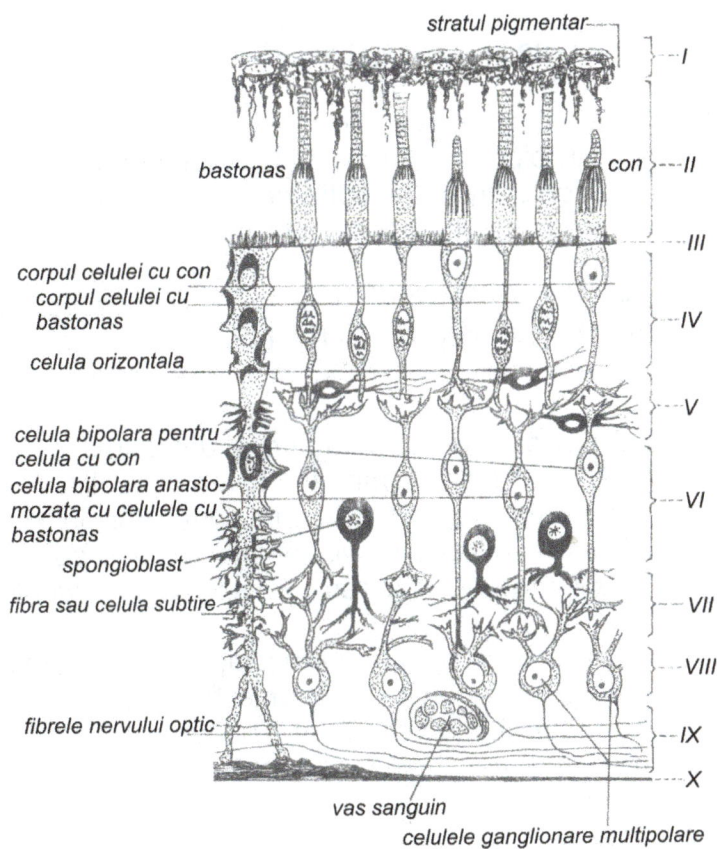

pupila
iris
process. ciliares
cornea
humor aqueus
lens cristallina
ora serrata
m.obliquus inf.
os zygomt.
chorioidea
corpus vitreum
macula lutea
retina
sclera
lam.papyr.oss.elhm.
m.rectus ext.
m.temporalis
n.opticus
n.opticus
ala magna oss. sphen.
m.rectus int.
for.opt.

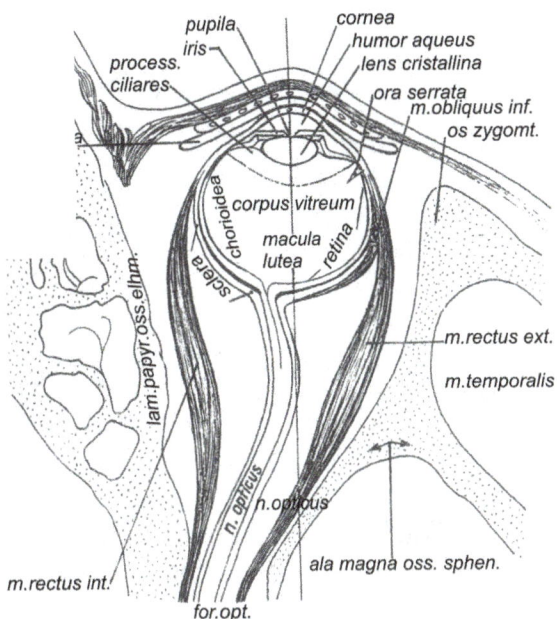

stratul pigmentar
bastonas
con
I
II
III
corpul celulei cu con
corpul celulei cu bastonas
celula orizontala
IV
V
celula bipolara pentru celula cu con
celula bipolara anasto-mozata cu celulele cu bastonas
spongioblast
fibra sau celula subtire
fibrele nervului optic
VI
VII
VIII
IX
X
vas sanguin
celulele ganglionare multipolare

Fig. 194 Histological structure of the retina

217

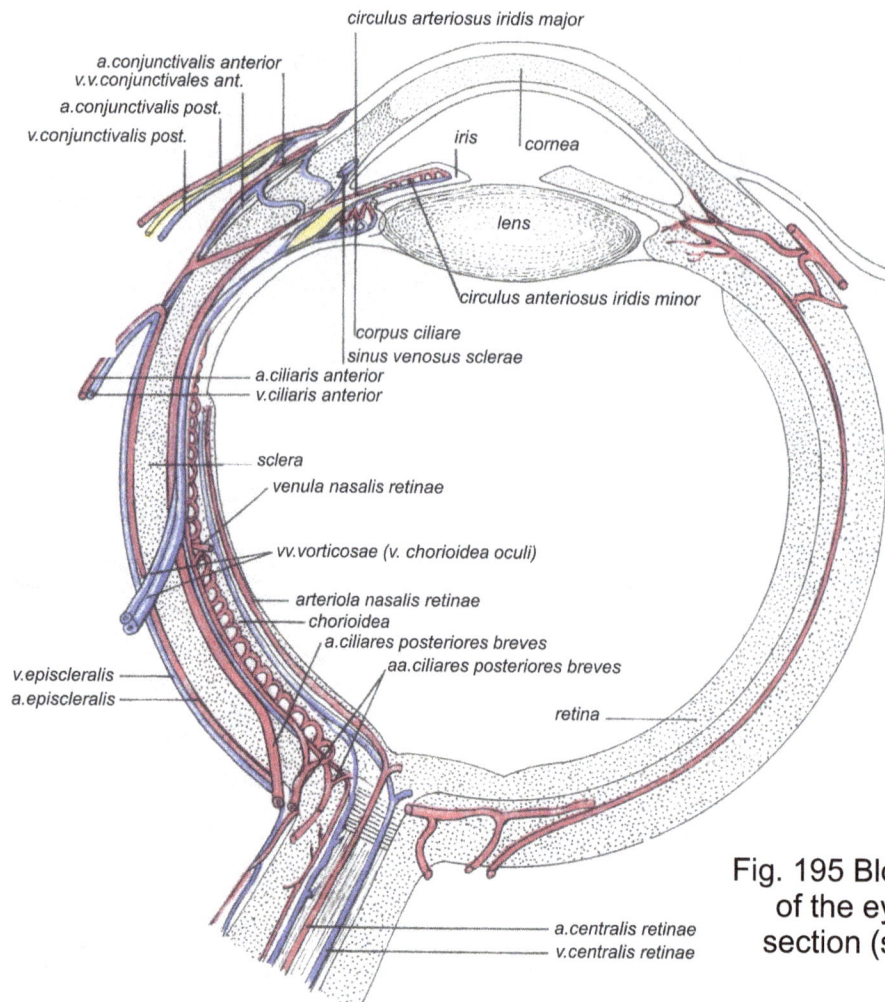

circulus arteriosus iridis major

a.conjunctivalis anterior
v.v.conjunctivales ant.
a.conjunctivalis post.
v.conjunctivalis post.

iris | cornea

lens

circulus anteriosus iridis minor

corpus ciliare
sinus venosus sclerae
a.ciliaris anterior
v.ciliaris anterior

sclera
venula nasalis retinae

vv.vorticosae (v. chorioidea oculi)

arteriola nasalis retinae
chorioidea
a.ciliares posteriores breves
aa.ciliares posteriores breves

v.episcleralis
a.episcleralis

retina

a.centralis retinae
v.centralis retinae

Fig. 195 Blood supply
of the eyeball –
section (scheme)

- The layer of multipolar neurons;
- The layer of optic fibers, formed of non-myelinated fibers with a parallel course to the surface of the retina;
- The internal limiting lamina, a thin membrane separating the retina from the vitreous body it resulted from the fusion of inner projections of Müller's cells.

Synthesizing, we mention that the visual sensory cells, with rods or cones (neuro-epithelial layer) are in connection (synapse) with bipolar neurons of the optic tracts, they anastomose with the multipolar neurons (ganglionic layers of optic nerves), the axons of which form the optic nerve; the latter leaves the eyeball, passing into the orbit and then, through the optic foramen, penetrating into the skull.

The papilla of the optic nerve, called also optic disc (discus n. optici), is a clear circular spot of 1.5 cm in diameter, functionally blind, situated medially to the posterior pole of the eyeball; it corresponds to the formation site of the optic nerve and of the opthalmic artery, hence to the vasculonervous hilum.

The macula lutea or yelow spot is an elliptic area of 1.5 by 3 mm, situated just in the posterior pole of the eyeball. It shows a central depression, termed the fovea centralis. The structure of the retina is modified at the level of the macula: only cells with cones specialized for the most precise and colored vision are present. Here, the visual impressions do not undergo diffusion and the most precise and clear images are formed ("we see with the whole retina, but we look with the macula").

Two axes perpendicular on each other passing through the macula, divide the optic retina into four quadrants: the superomedial and the inferomedial quadrants, forming together the nasal field (oriented laterally) and the superolateral and inferolateral quadrants, forming together the temporal field (oriented to the medial or nasal part); the light rays cross each other.

218

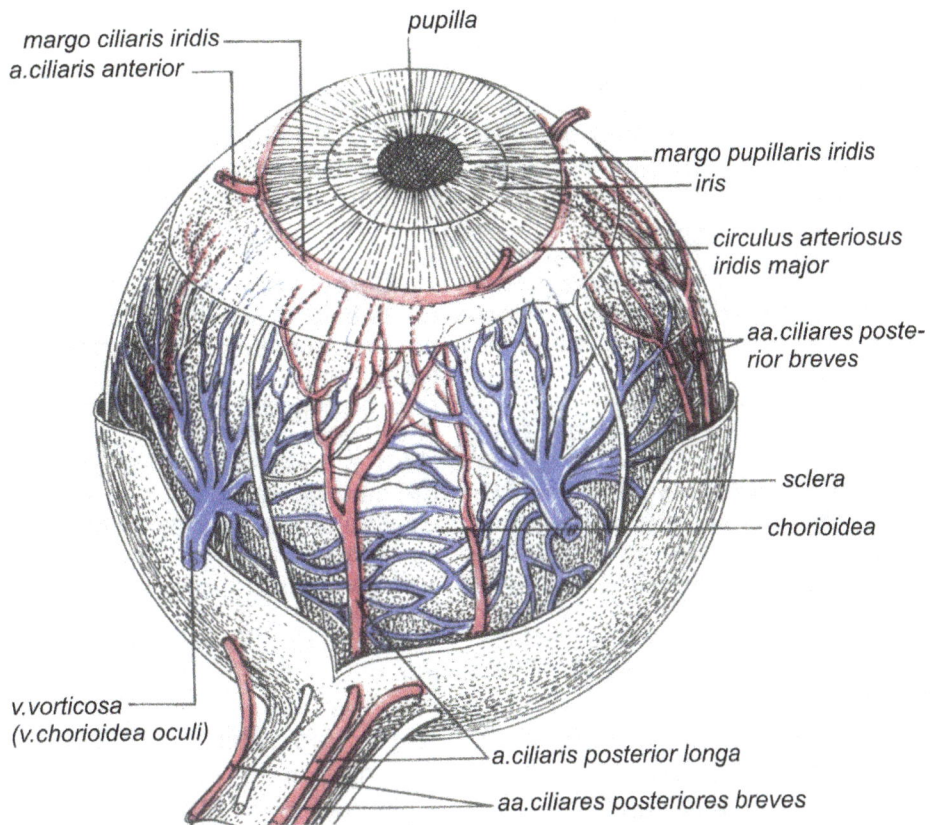

margo ciliaris iridis
a.ciliaris anterior

pupilla

margo pupillaris iridis
iris

circulus arteriosus
iridis major

aa.ciliares poste-
rior breves

sclera

chorioidea

v.vorticosa
(v.chorioidea oculi)

a.ciliaris posterior longa

aa.ciliares posteriores breves

Fig. 196 Eyeball – blood supply (scheme)

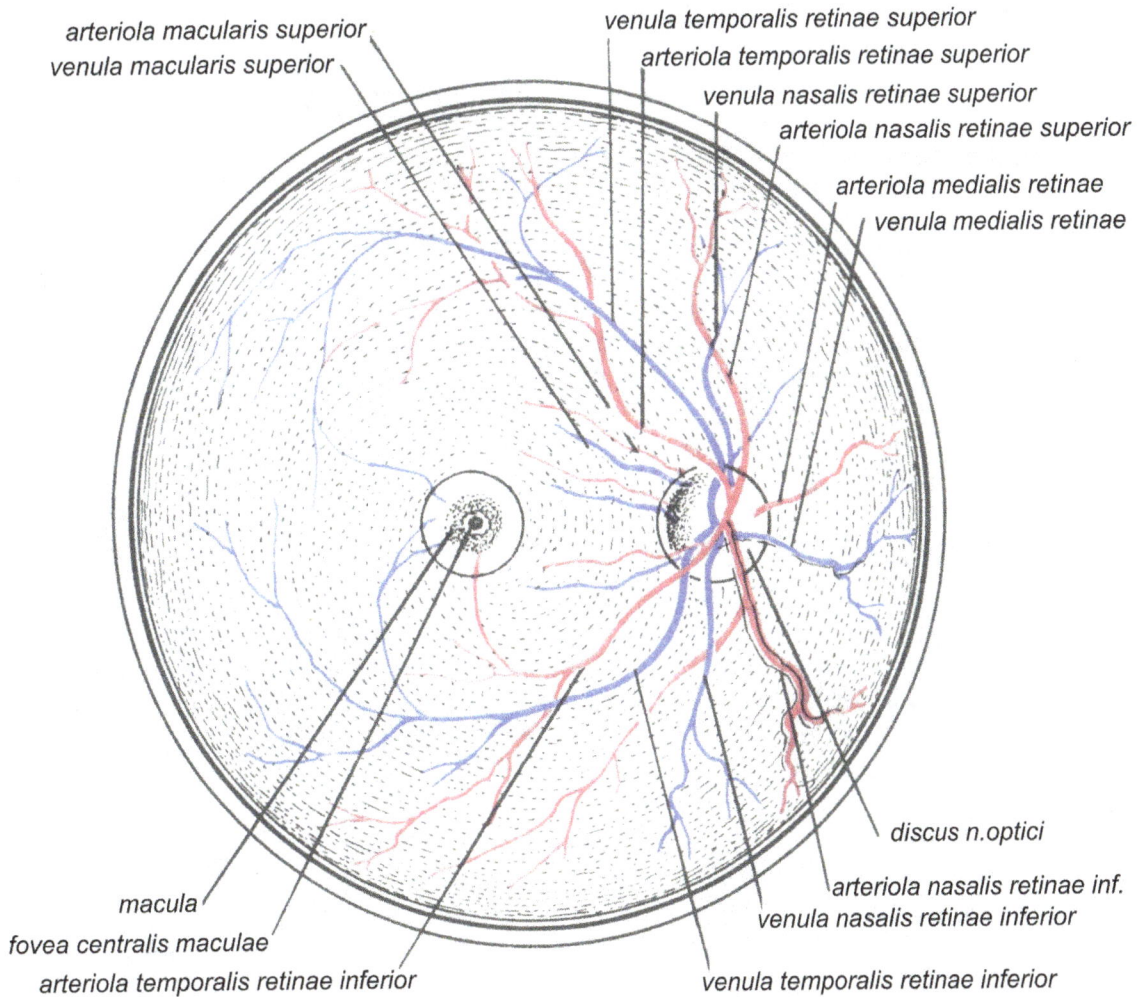

arteriola macularis superior
venula macularis superior

venula temporalis retinae superior
arteriola temporalis retinae superior

venula nasalis retinae superior
arteriola nasalis retinae superior

arteriola medialis retinae
venula medialis retinae

discus n.optici

arteriola nasalis retinae inf.
venula nasalis retinae inferior

macula
fovea centralis maculae
arteriola temporalis retinae inferior

venula temporalis retinae inferior

Fig. 197 Blood supply and macroscopic aspect of the retina

219

The **cilio-iridial retina** *(pars ciliare et pars iridica retinae)* is much thinner; it begins at *the ora serrata* and lines the internal surface of the ciliary body and the posterior surface of the iris; it is devoid of visual cells; this is the blind retina *(pars caeca retinae)*.

The ciliary portion of the retina extends from the ora serrata to the ciliary margin of the iris; it is formed of two layers of epithelial cells: the outer, pigmented layer, which continues in this area the pigmented layer of the visual retina, and the inner layer, formed of modified supporting cells and covered on the inner surface with a cuticle, which is a prolongation of the internal limiting lamina of retina.

The cells of the inner layer form a system of inextensible and transparent fibers inserted on the lens, in its equatorial region, constituting the fibers of Zinn's ciliary (the suspensory ligament of the lens). In the zone of the ciliary processes, these cells, possessing secretory properties, produce *the aqueous humour*.

The iridial part of the retina is a prolongation of the ciliary retina and lines the posterior surface of the iris; it is formed of two layers: an anterior layer, constituted of myo-epithelial cells (dilator muscle of the iris), and a posterior, pigmented layer, formed of cubic pigment cells (posterior epithelium of the iris)

Vessel supply. The vessels of the retina arise from a branch of the ophthalamic artery, termed the central retinal artery, which follows the course of the optic nerve; starting from the center of the papilla, it divides into an ascending and a descending branch, ending in a capillary network.

Three arterioles, a superior, an inferior and a medial one, supply the region of the macula. Anastomoses do not exist between the terminal branches of the central artery of the retina, neither between the latter and the choroidal arteries: they all these vessels are terminal arteries An embolus leads to the sudden and permanent loss of sight in the affected territory.

A central retinal vein drains the venules more or less satellites of the arterioles and empties into the ophthalmic vein.

The transparent media contained in the eyeball

The **aqueous humour** *(umor aquasus)*. It is a colorless, clear fluid, filling the anterior chamber of the eyeball, between the cornea and the lens. The iris divides this space into two "subchambers": anterior *(camera anterior bulbi)* and posterior *(camera posterior bulbi)*; which are communicating with each other through the pupillary aperture. *The aqueous humour* is secreted by the ciliary processes and drained through the venous iridocorneal sinus of the sclera (Schlemm's canal).

The **lens.** It has the shape of a biconvex lens and is situated behind the iris and in front of the vitreous body, at the level of the ciliary processes and of the ciliary muscle. The convexity of its posterior surface is more marked then that of its anterior surface. It is enclosed by a capsule *(capsula lentis)*, of which it may be extracted operatively in the case of opacification or cataract.

It is maintained in its position by a suspensory ligament (Zinn's zonule). It consists of transparent fibers running from the internal surface of the ciliary body to the periphery of the lens on the capsule of the lens.

At rest, the zonule is stretched and the lens is adjusted for the far sight. When the ciliary muscle contracts the zonule relaxes, the convexity increases and the near vision is possible; this is the mechanism of accommodation to variation in distance. In the older adults, the loss of elasticity of the lens produces hypermetropia (hyperopia) and the loss of transparency causes cataract.

The lens is devoid of blood vessels, as well as of nerves. The nutrition of the lens is achieved by diffusion from the level of the vessels of the ciliary processes.

The **vitreous body** *(corpus vitreum)*. It is situated behind the lens and the zonule and has the appearance of a viscous and transparent fluid, surrounded by the vitreous membrane *(membrana vitrae)*.

220

On the anterior surface there is a depression, the hyaloid fossa *(fossa hyaloidea)*, in which lies the posterior surface of the lens. Sometimes, the vitreous body is crossed, from behind forwards, by a canal called hyaloid canal *(canalis hyaloideus)*, which is the vestige of the hyaloid artery of the embryonic life (in pathological cases, in the vitreous body may appear flocks, leaving to the respective individual the impression of *muscae volitantes* (floaters).

Beside the fact that it represents a refractive medium, the vitreous body contributes to the maintenance of the lens in its position and prevents the retinal detachment; it has also a role in the nutrition of this nervous tunic of the eyeball.

The retrobulbar region

It is the space situated behind the peribulbar capsule. It contains retrobulbar fat (the orbital fat-part), enveloped by a connective fascia, a component of the peribulbar capsule.

On this fascia is inserted the orbital muscle, which has its origin in the bottom of the orbit, in the lateral part of the superior orbital fissure. Its action consists in the compression of the retrobulbar fat and, consequently in the production of exophthalamia, as this muscle is supplied by the cervical orthosympathetic nerve. The paralysis of the cervical orthosympathetic nerve causes the appearance of the Claude Bernard-Horner's syndrome, characterized by enophthalmia (consequence of the paralysis of the orbital muscle), myosis (consequence of the paralysis of the pupillar dilatory muscle), narrowing of the palpebral fissure (consequence of the paresis of the tarsal muscle).

In the retrobulbar region are also situated the muscles and neurovascular formations of the eyeball.

The muscles of the eyeball

They are six in number: four *recti* muscles – superior *(m. rectus superior)*, lateral *(m. rectus lateralis)*, medial *(m. rectus medialis)* and inferior *(m. rectus inferior)* and two oblique muscles: superior *(m. obliquus superior)* and inferior *(m. obliquus inferior)*, to which should be added the levator muscle of the superior eyelid *(m. levator palpebral superior)*, situated in the palpebral region (Fig. 199-200).

These muscles present a fascia which is a prolongation of the peribulbar capsule and is adherent to the body of the muscles, but which at the level of the tendons, inserted on the sclera, is separated from them through a loose connective tissue. Thus, the fascia becomes similar to a synovial membrane, permitting the slipping of tendons, which is very important in operation for strabismus.

These muscles are supplied by the following nerves:
- the superior rectus muscle and the levator muscle of the eyelid by the superior terminal branch of the oculomotor nerve (third cranial nerve);
- the inferior and medial *recti* muscles and the inferior oblique muscle by the inferior terminal branch of the oculomotor nerve (third cranial nerve pair);
- the lateral rectus muscle, by the m. abducent nerve (sixth cranial nerve);
- the superior oblique muscle by the trochlear nerve (fourth cranial nerve).

The action of these muscles may be systematized as follows: the *recti* muscles draw the eyeball each to its side, as they are antagonists; in addition, the superior and inferior *recti* muscles perform, agonistically, adduction movements; the oblique muscles produce together the abduction of the eyeball, but individually they are antagonistic in its rotation movement.

The vasculonervous formations
The arteries and veins

The ophthalmic artery *(arteries ophthalmica)*, it is a collateral artery of the internal carotid artery; it supplies all the structures contained in the orbital cavity.

During its passage through the petrous part of the temporal bone, the internal carotid artery (petrous part) describes a curve at a right angle, then it straightens and ascends the flank of the body of the sphenoid bone, passes through the cavernous sinus, describes a last curve and traverses the dura mater and the arachnoid; medially to the anterior clinoid process, it lies in the subarachnoid space.

Before dividing into its terminal branches (the anterior and middle or sylvian cerebral arteries), the internal carotid artery gives rise to the ophthalmic artery, a collateral branch situated below and medially to the anterior clinoid process.

The ophthalmic artery runs forwards, passes through the optic canal, inferolateral to the optic nerve, after leaving the canal, it traverses again the dura mater, penetrates into the orbital cavity and surrounds the external and then the superior surface of the optic nerve. It leaves the insertion cone of the *recti* muscles, and courses along the inferior border of the superior oblique muscle.

It ends in the internal angle of the orbit, forming an anastomosis with the facial artery: the angular artery.

Branches. In its course, the ophthalmic artery gives off numerous collaterals.

1. Optic branches:

- the central artery of the retina *(arteria centralis retinae)* penetrates into the optic nerve at 1 cm behind the eyeball;

- numerous ciliary arteries, which ramify in the choroid (posterior short and posterior long ciliary arteries) and in the iris (anterior ciliary arteries).

2. Branches to the adnexa of the eyeball:

- the lacrimal artery *(arteria lacrimalis)* runs along the supero-external angle of the orbit and supplies the lacrimal gland and the adjacent integuments;

- the supraorbital artery *(arteria supraorbitalis)*, symmetrical and medial to the lacrimal artery emerges from the supraorbital foramen and terminates in the frontal area;

- the inferior muscular artery supplies the inferior and lateral *recti* muscles and the inferior oblique muscle;

- the superior muscular artery supplies the superior and medial *recti* muscles, the levator muscle of the superior eyelid and the superior oblique muscle;

- the two ethmoidal arteries, anterior *(a. ethmoidalis anterior)* and posterior *(a. ethmoidalis posterior)*, pass through the ethmoidal foramina, traverse the cribriform plate of the ethmoid bone and reach *the nasal fossae*. The posterior ethmoidal artery can supply only the dura mater which covers the cribriform plate *(lamina cribrosa)*;

- the two palpebral arteries *(aa. palpebrales)*, superior and inferior;

- the supratrochlear (internal frontal) artery reaches the frontal integument;

- the facial artery supplies partly the eyelids;

- the suborbital artery *(a. infraorbitalis)*, a branch of the internal maxillary artery *(arteria maxillaris)*, penetrate into the orbit through the inferior orbital fissure and passes immediately into the suborbital canal.

The superior ophthalmic vein *(vena ophthalmica superior)* begins in the internal angle of the orbit and anastomoses with the facial vein, hence an infection of the face can be transmitted to the cavernous sinus, into which the ophthalmic vein empties. It is a satellite of the opthalmic artery and drains the veins of the superior half of the orbit.

The central vein of the retina *(vena centralis retinae)*, which is very thin, may empty into one of the ophthalmic veins or may enter the cavernous sinus *(sinus cavernosus)*, as all these veins pass through the sphenoidal fissure called also the superior orbital fissure *(fissura orbitalis superior)* in its widened part (outside the common tendinous ring) and terminates in the anterior part of the cavernous sinus.

The inferior ophthalmic vein *(vena ophthalmica inferior)* begins in the inferomedial angle of the orbit, collects the blood from the inferior rectus and inferior oblique muscles, from the lacrimal sac and from the eyelids and empties into the superior ophthalmic vein or directly into the cavernous sinus.

222

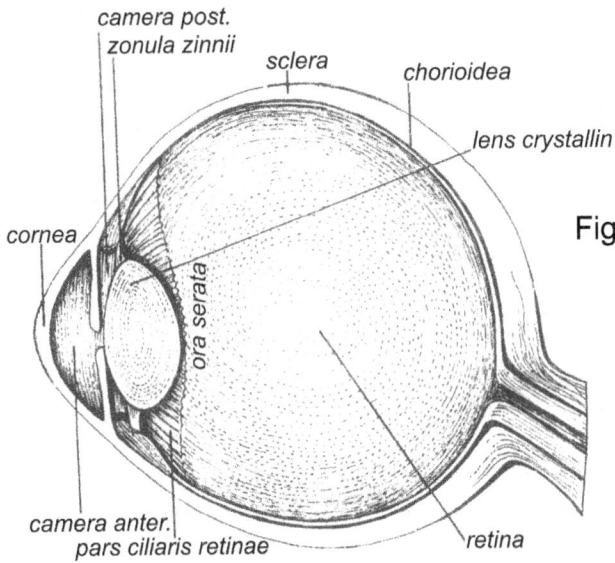

Fig. 198 Sagittal section through the eyeball, showing the refraction media

camera post.
zonula zinnii
sclera
chorioidea
lens crystallin
cornea
ora serata
camera anter.
pars ciliaris retinae
retina

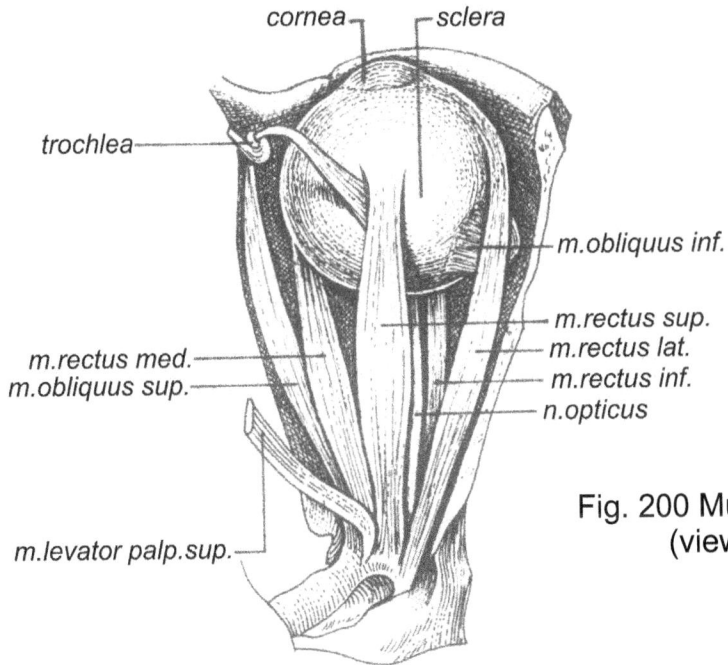

Fig. 199 Muscles of the eyeball (anterior view)

m. rectus lateralis
m.rectus superior
m.obliquus superior
capsula bulbi (Tenoni)
fascia bulbi (Tenoni)
m.rectus medialis
m.rectus inferior
m.obliquus inferior

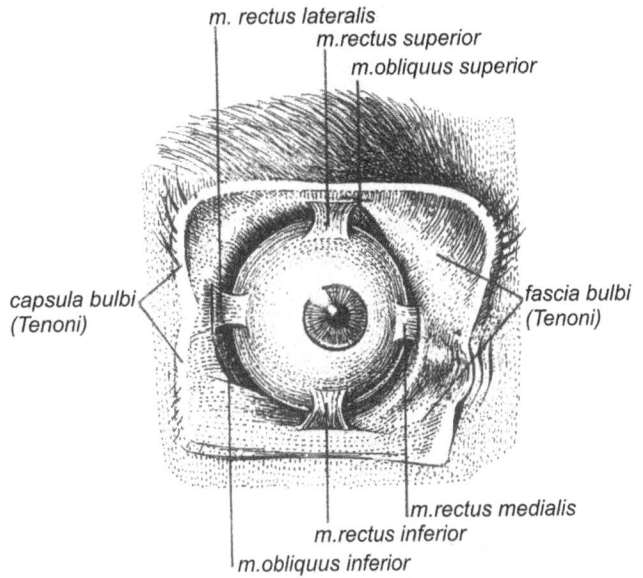

Fig. 200 Muscles of the eyeball (view from above)

cornea
sclera
trochlea
m.obliquus inf.
m.rectus sup.
m.rectus lat.
m.rectus inf.
n.opticus
m.rectus med.
m.obliquus sup.
m.levator palp.sup.

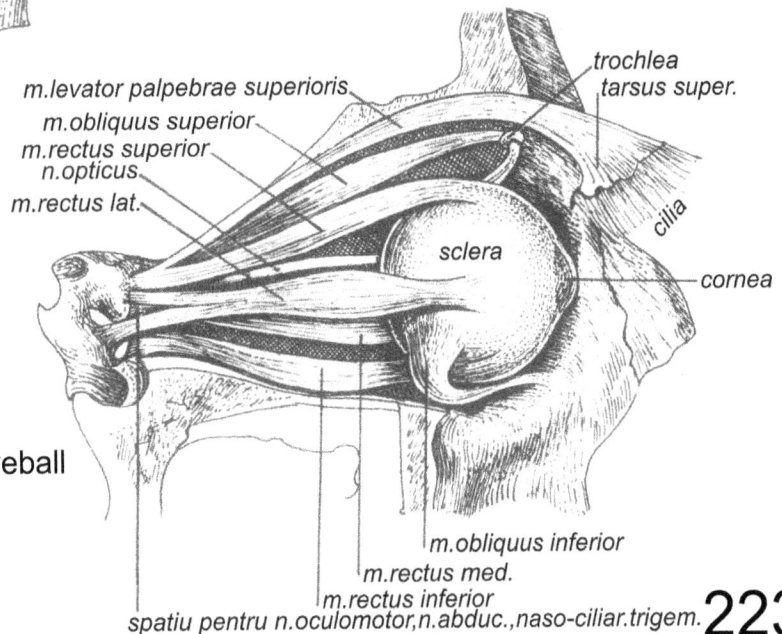

Fig. 201 Muscles of the eyeball (lateral view)

m.levator palpebrae superioris
m.obliquus superior
m.rectus superior
n.opticus
m.rectus lat.
trochlea
tarsus super.
cilia
sclera
cornea
m.obliquus inferior
m.rectus med.
m.rectus inferior
spatiu pentru n.oculomotor,n.abduc.,naso-ciliar.trigem.

223

The motor nerves of the eyeball

The oculomotor nerve (*nerve oculomotorius*, third cranial nerve) supplies all the extrinsic muscles of the eyeball (except the lateral rectus and the superior oblique muscles) and as the internal muscles of the eyeball, with the center in Westphal-Edinger's nuclei, among which lies the median nucleus, named Perlia's nucleus. Its fibers are responsible for the function of convergence. It arises from the periependymal motor nuclei of the mesencephalon, appears at the medial border of the eyelids respective cerebral peduncle (in the interpeduncular fossa), has an anterior and lateral course, enters the cavernous sinus and runs towards the orbit. It is contained in the deep lamina of the lateral wall of the sinus and divides into two terminal branches. In the wall of the sinus it is initially situated above, then it descends a little and is crossed by the trochlear nerve and by the terminal branches of the ophthalmic nerve.

Thus, it reaches the superior orbital fissure and enters into the orbit. It supplies the superior rectus, the levator of the upper eyelid (through the superior branch), the medial and inferior *recti* and the inferior oblique muscles through the inferior branch. Moreover, it gives off fibers to the ciliary ganglion (the short root), which conveys iridoconstrictor messages (the pupillar sphincter), and to the ciliary muscle; it receives an anastomosis of the sympathetic plexus, for the levator muscle of the superior eyelid.

The trochlear nerve (*nervus trochlearis*, fourth cranial nerve) is the only cranial nerve which arises from the dorsal side of the brainstem. Its fibers assure from a nucleus situated in the mesencephalon, too, but on the opposite side and appears laterally to the anterior medullary velum *(velum medullare anterius)* or Vieussens' valvula and surrounds the central peduncle, passing below the optic tract and the posterior cerebral artery.

Then it enters the cavernous sinus, inside which it runs along the deep lamina of the lateral wall. It crosses laterally the oculomotor nerve (third cranial nerve), which presents a reverse obliquity, then it leaves the cavernous sinus and passes into the orbit through the superior orbital fissure (its narrow part), lateral and above the common tendinous ring. It courses along the ceiling of the orbit, crosses the levator muscle of the superior eyelid and ends in the superior oblique muscle, which it supplies. Like the preceding nerve, it receives an anastomosis from the sympathetic plexus.

The abducent nerve (*nervus abducens*) (sixth cranial nerve) arises from the motor nervous in *the ponce*, emerges in the bulbopontine sulcus, courses forwards, laterally and upwards, traverses the dura mater in proximity to the petro-occipital sutura and adheres closely to the apex of the petrous part of the temporal bone (this explains its lesion in fractures of the base of the skull). It penetrate into the cavernous sinus, between the superior and inferior petrous sinuses; in this way it runs between the internal carotid artery and the deep lamina of the external wall of the sinus, lamina to which it adheres. It leaves the sinus and passes immediately through the superior orbital fissure (or sphenoidal fissure) and through the external part of Zinn's common tendinous ring. It terminates in the lateral rectus muscle which it supplies. It receives an anastomosis from the sympathetic plexus.

We mention at the same time that the nuclei of *the trochlear nerve* (fourth cranial nerve) and of *the abducent nerve* (sixth cranial nerve) are in a close connection with the nuclei of *the oculomotor nerve* (third cranial nerve) and form the primary motor centers of the muscles of the eyeball. They are the centers of voluntary and automatic movements of the eyeballs, movements which are conceivable only by a close synchronism between the motor nuclei on each side: the lateral rectus muscle, on the right side, contracts concomitantly with the left medial rectus, while the right medial rectus and the left lateral rectus relax in order to maintain the visual axes on the object looked at.

They are equally associated with the medullary nucleus of *the accessory nerve* (eleventh cranial nerve) and their association fibers form the intra-axial oculogyric system, starting point of reflex orientation movements (for example when a loud noise is heard, the head and the eye turn simultaneously to the respective direction).

224

The sensory nerves of the eyeball

The ophthalmic nerve (nervus ophthalmicus) is the superior branch of the trigeminal nerve; it is exclusively sensory. It divides into three terminal branches, which traverse the orbit: lacrimal, frontal and nasociliary. The ophthalmic nerve arises from the anteromedial part of the semilunar or gasserian ganglion of the trigeminal nerve.

The trunk of the ophthalmic nerves is situated in the deep lamina of the lateral wall of the cavernous sinus, in its inferior part, below the trochlear nerve. It gives off collateral meningeal branches, receives anastomoses from the internal carotid (sympathetic) plexus and from the three motor nerves of the eyeball, conveying also their proprioceptive sensibility. In the anterior part of the cavernous sinus it gives off the three terminal banches: lacrimal, frontal and nasociliary, which traverse the superior orbital fissure and penetrate into the orbit.

The lacrimal nerve (nervus lacrimalis). It passes through the lateral part of the superior orbital fissure, with the frontal nerve, then it courses along the lateral wall of the orbit above the lateral rectus muscle. It supplies the lacrimal gland and the lateral part of the skin of the superior eyelid. It receives a very important anastomosis from the zygomatico-orbital nerve, a branch of the maxillary nerve, which supplies the lacrimal gland, with parasympathetic secretory fibers.

The frontal nerve (n. frontalis). This nerve enters with the lacrimal nerve into the orbit, medially to the trochlear nerve in outside the common tendinous ring. It runs from behind forwards, between the root of the upper eyelid and the ceiling of the orbit. It ends through a medial supraorbital (internal frontal) branch, which passes above the trochlea of the superior oblique muscle and supplies the conjunctiva and the integument of the root of the nose and of the forehead, nearly to the vertex.

The nasociliary nerve (nervus nasociliaris). This last branch is the only which reaches the eyeball. It passes through the medial part of the common tendinous ring, follows the course of ophthalmic artery, runs like it above the optic nerve and then along the medial wall of the orbit. Its branches are:

- the long root of the ciliary (ophthalmic) ganglion, which supplies this ganglion with sympathetic fibres conveyed by the trigeminal nerve;

- the two long ciliary nerves, which run to the eyeball; they are sensory and sympathetic pathways;

- the posterior ethmoidal nerve (nervus ethmoidalis posterior) passes through the posterior ethmoidal foramen and supplies the mucosa of the ethmoidal labyrinth and of the sphenoidal sinus;

- the anterior ethmoidal nerve (nervus ethmoidalis anterior), a terminal branch which traverses the anterior ethmoidal canal, passes through the cribriform plate and the ethmoidal foramen; reaching the nasal fossae, where it divides into a medial and a lateral branch, its terminal branch, the external nasal nerve, innervate the skin of the dorsum of the nose up to its apex;

- the infratrochlear nerve (nervus infratrochlearis) is a terminal branch which accompanies the opthalmic artery up to the medial angle of the orbital opening and supplies the lacrimal ducts, the root of the nose and the medial parts of the eyelid.

The vegetative nerves

The orbit contains a very important vegetative center. The ophthalmic or ciliary ganglion. It is a transversally flattened quadrilateral corpuscle, situated on the lateral surface of the optic nerve, at the junction of its posterior third with the two anterior thirds.

It receives three roots:

- a short, thick motor or parasympathetic root, derived from the nerve of the inferior oblique muscle, a branch of the oculomotor nerve; it is iridoconstrictor;

- a long, thin sensory root, derived from the nasociliary nerve;

225

- a sympathetic, iridodilator root, issued from the internal carotid plexus and running along the Zinn's common tendinous ring.

The ganglion gives off numerous short ciliary nerves *(nervi ciliares breves)*, which penetrate into the sclera around the entrance of the optic nerve or even into the nerve and are distributed to the sclera, the choroid, the ciliary body, the iris and the central artery of the retina; these filaments assure the sensibility of the tunics of the eyeball and the motoricity of its intrinsic muscles (the sphincter muscle of the pupil and the ciliary muscle, through the fibers derived from the Westphal- Edinger's nucleus; the dilatory muscle of the pupil, through the postganglionic fibres issued from the superior sympathetic cranial ganglion; the protoneuron lies in a Budge's ciliospinal center in the spinal cord C8-T2).

As some of these muscles contain sympathetic and others parasympathetic fibers, their action is antagonistic.

The conduction pathway
(Fig. 202-203)

The visual impressions collected by the retina are initially received by the rod and cone cells, which actually do not belong to the nerve pathway but are neuro-epithelial sensory cells.

The nerve pathway contains as matter of fact, three neurons, which we shall describe from the very beginning in order to facilitate the understanding of the optic pathway (optic):

- the first neuron connects the visual cells (with cones or with rods) with the second neuron. It is the bipolar neuron completely situated inside the retina. The totality of bipolar retinal perikaryons is the equivalent of a spinal or sensory cranial ganglion;

- the second neuron has too its body and dendrites inside the retina and is the equivalent of the posterior horn of the spinal cord. The axon of the second neuron emerges from the eyeball at the level of the optic papilla (termed also optic disc and form the optic nerve (which actually is not a nerve, but a bundle, as it is constituted of central axons), the chiasma and the optic tract. It ends in *the metathalamus*, in the lateral geniculate body, where it forms a synapse with the third neuron;

- the third neuron is metathalamocortical, situated in the lateral geniculate body, this neuron arrives through its axon, in the visual area of the occipital lobe, at the level of *the calcarine sulcus*.

We mention that at the level of the retina may be distinguished:

- *the macula lutea* (yellow spot), which represents a zone where a cone cell or a rod cell synapses with a bipolar cell and, further away, the bipolar cell forms synapse with a single multipolar cell; in this area, owing to the above described anatomical disposition, the maximum clearness of vision is achieved,

- *the area around the macula lutea*, respectively the remaining visual retina is the site of a less clear vision, as anatomically it is observed that several cone and rod cells synapse with a single bipolar cell and several bipolar cells synapse with a single multipolar cell. This area may be divided by two axes, a vertical axis and a horizontal one, into four quadrants, respectively into two fields: nasal and temporal.

We mention that, as a consequence of the presence of the lens and of its function, the temporal retinal field perceives the bright rays coming from the nasal part and vice versa, which explains, in association with the manner of crossing of the fibers in the optic chiasm, the binocular vision;

- the existence, on the whole surface of the retina, of special cells named pupillary cells, responsible for the iridial and accomodation reflexes.

In addition, we mention the axons of the multipolar neurons in the temporal fields of retina remain homolateral and do not decussate in the optic chiasm; all those of the nasal field decussate (from the right nasal field they pass into the left tract and vice versa); the macular multipolar axons cross partially with each other; the arrangement of the fibers of the pupillary cells is similar to that of the macular multipolar axons.

226

n. opticus

chiasma opticum

substantia nigra

tractus opticus

nucleus ruber

tractus tectospinalis
(decussatio Meynert)

tractus longitudinalis medialis

colliculus superior

corpus geniculatum laterale et
mediale

radiatia optica

sulcus calcarinus

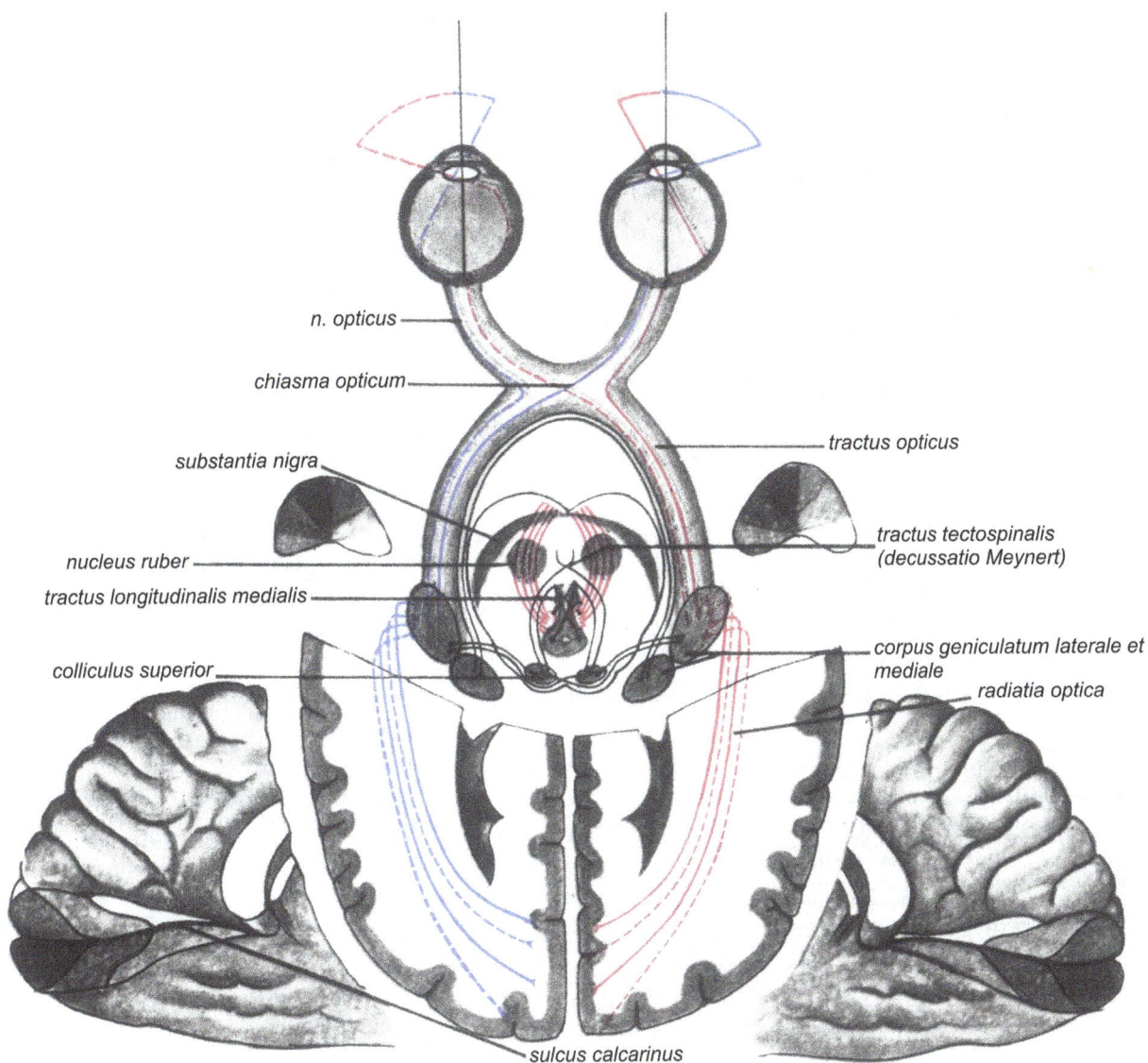

Fig. 202 Optic pathways

Fig. 203 Scheme of the main lesions of the optic
pathways
1. cortical lesion (crossed homonymous hemianopsia;
complete loss of vision; normal reflexes; consequence
of injuries, tumours, cysts, abscesses; 2. subcortical
lesion (optic radiations or primary optic centres; same
symptoms as mentioned above, but dim sight; 3. lesion
of the optic tracts (same symptoms as above
mentioned, but hemianoptic reaction; tumours); 4.
partial lesion of the chiasm (bitemporal hemianopia
and corresponding abolition of light reflexes; tumours
of the hypophysis, injuries of the sella turcica 5. total
lesion of the chiasm (bilateral blindness, bilateral
abolition of reflexes; tumours, injuries of the
hypophysial region) 6. lesion of the optic nerve (total or
partial unilateral blindness; scotoma or concentric
retraction of the visual field in case of partial lesion;
injuries, tumours, inflammatory diseases)

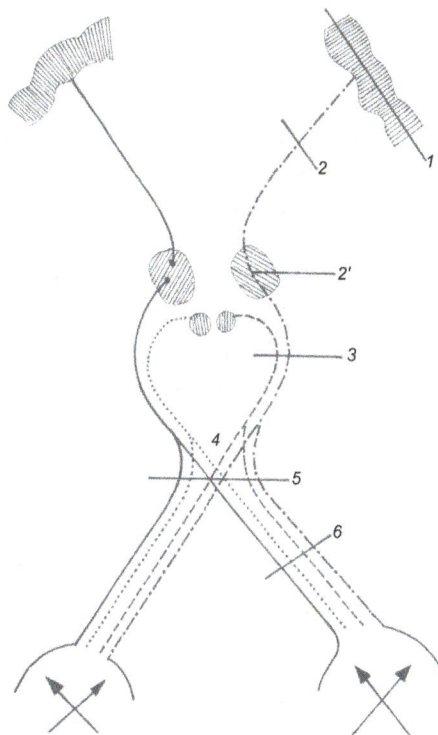

227

Hence, it ensues, for example, that in the left optic tract will be present: all the fibers of the temporal field of the left retina, all the fibers of the nasal field of the right retina, a part of the fibers of the right and left macular areas and a part of the pupillary fibers issued from the right and left retina.

After the description of the neurons of the optic tract and of the arrangement of the fibers arising from the level of the second retinal neuron, we shall deal again with some anatomical structures connecting the second reticular neuron to the third metathalamocortical neuron, respectively with the optic nerve, the optic chiasma and the optic tract.

The optic nerve *(n. opticus seu fasciculus opticus)* conveys the nerve inflow from the retina to the optic chiasm, from where the nerve inflow continues its course on the same axons up to the lateral geniculate body. It is not considered as a true nerve, but-owing to the optic invagination of the embryonic diencephalon, as the homologue of a bundle of the central nervous system. Macroscopically, it emerges from the eyeball at 3 mm medially to the posterior pole of the retina, under the form of a bulky, rounded, whitish cord, covered with meningeal tissue.

Its course comprises three parts:

The intraorbital segment forms the axis of the musculary-aponeurotic cone, constituted by the *recti* muscle of the eyeball; however, its course is not rectilinear, but slightly sinuous, preventing the nerve from stretching during the movements of the eyeball.

It is enclosed in three meningeal envelopes, which ensheath it *(vaginae nervi optici)*. It seems strangled in the orifice of the sclera; actually its apparently bulkier volume, is due to the myelin sheaths.

The optic nerve has connections with the orbital fat-pad or fat body *(corpus adiposum orbitae)*, with the orbital vein and nerves and especially with the ophthalmic artery; this artery surrounds it, passing laterally to it. Thus, it reaches the orifice of the tendon of origin of the *recti* muscles, then the anterior foramen of the optic canal.

The intracanalicular segment. The nerve follows the short course (5 mm) of the optic canal oriented, like the orbital axis, medially obliquely downwards and upwards, and is separated from the body walls of the canal through the meninges. It is in relation with the ophthalmic artery, adherent to it below and laterally. (A possible prolongation of the sphenoidal sinus may raise it and, in this case, it may be the source of an optic neuritis of local origin).

The intracranial segment. The nerve is flattened from above downwards and, after a course of 1 cm, reaches the optic chiasma. In this segment, the nerve is bathed by the cerebrospinal fluid of the perichiasmatic cistern. It corresponds: behind and laterally, to the terminal part of the internal carotid artery: above, to the anterior perforated substance of the respective hemisphere; below, to the *tentorium* of the hypophysis.

The optic chiasm is a superior inferiorly flattened horizontal X-shaped nervous lamina, situated between the optic nerves which reach it and the optic tracts which start from it. It is situated on *the tentorium*, of the hypophysis, in front of the pituitary stalk *(infundibulum)*, in the optic sulcus which unites transversally the two optic foramina.

The optic chiasm is not a commissure, but a partial decussation of fibers. From its posterior angles start the optic tracts, which surround the central peduncles and divide, behind the thalamus, into two roots; a chief lateral root, that terminates in the lateral geniculate body and in *the pulvinar* (according to some authors) and a medial root, which penetrates into *the superior colliculus* (superior quadrigeminal tubercle).

The efferent optic tracts in the chiasm contain uncrossed fibers from the temporal half of the ipsilateral side of the retina, crossed fibers from the nasal half of the contralateral side, as well as crossed and uncrossed fibers from both maculae. The optic tract winds round the lateral parts of the central peduncles and courses to the lateral geniculate body, from which some of the fibers run to the primary center of sight and to the superior colliculus of the quadrigeminal lamina, that constituted the reflex optic center, whereas other fibers form Gratiolet's optic radiation (the geniculocortical and the thalamocortical tracts, from *the pulvinar* to the cortex of the occipital lobe.

The optic striations pass on the most dorsal part of the posterior limit of the internal capsule, are continuous in a semicircular pathway with the medial concavity and terminate on the borders of *the calcarine sulcus*, the secondary optic tract, which is nevertheless, the center of optic perception. The cortical projection is achieved as follows: the superior half of the retina (the inferior part of the visual fields) is situated above the calcarine sulcus, while the inferior part (the superior part of the visual field) lies below the calcarine sulcus. The macula is situated exactly at the level of the scissure.

The cortical projection segment

The cortical segment of the visual analyzer has been described in the chapter dealing with the cerebral cortex.

Pathways of the optic reflexes

The optic reflexes have their origin in the pupillary cells distributed in the whole retina, the fibers of which follow the course of the muscle fibers, traverse the lateral geniculate body and through superior conjunctival limb, reach the superior colliculus, where they form a synapse with the pretectal nucleus, which is connected with the pupillary iridomotor nucleus (Westphal–Edinger's nucleus). From here starts the parasympathetic iridoconstrictor reflex pathway. The other nucleus is the ciliomotor or accommodation nucleus, it is under the control of the cerebral cortex and its peripheral pathway is coincident with the iridoconstrictor pathway.

The iridoconstrictor and ciliomotor pathway. It arises from the Westphal -Edinger's mesencephalic vegetative ganglion, in which lie the effector neurons of the pathway. Their axons unite with the oculomotor nerve (third cranial nerve), follow the course of the latter and reach the ciliary ganglion, where they synapse with the second neuron, after which, through the short ciliary muscles, the inflow arrives to the sphincter muscle of the iris and to the ciliary muscle.

The contraction of these two muscles diminishes the pupil (myosis) and respectively, relaxes the Zinn's zonule causing the bulging of the lens, as it happens in the near sight, when the photomotor reflex-through the sphincter of the iris – and the accommodation reflex – through the ciliary muscle – form a single physiological act. However, the actions may be dissociated. The abolition of the photomotor reflex with preservation of accommodation for distance variations constitutes the Argyll-Robertson sign. This sign appears in tabes, as a consequence of the lesion of the iridoconstrictor fibres which connect the oculomotor nucleus to the nuclei situated in the ceiling of the mesencephalon (in the superior colliculus or anterior quadrigeminal body).

The accommodation is maintained, since the ciliomotoricity lies under cortical control (ganglionic cells, whose axons from the optic nerve reach the pretectal nucleus, that is connected also with the opposite one, which explains the fact that the constrictor pupillary reflex is consensual.

The iridodilator pathway. The efferent tectospinal neuron descends its axons into the posterior longitudinal fasciculus, passes through the brainstem and reaches the ciliospinal (Budge's) center in the cervico-dorsal spinal cord (C8-D3). The nerve inflow passes from here through the communicating branches of the first and second spinal nerves, gains the stellate ganglion, ascends the cervical sympathetic chain and synapses in the superior cervical sympathetic ganglion.

Through the internal carotid nerve and through the fibres which it sends either to the gasserian ganglion or – rather – to the branches of the trigeminal nerve, it brings the nerve inflow into the ciliary ganglion – through which it passes and, through the short ciliary nerves gains the dilatory muscle of the pupil. The excitation of the sympathetic system produces mydriasis (dilation of the pupil) and exophthalmos or exophtalmia (protrusion of the eyeball).

The anesthesia or lesion of these pathways (at the level of the stellate ganglion) causes the Claude Bernard-Horner's syndrome, characterized by myosis, enophthalmia or *enophthalmos* (recession of the eyeball within the orbit), hemifacial vasodilation and ptosis of the upper eyelid.

To conclude, the ciliary ganglion constitutes the passage of two antagonistic nerve pathways, destined to the intrinsic musculature of the eyeball: the sympathetic iridodilator pathway and the parasympathetic iridoconstrictor and ciliomotor pathway (accomodative to distance variations); the parasympathetic pathway synapses in the ciliary ganglion, whereas the sympathetic pathway passes through this ganglion without synapsing, but it forms a synapse in the superior cervical sympathetic ganglion.

Moreover, the nuclei of origin of the nerves of the ocular muscles on both sides are interconnected through the medial longitudinal fasciculus *(fasciculus longitudinalis mediatis)* and through the ascending fibres of the lateral vestibular nucleus (Deters' nucleus). Likewise, they are traversed by collaterals from the olivary nucleus of the metencephalon. The two visual fields are coordinated by multiple internuclear connections. In normal developed ocular muscles innervation impairments may bring about a strabismus.

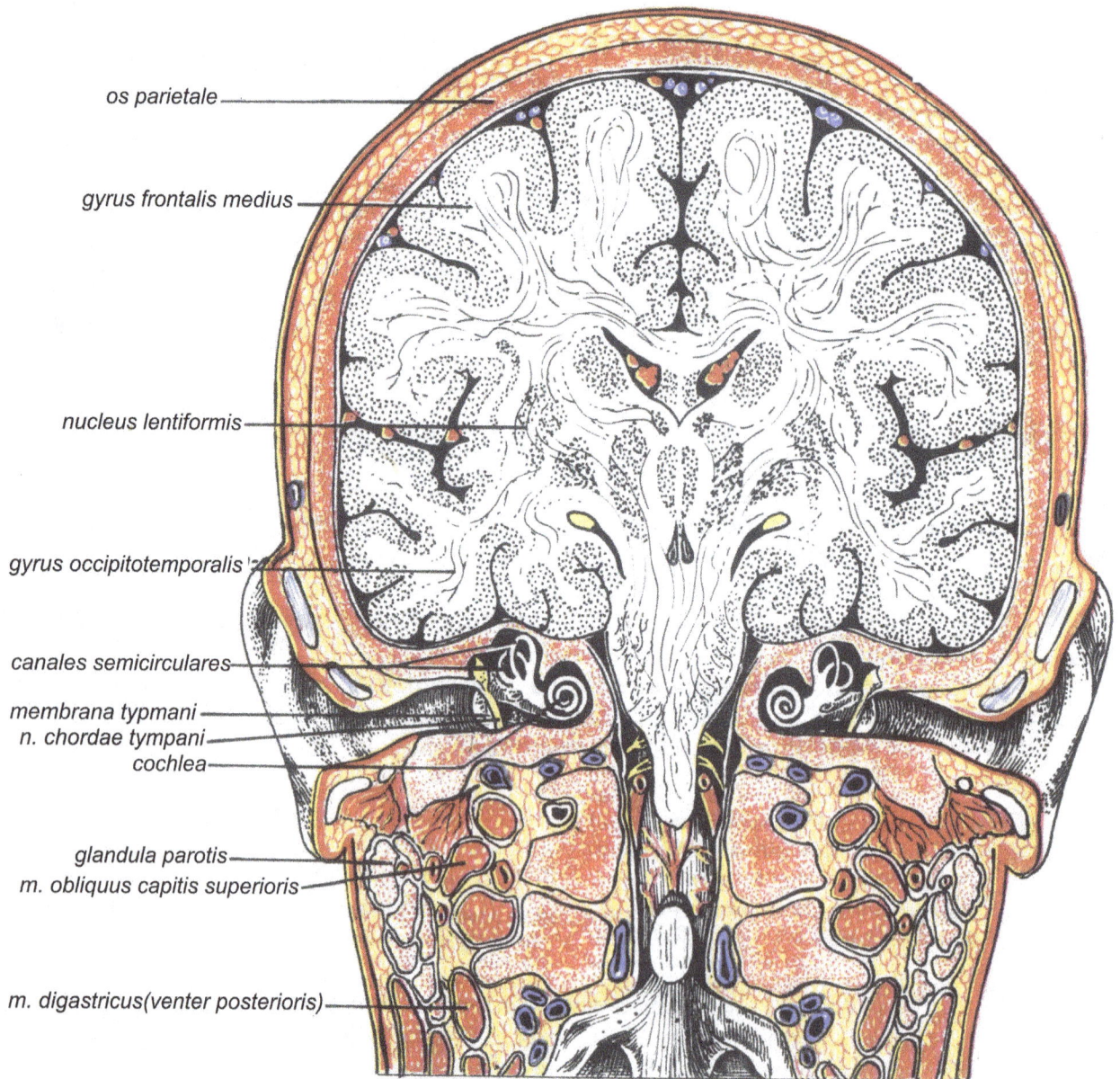

Fig. 204 Frontal section through the petrous portion of the temporal bone, showing the cochlea

THE COCHLEOVESTIBULAR ANALYZER

The topography, which is very intricate at the level of the receptor segment and, partially, at the level of the conduction pathway, compels us, for didactic reasons, to describe together the acoustic analyzer and the analyzer of equilibrium or static analyzer (Fig. 204-244).

The ear or stato-acoustic organ
(Auris seu organon statoacusticus)

Topographically, the receptor of this analyzer contains the auricular region constituted of the external ear *(auris externa)*, the region of the tympanic cavity or the middle ear *(auris media)* and the labyrinthine region or the internal ear *(auris interna)* which, on turn, contains, on the one hand, the Corti's organ specialized in the reception of sound waves, and on the other hand, receptor structures specialized in the captation of proprioceptive excitations serving to the achievement of the equilibrium of the organism.

For the unity of the work, we shall give a topographical description of these structures (Fig. 205-242).

The auricular region. The external ear
(Auris externa)

It consists of the auricle or pinna *(auricula)* the external acoustic meatus *(meatus acusticus externus)* and the tympanic membrane *(membrana tympani)* which separates the external acoustic meatus from the middle ear (Fig. 207-210).

The auricle or pinna
(Auricula)

The auricle is tegumental fold of varying form and size, formed of skin and of auricular cartilage *(cartilage auriculae)*, which constitutes its skeleton and of the auricular lobe *(lobulus auriculae)*.

The auricle presents two arcuate prominences, *the helix* and *the antihelix*, and two tubercular elevations, *the tragus* and *the antitragus*, as well as three depressions: *the scapha or scaphoid fossa, the triangular fossa and the concha.*

The cartilage of the ear is connected to the temporal bone by connective formations, which unite its perichondrium with the periosteum of the temporal bone.

Around the auricle lie the anterior auricular; posterior auricular and superior auricular muscles (respectively *musculi auricularis anterior, auricularis posterior* and *auricularis superior*), which impart to the ear very reduced movements. There are also rudimentary muscles: *m. helicis major, m. tragicus, m. helicis minor* and *m. antitragicus*, on the external surface of the auricle, and *m. obliques auriculae* and *m. transversus auricular*, on its internal surface; they are developed in quadrupedal mammals, but very reduced and devoid of any major practical significance in man.

Vessel supply. *The arteries* are branches of the external carotid artery. The anterior third of the auricle is supplied by the three anterior auricular branches, which arise from the posterior border of the superficial temporal artery. The two posterior thirds of the medial surface of the auricle are supplied by the branches of the posterior auricular artery. The perforating branches of the same artery supply the two posterior thirds of its lateral surface.

There are three groups of *veins*:

-the anterior group, which empties into the superficial temporal vein;

-the posterior group, which empties into the posterior auricular vein and into the mastoid emissary vein, and

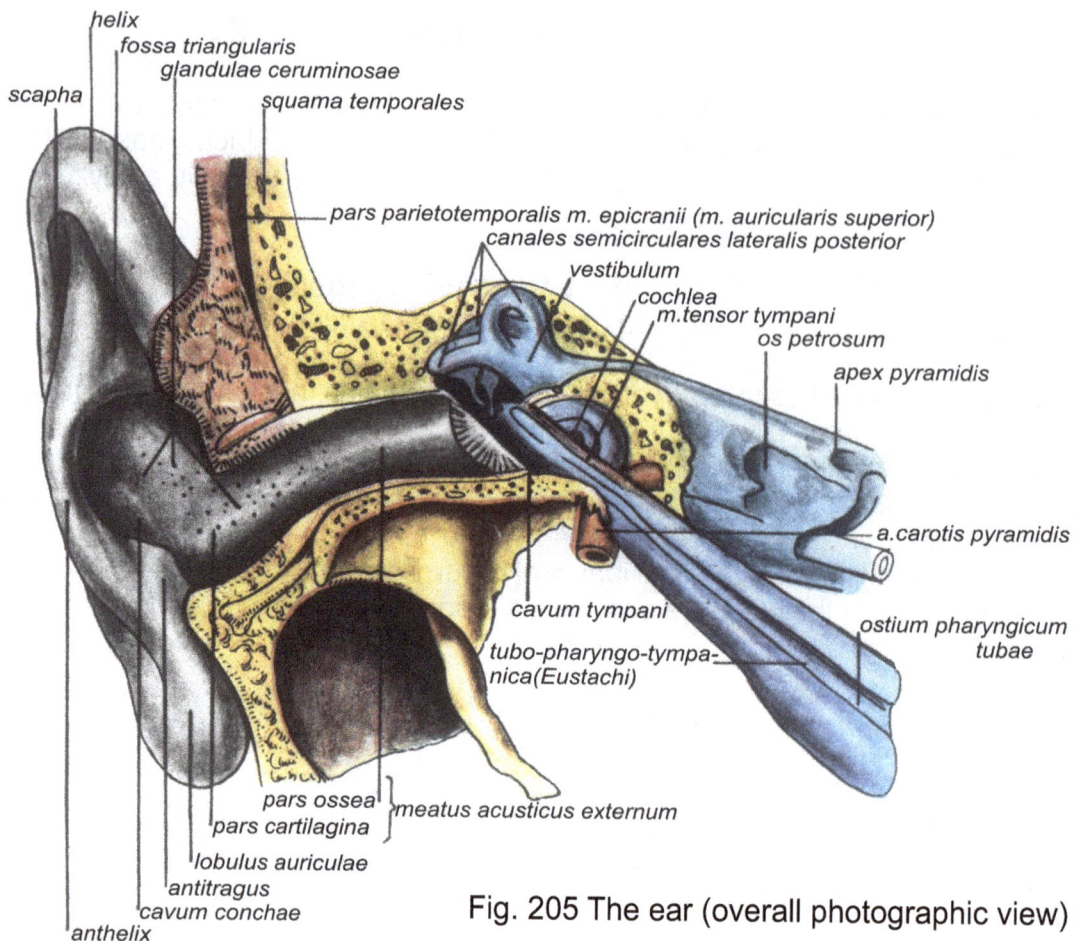

helix
fossa triangularis
glandulae ceruminosae
scapha
squama temporales

pars parietotemporalis m. epicranii (m. auricularis superior)
canales semicirculares lateralis posterior
vestibulum
cochlea
m.tensor tympani
os petrosum
apex pyramidis

a.carotis pyramidis

cavum tympani

ostium pharyngicum
tubae

tubo-pharyngo-tympa-
nica(Eustachi)

pars ossea
pars cartilagina
meatus acusticus externum

lobulus auriculae
antitragus
cavum conchae
anthelix

Fig. 205 The ear (overall photographic view)

-the inferior group, which empties into the external jugular vein or into the retromandibular vein.

The lymphatics of the anterior part of the helix and of the tragus reach the superficial (preauricular) and deep parotid nodes. The lymphatics of the medial surface and those arising from the lateral surface, behind the concha, are tributary to the retroauricular (mastoid) and parotid lymph nodes. The lymphatics of the region of the concha run towards the retroauricular and superficial (preauricular) parotid nodes.

Nerve supply. The auricular muscles are supplied by the facial nerve through the posterior auricular branch. This branch surrounds anteriorly the belly of the digastric muscle up to the anterior margin of the mastoid process; from this last portion of *the facial nerve* arise the branches to the auricular muscle. The occipital branch continues the posterior auricula nerve and supplies the occipital muscle and a part of the aponeurotic *galex*.

The sensory nerve supply is mainly assured by the auriculotemporal nerve (a branch of the mandibular nerve), the great auricular nerve (arising from the cervical plexus) and the lesser occipital nerve (arising from the posterior cervical plexus).

On the lateral surface of the auricle, *the tragus*, the ascending part of *the helix*, *the concha* and *the antihelix* are supplied by the auriculotemporal nerve. The concha or a part of it is supplied by the auricular branches of the vagus nerve. The remaining lateral surface is supplied by the auricular branch of the superficial cervical plexus. The auricular branch of the vagus anastomosis with the facial nerve, bringing about the Ramsay-Hunt's sensory innervation of the auricle.

The medial surface of the auricle is supplied by the auricular branch of the superficial cervical plexus. Some researches show the existence of an internal auricular branch of Wrisberg's intermediate nerve (cranial) (sensory branch of the external acoustic meatus), which would supply a more extensive area.

232

This branch arises at the level of the stylomastoid foramen, traverses the external acoustic meatus and supplies the tympanum, the posterior part of the external acoustic meatus, the region of the concha, the antihelix and the tragus. Moreover, it is stated that the Ramsay-Hunt area would correspond only to Wrisberg's intermediate nerve.

The external acoustic meatus

It is sinuous, oblique, anteromedial, with a terminal inclination downwards. At the junction of the external three quarters with the internal quarter, a constriction is present: the isthmus, during the otoscopic examination, this inflections are connected by drawing the auricle upwards and backwards (Fig. 209).

It measures 22-26 mm in length and is slightly flattened anteroposteriorly. It is constituted medially of the tympanic part of the temporal bone applied on the squama of the temporal bone, and laterally of the fibrocartilaginous canal (the cartilaginous external acoustic meatus), which is continuous with the auricular cartilage. The external acoustic meatus is lined with a thin tegument, which contains sebaceous glands, ceraminous glands and hairs called *tragi*.

The relations are the following:
- behind, with the mastoid process and the third portion of the facial nerve;
- above, with the middle cranial fossa, with the precentral cells which communicate, with the epitympanic recess, as well as with the retroauricular mastoid cells;
- below with the base of the parotid gland and
- anteriorly, with the temporomandibular joint.

Vessel supply. *The arteries* of the fibrocartilaginous segment have the same origin as those of the auricle, the anterior arteries arise from the superficial temporal artery, and the posterior arteries, from the posterior auricular artery. The bony segment is supplied by the anterior tympanic artery, which is a branch of the maxillary artery. The tympanic artery penetrates into the external acoustic meatus through the petrotympanic fissure (called also glasserian scissure or fissure).

The veins form a slender network, which drains the blood anteriorly into the superficial temporal vein and the maxillary vein, less frequently directly into the retromandibulary vein, and posteriorly into the posterior auricular veins.

The lymphatics form a network which opens in the same lymph nodes as the lymph vessels of the auricle.

Nerve supply. The anterior wall is supplied by the auriculotemporal nerve, a branch of the mandibular nerve and, according to some other authors, by the internal auricular branch of the intermediate nerve (Wrisberg's intermediate nerve). The posterior wall is supplied by an auricular branch, which derives from the vagus nerve and anastomoses with the facial nerve. The excitation of the skin at this level may induce vagal reactions (cough, vertigo etc) owing to the presence of the vago-auricular branch.

The tympanic membrane
(Membrana tympani)

The external acoustic meatus is separated from the middle ear by a membranous wall, *the tympanic membrane* (1/10 mm thick), constituted of two *epithelial laminae*, between which lie the proper fibroconnective tissue and *the manubrium mallei* (the handle of the hammer). The tympanic membrane is thin but resistent, fibrous but elastic, concave and almost circular (1 cm in diameter and a surface of 64 mm2), best with 450 against the horizontal line. It is attached downwards to the tympanic sulcus *(sulcus tympanicus)*, situated at the inner extremity of the external acoustic meatus and is fastened by means of a fibrocartilaginous ring (Gerlach's ring) *(annulus fibrocartilagineus)*.

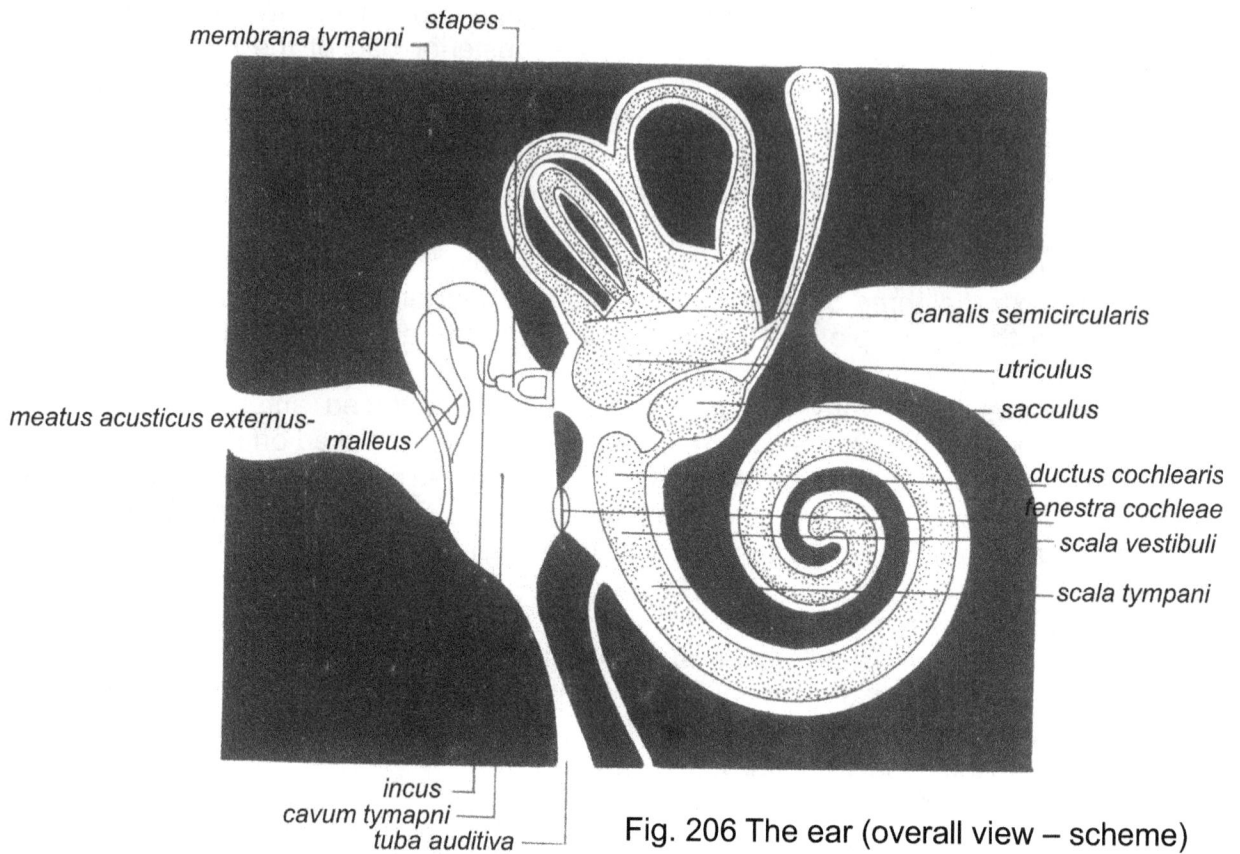

membrana tymapni
stapes
canalis semicircularis
utriculus
sacculus
meatus acusticus externus-
malleus
ductus cochlearis
fenestra cochleae
scala vestibuli
scala tympani
incus
cavum tymapni
tuba auditiva

Fig. 206 The ear (overall view – scheme)

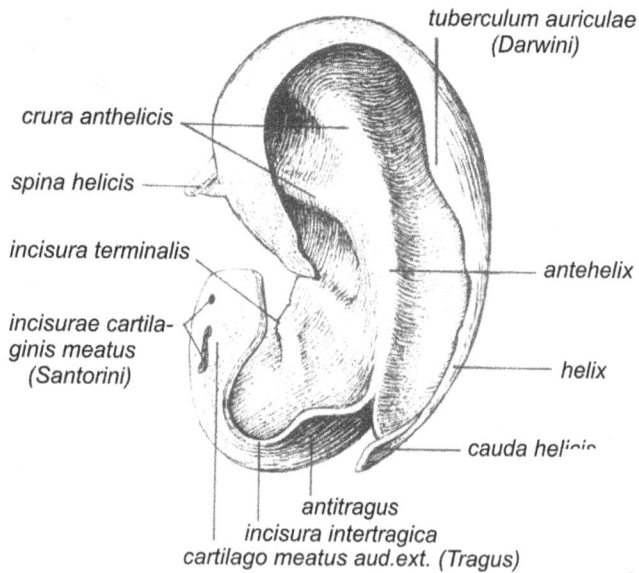

tuberculum auriculae
(Darwini)
crura anthelicis
spina helicis
incisura terminalis
incisurae cartila-
ginis meatus
(Santorini)
antehelix
helix
cauda helicis
antitragus
incisura intertragica
cartilago meatus aud.ext. (Tragus)

Fig. 207 Cartilages of the external ear

Fig. 208 The ear – external view

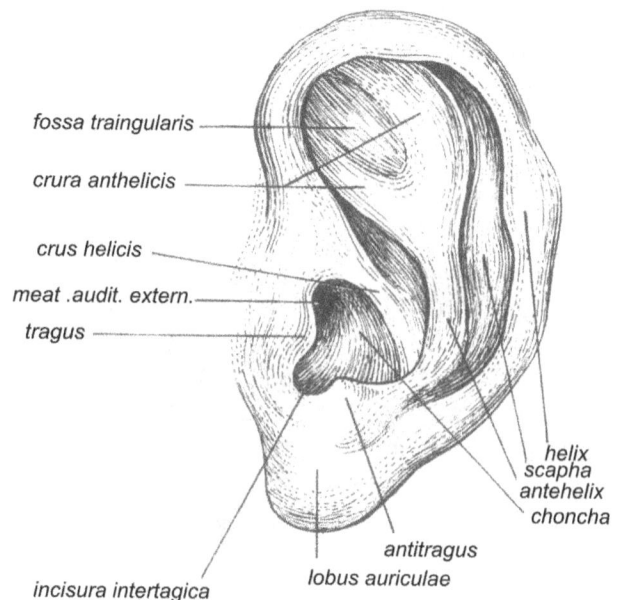

fossa traingularis
crura anthelicis
crus helicis
meat .audit. extern.
tragus
helix
scapha
antehelix
choncha
antitragus
incisura intertagica
lobus auriculae

234

This ring forms a thick peripheral zone, made up of radial and circular fibers, which insert below and behind the tympanic membrane into the tympanic sulcus. Its superior part is thinner and less tense (pars flaccida) it is attached on the squama of the temporal bone (pars squamosa), the remainder of the tympanic membrane is thicker and tant (pars tensa) (Fig. 213).

The superior half of the tympanic membrane is divided into two parts by the mallear prominence (prominentia malearis) (the prominence of the handle of the hammer) included in the tympanic membrane. The inferior extremity of this prominence lies in the center of the tympanic membrane and is called the umbo of the tympanic membrane (umbo membranae tympani). From here starts a triangular light reflex, visible otoscopically, called Politzer's cone of light or Wilde's luminous triangle. In the superior part, corresponding to the manubrium mallei may be seen the stria mallearis, an oblique stripe, running from the umbo upwards and forwards on the tympanic membrane his stripe gives to the tympanic membrane attachment to two fibrous ligaments, which lead to the formation of the anterior and posterior mallear folds (the anterior and posterior limiting folds). The anterior mallear stria is continuous with the fibrous ring from the tympanic sulcus to the apex of the lateral process of the malleus. The posterior mallear fold is formed of the posterior part of the ring, which is attached on the apex of the lateral process of the mallus. These folds separate the pars tensa of the tympanic membrane from the pars flaccida (Shrapnell's membrane).

Two lines perpendicular on each other, which cross at the level of the umbo, one of them being drawn into the axis of the mallear manubrium, divide the tympanic membrane, into four quadrants: anterosuperior, posterosuperior, antero-inferior and postero-inferior. Through the transparence of the posterosuperior quadrant may be seen the ossicles and the nerve of the chorda tympani. The paracentesis of the tympanum should be performed in the postero-inferior quadrant.

Structure. The tympanic membrane is made up of three layers: a proper fibrous tissue (lamina propria), covered on its outer surface by integument and called the cutaneous layer (stratum cutaneum), and on its inner surface by the tympanic mucosa, called the mucous layer (stratum mucosum). The fibrous layer, on turn, is formed of two layers of fibers: an outer layer, in which the fibers are oriented radially (stratum radiatum), with a central insertion point on the manubrium of the malleus, and an inner layer, with a circular and concentric arrangement of the fibers (stratum circulare). This system of fibers lacks at the level of the flaccid part, the tympanic membrane being formed here merely by the adherence of the cutaneous to the mucous layer.

Vessel and nerve supply. The arterial supply is assured for the external surface of the tympanic membrane, by the deep auricular artery, a branch of the maxillary artery, and for the internal surface of the tympanic membrane, by the anterior tympanic artery, which is also a branch of the maxillary artery and which penetrates the tympanic cavity through the petrotympanic (gloserian) fissure.

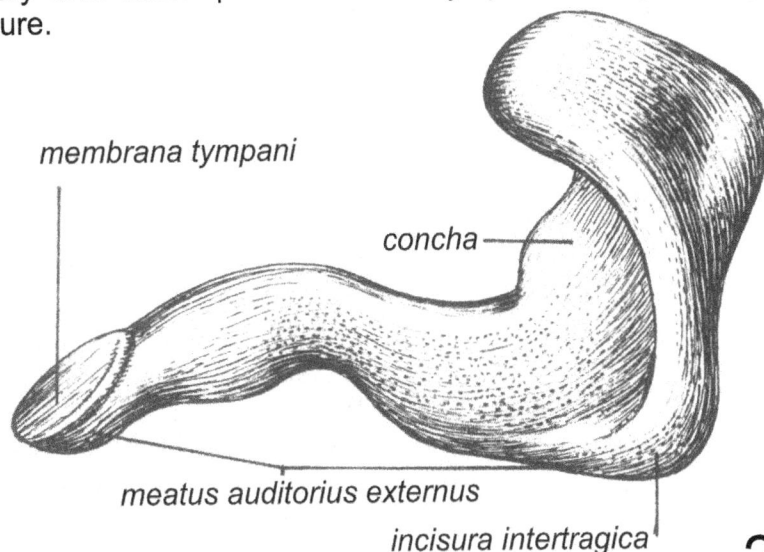

Fig. 209 External acoustic meatus

membrana tympani

concha

meatus auditorius externus

incisura intertragica

235

The veins drain either into the external jugular vein (those of the external surface) or into the veins of *the dura mater* and of the transverse sinus (those on the internal surface); *the lymph vessels* of the external zone of the tympanic membrane open into the pre- and retroauricular lymph nodes, while those of the internal zone of the membrane run into the lymph nodes of the tympanic cavity and respectively into the retropharyngeal lymph nodes. *The nerve* supply is assured by the great auricular nerve and an auricular branch of the vagus nerve.

The region of the tympanic cavity *(the middle ear)*
(Auris media)

The middle ear is situated in the petromastoid portion of the temporal bone and consists of three parts: the tympanic cavity *(cavum tympani)*, the auditory tube or eustachian tube *(tuba auditiva)* and *the mastoid antrum* and the mastoid cavities or sinuses or air cells *(antrum mastoideus* and *cellulae mastoideae)*. They are situated along an axis almost parallel to the great axis of the petrous portion of the temporal bone, they are air-filled cavities, their ventilation being achieved through the auditory tuba, which opens anteriorly in the rhinopharynx *(pars nasalis pharyngis)*. Infections of the pharynx may be also spread through this route towards the middle ear, which is lined with a mucosa continued, with changes, from the pharyngeal mucosa up to the mastoid cells (Fig. 211 - 233).

The tympanic cavity or middle ear
(Cavum tympani)

It has the shape of an irregular slit, of which six walls are described for didactic reasons; their thickness varies between 6 mm at the level of the posterior quadrant of the tympanum and 3 mm at the level of its anterior quadrant. It displays the same obliquity as the tympanic membrane and is constituted of three superposed parts:
- the epitympanic recess *(recessus epitympanicus)* which contains the most part of the chain of ossicles connecting the tympanic membrane to the oval window of the vestibule;
- the hypotympanic recess *(recessus hypotympanicus)*, situated below the tympanic membrane;
- the atrium, situated at the level of the tympanic membrane.
The six walls of the tympanic cavity have individual characteristics, in dependence on their structure or relations.
The membranous or lateral wall *(paries membranaceus)* is formed mainly by tympanic membrane, which is fastened in a bony frame formed: forwards, downwards and behind by the tympanic bone *(pars tympanica ossis temporalis)*; upwards, by that oblique part of the temporal bone which constitutes the roof of the medial segment of the external acoustic meatus. The tympanic membrane is attached to the tympanic sulcus by Gerlach's fibrocartilaginous ring *(annulus fibrocartilagineus)*, called also Gerlach's annular tendon. On the internal surface, the tympanic membrane is covered by a mucosa, which passes here over the handle of the hammer *(manubrium mallei)* and is in relation with *the chorda tympani nerve.*
The long frame to which the tympanic membrane is attached has a variable breadth. In front of the tympanic membrane, below and behind it, the bony wall is 1-2 mm in breadth. Above the tympanic membrane. It is broader, of 5-7 mm, and forms the lateral wall of the epitympanic recess (the wall of the recess of *the ossicles)*.
The labyrinthine or medial wall *(paries labyrinthicus)* corresponds to the internal ear and presents the formations described below.
The promontory, situated in the center, is an elevation caused by the first coil of the bony cochlea; it is furrowed by grooves in which pass branches of the tympanic or Jacobson's nerve *(n. tympanicus)* and of the tympanic artery. Behind the promontory lie several formations:

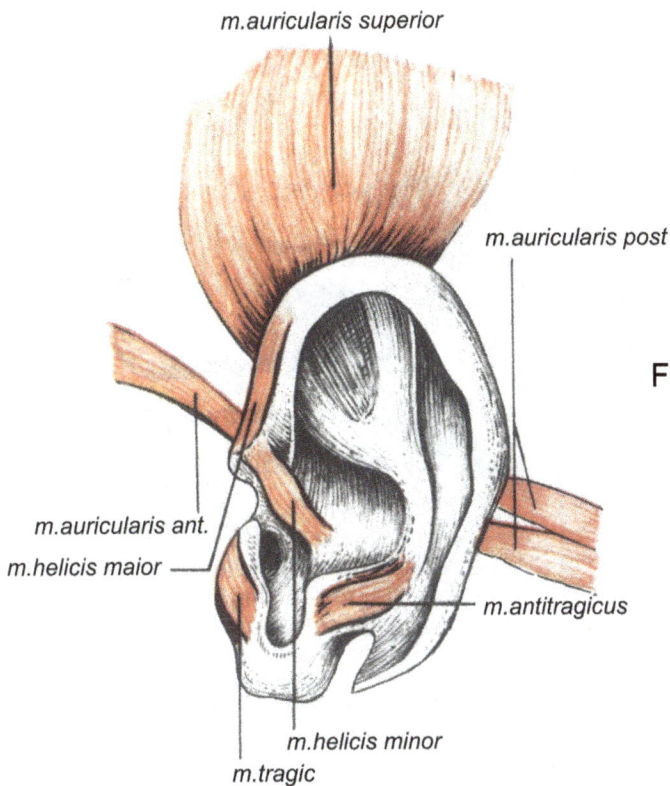

m.auricularis superior

m.auricularis post

m.auricularis ant.

m.helicis maior

m.antitragicus

m.helicis minor

m.tragic

Fig. 210 Muscles of the external ear

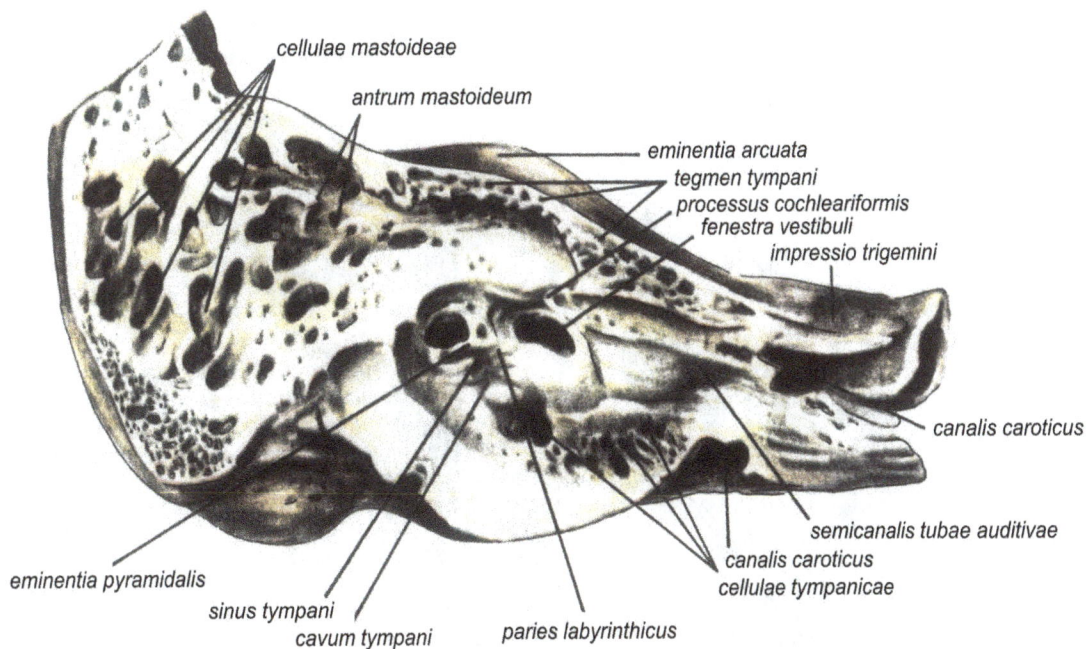

cellulae mastoideae

antrum mastoideum

eminentia arcuata
tegmen tympani
processus cochleariformis
fenestra vestibuli
impressio trigemini

canalis caroticus

semicanalis tubae auditivae
canalis caroticus
cellulae tympanicae

eminentia pyramidalis

sinus tympani

cavum tympani

paries labyrinthicus

Fig. 211 Bony lateral wall of the middle ear (tympanic cavity) and mastoid cells

- the oval window *(fenestra vestibuli)*, posterosuperior to the promontory corresponds to the vestibule of the internal ear; it is closed by a membrane, by the annular ligament of the stapes *(ligamentum annulare stapedis)* and by the base of the stapes *(basis stapedis)*. When the membrane of the stapes vibrates, the base of the stapes pushes the perilympha into the membranous labyrinth, transmitting to the auditory *ossicles* the sound vibrations;

- the round window *(fenestra cochleae)* is situated behind and below the promontory corresponding to the scala tympani of the cochlea, and is closed by the secondary tympanic membrane *(membrana tympani secundaria)*. Between the round and the oval window lies a deep depression, the tympanic sinus *(sinus tympani)*;

237

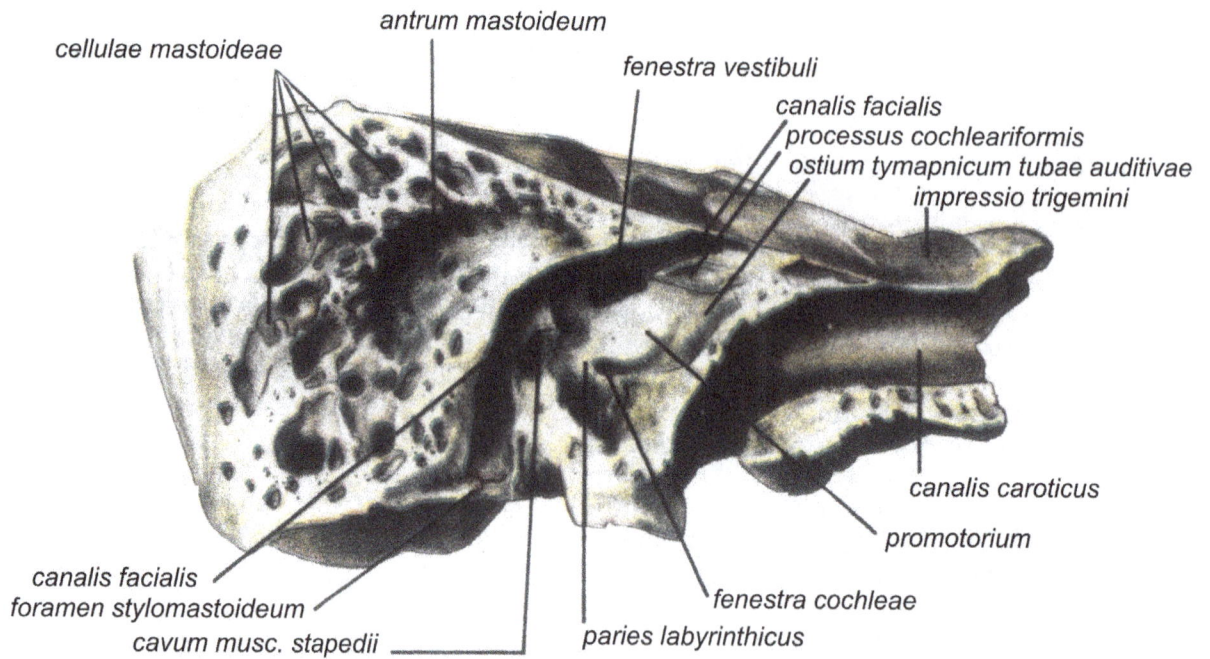

cellulae mastoideae

antrum mastoideum

fenestra vestibuli

canalis facialis
processus cochleariformis
ostium tymapnicum tubae auditivae
impressio trigemini

canalis caroticus

promotorium

canalis facialis
foramen stylomastoideum
cavum musc. stapedii

fenestra cochleae

paries labyrinthicus

Fig. 212 Tympanic cavity – bony walls – frontal section (the canal of the facial nerve is displayed)

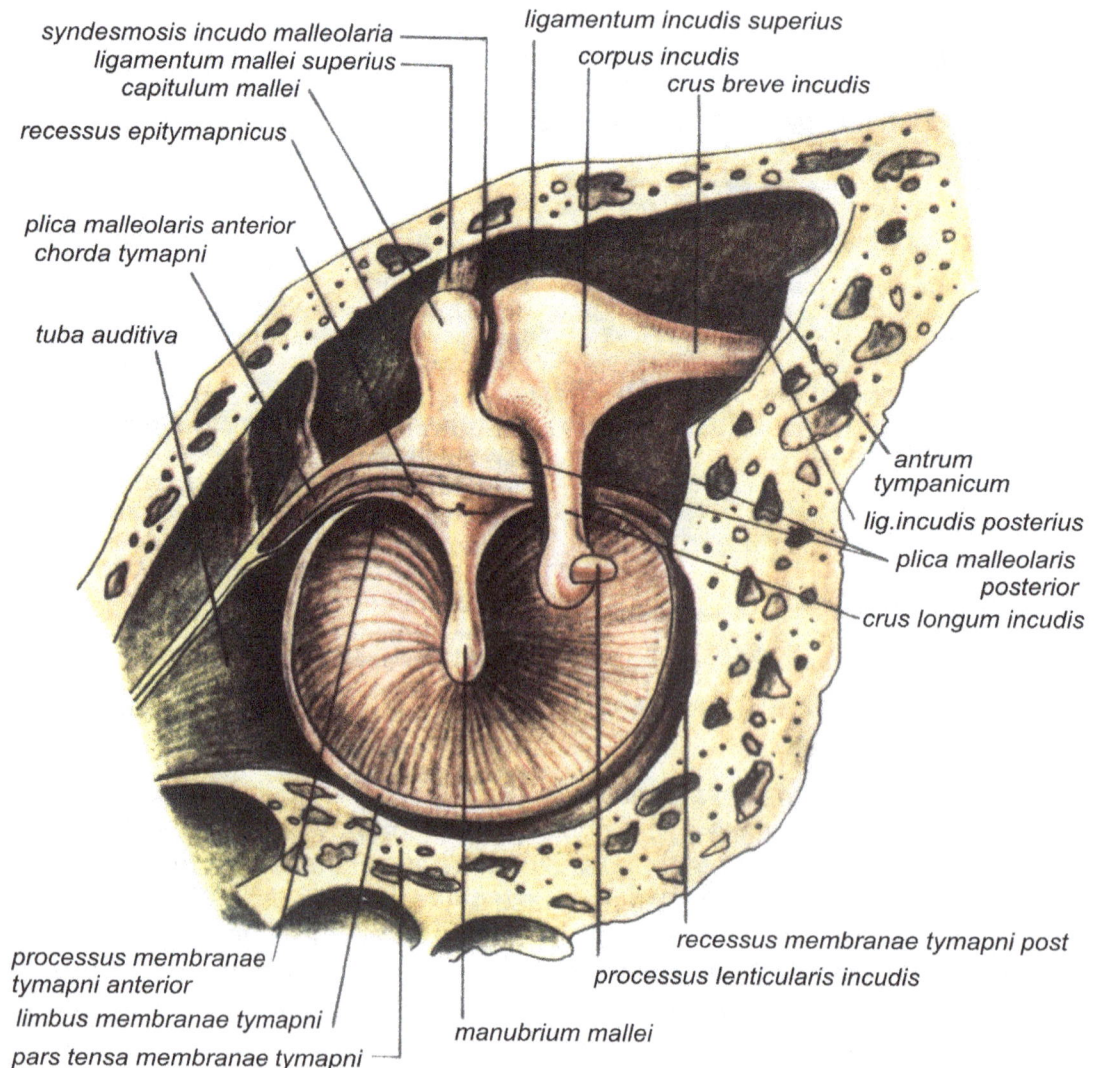

syndesmosis incudo malleolaria
ligamentum mallei superius
capitulum mallei

recessus epitymapnicus

plica malleolaris anterior
chorda tymapni

tuba auditiva

ligamentum incudis superius
corpus incudis
crus breve incudis

antrum tympanicum

lig.incudis posterius

plica malleolaris posterior

crus longum incudis

processus membranae tymapni anterior
limbus membranae tymapni
pars tensa membranae tymapni

manubrium mallei

recessus membranae tymapni post
processus lenticularis incudis

Fig. 213 Lateral wall of the tympanic cavity

238

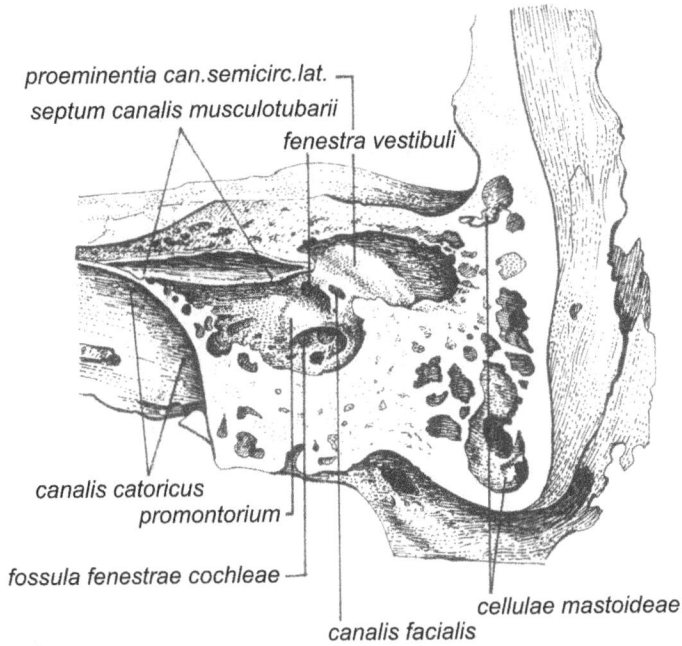

proeminentia can.semicirc.lat.
septum canalis musculotubarii
fenestra vestibuli

canalis catoricus
promontorium

fossula fenestrae cochleae

canalis facialis

cellulae mastoideae

Fig. 214 Osseous internal ear, right side,
posterior and medial walls, anterolateral view

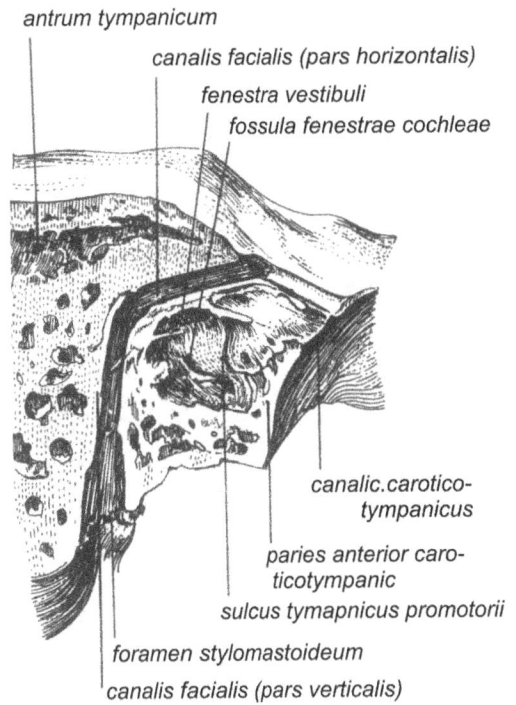

antrum tympanicum
canalis facialis (pars horizontalis)
fenestra vestibuli
fossula fenestrae cochleae

canalic.carotico-
tympanicus

paries anterior caro-
ticotympanic

sulcus tymapnicus promotorii

foramen stylomastoideum

canalis facialis (pars verticalis)

Fig. 215 Lateral wall of the osseous
internal ear, on the left side (left ear)

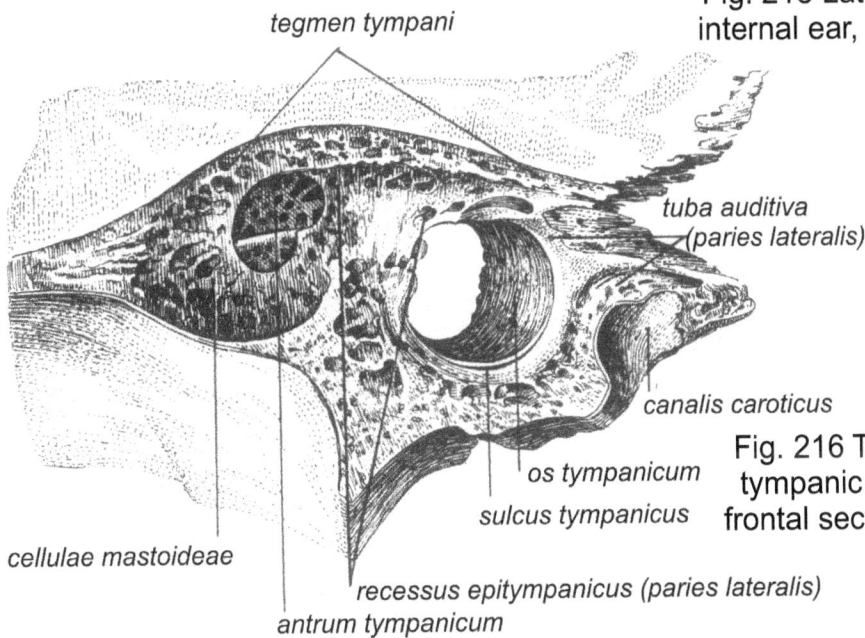

tegmen tympani

tuba auditiva
(paries lateralis)

canalis caroticus

os tympanicum

sulcus tympanicus

cellulae mastoideae

recessus epitympanicus (paries lateralis)

antrum tympanicum

Fig. 216 Tympanic membrane and
tympanic cavity, right side, after a
frontal section (performed in front of
the malleus)

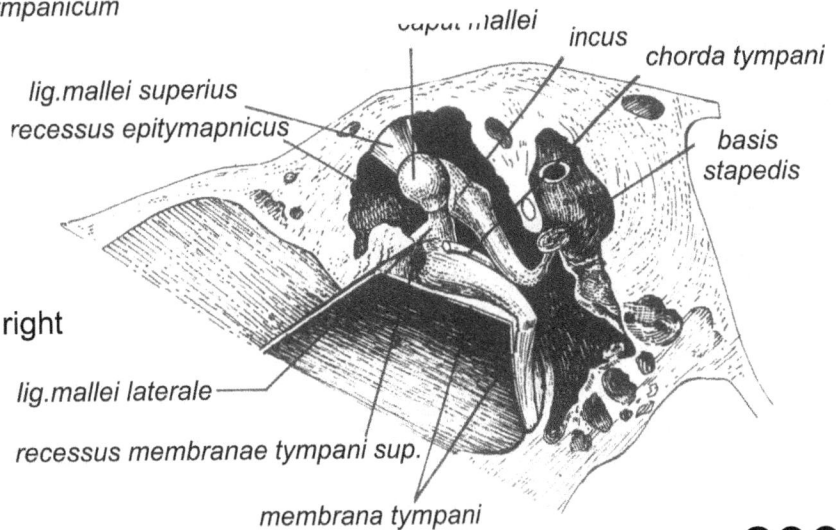

caput mallei
incus
chorda tympani

basis
stapedis

lig.mallei superius
recessus epitymapnicus

Fig. 217 Osseous internal, right
side, medial wall

lig.mallei laterale

recessus membranae tympani sup.

membrana tympani

239

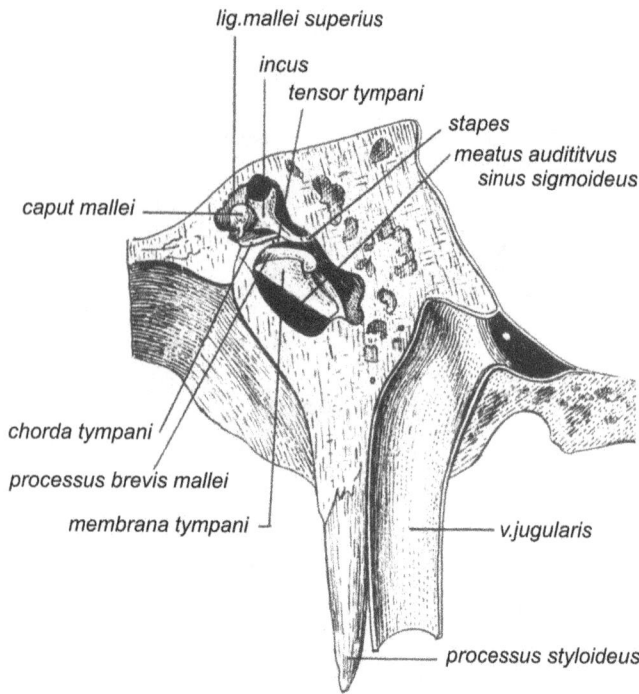

Fig. 218 Frontal section through the left internal ear

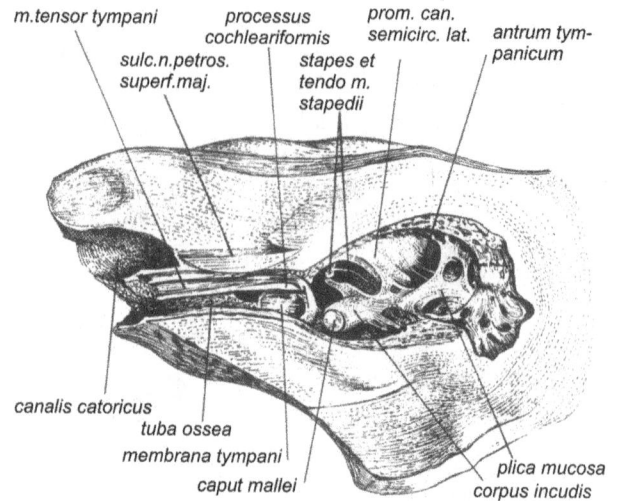

Fig. 219 Internal ear and osseous tube after removal of the roof of tympanum (tegmen tympani) – left side, seen from above

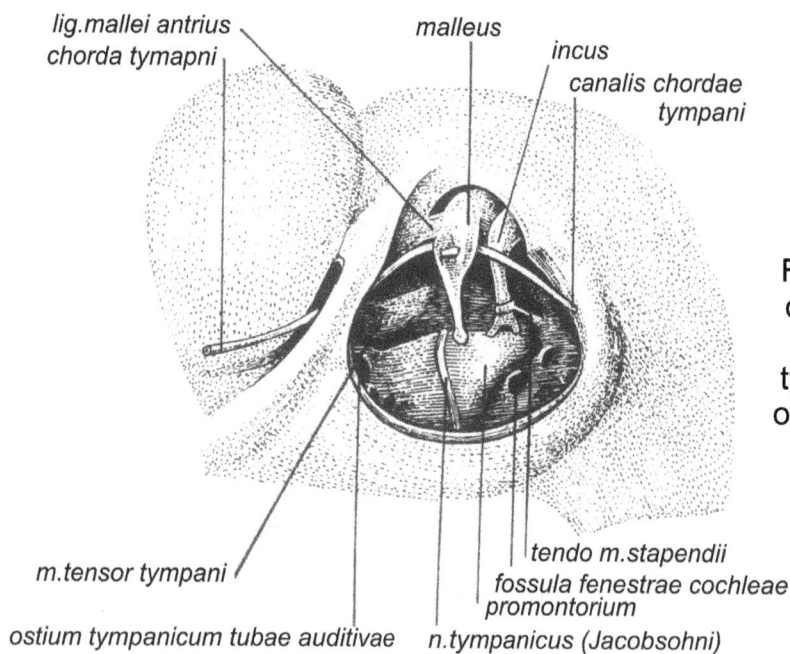

Fig. 220 Auditory ossicles, cord of tympanum (chorda tympani) and stapedius and tensor tympani muscles, after removal of the tympanic membrane – left side, seen from above

Fig. 221 Topography of the tympanic cavity and of the auditory tube at the level of the petrous portion of the temporal bone

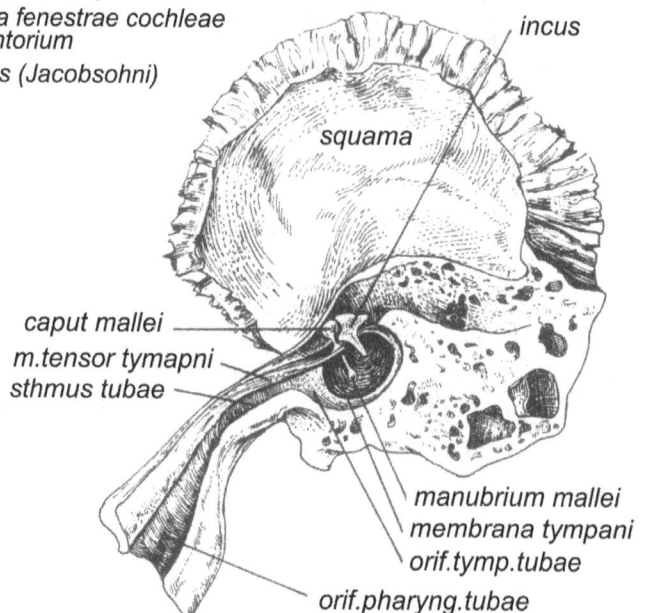

240

- the cochleariform process (*processus cochleariformis*, a bony canal ending by an outwards curved lamina, is situated anteroposteriorly to the promontory, the tensor muscle of the tympanic membrane *(m. tensor tympani)* passes along it;

- the prominence of the facial canal *(prominentia casalis facialis)* is the second portion of the fallopian aqueduct, situated above and behind the oval window. Its wall is thin and often dehiscent and the facial nerve, uncovered can be injured in the case of an otitis or during surgery.

- The superior wall *(tegmen tympani seu paries tegmentalis)* is a bony plate interposed between the middle ear and the middle portion of the skull, forming the roof of the tympanic cavity, it presents the internal petrosquamous suture, through which pass thin vessel, tributary of the meninges and of the superior petrous sinus. The presence of thin plate explains the possible meningeal and encephalic complications in otitis.

- The inferior wall or floor *(paries jugularis cavi tympani)* is an anfractuous groove situated below the tympanic membrane *(recessus hypotympanicus)* and below the promontory. On it is situated the opening for the passage of the tympanic nerve. The floor of the tympanic cavity is separated from the jugular fossa by a small bony plate, which explains the possibility of the occurrence of a jugular thrombophlebitis in an otitis media.

- The posterior wall *(paries mastoideus)* presents in its superior part, an aperture: *the aditus* to *the mastoid antrum (aditus ad antrum)*. In the middle part of *the aditus* is situated the pyramidal eminence *(eminentia pyramidali)*, a conic elevation from which emerges the tendon of the stapedius muscle *(m. stapedius)*. More lateral lies the opening for the entrance of *the chorda tympani nerve (apertura tympanica canaliculi chordae tympani)*. On this wall are also sited the prominence of the lateral semicircular canal *(prominentia canalis semicircularis lateralis)* and, more anteriorly, the prominence of the second portion of the facial canal *(prominentia canalis facialis)*; this canal bends and descends obliquely in the thickness of the posterior wall of the cavity (the third portion of the canal).

- The anterior wall *(paries caroticus)*. The inferior part is irregular, sometimes dehiscent; it corresponds to the carotid canal *(canalis caroticus)* (which explains the possibility of cataclysmic haemorrhages in *otitis media*) and is perforated by the caroticotympanic canaliculi *(canaliculi caroticotympanici)* and by small venules. The middle part of the wall is occupied by the tympanic opening of the auditory tube *(ostium tympanicum tubae auditivae)*. The fact that the pharyngotympanic tube does not open at the declive point in the tympanic cavity explains why, in the suppurative *otitis media*, the pus stagnates in the hypotympanic recess. Moreover, above the opening of the tube are situated the semicanal of the tensor tympanic muscle and the exit opening of *the chorda tympani nerve*, through the glaserian fissure.

The auditory ossicles
(Fig. 223-228)

The three ossicles, *the malleus, the incus* and *the stapes* are contained in the epitympanic recess. They articulate with each other, connecting the tympanic membrane to the oval window. The movements of this chain of *ossicles* are influenced by two muscles: the muscle of *the malleus (m. tensor tympani)* and the muscle of *the stapes (m. stapedius)*.

The malleus consists of a head *(caput mallei)*, a flattened neck *(collum mallei)*, a handle included in the tympanic membrane *(manubrium mallei)*, a short lateral process *(processus brevis)*, which produces on the tympanic membrane *the mallear* prominence *(prominentia mallearis)* and a long anterior process *(processus longus)*.

The incus is formed of a body *(corpus incudis)* articulated with *the malleus*, a short superior root *(crus breve)* in the fossa of the inferior angle of *aditus ad antrum* (aperture of the mastoid antrum) and an inferior long crus *(crus longum)*, with the extremity widened, the so-called lenticular process *(processus lenticularis)*, through which *the incus* articulates with *the stapes*.

241

The stapes consists of a head *(caput stapis)* articulated with *the incus* through the lenticular process of the latter, a flat base *(basis stapedis)*, applied on the oval window through the annular ligament of the stapes *(ligamentum annulare stapedis)* and two limbs: a shorter anterior limb *(crus anterius seu rectilinium)* and posterior one *(crus posterius seu curvilineum)*, which connect the head to the base of the stapes.

The connections between *the ossicles* are two complete articulations, with a capsule and a synovial membrane, one between the malleus and *the incus* called the incudomalleolar joint and the second between *the incus* and *the stapes*, called the incudostapedial joint.

The incudomalleolar joint is very tight and poorly movable; it derives from the embryonic articulation of the branchial arch 1. The vibration of the tympanic membrane displaces together these two *ossicles*, which make up a level with unequal arms, in which the movements have a swinging character and induce intensity changes.

Connection with the walls of the middle ear:

- *the malleus* is maintained in its position by three ligaments: superior, lateral and anterior *(lig. mallei superius, laterale et anterius)*;

- *the incus* is provided with two ligaments: superior and posterior *(lig. incudis superius et posterius)*,

- *the stapes* presents two ligaments: the annular ligament of the base of stapes *(ligamentum annulare stapedis)* and the obturator stapedial membrane *(membrana stapedis)*.

The muscles of the ossicles:

- the muscle of *the malleus (m. tensor tympani)* inserts on the sphenoidal spine and on the walls of the bony canal in which it is contained (musculotubal canal), its tendon bends laterally at the level of the cochleariform process and is attached to the handle of *the malleus*. The tensor tympani is supplied by a motor branch of the mandibular nerve. Its contraction stretches the tympanic membrane, which displaces *the ossicles* medially (accommodation to sounds of high intensity);

- *the stapedius* muscle *(m. stapedius)* is contained in the canal excavated in the posterior wall of the middle ear. Its tendon emerges from the orifice of the pyramid and is attached on the head of *the stapes*. It is supplied by a filament *(nervus stapedius)* of the facial nerve and is antagonistic to the tensor muscle of the tympanic membrane: it relaxes the tympanic membrane (accomodation to sounds of low intensity).

The tympanic mucosa behaves towards the contents of the middle ear like the peritoneum towards the abdominal organs. It invests the auditory *ossicles* the tendons and the ligaments and forms mesos which divide the tympanic cavity into compartments.

Owing to the presence of an interatriotympanic diaphragm, the epitympanic recess communicates with the atrium only through the narrow pass between the horizontal branch of *the incus* and the medial wall. The folds of the mucosa delineate the following cavities:

- the superior recess of the tympanic membrane, situated between the flaccid part of the tympanic membrane, the lateral ligament of the malleus and the neck of this ossicle;

- the anterior and posterior tympanic recesses, situated between the tympanic membrane and *the mallear* folds; they are separated by *the manubrium malleoli*;

- the superior recess situated between the wall of the niche of the ossicles, the body of *the stapes*, the head of *the malleus* and the lateral ligament of *the malleus*.

The presence of these cavities in the middle ear explains the passage to chronicity of an otitis in the case of an inadequate drainage.

The vessel supply of the tympanic cavity. *The arteries* supplying the tympanic cavity belong to the external carotid system. They are: the anterior tympanic artery, a branch of the maxillary artery; the inferior tympanic artery, a branch of the ascending pharyngeal artery; the posterior tympanic artery, a branch of the stylomastoid artery: the superior tympanic artery, a branch of the middle meningeal artery, the caroticotympanic artery, a branch of the internal carotid artery.

242

crista transversa

canalis facialis

area vestibularis superior

crus commune

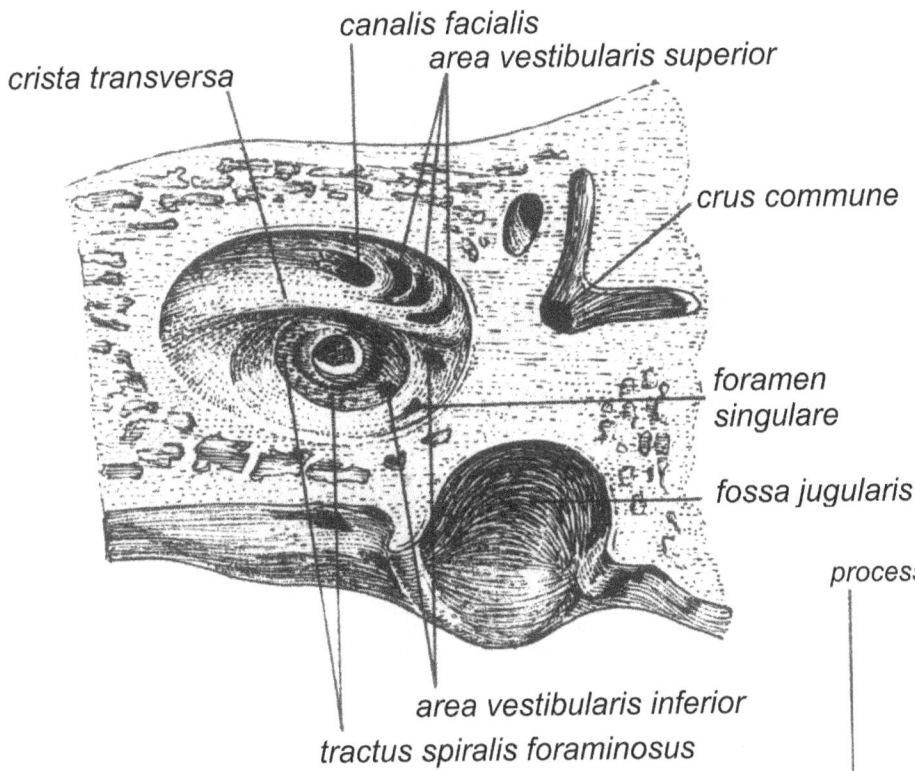

Fig. 222 Canal of the facial nerve – relationships with the osseous cavity of the middle ear in the petrous portion of the temporal bone

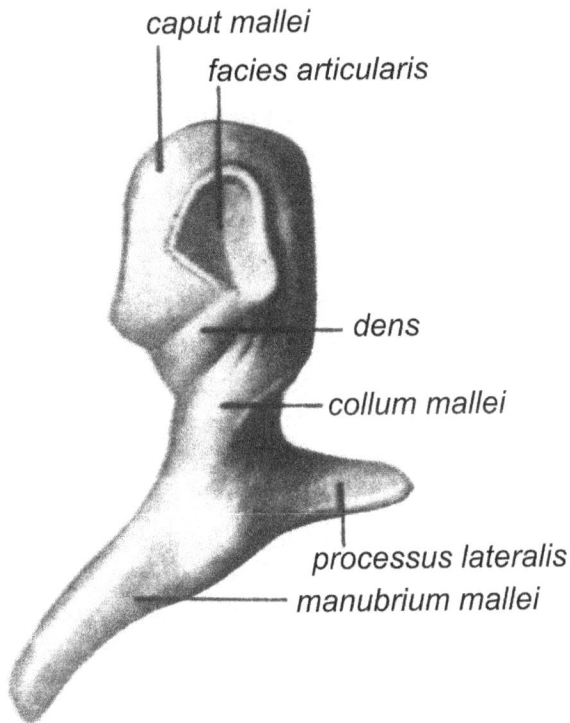

foramen singulare

fossa jugularis

area vestibularis inferior

tractus spiralis foraminosus

processus lateralis

collum mallei
caput mallei

caput mallei

facies articularis

dens

collum mallei

processus anterius
manubrium mallei

Fig. 223 The hammer (maleus) – posterior view

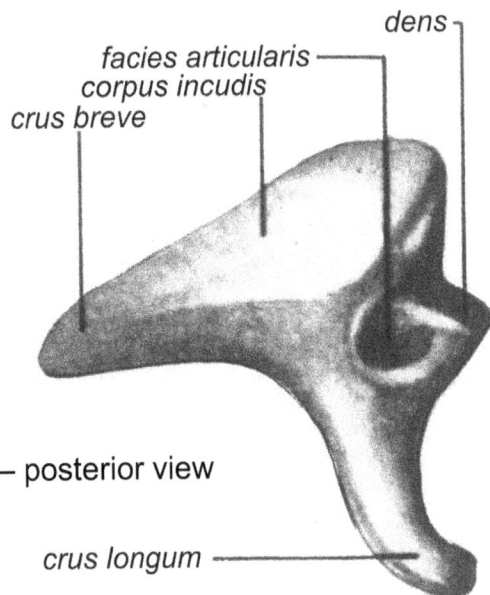

processus lateralis
manubrium mallei

Fig. 224 The hammer (malleus) – anterior view

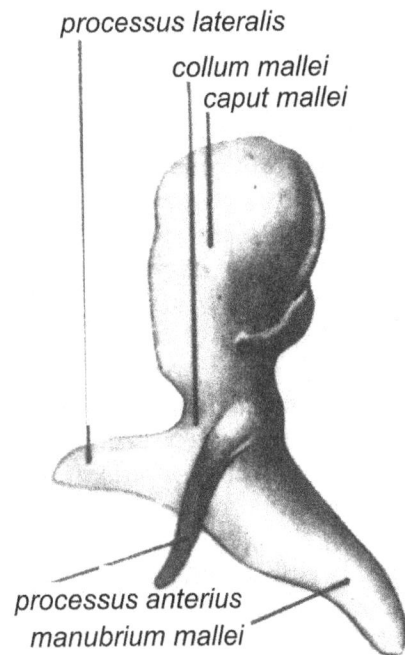

facies articularis
corpus incudis
crus breve

dens

Fig. 225 The anvil (incus) – posterior view

crus longum

243

dens
facies articularis
crus breve

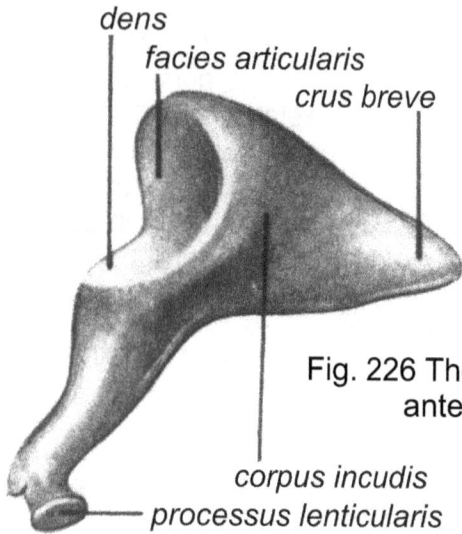

Fig. 226 The anvil (incus) –
anterior view

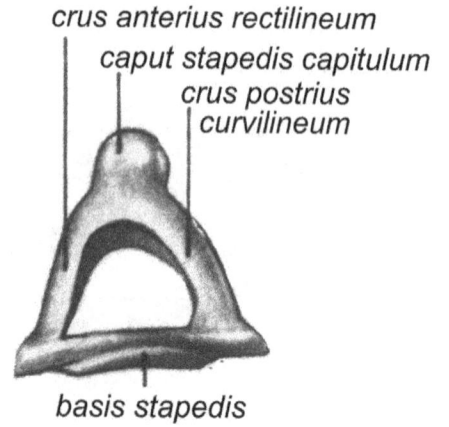

corpus incudis
processus lenticularis

crus anterius rectilineum
caput stapedis capitulum
crus postrius
curvilineum

basis stapedis

Fig. 227 The stirrup
(stapes)

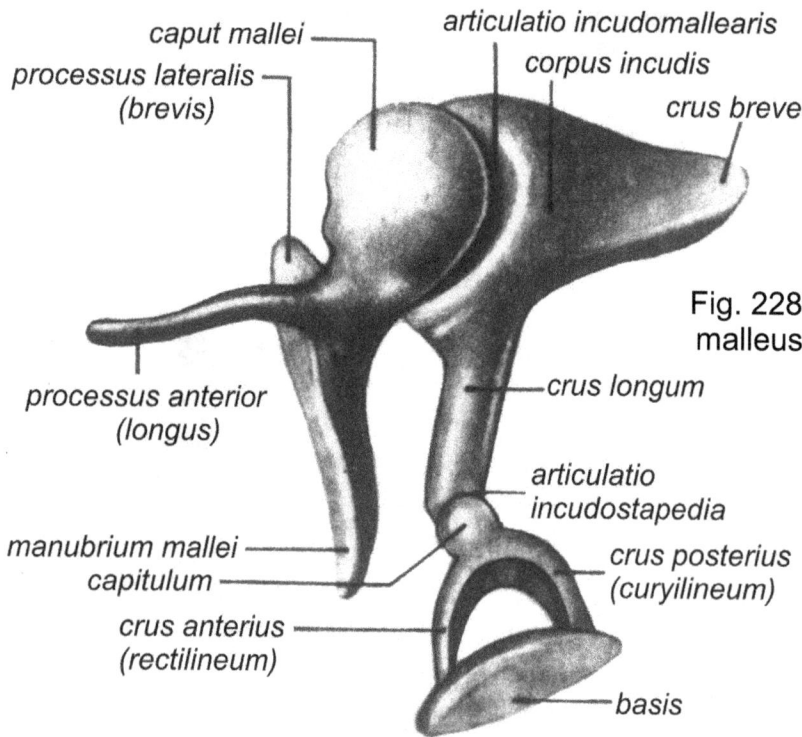

caput mallei
processus lateralis
(brevis)

articulatio incudomallearis
corpus incudis
crus breve

processus anterior
(longus)

crus longum

Fig. 228 Ossicles of the middle ear:
malleus, incus and stapes – overall
view

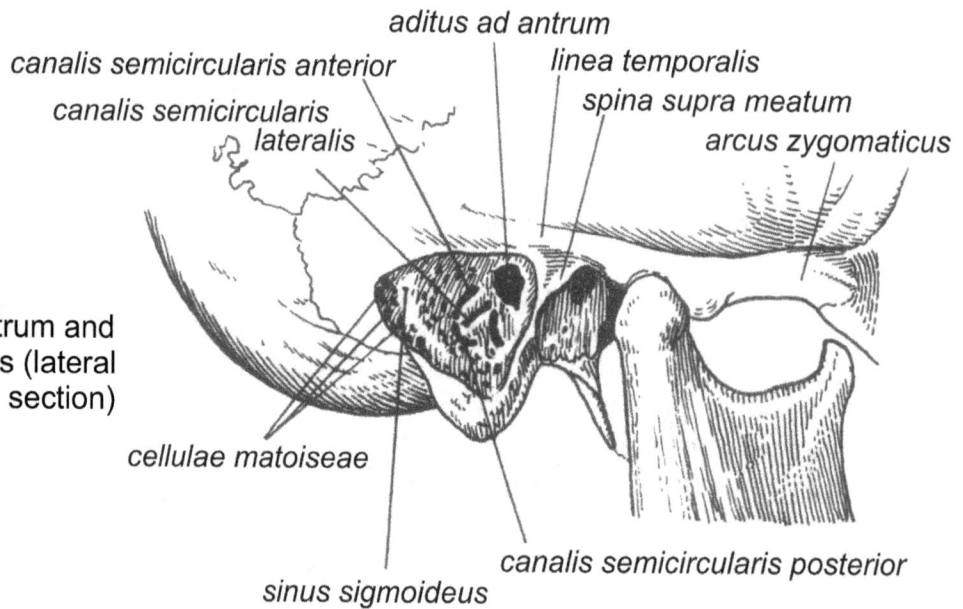

articulatio
incudostapedia
crus posterius
(curyilineum)

manubrium mallei
capitulum

crus anterius
(rectilineum)

basis

aditus ad antrum
canalis semicircularis anterior
canalis semicircularis
lateralis

linea temporalis
spina supra meatum
arcus zygomaticus

Fig. 229 Antrum and
mastoid cells (lateral
view, on the section)

cellulae matoiseae

244

sinus sigmoideus

canalis semicircularis posterior

The veins are drained through:
- affluents of the external jugular vein;
- the internal jugular vein;
- the inferior petrosal sinus;

The nerve supply of the tympanic cavity. The sensory nerve supply is assured by the tympanic nerve *(n. tympanicus)*, a branch of the glossopharyngeal nerve *(n. glossopharyngeus)*. It penetrates through the floor and ramifies on the promontory into its branches: one to the oval window, the second to the round window, one to the auditory tube, several carotidotympanic filaments to the internal pericarotid plexus, two petrosal nerves *(nn. petrosus major et minor)*, which traverse the tegmen.

The chorda tympani nerve *(n. chorda tympani)* represents a significant element of the region of the tympanic cavity; it arises from the facial nerve, shortly before its emergence from the petrous portion of the temporal bone.

It reaches the tympanic cavity via a posterior canal, passes along the internal surface of the tympanic membrane, emerges through an anterior canal (through which pass the tympanic artery and the anterior ligament of *the malleus*), reaches the sphenoidal spine and anastomoses with the lingual nerve (a branch of the maxillary nerve which arises from the trigeminal nerve).

In the tympanic cavity, the chorda tympani nerve runs laterally to the neck of the malleus, that lies on the mallear folds. Excited by the movements of the ossicles, the chorda tympani nerve induces the continuous salivary secretion; its reaction brings about gustatory and salivary impairments.

The mastoid cavities *(mastoid air cells)*
(Cellulae mastoideae)

The mastoid process is a bony prominence situated behind the external acoustic meatus. It belongs to the temporal bone and has a varying structure. In a fifth of cases it is compact and presents only a cavity, the mastoid antrum *(antrum mastoideum)*, connected to the tympanic cavity by the aditus and antrum *(aperture of mastoid antrum)*.

However, in other cases, the antrum is replaced by a great number of mastoid cavities achieving the pneumatic mastoid. The mastoid invasion by an infection originating in the middle ear is possible.

The auditus ad antrium is a bony, prismatic, triangular canal, that connects the middle ear to the mastoid antrum. Its superior wall is continuous with the roof of the tympanum *(tegmen tympani)*.

The medial wall presents the prominence of the lateral semicircular canal, which is an important surgical landmark. Its inferior angle corresponds to the hand of the canal of the facial nerve.

The mastoid antrum is a polymorphous cavity of a variable size, which exists constantly and is situated obliquely, behind and laterally to the tympanic cavity. The projection of the antrum on the surface of the temporal bone corresponds to a square, the topography of which is rather precise, like that of the aditus.

The mastoid cells, when they exist, are scattered around the antrum and we distinguish superior, inferior, posterior, lateral and medial cell groups. Moreover, aberrant, petrous, squanamous and zygomatic groups may exist, too, so that an infection starting from the middle ear can spread up to them (mastoiditis).

The narrow petromastoid canal connects the antrum to the subarcuate fossa (behind the internal acoustic meatus); it contains some slender vessels).

The vessels supply of the mastoid cavities is assured by branches of the stylomastoid artery, arissing from the posterior auricular artery, and of the middle meningeal artery, arising from the maxillary artery.

Relation of the mastoid region:

- with the sigmoid sinus *(sinus sigmoideus)*, of the dura mater which, normally, is remote from the antrum, but may get in a direct relation with its posterior wall;

- with the facial nerve, which traverses the petrous part of the temporal bone and bends twice in its course. The first segment, transverse on the axis of the petrous part of the temporal bone, passes in the neighborhood of the internal ear, between the vestibule and the cochlea. The second segment, oblique backwards laterally and downwards, is situated on the medial wall of the tympanic cavity, above and behind the oval window. The last segment is oblique downwards and slightly outwards, the nerve being contained in the bony, 5 mm thick wall, which separates, at first, the middle ear from the antrum, then the mastoid cells from the external acoustic meatus.

The facial nerve can be damaged in the two last portions either owing to the infection of the middle ear or in the course of surgery.

The auditory tube
(Tuba auditiva)

The auditory or eustachian tube is an osterocartilagionous channel connecting the tympanic cavity to the pharynx, and assuring the ventilation of the middle ear. It is oblique forwards, medially and downwards, forming an angle of 40° on the average. According to Guerrier, its length is 4 cm. The bony segment is contained in the temporal bone, while the cartilaginous segment lies below the base of the skull, in the lateral wall of the pharynx.

The bony part of the auditory tube *(pars ossea tubae auditivae)* is a 12-14 mm long-channel, situated between the petrous portion of the temporal bone and the tympanic bone, in the musculotubal canal below the semicanal of the tensor muscle of the canal, lateral to the carotid canal. Its caliber diminishes as it gets closer to the surface of the bone. It opens into the angle between the petrous portion of the temporal bone and the squama of the temporal bone, hence below the base of the skull.

The cartilaginous part of the auditory tube *(pars cartilaginea tubae auditivae)* is formed of a fibrocartilage in the form of a gutter with the concavity downwards. This gutter is transformed into a canal by a fibrous plate, which unites its edges. The canal unites with the opening of the bony portion, the caliber of the auditory tube being here the smallest. Initially, the two edges are at the same level, then the posteromedial edge becomes higher and thinner and the cartilage acquires the appearance of a hook.

The lateral cartilage is 24 mm long; it is more oblique downwards than the bony portion and its diameter increases gradually. It is finally attached at the level of the sphenopetrous suture, along the pharynx, then it diverges from the base of the skull and passes along the medial plate of the pterygoid process. The pharyngeal orifice *(ostium pharyngeum)* opens into the external wall of the nasopharynx, at 1cm behind the posterior margin of the inferior nasal concha, 1 cm above the soft palate *(velum palatinum)* and 1.5 cm anteriorly to the posterior wall of the pharynx.

This orifice of the tube is triangular, its anterior margin is formed by the salpingopalatine fold, its posterior margin, more prominent by the salpingopharyngeal fold and its inferior part by the levator muscle of the soft palate (the prominence of the internal peristaphyline muscle) *(torus tubarius)*. Above the orifice lies the supratubal fossa and behind the orifice is situated the pharyngeal recess or Rosenmüller's fossa, an important landmark in the catheterization of the tube.

Systematizing the main relations of the tube, we underline that in its superior part, it is adjacent to the middle meningeal vessels, posteromedially it is in relation with the internal carotid artery, which crosses it in its course in the carotid canal, and antero-externally with the mandibular nerve at its emergence from the oval foramen and with the tensor muscle of the soft palate.

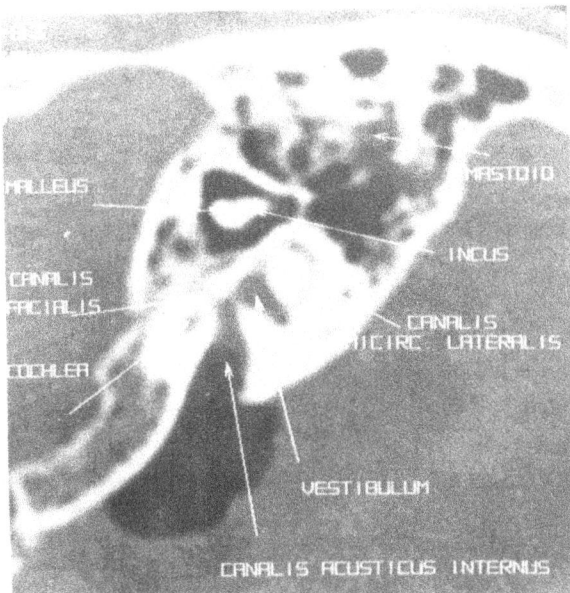

Fig. 230 Computed tomography on the petrous portion of the temporal bone (Siemens Collection)

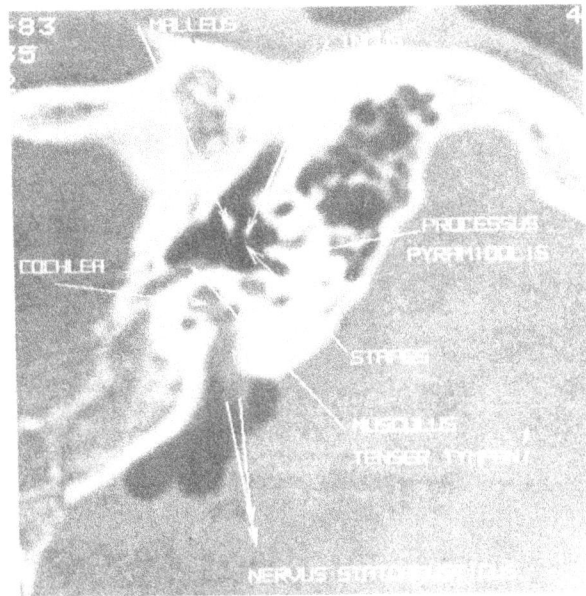

Fig. 231 Computed tomography of the middle ear (Siemens Collection)

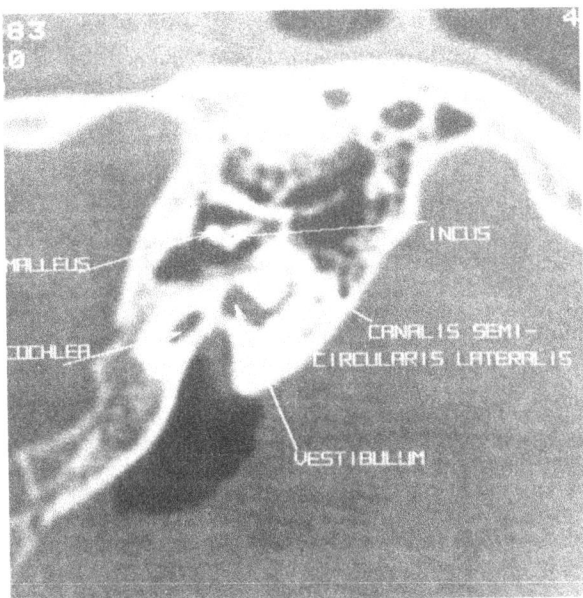

Fig. 232 Computed tomography of the auditory ossicles (Siemens Collection)

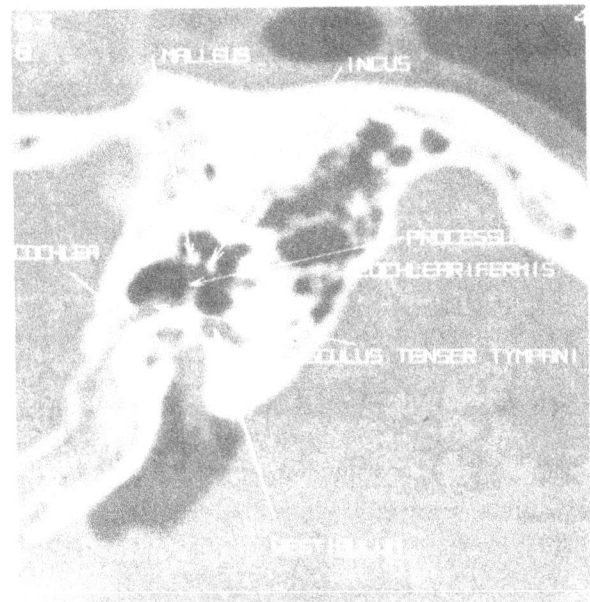

Fig. 233 Computed tomography of the petrous portion of the temporal bone at the level of the tensor tympani muscle

The tubal mucosa *(tunica mucosa tubae auditivae)* is thin in its bony portion and thickened in its cartilaginous portion. It contains lymphatic follicles and is innervated by the tubal branch of the tympanic nerve, which is a branch of the glossopharyngeal nerve, and by Bock's pharyngeal nerve, which arises from the pterygopalatine ganglion (Meckel's ganglion). The motor apparatus.

The tubal cartilage is subdivided by notches and mobilized by the tensor and levator muscles of the soft palate *(mm. tensor et levator veli palati)* (external and internal peristaphyline muscles). These two muscles course along the tube, from the base of the skull towards the soft palate. The first is supplied by the mandibular nerve, the second by the vagus nerve. Both are dilators of the auditory tube; when the soft palate is at rest, the cartilaginous portion is closed and during each deglutition movement, when the tube opens, the tympanic cavity is aerated.

247

The inflamation of the tubal mucosa prevents the opening of the tube and causes deafness through tubal catarrh.

The arterial supply of the auditory tube is assured by the ascending pharyngeal artery (branch of the external carotid artery), by the meningeal artery (branch of the maxillary artery) and by the vidian artery (branch of the maxillary artery). The satellite *veins* of the arteries drain into the pterygoid plexus and then into the jugular veins, forming anastomoses with the tympanic and pharyngeal veins. The *lymphatic vessels* drain into the retropharyngeal lymph nodes.

The labyrinthine region. The internal ear
(Auris interna)

The labyrinth consists of two anatomically and functionally completely different parts:
- *the vestibule* and the bony semicircular canals, which contain the receptors for equilibrium and spatial body orientation; the vestibular nerve forms a synapse with them;
- *the cochlea*, from the receptors of which the cochlear nerve, through slender neurofibrillar networks, collects the excitations, conveying them under the form of nervous inflows towards the cortical centres.

The internal ear or labyrinth consists of a series of cavities within the petrous part of the temporal bone, medially to the tympanic cavity (the middle ear). These cavities form the bony labyrinth *(labyrinthus osseus)*, inside which is situated the membraneous labyrinth *(labyrinthus membranaceus)*, that contains the auditory and vestibular receptors. The membranous labyrinth is separated from the bony labyrinth by a small space, filled with the fluid called perilymph. The membranous labyrinth contains a similar fluid, named endolymph (Fig. 234 - 242).

The bony (osseous) labyrinth
(Labyrinthus osseus)

The cavities of the bony labyrinth are delineated by a special bony tissue, derived from the embryonic auditory capsule. It consists of the vestibule and the cochlea situated in front of it. Behind it lie the bony semicircular canals.

To the bony labyrinth belongs also the internal acoustic meatus, which presents an orifice in the posterior wall of the pyramid of the temporal bone and thus penetrates laterally into the pyramid. Its bottom is similar to a sieve, through the holes of which pass the nerve fibres deriving from the cochlea and the vestibule.

The transverse crest *(crista transversa)* divides the fundus of the internal acoustic meatus into a superior and an inferior area.

The vestibule (vestibulum) is the central part of the labyrinth and constitutes the access space to the cochlea, situated anteriorly, and to the semicircular canals, situated behind; it has an irregulliar, lateromedially flattened ovoid shape and is oriented perpendicularly on the axis of the petrous portion of the temporal bone. It has the following delimiting walls:
- a lateral wall corresponding to the middle ear, to which it is connected above, through the oval window (covered by the base of the shapes) and below through the round window. In its superior half opens the lateral semicircular canal;
- a superior and a posterior wall, in which lie the orifices of the superior and posterior semicircular canals;
- an inferior wall, from which a bony plate detaches itself; from this plate starts the osseous spiral lamina of the cochlea;
- the anterior wall, corresponding upwards to the first portion of the canal of the facial nerve and downwards, to the cochlea;
- a medial wall corresponding, in its anterior part, to the posterior half of the internal acoustic meatus. It contains three recesses:

248

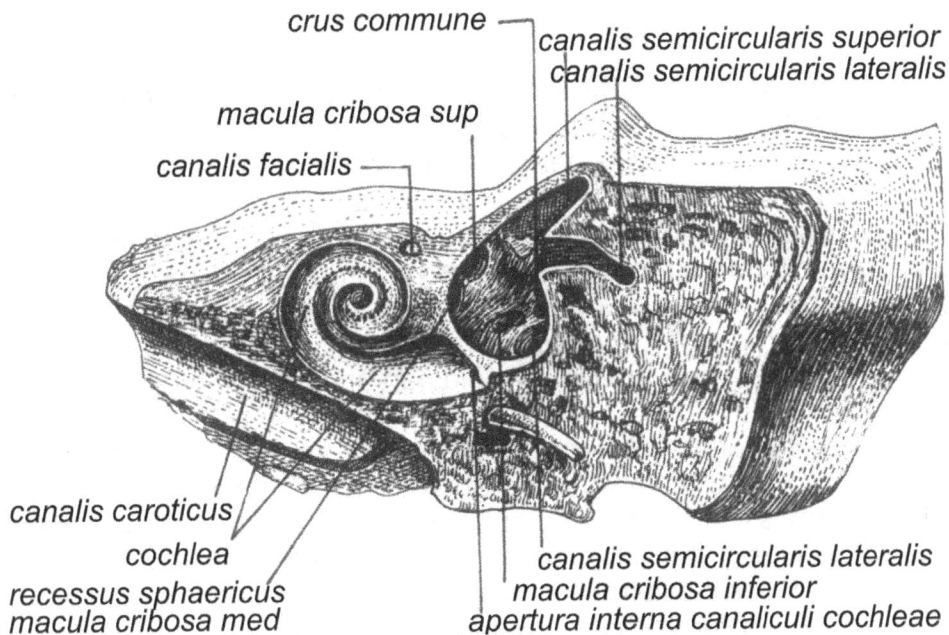

crus commune
canalis semicircularis superior
canalis semicircularis lateralis
macula cribosa sup
canalis facialis
canalis caroticus
cochlea
recessus sphaericus
macula cribosa med
canalis semicircularis lateralis
macula cribosa inferior
apertura interna canaliculi cochleae

Fig. 234 Semicircular canals and cochlea carved in the temporal bone

- the elliptical recess (recessus ellipticus), called also the semioval fovea, situated anterosuperiorly, in which the utricle (utricula) is lodged. It is covered with a long plate, through which passes the utriculo-ampullar nerve;

- the hemispherical recess (recessus haemisphericus), called also hemispherical fovea, situated antero-inferiorly and covered with a perforated plate, through which pass the fibres of the saccular nerve (n. saccularis); here is lodged the sacculus. In front of this recess lies the pyramid of the vestibule (pyramis vestibuli) and between them is situated the vestibular crest (crista vestibuli)

- the cochlear recess (recessus cochlearis), situated behind; above it lies the opening of the aqueduct of the vestibule, which contains the endolymphatic canal.

The semicircular canals (canales semicirculares ossei) are three cylindrical, curved canals, which open into the vestibule through their extremities. They are arranged in three planes of the space, approximately a sagittal, a frontal and a horizontal plane, and are bent from the sagittal median plane with 45° and from the horizontal plane with 45°.

The anterior or superior semicircular canal (canalis semicircularis superior) is vertical and perpendicular on the axis of the petrous part of the temporal bone. It begins in the anterior part of the superior wall of the vestibule and ends in the posteromedial part of the superior wall, by a portion which is common with the posterior semicircular canal. Its convexity forms the arcuate eminence (eminentia arcuata) on the endocranial surface of the petrous part of the temporal bone.

The posterior semicircular canal (canalis semicircularis posterior), is vertical, oriented nearly parallel to the axis of the petrous part of the temporal bone. The ampullar orifice is situated at the level of the posterior, inferior and lateral angle of the vestibule, while the other orifice is united with that of the superior canal.

The lateral semicircular canal (canalis semicircularis lateralis) is convex laterally and behind and oriented in a horizontal plane. The ampullar orifice is situated superoposteriorly to the oval window, on the lateral wall of the vestibule. Its non-ampullar orifice lies in the posterosuperior zone of the lateral wall. Through its convexity, this canal bulges on the medial wall of the aditus ad antrum.

The cochlea consists of a spiral canal, wound around a conical axis, called *columella or modiolus*. It is anterior to the vestibule and lies between the tympanic cavity and the internal acoustic meatus.

The columella (modiolus) is a bony cavity cone, the excavated base of which *(basis modioli)* forms the cochlear fovea *(cochlear fovea)*. The apex of the cone is oriented anterolaterally. The cochlear recess is perforated by foramina disposed along a spiral line; they lead into the spiral of the modiolus canal (*canalis spiralis modiolis*, Rosenthal's canal), which contains Corti's spiral ganglion or spiral cochlear ganglion. Numerous canaliculi arise from the spiral canal and penetrate into the thickness of the spiral lamina.

The spiral cochlear canal *(canalis spiralis cochleae)* winds around *the mediolus* (clockwise on the right and counterclockwise on the left), forming two turns and a half. It begins at the level of the vestibule by an unwound part and ends in cul-de-sac at the level of *the cupula* of the cochlea.

The osseous spiral lamina *(lamina spiralis ossea)* is a bony plate winding, like the cochlea, around *the modiolus*, alongside the spiral canal. It has a free edge in the spiral cochlear canal, which does not reach the outer lamina of the latter. At the level of *the cupula* of the cochlea *(cupula cochleae)*, the spiral lamina ends in a hook-shaped process: *the hamulus* of the spiral lamina. It divides the spiral cochlear canal into two parts: *the scala vestibuli*, so-called because it opens into the vestibule, and *the scala tympani*. The latter ends below the vestibule, through a more prominent part, which causes the formation of the promontory in the tympanic cavity (middle ear). The scala tympany communicates with the middle ear through the round window *(fenestra cochleae)* and with the cranial cavity through the cochlear canaliculus *(canaliculus cochleae)*, which opens into the petrosal fossula *(fossula petrosa)*, on the posterior margin of the petrous part of the temporal bone. The separation between the two *scalae* is completed by the presence of the membranous cochlear duct *(ductus cochlearis)*; the only place which remains open, permitting the communication between the two scalae, is situated at the level of the hook formed by the spiral lamina in *the cupula* of the cochlea. This opening is called *helicotrema*.

The membranous labyrinth
(Labyrinthus membranaceus)

The bony vestibule contains two membranous vesicles: a superior anteroposteriorly elongated vesicle, called the utricle *(utriculus)*, which corresponds to the semioval fovea, and a rounded, smaller inferior vesicle, the saccule *(sacculus)*, in the hemispherical recess. The two vesicles adhere to the corresponding recesses on the medial wall of the bony vestibule (elliptic and spherical recess). Both the utricle and the saccule present a sensory epithelial area of 3/2 mm, the maculae of the utricle and of the saccule; at the level of the sensory maculae arrive the dendrites of the Scarpa's neurons, whose axons form the vestibular nerve. From the medial wall of the saccule and of the utricle two canals slender are given off, which run backwards and upwards and unite to form the endolymphatic duct *(ductus endolymphaticus)*. This duct passes through the aqueduct of the vestibule *(aqueductus vestibuli)* and end below the cranial dura mater and in a dilated extremity, called the endolymphatic sac *(saccus endolymphaticus)*, situated between two leaflets of the dura mater on the posterior surface of the pyramid of the temporal bone (petrous pyramid).

The membranous semicircular ducts, contained in the bony semicircular canals, have the same shape, but a caliber reduced to one fourth of that of the bony canals. They are bathed in the perilymph and open into the utricle by five openings; each duct has a dilated ampullar end. From the medial wall of the ampullae of the membraneous semicircular ducts, at the level of the sensory area start nerve fibers (dendrites), which reach the interval acoustic meatus *(meatus acusticus internus)* through the orifices of the elliptical recess, except the fibers of the ampulla of the posterior duct, which pass through the solitary foramen *(foramen singulare)*.

250

The vestibular receptors. The utricle and the saccule contain *the maculae* which record the static position of the head. They consist of a sensory epithelium, formed of supporting cells and of sensory cells, called also *"hair cells"*, owing to their ciliated apical pole, while their basal pole is in contact with the dendrites of the neurons of Scarpa's ganglion. Above this sensory epithelium lies the gelatinous mass, *the statolithic* or *otolithic membrane*, which contains calcareous concretions called *otoliths (statoliths)*. The utricle contains receptors for horizontal movements, while those lying in the saccule are receptors for vertical movements.

The receptor zones in the semicircular canals are termed ampullary crests. Three in number, they are disposed in the ampullary region of the ducts . They are sensitive to gyratory motion and represent the starting point for the appreciation of the direction of motion; they are formed of a sensory epithelium containing supporting cells and sensory cells ciliated at the apical pole.

The cili are included in a dome-shaped gelatinous mass, called ampullary cupula *(cupula ampullaris)*, perpendicular on the cili in the sensory crests. The movements of the endolymph mobilize the cupulae or the gelatinous crests which, in turn, exert a traction on the cili of the sensory cells, exciting these cells.

The dendrites, which envelop the basal pole of the sensory cells, receive the movements of the cili and code them into a nerve inflow. These dendrites arise from the neurons of the Scarpa's ganglion. The axons of these neurons form the vestibular branch of the eighth cranial nerve (the vestibulocochlear nerve).

The cochlear (auditory) receptors. The membranous cochlea or cochlear duct *(ductus cochlearis)* has the shape of a small prismatic triangular tube extending from the osseous spiral lamina, which it prolongs, till the lateral wall of the bony cochlea. It ascends up to the apex of *the modiolus*, where it ends in a blind sac, called the cupular caecum or *lagaena* and separates the scala tympani from the scala vestibuli. It has three walls:

- the inferior tympanic wall *(lamina basilaris)* has a complex structure; it is formed of a bony part and of a membranous part, the basilar (or spiral) membrane, which inserts itself at the level of a prominence (the basilar crest) of the lateral spiral ligament. The bony portion consists of the osseous spiral lamina, of its periosteal thickening which forms the internal spiral ligament (the spiral crest), and of the epithelium which lines them. The triangular osseous spiral lamina, with the base attached on *the modiolus*, consists of two bony plates, between which the Corti's ganglion is situated.

The basilar membrane is formed of a great number of fibres (over 24,000 – according to Garven), embedded in a homogeneous ground substance; they are called auditory strings and their structure is different in the external and internal zone of the membrane

The inferior tympanic wall supports the spiral organ of Corti, constituted of supporting and sensory cells, to which arrive the dendrites of the neurons of Corti's ganglion, who's axons form the cochlear branch of the eighth cranial (vestibulocochlea) nerve.

The lateral (peripheral) wall, that makes up the base of the membranous cochlea, adheres to the periosteum of the bony cochlea, forming the external spiral ligament of cochlea (lig. spirale cochleae). At this level lies a pigment and vascular epithelium, which forms the stria vascularis that is involved in the production of endolymph and in the regulation of its ionic content, as well as in the endolymph resorption.

The vestibular wall *(paries vestibularis)* or Reissner's membrane, situated above and anteriorly, is in relation with the scala vestibuli. The vestibular membrane which forms this wall is involved in the active substance transfer from the vestibular perilymph into the endolymph of the cochlear duct.

The cochlear duct continues behind, lateral to the spiral lamina, on the floor of the vestibule, where it ends in a cul-de-sac *(caecum vestibulare)*. It is united with the saccule through a narrow canal, the uniting duct or Hensen's duct *(ductus reuniens)*. In this way, all the cavities of the membranous labyrinth communicate with each other.

Corti's organ or the auditory receptor is formed of a sensory epithelium with particular features, situated on the basilar membrane, on the membranous cochlea.

251

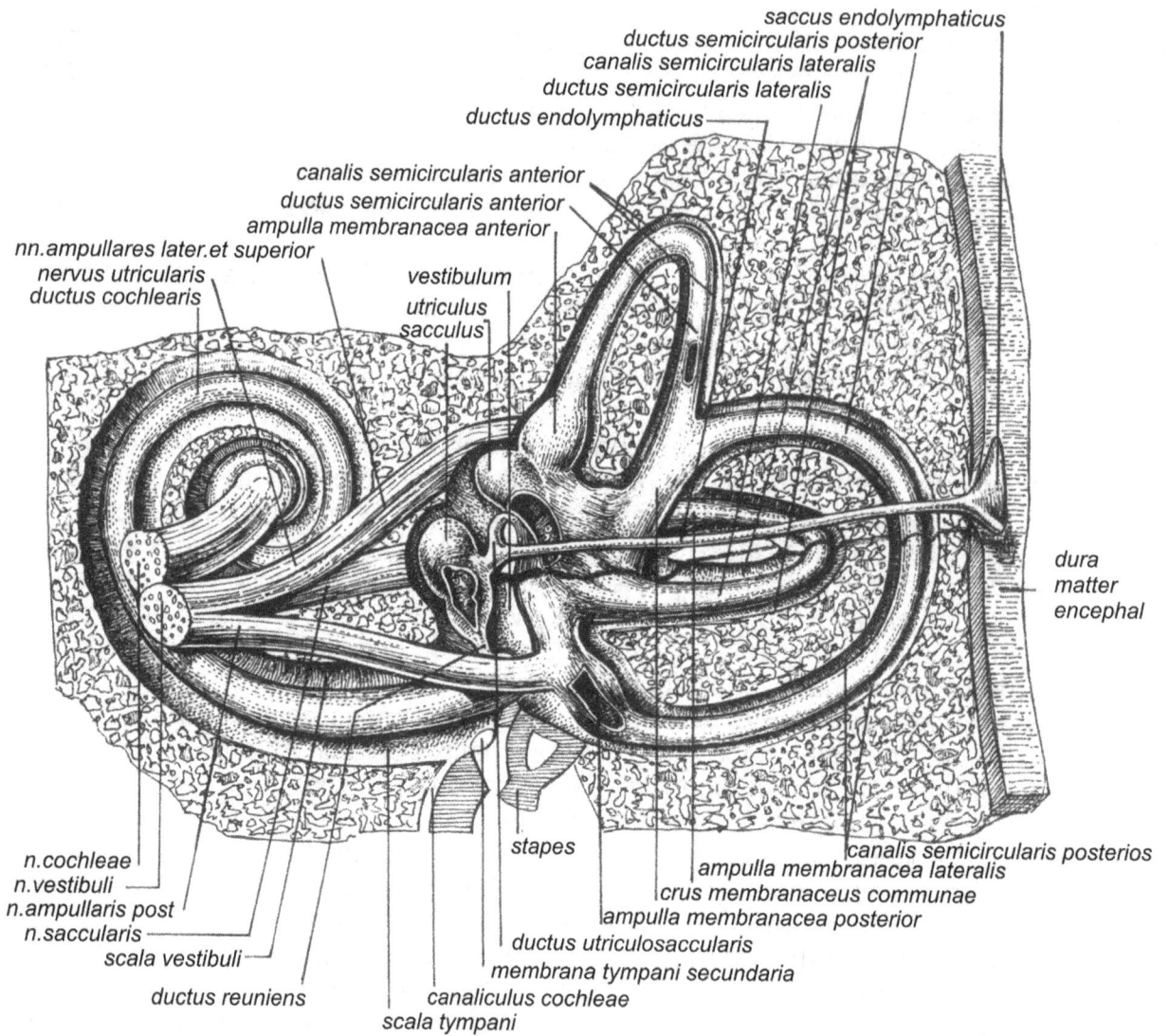

saccus endolymphaticus
ductus semicircularis posterior
canalis semicircularis lateralis
ductus semicircularis lateralis
ductus endolymphaticus

canalis semicircularis anterior
ductus semicircularis anterior
ampulla membranacea anterior

nn.ampullares later.et superior
nervus utricularis
ductus cochlearis

vestibulum
utriculus
sacculus

dura
matter
encephal

n.cochleae
n.vestibuli
n.ampullaris post
n.saccularis
scala vestibuli

ductus reuniens

scala tympani

stapes

canaliculus cochleae

ductus utriculosaccularis
membrana tympani secundaria

canalis semicircularis posterios
ampulla membranacea lateralis
crus membranaceus communae
ampulla membranacea posterior

Fig. 235 Membranous labyrinth – structure of the vestibulocochlear nerve

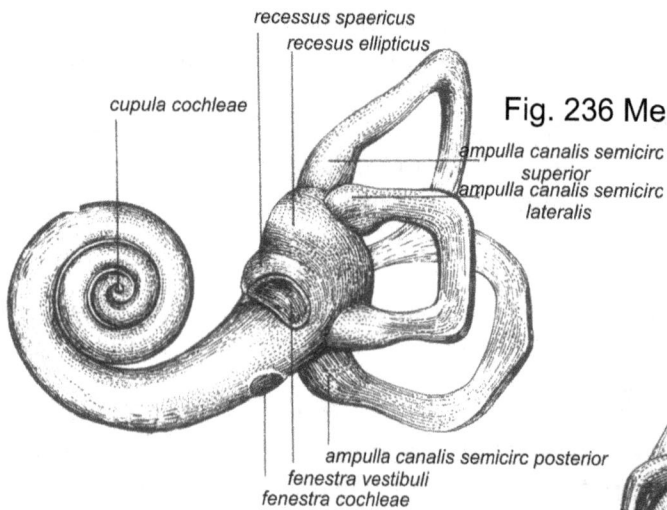

recessus spaericus
recesus ellipticus

cupula cochleae

Fig. 236 Membranous labyrith, left side (lateral view)

ampulla canalis semicirc
superior
ampulla canalis semicirc
lateralis

ampulla canalis semicirc posterior
fenestra vestibuli
fenestra cochleae

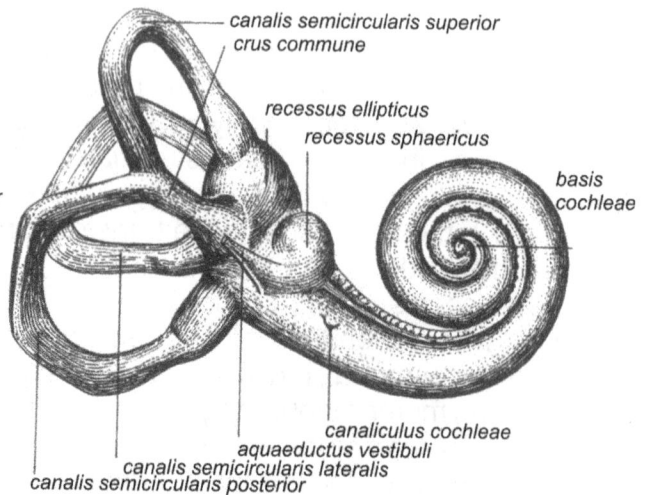

canalis semicircularis superior
crus commune

recessus ellipticus
recessus sphaericus

basis
cochleae

Fig. 237 Membranous labyrith,
left side (medial view)

canaliculus cochleae
aquaeductus vestibuli
canalis semicircularis lateralis
canalis semicircularis posterior

252

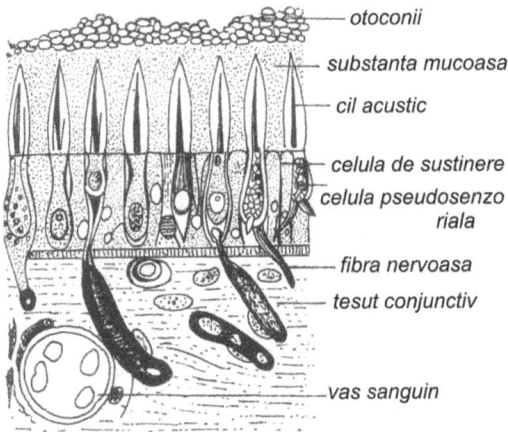

Fig. 240 Structure of the maculae

- otoconii
- substanta mucoasa
- cil acustic
- celula de sustinere
- celula pseudosenzoriala
- fibra nervoasa
- tesut conjunctiv
- vas sanguin

The sensory epithelium consists of two cell types: supporting and sensory cells, separated through their bases, but in a close contact through their apices; these cells form Corti's tunnel. The remaining supporting cells are scattered among the sensory cells. Medially to Corti's tunnel there are a row of sensory cells and the cells of the epithelium of the spiral sulcus, whereas laterally to the tunnel two, three and even four rows of sensory cells are present. This number increases as the basilar membrane gets closer to the apex of the cochlea. The sensory cells are ciliated at the apical pole and at this base lie dendritic processes which envelop them and which derive from the neurons of Corti's ganglion, situated in the cavity of the modiolus.

Fig. 238 Section at the level of the cochlea

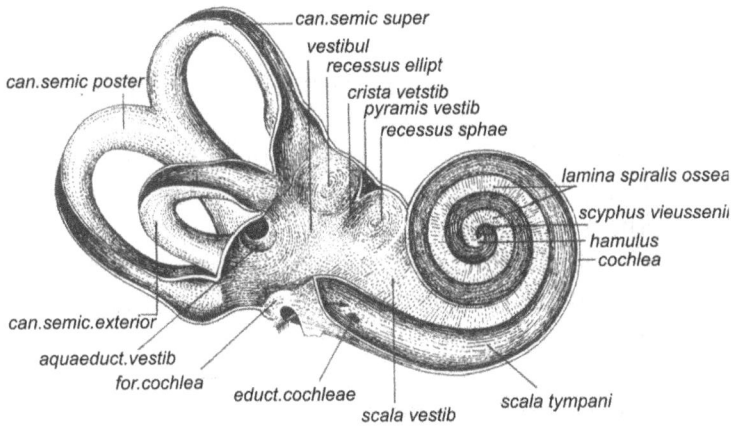

- can.semic super
- vestibul
- recessus ellipt
- crista vetstib
- pyramis vestib
- recessus sphae
- can.semic poster
- can.semic.exterior
- aquaeduct.vestib
- for.cochlea
- educt.cochleae
- scala vestib
- scala tympani
- lamina spiralis ossea
- scyphus vieussenii
- hamulus
- cochlea

Fig. 239 Histological structure of the vestibular (ampullary) crests

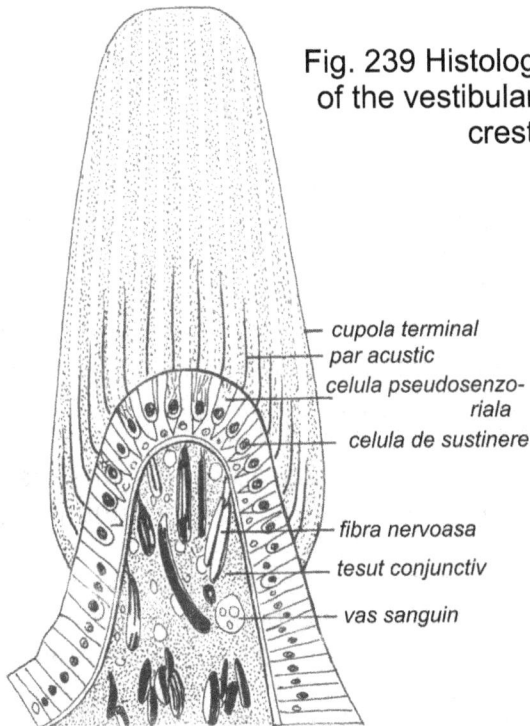

- paries externus ductus cochlearis
- ductus cochlearis
- scala vestibuli
- sulcus spiralis externus
- paries vestibularis ductus cochlearis (membrana vesti)
- membrana tectoria
- sulcus spiralis internus
- abium limbi vestibulare
- lamina spiralis ossea
- labium limbi tympanicum
- lamina basilaris
- proeminentia spiralis
- lig. spirale cochleae
- scala tympani
- ganglion spirale

Fig. 241 Section at the level of the cochlea, showing Corti's organ

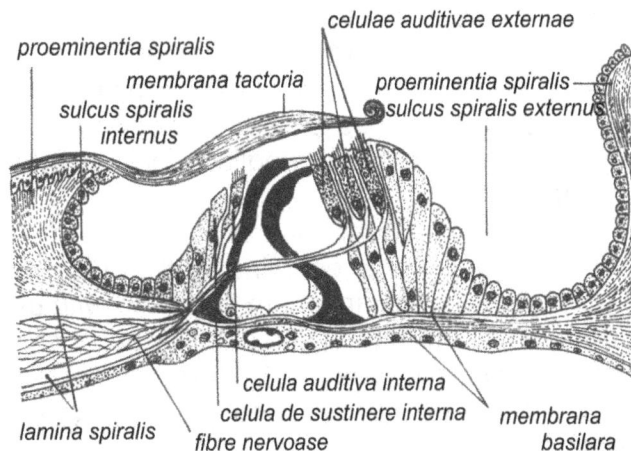

- cupola terminal
- par acustic
- celula pseudosenzoriala
- celula de sustinere
- fibra nervoasa
- tesut conjunctiv
- vas sanguin

Fig. 242 Histological structure of Corti's organ

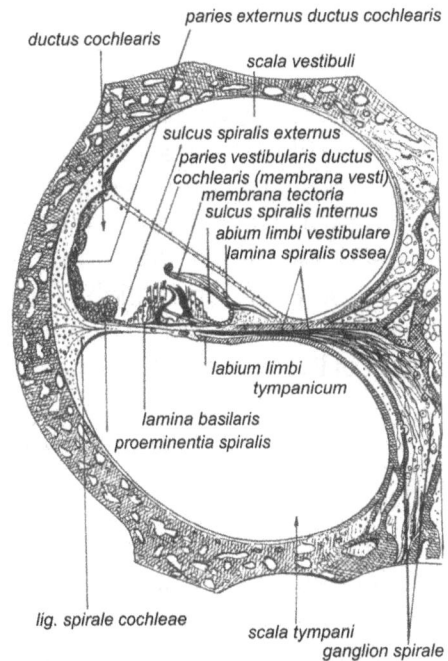

- proeminentia spiralis
- membrana tactoria
- sulcus spiralis internus
- celulae auditivae externae
- proeminentia spiralis
- sulcus spiralis externus
- celula auditiva interna
- celula de sustinere interna
- fibre nervoase
- membrana basilara
- lamina spiralis

253

The sensory cells are supported by Deiters' cells, which are laterally continuous with Hensen's cells and above the spiral ligament, with the epithelial, cubical cells of the cochlear duct.

Above Corti's organ lies the tectorial membrane (*membrana tectoria*), which has various contacts with *the cili* of the sensory cells, exciting them in dependence on the movements of the endolymph. The movements of the endolymph are synchronous with the movements of the perilymph in the cochlea and the latter, with the movements of *the ossicles* in the middle ear and with the movement of the tympanic membrane. High-pitched sounds are perceived at the base of the cochlea and low-pitched sounds, towards the apex.

The arterial supply of the labyrinthine region is assured by the posterior tympanic artery, a branch of the stylomastoid artery, and by the internal auditory (labyrinthine) artery, a branch of the basilar artery. The venous drainage is assured by internal auditory (labyrinthine) veins which empty mainly into the inferior petrosal sinus, by the vein of the vestibular aqueduct (*v. aqueductus vestibuli*), which opens into the superior petrosal sinus, and by the vein of the cochlear aqueduct (*v. canaliculi cochleae*), which empties into the internal jugular vein. Lymphatic vessels are not present at this level, as the ends and perilymph drain into the CSF.

The conduction pathways
The vestibular sensory pathways

The vestibular excitations arise from *the maculae*, as regards the position of the head and of the body at rest, and from the ampullary crest as regards the body equilibration during walking. They are conveyed from the first neuron (the receptor protoneuron situated at the level of Scarpa's ganglion) to the level of the brainstem, where the second neuron is sited, via the vestibular nerve that shapes its course with the cochlear nerve, forming the vestibulocochlear nerve (eighth cranial nerve), which, for didact and topographical reasons, will be described with the vestibular sensory pathways; merely references will be made on the occasion of the description of the acoustic sensory pathways.

The stato-acoustic or vestibulocochlear nerve
(Nervus octavus)

It is formed by the union of two nerves differing with respect to their origin, their function and their central connections: the cochlear nerve and the vestibular nerve, the fibers of which are connected in a single nerve.

The vestibular nerve (*pars vestibularis n. octavi*). From the sensory areas of the utricle, the saccule and the membranous semicircular ducts arise a series of nerve fibers (dendrites), which form the nerve of the saccule, the nerve of the utricle and the ampullar nerves of the superior and lateral semicircular ducts. The ampullar nerves unite with each other to form the utriculo-ampullar nerve, a superior branch of the vestibular nerve, which traverses the superior vestibular fovea. The saccular nerve, which passes through the inferior vestibular fovea, and the posterior ampullar nerve, which passes through *the foramen singulare*, unite with each other to form the inferior branch of the vestibular nerve.

The two branches join each other in the internal acoustic meatus; here is also the site of the vestibular (Scarpa's) ganglion, formed by the pericaryons of the first neurons of the vestibular pathway.

The further course and relations of the stato-acoustic nerve are those of the internal acoustic meatus which contains them. This meatus, situated in the petrous portion of the temporal bone is 1 cm in length and lined with *dura mater*, it is oblique anteroposteriorly and lateromedially. The lateral end of this meatus corresponds to the base of *the modiolus* and to the medial wall of the vestibule. A transverse crest divides it into two regions, each of which is, in turn, subdivided into two areas, so that we distinguish four areas in the internal acoustic meatus:

254

- the area of the facial nerve – anterosuperior;
- the cochlear area – anteroinferior;
- the superior (utricular) and inferior (saccular) vestibular areas.

The cochlear and vestibular areas are covered with bony laminae, perforated by minute orifices; the area of the facial nerve opens into the fallopian aqueduct.

The cochlear nerve *(pars cochlearis n. octavi)*. The acoustic excitations are received by Corti's spiral organ *(organum spirale)*, situated on the basilar membrane of the cochlear canal *(lamina basilaris ductus cochlearis)*. Here they are transformed into a nerve inflow and sent into the brainstem. The dendrites of the neurons from the spiral canal of *the modiolus* traverse the canals excavated in the depth of the spiral lamina and reach the sensory cells of Corti's organ. The axons of these neurons leave the spiral canal through a series of orifices which open through the cochlear recess into the internal acoustic meatus, where they join the axons of the vestibular nerve.

The vestibulocochlear nerve results from the junction of the cochlear and the vestibular fibres. An acoustico-facial anastomoses is present, through which the facial nerve receives fibres arising in the cochlea and in the semicircular canals.

The main relations of the vestibulocochlear are with the facial nerve. The first portion of the facial nerve canal is situated oblique anteriorly and inferiorly, perpendicular on the axis of the petrous portion of the temporal bone, between the vestibule and the cochlea. After 3-4 mm, the facial nerve forms the first bend, or genu, called *geniculum* of the facial nerve and runs behind, above and medially to the middle ear. On the anterior surface of *the geniculum* lies the geniculate ganglion *(ganglion geniculi)*, which is situated on Wrisberg's intermediate nerve *(nervus intermedius)*. From this ganglion arise two superficial petrosal nerves (the greater and the lesser superficial petrosal nerves), exiting from the petrous portion of the temporal bone via the fallopian hiatus (hiatus of the facial canal). Then the facial nerve gives off the nerve to the stapedius muscle *(n. stapedius)* and then the nerve called chorda tympani. The complexity of the study of the labyrinthine region is due not only to its functional duality, but also to the intrication of the relationships with the facial nerve, the internal carotid artery and the sigmoid sinus of the dura mater.

The second neuron lies in the bulbopontine vestibular nuclei (Bechterew's dorsal nucleus, Schwalbe's medial nucleus and Deiters' lateral nucleus); the axons penetrate the inferior cerebellar peduncle, forming the vestibulocerebellar tract and conveying the nerve inflow at the level of the cerebellar cortex of the archeocerebellar flocculonodular zone.

Moreover, it is believed that some axons of the cells of Scarpa's ganglion would not synapse at the level of the second neuron and course directly, without relay, to the cerebellar cortex.

In addition, we mention that from the vestibular nuclei (Deiters) arise also descending fibres running towards the spinal cord, which forming the vestibulospinal tract, and ascending fibres, some of which travel to the nuclei of the oculomotor nerve, eliciting the oculocephalogyric reflexes of the labyrinthine origin, to the centromedian nucleus in the thalamus which, connected with the lenticular nucleus, achieves postural oculogyric reflexes, and to the temporal and frontal cerebral cortex, assuring the awareness of the position of the body in the space (these cortical projection fibers run through the medial lemniscus and synapse with the lateral thalamic nucleus).

Sensory auditory pathways

The first neuron of the auditory pathway is situated at the level of the Corti's ganglion. The axons of the cells of this ganglion form the cochlear branch of the vestibulocochlear nerve (previously described).

The second neuron lies in the anterior and posterior cochlear nuclei. Recent researches (Eyries) demonstrate that all the fibers of the protoneuron reach the anterior nucleus, the posterior nucleus being merely a secondary nucleus.

The axons of the deutoneuron (the second neuron) constitute an *anterior fasciculus*, the trapezoid body, the fibers of which cross on the midline, forming on each side the lateral lemniscus. Passing through the pons and the mesencephalon, the lateral lemniscus reaches the third thalamocortical neuron, represented by the medial geniculate body *(metathalamus)*, after which the axons of these cells, via the acoustic radiations, reach the temporal cerebral cortex. There are commissural fibers between the medial geniculate bodies and the inferior *quadrigeminal colliculi*. We mention also that a group of fibers arising from the anterior cochlear nucleus do not cross and remain homolateral, so that the two *lemnisci,* right and left, contain both homo – and heterolateral fibers; in the superior area of the pons there are also association fibers between *the lemnisci.*

In addition, fibres from the anterior cochlear nucleus run to the pontine oliva, after which they travel to the nucleus of the facial nerve and to that of the oculomotor nerve.

At the same time, from the anterior cochlear nucleus start a series of fibers towards the posterior cochlear nucleus, where they synapse, after which they form a superficial bundle, *the cochleoreticular fasciculus* situated on the floor of the fourth ventricle, *the acoustic striae*, which after crossing on the midline, runs to the central reticular nuclei.

From the description of these structures ensues the complexity of the association pathways represented by connections with the facial nerve which, supplying *the stapedius muscle*, achieves auditory accomodation reflexes, with the oculomotor nerve and with the abducens, through *the medial longitudinal fasciculus*, for the auditory-oculogyric reflexes, with the nuclei of the reticular substance, which has numerous regulation functions, as well as with the mesencephalic nuclei, respectively with the inferior quadrigeminal colliculi, that are involved in the muscle tonicity regulation.

The cortical segment

The cortical segment of the auditory and vestibular analyzers is situated at the level of the cerebral cortex – see the subsection *"The Auditory Analyzer Area"*, in which the problem is already approached. We mention here only some elements.

The auditory area is situated in the superior part of the first temporal gyrus, at the level of Heschl's transverse gyri, which represent, on the sylvian side of the temporal lobe, the retroinsular region. It is considered that, at this level, the high-pitched sounds are recorded in the internal part of the transverse gyri and the low-pitched sounds, in their external part and on the inferior lip of the sylvian scissure, a true cortical cochlea being thus achieved.

We mention that the auditory cortical centre corresponds to Brodmann's areas 41 and 42, representing, as a structure, a granular type cortex (auditory coniscortex). The unilateral destruction of the zone does not lead to total deafness, but only to the lowering of the auditory acuity, while the bilateral destruction causes a total deafness.

By comparison with other projection centers, it is admitted that this area, which records the sounds without interpreting them, is surrounded by a perception zone and by a gnosia zone (auditory-psychic area), the latter accomplishing the role of integrating and analyzing the auditory sensations; it corresponds to the first temporal gyrus and to the superior part of the second temporal gyrus(at this level, in the superior part of the first temporal gyri, is situated Wernicke's verbal audition center). The existence of these areas explains why, in dependence on the site of the possible lesions, impairments occur in the recording, identification and, finally, interpretation of sounds, impairments which may appear dissociated. Moreover, the presence of these areas explains also why a subject with a normal auditory acuity is unable to distinguish several sounds between each other or to interpret them as a performer or as a hearer, with respect to the duration or the rhythm, or to understand the symbolic significance of the music.

The vestibular area is still a controversial element. Some authors consider that, as the sense of equilibrium and of body positions does not belong to the field of consciousness, the vestibular pathway is only of a reflex nature.

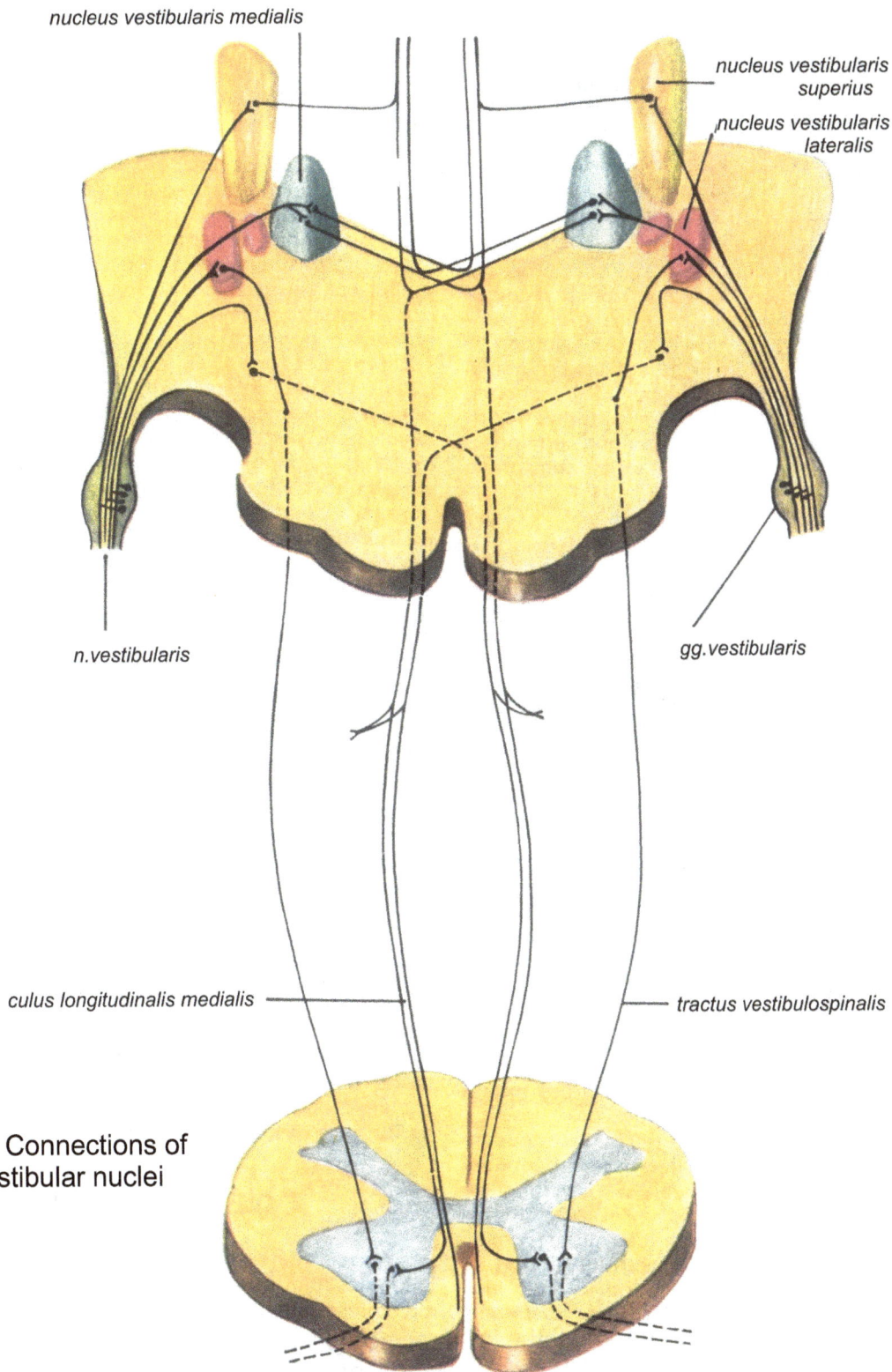

nucleus vestibularis medialis

nucleus vestibularis superius

nucleus vestibularis lateralis

n.vestibularis

gg.vestibularis

culus longitudinalis medialis

tractus vestibulospinalis

Fig. 243 Connections of the vestibular nuclei

Other authors, starting from the fact that the vestibular lesions are associated with spontaneous vertigo and conjugate deviations of the head and of the eyes, admit the existence of a cortical control area of the vestibular system. Another argument is the existence of a temporopontine tract, which plays a role in the cerebellar control of motoricity. Spiegel describes the presence of a vestibulocortical tract, the direct and uncrossed fibres of which seem to follow the course of the Reil's band, with or without thalamic relay, connecting the vestibular receptor to the cerebral cortex.

257

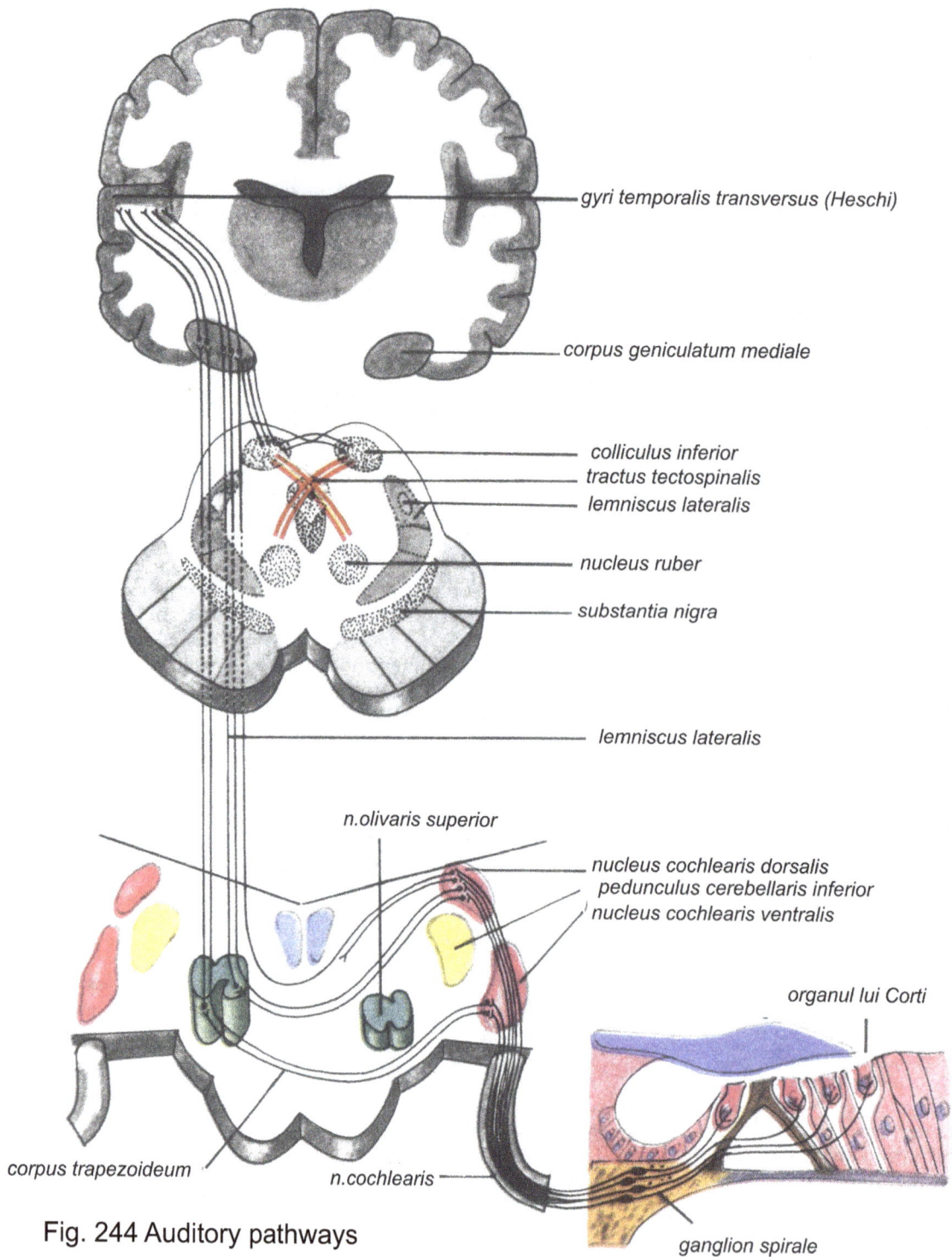

Fig. 244 Auditory pathways

Labels on figure:
- gyri temporalis transversus (Heschi)
- corpus geniculatum mediale
- colliculus inferior
- tractus tectospinalis
- lemniscus lateralis
- nucleus ruber
- substantia nigra
- lemniscus lateralis
- n.olivaris superior
- nucleus cochlearis dorsalis
- pedunculus cerebellaris inferior
- nucleus cochlearis ventralis
- organul lui Corti
- corpus trapezoideum
- n.cochlearis
- ganglion spirale

The cortical center of the vestibular analyzer is located, according to some authors (Spiegel) at the level of the first temporal gyrus, according to others at the level of the ascending parietal gyrus – a region which receives proprioceptive afferents, too, giving rise also to extrapyramidal fibers (the lesions occurring in this area lead to the loss of the motion of body position – gait apraxia) and, finally, there are authors (Marsalet) who situate the vestibular projection in the prefrontal area, explaining in this way the frontal ataxia.

We mention that the cerebellar centers are well-known; they are represented by the flocculus, connected to the vestibular nuclei, and by the fastigial nuclei, both belonging to the archicerebellum, the cerebellum being the regulating element of the static and dynamic equilibrium.

258

THE GUSTATORY AND OLFACTORY ANALYZERS

The gustatory analyzer
The gustatory receptors

They are represented by corpuscular structures called taste buds or *gustatory caliculi*, distributed at the level of the lingual mucosa and on the epiglottis (Fig.246).

The taste buds are ovoid in shape, with the basal pole sited on the basement membrane of the epithelium, while the apical extremity is continuous with the gustatory canal – also situated inside the epithelium -, which opens at the surface of the mucosa through an aperture named gustatory pore. The taste buds are constituted of gustatory, supporting and basal cells.

The gustatory cells, situated centrally, are elongated and have a hyperchromic nucleus. At the apical extremity are present fixed ciliary formations, the gustatory hairs, while the basal extremity has the appearance of a projection which inserts on the basement membrane.

The taste bud is surrounded by a rich network of sensory nerve fibers, belonging to the seventh, ninth and tenth pair of cranial nerves, which form a nerve plexus.

The gustatory area, represented especially by the superior surface of the tongue, comprises three important nerve territories:

1) The territory of the lingual nerve (the lingual branch of the facial nerve, which assures the gustatory sensibility in a limited zone at the lateral side of the base of the tongue and at the level of the anterior pillar of the soft palate), of the chorda tympani and of Wrisberg's intermediate nerve (in the anterior two-thirds of the tongue in front of the V-shaped sulcus terminalis some authors consider that this territory belongs only to Wrisberg's intermediate nerve).

2) The territory of the glossopharyngeal nerve, constituted of t*he lingual sulcus terminalis* – a region rich in *caliciform papillae* – and the region situated behind this sulcus.

3) The territory of the vagus nerve, through its superior laryngeal branch, which supplies a small territory in the region of the glosso-epiglottic fields, of *the valleculae* and of the base of the epiglottis. It is considered that there is a selectivity in the reception of the various tastes, respectively bitter, sour (acid), salt and sweet, in the different zones of the lingual gustatory area. Thus, for example, the territory of the lingual nerve, of the chorda tympani and of the Wrisberg's intermediate nerve receives the sour sensations, whereas the territory of the glossopharyngeal nerve is sensitive to the bitter taste.

The conduction pathways

The first neuron (protoneuron) is represented by the nerve cells situated at the level of Andersch's ganglion for the territory of the glossopharyngeal nerve, of the geniculate ganglion for the territory supplied by the system of branches of the facial nerve (lingual nerve, chorda tympani and Wrisberg's intermediate nerve) and of the plexiform ganglion for the territory innervated by the vagus nerve. The cellulifugal fibers of the protoneuron reach the medulla oblongata and terminate at the level of the middle third of the nucleus of the solitary tract *(nucleus tractus solitarii)*, which represents the relay of the gustatory pathway. Therefore, the area of the solitary nucleus is termed also Nageotte's gustatory nucleus (some authors mention as relay of the gustatory pathway, instead of the nucleus of the solitary tract the sensory nucleus of the trigeminal nerve, considering that the gustatory fibers follow the course of the mandibular branch of the trigeminal nerve).

The deutoneuron at the level of the gustatory nucleus gives off fibers via Reil's band, where their crossing takes place (the gustatory chiasm); on reaching the thalamus, they establish a synapse in the arcuate nucleus *(nucleus arcuatus)*, where the terminal diencephalocortical neuron is situated.

The cortical segment

The gustatory cortical area lies at the level of the anterior part of the fifth temporal gyrus, corresponding to Brodmann's area 38. Some authors consider that there is also a mixed, olfactory and gustatory area at the level of *the uncus*. Moreover, a gustatory area above the sylvian scissure, associated to the sensory lingual area. The general sensitivity of the tongue, in the anterior two-thirds, is imparted by the lingual nerve, its fibers reaching, via the mandibular nerve, the nucleus of the descending root of the trigeminus, where the second neuron is sited.

From here, via Reil's band, they reach the arcuate nucleus of the thalamus, from which the third neuron sends its fibers to the ascending parietal convolution; here lies the sensory lingual area, which achieves the integration of the general sensibility of the tongue, vehiculated through the trigeminus (fifth cranial nerve pair), with the gustatory sensitivity proper, vehiculated mainly through the glossopharyngeal (ninth cranial nerve pair), the vagus (tenth pair) and Wrisberg's intermediate nerves.

THE OLFACTORY ANALYZER
The olfactory receptor

It is represented by the olfactory mucosa, located in the superior region of *the nasal fossae*, the area of each fossa being of about 2.5 cm^2. It is yellow-brown and formed of a corium and a two-layered epithelium, separated by a basement membrane. The corium, formed of loose connective tissue and elastic fibers, contains serous tubular glands, called Bowman's glands, the secretion of which maintains the moisture of the mucosa (Fig.247).

The olfactory epithelium contains basal, supporting and sensory olfactory cells; the latter are less numerous than the supporting cells. The olfactory cells are unipolar neurons with an apical dendritic and axonic basal processes.

The apical process ends through a globulous structure, the olfactory vesicle, provided with short and thin cili, the olfactory cili. They lie on the surface of the olfactory mucosa and are moistened by secretion of Bowman's glands. The basal process passes through the cribriform plate of the ethmoid bone and ends at the level of the mitral cells of the olfactory bulb, with which the apical process establishes a synapse.

The conduction pathways

The sensory olfactory pathways are corticopetal. The protoneuron represented, according to some authors, by the sensory cells of the olfactory mucosa (other authors consider the olfactory cells of the olfactory mucosa as simple sensory cells, the protoneuron being actually the mitral cell), establishes a synapse with the deutoneuron, represented by the mitral cells in the olfactory bulbs, from which start three pathways towards the corresponding secondary olfactory centers:

-through the lateral olfactory stria to the cortex of the hippocampus;

-through the medial olfactory stria via the subcallosal gyrus, the medial and lateral longitudinal (Lancisi) olfactory striae, the fasciolar gyrus, the dentate gyrus (called *"corps goudronssé"* by the French authors) and *the hippocampal uncus*, also to the cortex of the hippocampus.

The cortical segment

Although the pathways of conduction and the cortical projection have been described in the section referring to the cortex, we mention again some elements. The part of the encephalon constituted of the olfactory formations is the rhinencephalon.

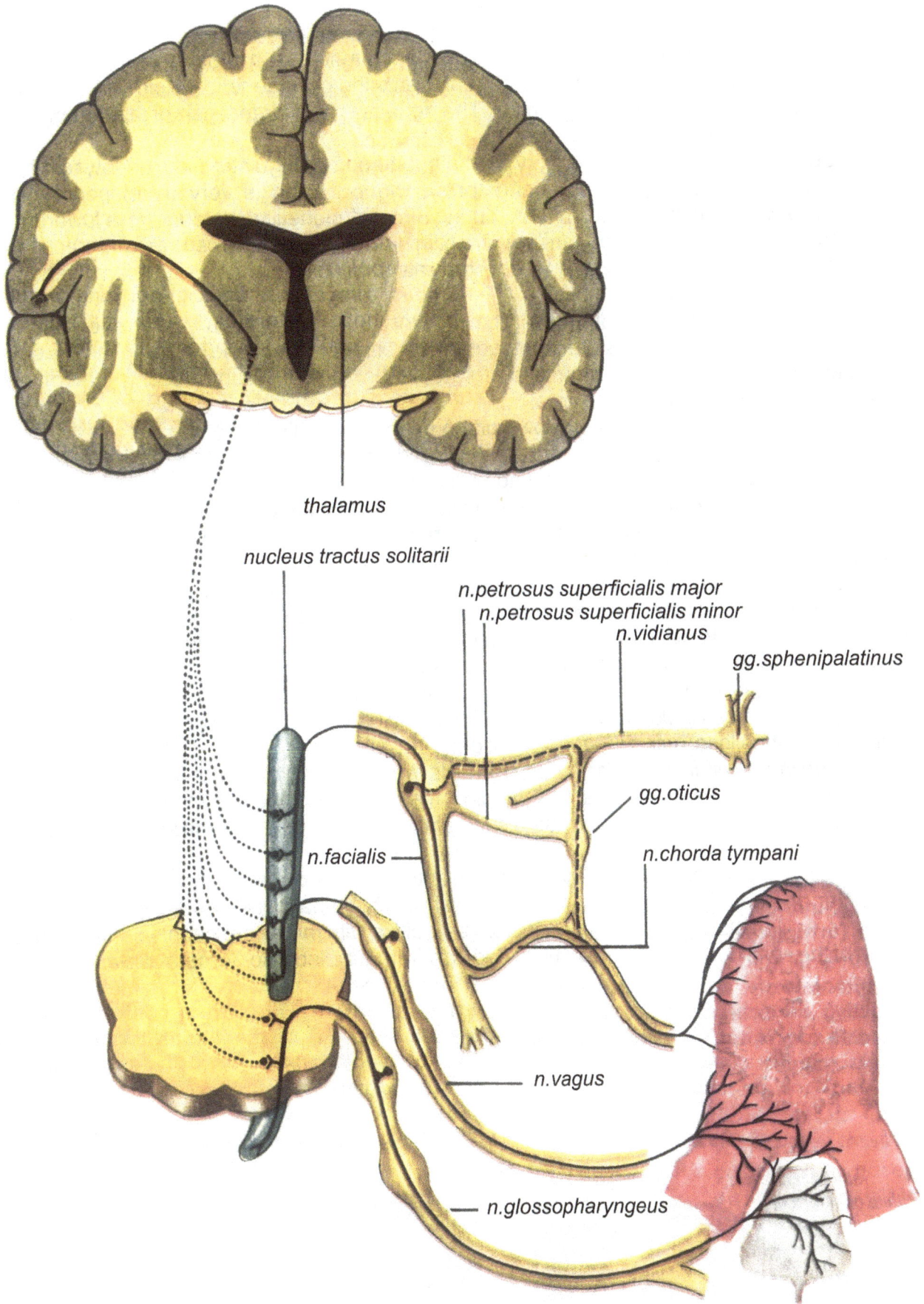

thalamus

nucleus tractus solitarii

n.petrosus superficialis major
n.petrosus superficialis minor
n.vidianus

gg.sphenipalatinus

gg.oticus

n.facialis

n.chorda tympani

n.vagus

n.glossopharyngeus

Fig. 246 Gustatory pathways

The rhinencephalon

The rhinencephalon or the limbic system is situated at the boundary between the telencephalon and the diencephalon and represents the phylogenetically oldest constituent of the cerebral hemispheres *(archicortex)*. The remaining and largest part of the cortex is the neocortex (Fig.245).

Some animals (dogs, predatory animals, and some ungulates) are endowed with a particularly high olfactory capacity and their *rhinencephalon* is very developed. These animals are called macrosmatic. In man, during the phylogenetic development, a lowering of the sense of smell and concomitantly a diminishing of the olfactory centres and tracts has taken place. In comparison with the other vertebrate animals, man has a significantly reduced olfactory capacity, he is microsmatic.

The rhinencephalon comprises: the olfactory bulb, the olfactory tract, the olfactory trigone (these three structures forming together the olfactory lobe), the medial olfactory stria, the lateral olfactory stria, the anterior perforated substance, the subcallosal (paraterminal) gyrus and the parahippocampal gyrus.

The olfactory lobe. It is a narrow portion of the brain, situated at the base of the brain and directed rostrally. It begins with an enlarged portion, termed olfactory bulb *(bulbus olfactorius)*, which represents the ending of the olfactory nerves, and is situated above the cribriform plate of the ethmoid bone *(lamina cribrosa ossis ethmoidalis)*. The olfactory bulb is continuous with the olfactory tract *(tractus olfactorius)*, which is thin and situated in the olfactory sulcus *(sulcus olfactorius)* of the frontal lobe. It ends with the olfactory trigone *(trigonum olfactorium)*. The olfactory trigone is a triangular elevation, situated laterally to the optic chiasma. From here starts the internal olfactory gyrus, which is connected to the callosal gyrus and to the external olfactory gyrus. It is limited laterally by a bundle of white fibers, the medial and lateral olfactory *striae (stria olfactoria medialis et lateralis)*.

The medial olfactory stria runs to the paraterminal gyrus, which is situated below the rostrum of *the corpus callosum* in the frontal lobe, and continues its course to the gyrus in front of it, called subcallosal gyrus or area subcallosa.

The lateral olfactory stria courses towards *the uncus* and the paraterminal gyrus of the temporal lobe. Between the olfactory trigone, the optic tract and the medial border of the temporal lobe lies the anterior perforated substance, the olfactory area, the surface of which is perforated by vessels.

The subcallosal area is delineated by an anterior and a posterior groove.

Its afferent and efferent connections with the hypothalamus, the activity of which is partly controlled also by the *rhinencephalon*, as well as with the mesencephalon, accomplish a part of the control of the visceral activity and is therefore called also *"visceral brain"*.

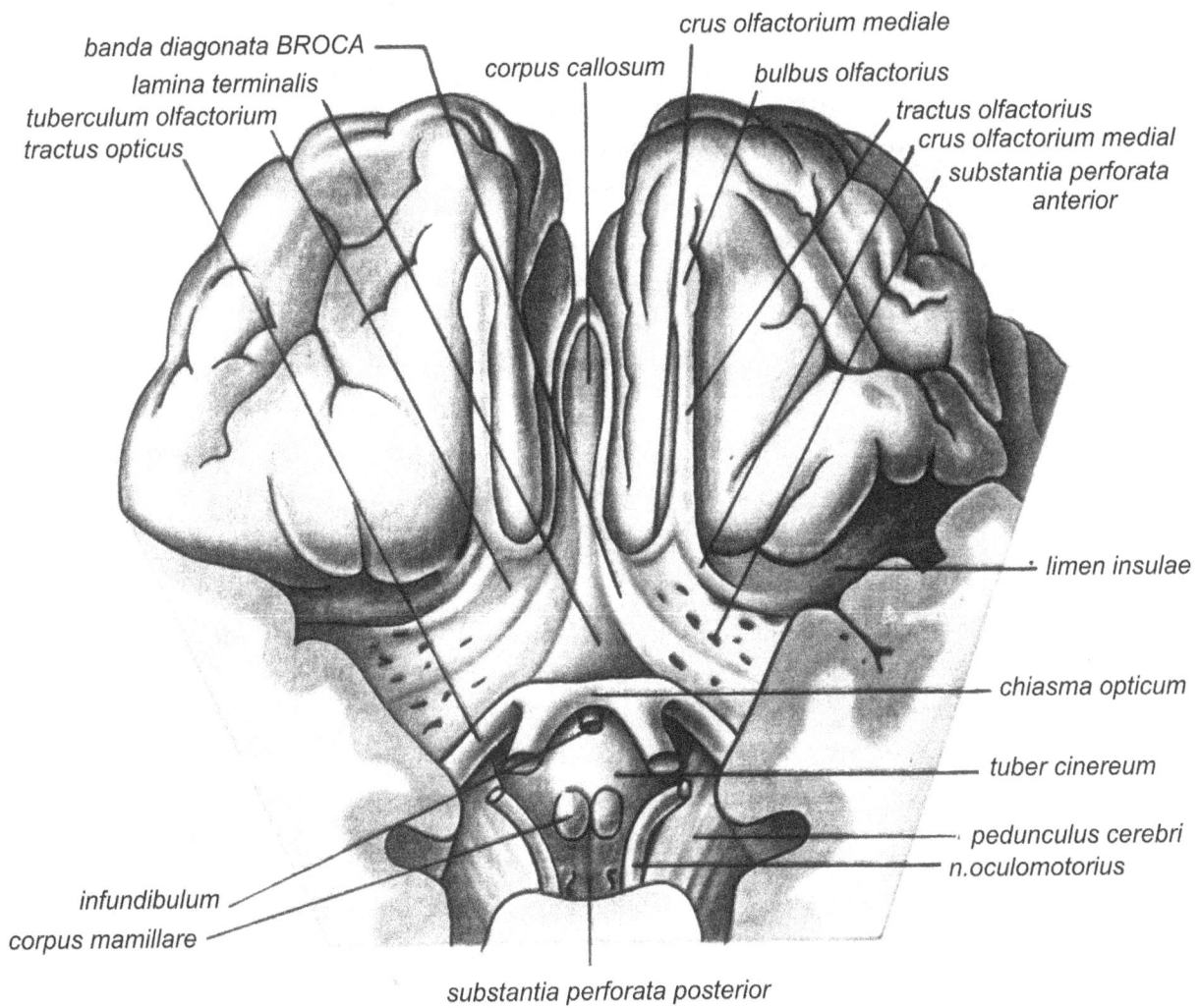

banda diagonata BROCA
lamina terminalis
tuberculum olfactorium
tractus opticus

corpus callosum

crus olfactorium mediale
bulbus olfactorius
tractus olfactorius
crus olfactorium medial
substantia perforata anterior

limen insulae

chiasma opticum

tuber cinereum

pedunculus cerebri
n.oculomotorius

infundibulum
corpus mamillare

substantia perforata posterior

Fig. 245 Rhinencephalon – olfactory bulb, perforated substance and lateral, medial and intermediate striae

263

THE CUTANEOUS ANALYZER

The skin

The skin is an uninterrupted covering which at the level of the large orifices (mouth, nose etc.) is continuous with a semimucosa (partially keratinized) and which, inside the respective cavities, becomes a true mucous membrane. *The semimucosae* and *the mucosae* are of an embryonic origin, identical with that of the skin, but their macroscopic and microscopic structure is different.

The surface of the skin is not uniform, it is marked by orifices, folds and prominences.

The orifices are of two types: some of them are *large*, leading to the natural cavities (mouth, nose etc.), others are *small*, hardly visible to the naked eye, but well visible with the magnifying glass. The latter correspond either to the hair follicles (from which the hair emerges) or to the eccrine sweat glands (pores). All the openings, but especially the large ones, as well as the follicular openings, are intensely populated by microbes, a phenomenon which explains the high frequency of folliculitides. The follicular openings represent also the place where the percutaneous absorption of water, electrolytes, drugs (ointments, creams etc.) and other substances is maximum.

The folds of the skin are also of two types: *congenital* (or structural) and *functional*, the latter occurring with ageing and elasticity lowering.

The structural folds are either *large* (axillary fold, inguinal fold etc.) or *small* (microfolds).

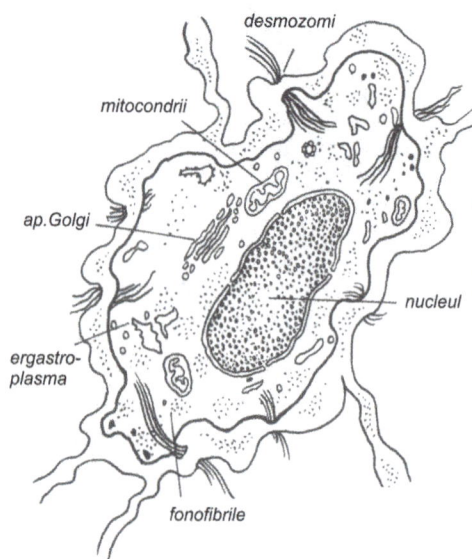

Fig. 248 Electron microscopic structure of an epidermal cell (C. Orfanos)

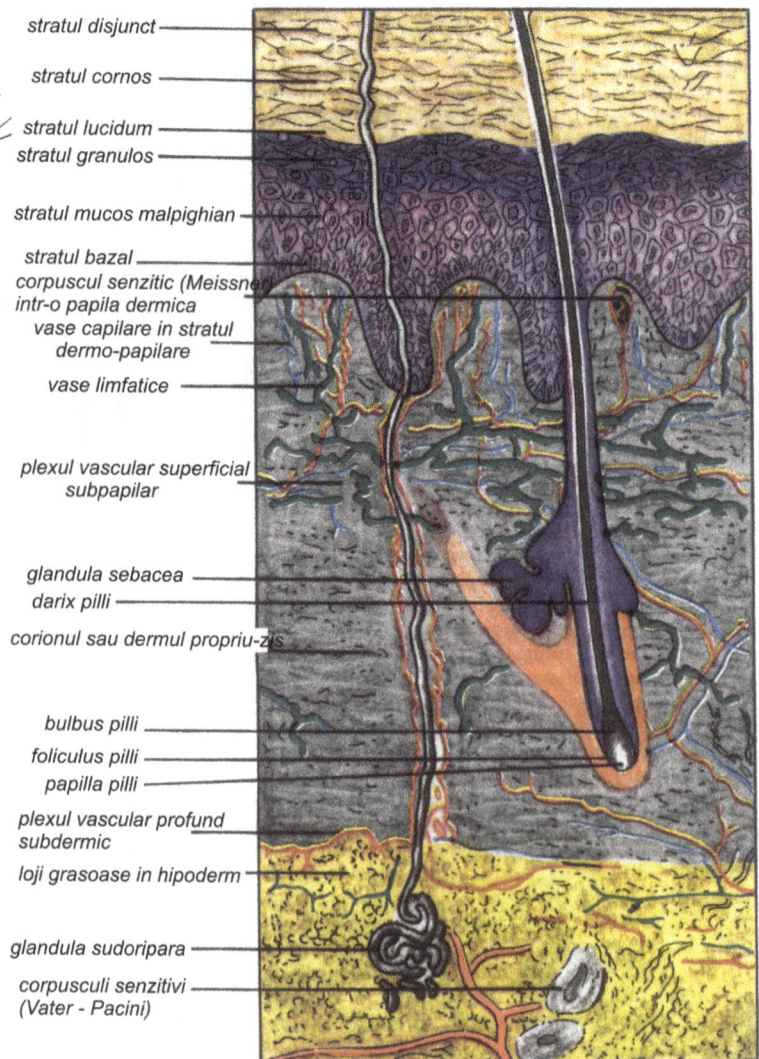

Fig. 249 Structure of the integument

264

The large folds exhibit certain pathophysiological peculiarities, such as: more marked moisture as compared with the rest of the skin, an alkaline or neutral pH, a more marked pilosity. Owing to these characteristics, they may be affected by some specific diseases, such as mycoses, fissures, intertrigo etc. Small folds (microfolds) are present on the whole surface of the skin, they unite the openings of the pores; in this way they bring about the formation of small rhomboidal surfaces, which are the expression of a normal elasticity. These small folds disappear at the level of scars, in states of epidermal atrophy or dermal sclerosis (scleroderma). At the level of the palms, soles and digits, the microfolds are disposed in parallel arcuate lines, achieving the fingerprints, with characters transmissible by inheritance; they are important for the juridical identification of the individuals. On the ridges between the folds, also disposed in parallel lines, are present the openings of the sweat pores.

The functional folds appear as a consequence of the decrease of the cutaneous elasticity, as well as following muscular contractions (wrinkles).

The color of the skin depends on the amount of melanotic pigment, varying from white (lack of pigment) to black (excess of melanin).

The amount of melanin is determined genetically, but the color variations of the skin in dependence on the geographical latitude (from the pole to the equator) shows also an adaptation. Melanin variations occur within certain limits also in dependence on the exposure to ultraviolet rays.

The degree of capillary vessel supply determines the pink-red hue of the skin. A more abundant vessel supply of the face produces also regional morbid peculiarities – the congestive diseases of the face are more numerous. The color of the skin depends also on the amount of hemoglobin (pallor in anemias). The skin of infants is characterized by a rich vessel supply; therefore it is pink-coloured.

The thickness of the skin influences its color: the skin of the infants is thinner and, as mentioned above, pink; the skin of the palms and the soles is yellowish owing to the horny layer (keratin), especially in a case of hyperkeratosis. The abundance of keratohyalin (granular layer) imparts to the skin a white color.

The mucous membranes are thinner than the epidermis: normally, keratin, keratohyalin and melanin are absent at the level of *mucosae*, so that they are transparent, permitting to perceive the blood supply in their depth, which explains their red color.

In leukoplastic states (leukos: white) appear keratin and keratohyalin and the damaged surfaces become white and more rugose.

In children under 1 year of age the pigmelogenesis is deficient and their accomodation to the sunlight is deficient, actinic burns after exposures to the sun or to ultraviolet rays appearing easier in them.

The surface of the skin in a mature, middle-sized man is of 1.5 - 1.8 m^2.

Its total weight corresponds to about 20% of the body weight (on the average 14 – 16 kg), of which 15% represent the hypodermis (with large variations according to the corpulence), approximately 5% the dermis and less than 1% the epidermis.

The thickness of the skin varies according to the regions: thin on the face, the forehead and the genitalia and thicker on the anterior thorax, the abdomen and the large folds. The thickness increases on the back and on the extensor surfaces of the limbs, the maximum thickness being attained on the palms and the soles. Among the layers of the skin, the epidermis is the thinnest (between 0.03 and 1mm), the dermis is thicker (between 0.5 and 0.8 mm on the face, and 2 mm on the thorax) and the hypodermis is marked by large regional variations.

The elasticity of the skin is due, on the first place, to the dermal fibrillar system, especially to the elastic fibers, owing to which the skin is depressible. To its elasticity contributes also *the panniculus adiposus*, the fat lobules of which, enveloped in a connective-elastic tissue, function as minute rubber balls that are depressed by compression, after which they return to their original form.

The elasticity diminishes with increasing age and disappears in *oedematous* states or in cutaneous sclerosis. Owing to the elasticity, the wounds become larger than the sectioned surface and the excised surface (grafts) and the excised skin portions (the grafts) are smaller than the removed surface.

The mobility of the skin is variable: in relation with the deep planes, it is easily mobilizable on the face, the thorax, the limbs, and the penis, and poorly mobilizable on the palms, the soles, *the pinnae* and *the alae nasi*. It diminishes in sclerosis processes (scars, secondary sclerosis, essential scleroderma).

The microscopic structure of the skin

The skin is composed of three layers: the epidermis of ectodermal origin; the dermis or corium and the hypodermis (the subcutaneous cell layer) of mesodermal (mesenchymal) origin (Fig. 248 - 255).

The epidermis

The epidermis is composed of a stratified pavement, squamous cornified epithelium, its cells being in a continuous regeneration. Blood vessels are absent and the nutrition takes place by diffusion of the interstitial lymph from the dermis; through the basement membrane and the narrow spaces (of about 10 millimicrons), which separate the vital cells of this layer from each other. The solutions of continuity which are limited to the epidermis (erosions) do not cause hemorrhages, but only a lymph exoserosis.

According to their origin, to their microscopic aspect and to their functions the cells of the epidermis are of two distinct types: *keratinocytes*, which make up the greatest part of the cell mass, and melanocytes, significantly less numerous.

The keratinocytes derive from the cells of the basal layer, which undergo a permanent division. As the daughter cells are continuously pushed towards the surface, a slow ascending movement is achieved in the epidermis, during which the cells are progressively loaded with keratin. In the course of this movement each cell traverses the whole depth of this layer.

The melanocytes elaborate and release the pigment melanin, which is stored both in the epidermal cells (especially in the basal layer) and in the dermal macrophages, that become in this way melanophorous.

This *"epidermal renewal time"* lasts normally 25-30 days (studies with labeled isotopes), but in *parakeratosis*, characterized by an accelerated cell multiplication (for example psoriazis) the renewal time may be reduced to 3-4 days.

As epidermis is devoid of blood vessels, the nutrition of the cells is assured by the transfer of the fluids from the ground substance of the dermis, through the basement membrane.

pigment

filamente Herxhiemer

membrana bazala

celula bazala in mitoza

Fig. 250 Cells of the basal germinative layer

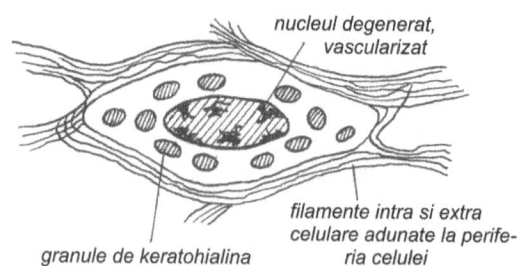

nucleul degenerat, vascularizat

granule de keratohialina

filamente intra si extra celulare adunate la periferia celulei

Fig. 251 Granular cell (after Policard)

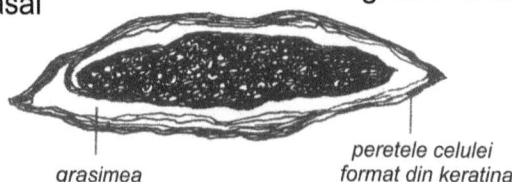

grasimea

peretele celulei format din keratina

Fig. 252 Horny cell (after Policard)

266

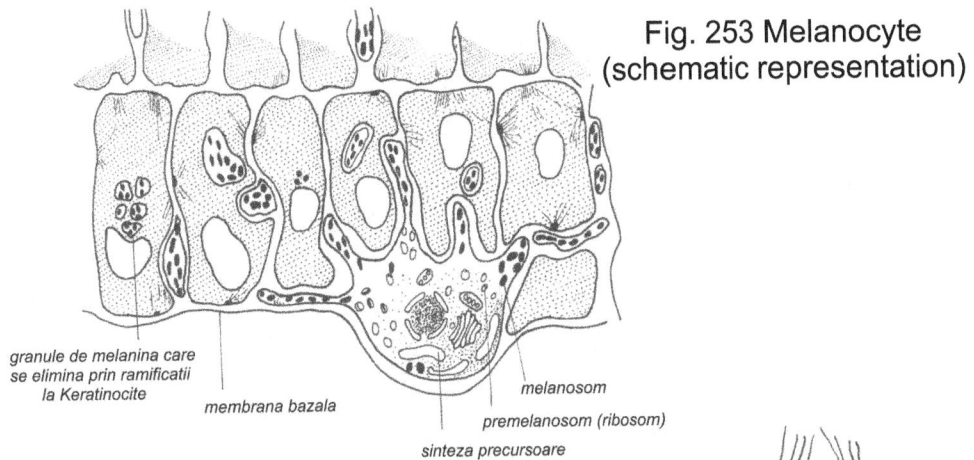

Fig. 253 Melanocyte (schematic representation)

granule de melanina care
se elimina prin ramificatii
la Keratinocite

membrana bazala

melanosom

premelanosom (ribosom)

sinteza precursoare

Fig. 254 Lacunar epidermal system (cells separated by spaces and joined by cell-membrane tonofibrils)

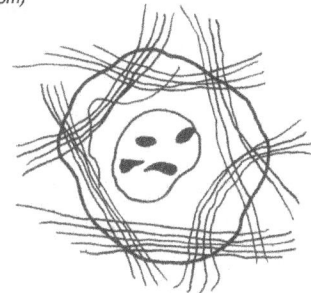

The nutritional impairments (vascular or of permeability) of the dermis, as well as of the permeability of the basement membrane have a significant repercussion on the epidermal metabolism. Even under normal conditions, the gradual receding of the cells from the basement membrane brings about significant metabolic cell changes, which are reflected in morphological cell changes. These modifications, as they are similar in the cells which are at an equal distance from this membrane, have as consequence the appearance of *a functional stratification*, the last stage of which is represented by the horny, devitalized cell, devoid of nucleus. The epidermis is composed of the layers enumerated below:

- **the basal cell layer** (germinative) *(stratum basale)* is the deepest layer, in contact with the basement membrane. Its cells are 5-6 micron-sized and have an ovoid shape, with the great axis perpendicular on the basement membrane. They exhibit all the signs of an intense biological activity: a large nucleus situated at the apical pole, rich in chromatin, and a cytoplasm with numerous organelles. At the apical pole are disposed melanin granules like an umbrella above the nucleus. Melanin plays a photoprotective role, namely it protects the nucleic acids (especially DNA) from the ultraviolet rays and exerts an inhibitory action on them. The cytoplasm contains filaments parallel to the great axis of the cell (Herxheimer's fibers), which attach at the level of the basal pole of the cell on nodular formations of the membrane, called half-dermosomes. About one of 200-600 basal cells undergoes mitosis. The ratio of these cells represents the mitotic index. Paradoxically, this index increases with ageing and is probably one of the sources of malignancy at an advanced age (basocellular epithelioma). Among the basal cells are distributed the melanocytes and the sensory Merckel-Ranvier's corpuscles.

- **the spinous layer** *(stratum spinosum)* is termed together with the basal one, malpighian layer. It lies immediately above the basal layer from which it derives. Normally it is formed of 6-15 rows of polyhedral cells which, as they ascend towards the surface, become increasingly flatter and less vital. They are larger than the basal cells, measuring 10-15 microns, but their nucleus does not increase in size. They are more acidophilic than the basal cell, but are intensely vital, this layer being in eczema or in *metaplezias* (spinocelular epithelioma) and in numerous other diseases the site of significant transformations. The cells are separated by narrow spaces of subcapillary dimensions, namely of about 10 millimicrons, through which nutritive interstitial lymph circulates and which contain unfrequent lymphocytes or polynuclear cells and amyelinic nerve endings.

267

These spaces form together the *"epidermal lacunar system"*, in which the cell cohesion is maintained by intercellular bridges. The cell cytoplasma is characterized, in addition to the usual formations, by filaments arranged in bundles, called tonofibrils, under the optic microscope these tonofibrils appear with an intracellular portion and an extracellular part, owing to the fixation techniques under the electron microscope, only intracellulary filaments are observed, which fix themselves on the internal leaflet of the cell membrane, on a thickening termed desmosome. The desmosomes of the neighbouring cells are in contact, so that the tonofilaments are only intracellular. It seems that the synthezis site of the tonofilaments is represented by the desmosomes. The tonofibrils and the tonofilaments play a significant role in the keratin synthesis.

- the granular layer lies above the preceding layer; it is composed of 1-5 rows of flat cells. Their characteristic is the abundance of cytoplasmic of *keratohyalin* granulations, intensely a basophilic substance, resulted from the degeneration of the tonofibrils, which intermingle with residues of nuclear and cytoplasmatic material. The remaining tonofibrils lie at the periphery of the cell and are to a great extent fragmented. The nucleus presents a pyknotic degeneration. The intercellular spaces are narrower, the nerve filaments stop at this level. This layer is absent in accelerated keratinization processes (parakeratosis, for example in psoriazis) and is thickened in lichen planus. It is normally absent at the level of the mucosae and, when it appears,the respective portion of mucosa becomes white owing to the white colour of keratohyalin and the transparency disappears (leucoplasia).

- the stratum lucidum or clear layer, called also basal horny layer because it is sited at the depart level of the horny layer, appears in the usual slaining techniques homogeneous and acellular owing to a prehorny fat, present in the cells. Special slainings permit to observe that it is composed of cells rich in glycogen, eleidin and fats. This layer is visible especially at the level of the palmar and plantar skin, where also the horny layer is thicker. The presence of glycogen attests the existence of vital processes necessary to *the final stages in keratin synthezis*. This stratum is the last vital layer of the epidermis which, with the deep horny layer, constitutes the so-called *"epidermal"* barrier (barrier to water, chemical substances and microorganisms).

- the horny layer (*stratum corneum*) is the most superficial. It is formed of two substrata: the deep or conjoined horny layer and the superficial or disjoined layer, called also exfoliative stratum. The horny cells of the deep layer adhere strongly one to another, while those of the superficial layer have loose connections and flack away at the surface. The normal horny cells are scale-shaped, the nucleus as well as the cell organelles have disappeared and the cell has the appearance of a sac formed of an envelope of keratin and a content rich in *osmiophilic fats* (lipids and cholesterol). Above the horny layer and intermingled with the cells of the disjoined layer lies *a functional* (physiological) *stratum*, resulted from the trickling of sweat and of sebaceous secretion and from the remnants of the horny cells and of the intercellular substance. This layer, called *the acid lipoproteic film or pellicle of the skin* (pH=4.5-5.5), confers protection against microorganisms and chemical substances. On the surface of the skin and among the cells of the disjoined layer are microorganism of *the saprophyte flora*. The number of these germs diminishes gradually towards the depth, as they are arrested at the level of the conjoined layer.

Under *the electron microscope* the keratinocytes have a very irregular polyhedral shape. hey are separated through intercellular spaces and the connection between the cells is achieved by the desmosomes. The cells of the horny layer are flat, of 3-4 microns in thickness and 30-40 microns in width. Any contact between them has disappeared, as the desmosomes are degenerated. The malpighian cell has a membrane formed of three leaflets, of which the external and internal are of a proteic nature, while the middle one is lipidic. The membrane of the horny cells is unique, very sinuous and presents thicknesses corresponding to the desmosomes, a level at which the cells are fixed to one another and from which the tonofilaments start towards the inside of the cell. In the cytoplasm, the characteristic elements observed are the tonofilaments.

268

They are thin, of about 5 millimicrons in the deep layers, whereas in the superficial layers they thicken under the form of bundles, becoming tonofibrils. The tonofilaments synthesize at their desmosomal extremity through which they attach themselves to the membrane. They are composed of a fibrillary protein twisted in a triple helix with a high content of biologically active sulfhydryl groups (of SH type). Concomitantly with their thickening and keratinization, they enrich themselves in biologically inactive disulphide groups (of –S-S– type). In the horny cells the tonofibrils form a dense and homogeneous network, as they are enveloped in a covering made up of residues of nuclear and cytoplasmic material.

Cell organelles. The ribozomes responsible for the protein synthesis appear often in the neighborhood of the tonofibrils. They are present up to the granular layer. *The endoplasmic reticulum* and *the Golgi apparatus* are reduced, their presence being confined to the deep vital layers. *The mitochondria*, which play a role in the aerobic glycolisis, are small and unfrequent, aspects which correspond to an anaerobic cell metabolism (explainable by the lack of vessels). *Melanin granules* included in the cells and partly in the lysosomes, in which they undergo decomposition, are also observed. *The lysosomes* (enzyme concentrates) are increased in inflammatory states and especially in eczema.

In the basal and in the filamentous layer, *the nucleus* contain one or several nucleoli. In the granular layer begins the decomposition of the nucleus and the nucleoli disappear. The latter disappear also in the neighbourhood of the vehicles in eczema.

The melanocytes, which originate from the neural crest, have a morphological appearance and tinctorial affinities (they impregnate with silver nitrate, stain with methylene blue etc.), resembling those of nerve cells. Embryologically, they derive from the neural crest under the form of melanoblasts which, during the first months of foetal life, migrate towards some regions of the central nervous system (*tuber cinereum, locus niger* etc.), into the peritoneum and into the skin. In the skin they arrange themselves among the basal cells. Through their processes they get in contact with the adjacent keratinocytes, as well as with one another, forming a flat unicellular network.

The melanic pigment is synthesized in specialized ribosomes (premelanosomes and melanosomes) which, by polymerizetion and coupling an a protein support, become melanin granules. These, then, are transfered through the cell ramifications to the neighbouring keratinocytes. In this way, a true cell cooperation is achieved, making up the "epidermal melanin unit", visible under the electron microscope.

The dermis

The dermis constitutes the resistant connective-fibrous skeleton of the skin. It is separated from (and at the same time united with) the epidermis through the basement membrane.

The basement membrane consists of interwoven epidermal and dermal fibers. The epidermal elements continue the filaments of the basal layer (Herxheimer's filaments), while the dermal elements are of a reticular nature. The meshes of this network are filled with the dermal ground substance, which at this level contains highly polymerized (thus viscous and poorly permeable) mucopolysaccharides. In this way, the basement membrane accomplishes a function of selective filter for substances coming from the dermis and serving to the nutrition of the epidermis, but it represents also a second "barrier" to the substances which could penetrate from the epidermis.

The permeability of the basement membrane is increased in eczema and psoriazis and lowered (the membrane is thickened) in lupus erythematosus. In malignant tumors, the basement membrane may play a role in stopping the spread of the process to the depth: in the basal cell epithelioma (a non-metastasing neoplasm), the perforation of the basement membrane does not occur, whereas in the spindle cell epithelioma and especially in the malignant melanoma the process extends beyond the membrane, facilitating the appearance of metastases.

269

The separation plane between the epidermis and the dermis corresponding to the basement membrane is very sinuous, owing to the interpenetration of the dermal papillae, which are directed to the surface, with the epidermal papillae, bulging towards the depth. Owing to this configuration, the contact surface between the two layers increases 10-15 times in comparison with a plane surface, augmenting in this way the adhesion between them (layers of different embryonic origin) and assuring a sufficient nutrition to the epidermal cells. In psoriazis this surface increases 3-5 times in comparison with the normal size, while in states of epidermal atrophy or/and dermal sclerosis it diminishes and becomes plane.

Layers. The dermis is composed of two layers, both exhibiting the characteristic features of the connective tissue. The superficial subepidermal layer comprises the dermal papillae and a thin area below them. It is called the subpapillary layer and is characterized by gracile fibrillary elements, more numerous cell elements, a more abundant ground substance and a rich vessel supply (subpapillary plexuses). The deep layer, called dermis proper or corium, is much thicker (4/5 of the thickness of the dermis), much more resistant and predominantly composed of fibers. Morphologically, the dermis is composed of cells, fibers and ground substances, the latter separating the figurated elements from one another.

The cells are of two types: fixed and mobile. But this classification is not categorical.

The mobile cells may be "fixed" and the fixed ones can spread by proliferation. Among the "fixed" cells rank the slightly elongated *fibroblasts*, with a comma-shaped or round nucleus and with characters of young cells, and the marked by elongated, spindle-shaped *fibrocytes* with a linear nucleus and characters of mature cells. In the cytoplasm of the fibroblasts the collagen is synthesized under the form of *precollagen*, from which arise connective, reticular and elastic fibres, which polymerize after their elimination from the cell and acquire the well-known fibrillary appearance.

The histocytes are characterized by their ovoid shape, but especially by their clear, loaf-shaped nucleus. They are involved in phagocytosis processes (macrophages), their role being to engulf foreign particles (acid-fast bacilli, inert particles, immune precipitates). They hold an intermediate place between the fixed and the mobile cells. They contain numerous lysosomes. They belong to the reticulohistiocytary system.

Mastocytes or mast-cells, sited especially around blood vessels, exhibit as a characteristic feature the presence of mucopolysaccharide cytoplasmic granulations, containing heparin and histamine, which are released through the *"degranulation"* phenomenon; they function as true unicellular glands. They play a significant role in inflammatory states and especially in the anaphylactic type allergic reactions, in which, owing to the histamine (and other substances) released around them, they produce vasodilatation and oedema (for example, like in urticaria).

The tissular lymphocytes and plasma cells are mobile cells, sited around the blood vessels. Their role is significant especially in immunologic humoral (plasma cells) or tissular (lymphocytes) processes.

The above listed elements may proliferate, causing in the fibrous malignity order histiocytomas (or diffuse histicytosis), mastocytosis (called also urticaria pigmentosa), lymphomes and plasmocytomas.

The fibers are of three types:

1) *Collagen fibers* composed of a fibrillary protein, which is converted by boiling (from Latin: colla=glue). They consist of protofibrils composed of linear protein macromolecule chains twisted in a triple helix, with a specific content rich in proline and hydroxiproline. They show under the electron microscope a 640Å periodicity, formed of transparent and dense bands, produced by the alternating arrangement of the polar and apolar amino acids. These protofibrils join into bundles forming collagen fibrils and then 10-20 micron-thick fibers, ramified but not anastomosed. Their number increases in fibrosis and sclerosis processes, their synthesis and polymerization being favored by states of tissular anoxia (scleroderma, hypertrophic scars etc.).

2) *The reticular fibers* are thinner than the collagen fibers and argentaffin; they are branching and anastomosing fibers. They are in contact with the connective cells (fibroblasts) from which they derive being formed of precollagen. They do not exhibit the periodic structure of collagen fibers. They are the first which appear in the intrauterine life and during regeneration processes after wounds.

3) *The elastic fibers* are thinner than collagen fibers, but may aggregate in thin bundles, especially in degenerative states, in which they break and become irregular. They are formed of an axial elastic filament and a polysaccharide covering (elastomycin), with affinity for metal cations (calcium, iron) or fats (cholesterol). They undergo regeneration under the action of sun rays and in aging processes, causing wrinkling of the skin.

The ground substance *(substantia fundamentalis)* is partly of blood origin; it is secreted by cell elements. It has the structure of a colloidal gel, the fluidity of which depends on the polymerization state of the acid mucopolysaccharides: hyaluronic acid (molecular weight of 2-100 million daltons) and chondroitin-sulfate B (dermatan sulfate), the latter functioning as a "ion exchanger". The higher the polymerization degree, the more increased the viscosity. The polymerization state depends on the action of some enzymes and antienzymes (for example the endogenous or microbial hyaluronidase causes the lowering of polymerization, and fluidifies the ground substance, while the antihyaluronidase increases the polymerization). The ground substance is rich in water and contains salts (especially sodium and calcium), proteins, glycoproteins and lipoproteins, glucose, the concentration of which at this level is identical with that in the plasma, while in the epidermis it is reduced to one-third).

The hypodermis *(superficial fascia)*

The hypodermis is the layer which separates the skin from the underlying structures. It consists of lobules of fat cells (lipocytes), containing triglycerides, which have the role of accumulating nutritional reserve and of acting as a thermal and mechanical insulator. These lobules are separated by connective septa, in which are situated vessels and nerves. The hypodermis has a variable thickness, in dependence on endocrine and metabolic influences. Its hypertropohy causes obesity, which implies a certain cutaneous pathology. The pathology of the hypodermis is to a great extent related to vessels and to perivascular connective tissue (interlobular septa), where inflammatory nodular hypodermitides develop, situated especially at the level of the leg. Lipomas (benign tumors) develop at this level, too. The dermis and the hypodermis are the site of oedema.

A special cutaneous structure is the apocrine line. It starts from the axilla, in the mamillary region, and descends converging laterally towards the perineum. It consists of clear cells agglomerations which structurally resemble the cells of the mammary gland. Accordingly the mammary gland may be considered an enormous apocrine gland, the structure of which is correlated with its secretory function.

We mention that between the secretory activity of the mammary gland and that of the apocrine glands there are close resemblances, namely:
- the cells secrete a significant amount of proteins;
- the secretory cell type is characterized by the "decapitation" of the secretory pole of the glandular cells.

Anatomically the apocrine line is visible with the naked eye in cases of polymastia and explains the possibility of the appearance of aberrant mammary tumours and of the extramammary Paget's disease.

Vascularization and innervation

The blood and lymph vessels form two plexuses parallel to the surface of the skin: *a superficial, subpapillary plexus* and a deep, *subdermal* one. The corium, of dense collagen fibers, is poorly vascularized; it is traversed by the communicating vessels between the two above-mentioned plexuses and by appendages (follicles, sudoriferous glands) which reach with their apex the hypodermis.

From these plexuses arise vessels with a precapillary calibre. For the external pathology are important the arterioles which run towards the surface of the dermis (in the papillae), forming "vascular cones" of a terminal character (between the directly supplied surfaces are present only capillary type anastomoses). This arrangement explains the circumscribed character, on small areas, of some roseolous or popular rashes.

The subcapillary plexus is formed of thinner vessels, while the subdermal plexus is made up of the thicker vessels. Between them, beside those described, are present arteriovenous shunts, called glomera (myo-arterio-venous glomera), which play a role in the control of the arterial tension (they have a role in shock states) (Fig.260).

The lymphatic vessels are present under the form of capillaries and lymphatic plexus, disposed similarly to the blood vessels, but without a muscle layer. Their role consists in the drainage of the excess of fluids ultrafiltrated, through the capillaries, inclusively of macromolecular substances, microbes, immunoglobulins etc.

The cutaneous nerves are sensory (connected to the cerebrospinal system), motor (of vegetative nature) and secretory. Hence, the skin has a somatic innervation, mainly represented by sensory (afferent) fibers, and a neurovegetative innervation, of the effector (motor) type, represented by efferent fibers (Fig. 261, 262).

The sensory, free of encapsulated elements of the skin represent the peripheral endings of the sensory nerve fibers. They are present under three main aspects: 1) free nerve endings; 2) corpuscular or encapsulated nerve endings and 3) peritrichal nerve endings (around hairs).

-The free endings are of two types: the intradermal (Langerhans) network and hederiform endings.

The intradermal network is represented by a fine plexus formed of unmyelinated fibers and situated in the depth of the epidermis, while the hederiform endings are represented by unmyelinated fibres derived from the superficial dermal plexus, terminating under the form of a basket around large, clear, horizontally arranged epithelial cells, considered as tactile sensory cells.

-The encapsulated nerve endings, called sensory corpuscles, are structures with a more complex morphological organization and are sited only in the dermis and hypodermis.

These corpuscles consist of a capsule and an axial portion. The capsule, of a connective nature, is formed of concentric lamellae consisting of collagen fibers and lined with a layer of flattened connective cells. The innermost lamella establishes contacts with the intracorpuscular nerve ending. The axial portion is represented by one or several unmyelinated nerve fibers, constituting Timofeev-Dogiel's neurofibrillary apparatus.

According to their structure, the sensory corpuscles are lamellar or non-lamellar, bulb-shaped or helicoid. In the hypodermis are sited lamellar corpuscles, some of which are adapted to the tactile sensibility, respectively Vater-Pacini's and Golgi-Mazzoni's organs, and others to the thermal sensibility to heat, namely Ruffini's organ. In the depth of the dermis lie, on the one hand, non-lamellar corpuscles, adapted to the tactile sensibility, named Wagner-Meissner's corpuscles, and on the other hand, bulb shaped corpuscles, adapted to the thermal sensibility to cold, termed Krause's corpuscles.

-Peritrichal nerve endings. Hairs accomplish also the function of tactile receptors. Around the hair follicles there are two "tactile" rings, formed of sensory unmyelinated fibers, among which lie also vegetative nerve fibers. One of the tactile rings is external, the other is internal. From the internal tactile ring arise slender nerve fibers, which end around the cells of the external sheet under the form of tactile menisci or baskets.

The number of nerve receptors varies according to the regions: their maximum density is attained at the level of the pulp of fingers, at the face (lips, nose, eyelids) and in the genital region.

272

Their number is the lowest on the scalp (attached especially to the hair follicles) and on the back. Variations in the density of receptors are also observed according to the excitation types: the most numerous are the pain and pruritus receptors (free endings), followed in decreasing order by touch-pressure and cold-heat receptors.

The vegetative endings act upon the blood vessels, producing through the acetylcholinergic fibers vasodilation and through the adrenergic fibres, vasoconstriction. The sweat secretion is acted upon by the acetylcholinergic fibers, which stimulate the secretion. The adrenergic fibers produce the contraction of the sudoriferous glomerulus (evacuation of "cold" – emotional sweat) and of the erector muscles of hairs *(arrector pili muscles)*.

At the level of the terminal plexuses, all the nerve fibers are unmyelinated, so that the fibers of the cerebrospinal nerves (myelinated fibers) cannot be distinguished of the vegetative (unmyelinated) fibers.

The close correlation in the terminal plexus between the two systems explains the phenomenon of emotional (pudency) erythema and of antidromic, reflexes, in which, the nerve inflow received by the cutaneous surface, running along the axonal fibers, partly travels towards the spinal ganglion and partly to the capillary, producing vasodilation.

Appendages of the skin

The appendages of the skin are represented by the superficial structures and the glands of the skin, formations specially differentiated for its protection (Fig.256-259).

Superficial structures

The superficial structures are appendage organs of the skin differentiated on its surface and accomplishing the function of protection of the organism. They are present in man under the form of nails and hairs, in mammals under the form of fur, claws, hoofs and horns and in birds of feathers, claws, spurs, breaks etc.

The nail is formed of a distal, hard, horny plate, termed *the body of the nail*, constituted of the limb and the nail bed, and a *root or radix*, situated proximally, corresponding to the part covered by a fold of skin, called supraungual fold *(plica supraungualis)* which is prolonged on the sides of the nail (latero-ungual fold). The fold covers the lunula, the pale, pink, crescentic portion of the nail body, which in the depth is continuous with the root.

The nail plate is formed of a superficial hard portion and a deep soft layer. The hard layer is generated by *the nail matrix* (the deepest part of the root), while the soft layer grows through the cornification of the cells of the nail bed on which the nail rests. Beneath the free border of the nail lies the nail groove, at the level of which the epidermis, with its horny layer, is continuous with the nail, forming the hyponychium.

Impurities and microorganisms accumulate in the nail groove, the level at which appear the mycoses of the nail. The nail grows about 1mm a week. The limb of the nail is the actual nail and it consists of keratinized squamous (scala-like) cells. Those from the surface of the limb are continuous, at the level of the supraunguinal fold, with the horny layer of the epidermis, forming the eponychium (perionyx).

The nails may be involved in pathological processes of the whole organism and in this case the disease starts at the level of the root, the consequence being the appearance of longitudinal or transverse striae (eczema, erythroderma, pemphigus), or of punctiform depressions (psoriazis).

A damage to the nail bed causes the thickening of the nail, either compact (onychogryphosis), or stratified (nail parakeratosis).

The chronic inflammation (due to staphylococci or Candida) of the supraungual fold leads to changes of the nail, too.

The hairs are anatomical formations proper to man and mammals. A hair is generally formed of two parts: an external, visible, part, which is the shaft, and a part deeply implanted in the dermis, the pilosebaceous follicle or the root.

273

The shaft develops from the epidermis and is horny, flexible, elastic and 0.006-0.6 mm thick; its length varies from a few millimeters to more than a meter. In mammals, the hair shafts form the fur, while in man they may cover the whole body (in hirsute individuals); the length and density vary according to the region of the body and to the sex, the pillosity being more marked in the region of the head (where the hairs are also longer), on the face in males, in *the axillae*), in the pubic region, on the thorax and on the abdomen (in males).

The pillosebaceous follicle represents an invagination in the depth of the skin, which with its tip reaches the hypodermis. It contains the hair and to it are attached a sebaceous gland and a fibre of the smooth: *arrector pili muscles*.

The follicle exhibits an orifice, called *ostium*, which is continuous with a funnel-shaped duct, termed *infundibulum*, in which the hair moves freely and where the cell debris of epithelial sheaths, sebum, impurities and microorganisms accumulate. It is beneath the infundibulum that the sebaceous gland opens and here commences the follicle sac proper, formed of four sheaths. The most peripheral is *the fibrous outer sheath*, which represents a continuation of the dermis.   Below it lies *the vitreous sheath*, which continues the basement membrane, then follows *the outer epithelial sheath* (in continuation of the epidermis), the innermost being the inner epithelial sheath, which is in contact with the hair through its internal lamina, the cuticle. *The cuticle* consists of brick-like overlapping cells, but in an opposite direction to those of the epidermicula of the hair, the adherence of the hair in the follicle being thus increased. The cells of the internal epithelial sheath (composed of three layers) have the same origin as the hairs, arising from the lateral zone of the hair matrix.

They grow towards the surface concomitantly with the hair and empty into the infundibulum.

The hair has a free part, called shaft, and an intrafollicular part named *root*. The latter ends by a club-shaped enlargement, *the bulb*. In its deepest part, the bulb presents a depression in which the intensely vascularized nutrient dermal hair papilla. In contact with the papilla are the cells of the hair matrix, which are in an intense multiplication. This multiplication continues also in the bulb, whereas in the area above the bulb the cells become elongated (elongation zone) and then keratinized (keratogenous zone). The keratophilic fungi penetrate only up to this zone.

As regards the structure, the shaft of the hair is composed of a *central medulla*, formed of large cells filled with fat droplets, of a thick cortex consisting of keratinized and variously pigmented scaly cells (which confer the color upon the hair) and an *external epidermicula*, in which the cells overlap brick-like with the free extremity directed towards the tip of the hair.

The hairs are of several types: lanugal hairs (downy), hairs of the scalp, of the eyelashes and eyebrows and, finally, those which belong to the secondary sexual characters (of the axillae, the genitalia, the breast, the back etc.). The colour of the hair is various: fair, red, brown, black. It is conferred by a granular brown or reddish pigment, which is formed in the bulb. The white hair of the old age is due to the penetration of air bubbles in the shaft. The shape of the hairs varies, also, to a great extent: crimp hair (relotrichous) encountered in Negroes, straight hair (lissotrichous) characteristic of Mongolians and slightly wavy hair (hymotrichous) characteristic of Europeans.

Hairs are in a continuous renewal, with a three-phase evolution: *anagen phase* (anabolism), *catagen phase* (interruption of growth) and *telogen phase* (resting phase with shedding of the old hair and appearance of a new hair bud).

Normally, 85-90% of the hairs are in the anagen phase, 9-15% in the catagen phase and 1% in the telogen phase. In the accute (massive) hair loss, the hairs enter into the catagen phase (for example, after roentgenotherapy or infectious diseases).

In chronic alopecias, hairs enter into the telogen phase. Hairs grow on the average by 0.3 mm/day.

274

The glands of the skin

The sebaceous gland is an acinous gland. It is a *holocrine gland*, the secreted sebum resulting from the fatty degeneration of the cells which line the walls of the gland. In the so-called seborrheic regions (nose, forehead, chin, ear, mediosternal region etc.) these glands are hypertrophial, conferring upon the respective zones an unctuosness and a peculiar reactivity (seborrheic diseases, for example juvenile acne). The function of these glands is endocrinodependent hypophyseosteroid corticosuprarenal and sexual system.

The sudoriferous or sweat glands are tubular, ending by a secretory glomerulus. They are classified into two types: *eccrine glands*, which are smaller, distributed almost on the whole surface of the body and open directly at the surface of the epidermis through *pores*, and *apocrine glands*, much larger, sited only at the level of the axillae, around the nipple of the breast and on the perineum.

The eccrine glands release the product without modifying the structure of the cells, which remain intact: the sweat elaborated by these glands is aqueous and rich in salts, with an acid pH, and does not contain proteins or parts of the secretory cells. In the epidermis, the duct of the gland is transformed in *a simple coil*, devoid of proper cells, which by gliding of the horny cells may be easily obliterated, the consequence being the retention of secretion and the appearance of a typical rash characterized by intraepidermal microvesicles, with a clear content, occurring especially in summer, when perspiration is more abundant, and leading to the clinical picture of *sudamina* (children, obesity, vagotonia).

The apocrine glands are *merocrine*: their secretion results partly from the elimination of some of the secretory cells. Their function starts after puberty, hence they are endocrinodependent.

The pathology of these glands has a post puberal onset (hydrosadenitis, Fordyce's disease etc.). They open on the follicular infundibulum, the secreted sweat is more, viscous, richer in proteins and has a neutral pH, which explains their frequent infection.

The sweat glands have a neurodependent secretion, the excitation the parasympathetic system by pilocarpine produces increases in the secretion, its inhibition by pilocarpine decreases the secretion. The states of vagotonia are characterized by profuse sweating. The "cold" sweat caused by emotions is due to adrenergic influences, through the evacuations of the glomerulus (contraction of the myoepithelial cells).

The moisture of the cutaneous surface is due to sweat (sensible perspiration), to which the epidermal evaporation of water (insensible respiration) asociates.

The sebum and sweat contribute to the formation of the biological lipoproteic and acid pellicle on the surface of the epidermis (pH 4-5), which has a significant role in the biological protection of the skin. This layer is frequently removed by using soap, solvents and detergents, the skin becomes in this way dryer and more vulnerable to microorganisms and chemical substances.

The mucosae

There are numerous similarities between the skin and the mucosa which lines the natural cavities, such as: their common embryonic origin (from the ectoderm), their stratification and the exfoliation of the superficial layer. The differences consist mainly in the lack of keratinization of *the mucosae* and in the absence of superficial structures at this level. The cells of mucous epithelium do not undergo keratinization, their degeneration is of vacuolar type, appear clear fluid droplets inside the cytoplasm, the nucleus does not disappear.

The cells contain no pigment, although melanocytes are present at this level, but they are inactive. In some pathological conditions (for example in Addison's disease), they may become activated and pigmentations may appear also at this level.

suprafata unghiei

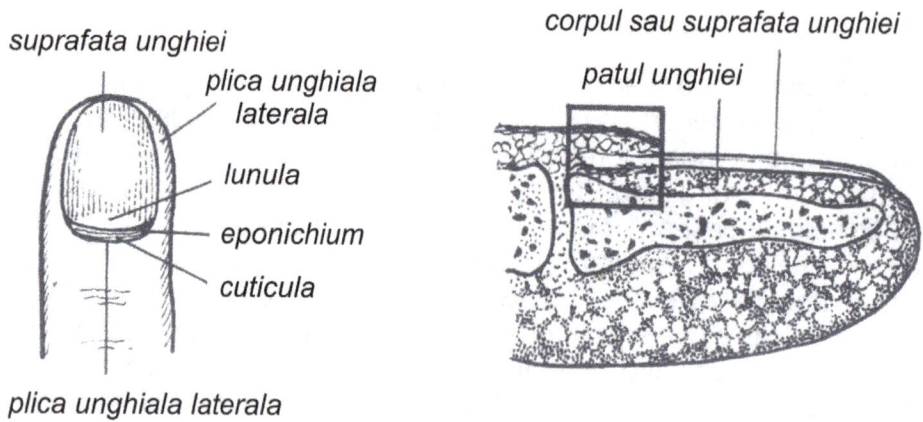

plica unghiala
laterala

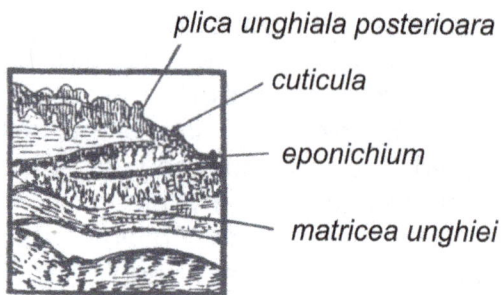

lunula

eponichium

cuticula

plica unghiala laterala

corpul sau suprafata unghiei

patul unghiei

plica unghiala posterioara

cuticula

eponichium

matricea unghiei

Fig. 256 Nail (after Anghelescu)

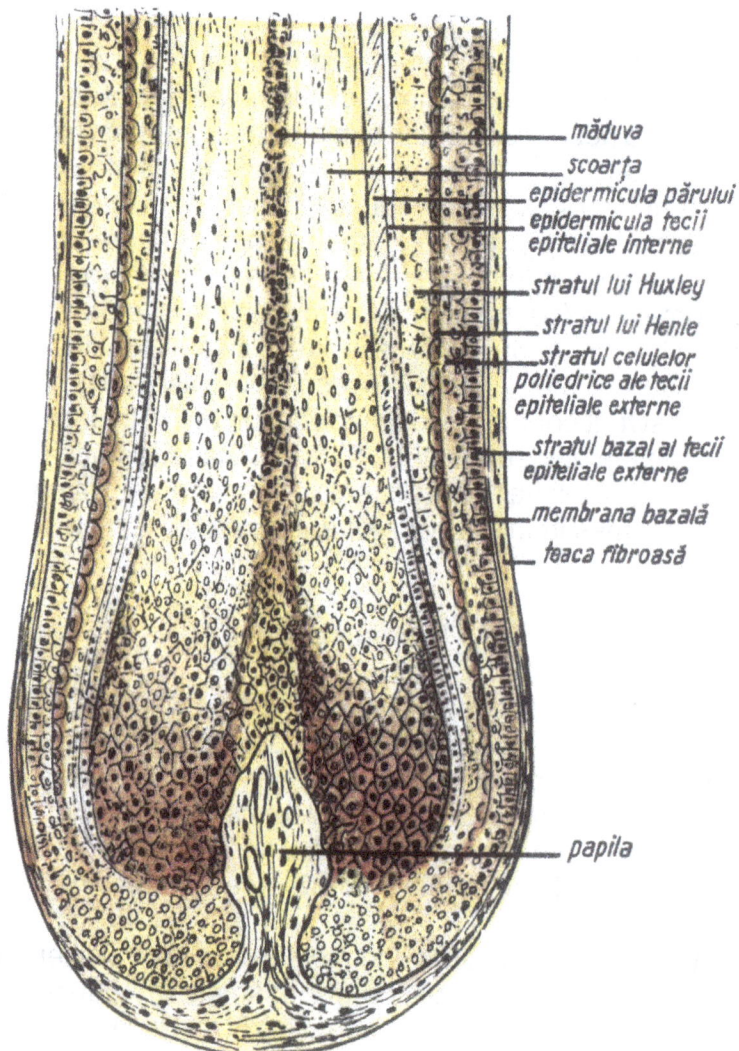

măduva

scoarţa

epidermicula părului

epidermicula tecii
epiteliale interne

stratul lui Huxley

stratul lui Henle

stratul celulelor
poliedrice ale tecii
epiteliale externe

stratul bazal al tecii
epiteliale externe

membrana bazală

teaca fibroasă

papila

276

Fig. 257 Hair follicle

As keratinization does not occur, the "prehorny" layers (*stratum granulosum* and *stratum lucidum*) are also absent; therefore, with the exception of some parts of the tongue and of the palate, the epithelium is transparent and the red color of the underlying corium may be seen through it. At the level of the *semimucosae* (free border of the lips, glans of penis etc.) a discreet keratinization takes place, which becomes exaggerated in frequently sun exposed individuals (farmers etc.), leadind to keratotic-actinic cheilitis. In leucoplasias (premalignant states) appear the granular and the horny layers and the mucosa acquires a white color.

Another difference between the mucosae and the skin consists in the absence of appendages (sweat glands and hair follicles). It should be mentioned that on the free border and the vestibular surface of the lips are present small white-yellowish spots, without a malignant potential (Fox-Fordyce's disease), but often causing cancerophobia. Their treatment consists in punctiform diathermocoagulation. Moreover, a special mention deserves the fact that there are some analogies between the formation of the tooth buds and of the epithelial buds of the sudoriferous glands and of the pilosebaceous follicles. Therefore, in genodermatoses with epidermal impairments and disorders of the superficial structures of the skin (ichthyosis a.o.), dentition impairments are often observed, which in some more severe syndromes (for example, in the complex ectodermal dysplasia) can lead to the triad hypotrichosis-anhydrosis-anodontia.

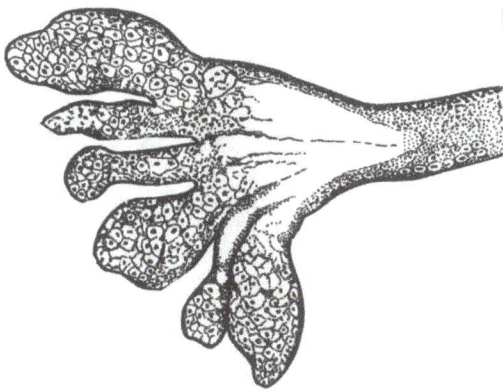

Fig. 258 Sebaceous gland (microscopic structure)

Fig. 259 Comparative structure of the apocrine and eccrine gland

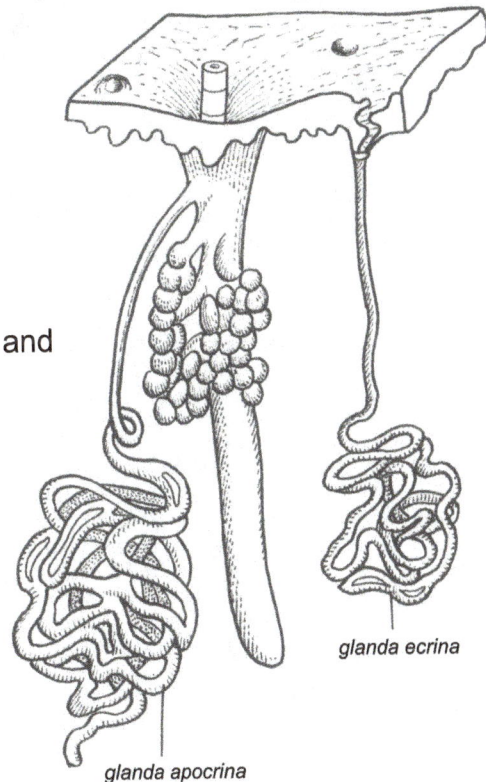

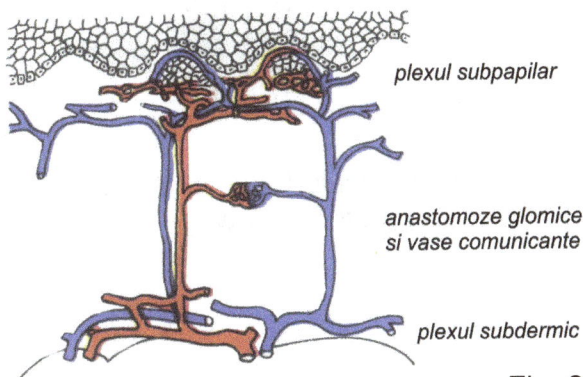

glanda ecrina

glanda apocrina

plexul subpapilar

anastomoze glomice si vase comunicante

plexul subdermic

Fig. 260 Blood and lymph vessels

277

Fig. 261 Medulloradiculo-tegumental topography
(anterior view)

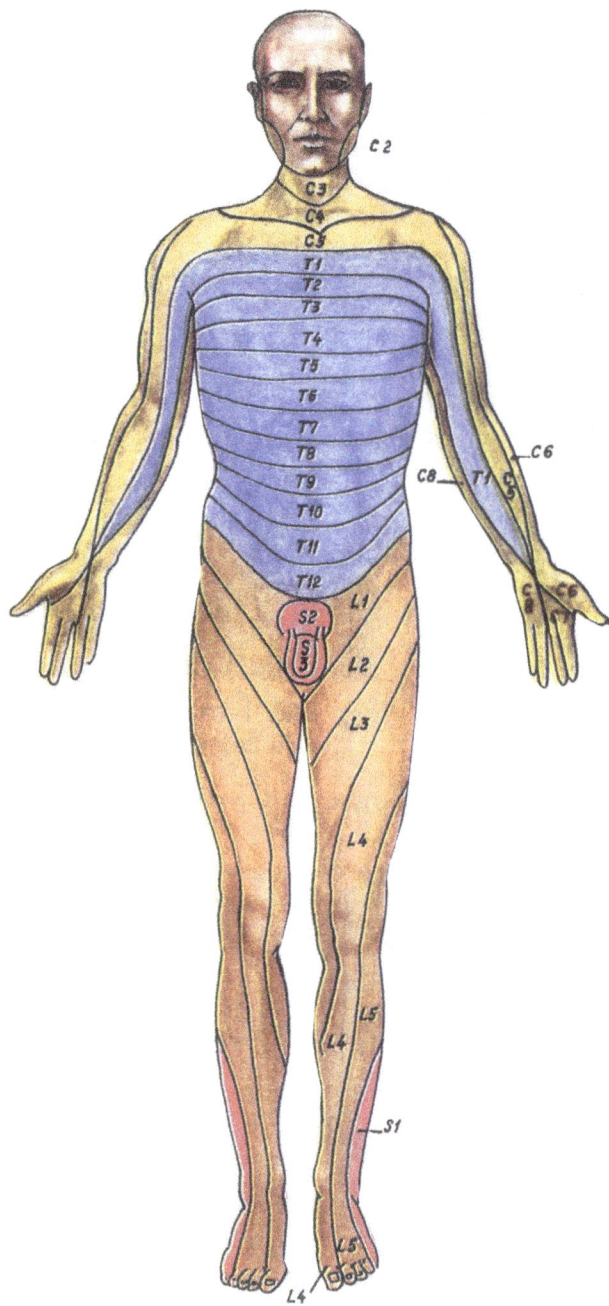

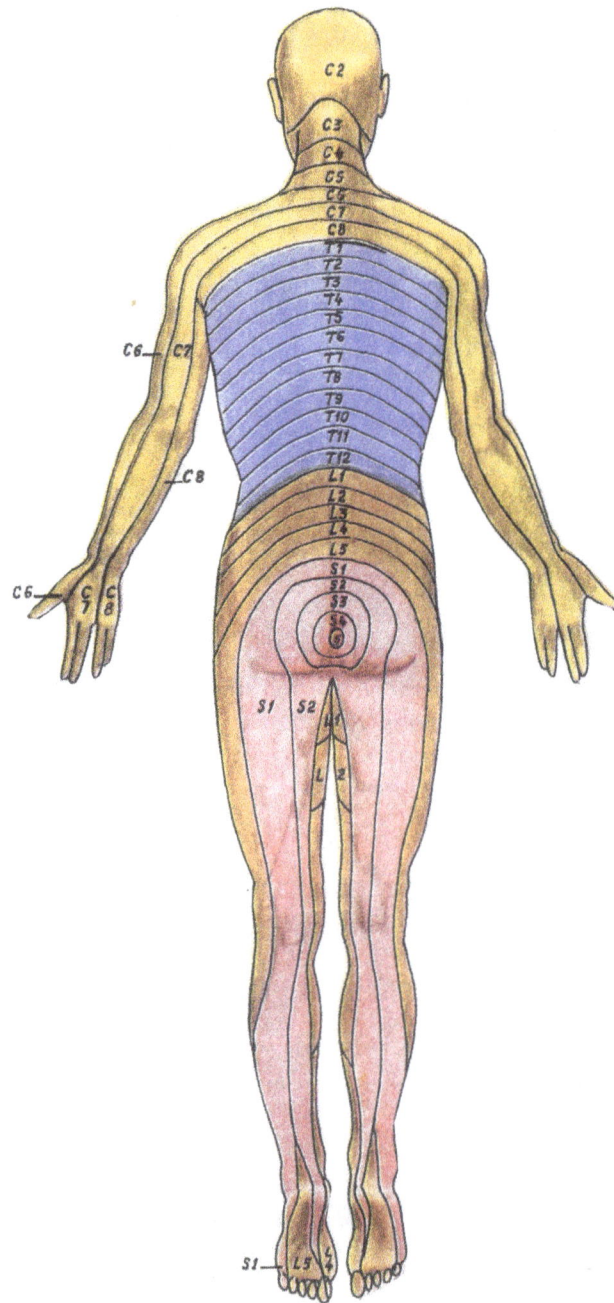

Fig. 262 Medulloradiculo-tegumental topography
(posterior view)

Tegumental thermographies

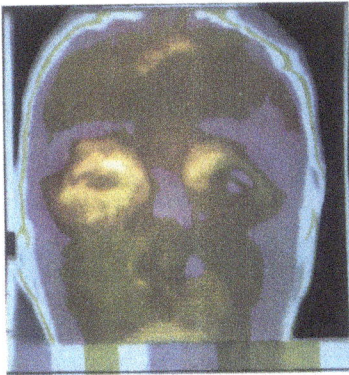
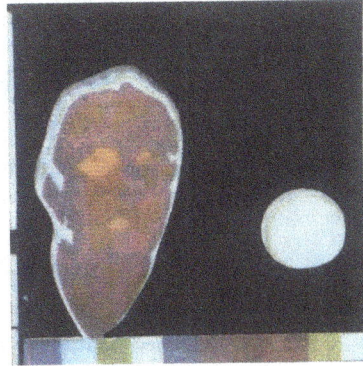
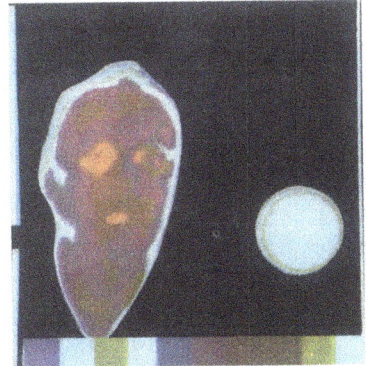

Fig. 263. 264. 265. Tegumental thermography at the level of the eye. Recurrent conjunctival melanoma

Fig. 266 Cranial tegumental thermography. Blastocytoma of the left frontal lobe

Fig. 267 Tegumental thermography of the eye. Tumour of the right eye

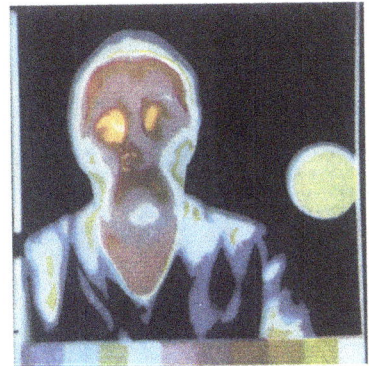

Fig. 268 Normal mammary tegumental thermography

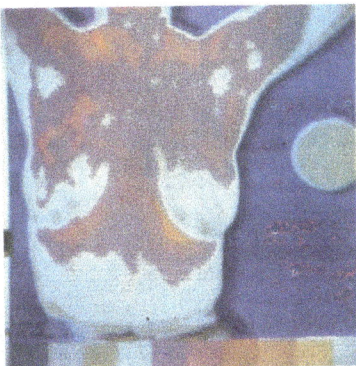

Fig. 269 Mammary tegumental thermography. Carcinomatous mastitis

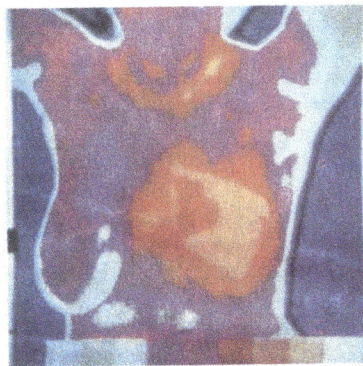

Fig. 270 Mammary tegumental thermography. Mammary carcinoma with supra- and subclavicular adenopathy

Fig. 271 Mammary tegumental thermography. Left mammary adenofibroma

279

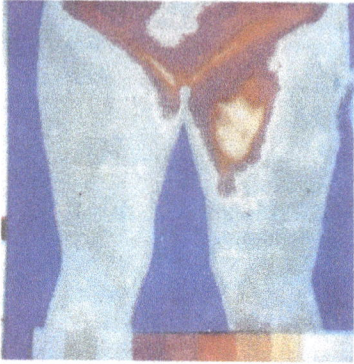

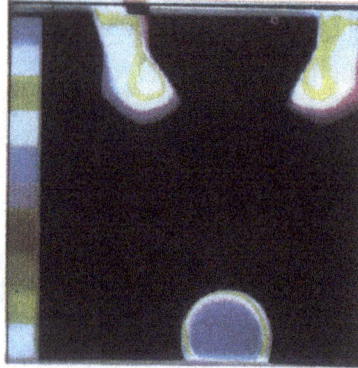

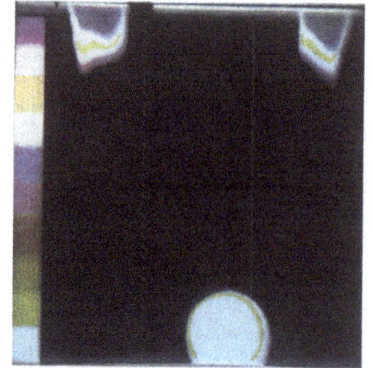

Fig. 272. 273. 274. Tegumental thermography of the hands
(Raynaud's syndrome)

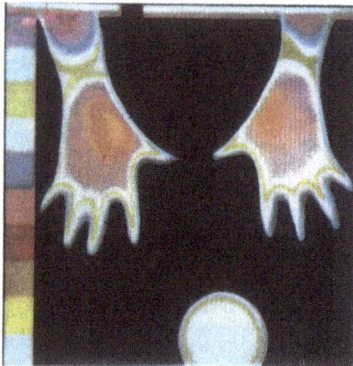

Fig. 275 Tegumental thermography of
the hand. Operated melanoma

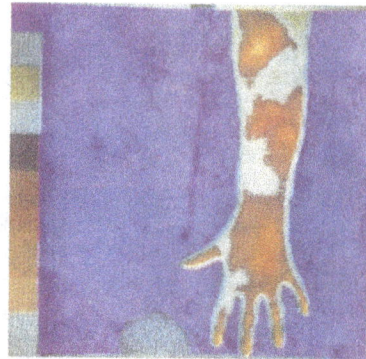

Fig. 276 Tegumental
thermography of the lower
limbs. Osteosarcoma of the
internal condyle of the left femur

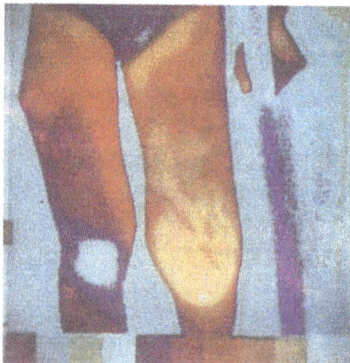

Fig. 277 Tegumental thermography
of the lower limbs.
Thrombophlebitis of the left lower
limb

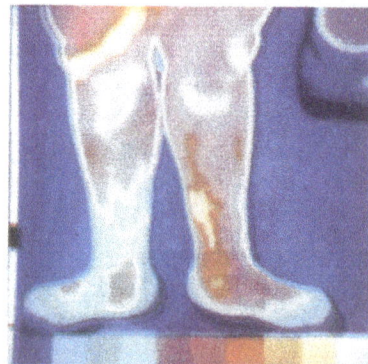

Fig. 278 Tegumental thermography of the
lower limbs. Melanosarcoma of the left leg;
metastasis after operated melanoma

280

Topographical images, computerized radiographs and histological structures

In order to facilitate the comprehension of the topography of the somatic peripheral nervous system (presented in vol. I) and of some peripheral thoraco-abdominal segments of the vegetative (autonomic) nervous system, we present in this chapter a series of topographical sections, performed both classically (Fig. 279 - 299) and by computerized radiographs (Fig. 300 - 316). These sections have the role of making up useful working tools for the clinicians, in establishing the topographical lesional diagnosis, of absolute necessity both in neurological diseases and in the general pathology.

The explanation of the images in the tomographies achieved at the level of the thorax and of the abdomen results from the correlation with the similar drawings contained in vol.II.

The histological images (Fig. 317 - 332) complete the informations on the nervous system.

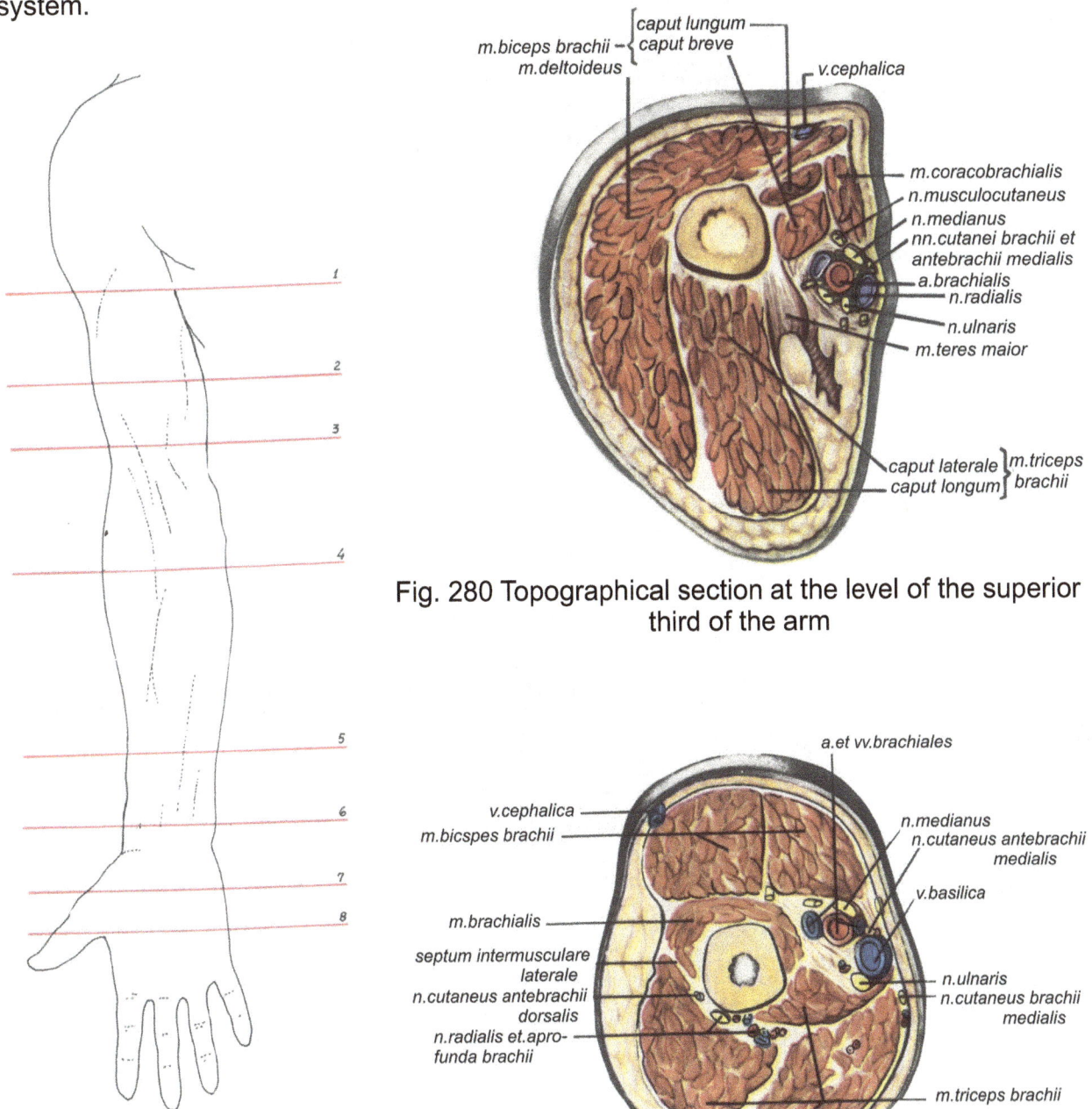

Fig. 280 Topographical section at the level of the superior third of the arm

Fig. 279 Section levels through the upper limb

Fig. 281 Topographical section at the level of the middle third of the arm

281

Fig. 282 Topographical section at the level of the inferior third
of the arm

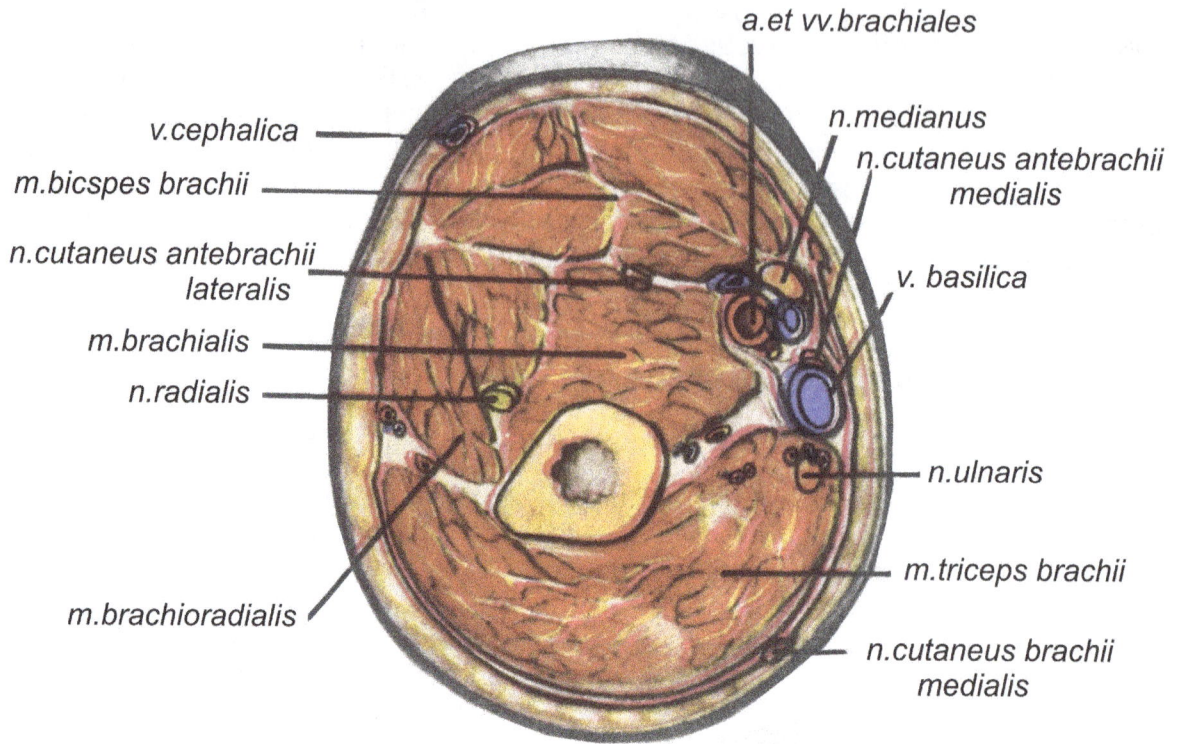

a.et vv.brachiales

v.cephalica

m.bicspes brachii

n.cutaneus antebrachii
lateralis

m.brachialis

n.radialis

n.medianus

n.cutaneus antebrachii
medialis

v. basilica

n.ulnaris

m.triceps brachii

n.cutaneus brachii
medialis

m.brachioradialis

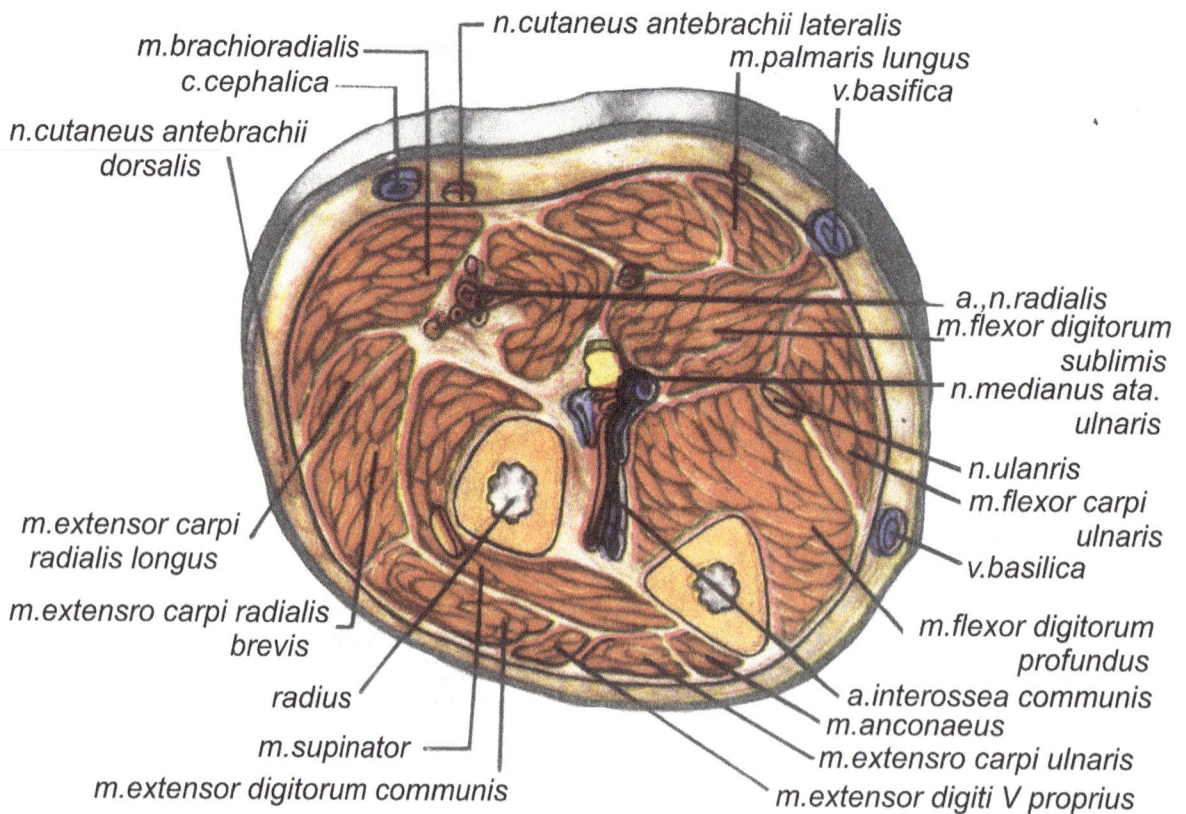

m.brachioradialis
c.cephalica

n.cutaneus antebrachii
dorsalis

n.cutaneus antebrachii lateralis
m.palmaris lungus
v.basifica

a.,n.radialis
m.flexor digitorum
sublimis
n.medianus ata.
ulnaris
n.ulanris
m.flexor carpi
ulnaris
v.basilica

m.extensor carpi
radialis longus

m.extensro carpi radialis
brevis

radius

m.supinator

m.extensor digitorum communis

m.flexor digitorum
profundus
a.interossea communis
m.anconaeus
m.extensro carpi ulnaris
m.extensor digiti V proprius

Fig. 283 Topographical section at the level of the superior third of
the arm

282

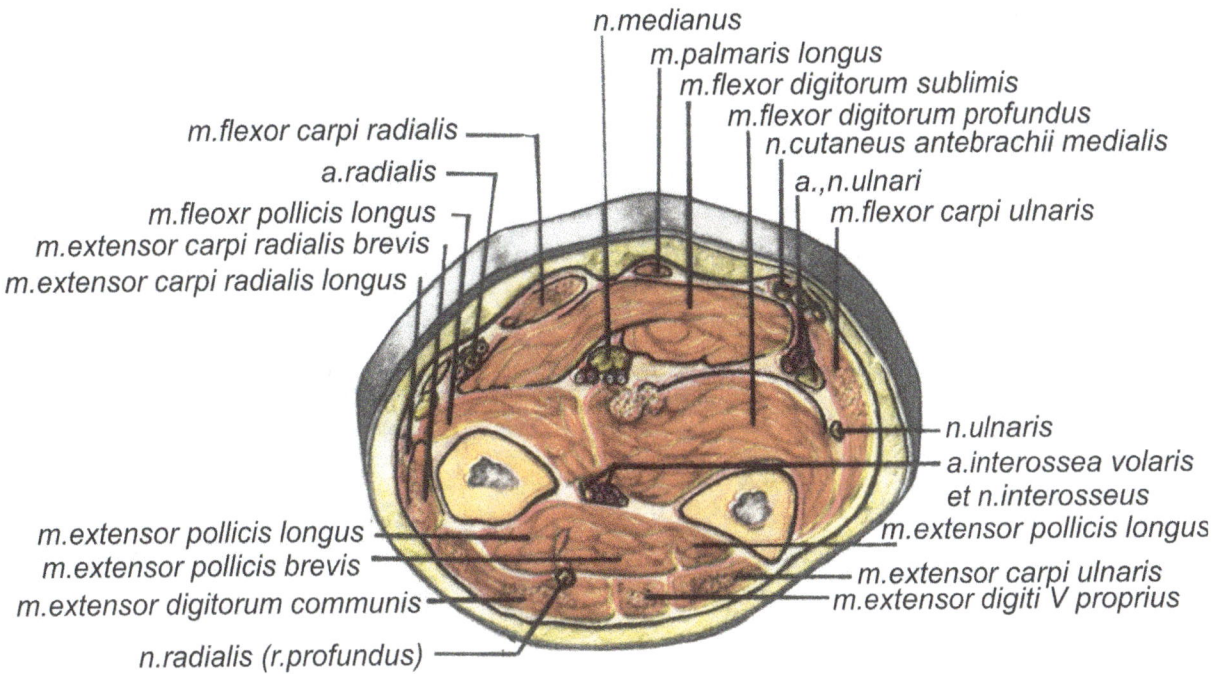

Fig. 284 Topographical section at the level of the middle third of the forearm

n.medianus
m.palmaris longus
m.flexor digitorum sublimis
m.flexor digitorum profundus
n.cutaneus antebrachii medialis
a.,n.ulnari
m.flexor carpi ulnaris

m.flexor carpi radialis
a.radialis
m.fleoxr pollicis longus
m.extensor carpi radialis brevis
m.extensor carpi radialis longus

n.ulnaris
a.interossea volaris
et n.interosseus
m.extensor pollicis longus

m.extensor pollicis longus
m.extensor pollicis brevis
m.extensor digitorum communis

m.extensor carpi ulnaris
m.extensor digiti V proprius

n.radialis (r.profundus)

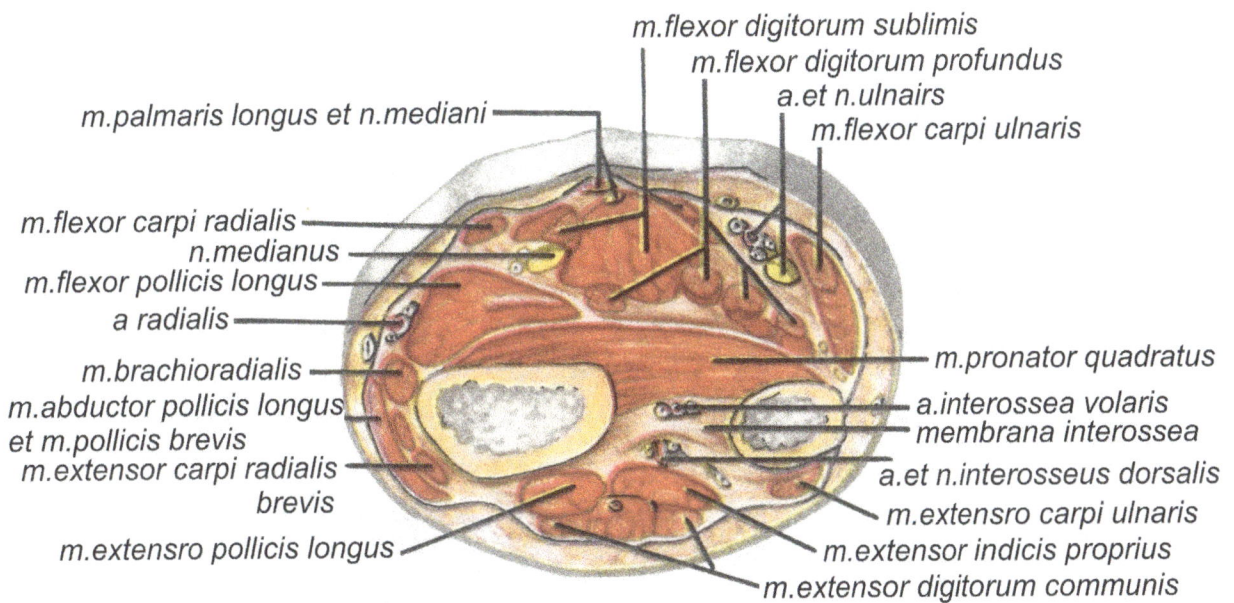

m.flexor digitorum sublimis
m.flexor digitorum profundus
a.et n.ulnairs
m.flexor carpi ulnaris

m.palmaris longus et n.mediani

m.flexor carpi radialis
n.medianus
m.flexor pollicis longus
a radialis

m.brachioradialis
m.abductor pollicis longus
et m.pollicis brevis
m.extensor carpi radialis
brevis

m.pronator quadratus
a.interossea volaris
membrana interossea
a.et n.interosseus dorsalis
m.extensro carpi ulnaris
m.extensor indicis proprius
m.extensor digitorum communis

m.extensro pollicis longus

Fig. 285 Topographical section at the level of the inferior third of the forearm

Fig. 286 Topographical section at the level of the superior third of the palmar region

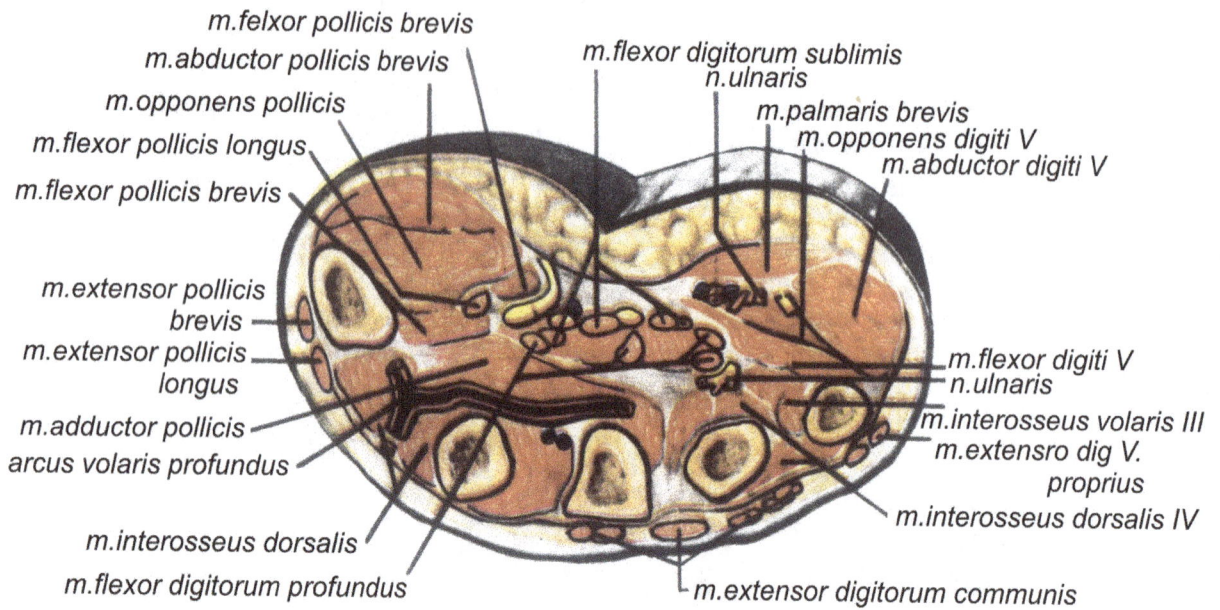

m.felxor pollicis brevis
m.abductor pollicis brevis
m.opponens pollicis
m.flexor pollicis longus
m.flexor pollicis brevis

m.extensor pollicis brevis
m.extensor pollicis longus
m.adductor pollicis
arcus volaris profundus

m.interosseus dorsalis
m.flexor digitorum profundus

m.flexor digitorum sublimis
n.ulnaris
m.palmaris brevis
m.opponens digiti V
m.abductor digiti V

m.flexor digiti V
n.ulnaris
m.interosseus volaris III
m.extensro dig V. proprius
m.interosseus dorsalis IV

m.extensor digitorum communis

m.abductor pollicis brecis
m.flexor pollicis brevis
m.flexor pollicis longus
aponeurosis palmaris
a.et n.digitales volares communes
m.abductor digiti V

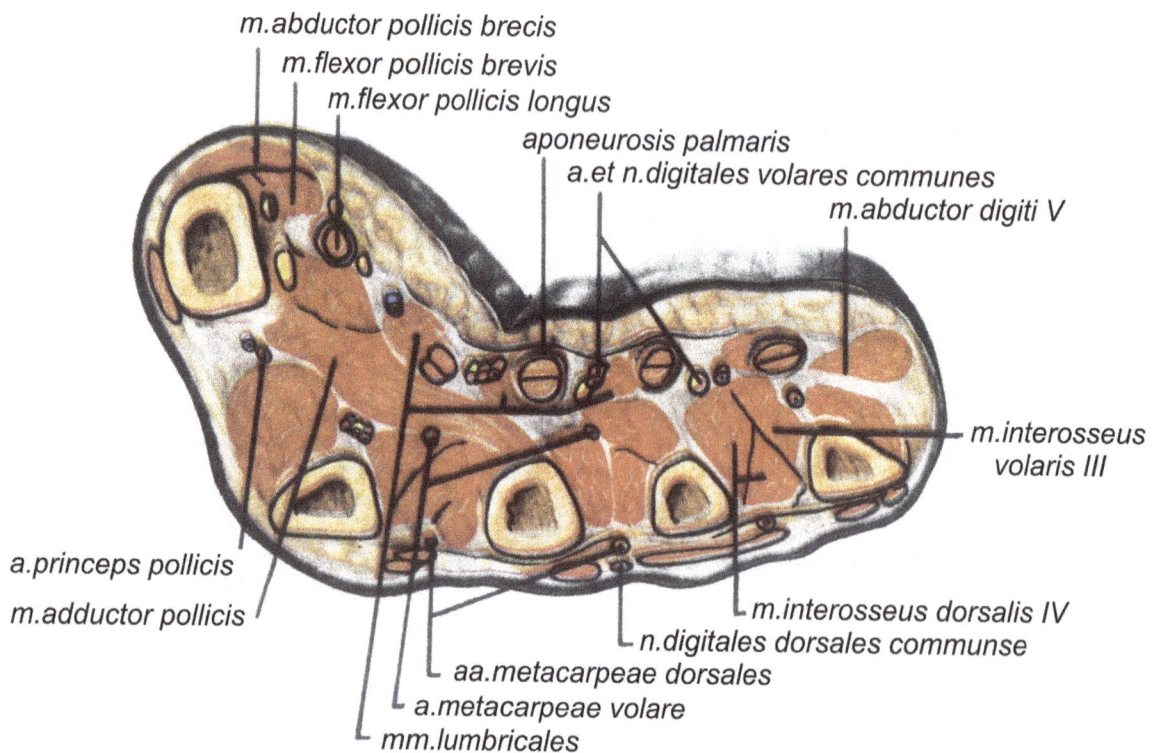

a.princeps pollicis
m.adductor pollicis

m.interosseus volaris III

m.interosseus dorsalis IV
n.digitales dorsales communse
aa.metacarpeae dorsales
a.metacarpeae volare
mm.lumbricales

Fig. 287 Topographical section at the level of the middle third of the palmar region

284

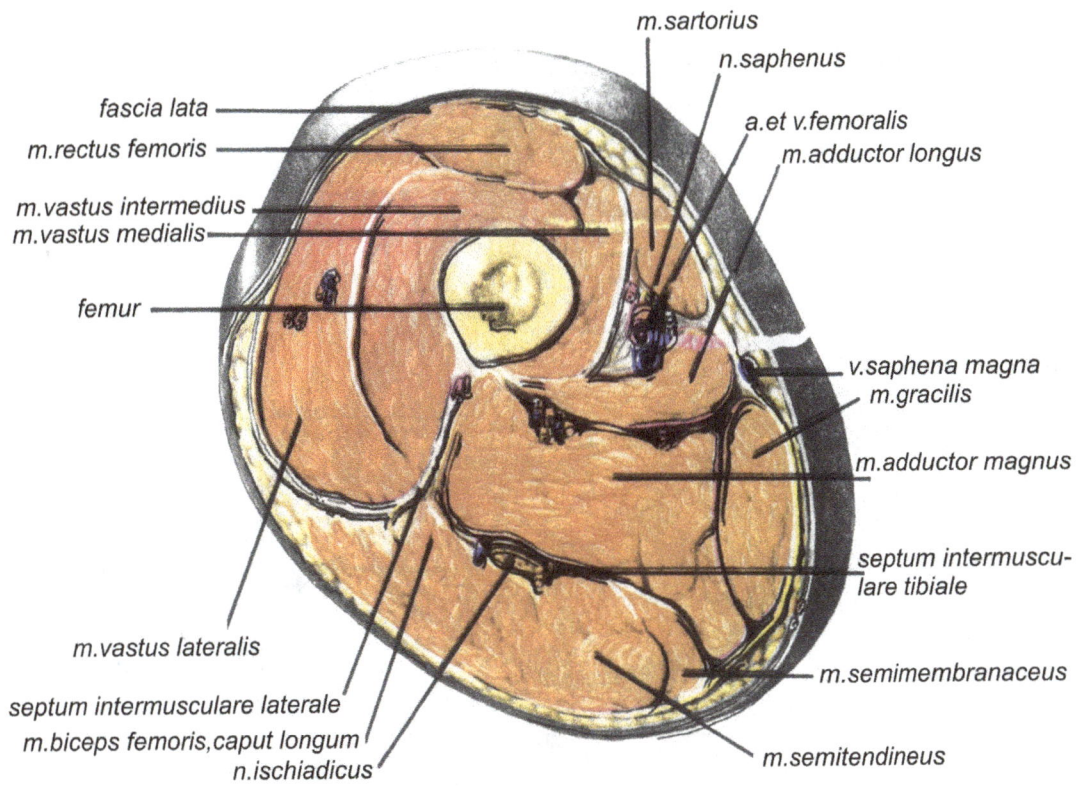

m.sartorius
n.saphenus
a.et v.femoralis
m.adductor longus
fascia lata
m.rectus femoris
m.vastus intermedius
m.vastus medialis
femur
v.saphena magna
m.gracilis
m.adductor magnus
septum intermuscu-
lare tibiale
m.vastus lateralis
septum intermusculare laterale
m.biceps femoris,caput longum
n.ischiadicus
m.semimembranaceus
m.semitendineus

Fig. 289 Topographical section
at the level of the superior third
of the thigh

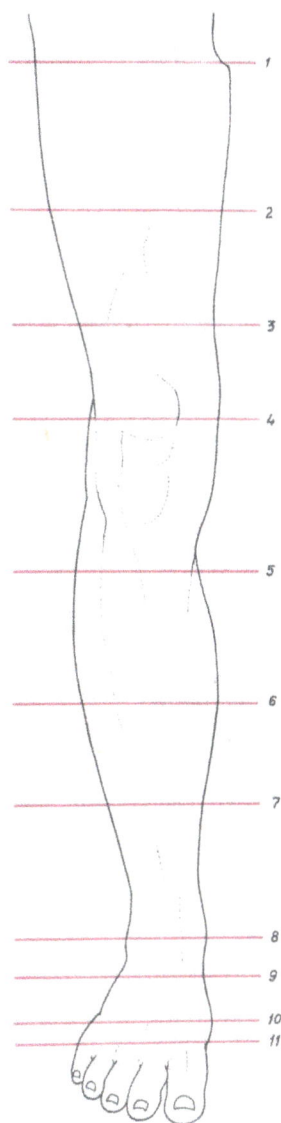

1
2
3
4
5
6
7
8
9
10
11

Fig. 288 Section levels
through the limb

285

Fig. 290 Topographical section at the level of the middle third of the thigh

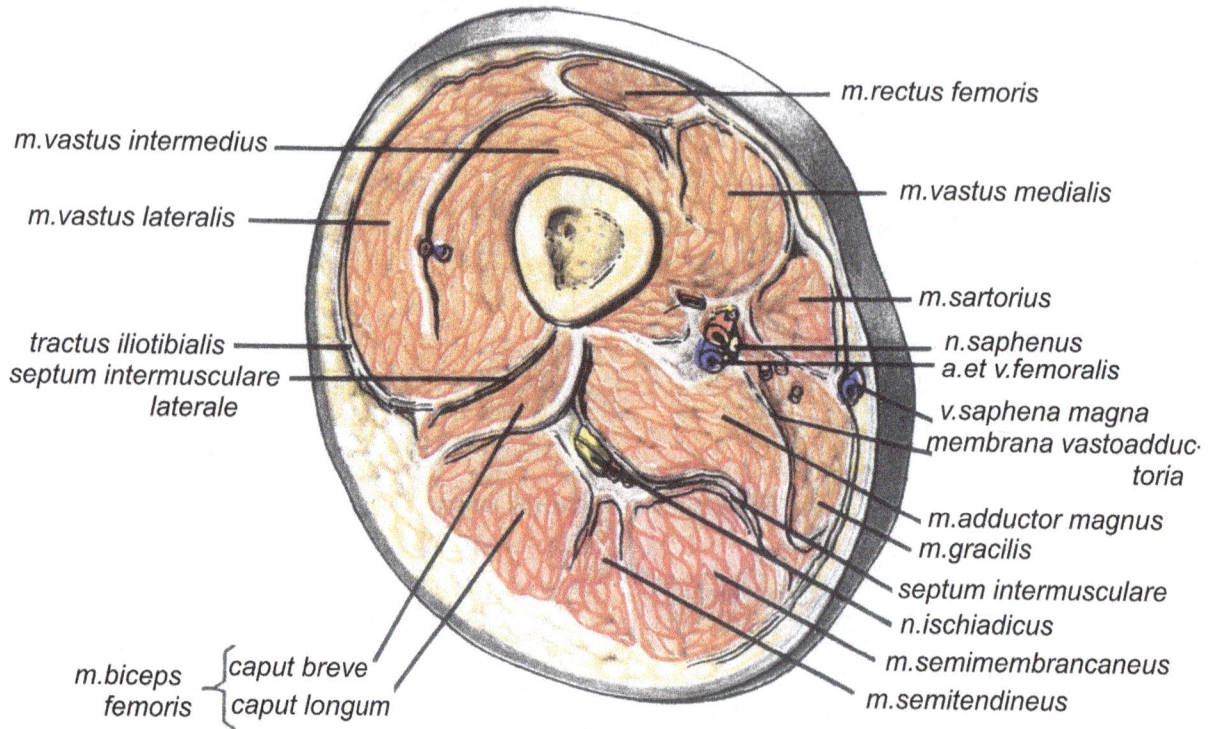

m.vastus intermedius

m.vastus lateralis

tractus iliotibialis
septum intermusculare
laterale

m.biceps {caput breve
femoris {caput longum

m.rectus femoris

m.vastus medialis

m.sartorius

n.saphenus
a.et v.femoralis

v.saphena magna
membrana vastoadduc·
toria

m.adductor magnus
m.gracilis

septum intermusculare

n.ischiadicus

m.semimembrancaneus

m.semitendineus

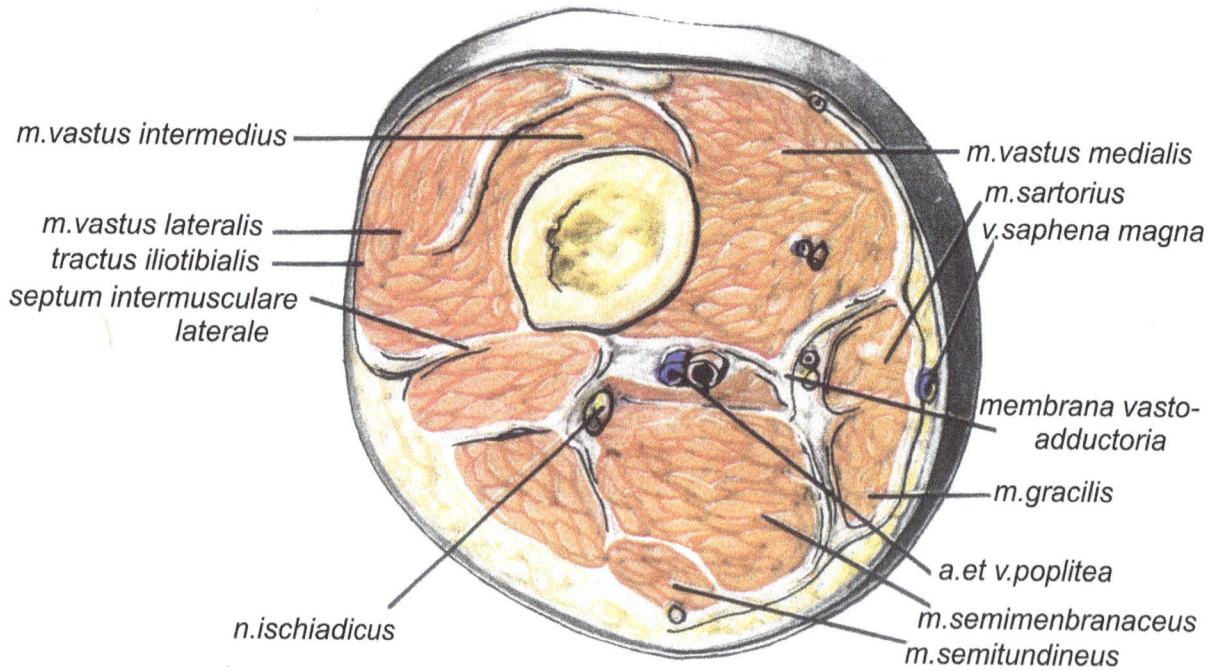

m.vastus intermedius

m.vastus lateralis
tractus iliotibialis
septum intermusculare
laterale

n.ischiadicus

m.vastus medialis
m.sartorius
v.saphena magna

membrana vasto-
adductoria

m.gracilis

a.et v.poplitea

m.semimenbranaceus
m.semitundineus

Fig. 291 Topographical section at the level of the inferior third of the thigh

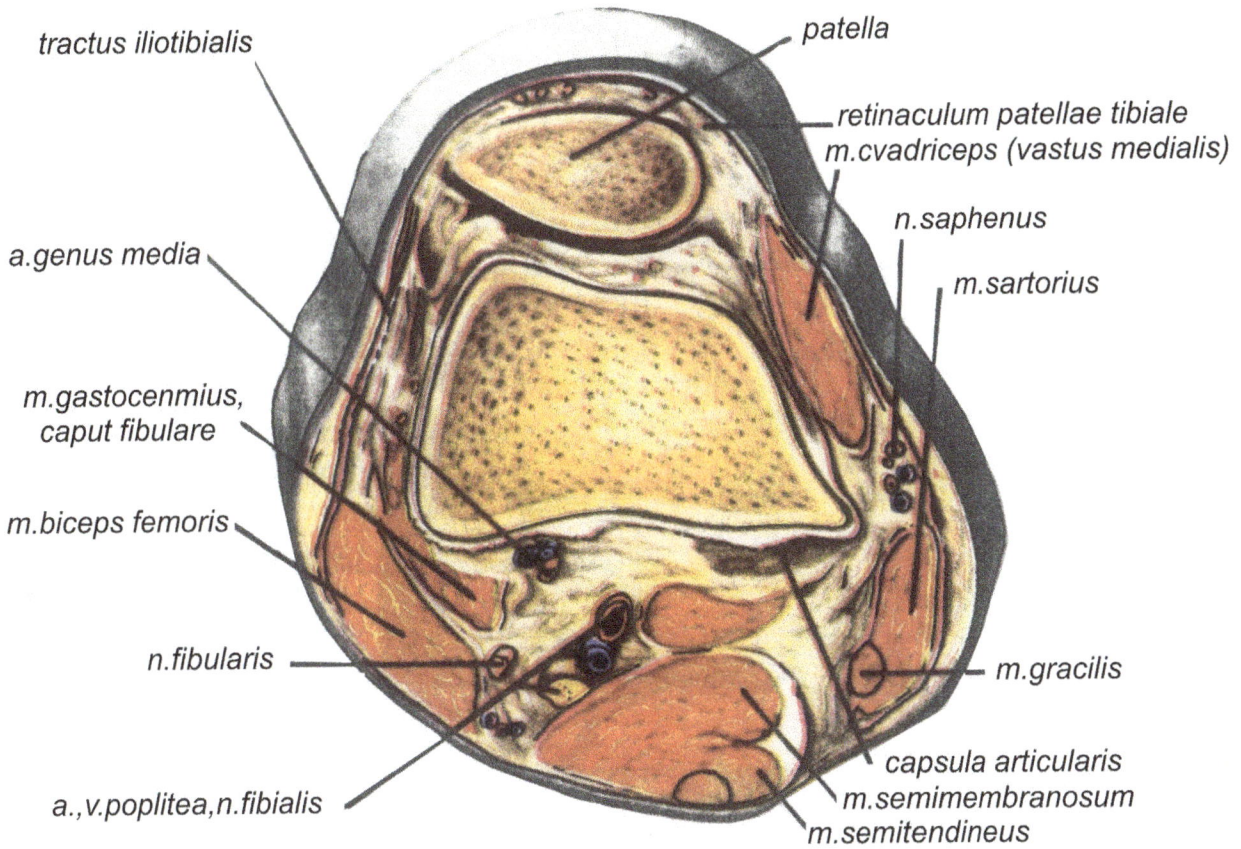

Fig. 292 Topographical section at the level of the superior portion of the region of the knee

tractus iliotibialis

patella

retinaculum patellae tibiale
m.cvadriceps (vastus medialis)

n.saphenus

m.sartorius

a.genus media

m.gastocenmius,
caput fibulare

m.biceps femoris

n.fibularis

m.gracilis

a.,v.poplitea,n.fibialis

capsula articularis
m.semimembranosum
m.semitendineus

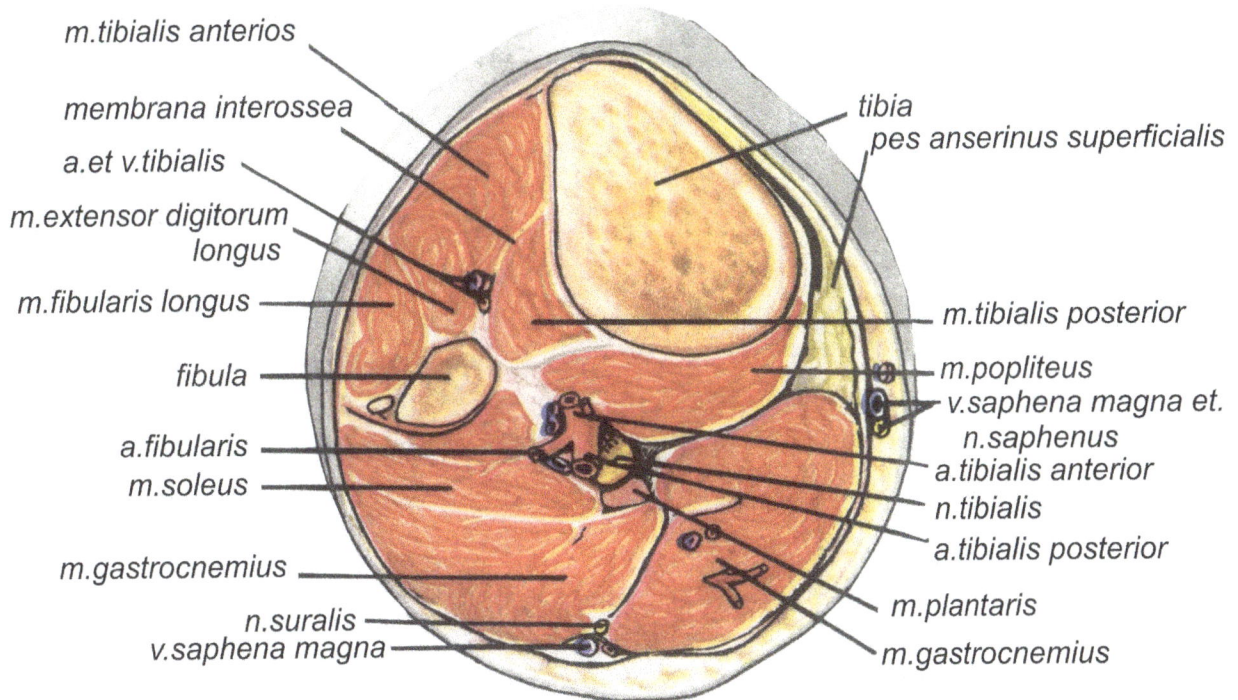

m.tibialis anterios

membrana interossea

a.et v.tibialis

m.extensor digitorum
longus

m.fibularis longus

fibula

a.fibularis

m.soleus

m.gastrocnemius

n.suralis

v.saphena magna

tibia
pes anserinus superficialis

m.tibialis posterior

m.popliteus
v.saphena magna et.
n.saphenus
a.tibialis anterior
n.tibialis
a.tibialis posterior

m.plantaris

m.gastrocnemius

Fig. 293 Topographical section at the level of the superior third of
the leg

287

Fig. 294 Topographical section at the level of the middle
third of the leg

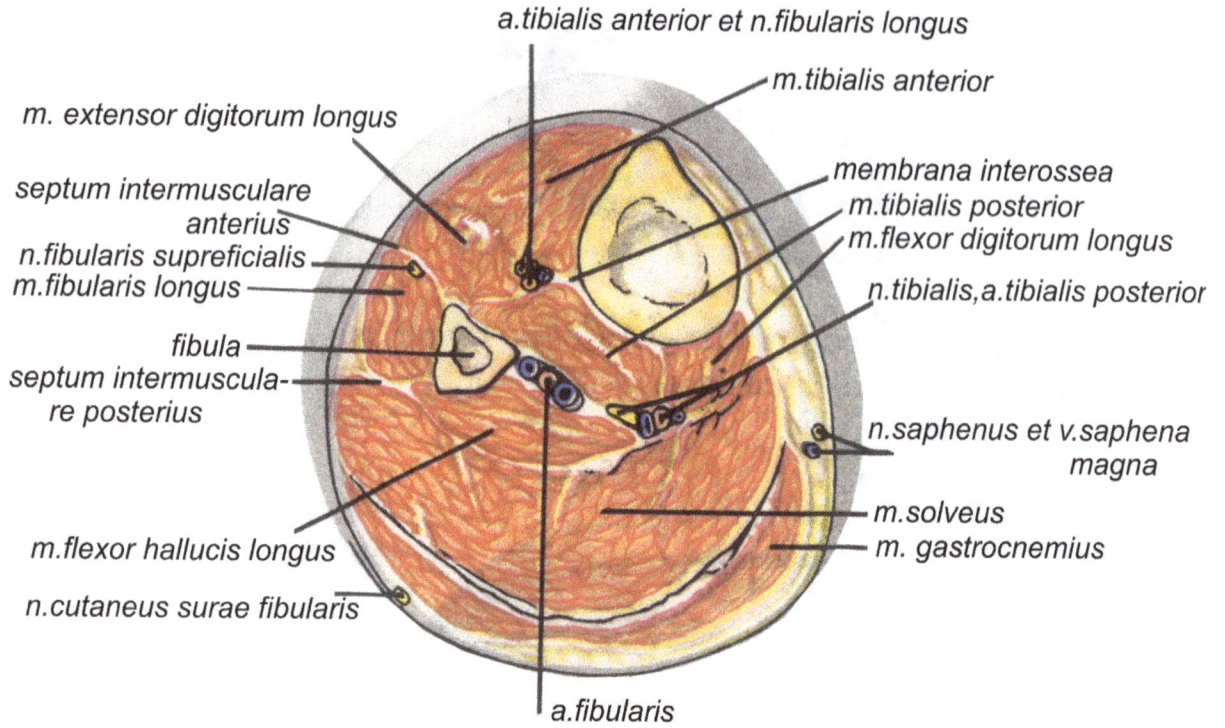

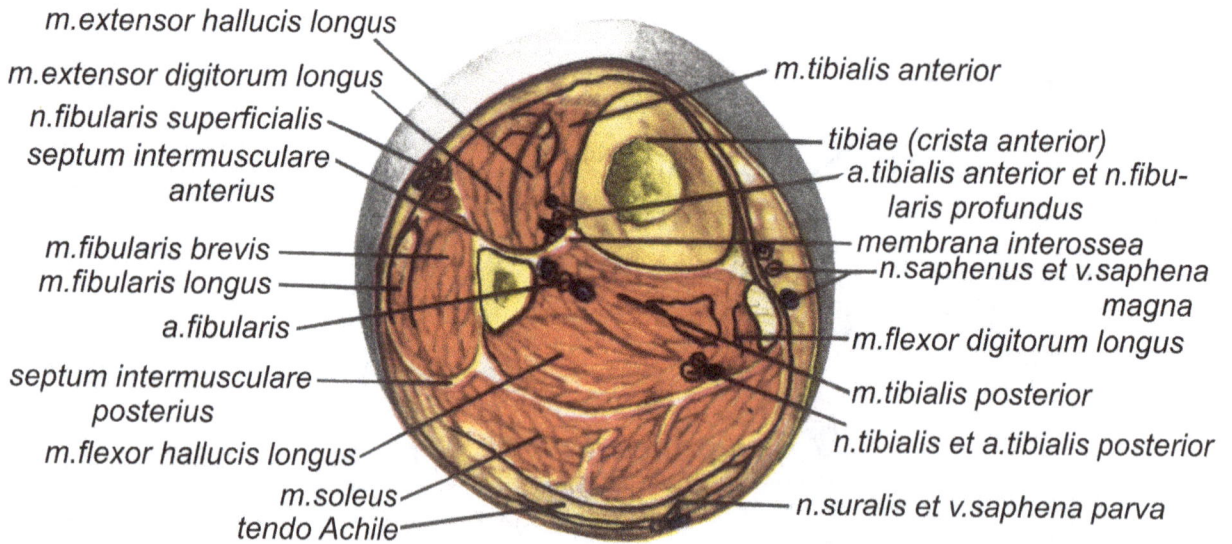

a.tibialis anterior et n.fibularis longus

m.tibialis anterior

m. extensor digitorum longus

septum intermusculare
anterius

n.fibularis supreficialis

m.fibularis longus

membrana interossea
m.tibialis posterior
m.flexor digitorum longus

n.tibialis,a.tibialis posterior

fibula

septum intermuscula-
re posterius

n.saphenus et v.saphena
magna

m.solveus
m. gastrocnemius

m.flexor hallucis longus

n.cutaneus surae fibularis

a.fibularis

m.extensor hallucis longus

m.extensor digitorum longus

n.fibularis superficialis

septum intermusculare
anterius

m.tibialis anterior

tibiae (crista anterior)
a.tibialis anterior et n.fibu-
laris profundus
membrana interossea
n.saphenus et v.saphena
magna

m.fibularis brevis

m.fibularis longus

a.fibularis

m.flexor digitorum longus

septum intermusculare
posterius

m.flexor hallucis longus

m.soleus

tendo Achile

m.tibialis posterior

n.tibialis et a.tibialis posterior

n.suralis et v.saphena parva

Fig. 295 Topographical section at the level of the inferior third
of the leg

288

Fig. 296 Topographical section at the level of the talocrural joint

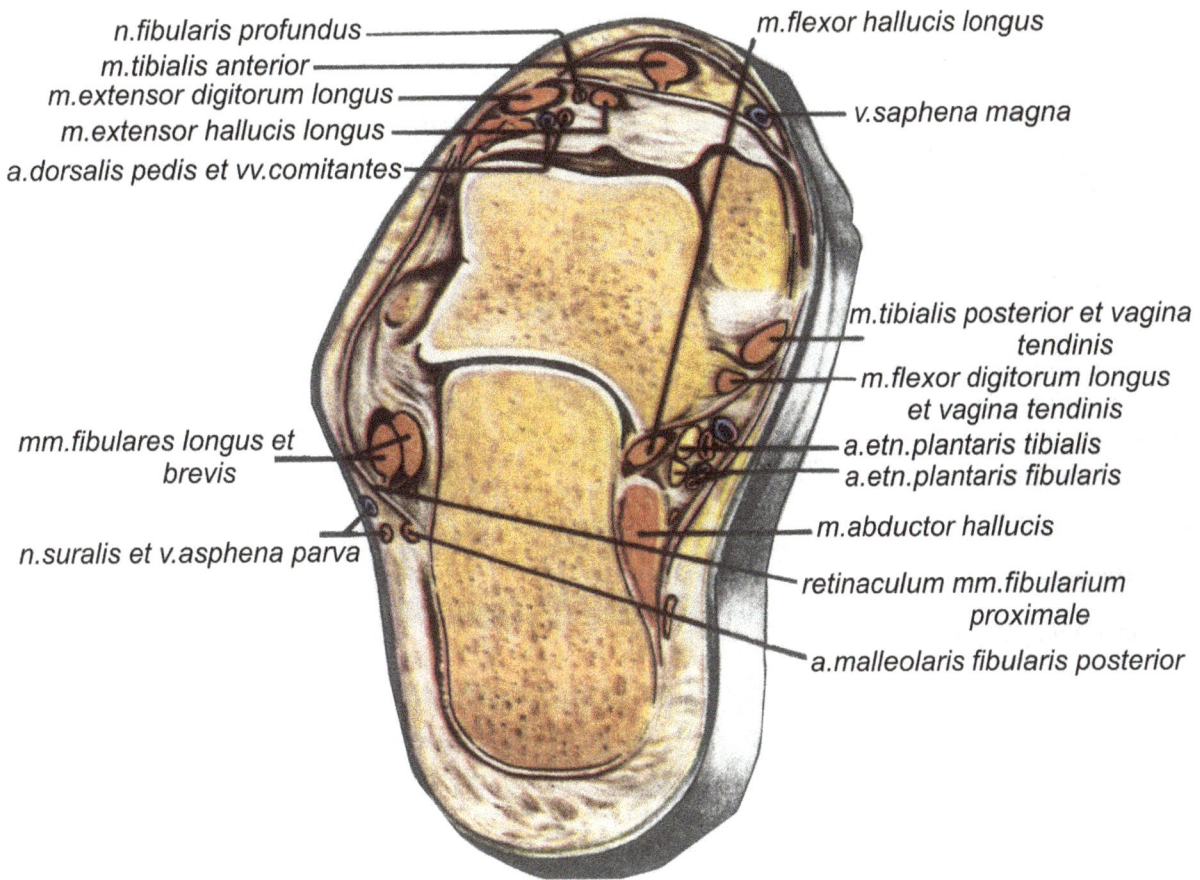

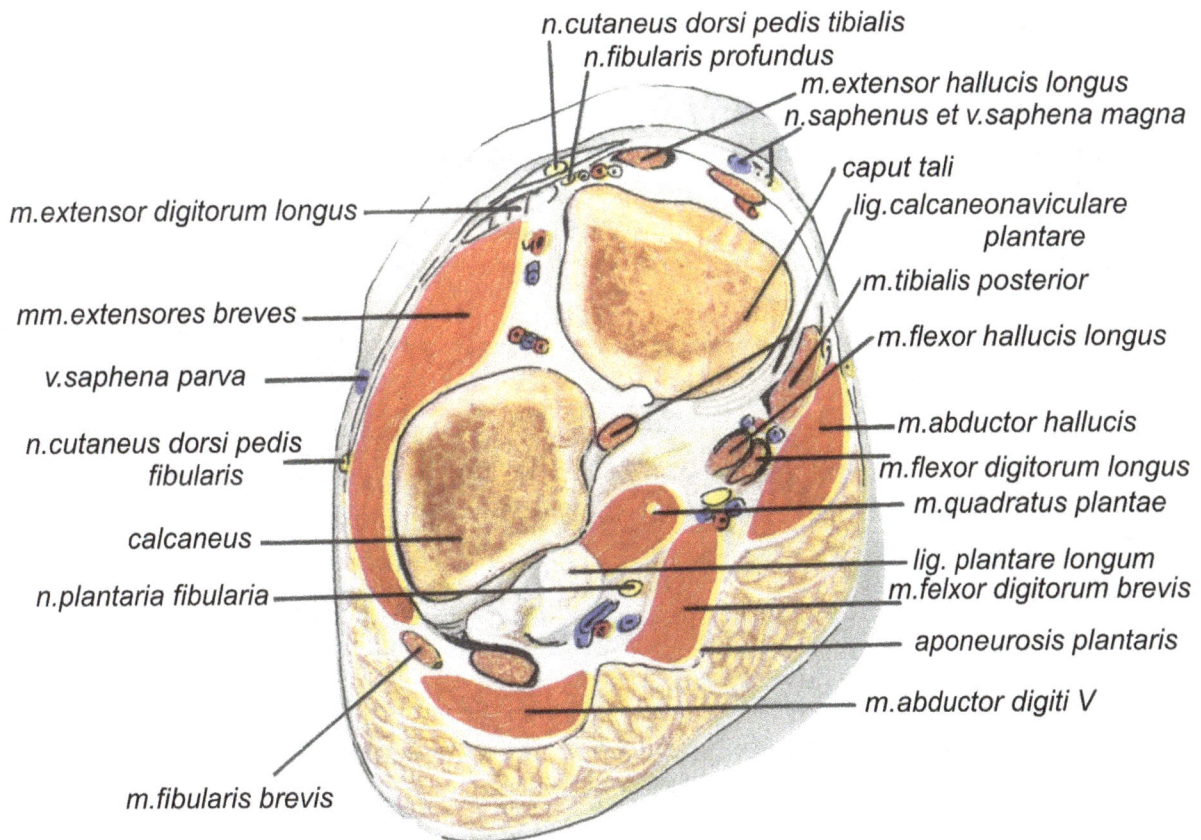

n.fibularis profundus
m.tibialis anterior
m.extensor digitorum longus
m.extensor hallucis longus
a.dorsalis pedis et vv.comitantes

m.flexor hallucis longus

v.saphena magna

m.tibialis posterior et vagina tendinis
m.flexor digitorum longus et vagina tendinis
a.etn.plantaris tibialis
a.etn.plantaris fibularis
m.abductor hallucis

mm.fibulares longus et brevis

n.suralis et v.asphena parva

retinaculum mm.fibularium proximale

a.malleolaris fibularis posterior

n.cutaneus dorsi pedis tibialis
n.fibularis profundus
m.extensor hallucis longus
n.saphenus et v.saphena magna
caput tali
lig.calcaneonaviculare plantare
m.tibialis posterior
m.flexor hallucis longus

m.extensor digitorum longus

mm.extensores breves

v.saphena parva

n.cutaneus dorsi pedis fibularis

calcaneus

n.plantaria fibularia

m.abductor hallucis
m.flexor digitorum longus
m.quadratus plantae
lig. plantare longum
m.felxor digitorum brevis
aponeurosis plantaris
m.abductor digiti V

m.fibularis brevis

Fig. 297 Topographical section at the level of the proximal region of the sole

289

Fig. 298 Topographical section at the level of the middle
region of the sole

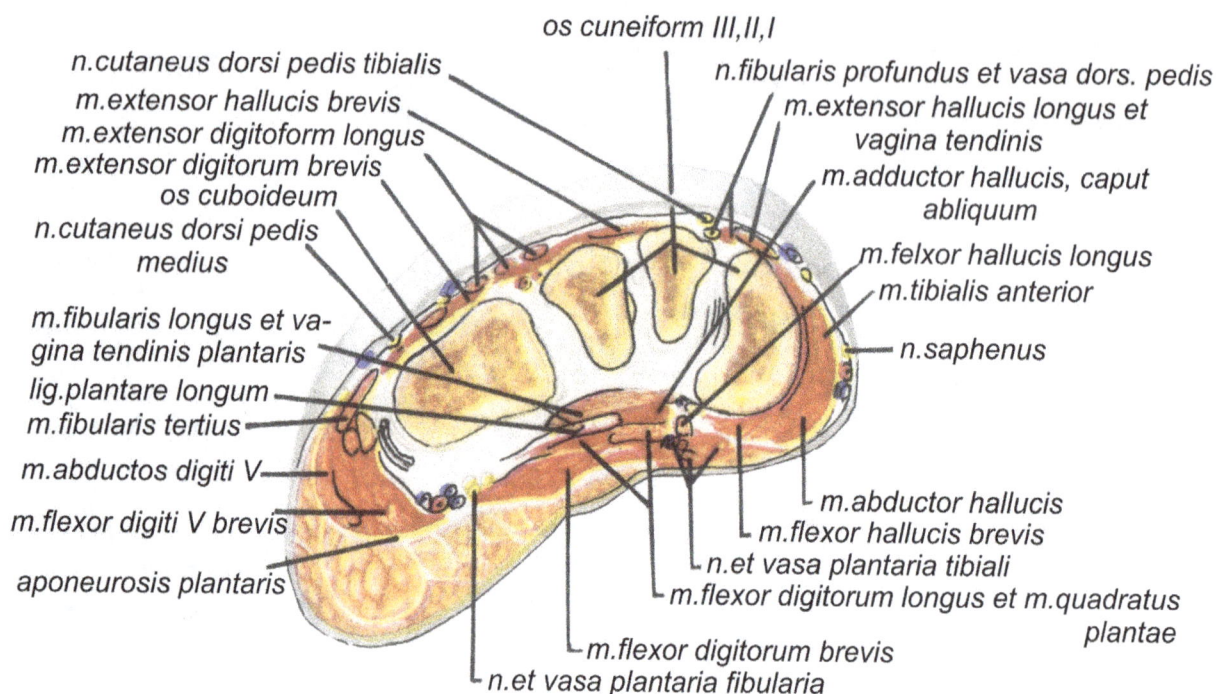

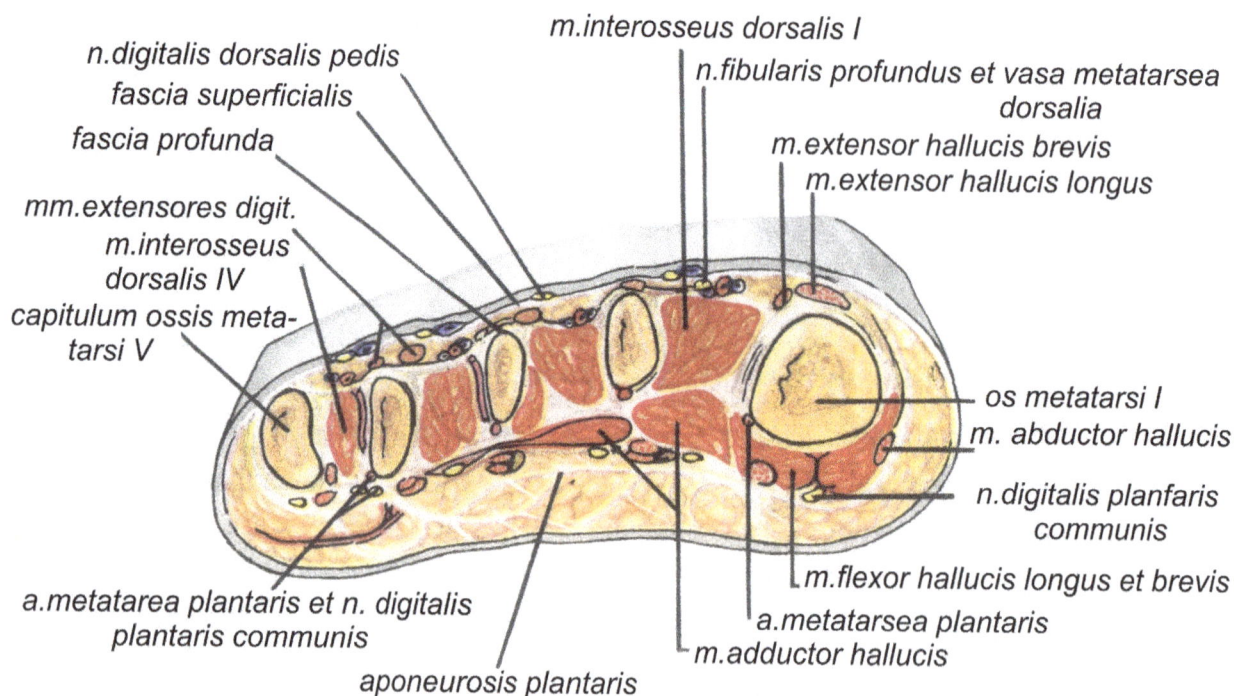

os cuneiform III,II,I

n.cutaneus dorsi pedis tibialis

m.extensor hallucis brevis

m.extensor digitoform longus

m.extensor digitorum brevis

os cuboideum

n.cutaneus dorsi pedis
medius

m.fibularis longus et va-
gina tendinis plantaris

lig.plantare longum

m.fibularis tertius

m.abductos digiti V

m.flexor digiti V brevis

aponeurosis plantaris

n.fibularis profundus et vasa dors. pedis

m.extensor hallucis longus et
vagina tendinis

m.adductor hallucis, caput
abliquum

m.felxor hallucis longus

m.tibialis anterior

n.saphenus

m.abductor hallucis

m.flexor hallucis brevis

n.et vasa plantaria tibiali

m.flexor digitorum longus et m.quadratus
plantae

m.flexor digitorum brevis

n.et vasa plantaria fibularia

m.interosseus dorsalis I

n.digitalis dorsalis pedis

fascia superficialis

fascia profunda

mm.extensores digit.

m.interosseus
dorsalis IV

capitulum ossis meta-
tarsi V

n.fibularis profundus et vasa metatarsea
dorsalia

m.extensor hallucis brevis

m.extensor hallucis longus

os metatarsi I

m. abductor hallucis

n.digitalis planfaris
communis

m.flexor hallucis longus et brevis

a.metatarsea plantaris

m.adductor hallucis

a.metatarea plantaris et n. digitalis
plantaris communis

aponeurosis plantaris

Fig. 299 Topographical section at the level of the distal
region of the sole

290

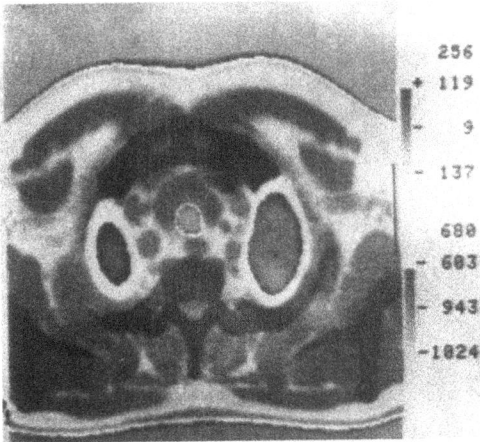

Fig. 300 Computed tomography at the level of a horizontal section through T1 (Collection Dr. Şerban Georgescu – "Fundeni" Hospital)

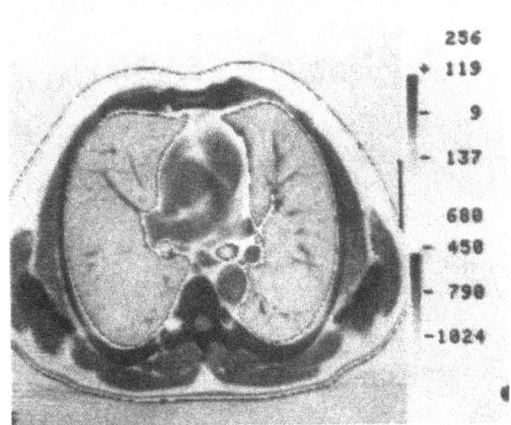

Fig. 301 Computed tomography at the level of a horizontal section through T6 (Collection Dr. Şerban Georgescu – "Fundeni" Hospital)

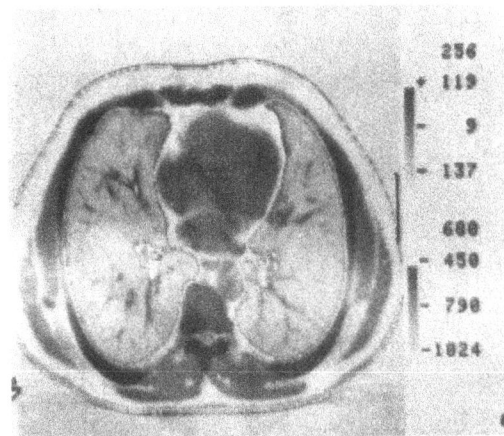

Fig. 302 Computed tomography at the level of a horizontal section through T7 (Collection Dr. Şerban Georgescu – "Fundeni" Hospital)

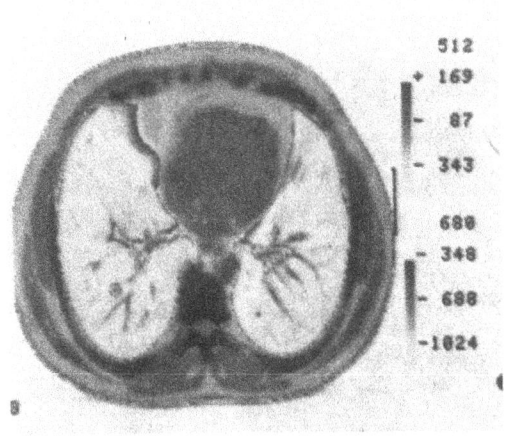

Fig. 303 Computed tomography at the level of a horizontal section through T8 (Collection Dr. Şerban Georgescu – "Fundeni" Hospital)

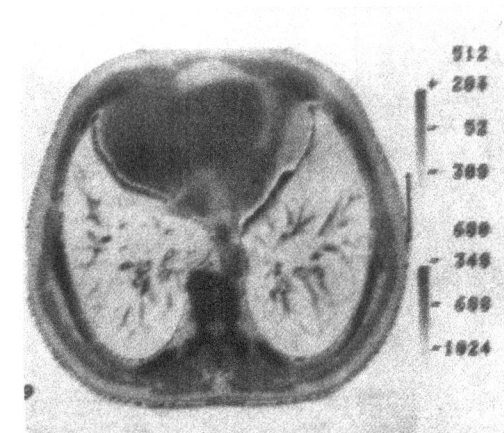

Fig. 304 Computed tomography at the level of a horizontal section through T9 (Collection Dr. Şerban Georgescu – "Fundeni" Hospital)

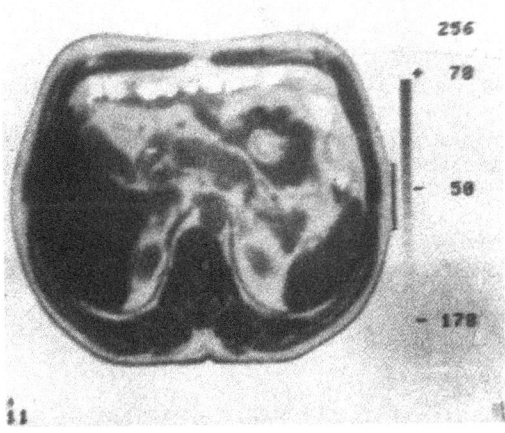

Fig. 305 Computed tomography at the level of a horizontal section through T11 (Collection Dr. Şerban Georgescu – "Fundeni" Hospital)

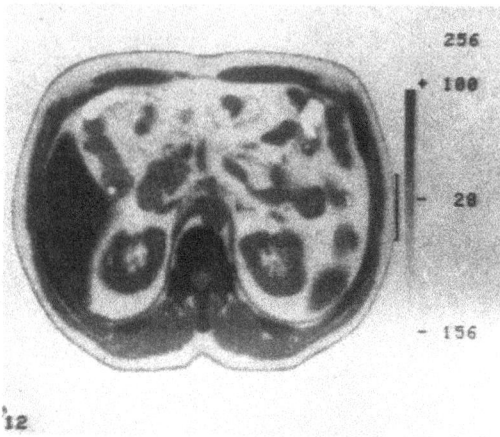

Fig. 306 Computed tomography at the level of a horizontal section through T12 (Collection Dr. Şerban Georgescu – "Fundeni" Hospital)

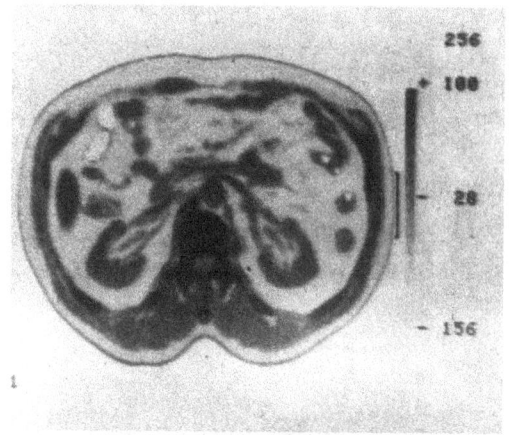

Fig. 307 Computed tomography at the level of a horizontal section through L1 (Collection Dr. Şerban Georgescu – "Fundeni" Hospital)

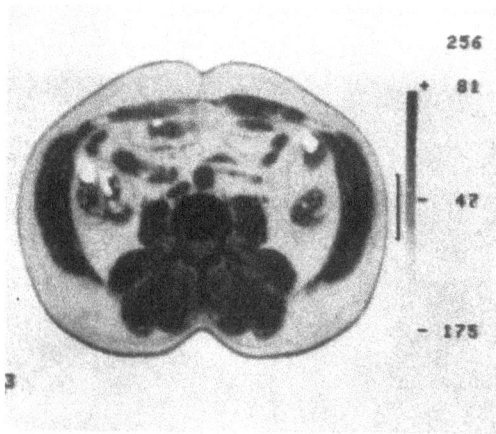

Fig. 308 Computed tomography at the level of a horizontal section through L3 (Collection Dr. Şerban Georgescu – "Fundeni" Hospital)

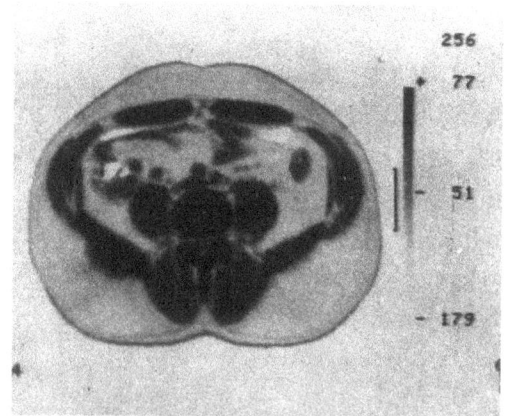

Fig. 309 Computed tomography at the level of a horizontal section through L4 (Collection Dr. Şerban Georgescu – "Fundeni" Hospital)

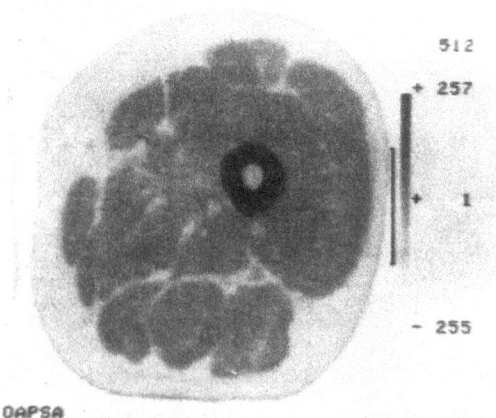

Fig. 310 Computed tomography at the level of the middle region of the thigh (Collection Dr. Şerban Georgescu – "Fundeni" Hospital)

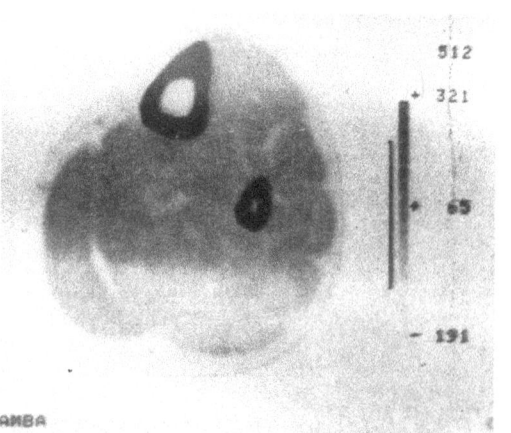

Fig. 311 Computed tomography at the level of the middle region of the thigh (Collection Dr. Şerban Georgescu – "Fundeni" Hospital)

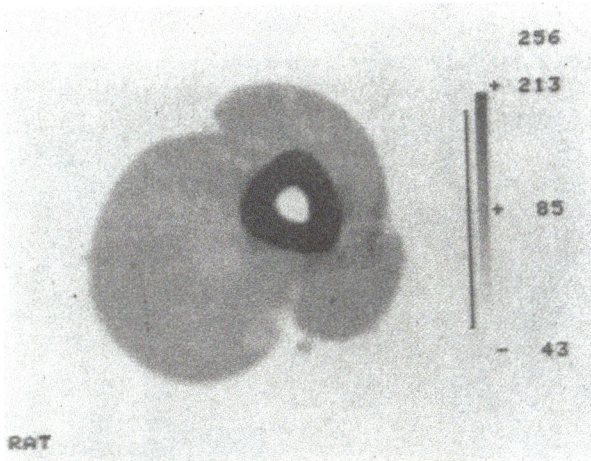

Fig. 312 Computed tomography at the level of the middle region of the arm (Collection Dr. Şerban Georgescu – "Fundeni" Hospital)

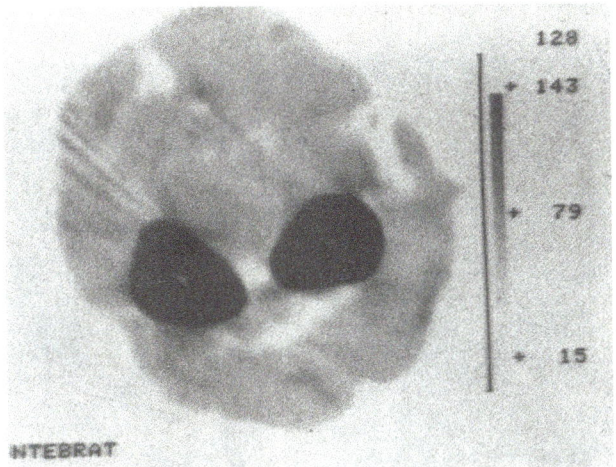

Fig. 313 Computed tomography at the level of the middle region of the forearm (Collection Dr. Şerban Georgescu – "Fundeni" Hospital)

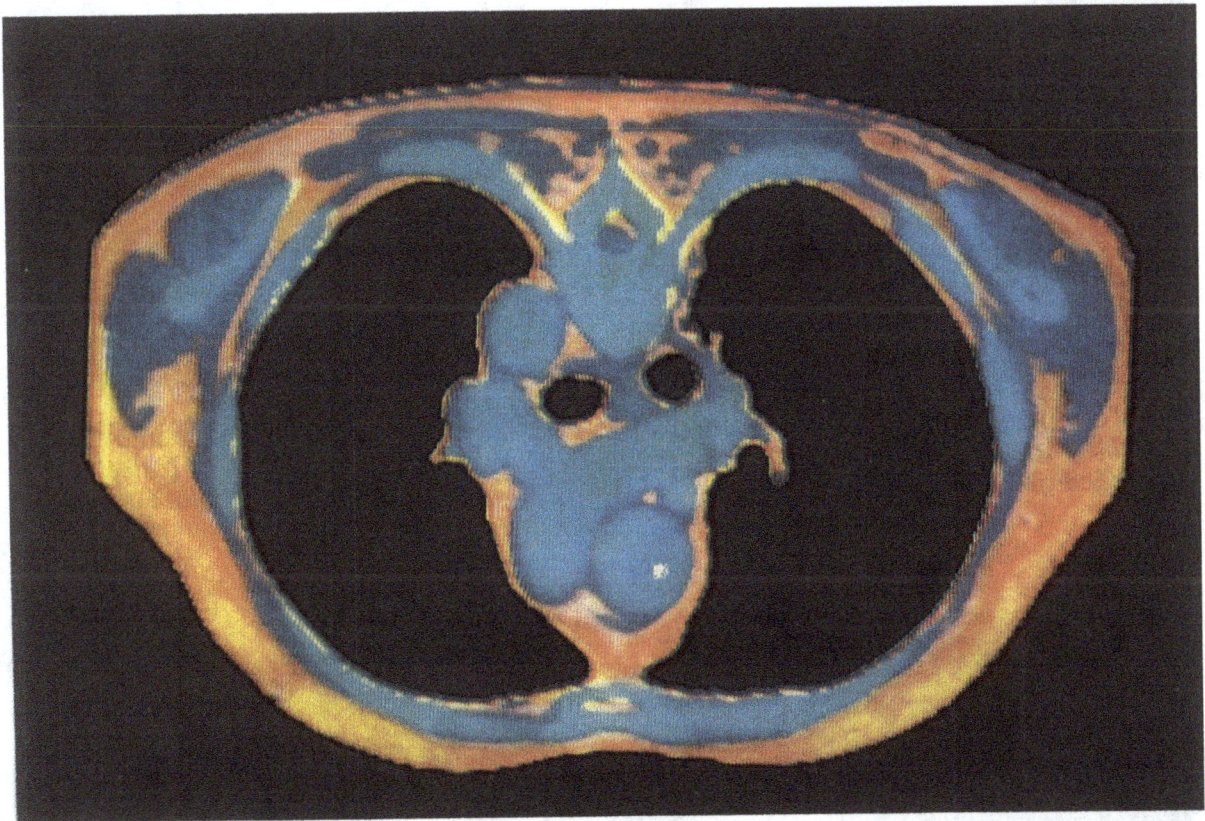

Fig. 314 Computed tomography (colour) at the level of a horizontal section through T6 (Siemens Collection)

293

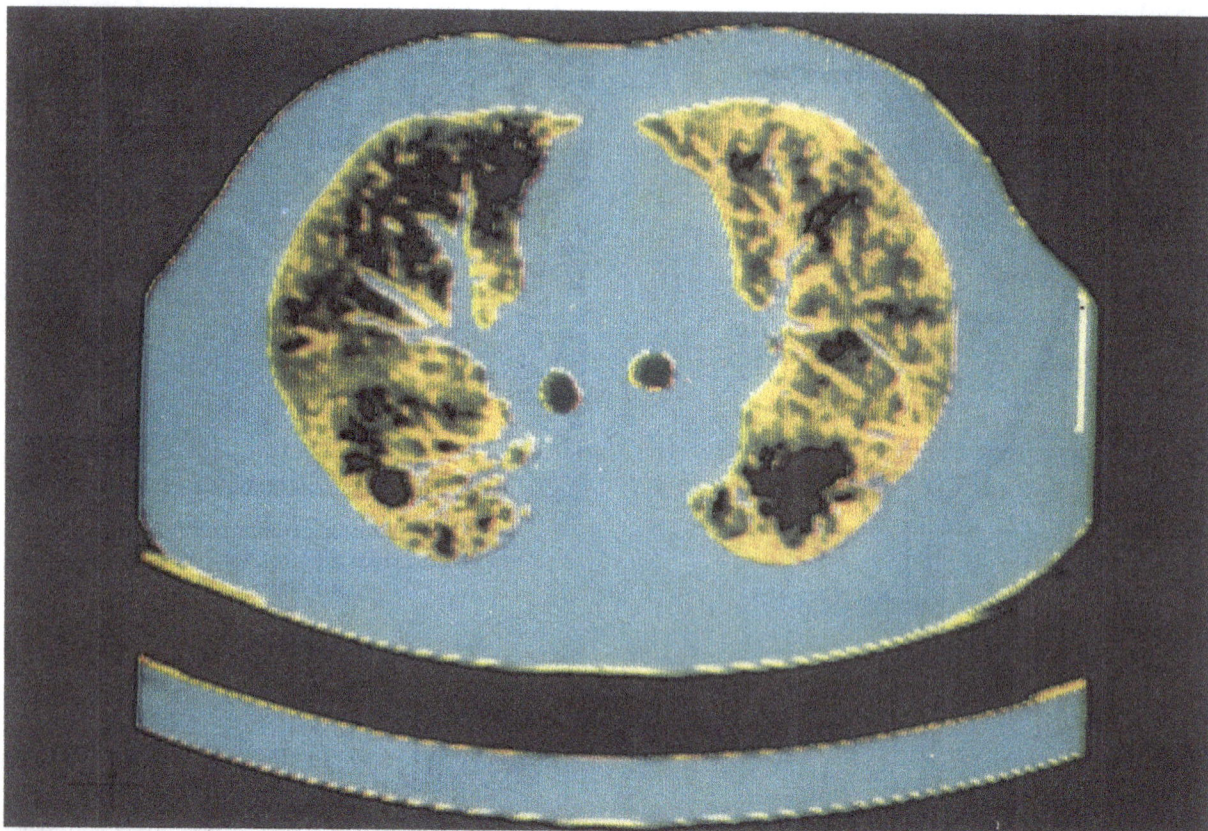

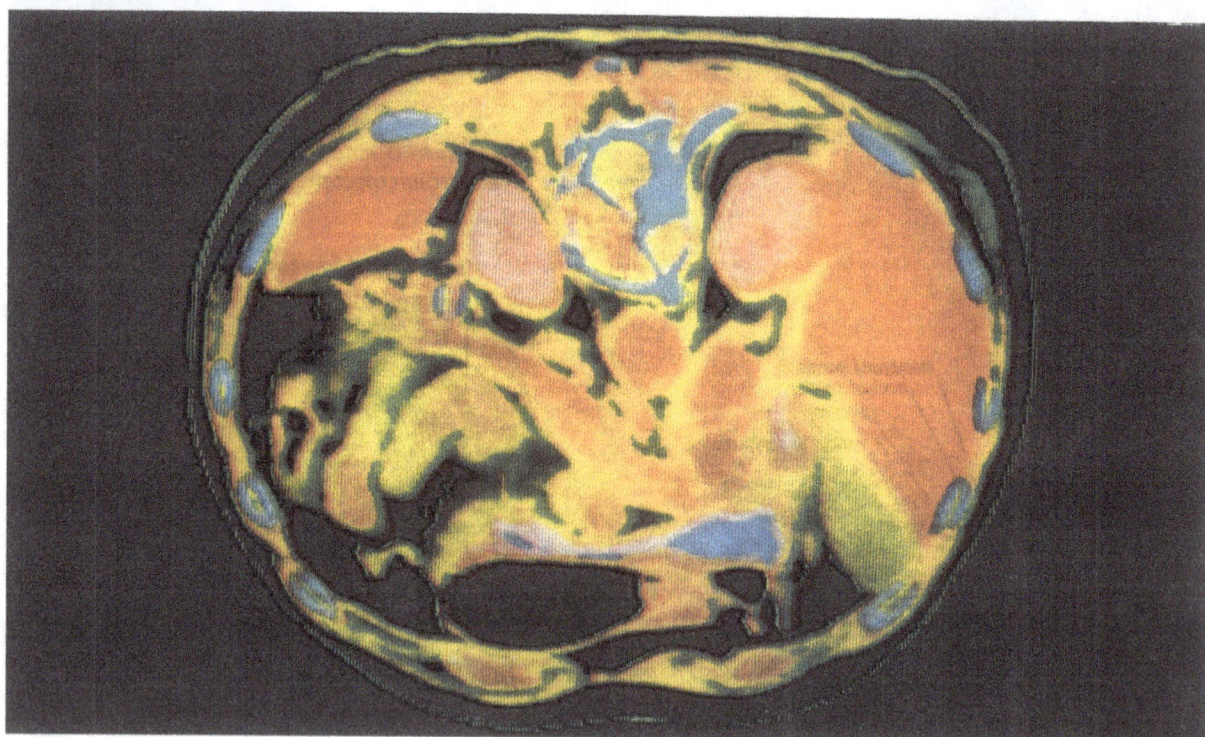

Fig. 316 Computed tomography (colour) at the level of a horizontal section in
the area between T11 and T12 (Siemens Collection)

294

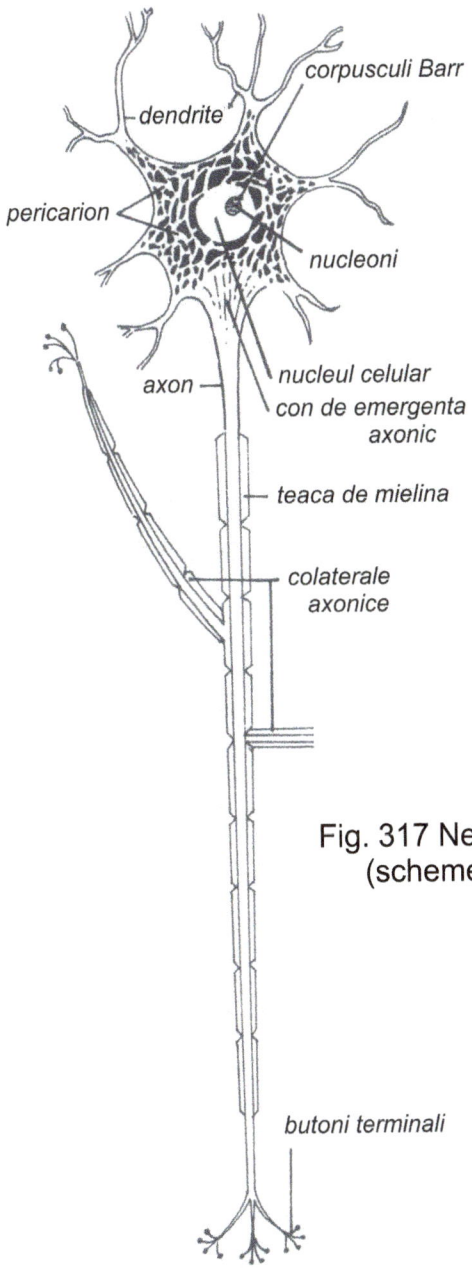

Fig. 317 Neuron (scheme)

- dendrite
- corpusculi Barr
- pericarion
- nucleoni
- axon
- nucleul celular
- con de emergenta axonic
- teaca de mielina
- colaterale axonice
- butoni terminali

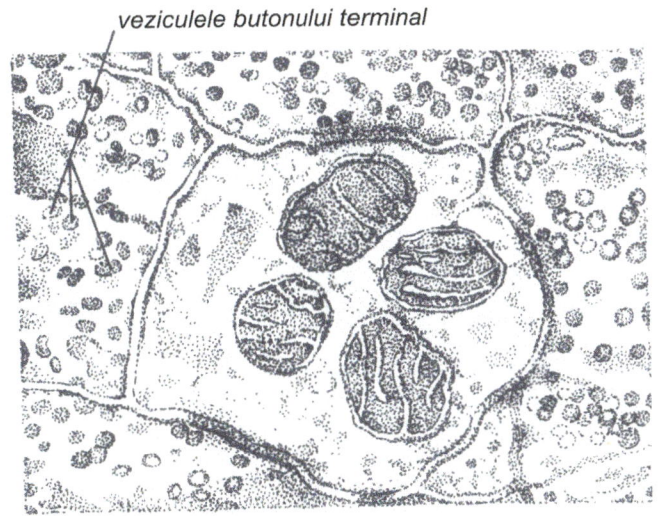

veziculele butonului terminal

Fig. 319 Neuron with synapses – electron microscopic aspect (scheme after Bek)

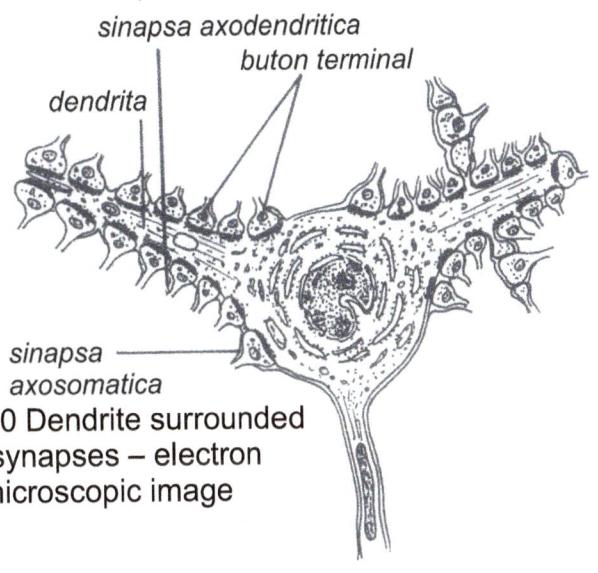

sinapsa axodendritica
buton terminal
dendrita
sinapsa axosomatica

Fig. 320 Dendrite surrounded by synapses – electron microscopic image

Fig. 318 Ultrastructure of the nerve cell

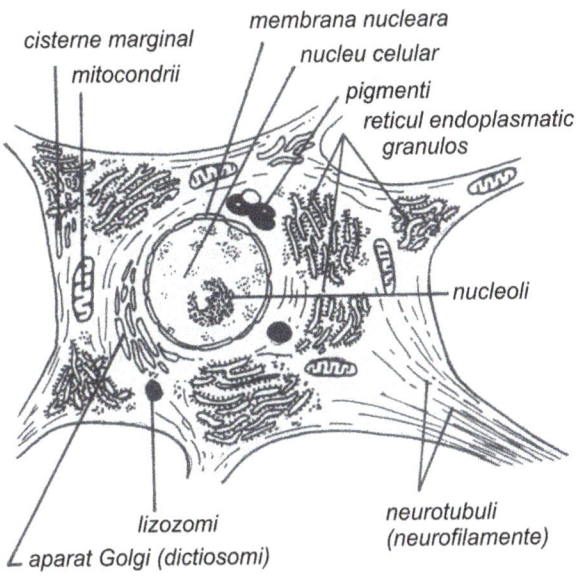

- cisterne marginal
- mitocondrii
- membrana nucleara
- nucleu celular
- pigmenti
- reticul endoplasmatic granulos
- nucleoli
- lizozomi
- aparat Golgi (dictiosomi)
- neurotubuli (neurofilamente)

Fig. 321 Peripheral nerve fibre (electron microscopic aspect) (after M. Schröder)

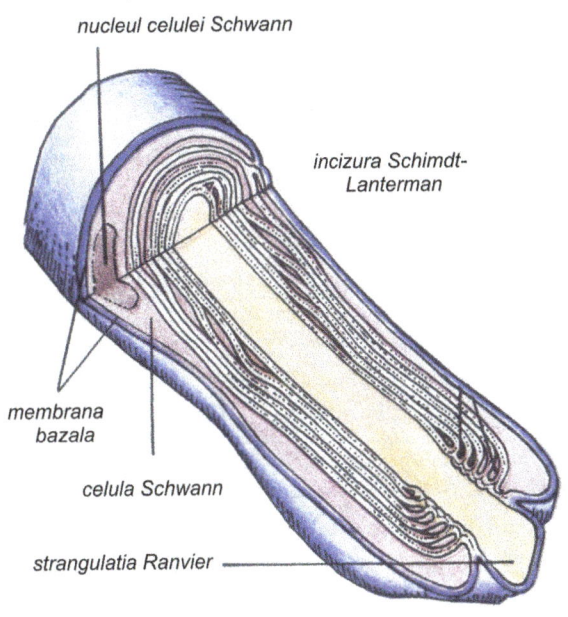

- nucleul celulei Schwann
- incizura Schimdt-Lanterman
- membrana bazala
- celula Schwann
- strangulatia Ranvier

295

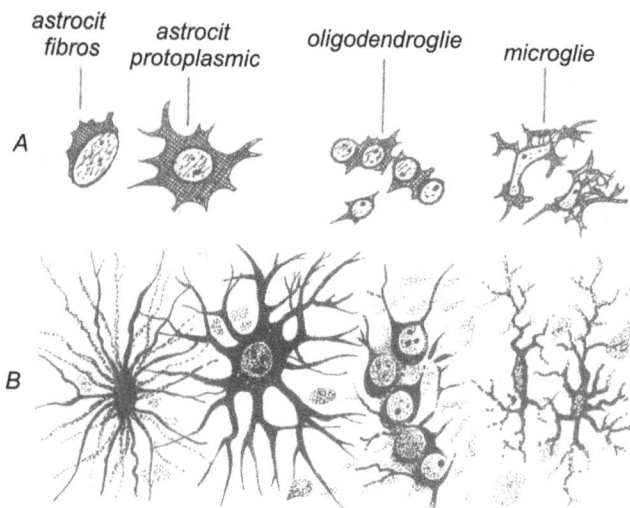

Fig. 322 Neuroglia – types
A – Nissl staining; B – silver
impregnation

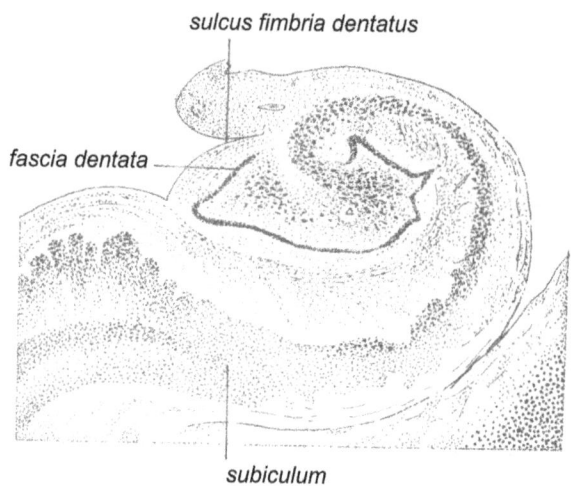

Fig. 323 Frontal section through the
hippocampus (microscopic structure)

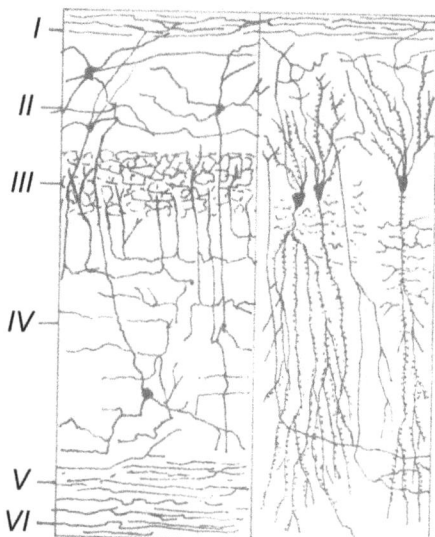

Fig. 324 Structure of the
hippocampal cortex – silver
impregnation (after Cajal)

Fig. 325 Cortical layers – microscopic
structure
A – silver impregnation; B – cell staining;
C – staining for myelin

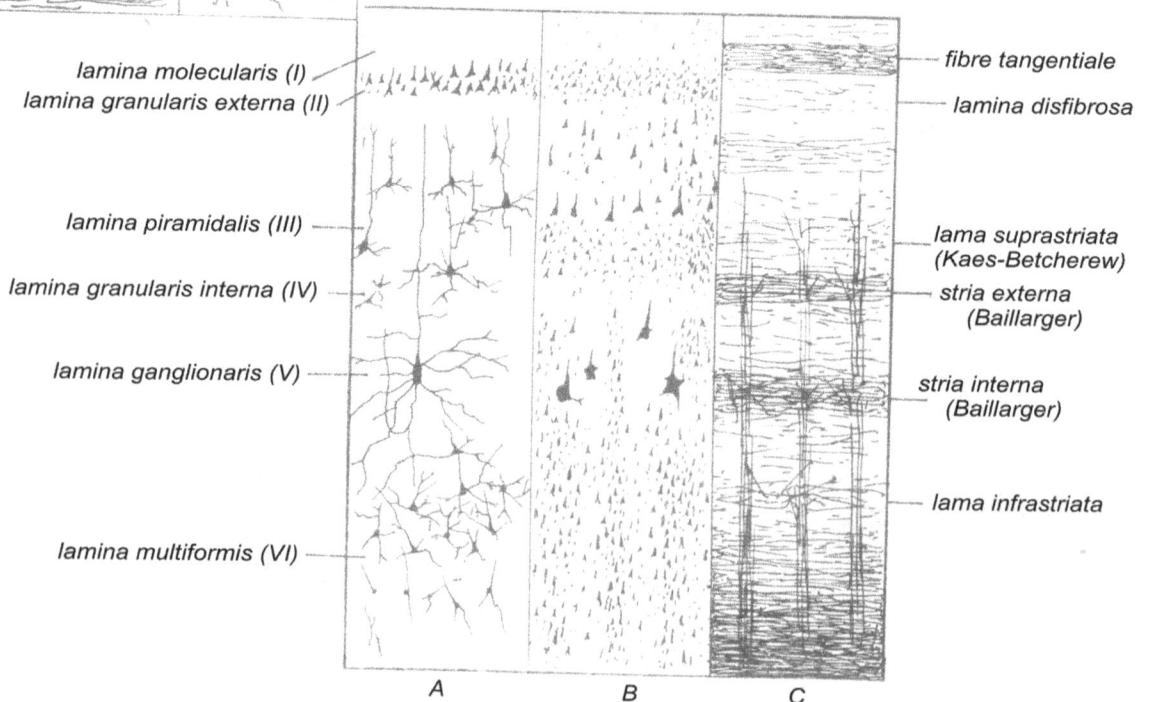

296

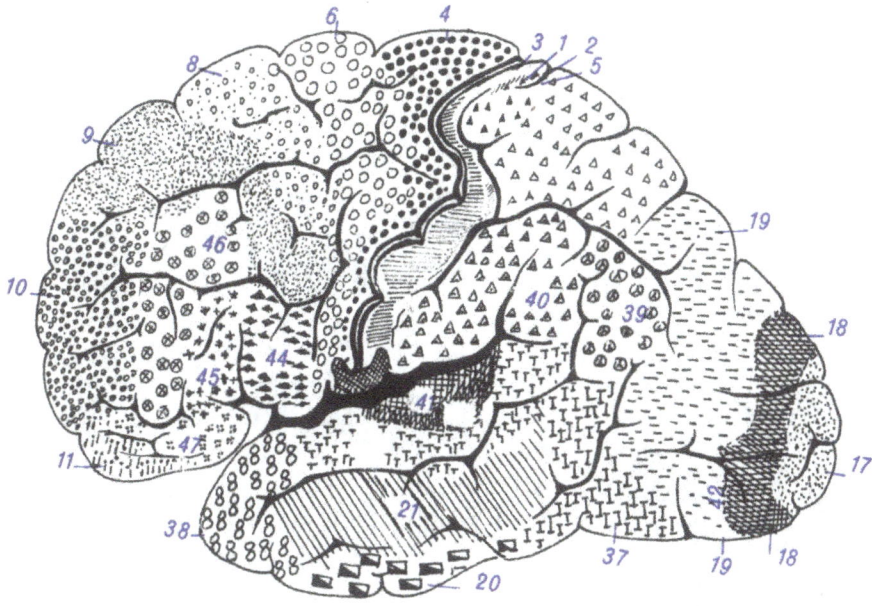

Fig. 326 Brodmann's cortical areas (lateral view)

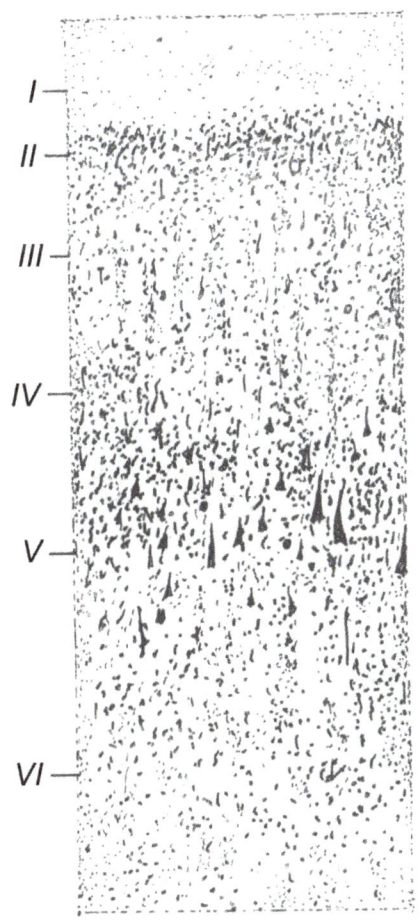

Fig. 327 Cortical areas, after Brodmann (medial view)

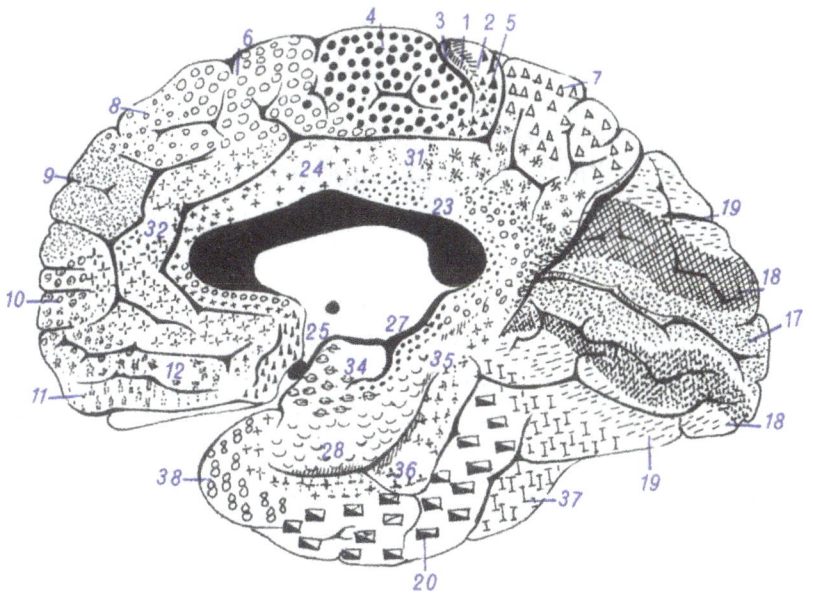

I
II
III
IV
V
VI

Fig. 328 4th and 6th precentral cortical areas
(microscopic structure)

297

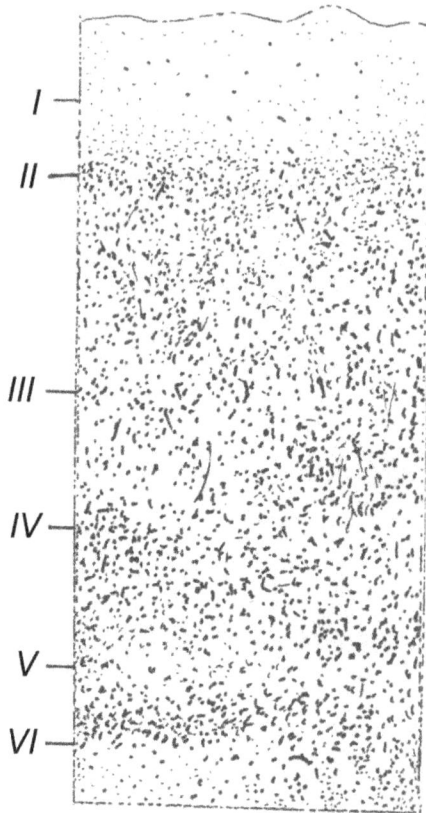

Fig. 329 Sensory cortical area
(area 3), the microscopic structure

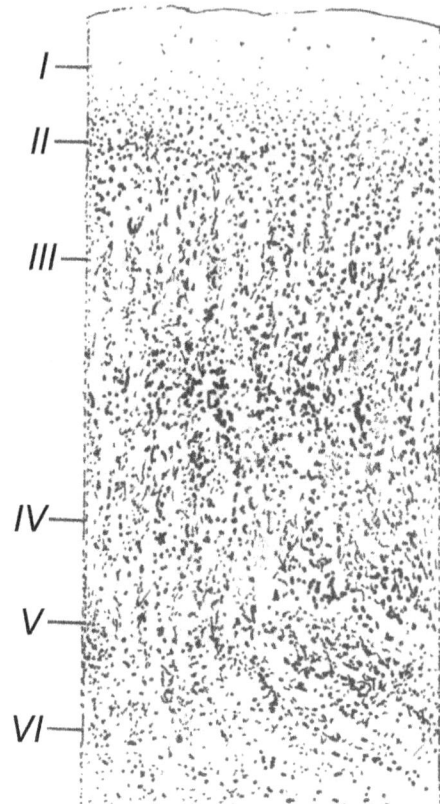

Fig. 330 Auditory cortical area
(area 41, the microscopic
structure)

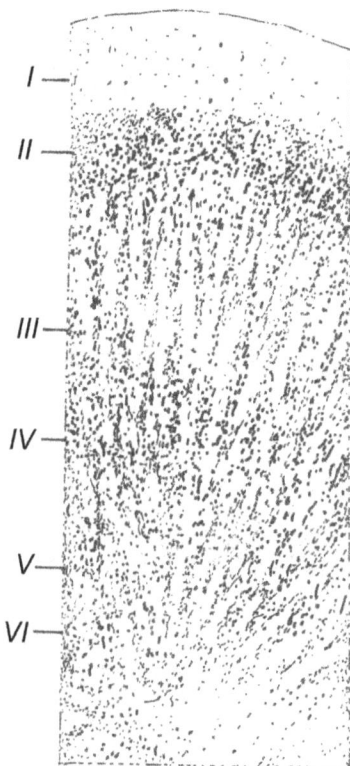

Fig. 331 Auditory cortical area
(area 21, the microscopic
structure)

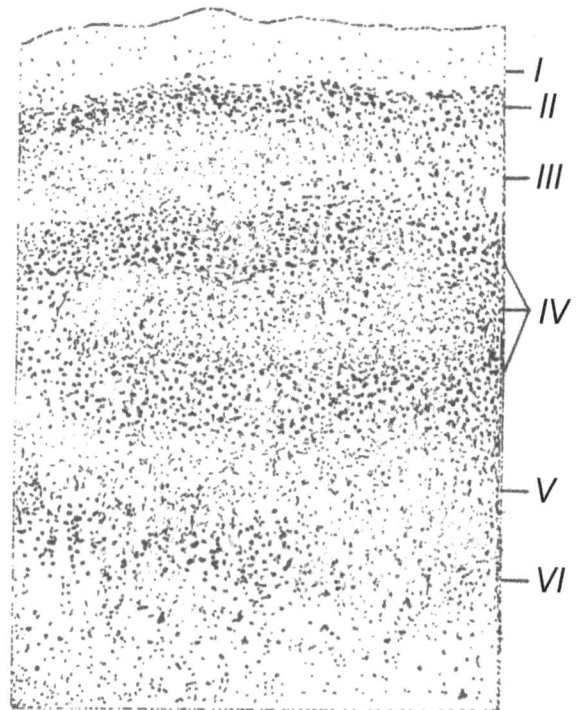

Fig. 332 Visual cortical area (area 17,
the microscopic structure)

Notion of clinical anatomy in vivo

The vertebral column and the spinal cord

The vertebral column *(columna vertebralis)* is the resistance and support axis of the body, which assures and maintains the bipedal position and which, by its biomechanical structure, assures the flexibility required for movements. Moreover, it affords protection to spinal cord. The resistance of the vertebral column is due to the existence of the two bony systems: one anterior, formed of the succession of vertebral bodies and one posterior, constituted by the superimposition of the vertebral arches, laminae and spinous processes. The elasticity is due to the presence of the intervertebral discs. These elements make up the structure called by Rainer the ventral pillar, to which is added the dorsal plate of the vertebral arches united by articulations of the respective vertebral processes and by the yellow ligaments *(ligamentum flavum)*.

The main part bears the weight of the trunk, a fact materialized in the increase of the section surface downwards up to the sacrum.

The dorsal plate has the role to maintain the curvatures of the vertebral column by means of the yellow ligaments and, moreover, to guide the direction of movements and the orientation of the articular processes.

The vertebral bodies, are adapted, by their structure, to resist to overloading, while the discs afford flexibility to the column, without diminishing its resistance capacity to compression.

As regards the materialization of a functional structure the fibrous apparatus assures the unity of the segments of the column, while the vertebrae may be considered as strengthening zones and insertion points of the local anatomical formations. Owing to this fact, Kapandji (1972) compares this structure to the mast of a ship.

The vertebral column has a sinuous course, with four curvatures: two sagittal and two frontal curvatures.

The examination of the vertebral column is performed with the subject undressed, in upright posture, viewed from the front, from the back and from profile, a symmetrical aspect being observed in the sagittal plane, on the basis of the anatomical landmarks (Fig.333).

On the posterior median line, the C7 spinuous process and, the vertebrae to which correspond various other landmarks, may be palpated (Fig.334).

The vertebral column presents physiological curvatures which are more obvious in the sagittal than in the frontal plane. Those in the frontal plane are termed scolioses and those in the sagittal plane, *kyphoses* (Fig.335).

According to Staffel's classification, there are five types of normal aspect of the spine (Fig.336).

The scoliotic atitude may be either vicious or real, it disappears in flexion, in the first case, or it becomes more marked in the case of a constituted lesion (Fig.337).

The range of movements is assessed both on the whole and by segments. To end, the Ionitzki method for the appreciation of the flexion, lateral bending, extension and rotation movements is used in the clinic (Fig.338-341).

For the numbering of the thoracic spinous processes it is advisable to perform their palpation from below upwards – the *"stairs sign"* – and in this way the less obvious unevenness may be identified (Fig.342).

According to Brueger the following vertebral functional segments may be differentiated: an anterior pillar, which plays the role of support (static role) and a posterior, articular pillar (dynamic role).

Schmorl considers that there is a passive segment – the vertebral body – and a motor segment formed of the intervertebral disc, the conjugate foramen, the interspinous articulations, the yellow ligament and the interspinous ligament.

It is believed (Kapandji a.o.) that between *the interior and the posterior pillar there is a functional relationship*, assured by the vertebral pedicles and the interspinous articulation, which permits the absorption of the axial compression stress exerted upon the vertebral column: *a direct and passive absorption* at the level of the intervertebral disc and *an indirect and active absorption* at the level of the muscles of the intervertebral grooves.

The intervertebral disc plays a significant role in the performance of the movements of the vertebral column. In the normal state, a pressure of the fibrous ring is exerted on the nucleus of the disc. The axial elongation diminishes the pressure inside the nucleus (basis of the treatment by elongation applied in herniations of the intervertebral discs) and, consequently the gelatinous substance permits the nucleus to resume its spherical form. The axial compression produces the crushing and widening of the disc and to the flattening of the nucleus.

The various movements of the vertebral column bring about changes of the disc, as follows: the flexion produces the migration of the disc backwards, while the axial rotation tears the fibrous ring, increases the pressure and injuries the nucleus.

The normal freedom degree of the movements of the vertebral column is assessed by radiological examination. Both profile and front view radiographies are used. We reproduce the data presented by Kapandji.

The profile radiograph shows the following values:

The segmental range:

-at the level of the lumbar column:

-flexion 60°;

-extension 35°;

-at the level of the thoracolumbar column:

-flexion 105°;

-extension 60°;

-at the level of the cervical column:

-flexion 40°;

-extension 75°.

Total range: flexion of the vertebral column is 110° and its extension is 140°.

The front radiograph shows the total segmental range of the lateral inflexion:

- lateral inflexion of the lumbar vertebral column = 20°;

- lateral inflexion of the thoracic vertebral column = 20°;

- lateral inflexion of the cervical vertebral column = 35°-45°;

- lateral inflexion on the whole, including the sacrum and the skull: 75°-85°.

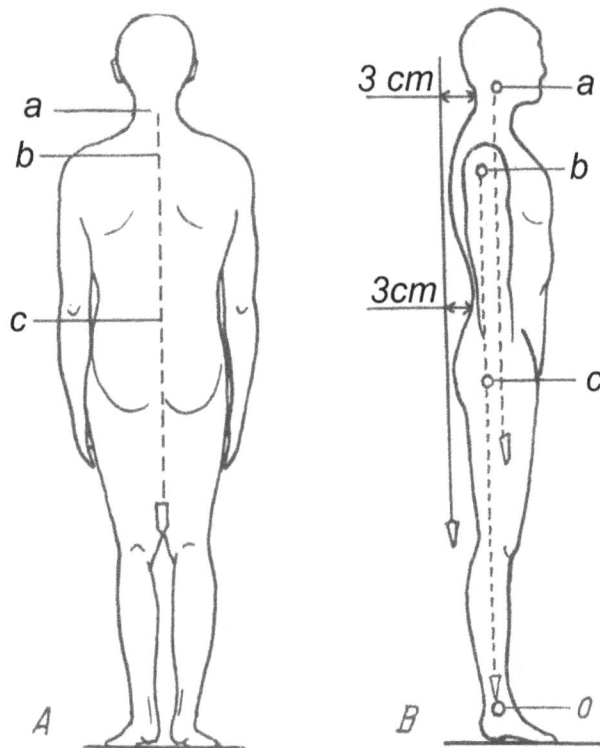

Fig. 333 Normal axis of the spine: A) The posterior (dorsal) visible aspect; a) external occipital protuberance, b) C7 vertebra, prominent (vertebra proeminens) inter-fesier fold. B) The profile visible aspect; a) tragus, b) the mean of external face of the arm, The b-c-o line joining the shoulder, the greater trochanter and the external malleolus.

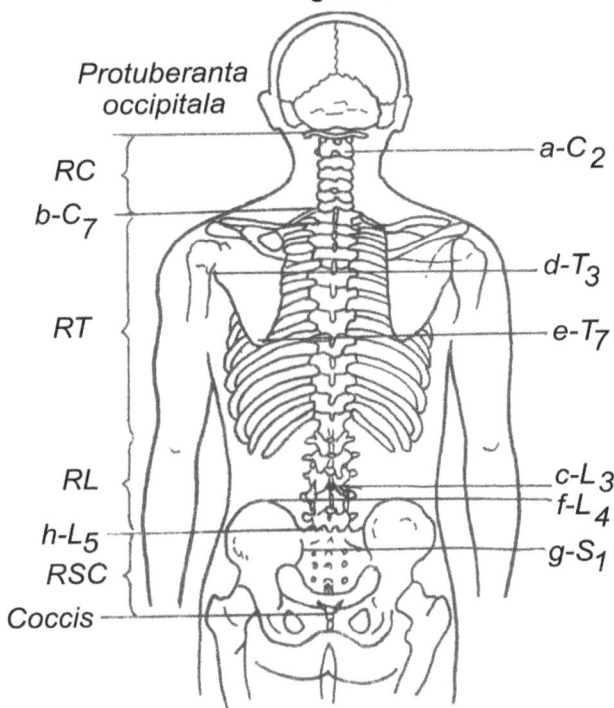

Fig. 334 Landmarks and topographical distribution of the spine -posterior view:
a- C2 spinal apophyses of axis, b-C7 vertebra protruding, c-L3 spinal apophyses of L3 (lumbar prominence), d-T3, T3 corresponds to the scapular bispinal line, e-T7, biangular scapular line corresponds to T7, f-L4, L4 corresponds to the bicreastal line, g-S1, S1corresponds to the SIPS's bispinal line, h-L5 - The previous superjacent line corresponds to L5.

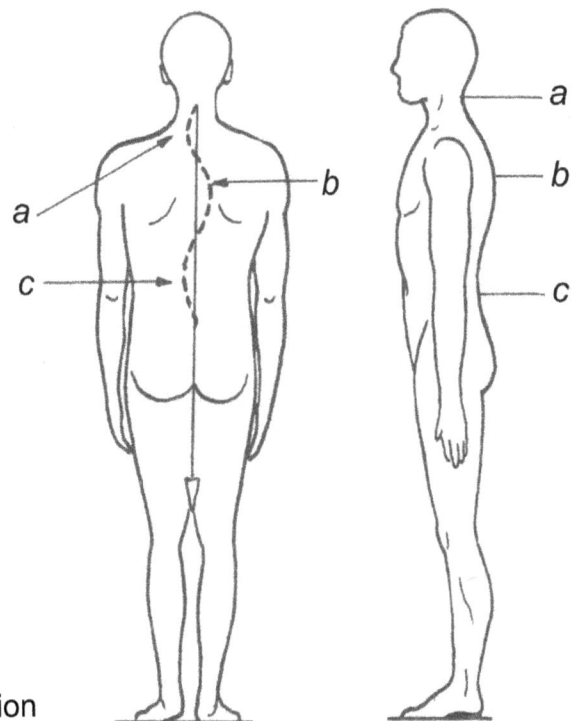

Fig. 335 Normal curvatures of the vertebral column in the frontal (A) and sagittal (B) plane
a – cervical curvature;
b – dorsal curvature;
c – lumbar curvature;

301

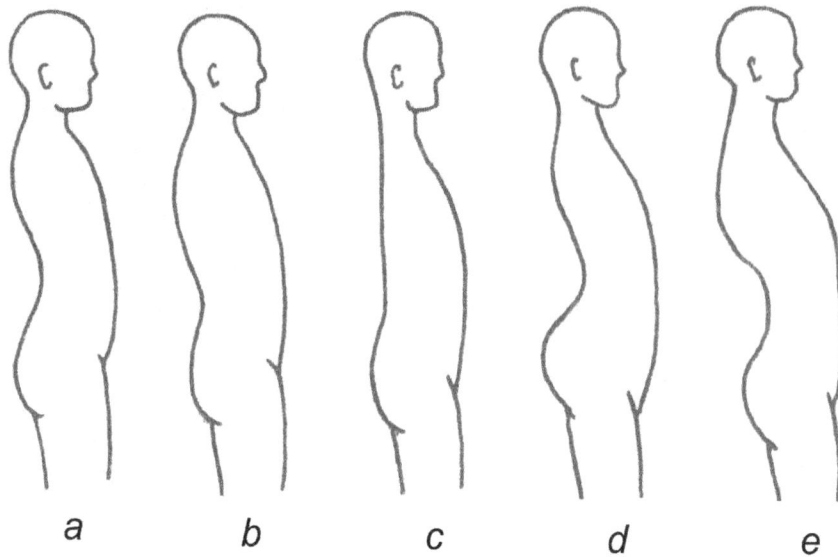

Fig. 336 Curvature types of the vertebral column
a – normal curvature; b – round back; c – flat back; d – flat-concave back;
e – round-concave back;

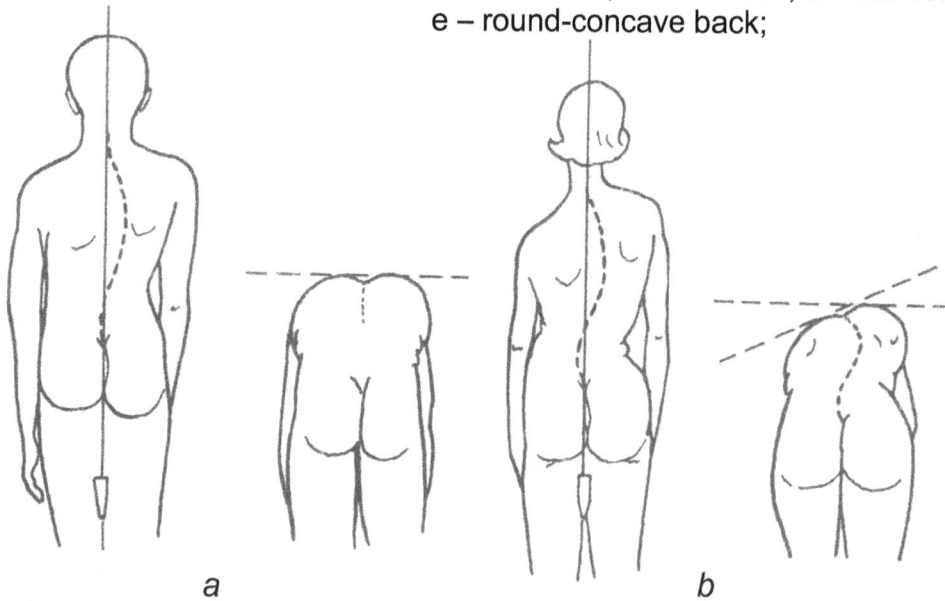

Fig. 337 Scoliosis
a- vicious attitude; b- fixed attitude

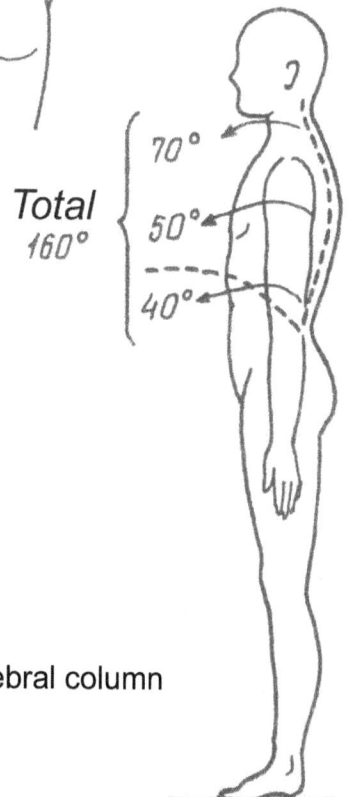

Total
160°

70°

50°

40°

Fig. 338 Amplitude of the flexion movement of the vertebral column

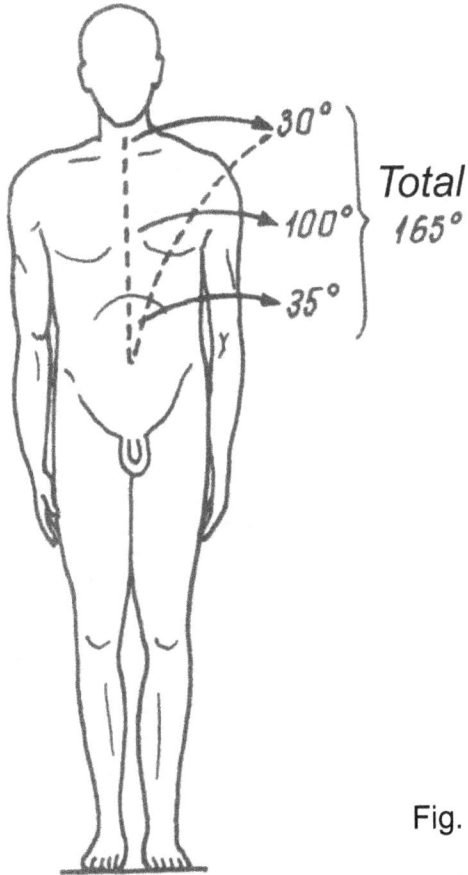

Fig. 339 Amplitude of the lateral bending movement

Fig. 342 Examination of the spinous processes – ladder-step sign

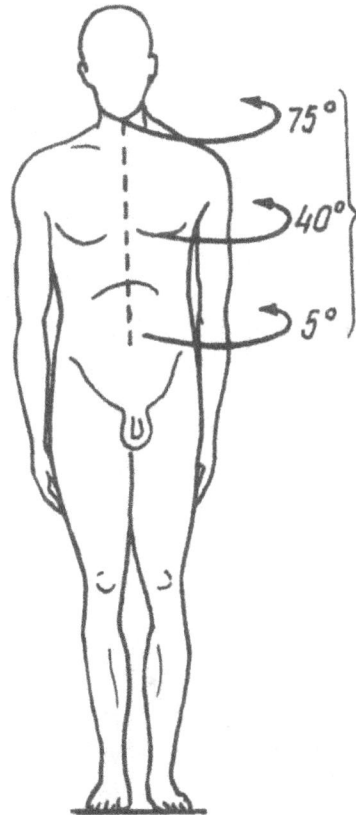

Fig. 340 Amplitude of the extension movement

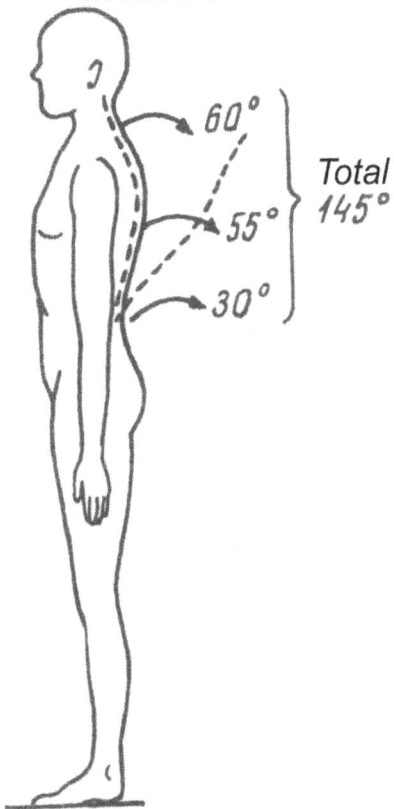

Fig. 341 Amplitude of the rotation movement

303

Investigation on the motility

Owing to the fact that the dissociation by segments of the cerebrospinal axis permits a clear analysis of the motility, we shall refer both to the voluntary and to the involuntary motility, as the anatomical in vivo study includes the motility in general.

Voluntary motility

The determination of the voluntary or active motility involves a complex functional system, directed by the cerebral cortex, which has been previously described. It comprises all the physiological movements (flexion, extension, abduction, adduction etc.) which are performed under voluntary control, both at the level of thoracic and pelvic limbs. Movements will be performed concomitantly (by both limbs), by segments, at a different rhythm, and their strength and range will be assessed.

The muscle force is evaluated by segments, comparatively and systematically, the examiner opposing resistance to the performance of certain movements which the patient is requested to carry out. In the framework of this examination are included several tests:

- *The test of stretched out arms*: the subject stretches forward both thoracic limbs, with the palms upwards, and closes the eyes. Normally, the limbs remain symmetrical, in the required posture. Pathologically (paralysis), the respective limb will descend;

- *The Grasset test*: the subject lies in dorsal decubitus position, with the pelvic limbs lifted and far apart, forming a 45° angle with the plane of the bed. Normally, the limbs remain in the respective position for some minutes; pathologically, the affectted limb falls towards the bed;

- *The Mingazzini test*: the subject lies in dorsal decubitus, with closed eyes and the limbs stretched aut and apart from each other by about 10 cm; he is requested to perform a straight angle flexion of the leg on the thigh and of the thigh on the abdomen. Normally, the position remains unmodified; pathologically. The affected limb will descend;

- *The Barré test*: the subject lies in ventral decubitus, with the limbs held in the same position as in the former test. He is requested to perform the flexion of the leg on the thigh up to the obtention of an angle of about 60o. Normally, this position is achieved; pathologically, the leg descends towards the plane of the bed.

The muscle tone represents the state of permanent muscle tension; it is objictified by its resistance to passive movements, as well as by its characteristic extensibility and consistency, and is controlled by the myotatic reflex.

The myotatic reflex has as anatomical substrate the neuromuscular spindle, the afferent pathway, the spinal cord and the efferent pathway.

The neuromuscular spindle is formed of 8-10 muscle fibers, provided with sensory annulospinal or flower spray formations, and disposed parallel to the muscle fibers of the respective segment.

The afferent pathway is made up of the fibers arising from the muscular annulospinal formations and establish a synapse in the alpha motor neuron situated in the anterior horn of the spinal cord. The fibers arising from the flower spray endings establish a synapse in a neuron lying in the posterior horns and, through an intercalary neuron, run also the alpha motor neuron.

The efferent pathway starts from the alpha motor neurons situated in the anterior horns and penetrates in the muscle fibers, outside the neuromuscular spindles.

The muscle tone is maintained by two mechanisms: mainly, by the variations of passive stretches at the level of the striated muscles, which excite the annulospinal and flower spray endings and activate permanently the alpha motoneuron lying in the anterior horns; secondarily, by the feedback achieved by the collaterals of the alpha cells ending at the level of Renshaw cells: they are the gamma neurons situated in the anterior horns of the spinal cord, controlled by the supraspinal formations, send - through the gamma fibers – impulses to 304 the intrafusal fibres, increasing the intensity of discharges from the afferent pathways.

The Renshaw neuron sends back fibers to the alpha motoneurons and exerts an inhibitory action.

The resting tone is assessed by examining the muscle relief at rest, especially the gluteal folds which should have the same arrangement and depth. The following tests may be used:

- *The Stewart-Holmes test:* the subject performs a strong flexion of the forearm on the arm and the examiner opposes resistance for a few seconds, after which he releases suddenly the forearm of the subject. Normally, after 1-2 oscillations, the forearm enters in a resting state; in the case of hypotonia, the forearm will knock against the shoulder, while in hypertonia the forearm will not perform any oscillation;

- *The Foix-Thévenard test:* the subject, in upright posture, is oriented towards the examiner, who pushes him slightly backwards, permitting in this way the visualization of the anterior tibial muscle. The absence of this sign shows muscle hypotonia.

In the clinic, *the voluntary motility* is assessed by the exploration of the osteotendinous and cutaneous reflexes, as follows:

- *The osteotendinous reflexes:* the tapping of the tendon of the muscle to be explored, is followed either by a simple muscle contraction or by the displacement of the corresponding segment. During the examination the musculature should be relaxed (conversation, tightly clinching of the fists if this territory is not examined etc.).

- *The biceps (C5-C6) reflex:* the tapping of the tendon of the biceps at the level of the bend of the elbow, with the forearm flexed, causes the flexion of the forearm on the arm.

- *The triceps (C6-C7) reflex:* the tapping of the tendon of the triceps above the olecranon, with the forearm in a 90o flexion on the arm, causes the extension of the forearm.

- *The styloradial (C5-C6) reflex:* the tapping of the styloid process of the radius, with the forearm in a slight pronation and flexion, produces the flexion of the forearm on the arm.

- *The ulnopronator (C7-C8) reflex:* the tapping of the styloid process of the ulna, with the forearm in a slight pronation and flexion, causes the pronation of the forearm.

- *The pectoral (C5-T1) reflex:* the tapping of the tendon of the greater pectoral muscle causes the abduction and inner rotation of the arm.

- *The mediopubic (T6-T12; L2-L3) reflex:* the tapping of the pubic bone, with the subject in dorsal decubitus, causes the abduction of the thighs and the contraction of the rectus abdominis muscle.

- *The patellar (L2-L4) reflex:* the tapping of the patellar tendon, with the legs slightly flexed (the subject is sitting on a chair), causes the extension of the leg on the thigh through the contraction of the quadriceps.

- *The Achilles (S1-S2) reflex (tendo Achillis reflex):* the tapping the Achilles tendon (tendo calcaneus), with the leg flexed on the thigh, and the foot on the leg (the subject is kneeling on a chair), causes the plantar flexion on the foot.

- *The medioplantar (S1-S2) reflex:* the same position and the same response as to the Achilles reflex, but the tap is applied on the middle region of the sole.

- *The cutaneous reflexes* are obtained by the excitation of the skin with a needle, manoeuvre followed by the muscle contraction. Among the most frequent we mention:

- *The abdominal cutaneous reflexes:* the subject lies in dorsal decubitus, with the pelvic limbs in semiflexion and the abdominal musculature relaxed; the skin is excited with a needle from outside-inwards, at various levels: below the costal arch for the superior abdominal reflex (T6-T7); at the level of the umbilicus for the medio-abdominal reflex (T8-T9) and at the level of the inguinal fold (plica inguinalis) for the inferior abdominal reflex (T10-T12);

- *The cremasteric (L1-L2) reflex:* the excitation of the integument on the supero-internal surface of the thigh causes the elevation of the homolateral testis; the subject lies in dorsal decubitus, with a slight abduction and external rotation of the thighs;

- *The plantar cutaneous (S1-S2) reflex:* the excitation of the external margin of the sole, from the heel to the digits, produces the plantar flexion of the digits. Pathologically – Babinski's sign -, extension of the hallux.

305

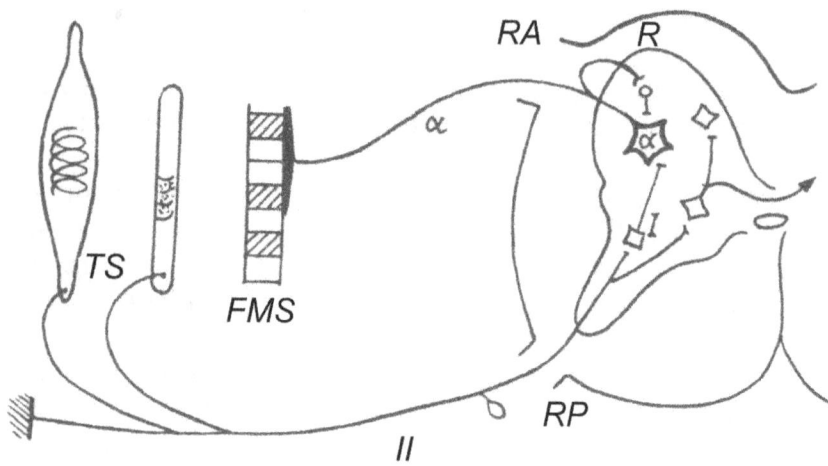

Fig. 343 Scheme of the motor unit (after F. Coutamine and O. Sabouraud)
AH – anterior horn; alfa-motoneuron; AR – anterior root; PR – posterior root; MF – motor fibre; MP – motor plate; SMF – striate motor fibre

Fig. 344. Motor unit scheme (after F. Coutamine and O. Sabouraud):
CA - anterior horn; α - alpha motoneurons, RA - anterior root, RP - posterior root; FM - motor fiber; PM - motor plate, FMS - striated muscle fiber

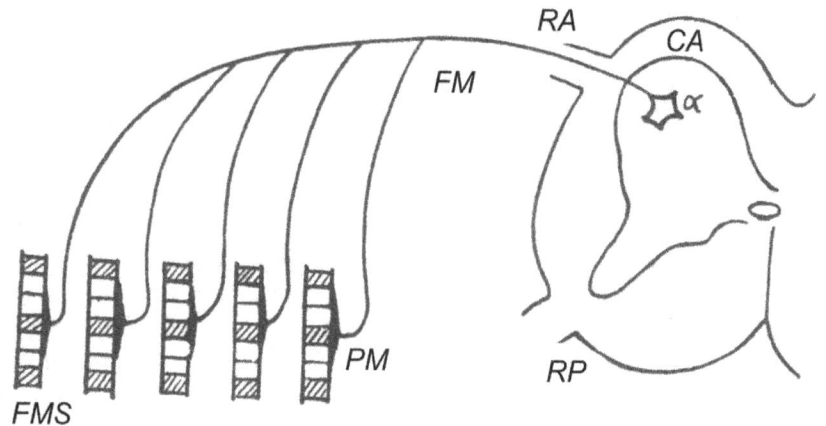

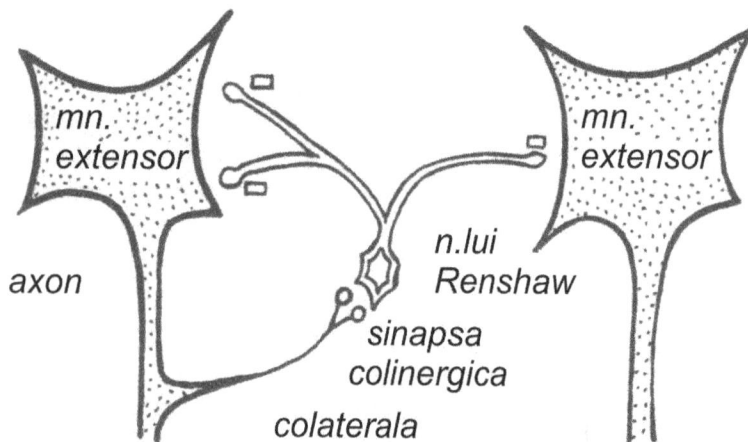

Fig. 345 Renshaw's neuron (after I. Cambier and M. Masson)

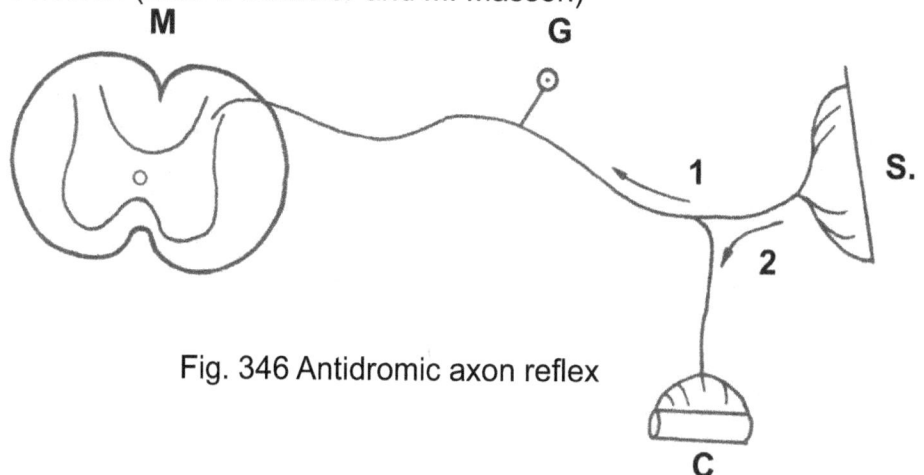

Fig. 346 Antidromic axon reflex

Involuntary motility

It appears independently of the subjects will, objectifies a lesion of the extrapyramidal system and is a pathological sign.

Investigation of the sensibility

It is an objective test, by means of which the state of the tegumental radicular innervation is established. It requires an attentive, patient and detailed examination; the eventual sensibility limits should be inscribed with the dermatographic pencil.

The superficial sensibility is examined comparatively on all the segments of the body, both on the posterior and the anterior surface, including the perineal region.

The tactile sensibility is examined qualitatively and quantitatively with a woollen pad or by investigation of Weber's fields or von Frey's zones (Fig.347).

The subject, with closed eyes, counts how many times he feels the touch with the woollen pad.

The Weber's test consists in the concomitant application of two tactile or painful (needle points) stimuli, of the same size and intensity, on the area to be explored, and in the assessment of the minimum distance up to which the subject perceives two touches or two pricks. At a lesser distance, he perceives only one touch or one prick. The respective distance represents the density of the sensory corpuscles and it is appreciated on the basis of a table.

The von Frey's test is performed with horse hairs of a various thickness and hardness. The area to be explored is touched with these hairs and the tactile sensibility is assessed.

Very reliable is the topography of analgesia to prick and deep pressure, which may detect a complete lesion of the respective nerve trunk.

Beside the painful sensibility (tested with the needle or the compass), the thermal sensibility is also investigated, by means of two test-tubes: one with warm water and the other with cold water. Throughout the examination, the subject has the eyes closed and counts how many times he perceives the thermal or painful sensation.

In the case of a sensibility dissociation, namely if the tactile sensibility disappears, whereas the painful sensibility is maintained, the nervous lesion is incomplete.

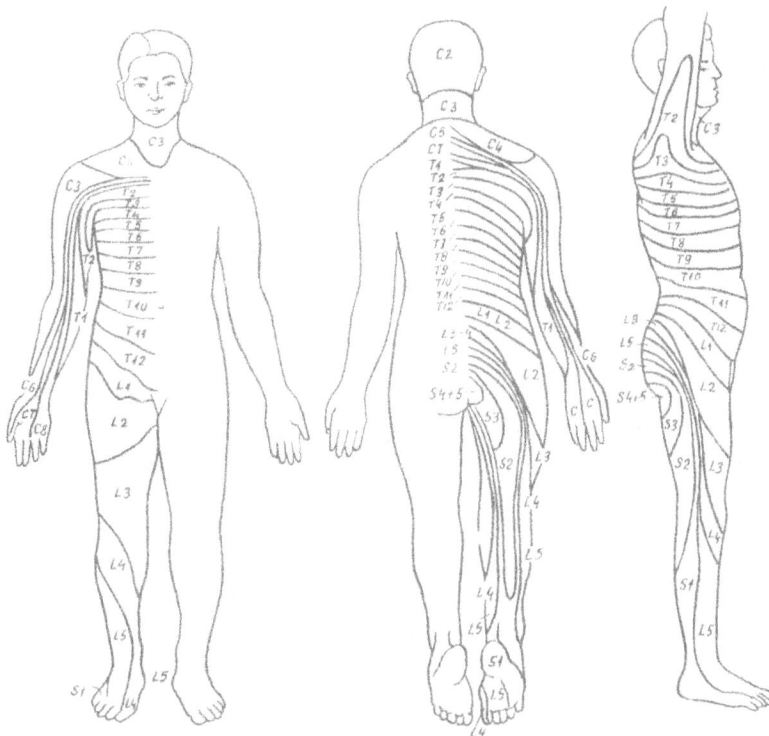

Fig. 347 Sensibility scheme showing the dermatomal distribution

307

The deep sensibility is assessed by means of a tuning fork applied at the level of bone prominences. Normally, the vibrations of the tuning fork are felt, but they may be absent, reduced or even stronger.

The sensibility reappears in the following succession: pain, sensibility to cold, then to heat and, finally, the tactile sensibility.

Investigation of the vegetative system

It permits, by the examination of the pilomotor reflex, of the state of local perspiration and of vasomotoricity, to assess its integrity or the presence of some impairments.

- *The pilomotor reflex* consists in hair erection, the skin acquiring the aspect of *"gooseflesh"* after the application of various cutaneous stimuli, either disagreeable - either pad, cold water, ice etc. – or painful – electric, galvanic, faradic current.

The reflex is limited to the respective dermatomas and stops, usually, at the midline.

- *The sweating reflex* is studied by injecting pilocarpine and observing the limit between hyperhidrosis and anhidrosis (lack of perspiration). The light bath may be also used, during which the examiners hand appreciates the area in which the skin begins to become moist (it corresponds to the site of the lesion).

The minor test : the body is painted with Lugol's solution or dilute iodinate alcohol; after drying, starch powder is applied.

The subject is introduced into a light bath, in order to induce perspiration and in the area where sweat is present a chestnut-brownish colour appears, whereas the anhidrotic area remains white.

- *The vasomotor reflex* (dermatographism or vasomotor stria): along the tegumental line traversed with a blurt-shaped device appears a white streak, accompanied on each side by erection of the hairs. After 30 seconds the white streak becomes red-coloured, but after 30 minutes the redness becomes diffuse and the skin bulges.

Plain radiological examination of the vertebral column

It is aimed at identifying the bone structures or the eventual congenital or acquired lesions. Usually, front and profile radiographs, centred the region to be examined, are performed. The data obtained are integrated in the general context of changes of the vertebral statics and dynamics.

The shape, the outlines and the intimate structure of bones will be examined.

The localized deformities may acquire various forms, from the cuneiform aspect, specific to the vertebral kyphoses or consecutive to injuries, to caries (Pott's disease – vertebral tuberculosis) and to the impressions of the pulpous in the framework of vertebral herniations, up to vertebral or polyvertebral blocks, caused by congenital diseases (Sprengel-Klipel-Feil syndrome: plurivertebral block, spina bifida, elevation of the scapula), by injuries or by injections.

The wearing away of the outlines of the vertebral bodies may occur after infectious spondylodiscal or tumoral, osteolytic or per contiguum processes.

The intimate bony structure may be modified as a consequence either of the resorption of bony tissue (osteolysis) or of a limited (Paget's disease) or diffuse (opaque vertebra – osteoid osteoma, prostate metastasis etc.) condensation. Often appear *osteolytic lacunae*, which impart the aspect of *"spotted"* vertebra (myeloma).

As regards the anomalies of the vertebral bodies, we mention somatoschisis (the vertical division of the vertebra into two parts, a butterfly-like aspect), rachischisis – spina bifida (cleft of the posterior neural vertebral arch, which may be partial or total, uni- or plurivertebral, occult or accompanied by clinical signs), spondylolytis (vertebral gliding), ranging from spondylolisthesis to spondylosteosis and discal agenesis (congenital vertebral block).

Radiological examination of the vertebral canal with contrast media

When the plain radiological examination does not permit the detection of the lesion of the vertebral column, especially in medullary or radicular compressions, it is necessary to have recourse to gas myelography, myelography with contrast media, medullary angiography and vertebral phebography.

- *The gas myelography* is used for the detection of disc herniations (cervical, thoracic or lumbar) and vertebromedullary myelopathies. After the extraction, by lumbar puncture, of 25-30 ml of CSF, an equivalent amount of air is introduced with the same needle and front, ¾ profile radiological images in Trendelenburg's position (the head lower then the pelvic limbs) are obtained. Although the radiographic signs are more evident following the introduction of a larger amount of air, it is preferable to avoid this technique, since it distends the dural sac, irritates the meningeal membranes and the appearance of contractures and major seizures is possible.

- *The myelography with contrast media* allows the examination of the subarachnoid spaces with Lipiodol and is indicated only when the other means do not permit to make the diagnosis. Through lumbar or suboccipital puncture, 3-5 ml of contrast medium are introduced, its supero-inferior migration is followed-up and, with the patient in upright posture, front and profile radiological images are taken. The conic aspect of the cul-de-sac varies in shape and length. This technique involves the risc of late complications (arachnoiditis) or of aggravation of pre-existant inflammatory processes. The same dangers are mentioned also in the case of the use of water-soluble contrast media (Fig.348-351).

- *The medullary angiography* is a method of fineness, which requires a special equipment and a highly skilled medical staff.

- *The medullary flebography* is performed with water-soluble radiopaque substances, which are injected into the vertebral body or into the tip of the spinous process (after a previous locating). Front and profile radiographs are performed, the local peridural plexus being also visualized.

- *The discography,* a modern method of investigation, is used for the detection of the diseases of the intervertebral pulpous nucleus (Fig.352).

The motor neuromuscular activity

The electromyography is the method by means of which the motor neuromuscular activity is investigated.

Its principle consists in the study of the biocurrents which appear during the muscle contraction and which are recorded by arciform electrodes introduced in the muscle to be studied.

The biocurrents appearing during the muscular activity acquire the aspect of the so-called "motor unit potentials".

Muscle denervation causes the appearance of a succession of small deflections, of a 1-2 millisecond-duration and a 1-10 second-frequency, called "fibrillation potentials" that occur early (15 days) after the section or crushing of the nerve and disappear gradually during the reinnervation, when the voluntary action potentials reappear.

The muscle action potential has its origin at the level of the motor end-plate and crosses the muscle fibre in both directions with a velocity of 4 m/s; it is recorded with the electromyograph.

In the resting state, the muscle does not display any electrical activity.

The voluntary contraction brings about the appearance of the action potential of a frequency varying from 5 to 50/s, in dependence on the intensity of the contraction.

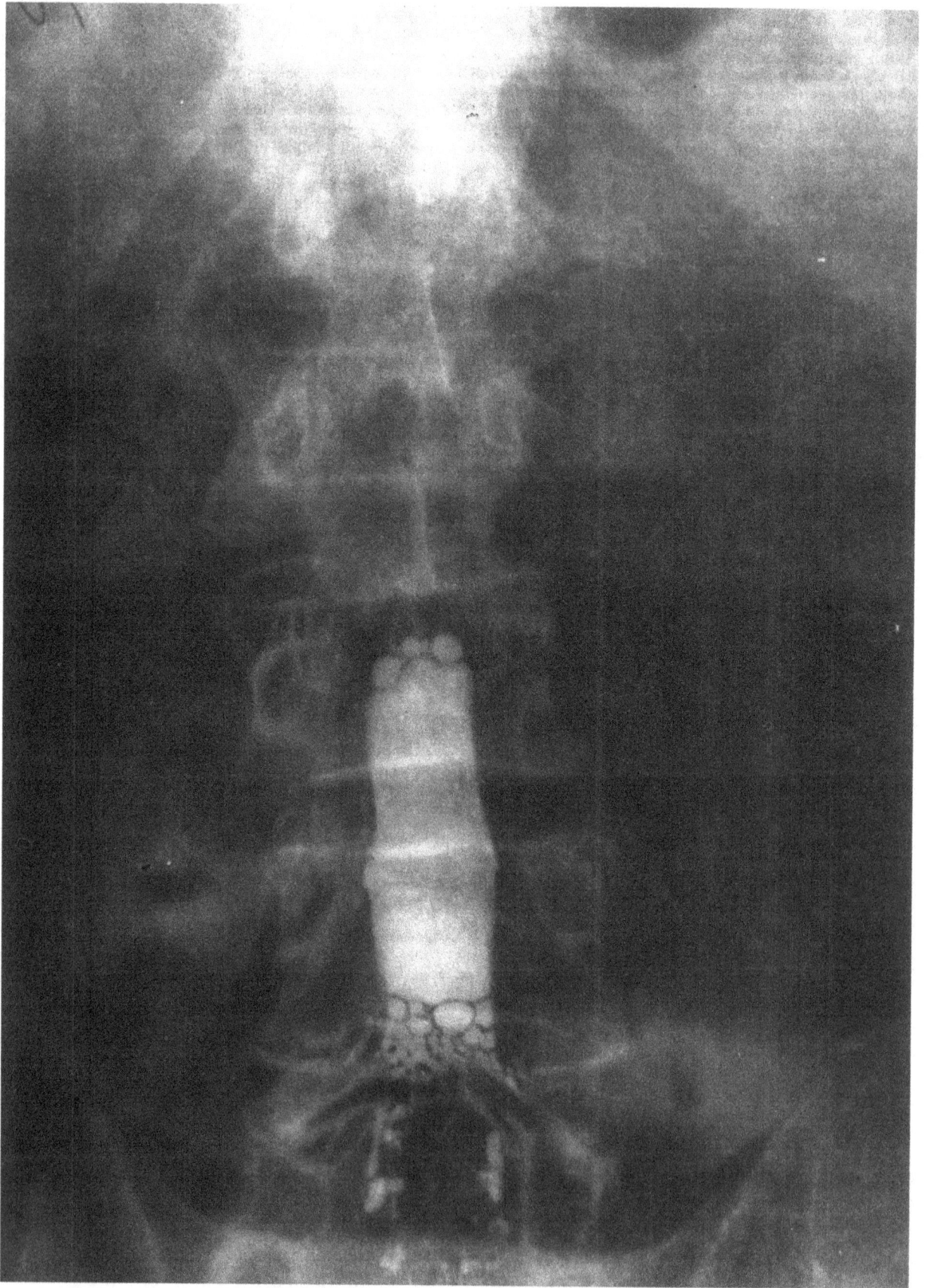

Fig. 348 Lumbar myelogram – anterior view
(Collection Dr. Iacob Mircea – Central
Military Hospital)

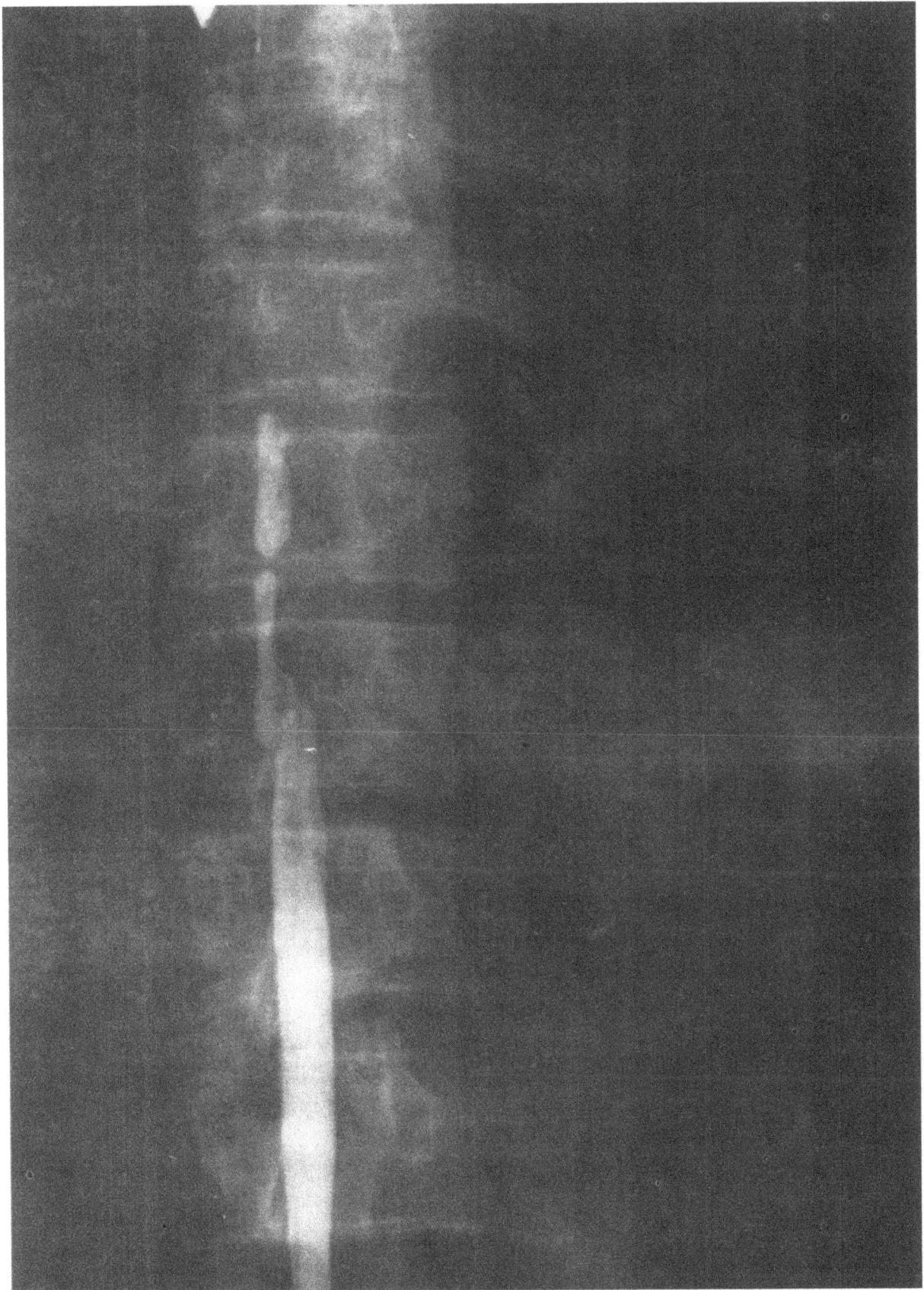

Fig. 349 Thoracolumbar myelogram – anterior view
(Collection Dr. Iacob Mircea – Central Military Hospital)

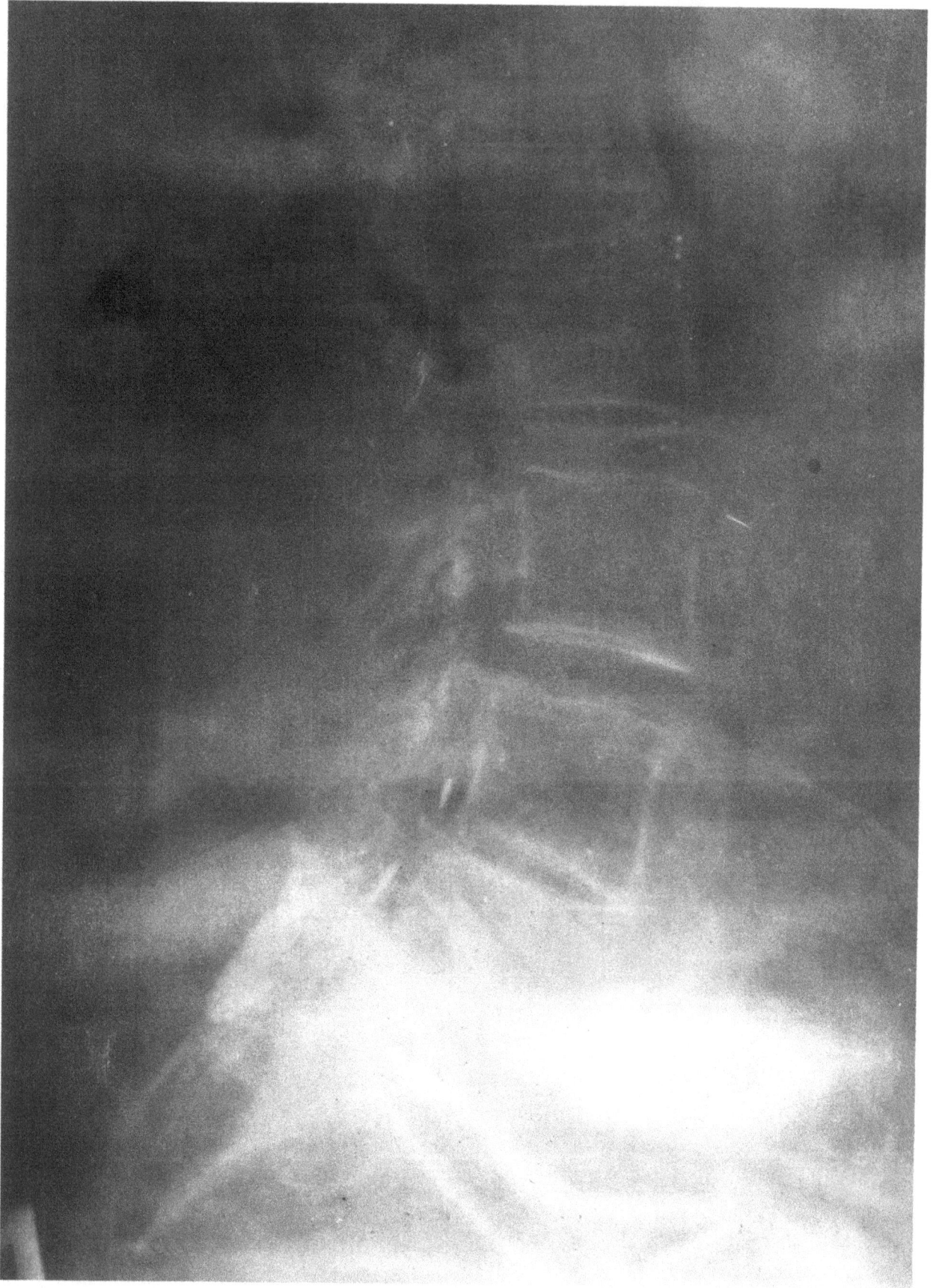

Fig. 350 Lumbosacral myelogram – lateral view
(Collection Dr. Iacob Mircea – Central Military
Hospital)

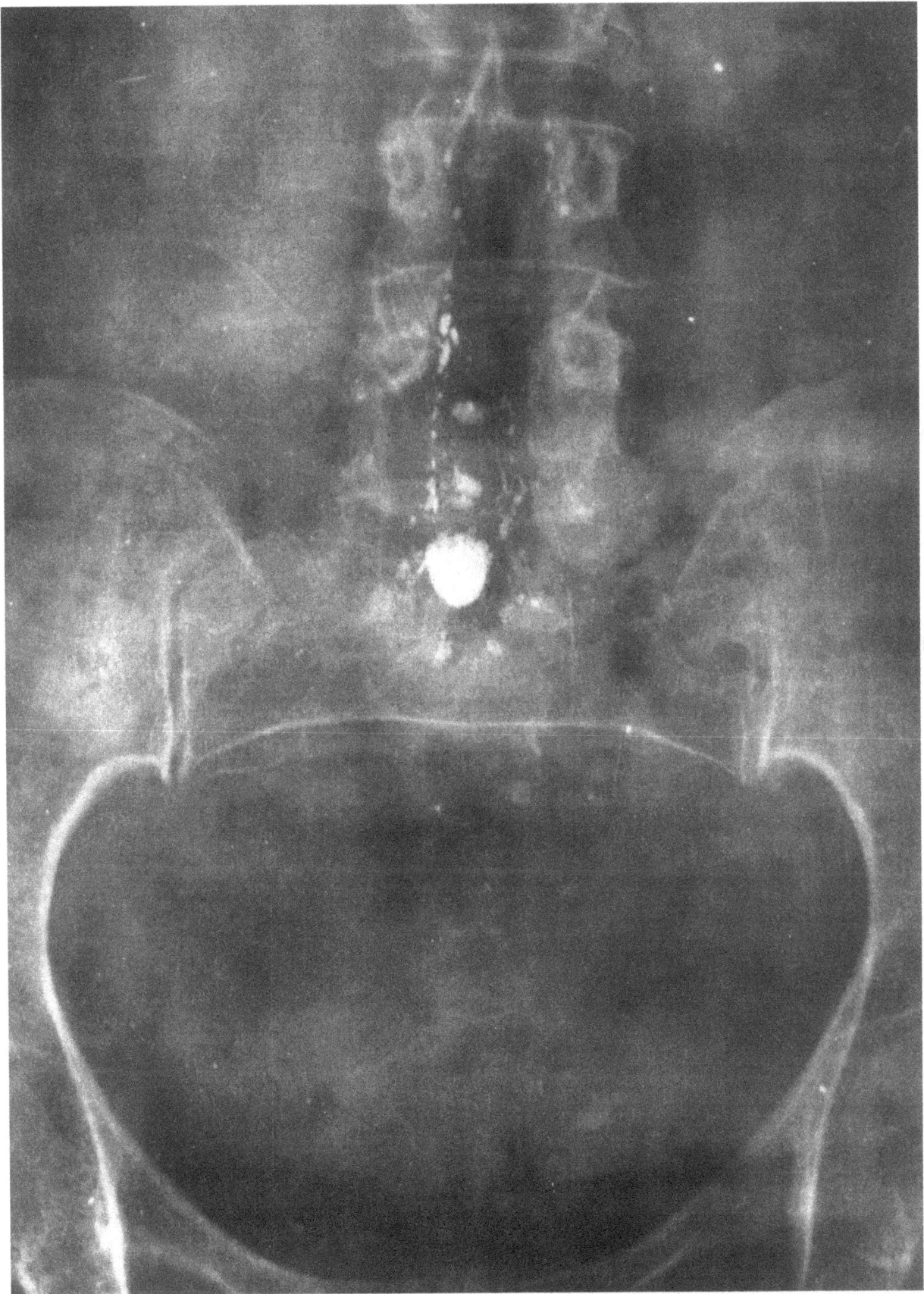

Fig. 351 Lumbosacral myelogram – anterior view
(Collection Dr. Iacob Mircea – Central Military
Hospital)

313

The skull and the encephalon

The neurocranium, formed of 8 bones, lodges the encephalon and some sense organs. For the establishment of the topographical cranio-encephalic relationships, *"craniometric points"* are used as landmarks, some of them (odd) situated on the midline and others (even) on the lateral line.

The splanchnocranium or viscerocranium is the bucco-maxillofacial region, formed of 14 bones, which lodges segments of olfactory and optic sense organs, as well as the initial segments of the respiratory and digestive systems.

The craniometric landmarks are represented by the fixed points situated on the midline or on the lateral line.

The exploration of the nasal region and of the oral cavity is described in vol.II.

The craniocerebral topography

The projection of the main scissures, gyri and cerebral vessels on the splanchno- and neurocranian permits the determination of the trepanation site on the braincase in view of the performance of surgical procedures. Various topographical schedules were proposed for this purpose.

The simplest is *the scheme established by Krönlein*, which has a wide applicability. In order to obtain the desired landmarks, the following five main lines are drawn on the tegument of the skull (Fig.359):

- The inferior horizontal line (1), uniting the inferior border of the orbit with the superior margin of the external acoustic meatus (the Frankfurt horizontal);
- The superior horizontal line (2), drawn along the superior border of the orbit, parallel to the preceding line;
- The anterior vertical line (3), drawn through the middle of the zygomatic arc;
- The middle vertical line (4), which passes through the middle of the mandibular condyle;
- The posterior vertical line (5), which passes through the posterior margin of the mastoid process, parallel to the preceding line.

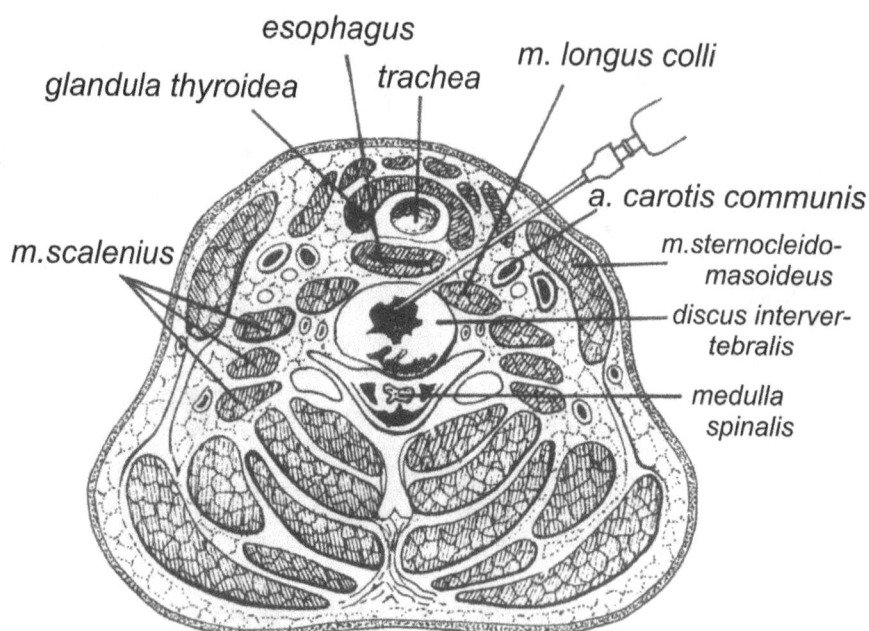

Fig. 352 Discography

The central (rolandic) fissure is represented by a line (6), which unites the intersection point between the anterior vertical and the superior horizontal lines with the highest-sited point of the posterior vertical line. The segment between the middle vertical and the posterior vertical line determines length of the central fissure.

The sylvian fissure (7) is represented by a line which divides into two parts the angle formed by the superior horizontal line and the line of the central fissure (6).

The main trunk of the middle meningeal artery (a frequent cause of extradural hematomas) is sited at the crossing of the superior margin of the zygomatic arch with the anterior vertical line.

The branches of the middle meningeal artery are situated on the superior horizontal line, at the level where it crosses the anterior vertical line (the anterior branch), and on the posterior vertical line (the posterior branch).

Briusova's scheme completes the Krönlein scheme and makes possible a better topographical determination of the cerebral arteries, by drawing a third horizontal line, which begins at the intersection of the posterior vertical line with the projection line of the sylvian fissure.

Hence, in the Krönlein-Briusova scheme there are three horizontal lines: inferior (a-c), middle (d-f) and superior (g-i), and three vertical lines: anterior (a-g), middle (b-h) and posterior (c-i) (Fig.360).

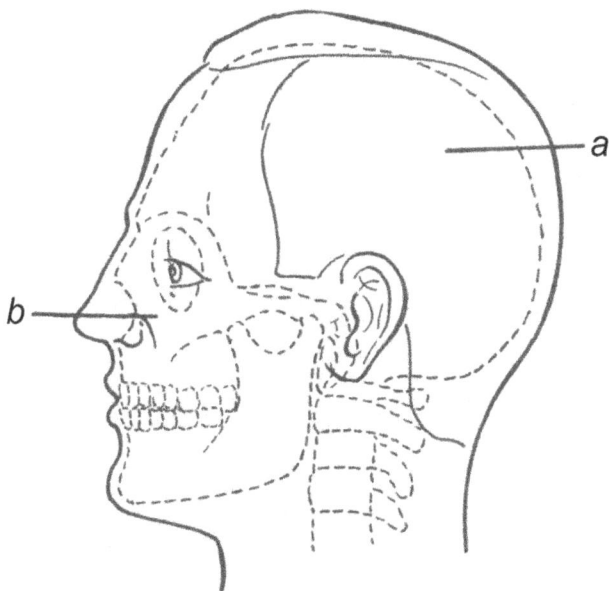

Fig. 353 Division of the cranium
a- cerebral cranium (cerebrocranium or neurocranium); b – visceral cranium (viscerocranium or buccomaxillofacial segment)

Fig. 354 Craniometric points
a - glabellab - ophryon; c - bregma;
d - obelion; e - lambda; f - inion;
g - opisthion; h - basion;
1 - stephanion; 2 - pterion;
3 - asterion.

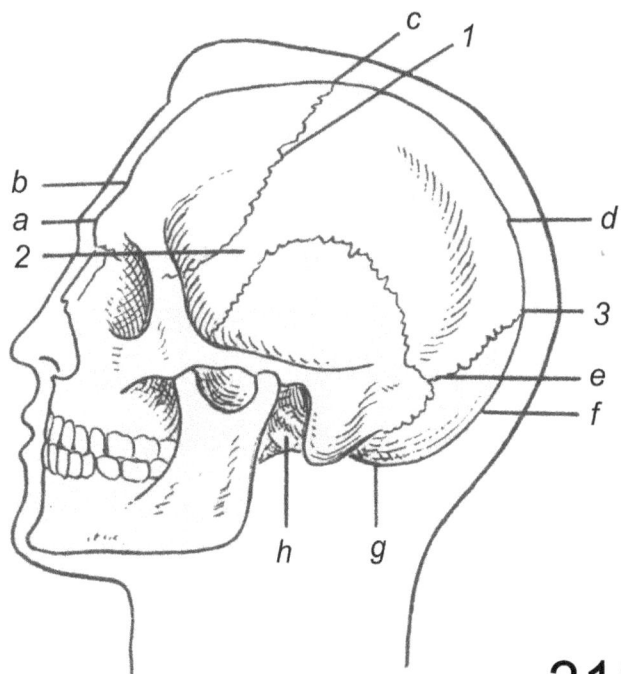

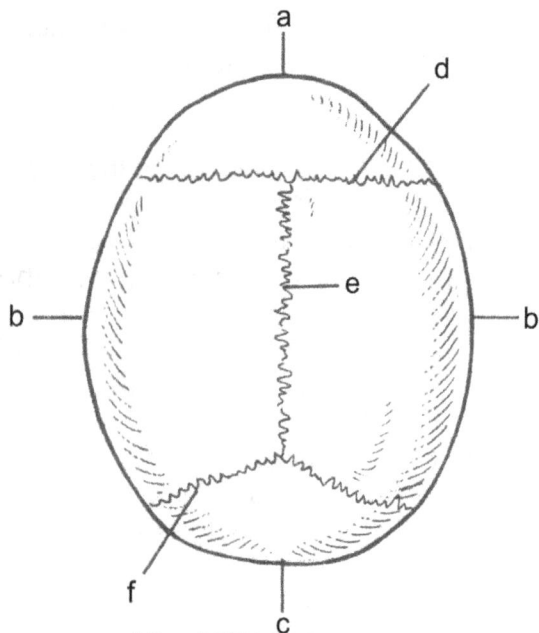

Fig. 355 Vault of cranium and bony suture
a – frontal bone; b – parietal bones; c – occipital bone;
d – frontoparietal suture; e – sagittal suture; f – lambdoid
suture

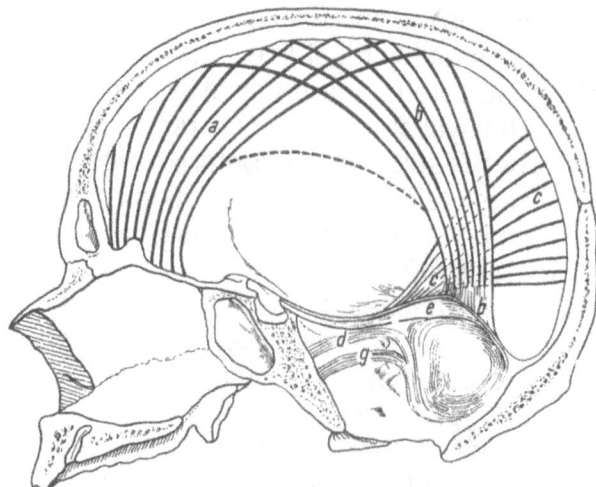

Fig. 358 Main fibre groups of the dural
septa (a-g)

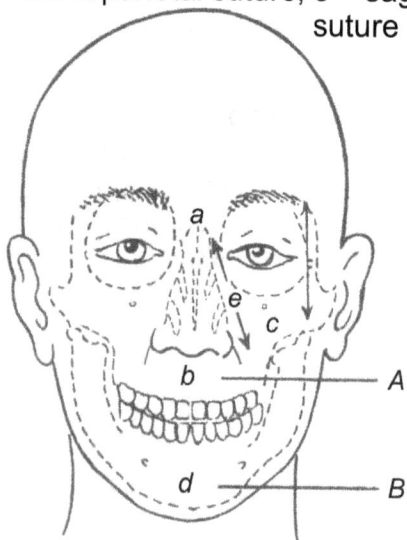

Fig. 356 Landmarks
situated on the midline
A – maxilla; B –
mandible
a – nasal print (nasion);
b – subnasal point;
c – alveolar point
(prosthion); d – mintal
point (pogonion);

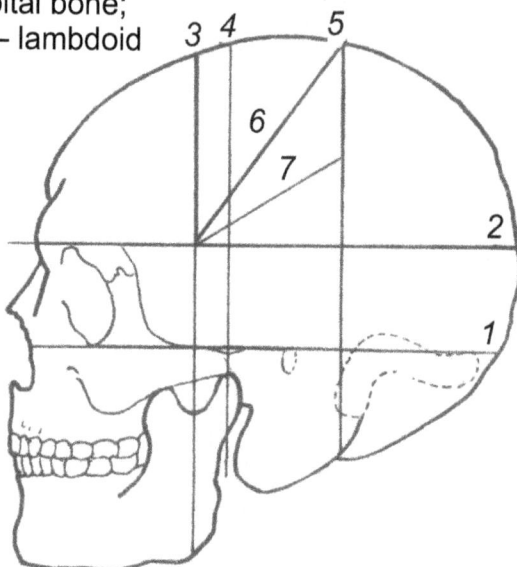

Fig. 359 Craniocerebral topography (after
Krönlein)

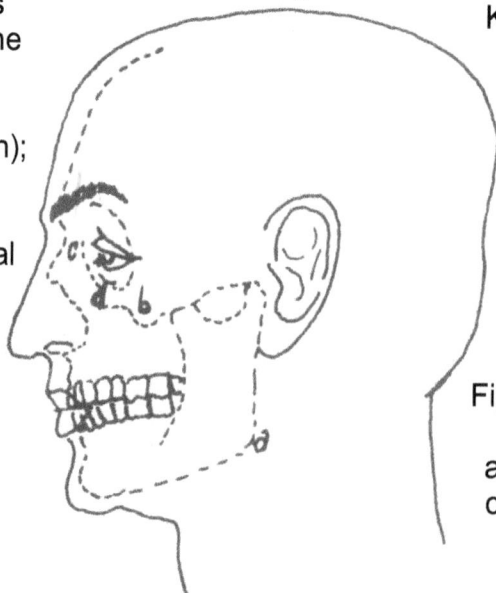

Fig. 357 Landmarks situated
on the lateral line
a - gonion; b – malar print;
c – dacrion; d – zygomatic
arch;

316

Fig. 361 Topographical landmarks for the appreciation of the platybasia

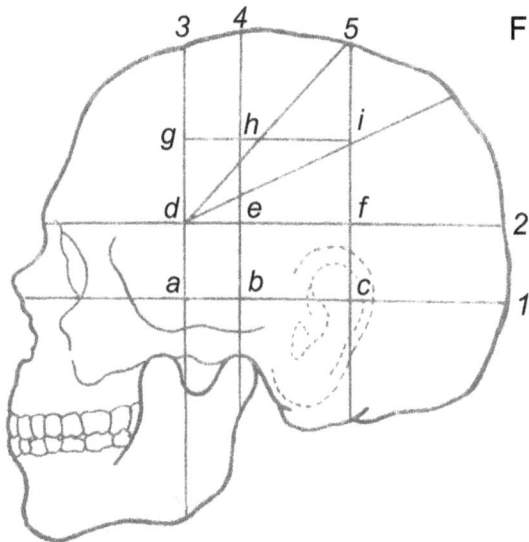

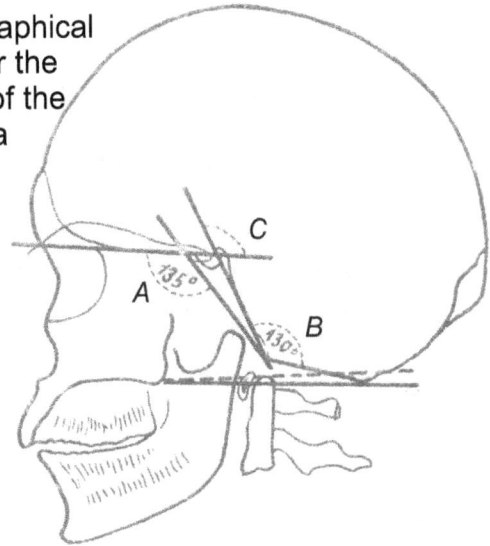

Fig. 360 Krönlein-Briusova scheme

The sinuosities of the internal carotid artery – at the exit from the region of the cavernous sinus – are projected in the antero-internal quadrant; the anterior cerebral artery, at the level of the superior horizontal line (g-i); the division site of the middle cerebral artery corresponds to the intersection point of the anterior vertical line with the middle horizontal line and the posterior cerebral artery is projected above the middle horizontal line, in its posterior part. The platybasia, a congenital anomaly at the level of the occipitocervical region, is represented by two angles.

-*The Welker angle* is formed by a line starting from *the nasion* and running up to *the sellar center* and by a second line running from *the sellar center* up to the anterior border of *the foramen magnum*. The respective angle has 135°.

-*The Boogard angle* is formed by a line running from *the nasion* to the tubercle of *the sella turcica* and by a second line running from the sellar tubercle to the anterior border of *the foramen magnum*. The rang of its normal values is between 115° and 145°. In the case of a large base angle appears the *platybasia*, which has a greater anthropological than clinical value (Fig.361).

For the identification of the basilar impression, Chamberlain has described a line drawn from the hard palate to the posterior border of the foramen magnum; the apex of the odontoid does not pass beyond this line. McGregor uses the base line starting from the border of the hard palate, tangential to the lowest part of the occiput; the apex of the odontoid does not exceed it by more than 4-5 mm.

Investigation of the integrity of the cranial nerves

The twelve pairs of cranial nerves are involved in the motor, sensitive, sensory and vegetative innervation of the structures of the cephalic extremity, as well as in the innervation of the viscera.

The olfactory nerve. The sense of smell is investigated, by asking the subject to smell well-known substances (flowers, perfumes, condiments) or by using the Elsberg olfactometer. Normally, the subject recognizes the odors, but a lesion of the olfactory mucosa or of the olfactory analyzer may induce: anosmia (complete loss of the sense of smell), hyposmia (diminution of the sense of smell), hyperosmia (exaggeration of the sense of smell), parosmia or olfactory daltonism (agreeable odors are confounded with disagreeable odors, and, rarely, vice versa), cacosmia (perception as disagreeable of all the odors), colored olfaction (perception of colours under the influence of odorant substances) and olfactory hallucinations (description of non-existent unpleasant smells, in case of schizophrenia, epilepsy etc.).

317

The optic nerve conveys to the integration centre the inflow produced by the excitation of the sensory cone and rod cells of the retina. For each eye, separately are examined the far vision (lecture of a small-letter-text) and the near vision (appreciation of the details in the room).

In clinics of ophthalmology is used the optotype, a table containing several rows of letters of decreasing size, at the level of each row being recorded the distance from which a subject with a normal sight, standing at 5 m from the table, can read the respective letters. If he does not distinguish the letters which a subject with a normal sight reads from 10 m, its visual acuity will be 5/10=1/2. In the case of marked impairments, the distance from which the fingers may be counted will be appreciated and in more severe cases the movements of the hand, whether he distinguishes light from darkness or whether he does not at all perceive the light are evoluated.

The visual field is, according to Traquair's definition, *"an island of light in a sea of darkness"*; one distinguishes the monocular visual field and the binocular visual field, which results from the superposition of the nasal portions of the two monocular visual fields, to which are added, on either side, the temporal portions (temporal crescents) (Fig.362, 363).

Colors are not perceived at the periphery of the visual field; from here up to the centre, colours are perceived in the following order: blue, yellow, red, green.

The visual field is examined either by comparison or with the Goldmann device. Beside the physiologic scotoma, which represents the projection of the blind spot or Mariotte's spot on the surface of the visual field, pathological scotoma are also described.

Among the visual deficiencies we mention hemianopsia (loss of a half of the visual field) and cortical cecity (lack of any visual perception).

The examination of the fundus oculi permits to appreciate the plate of the pupil eye, of the retinal vessels and of the macula. The examination is performed bilaterally with the ophthalmoscope, an apparatus constituted of a system of lenses and a source of light, permitting the magnification of the examined elements. For the dilation of the pupil may be used 1-2 drops of 1% atropine. Normally, t*he fundus oculi* is reddish-colored; the central artery of the retina, which starts from the centre of the pupil, gives off two branches – one superior and the other inferior -, each divided into a temporal trunk and a nasal trunk. The venous system is disposed in the same manner as the arterial system. The macula, which occupies the posterior pole of the retina, is vascular and vividly coloured.

The oculomotor, trochlear and abducent nerves are examined by investigating the mobility of the eyeballs, of the eyelids and of the pupils. The normal position of the eyeballs in the orbit, the presence of the exophthalmia or enophthalmia, as well as the ocular motility are examined (Fig.365).

The factors which maintain the eyeballs in their normal position are:
- The connective tissue of the orbit (Tenon's capsule and orbital septum) and especially the fat mass;
- The venous oculo-orbital drainage system (the superior and inferior ophthalmic vessels are avalvular and form an anastomotic system);
- The tonus of the extrinsic musculature of the eyeballs: the oblique muscles are protractors, the rectus muscle is retractor of the eyeball, the orbicular muscle is levator of the eyelid and Müller's smooth are retropulsors. They may act either isolated or synergistically;
- The optic nerve.

Exophthalmy (or exophthalmos) is the protrusion of the eyeball, induced by oculo-orbital, paraorbital, cranio-encephalic or general causes. The result is a quantitative disproportion between an inextensible container (the orbit) and a content which may modify either its size (inflammations, phlebitis, glaucoma, intra-ocular hypertension) or its components (tumours).

The motor ocular activity is performed by an *extrinsic* muscle group, which induces the displacements of the eyeball, and by an *intrinsic* muscle group, that acts upon the pupil and the lens.

318

The extrinsic mobility is due to the six motor ocular muscles, their action being conditioned by their innervation and by the position of the eyeball at the movement of the examination.

In the verticality movements are concomitantly involved the superior or inferior recti muscles of both eyes; in the laterality movements are acting simultaneously the lateral rectus muscle of one eye and the medial rectus muscle of the other eye. The muscles of the eyeballs exert their action according to Sherrington's reciprocal innervation principle: while the agonistic muscles contract, the antagonistic muscles relax.

The associated movements of the eyeballs are under the control of the cerebral cortex, they may be voluntary, automatic and reflex. The lesion of these circuits brings about dissociated, voluntary, automatic or reflex paralyses of the eyeballs.

The intrinsic motility is characteristic to the pupil and the lens. The range of the contraction and dilatation movements of the pupil does not exceed 0.5 mm. They are visually imperceptible and have a frequency of 70-120/minute.

Stronger stimuli induce *myosis* (contraction of the pupil) or *mydriasis* (dilation of the pupil), under the action of two antagonistic muscle: the constrictor (innervation dependent on the Edinger-Westphal nucleus) and the dilator muscle (sympathetic innervation, with the Budge's centre in the inferior cervical and superior thoracic spinal chord).

The ciliary muscles give rise to changes in the shape, sphericity or transparency of the lens, followed by the modification of the refractive index, in relation with the near or far vision. Two types of fibers are present for this purpose: circular fibers – Rouget-Müller's muscle or ciliary sphincter (involved in the accommodation for near vision, with parasympathetic innervation) – and radial fibers – Brücke's muscle (involved in the accommodation for far vision, with sympathetic innervation).

This explains the synergistic action to accommodation of the lens and the pupil.

The change of the refractive index between the various regions of the lens – like in evolutive cataract – may lead to monocular *diplopia*.

The trigeminal nerve is explored by the investigation of the sensibility of the face, the eyes, the mouth and the tongue (the sensory root – *radix sensoria n. trigemini*) and of the contraction force of the temporal, the masseter and the pterygoid muscles (the motor root – radix motoria n. trigemini)(Fig.366).

As regards the sensory branches for the ophthalmic branch *(n. ophthalmicus)* are explored the teguments of the forehead, for the maxillary branch (n. maxillaris) the upper lip and for the mandibular branch *(n. mandibularis)* the region of the chin. The tegument at the level of the mandibular angle are supplied by a second cervical root. Pain may be induced by the compression of the Valleix points: supraorbital, suborbital and mintal. Sometimes appears the *"painful facial tic"*(Trousseau).

The muscle tone is explored by asking the subject to clench the teeth, while the examiner palpates the temporal muscles and the muscles of mastication.

In the unilateral paralyses, the subject is asked to open the mouth, which assumes an asymmetric form, deviated on the affected side; in bilateral paralyses, the mandible prolapses, and its upwards, downwards and lateral movements are markedly impaired.

In addition should be examined:

The corneal reflex: the excitation of lateral portion of the cornea with cotton-wool elicits the bilateral blinking; the afferent pathway is represented by the ophthalmic branch of the trigeminal nerve and the efferent pathway by the facial nerve (motor innervation of the orbicular muscle of the eyelids);

The masseter reflex: the percussion of the mental region, while the mouth is half-open, elicits the elevation of the mandible and the contraction of the masseter muscles.

The facial nerve is explored by the appreciation of the movements of the eyelids, as well as of the facial musculature by wrinkling of the forehead etc.(Fig.367).

The gustatory sensibility of the tongue is examined by applying on the tongue sodium chloride, citric acid, glucose or quinine.

punct de fixație — Pata oarba Mariotte

Porțiune binoculară — a cîmpului vizual

90° — 60°

Semiluna temporala

C.V.monocular

Macula
Papila

O.S. — O.D.

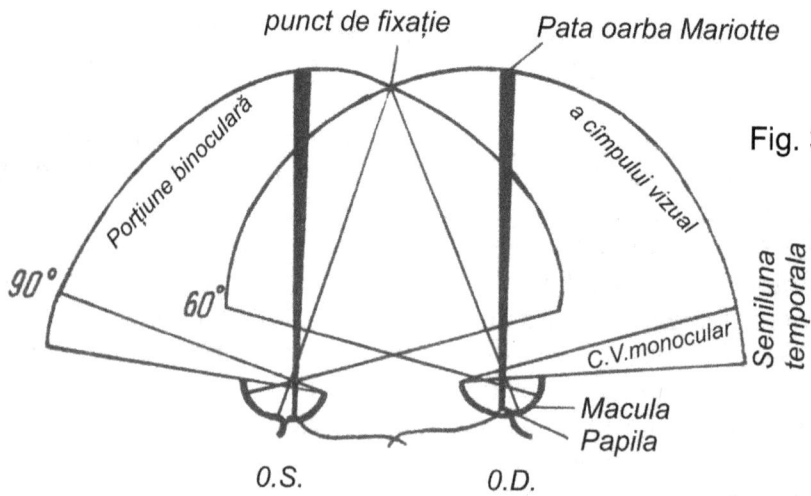

Fig. 362 Binocular visual field

Fig. 363 Normal visual field

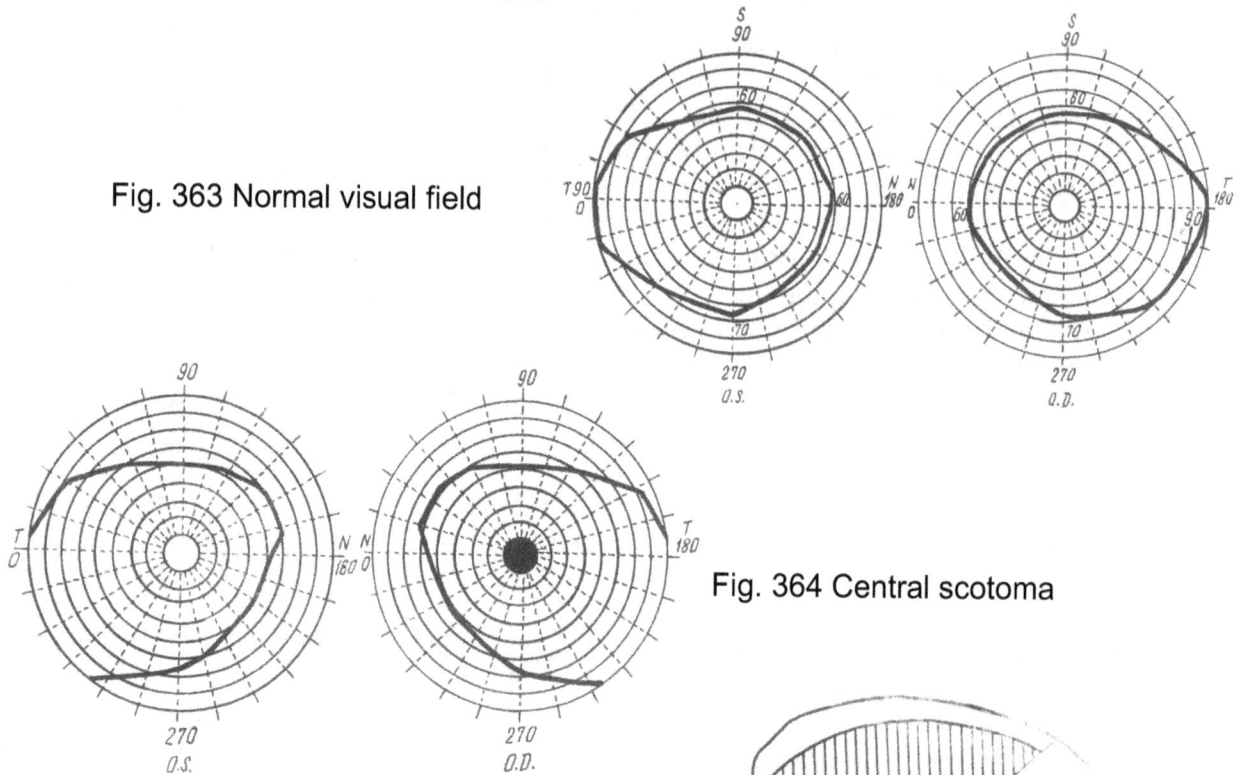

Fig. 364 Central scotoma

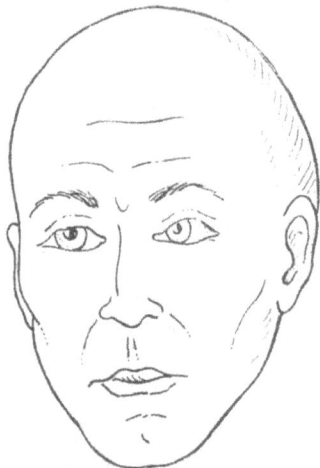

Fig. 365 The abducent nerve
paralysis manifests itself by a
convergent strabismus (deviation of
the eye in the direction opposite to
the damaged nerve)

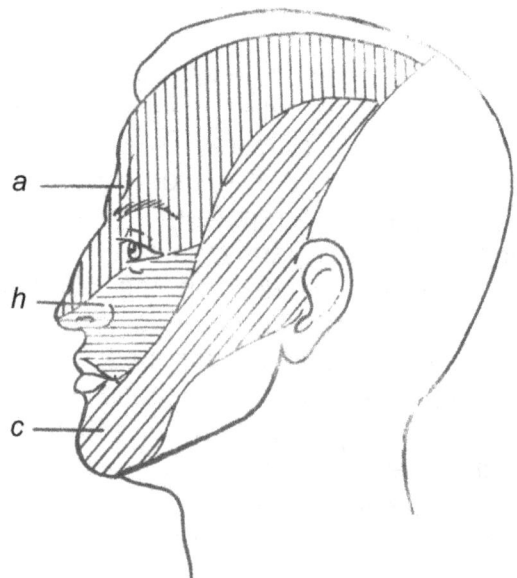

Fig. 366 Sensory areas of the
ophthalmic (a), maxillary (b) and
mandibular (c) branches of the
trigeminal nerve

320

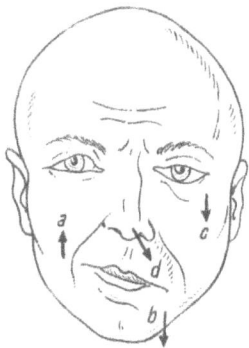

Fig. 367 Manifestations of the facial nerve paralysis:
a- deviation of the mouth towards the healthy side ; b- descent of the labial commissure towards the damaged side; c- enlargement of the palpebral fissure on the damaged side; d- deviation of the tip of the nose towards the damaged side.

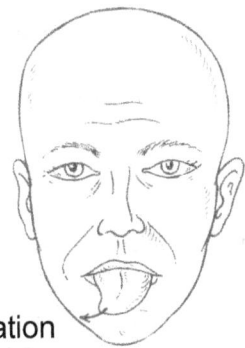

Fig. 368 Paralysis of the hypoglossal nerve: deviation of the apex of the tongue towards the damaged side

Usually, the facial paralysis is accompanied by a loss of taste *(aguesia)*, the latter symptom being unique in the case of the lesion of the chorda tympani.

In the facial paralysis, the impossibility of whistling and blowing is observed; speech and laughter increase the asymmetry and the Bell's phenomenon (an upward and downward deviation of the eyeball, when the patient is requested to close the eyes).

The cochleovestibular nerve is explored by examining the hearing (audiometry and by equilibrium tests (the vestibular nerve).

- *For the cochlear nerve,* the simplest test is *the phonic acumetry*; the whispered voice is perceived from 6 m and the loud voice from 40 m.

In the clinic, audiometry with the audiometer is used. The average value of the human sonic perception is 32-16.000 Hz, the conventional frequencies ranging on the average between 250 and 2,000 Hz. The maximum sensibility for the differentiation of tonalities varies between 50 and 400 Hz. Beyond the average limits the vibration may be perceived only by the cutaneous receptors and when the limits are significantly exceeded, the acoustic excitations becom painful for the ear. The lowering of hearing is appreciated in decibels.

Moreover, the following tests may be used:

- *the Rinne's test:* the tuning fork is first applied on the mastoid process, then in front of the auricle and the duration of the perception of vibrations is recorded. Normally, the vibration is longer perceived by air conduction than by bone conduction; the reverse situation shows that the text is negative.

- *the Weber's test:* the tuning fork is applied on the vertex and the vibration are perceived normally on both sides. In the conductive deafness the vibrations are perceived on the side of the affected ear, whereas in the perceptive deafness, the vibrations are perceived on the healthy side.

- *The vestibular nerve* is explored by tests which confirm impairments.

- *the "Plumb-line" test* – in upright posture, the head and the body of the patient deviate laterally, sagittally or obliquely.

- *the Romberg's test:* in upright posture, with the feet approximated, when the patient closes the eyes, his body bends slowly, gradually.

- *the test of stretched arms:* patient sits on a chair, the arms stretched out, in contact with the examiner's fingers; on closure of the eyes, the arms deviate towards the damaged area.

The glossopharyngeal vagus and spinal nerves are investigated by examination of the pharyngeal and palatine reflexes, by appreciation of the tone and form of the sternocleidomastoid and the trapezius muscles and by testing the taste.

The hypoglossal nerve is examined by the study of the movements of the tongue and of its trophicity (Fig.368).

The plain radiological examination of the braincase and of the encephalon

This examination reveals both the lesions of the container (the braincase) and of the contents (the encephalon with its coverings).

To this end, an overall study of the skull in the front and right or left lateral position or a radiograph of certain regions, such as the basis of the skull and the orbital region, by *"Hirtz's axial position"*, or the posterior fossa and especially *the sella turcica*, by the *"semi-axial fronto-suboccipital position"* are performed.

During the examination, the following guide marks are taken into account:

- the basal (Virchow's) plane, formed by the so-called German horizontal (Frankfurt horizontal), which is made up by the transverse line which passes through the inferior border of the orbit and of the external acoustic meatus;

- the Reid plane, represented by the line which unites the external angle of the orbit with the external acoustic meatus;

- the frontal plane and the sagittal plane;

- the Boogard angle.

The shape of the skull is appreciated by the Retzius formula:

$$I = \frac{\text{maximum bitemporal endocranial distance x 100}}{\text{maximum anteroposterior endocranial distance}}$$

According to this formula may be distinguished:

- the mesocephalic skull with I=75-80;
- the dolichocephalic skull with I=75 or less;
- the brachycephalic skull with I=80 or more.

Moreover, plain radiographs permit to investigate: the state of bone sutures, the vascular imprints, the presence of intracranial calcifications, the presence of osteitic inflammatory, parasitic or neoplastic lesions, of hyperostoses etc. The optic canal, projected into the infero-external quadrant of the orbit has normally a diameter of 2.8-6 mm, but is dilated bilaterally in tumours of the optic nerve and is destroyed in sarcomas of the base of the skull and in pituitary adenomas.

The radiological examination of intracranial lesions by contrast media

In the framework of these examinations are included: the gas encephalography, the iodine ventricolography, the cerebral arteriography, the computed axiel tomography etc.

The gas encephalography may be performed in two ways: fractional gas encephalography and total gas encephalography.

The fractional gas encephalography consists in a lumbar puncture, followed by the successive extraction of 2-3 ml of CSF, after which the fractional introduction of equal amounts of air is performed. Radiographs are taken as air migrates towards the encephalon and then to the lateral ventricles through the fourth ventricle, the cerebral aqueduct and the third ventricle. As a rule, the amount of air introduced does not exceed 10-15 ml and no more than 10-12 radiographs, in various positions of the skull, are performed.

The total gas encephalography is made after a previous suboccipital or lumbar puncture. Twenty-five – thirty ml of CSF are extracted, and, through the same needle, a similar amount of air is introduced, after which the penetration of air, the deformation of the ventricular system, the enlargement of the subarachnoid spaces and of the basal cisterns etc – are followed-up.

The gas ventriculography consists in a ventricular puncture, followed by the introduction of a certain amount of air directly into the ventricular system, after which front and profile radiographs are taken (Fig.369-372).

The iodine ventriculography is aimed at the following-up, under radiological control, of the migration of the ultrafluid Lipiodol introduced by ventricular puncture. This method permits to detect the presence of possible expansive processes occurring in the third ventricle and the permeability of the cerebral (sylvian) aqueduct and to locate some anatomical structures, in view of stereotaxic interventions (introduction of radioactive isotopes).

The cerebral arteriography is the most significant paraclinical method by means of which the diagnosis of intracranial tumoral formations, vascular cerebral malformations or diseases and traumatic craniocerebral lesions is made.

The common carotid artery or the internal carotid artery is punctured at the level of the neck and an amount of 4-6 ml of tri iodinated substance is injected under a pressure of about 4 atmospheres. Simultaneously, serial front and profile radiographies are made.

As regards the evolution, three phases are described:

- *the arterial phase*, immediately after the injection of the substance, in which the cervical and cerebral arterial network is visualized;

- *the capillary phase*, in which a diffuse and homogeneous shading of the cerebral tissue is obtained;

- *the venous phase (phlebogram)*, in which the cerebral veins and the dural venous sinuses are viewed.

The best images are achieved by the biplane angiography, permitting to obtain 3-6 images per second for the arterial phase, 2 per second for the capillary phase and 1 per second during 7 seconds to visualize also the venous phase.

The contrast medium is introduced rapidly, by means of an automatic mechanical injector, coupled to a radiographic apparatus.

The cerebral arteriography may be performed also by catheterization of the femoral or brachial artery, according to Seldinger's method; very useful data are obtained in this way regarding: the arterial circulation tissue (the normal value for the internal carotid artery is 4 seconds), the caliber and sinuosities of the cerebral vessels, the integrity of the cerebral vascular tree, the presence of vascular malformations, the existence of possible avascular areas (hematoma, empyema, serous meningitis etc.), the displacement of normal vascular tracings.

The cranial computed tomography represented a true revolution in medicine, its deviser being the electronics engineer Godfrey Newbold Hounsfield. The image is not obtained with a film or a fluorescent screen, but with very sensitive detectors, which may reveal even the smallest variations of the absorption coefficient of the X rays that have traversed the explored skull. Without the use of contrast media, details are obtained, regarding the state of fluid media of the encephalon, the presence of intracerebral tumors, abscesses, cysts, or hematomas (Fig.373-374).

The image may be read immediately from the screen of the TV monitor or may be fixed definitively with the photographic apparatus.

The patient lies in dorsal decubitus and the roentgen tube is rotated around the head or the neck, transmitting the image to the video monitor, after a previous automatic processing and amplification.

The radioisotope methods are modern and efficient techniques, which permit the detection of pathological processes by the two-dimensional representation of the gamma rays emitted by a radioisotope and they are used when the other methods do not furnish conclusive data.

By means of scintiphotography from various incidences (anteroposterior, right and left profile, postero-anterior etc.), particularly useful informations are obtained regarding the vascular changes consecutive to various craniocerebral diseases (Fig.375-380).

In dependence on the rapidity and duration of the concentration of these radioisotopes, the warm areas of hyperactivity are revealed, permitting to draw conclusions about the site, the size, the form and the number of lesional processes at the level of the encephalon. The radioisotopes ^{131}I (under the form of labeled albumin), ^{99}Tc etc. are used (Fig.381-388).

By means of radioisotopes, significant data regarding the cerebral blood output and the energy metabolism are also obtained.

The electroencephalography is a functional exploration method, by which the cerebral bioelectrical activity is recorded by means of electrodes attached to the scalp and connected to the electroencephalograph.

With their help, the cortical waves produced by the variations of the postsynaptic somatic and dendritic neuronal membrane potential are detected, amplified and recorded, they are expressed under the form of the following rhythms:

- *alpha rhythm*: formed of continuous, by the amplitude modulated oscillations; it appears in the period of wakefulness, in the state of physical and psychic relaxation, and is in close relations with the visual function;

- *beta rhythm*: it appears especially in children, in the motor and frontal areas, and is blocked by an involuntary effort;

- *theta rhythm*: it appears under the form of short volleys in children with behaviour disorders;

- *delta rhythm*: it occurs in the infant and during deep sleep in the adult; its presence in the awake adult is indicative of a pathological state.

The stereoencephalography is the recording of the electrical activity of deep structures of the brain; the electrocorticogram is obtained by recording the electrical activity derived directly from the cerebral cortex, during surgery on the brain.

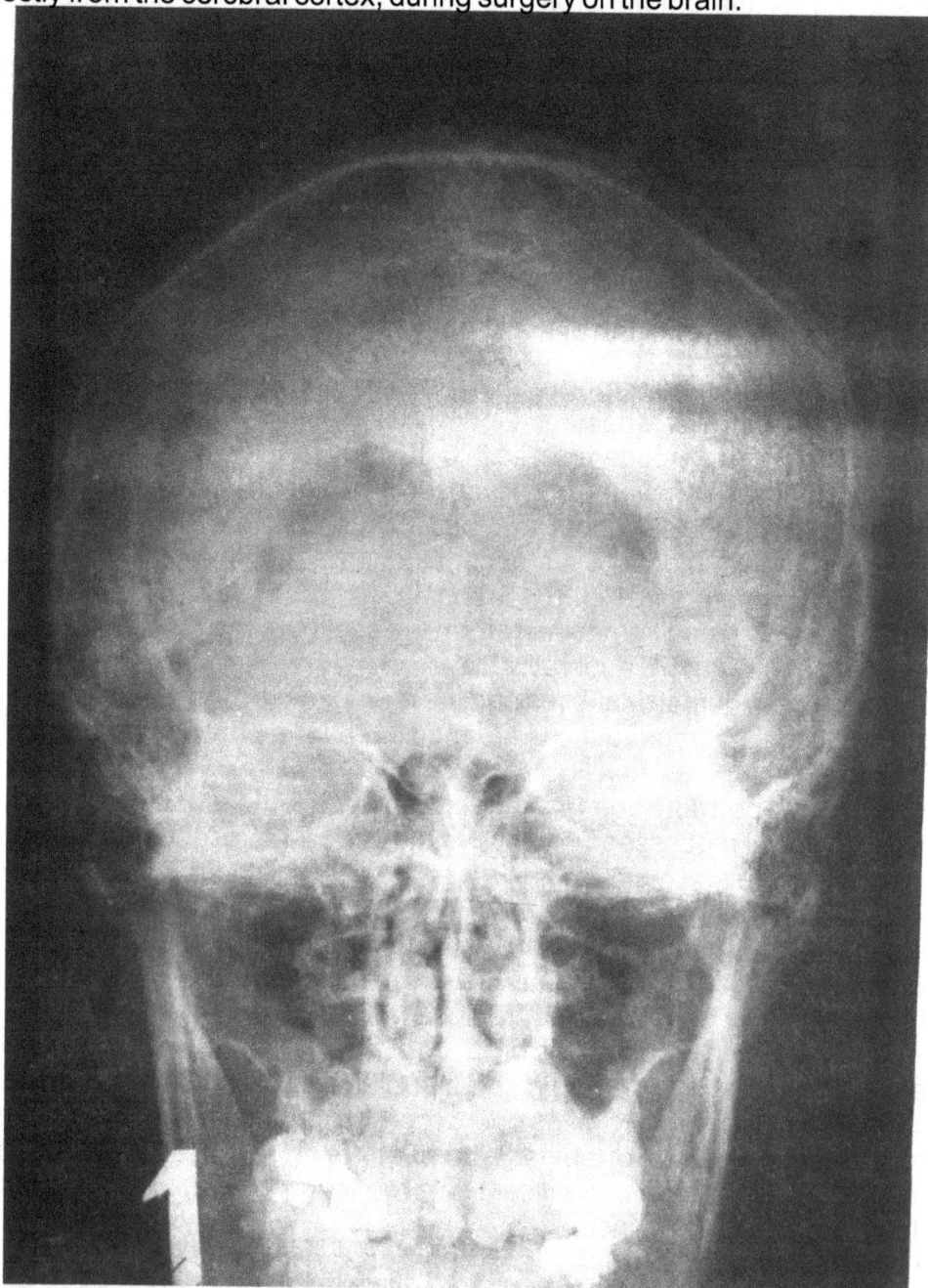

Fig. 369 Pneumoventriculography – anterior view (Collection Dr. Iacob Mircea – Central Military Hospital)

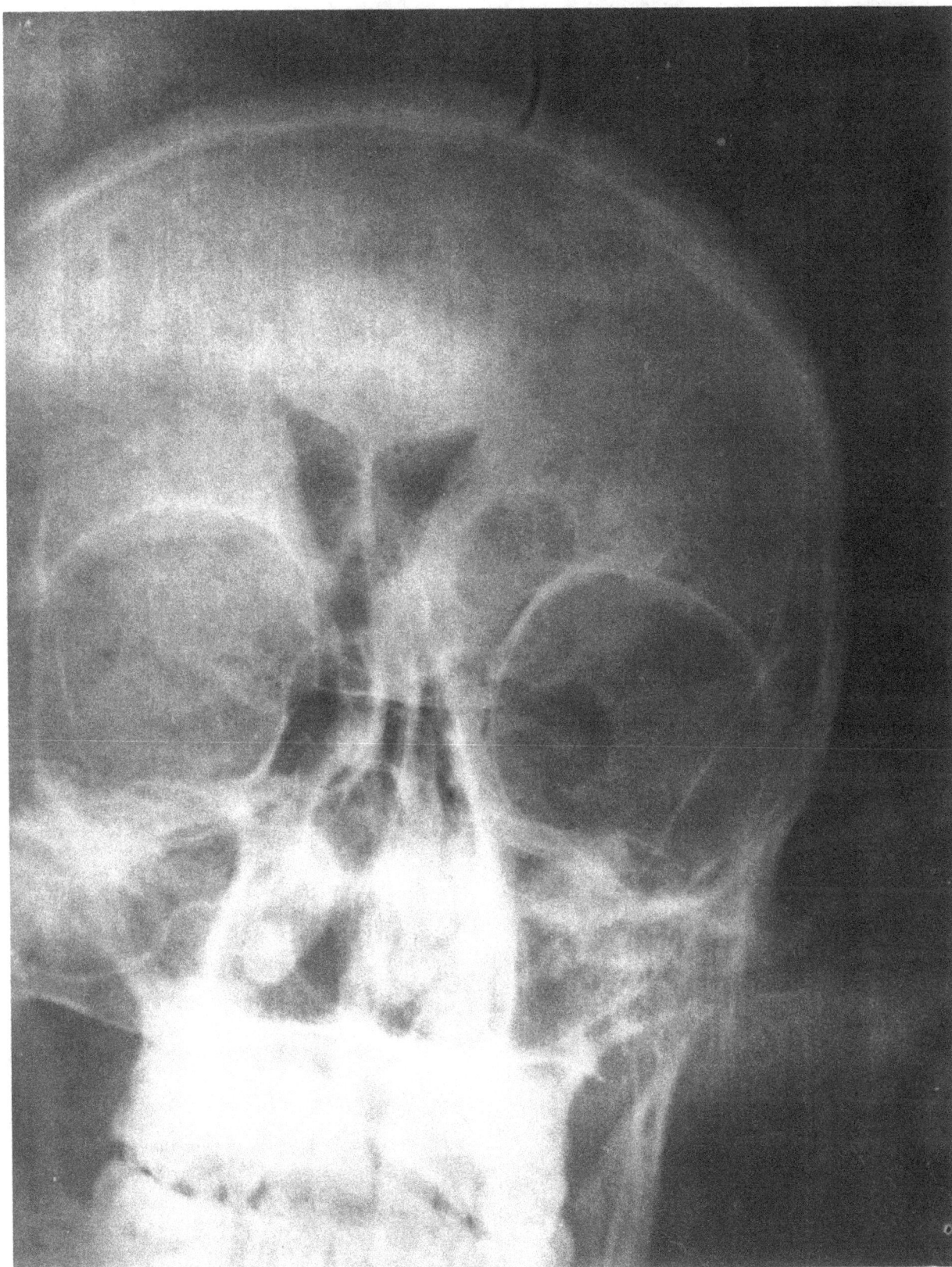

Fig. 370 Pneumoventriculography – anterior view (Collection
Dr. Iacob Mircea – Central Military Hospital)

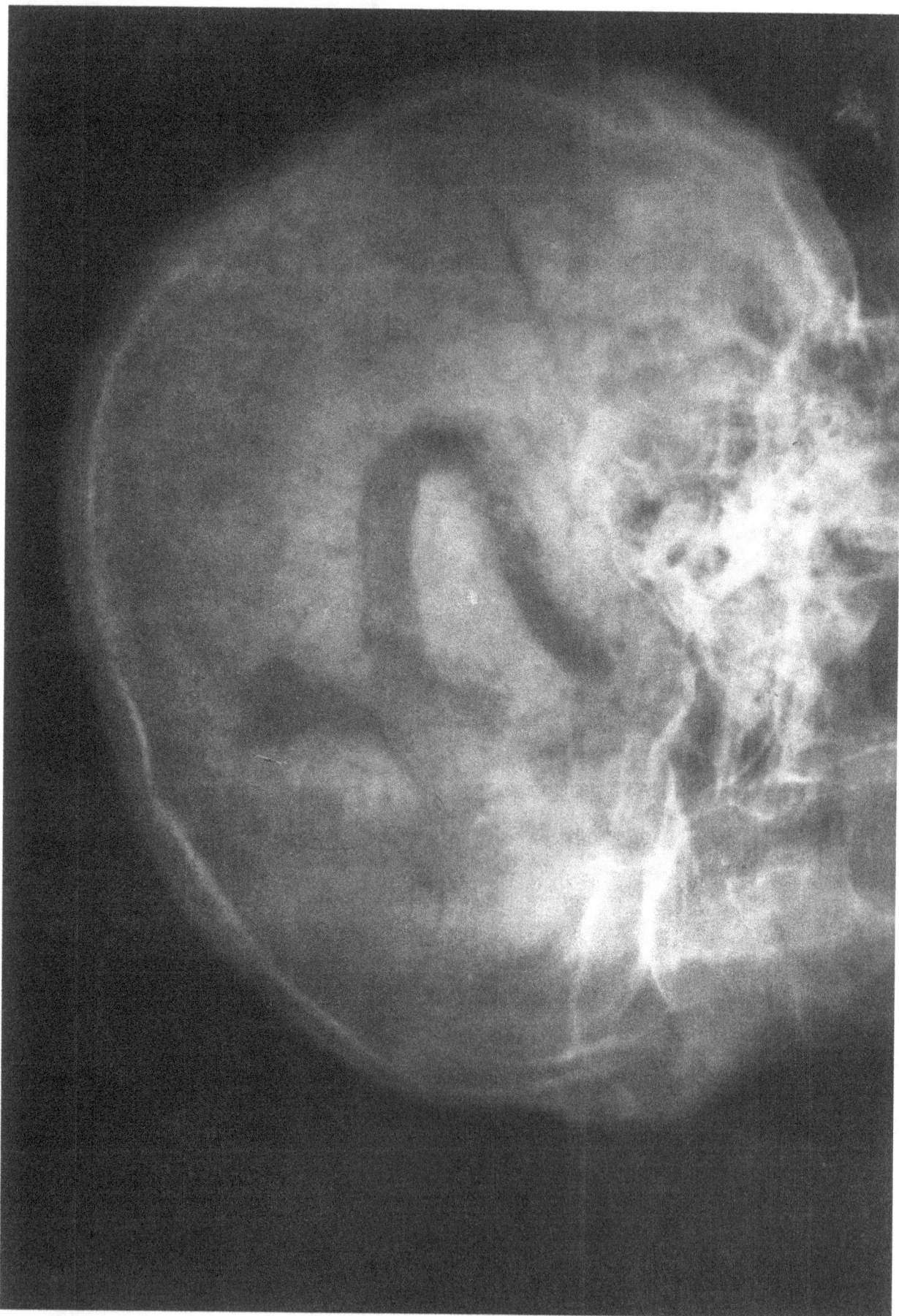

Fig. 371 Pneumoventriculography – lateral view (Collection Dr. Iacob Mircea – Central Military Hospital)

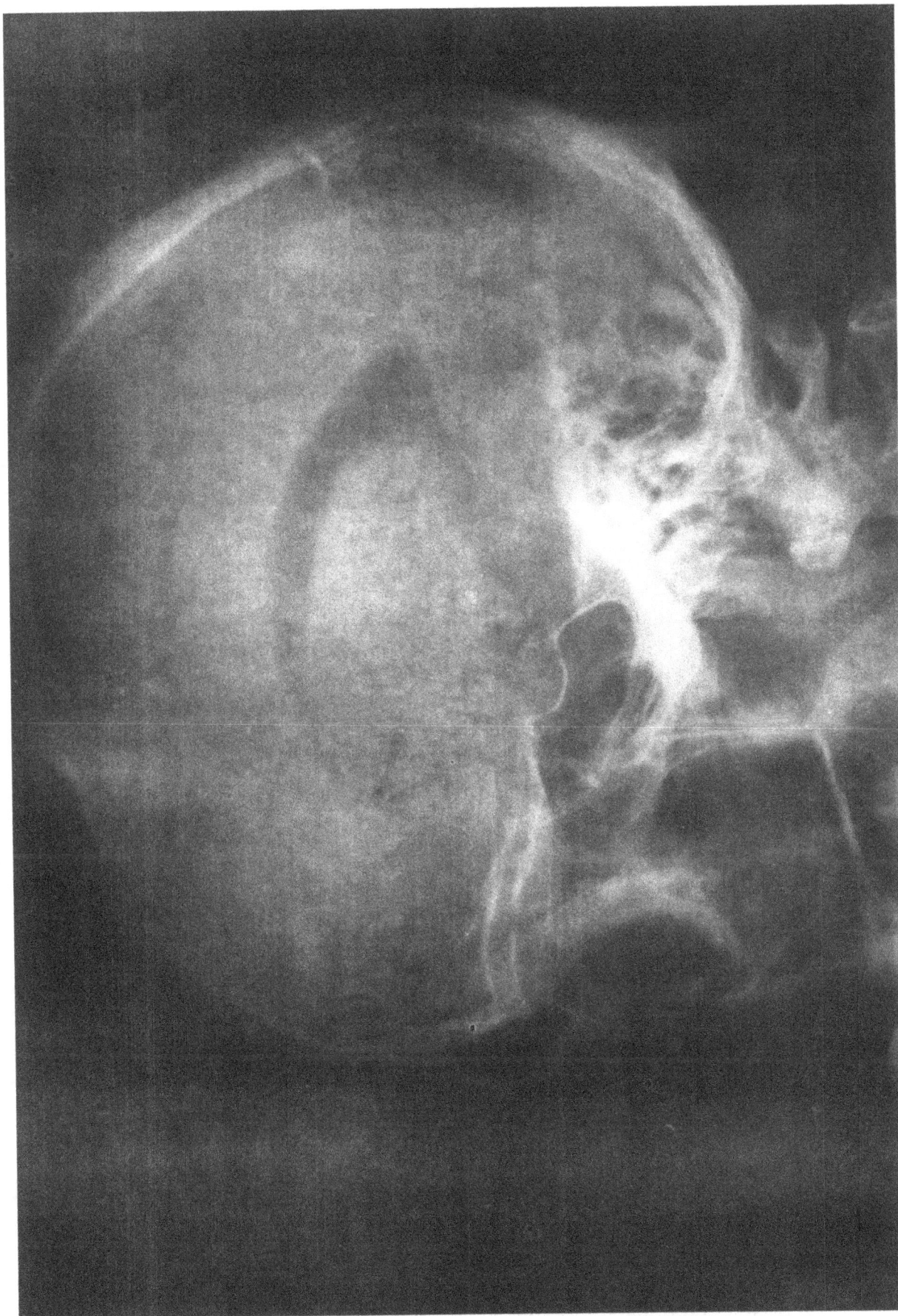

Fig. 372 Pneumoventriculography – lateral view
(Collection Dr. Iacob Mircea – Central Military Hospital)

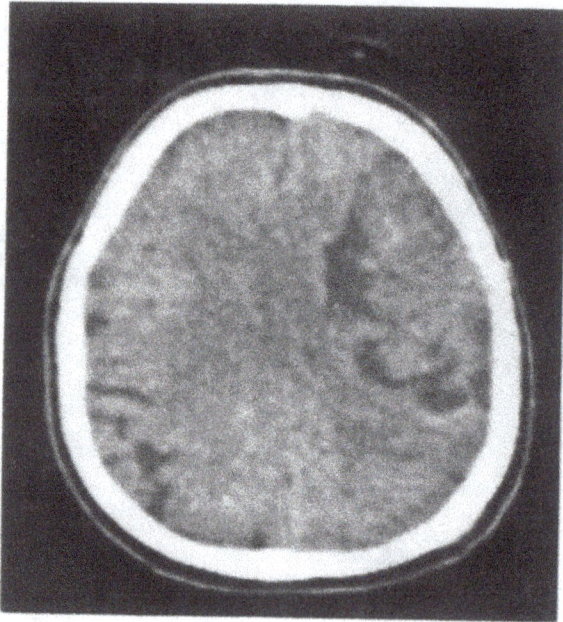

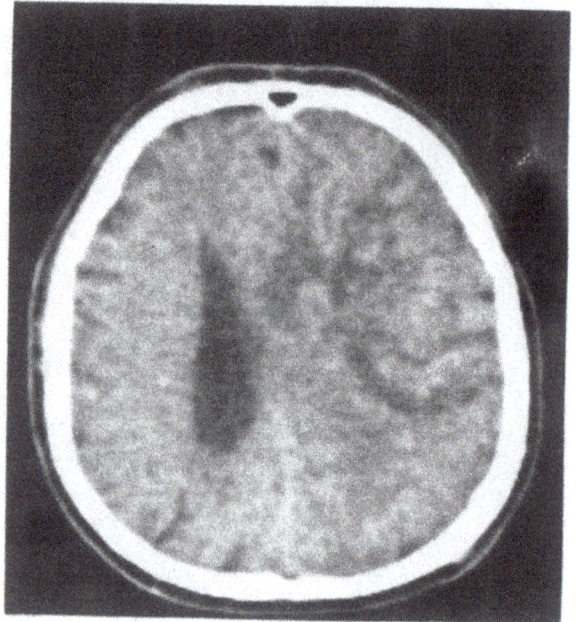

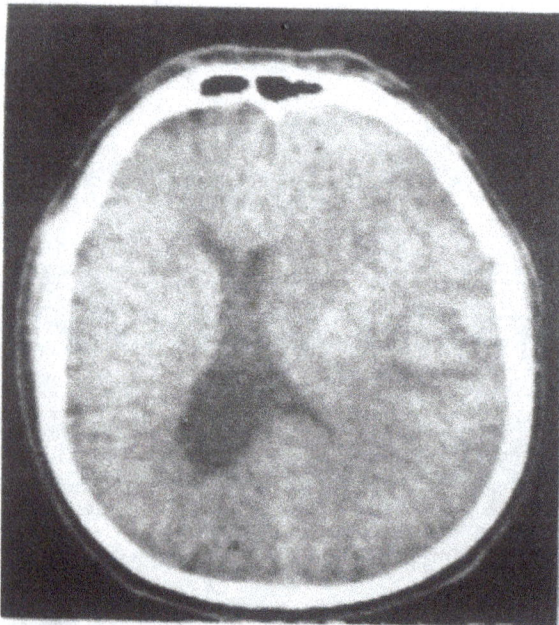

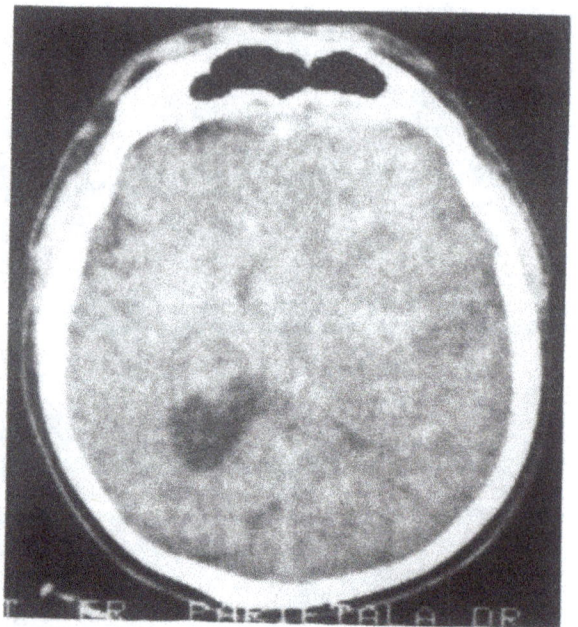

Fig. 373 Computed tomography in a case of frontoparietal tumour (A-D, consecutive sections at various levels) (Collection Dr. Şerban Georgescu – "Fundeni" Clinical Hospital)

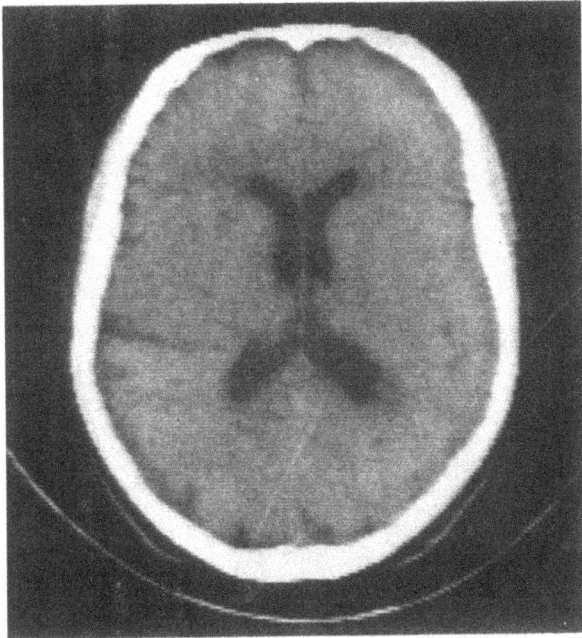
A

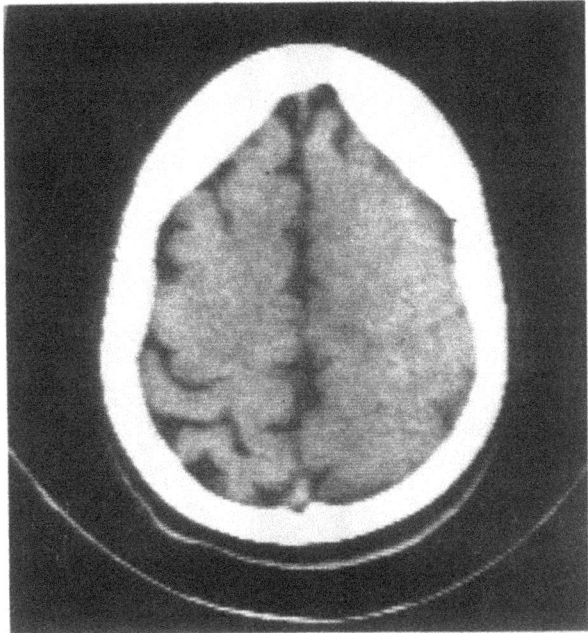
B

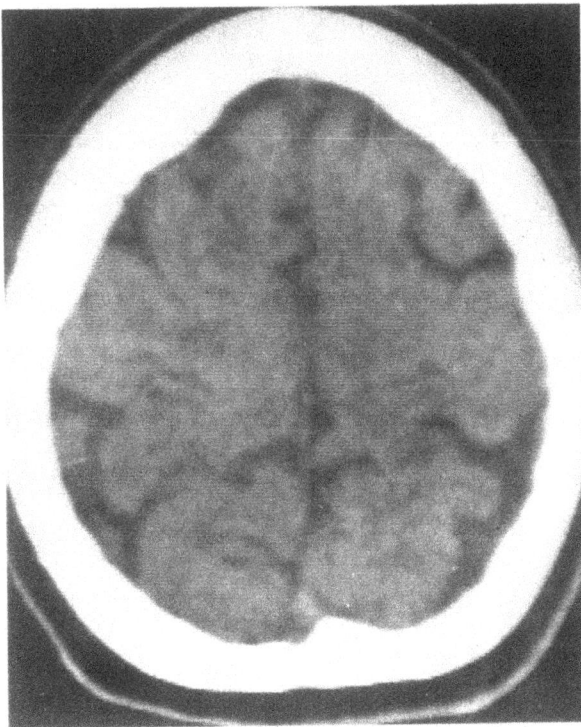
C

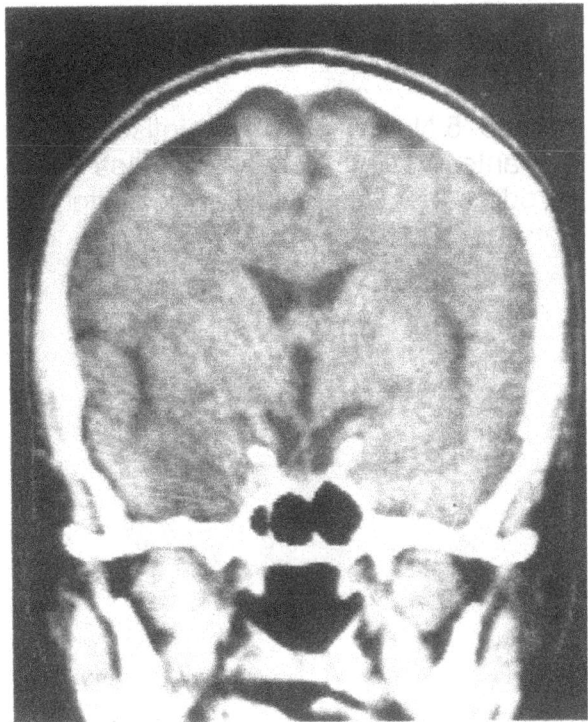
D

Fig. 374 Computed tomography in a case of posterior fossa syndrome (A-D, consecutive sections at various levels) (Collection Dr. Şerban Georgescu – "Fundeni" Clinical Hospital)

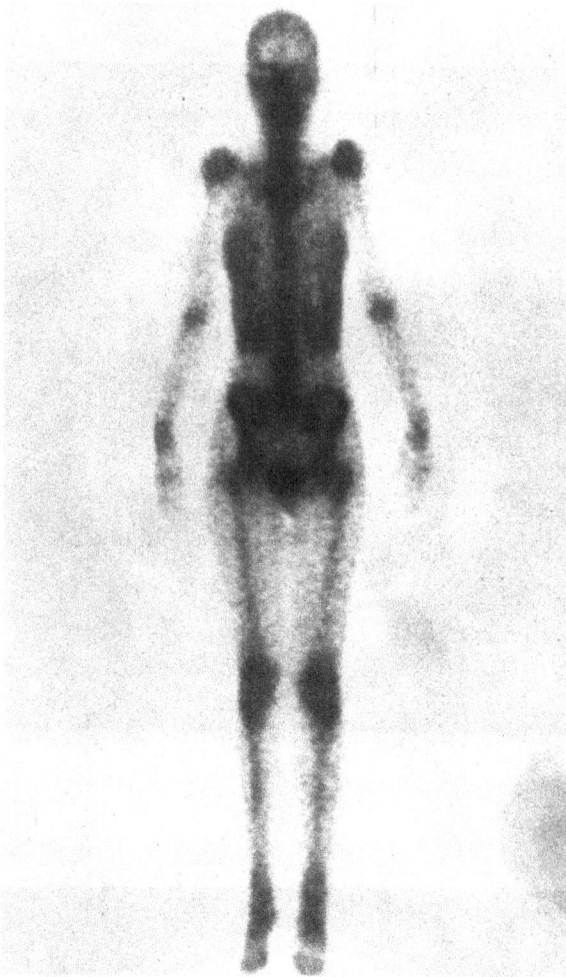

Fig. 375 Normal skeletal scintiscan –
anterior view (Collection Dr. Ion
Codorean – Central Military Hospital)

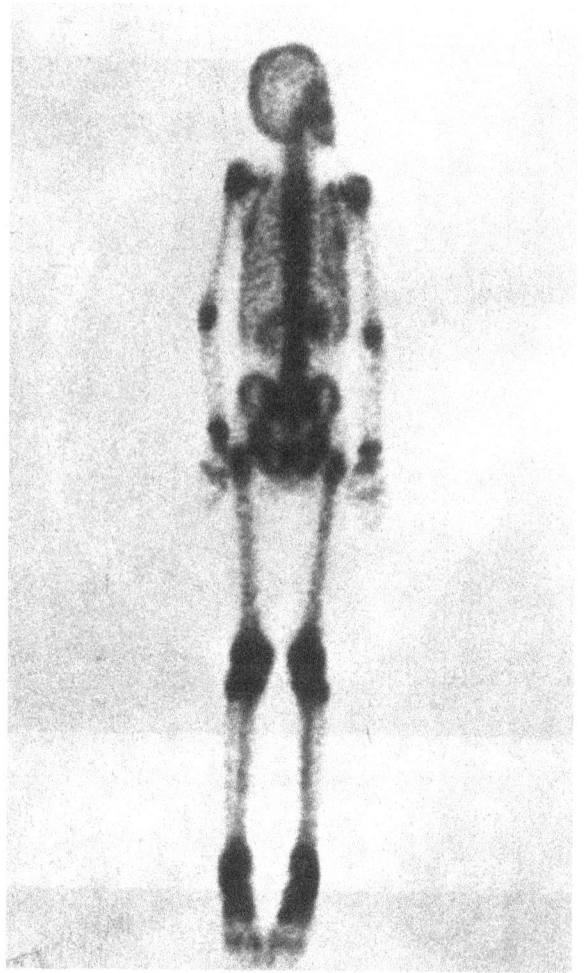

Fig. 376 Normal skeletal scintiscan –
posterior view (Collection Dr. Ion
Codorean – Central Military Hospital)

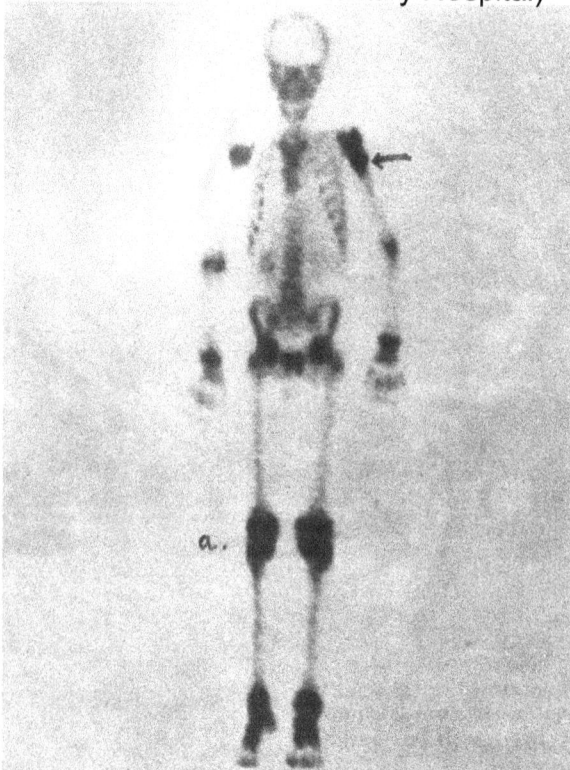

Fig. 377 Skeletal scintiscan, humeral
metastasis – anterior view (Collection Dr.
Ion Codorean – Central Military Hospital)

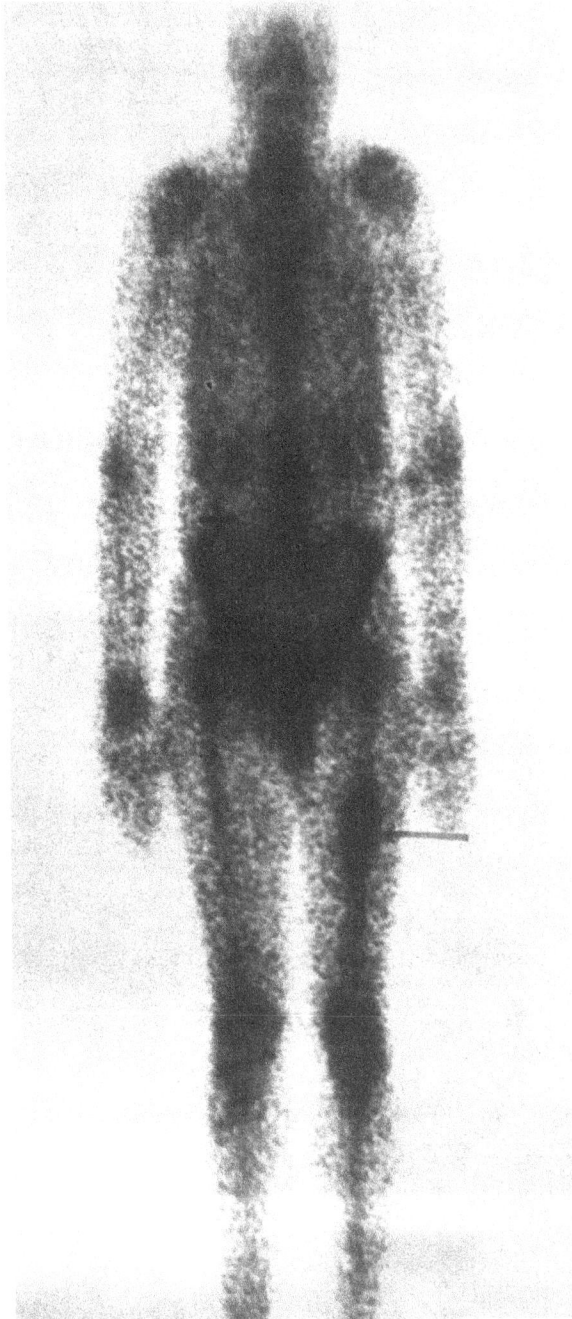

Fig. 378 Skeletal scintiscan – anterior view. Vertebral tumour with metastasis on the left femur (Collection Dr. Ion Codorean – Central Military Hospital)

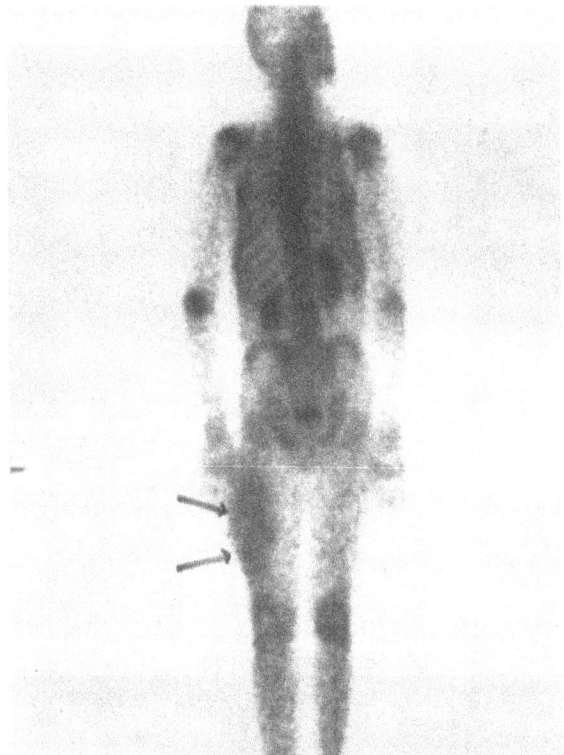

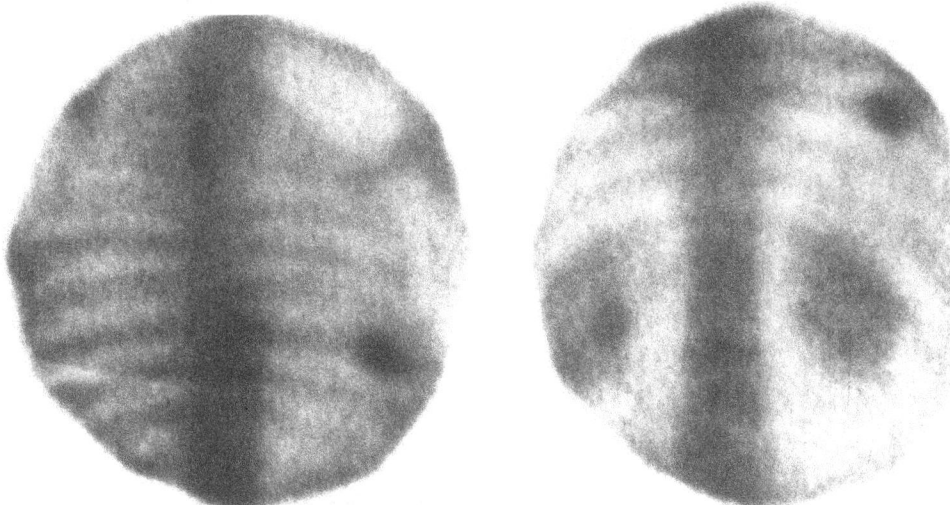

Fig. 379 Skeletal scintiscan – posterior view. Vertebral tumour with metastasis on the left femur (Collection Dr. Ion Codorean – Central Military Hospital)

Fig. 380 Skeletal scintiscan – anterior view. Metastasized costal tumour (Collection Dr. Ion Codorean – Central Military Hospital)

331

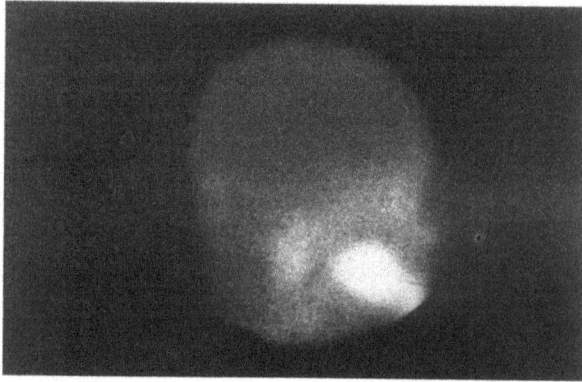

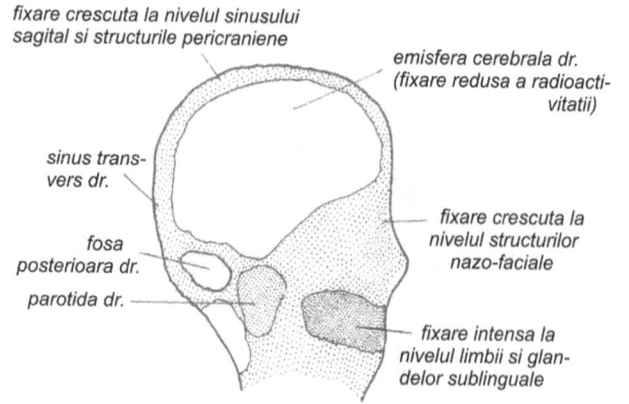

Fig. 381 Normal scintigrafic image (and scheme) in right lateral incidence, characterized by a low radioactivity fixation at the level of the right cerebral hemisphere, delimited through the increased fixation at the level of the pericerebral structures (Collection Dr. Ion Codorean – Central Military Hospital)

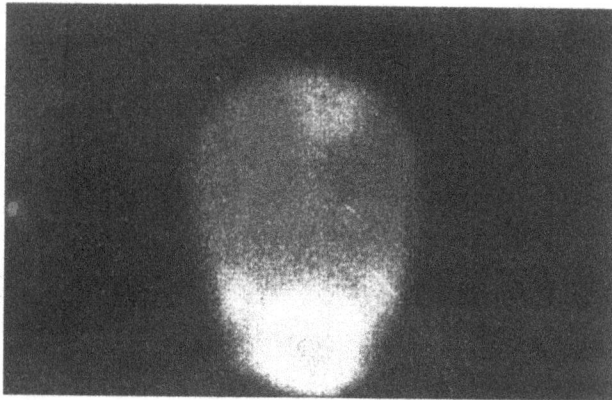

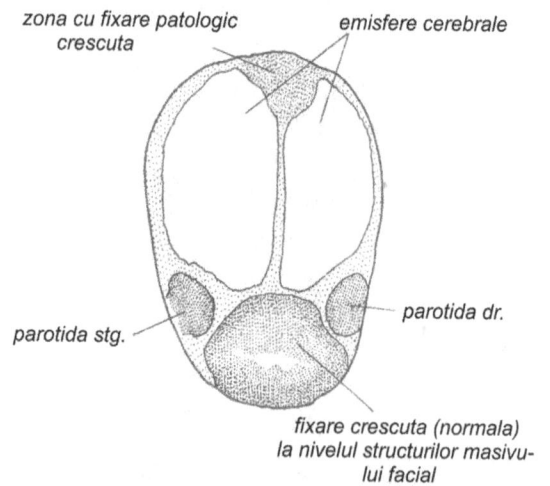

Fig. 382 Central scintiscan (and scheme) in posterior incidence, showing the presence of an area characterized by a pathological fixation of the radioactive substance, rather well delineated, situated at the level of the right cerebral hemisphere, in the internal superior third (right parietal, operatively confirmed meningioma) (Collection Dr. Ion Codorean – Central Military Hospital)

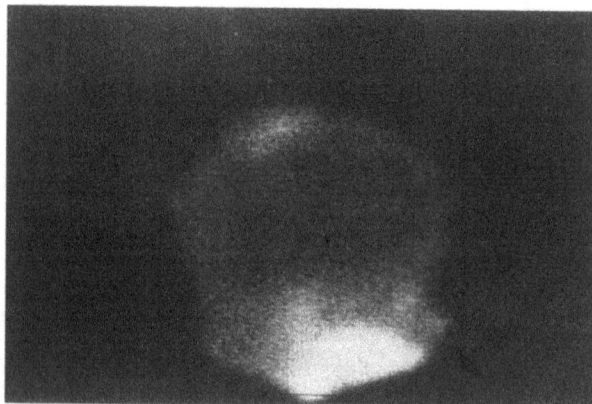

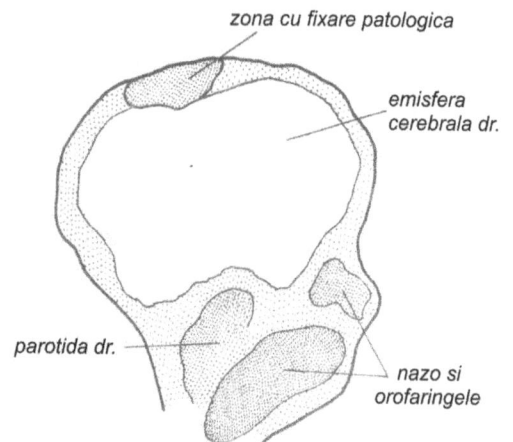

Fig. 383 cerebral scintiscan (and scheme) in right lateral incidence, showing the presence of an area characterized by a pathologically increased radioactivity fixation at the level of the right parietal lobe, in the upper position (right parietal, operatively confirmed meningioma) (Collection Dr. Ion Codorean – Central Military Hospital)

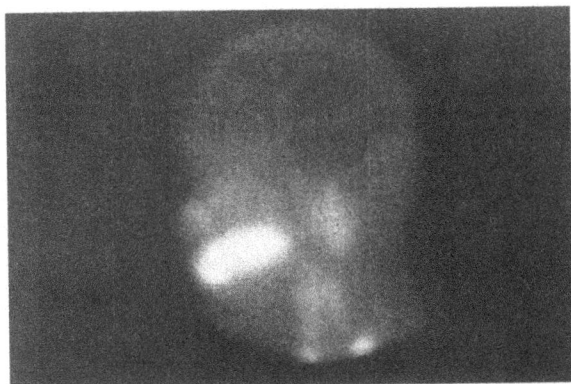

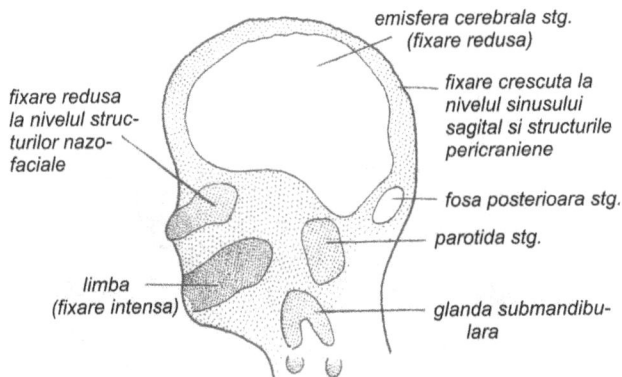

Fig. 384 Normal scintigraphic image (and scheme) in left lateral incidence, characterized by a low radioactivity fixation at the level of the left cerebral hemisphere and by an increased fixation at the level of the pericerebral structures (Collection Dr. Ion Codorean – Central Military Hospital)

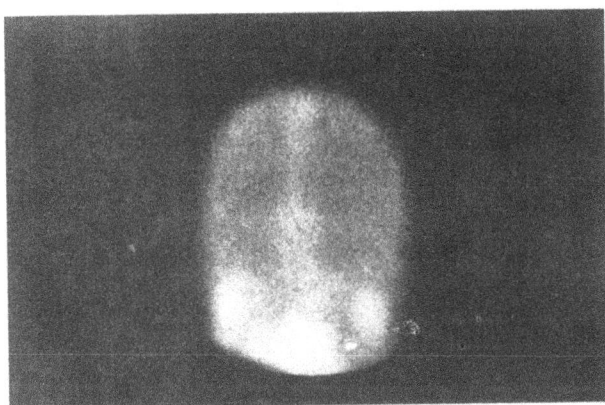

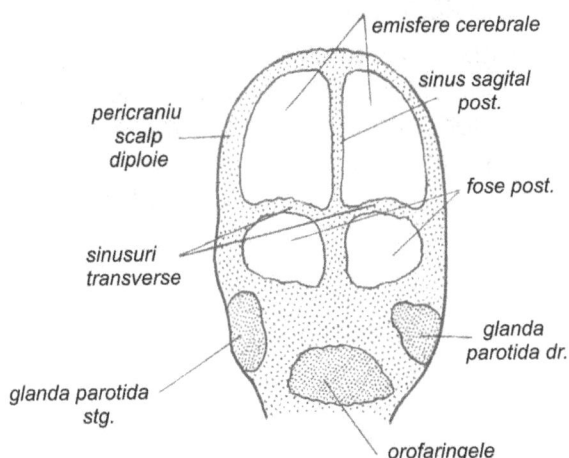

Fig. 385 Normal scintigraphic image (and scheme) in posterior incidence, characterized by a low radioactivity at the level of the posterior fossa of the cerebral hemispheres. It is the elective incidence for the visualization of the posterior fossa (Collection Dr. Ion Codorean – Central Military Hospital)

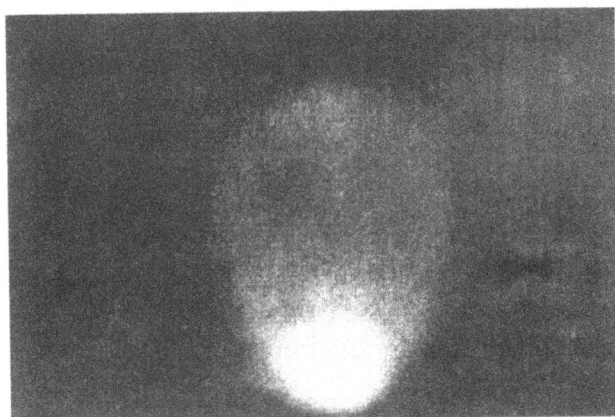

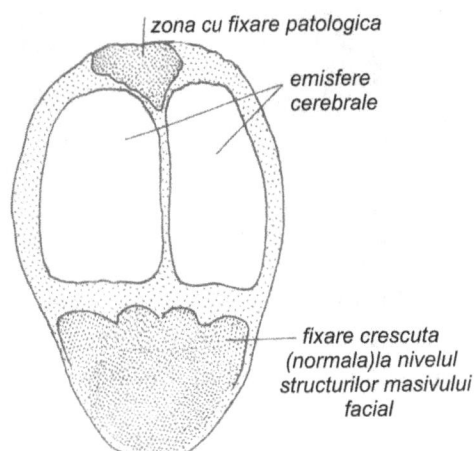

Fig. 386 Cerebral scintiscan (and scheme) in anterior incidence, showing an increased fixation of a pathological significance in the superior third of the right cerebral hemisphere, in the internal half (operatively confirmed right parietal meningioma) (Collection Dr. Ion Codorean – Central Military Hospital)

333

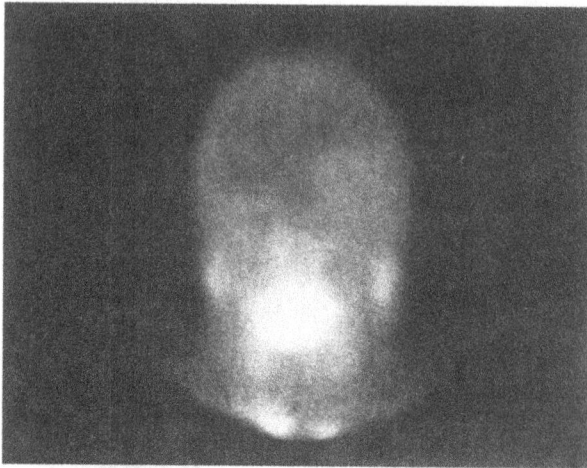

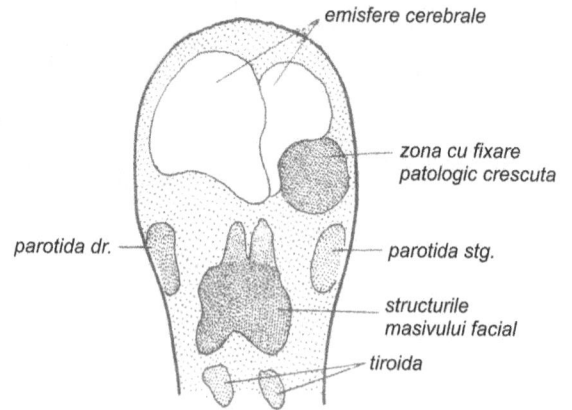

Fig. 387 Cerebral scintiscan (and scheme) in anterior incidence, showing an area characterized by a pathological fixation of the radioactive substance, unprecisely delimited, sited at the level of the inferior half of the left cerebral hemisphere (operatively and anatomicopatologically confirmed left temporo-occipital glioblastoma) (Collection Dr. Ion Codorean – Central Military Hospital)

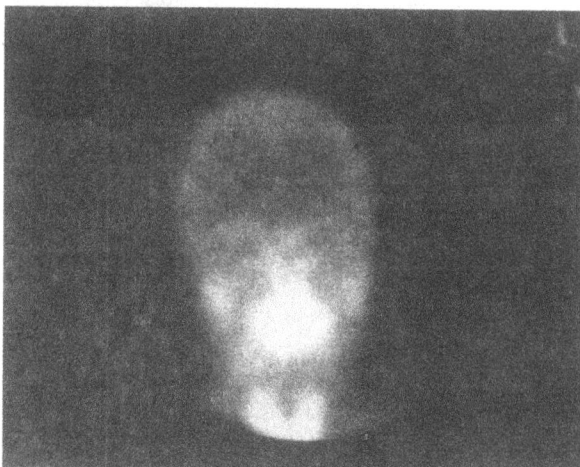

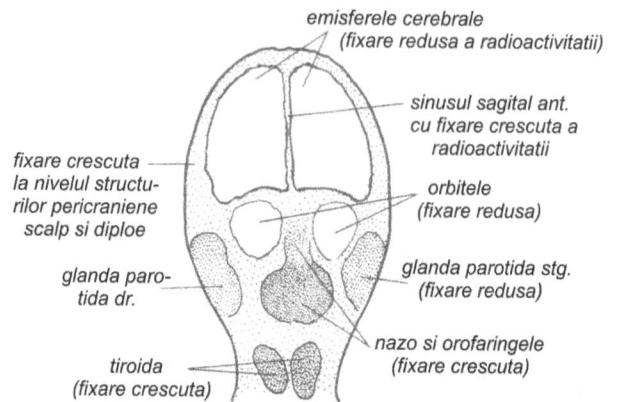

Fig. 388 Normal scintigrafic aspect (and scheme) in anterior incidence, characterized by a radioactivity fixation of low intensity at the level of cerebral hemispheres, delimited by an increased fixation at the level of the pericerebral structures (Collection Dr. Ion Codorean – Central Military Hospital)

334

SOME MEDICOSURGICAL APPLICATIONS

The vertebral column and the spinal cord

The vertebral column may be the site of various congenital, traumatic or orthopedic diseases, manifested by painful syndromes or motility impairments, explained by the previously mentioned anatomical data (Fig.389).

Pain of medullary origin

Pain is a very significant alarm symptom, of a primary value in the detection of a neurological disease. It may be radicular, truncal or plexal.

To understand the peripheral disposition of the radicular and truncal innervation, it is necessary to recall briefly some embriology data. The body of the embryo is formed by the orderly juxtaposition of anatomic units, called metaseres, each formed of three segments: a neural segment, termed *neuromere* or neurotome, a cutaneous segment, called *dermatomere* or *dermatome*, and a muscular segment, called *myomere* or *myotome*. Each metameric unit corresponds to a pair of spinal roots, the radicular innervation being identical itself with the truncal innervation.

As a consequence of the development of the limbs and of the formation of the nerve plexuses, the metamerism is maintained only at the level of the trunk. As at the level of the limbs the plexuses are formed by interjoining of branches of the spinal roots, the metamerism disappear and the radicular innervation differs from the truncal innervation.

Radicular pain *(radiculalgia)*. On their course, between the spinal cord and *the dura mater*, the roots maintain their pial covering, then – after perforating *the dura mater* – the arachnoid adheres to the root, forming the interradicular septum, while *the dura* forms a sheath both for the motor root and the sensory root. Then, the arachnoid penetrates for a small distance into the intervertebral foramen and fuses with the perineurium, while the dura passes gradually into the epineurium. The compressive or degenerative processes may damage this structure of the radicular root, causing *the radicular sheath fibrosis syndrome*. The radicular pain spares the distribution of dermatomes and is located at the trunk as horizontal bands – *"girdle pain"* – and at the level of the limbs as longitudinal bands. It is induced or intensified by any increase of the CSF pressure (cough, sneezing, defecation effort, compression of jugular veins).

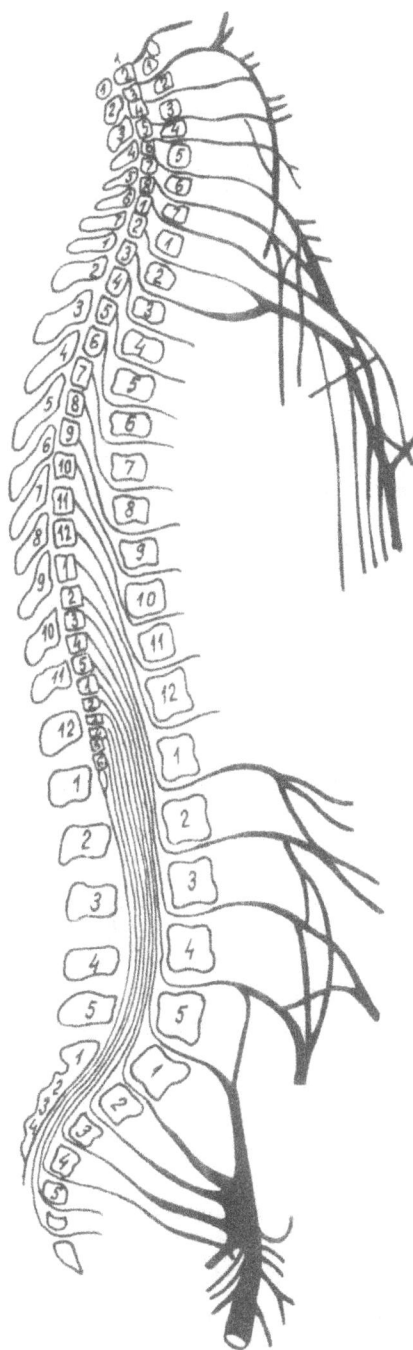

Fig. 389 Topographical relation between the segments of the spinal cord, the vertebral bodies and the emergence site of the nerve roots

335

It may advance towards pareses and paresthesias. In traumatic lesions, amyotrophies and abolition of the osteotendinous reflexes are recorded.

Beside the establishment of the motor deficit, it is necessary, by evaluation of the articular movements, to assess by the electromyography (EMG) the denervation degree, which may range from potential anomalies to a total sideration of the physiological functions of the nerve root.

Sicard and Forester distinguish:
- the pure radicular syndrome, consecutive to the exclusively radicular involvement;
- the ganglionic and funicular syndrome of bony etiology.

The truncal pain (plexalgia) or the neuralgia proper differs from the radicular pain, because it objectifies the truncal or plexal innervation.

The medullary painful syndromes

Clinically, two main syndromes are distinguished:
- *the neuralgic syndrome or neuralgia*, which is a peripheral irritative syndrome, and
- *the neuropathic syndrome or neuropathy*, which is a peripheral deficient syndrome.

The clinical manifestations of the two syndromes are often intricated, so that a clear-cut differentiation is not possible.

The neuralgic syndrome

It is characterized by paroxysmal pain in the distribution area of the sensory protoneuron and is caused by various pathological processes the site of which may be:
- intraspinal: meningoradicular tumours, arachnoiditis, epiduritis, herniation of the intervertebral disc etc;
- spinal: herniation of the intervertebral disc, spondylosis, vertebral injuries, Pott's disease, vertebral metastases etc.;
- extraspinal: tumours, vascular aneurysms, postfracture calluses etc.

We mention below some of the neuralgic syndromes.

The Arnold neuralgia (cervico-occipital neuralgia) is characterized by pain in the superior nuchal and occipital regions, the distribution area of *the Arnold's nerve* (dorsal branch of the C2 nerve) which, at the level of the vertex, establishes an anastomosis with the ophthalmic nerve, which explains the irradiation of pain.

The brachyalgia or cervicobrachial neuralgia manifests itself by pain in the inferior laterocervical region and in the corresponding thoracic limb.

The lesion involves the vasculonervous elements contained in the anatomic region called *cervicobrachial junction* (Sunderland), *ilar area of the upper limb* (Feringa) or *thoracic* (Rob) and expresses itself under the form of various syndromes.

The thoracic aperture syndrome ("cervicothoracic outlet syndrome"- Rob) appears following the compression, obstruction or irritation of the vasculonervous formations from the base of the neck (jugulosubclavicular confluence, vagus, phrenic nerve, subclavian artery, carotid artery, recurrent nerve, C8-T1 inferior trunk of the brachial plexus, sympathetic chain of the stellate ganglion).

It is characterized by a symptomatic polymorphism: pain, paresthesias, diminution of force, colour change of the integuments, oedema, sometimes ulcerations (of Raynaud's disease type) etc.

In the etiology of these syndromes may be also incriminated:
- congenital factors: mega-transverse process, cervical rib, bifid clavicle, intercostal synostosis, aberrant insertion of the middle scalene muscle on the first rib, etc.;
- dynamic factors: hyperabduction of the arm followed by the extension of the axillary vasculonervous bundle on the coracoid, like on a pulley;

- traumatic factors: fracture of the clavicle, scapulohumeral dislocation – of the subacromial type -, vicious callus of the clavicle, etc.

We enumerate below the most frequently encountered syndromes in this case:

1) The cervical rib syndrome. It is encountered in 2% of normal individuals and is bilateral in 80% of cases. It remains often asymptomatic, clinical manifestations appearing only in 10% of cases: nuchal pain of a variable intensity, without contracture and irradiation towards the extremities; paresthesias, especially of the hands and mainly on the territory of the ulnar nerve; compression phenomena of the subclavian artery (cold sensations, cramps, lowering of the amplitude of the pulse etc.); pareses, atrophies.

2) The scalenus anterior syndrome. The variation of the form and insertion of the scalene muscles or the existence of the fibrous tracts between the transverse process of the C7 vertebra and the floor of the interscalene space may produce the compression of the vasculonervous bundle. It is considered that the hypertonia of the anterior scalene muscle would cause the ascent of the first rib, compressing in this way the brachial plexus. Therefore the anterior scalenotomy is recommended not only in the treatment of this syndrome , but also in that of the cervical rib syndrome.

Significant in this case is the Adson and Coffey manoeuvre: the hyperextension of the cervical column, with rotation of the chin to the examined side and respiratory arrest after a deep inspiration, brings about the diminution or depression of the radial pulse.

3) The costoclavicular syndrome. It was described by Falconer and Weddell in soldiers carrying the haversack, their shoulders being drawn downwards and backwards, which leads to the narrowing of the costoclavicular space and to the compression of the vasculonervous bundle.

4) The hyperabduction syndrome. It appears after a long and repeated hyperabduction of the arms in house painters, mechanics, gymnasts etc.

It is recognized by evaluation of the radial pulse changes, with the upper limb in maximum abduction.

Furthermore, the maximum abduction of the arm with the forearm in extension induces pain – the Lasègue test of the upper limb.

5) The Pancoast-Tobias syndrome (superior thoracic aperture syndrome). It appears following the development of a malignant tumor at the level of the thoracic apex. It may be associated with the Claude Bernard-Horner's or Pourfour du Petit syndrome.

The intercostal neuralgia (*thoraco-abdominal neuralgia*) is characterized by *"belt pain"* or *"grindle pain"*, either at the level of the thoracic wall or on the anterolateral wall of the abdomen, an area supplied by the T1-T12 thoracic nerves. It occurs often after costal fractures or vicious calluses or after viral infections (herpes zoster).

Pain is elicited by compression of the Valleix points for the thoracic nerves: the parasternal point (intersection of the anterior axillary line wit the intercostal space); paravertebral point and iliac point (at the level of the anterosuperior iliac spine).

The sciatic neuralgia in the course of *the lumbar discopathy* is caused by the posterior displacement of the lumbar vertebral disc into the spinal canal and is characterized by well defined neurological and radiological signs.

The intervertebral disc is formed of the pulpous nucleus and the fibrous or lamellar ring, both situated between two cartilage plates which cover the surfaces of the vertebrae.

The pulpous nucleus is the element which governs the mobility between the vertebral bodies. According to Armstrong, it would act as a pivot, according to other authors it would move in the so-called cavity which surrounds it, shifting during extension and effecting a backward movement during flexion, Roaf comparing it, therefore, with a *"well-inflated tyre"*. It contains neither vessels nor nerves and is spatted by fibrous tracts starting from the periphery.

Around the pulpous nucleus are disposed concentric *lamellae* of colagenous fibres, which form the fibrous ring *(annulus fibrosus)*. Laterally the respective fibers are anchored to the bone, forming the Sharpey's fibres and designated under the name of marginal ring.

During the movements of the column, the pulpous nucleus shifts into the mass of the fibrous ring towards its convexity.

If the disc is submitted to a symmetrical axial pressure, the ring and the nucleus form a functional couple, which withstands the displacements. The excessive hydrophilia of the disc (88% water) shoul be emphasized, which under such conditions increases about four times the initial pressure, that is higher in young subjects. In the upright posture, the disc loses a certain amount of the included water and, according to Kapandji, its thickness diminishes, the subject decreasing in height by about 2 cm.

In dorsal decubitus the hydrophilia intervenes again as the axial pressure is no more exerted upon the vertebral bodies and the height returns to its former value.

The inner tension of the disc, its explosive force and its elastic damping capacity spread uniformly the locally exerted pressure. If the fibers of the fibrous ring are strong, the sinking of the vertebral plate, respectively the so-called "intraspongious hernia", occurs.

The mobility, the flexibility and the resistance to strain of the vertebral column are due to the algebraic summation of the functional capacity of the motor segments, the ligaments and the somatic musculature.

The most frequent is the lumbar location, followed by the cervical, which is predominantly observed in males, as a result of repeated microtraumatisms, of accidental injuries or of excessive strains exerted on the vertebral column.

The degenerative phenomena (dehydration of the nucleus, disappearance of the fibrillary elasticity, disorganization and cleavage of the fibers) at the level of the intervertebral disc, which occur already at the age of about 20 years, explain the high frequency of this disease. The gelatinous substance of the nucleus may diffuse along the fibers of the ring, the fusion being concentric or, most frequently, radial, seldom anterior and more often posterolateral.

Most authors consider that whenever a free physiological flexion or extension occurs, unacompanied by the contraction of the antagonistic muscles, the pulpous nucleus exhibits a tendency to shift.

In dependence on the topographical site, the discal hernias are classified as follows:
- intraspinal discal hernia;
- infraforaminal discal hernia – compression of the corresponding root by the external portion of the disc;
- lateral discal hernia;
- anterior discal hernia.

As a result of the compression or traction of the nerve roots, three stages are distinguished in the course of the disease:
- the first stage, characterized by paresthesias and radicular type pain;
- the second stage (radicular compression syndrome), manifested through deficit phenomena in the area of the affected root: hypesthesia, force deficit, changes of reflexes etc.;
- the third stage (radicular paralysis) characterized by sensorymotor deficit in the radicular distribution area.

As regards the migration of the gelatinous substance towards the common vertebral ligament, Kapandji distinguishes the following stages: initially it is attached to the nucleus and blocked under the ligament, but it may be reintegrated by vertebral tractions, but most often it dislocates the posterior common vertebral ligament, producing the so-called "free" discal hernia, or remains blocked under the ligament, preventing thus any return to its normal site; afterwards it may slip upwards or downwards, giving rise to the subligamentous migratory hernia.

As a consequence appear either lumbar pain, called lumbalgia (the nerve fibres are put under pressure), or radicular pain, called radiculalgia (the hernia compresses the spinal nerve).

338

Hugo A.Keim considers that there is a correlation between the site of hernia, pain, sensibility, muscular insufficiency, atrophy and reflexes.

We present some data, according to Kapandji, regarding the disposition of the spinal nerve.

The spinal nerve emerges through the conjugate foramen, bounded anteriorly by the posterior part of the intervertebral disc and the adjacent part of the vertebral body, inferiorly by the pedicle of the underlying vertebra, above by the pedicle of the overlying vertebra and behind by the interapophysial joints. Initially it is situated inside the dural sac; it approaches the internal wall of the latter, perforating it at the level of the radicular neck, which is the point, the passage point of the nerve, maintained by the dural sac.

The relations are revealed by a transverse section on the upper surface of the vertebra, which shows: the spinal cord, surrounded by the dural sac, is situated in the vertebral canal and bounded anteriorly by the posterior common vertebral ligament and behind the yellow ligament; the anterior surface of the interapophysial joints contains the capsule strengthened by a ligament, which is a prolongation of the yellow ligament.

In this way, the nerve passes through a tunnel, which has in front the disc, doubled by the posterior common vertebral ligament, and behind the interapophysial joint doubled by a prolongation of the yellow ligament.

In the pathogenetic mechanism of discal herniation, the radicular herniation occurs, according to Kapandji, in three stages: anteflexion of the trunk, onset of the lifting effort and, finally, redressement, straightening of the trunk, which implies the appearance of the radicular pain due to the compression of the roots of the sciatic nerve.

Afterwards occur changes of the osteo-articular reflexes and, finally, motor disturbances at the level of the damaged sciatic nerve.

The elongation tests of the sciatic nerve permit to establish an early diagnosis:

- *the Lasègue test:* patient in dorsal decubitus, the examiner raises the patient's lower limb with the leg in extension and pain appears along the sciatic nerve. Charley's works after an explanation of the Lasègue sign, which appears as the result of the tension exerted upon the sciatic nerve or its roots, which may perform a free run of 12 mm at the L3 level, permitting, normally, to raise the pelvic limb with the knee in extension;

- *the Bonnet test:* patient in dorsal decubitus, the examiner flexes the patient' leg on the thigh and the thigh on the abdomen, followed by the maximum adduction of the thigh, moment at which pain appears;

- *the Neri test:* patient in upright posture, with the lower limb in extension, is asked to effect the anteflexion of the trunk, touching the ground with the hands; pain appears on flexion.

In addition, *the painful Valleix points for the sciatic nerve are perceived:* the middle of the posterior region of the thigh (the major sciatic incisura), the area behind the greater trochanter.

The local circulatory disturbances intensify the neurological phenomena. Therefore, cold and moisture, by congestion of the regional nerve plexuses, increase the pain, while rest, warmth and antiphlogistic medication reduce it, as a consequence of diminution of the congestion.

The disappearance of pain is not always indicative of the return of the disc in the intervertebral space (which occurs infrequently) or of the accommodation of the roots to the decrease of the volume of the hernia. Most often is due to the functional interruption of the sensory root or to the appearance of a space in the vertebral body in which the hernia retires itself.

Surgery, consisting in laminectomy with excision of the herniated disc, is often indicated.

The neuropathic syndrome

It is characterized by sensory, trophic, vegetative and motor (paralyses) disturbances, due to the interruption of the anatomical and functional continuity of the peripheral motor neuron and of the sensory protoneuron (Fig.390-406).

Several clinical aspects may be distinguished in the framework of this syndrome.

The radiculopathy. The most clear-cut clinical form is *the cauda equina syndrome*.

As a consequence of injuries (T11-T12), of disc herniation, of a hypertrophic callus or of an arachnoiditis, a partial lesion of the L1-S5 roots may appear, which brings about both motor phenomena and a sensory deficit along several nerve roots.

According to the lesional site, several variants may be described:

- the superior lumbar type, involving the L1-L4 roots. It is characterized by the abolition of the patellar reflex and maintenance of the Achille's reflex, anaesthesia on the antero-external surface of the thigh, gait and upright posture impossible.

- the middle lumbosacrate type, involving the L5-S1-S2 roots. It gives rise on the clinical plane to the abolition of the Achilles reflex with the maintenance of the patellar reflex, presence of sensibility on the scrotum and on the internal part of the buttocks; sphincterial or genital disorders are lacking and the anal reflex is present;

- the inferior type, involving the S2-S5 roots, is characterized by the presence of the sphincterial and genital disturbances, as well as by the absence of motor disorders, with abolition of the anal and bulbocavernous reflexes.

In the current practice, so clear-cut differentiated syndromes are unfrequently encoutered. Usually, concomitantly with the spinal segment, the injury destroys also the spinal cord and the roots, causing a flaccid paraplegia, total anaesthesia, irreversible sphincterial, trophic, vascular and genital impairments.

The plexopathy or plexus paralysis may assume various clinical forms (cervical, brachial, lumbar and sacropudendal), but the best individualized clinically and of a real practical significance is the paralysis of the brachial plexus or the inferior cervical syndrome.

The inferior cervical syndrome is caused by compressions (scapulohumeral dislocations, tumours), sections (wounds caused by cutting and thrusting weapons and gunshot wounds) or elongations and avulsions (brutal tractions).

It is characterized by symptoms caused by the involvement of the roots forming the brachial plexus and manifests itself in three ways:

- *the superior type, Duchenne-Erb* (C5,C6 and C7), manifested by paralysis and atrophy of the deltoid, biceps, anterior brachial, brachioradial, supraspinatus and infraspinatus muscles, of the clavicular fasciculus of the greater pectoral muscle, of the triceps muscle and inversion of the styloradial reflex. The involvement of the C6 roots leads to the characteristic Thornburg position – arms in abduction, forearm and wrist in supination and flexion and fingers in flexion, position due to the paralysis of the adductors of the arm and of the extensors of the forearm, the hand and the fingers;

- *the inferior type, Dejerine-Klumpke syndrome* (C8,T1 and T2), is a condition similar to the Aran-Duchenne syndrome, but characterized in addition also by oculopapillary impairments (narrowing of the palpebral fissure, paralytic miosis, enophthalmia), induced by the involvement of the Budge's ciliospinal centre (situated in C8 and in the first two thoracic segments of the spinal cord), vasomotor disturbances in the respective hemiface (redness, warmth, cyanosis with anhidrosis or hypohidrosis) and cardiac disorders (bradycardia). The Claude Bernard-Horner syndrome (miosis, palpebral ptosis and enophthalmia) may be sometimes associated;

- *the middle type, Remak* (only C7 is injured), is characterized by paralysis of the brachial triceps (the extension of the forearm is not possible) and of the posterior muscles of the forearm (patient cannot effect the extension of the hand and of the first phalanx of the fingers). The Froment's test is characteristic: if a traction is exerted on the hand, with the forearm flexed from the elbow, the deltoid muscle must not contract;

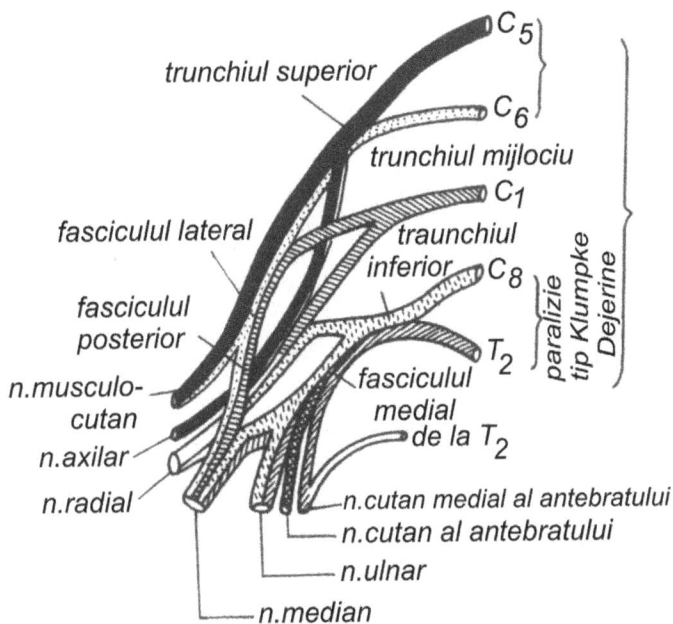

Fig. 390 Scheme of the structure of the
brachial plexus

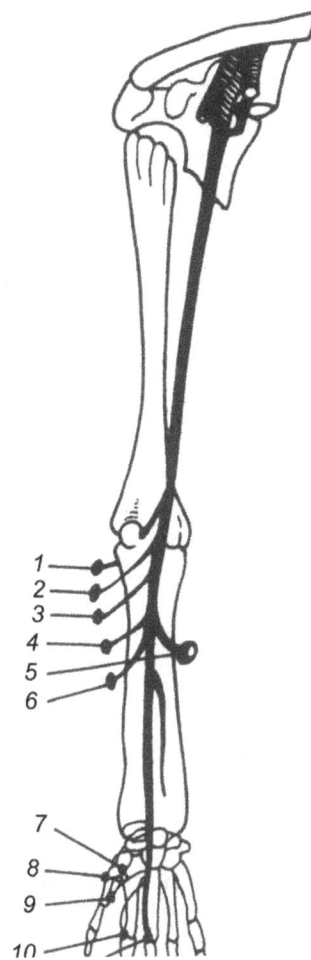

Fig. 393 Median nerve – motor distribution
1 - m. pronator teres; 2 - m. flexor carpiradialis;
3 - m. palmaris longus; 4 - m. flexor digitorum
superficialis; 5 - m. flexor digitorum profundis;
6 - m. flexor pollicis longus; 7 - m. abductor pollicis
brevis; 8 - m. flexor pollicis brevis; 9 - m. opponeus
pollicis; 10 - m. Lumbricalis I; 11 - m. lumbricalis II

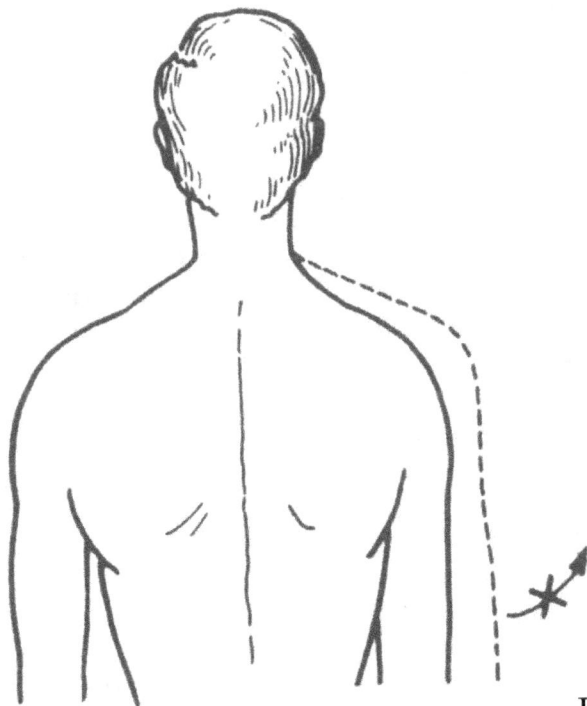

Fig. 391 Paralysis of the circumflex
(axillary) nerve: atrophy of the
deltoid muscle; the abduction of the
arm is impossible

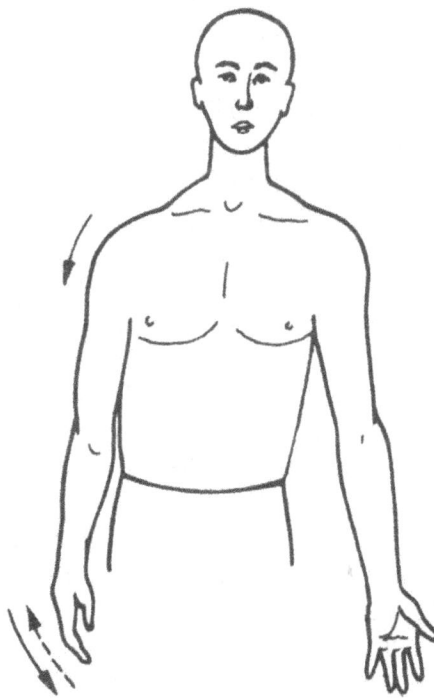

Fig. 392 Prissman type total paralysis of
the brachial plexus: drop shoulder, "bell-
clapper-like" hanging thoracic limb

341

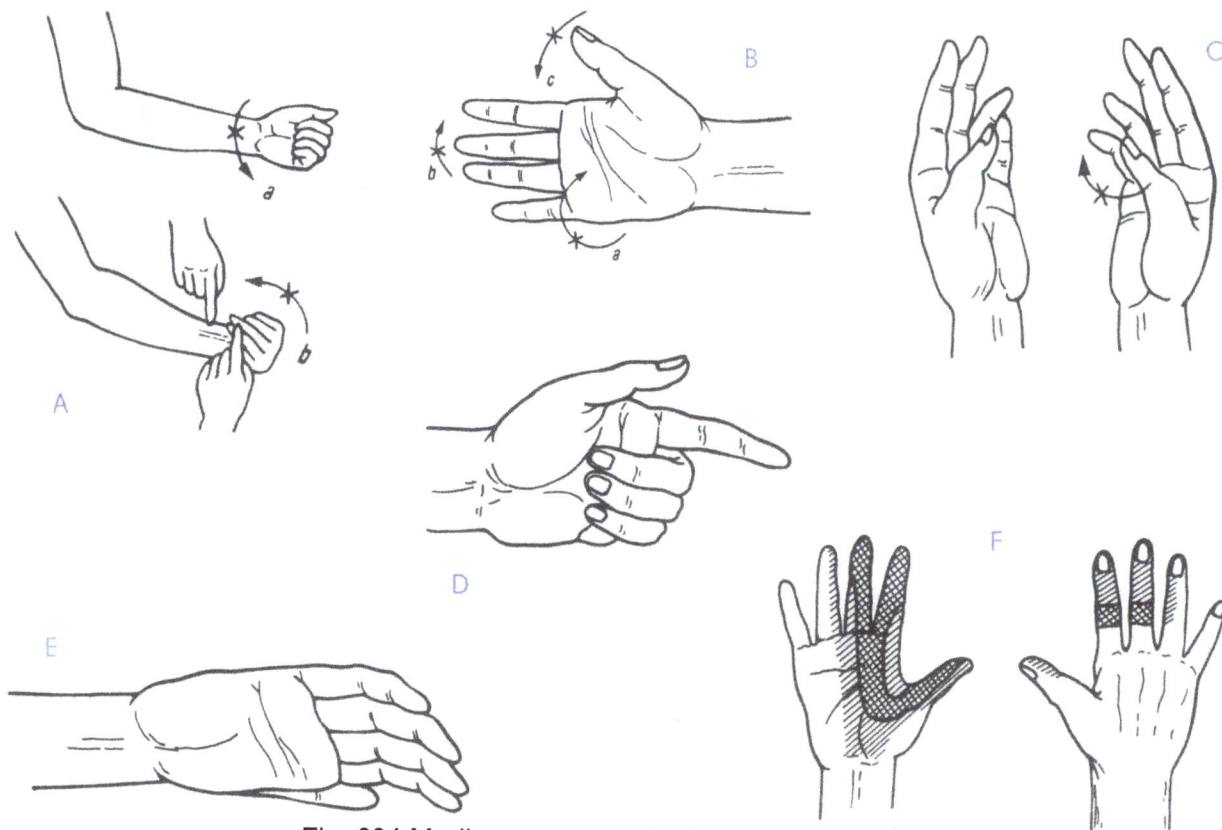

Fig. 394 Medium nerve paralysis – clinical aspects
A – abolition of pronation movements (a) and of the movements of the first three fingers (b); B – abolition of flexion and opposition movements of the thumb (c); C – abolition of the "digital clamp"; D – "indicator hand", through the possible partial flexion of fingers III, IV and V; E - "monkey hand (simian hand), due to paralysis of the thenar eminence; F – injury to the sensory territory

- *Prissman's total paralysis* involves all the spinal roots at this level.

The mononeuritis represents the inflammation of a single nerve trunk and is known in the clinic under the name of paralysis of the median, ulnar, radial, femoral, ischial, common fibular (external poplited sciatic) and tibial (internal popliteal sciatic) nerves, which are the most frequently affected by paralysis.

The polyneuritis or polyneuropathy is a syndrome marked by the exclusive and systematized involvement of peripheral nerve trunks. It is usually located only at the nerves of the lower limbs and exceptionally only at those of the upper limbs; it is characterized by a motor and sensory deficit, better visible in the distal segments. This group includes the infectious (diphteric, botulinic, typhic etc.), toxic (saturnine, alcoholic, arsenical, sulfonamidic), deficiency (beri beri, pregnancy), metabolic (diabetic), hereditary (Dejerine-Sottas disease, Charcot-Marie-Tooth disease) polyneuritis, Besnier-Boech-Schaumann sarcoidosis etc.

The polyradiculoneuropathy is characterized by the extensive, systematized and concomitant involvement both of the spinal roots and of the peripheral nerves, the cranial nerves inclusively. The primary form occurs after a short influenza episode, while the secondary form appears either during the development of the causal disease (sarcoidosis, septicemia) or at about 10 days after the onset of the eruptive infectious diseases or after vaccinoserotherapy. The patient is completely inert (flaccid tetraplegia), the osteotendinous reflexes are abolished, paresthesias and spontaneous pain are present, followed by trophic disorders. In addition to the strictly located forms (cranial nerve tetra- or paraplegia), Landry type ascending polyradiculopathies from the lower to the upper limbs and to the cephalic extremity) and descending polyradiculopathies (located initially at the cranial nerves), they descend afterwards to the upper and lower limbs.

342

Fig. 395 Ulnar nerve – motor area
1 - m. adductor pollicis; 2 - m. flexor pollicis brevis; 3 - m. osseus dorsalis 1 ;
4 - m. interosseus valoris 1 ; 5 - m. Flexor carpatiulnaris; 6 - m. flexor
digitorum profundis; 7 - m. palmaris brevis; 8 - m. abductor digiti V; 9 - m.
opponens digiti V; 10 - m. flexor digiti V; 11 - m. lumbricalis IV; 12 - m.
lumbricalis III

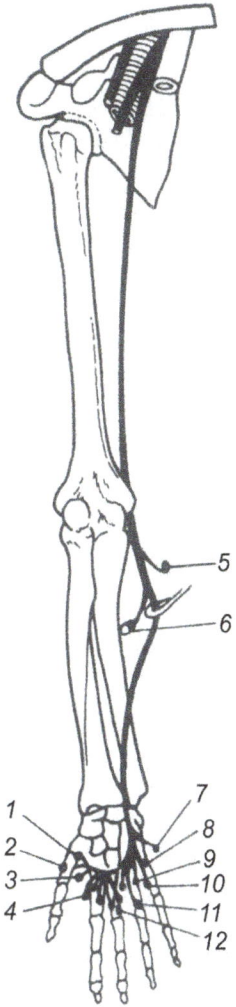

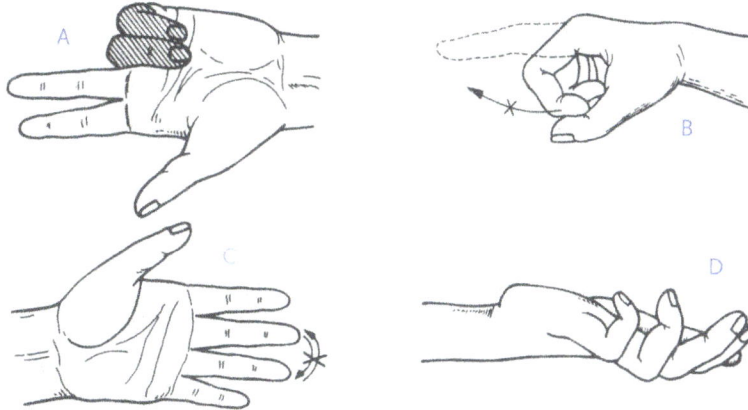

Fig. 396 Ulnar nerve paralysis –
clinical aspects
A- impossibility of flexion of the first
phalanx and of extension of the
second and third phalanges, more
obvious at the auricular (little) and
ring fingers and less marked at the
index and medius (middle) fingers; B
– the impairment of the extension of
the index precludes the "flock"
movements; C – the lesion of the
interosseous muscle renders difficult
the abduction and adduction
movements of the fingers; D – aspect
of the hand due to the damaging of
the interosseous muscles

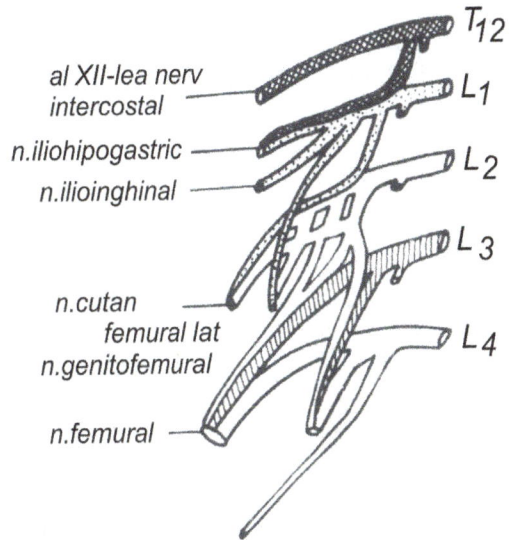

Fig. 398 Scheme of the structure of the
lumbar plexus

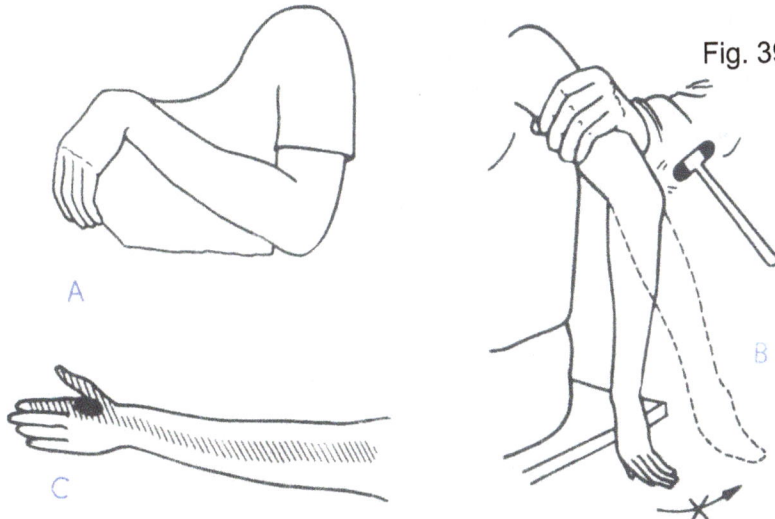

Fig. 397 Radial nerve paralysis – clinical aspects
A – flexion of the forearm on the arm, the hand in
pronation, the fingers in semiflexion, the palm in
abduction; B – abolition of the triceps reflex;
C – anesthesia area.

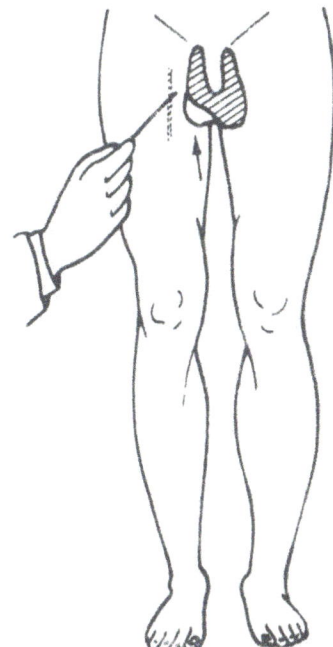

Fig. 399 Genitofemoral nerve paralysis: abolition of the cremasteric reflex

343

Fig. 400 Gluteal nerve paralysis – clinical aspects
A – superior gluteal nerve: imposibility of abduction and inner rotation of the thigh; B – partial paralysis of the superior gluteal nerve; lateral flexion of the vertebral column on the opposite side (Duchenne-Hirker sign); C – lesion of the superior gluteal nerve: tilting the pelvis on the damaged side during walking (b) (Trendelenburg sign), in comparison with the normal gait (a); D – inferior gluteal nerve: the extension of the thigh on the pelvis is impossible and raising from a chair, as well as mounting the stairs are difficult

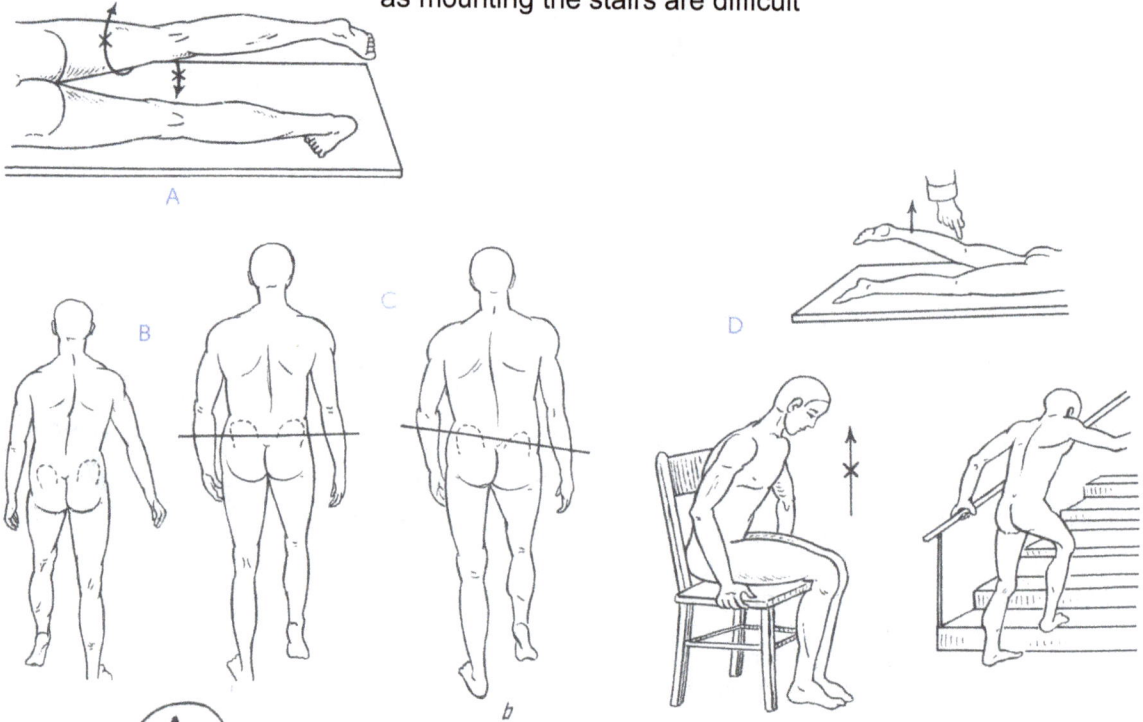

Fig. 401 Sciatic and tibial nerve paralysis – motor distribution
1 - n. ischiaticus; 2 - n., m. semitendinous; 3 - m. biceps; 4 - m. semimembranosus; 5 - m. biceps; 6 - m. adductor magnus; 7 - artera poplitee; 8 - vena poplitee; 9 - n. peronaeus; 10 - n. tibialis; 11 - m. popliteus; 12, 13 - m. gastrocnemianus; 14 - m. plantares; 15 - m. soleus; 16 - m. tibialis posterior; 17 - m. flexor digitorum longus; 18 - m. hallucis longus; 19 - n. plantaris lateralis; 20 - n. plantaris medialis; 21 - m. abductor digiti V; 22 - abductor hallucis; 23 - m. interosseus dorsalis I; 24 - m. interosseus dorsalis IV.

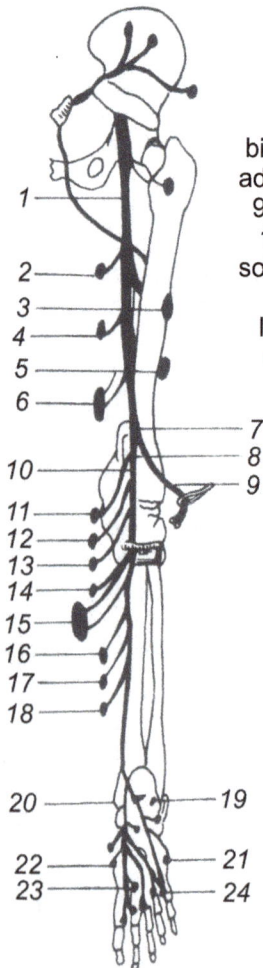

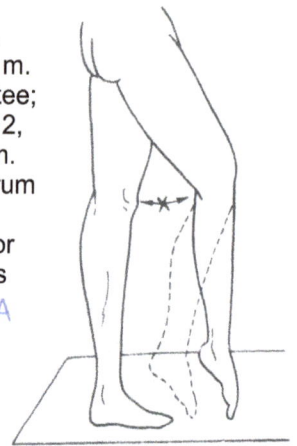

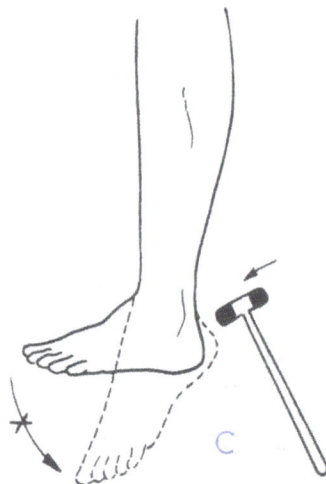

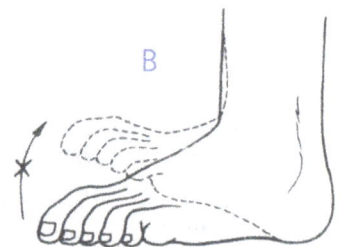

Fig. 402 Sciatic nerve paralysis – clinical aspects
A – imposibility of flexion of the leg; B – spastic flexion of the toes; C – abolition of the Achilles tendon reflex.

344

Fig. 403 Femoral, obturator and peroneal nerves – motor distribution

1-m. iliacus; 2-m. psoas; 3-m. sartorius; 4-m. rectus femoralis; 5- m. vastus medianus; 6-m. vastus lateralis; 7- m. vastus intermedius; 8- m. obturator extremis; 9- m. pectineus; 10- m. gracilis; 11- m. adductor brevis; 12- m. adductor longus; 13- m.adductor magnus; 14- m. tibialis anterior; 15- m. peronaeus longus; 16-m. extensor digitorum longus; 17- m. peronaeus brevis; 18- m. extensor hallices longus; 19-m. peronaeus tertius.

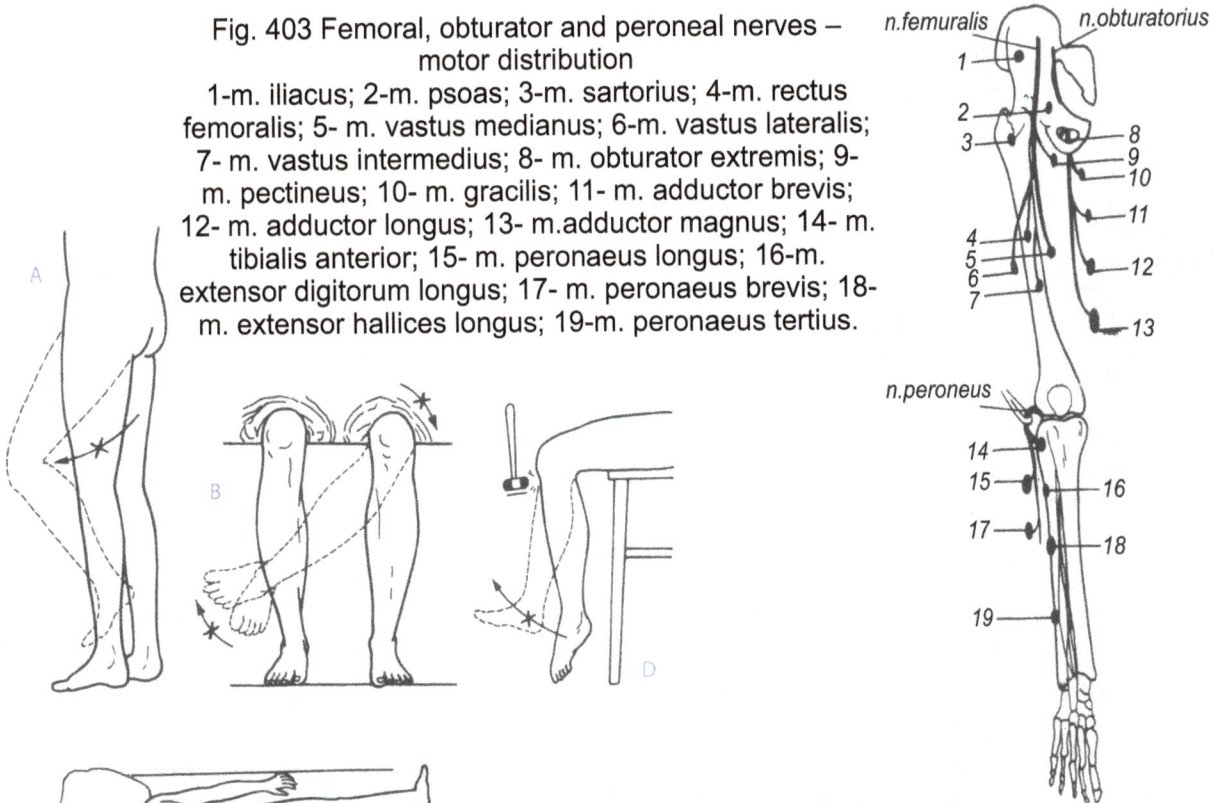

n.femuralis   n.obturatorius

n.peroneus

Fig. 404 Femoral nerve paralysis – clinical aspects

A - femoral quadriceps muscle paralysis makes impossible the active extension of the leg on the thigh; B - tailor (sartorius) muscle paralysis does not allow execution of opposition, flexion and external rotation of the knee; C - pectineus muscle paralysis reduces the adduction movement of the thigh; D - partial abolition of the patellar reflex .

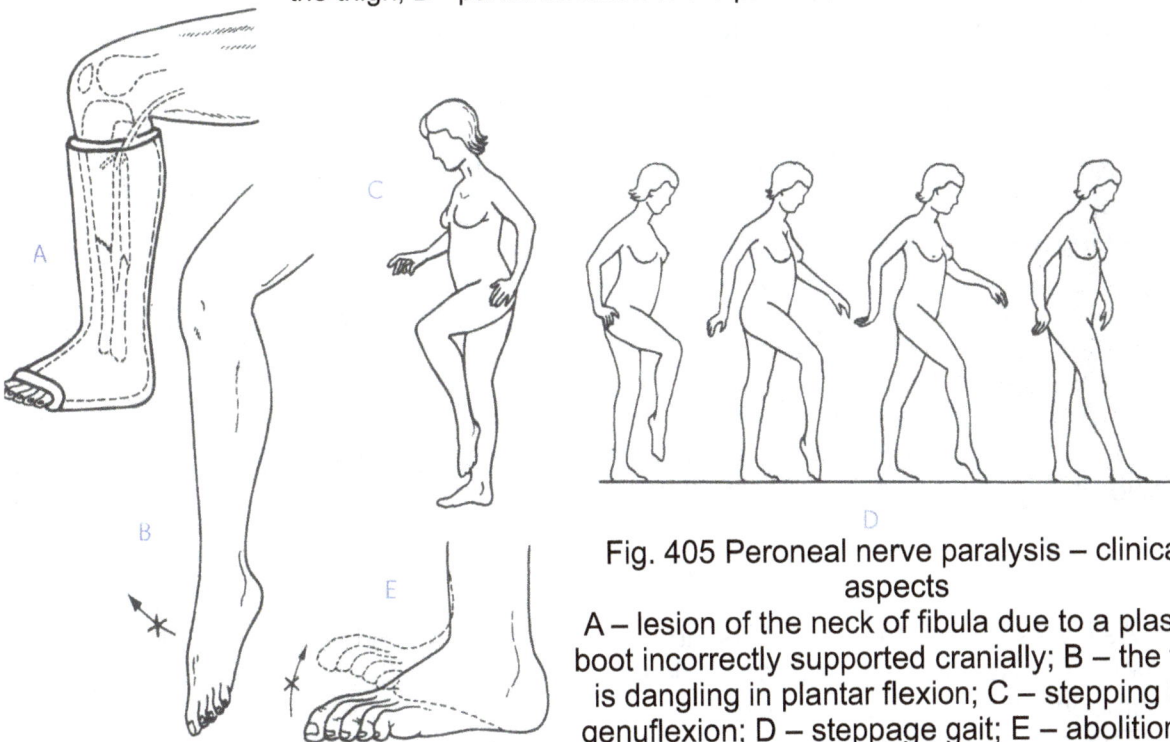

Fig. 405 Peroneal nerve paralysis – clinical aspects

A – lesion of the neck of fibula due to a plaster boot incorrectly supported cranially; B – the foot is dangling in plantar flexion; C – stepping by genuflexion; D – steppage gait; E – abolition of the dorsiflexion movement of the foot

345

Fig. 407 Diagram of the injured areas in the diseases of the motor unit + I sudden destruction (anterior acute poliomyelitis) or slow distruction (progressive muscular atrophy) of the cell body; II sudden destruction of the axon (peripheral nerve lesions); III impairment of the neuromotor synapse (myasthenia gravis), four mechanisms being suggested in this respect: the deficient acetylcholine production or elimination, an overactive cholinesterase system, a diminished sensibility to acetylcholine of the terminal plate or, more surely, the release of a curara-like substance into circulation; IV lesion of the muscle fibre membrane (myotonia); V diseases of the contractile mechanism

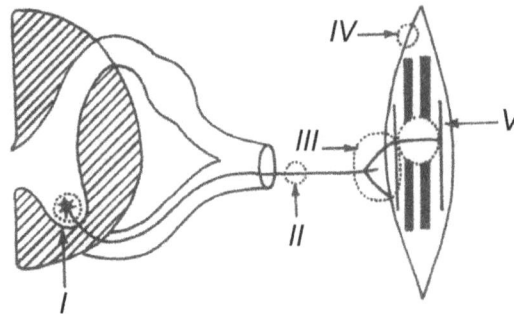

Fig. 406 Tibial nerve paralysis, impossibility of standing tiptoe on the damaged side

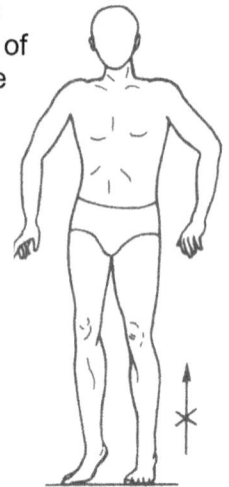

Motor disturbances

Motor disturbances are characterized by *pareses* or *paralyses* in dependence on the severity of the lesional process and they appear as a consequence of the involvement either of *the central motor neuron* – from the cerebral cortex up to the spinal cord -, or of *the peripheral motor neuron* – from the spinal cord up to the muscles (Fig.407).

For the detection of these pathological phenomena it is necessary to study the osteotendinous reflexes, which may be diminished, abolished, pendular or reversed.

In some cases, an excitation may cause successive muscle contractions, known under the name of polykinetic osteotendinous reflexes.

In the case of lesions of the pyramidal tracts, the following signs are investigated:

- *Babinski's sign* – the reflex is conducted through the delta fibres of the painful sensibility and indicates the lesion of the area 4 and of the pyramidal tract. The patient lies in dorsal decubitus, with the lower limbs in extension and the musculature relaxed; the excitation of the internal border of the foot with a pointed object, from the heel towards the toes, produces the dorsal extension of the hallux;

- *Oppenheim's sign:* the excitation of the tegument at the level of the anterior surface of the tibia elicits the same reaction;

- *Gordon's sign:* the compression of the posterior musculature of the leg induces the same response;

- *Schäffer's sign:* the vigorous pinching of the Achilles tendon is followed by a similar response;

- *Grigorescu's sign:* the sudden plantar flexion of the four last toes elicits the same response;

- *Bing's sign:* by pricking with the point of a needle the posterior surface of the lower limb, an identical response is obtained;

- *Noica's sign:* the flexion of the second toe, elicits a similar response;

- *Rossolimo-Hoffmann's sign:* it is characteristic of the lesion of the cortical area 6. The percussion with the plexor for reflexes of the toes of the affected lower limb produces promptly the flexion of the toes.

Medullary vascular accidents

As it was shown, the vascularisation of the spinal cord during the embryonic period is metameric. Afterwards it undergoes significant changes and in the adult a vascularisation with poor compensation resources is present.

The areas with the best vessel supply are those situated at the level of the two intumescences, cervical and lumbar, the first being better visualized.

The thoracic spinal cord receives a few branches through the conjugate foramina, two critical areas being present at the level of the fourth and ninth thoracic vertebrae.

Hence, Schneider compares the spinal cord with an irrigated field and the thoracic area with the terminal portion of the irrigation system starting from the cervical and lumbar areas.

Any compression at this level causes ischaemia. Since these areas are rich in motor cells, which display avidity for oxygen, they will be the first to be affected by ischaemia.

The anterior spinal artery, as well as the anterior radicular arteries which reach the cervical region, supply only the anterior two-thirds of the spinal cord. The posterior branches supply the posterior horns and the posterior cords. Between them lies a *limitrophe zone*, which is also damaged in the case of a circulatory deficit.

As a consequence appear paralyses of a spastic character, due to the lesion of the crossed pyramidal tract, as well as impairments of the superficial sensibility, especially thermalgesic disturbances, due to the involvement to the short and medium fibres of the sensory protoneuron and of Clarke's column.

The acute ischaemia process sited in the anterior two-thirds of the spinal cord produces a flaccid paralysis, accompanied by severe sensibility impairments.

The lesion of the great radicular artery of Adamkiewicz (arteria radicularis magna) results in the necrosis and softening of the whole lumbosacral cord, leading to severe and irreversible paralyses.

Particularly severe is the transverse medullary softening, situated preferentially in the middle thoracic region (T4-T6) and characterized by *the transverse medullary infarct*, which extends above and below under the form of softenings described by Zülch.

The selective angiography permitted to detect, beside the venous dilatations consecutive to compressive medullary processes, also arterial anomalies associated with congenital cardiopathies of the angioma and aneurysm type.

Medullary angiomas have the aspect of tufts of vessels (localized vascular proliferations) and manifest themselves under the form of spinal meningeal hemorrhages, accompanied by a medulloradicular syndrome: unilateral or bilateral radicular type pain, modifications of reflexes, diminution of muscle strength etc. The treatment is exclusively surgical.

The arteriovenous aneurysms are usually intra- and extradural; their treatment, too, is surgical.

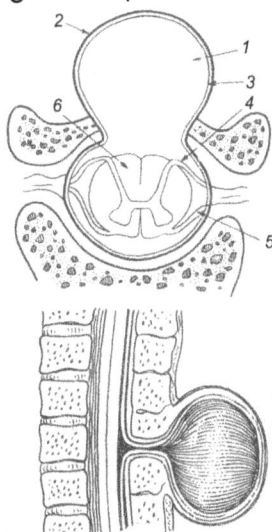

Fig. 409 Spina bifida with meningocele
a- dura mater; b- arachnoid; c- vertebral lamina; d- vertebral body; e- meningocele

Fig. 408 Spina bifida cystica

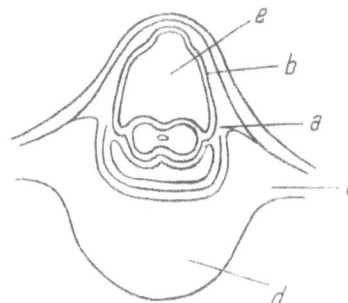

Fig. 410 Various clinical forms of spina bifida aperta

Myasthenia gravis

Myasthenia gravis or Erb-Goldflam's disease is an autoimmune disease, characterized by an excessive fatigability and deficit of the striated musculature (diminution of the power of contraction). The following forms are distinguished:
- the superior bulbar form, in which the deficit is sited at the cranial nerves (ocular muscles, muscles of mastication, of deglutition, of phonation etc.);
- the inferior or spinal form, affecting at first the musculature of the trunk and afterwards that of the limbs;
- the generalized, severest form, which may involve all the muscles of the body.

The statistics on the world-wide scale show that the thymectomy (ablation of the thymus) improves the course of disease.

Hysterical paralysis

The hysterical paralysis appears in tainted individuals, usually after a convulsive seizure of the same nature, and has a sudden onset, by a generalized or localized motor deficit, accompanied by sensitive-sensory anesthesia, with the same distribution. The paralisis is flaccid, spastic and does not imply changes of the osteotendinous, cutaneous and electrical reflexes, neither trophic disturbances. The treatment is performed by suggestion.

Spina bifida

Spina bifida is a congenital malformation, characterized by a closure defect, on a variable length, of the vertebral canal, through which the meningeal envelopes and nerve structures may or may not herniate (Fig.408-410).

If the whole vertebral canal is involved, the defect is called *rachischisis* or *holorachischisis*.

Usually, it is located in the cervico-occipital or lumbosacral region, showing closure impairments of the vertebral canal.

The following clinical forms are described:
- spina bifida occulta, usually detected occasionally, as it may remain asymptomatic throughout the life;
- spina bifida aperta (cystic or tumoral) with meningocele; it is the severest form.

The meningocele is usually associated with hydrocephalus and surgery at the level of the spinal cord may aggravate the hydrocephalus.

Pathological deviations of the vertebral column
(kyphosis, lordosis, scoliosis)
(Fig.411-418)

Kyphosis consists in the exaggeration by at least 20o of the curvature of the vertebral column in its flexion area.

Lordosis consists in the increase of the lumbar curvature, as a consequence of the tendency to counterbalance the anterior weight of the abdomen (obesity, pregnancy). In these cases the centre of gravity is displaced forwards.

Scoliosis is a permanent deformity of the vertebral column in the frontal and lateral plane, associated with a costoscapular prominence on the convexity side of the column, consecutive to the vertebral rotation.

The vertebral body rotates towards the convexity of the curvature, the spinous process towards the concavity, the intervertebral discs are widened towards the convexity and the vertebral pedicles are thinner towards the concave side.

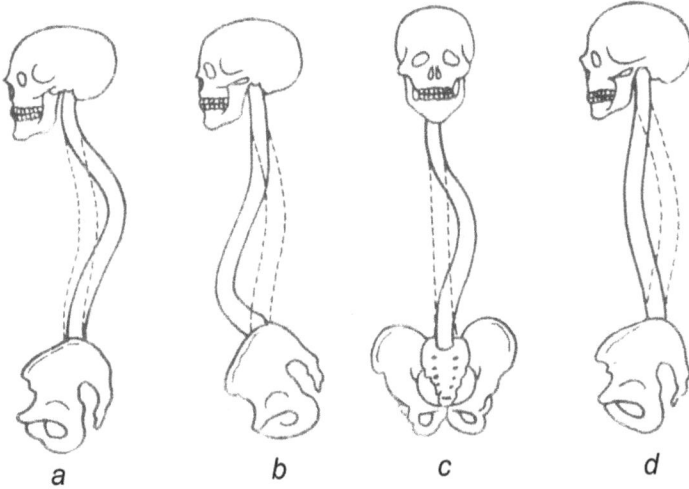

Fig. 411 Pathological deviations of the vertebral column
a- kyphosis; b- lordosis; c- scoliosis; d- vertebral inversion

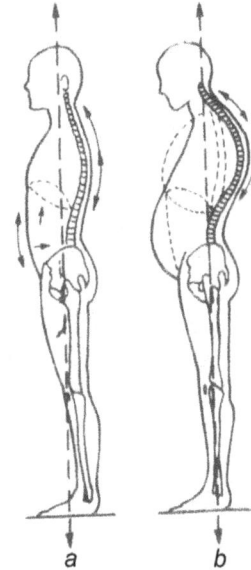

Fig. 412 Kyphotic aspect of the thoracic column a- normal aspect in the sagittal plane; b- kyphotic aspect in the sagittal plane

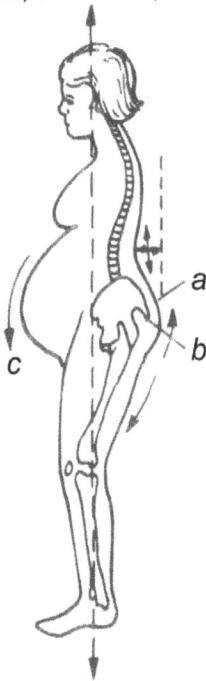

Fig. 413 Lordotic aspect of the vertebral column
a- the gluteal region is pushed upwards and dorsally;
b- the sacrum is more prominent; the centre of gravity of the organism is shifted forwards

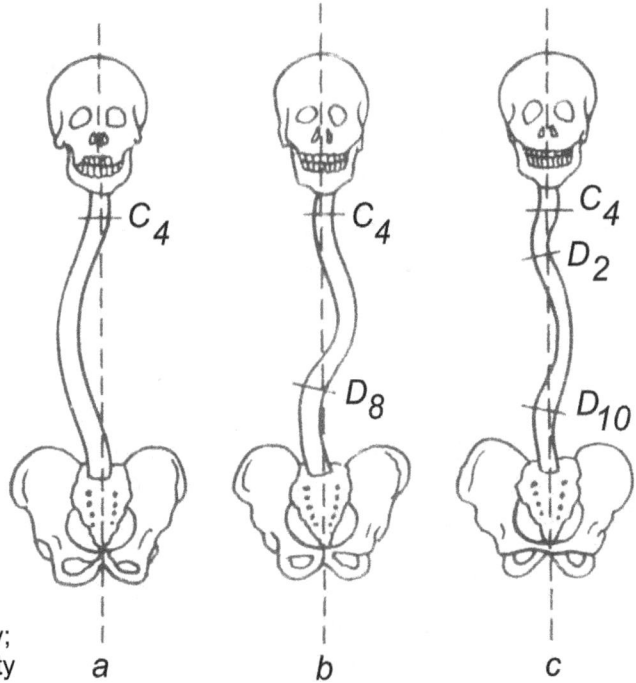

Fig. 414 Scoliosis is a permanent deviation of the vertebral column, in the lateral plane, associated with a costoscapular prominence on the convexity side of the column. Scoliotic aspect of the vertebral column
a- unique curvature;
b- double curvature;
c- triple curvature

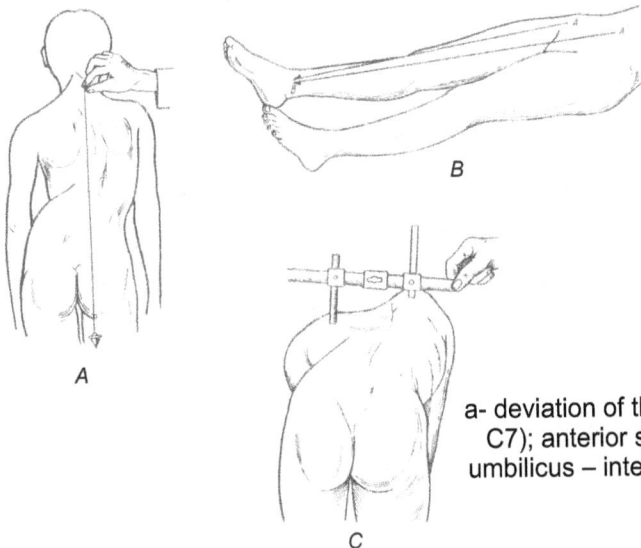

Fig. 415 Diagnosis of scoliosis
a- deviation of the spinous processes (the plumb line is fixed on C7); anterior superior iliac spine-internal malleolus (AB) and umbilicus – internal malleolus (A'B) lines; C- assessment of the gibbosity.

349

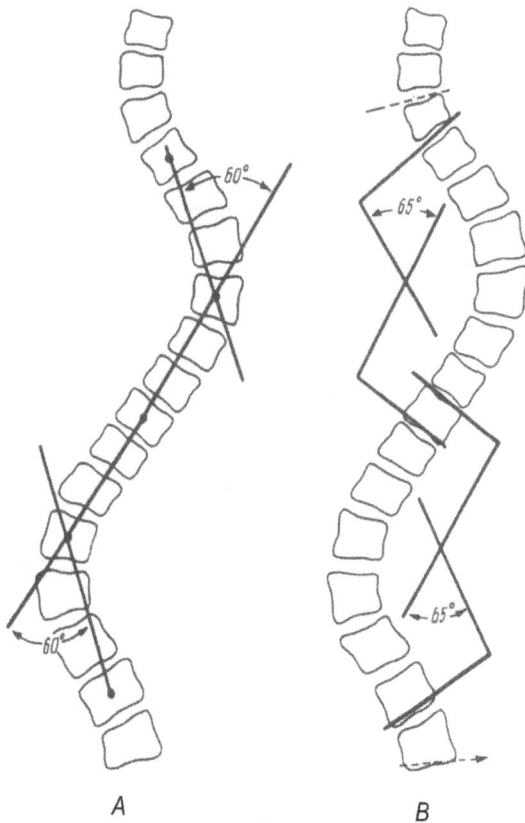

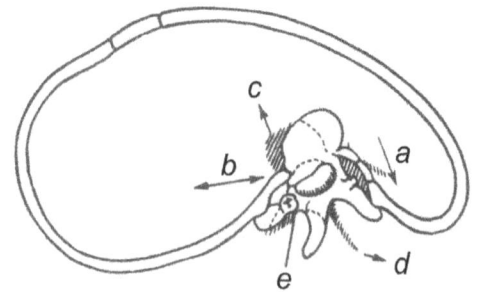

Fig. 417 Prominence of the costovertebral complex on the scoliotic convexity side
a- costal deformation; b- displacement of the costovertebral pint; c, d, e- normal aspect of the vertebra

A   B

Fig. 416 Measurment of the curvature of the vertebral column

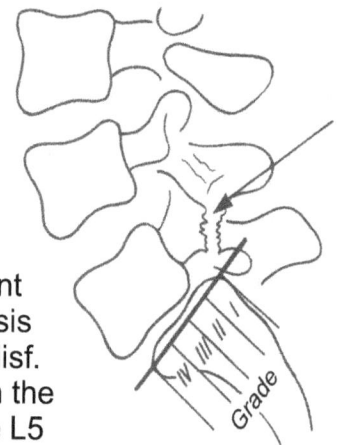

Fig. 418 Assessment method of the listhesis degree after Meyer disf. The distance between the posterior angle of the L5 vertebra and the posterior angle of the S1 vertebra shows the slipping

As a consequence of the vertebral imbalance, the vertical drawn from the apex of the spinous process of the seventh cervical vertebra will not fall on the spinous process S1. The appreciated value of the gibbosity angle corresponds rather exactly to the vertebral rotation degree.

For the appreciation of the severity of scoliosis and for the follow-up of its course – treated or untreated -, Risser-Ferguson's method or Cobb's method are the procedures used.

The rehabilitation treatment of scoliosis should be initiated as early as possible and consists in physiokinesitherapy, orthopedic treatment (Milwaukee brace) or surgical treatment (rachisynthesis after the orthopedic reduction and fixation with Harrington's nails).

Spondylolisthesis is a severe disease of the vertebral column and appears – as Kapandji shows – when the lumbosacral muscle mass can no more assure to the lumbosacral region an efficient girth.

We mention again, according to Kapandji, that the mechanism of its development is the following: as a consequence of the bending of the upper plate of the S1 vertebra, the body of the L5 vertebra has the tendency to glide downwards and backwards; the weight of the body (wg) is decomposed into two forces: a force perpendicular on the upper plate of the sacrum (N) and a force parallel to the upper plate of the sacrom (M) which draws upwards the body of the L5 vertebra.

The slipping is stopped by the firm implantation of the L5 posterior arch. The inferior articular processes of the L5 vertebra are imparted between the superior articular processes of the S1 vertebra, and consequently the slipping force applies the L5 processes on the superior processes of the sacrum, which whistand on each side of the direction of the force.

The transmission occurs through the vertebral arch. The rupture or distruction of the arch, leads to *spondylolysis*. The slipping downwards and forwards of the body of the L5 vertebra results in *spondylolisthesis*. The single elements which still can maintain the L5 vertebra on the sacrum are the muscles of the vertebral grooves and the lumbosacral disc.

Anterior acute poliomyelitis or *Heine-Medin's disease*, of viral origin, appears under the form of epidemics with digestive phenomena. These are followed by a meningeal syndrome and after 4-17 days occur flaccid paralyses; the osteotendinous reflexes are abolished.

After other 6-7 days the paralyses retrocede partially or totally, leaving paralytic sequelae and vicious attitudes: scoliosis, deviations of the lower limb (*pes varus, valgus, equinus, cavus* etc.) etc.

Fractures of the vertical column

They represent approximately 50% of all the lesions of the vertebral column.

According to Arseni, these lesions imply a potential medullary risk, the medullary complications occurring both at the moment of the accident and afterwards, during the transport or the hospital treatment.

In most cases, the fractures occur in the thoracolumbar transition zone (T12-L1), a region of maximum mobility, known under the name *"Lambert's inflexion zone"* (Fig.419).

Fractures may assume various aspects, from the partial fracture, which involves the transverse process, up to the complete fractures, with or without medullary lesions (amyelic fractures or myelic fractures accompanied by medullary shock (Fig.420-422).

In the comminuted fractures, the vertebral body is crushed, and, in addition to the anterior narrowing (specific to compression fractures), also lateral angulation is observed (Fig.422,424). In the case of association with rotation, the ligaments become vulnerable and, if the action continues, they will rupture in the following order: the posterior ligaments, the articular capsules and the posterior longitudinal ligament (Holdsworth's posterior ligamentous complex), resulting in dislocation or dislocation-fracture (unstable fracture)(Fig.425).

In the cervical region, two specific fracture types occur: the fracture of the atlas or the Jefferson type fracture and the fracture of the axis or the James wood type fracture (the fracture occurring in cases of in cases of execution by hanging)(Fig.426, 427).

The severity of these injuries depends on the site of the lesion. As the medullary segments do not correspond to the vertebral ones, it is necessary to appeal to Chipault's scheme (Fig.428).

Each of these lesions involves the reflexogenic centers, which explains the symptomatology of the various lesions, in dependence on the site.

As regards the treatment of fractures of the vertebral column, it is nowadays considered that in stable fractures immobilization is useless, the most rapid restoration of a good musculature (representing the protective cast, being more efficacious than *the best protective cast*)(Merle d'Aubigné, Nicoll).

In the case of unstable fractures (40% of all the fractures), the therapeutic attitude is different: in the fracture of a vertebral body with distruction of the intervertebral disc – plaster cast; in the dislocation fracture – emergency surgery.

In fractures of the cervical region, in order to present the dislocation accompanied by the possible section of the spinal cord, traction according to Crutschfield's method, followed by immobilization with the Schanz type collar or the Minerva type collar, is used. In the thoracolumbar fractures the plaster cast is used, after a previous reduction.

Among the most frequent complications after fractures of the vertebral column rank the Kümmel-Verneuil syndrome and the Brown-Ségnard syndrome.

The Kümmel-Verneuil syndrome: after a minor vertebral injury and a free interval of apparent health, occurs the progressive deformation of the vertebral bodies, followed by the appearance of significant neurological disorders, sited preferentially at the T11-L2 level. Vertebral rachisynthesis with bone graft is used.

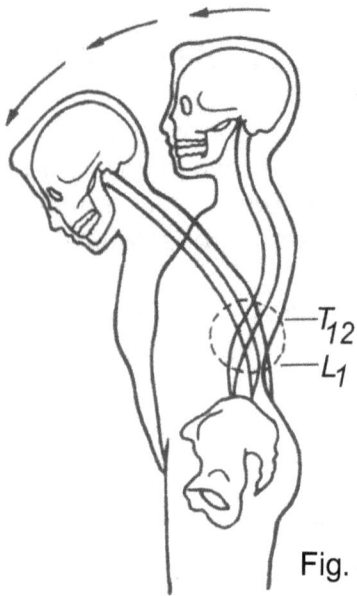

Fig. 419 Lambert's inflexion zone

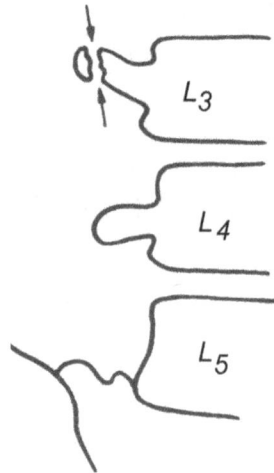

Fig. 420 Fracture of the transverse process of the L3 vertebra

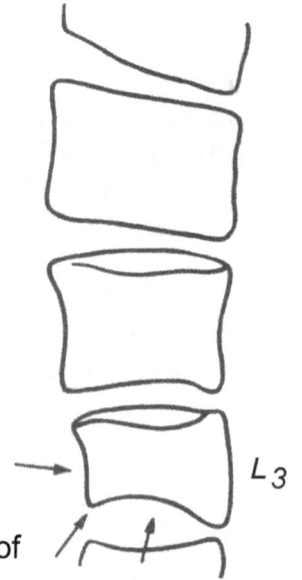

Fig. 421 Fracture of the vertebral body (compression). The severity of the medullary or radicular lesion is appreciated in dependence on its site

Fig. 422 Fracture of the vertebral body with fragmentary displacement and concomitant lesion of the spinal cord

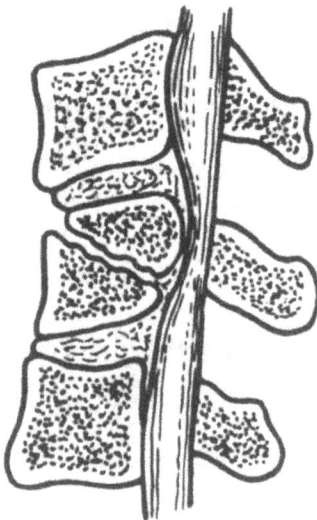

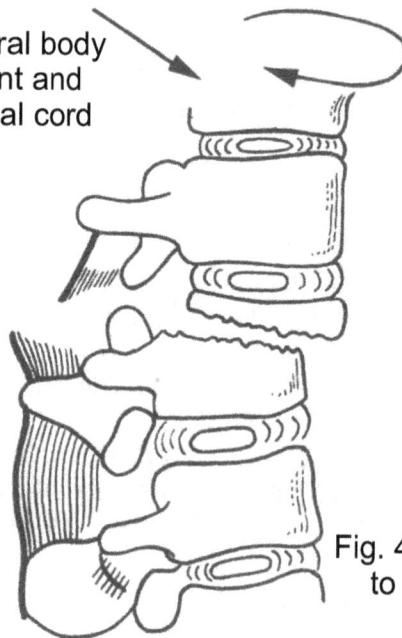

Fig. 424 Wedged fracture type due to compression, described by Watson Jones

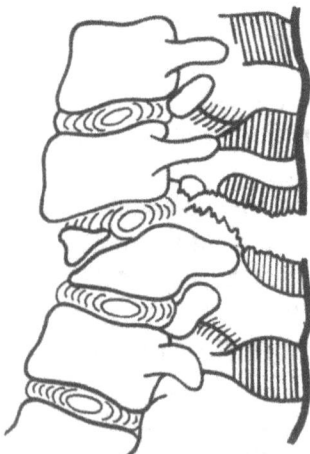

Fig. 423 Unstable fracture type. In the production of fractures the twisting forces play a prevalent role, since they rupture in the first place the ligaments

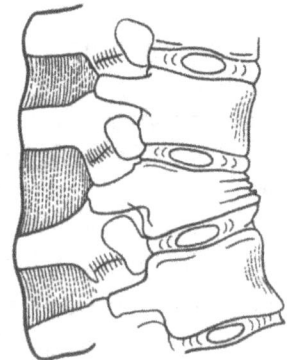

Fig. 425 The dislocation-fracture implies, according to Watson-Jones, the rupture of the posterior ligaments, the fracture of the articular capsules, the fracture of the anteroposterior angle of the underlying vertebral body and the haustation of the overlying column

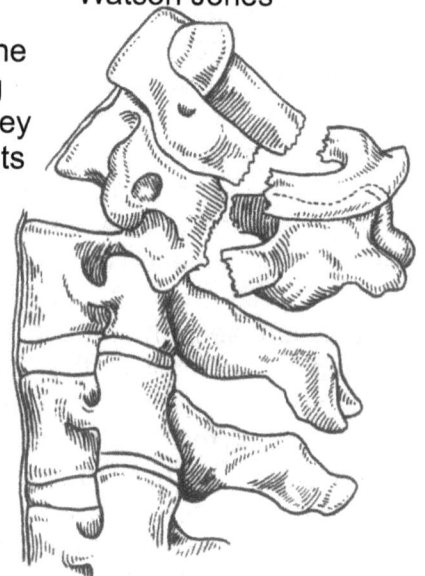

352   Fig. 426 Scheme of the haumatic lesion, called "fracture of the hanged"

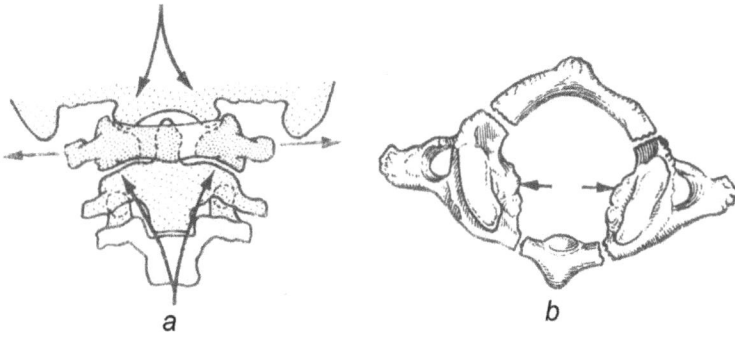

Fig. 427 Production mechanism of the Jefferson type fracture

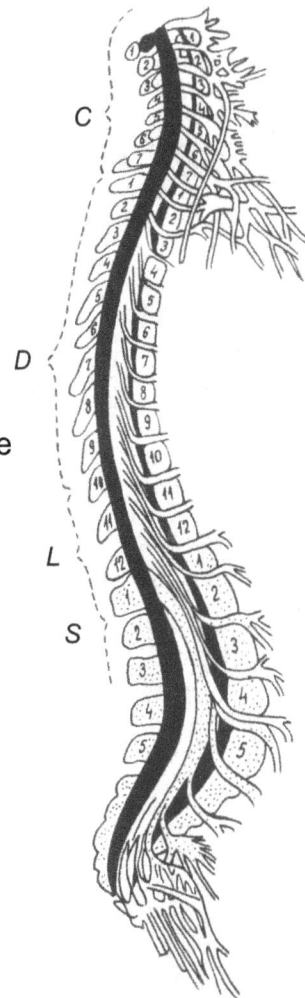

Fig. 428 Medullary vertebral topography.
As the medullary segments do not
correspond to the vertebral ones, the
Chipault's scheme is next

The Brown-Ségnard syndrome appears after cervical and thoracic medullary injuries and is characterized by:

- on the side of the lesion: motor paralysis through lesion of the crossed corticospinal tract, impairments of the deep sensibility by lesion of Goll's and Burdach's fasciculus (their fibres do not cross in the spinal cord) and vasomotor paralysis through interruption of the vasoconstrictive fibres of the damaged hemicord;

- on the side opposite to the lesion: impairments of the superficial sensibility due to the lesion of the fibres of the spinothalamic tract which cross in the spinal cord.

Tuberculosis of the vertebral column

The tuberculosis of the vertebral column or *Pott's disease* occurs especially at the young age (15-30 years) and is sited in areas which are more active from the circulatory viewpoint and mechanically more strained (Fig.431).

The lesion of the vertebral body by hematogenic route results in necrosis, sequestration and lysis of the bone tissue, with the formation of ossifluent abscesses along the tendinomuscular sheaths. In the absence of any treatment, the damaged vertebra may collapse and neurological disorders may appear, but the latter are, nowadays, fortunately, exceptional.

The treatment is medicosurgical: specific medication (antibiotics, hyadrazide), immobilization of the damaged segment and surgical intervention in the focus or posterior rachisynthezis.

The epidural or paraspinal abscess is often attended with neurological disorders (Fig.432).

353

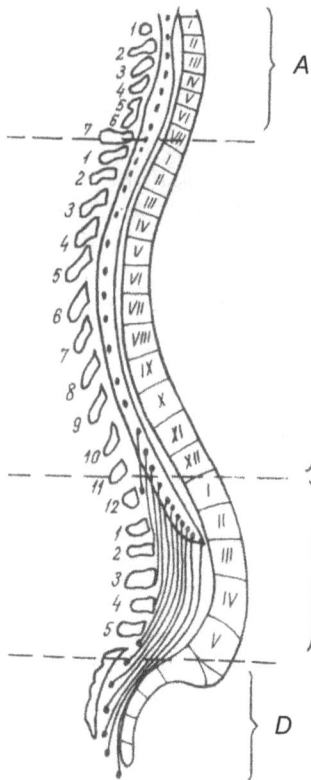

Fig. 429 Chipault's scheme for the determination of the convergence level of the nerve roots A- cervical region; B- thoracic region; C- lumbar region; D- sacrococcygeal region

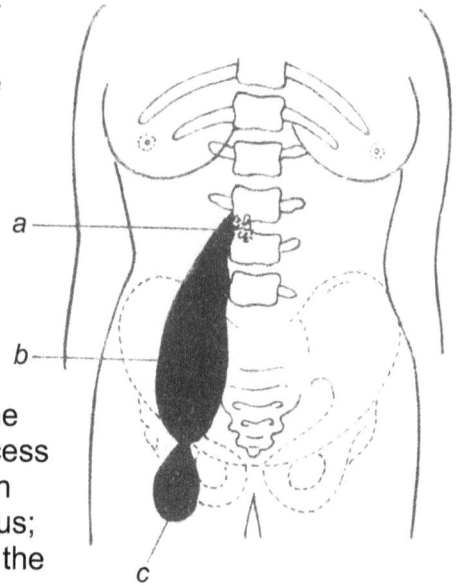

Fig. 431 Tuberculosis of the vertebral column – cold abscess fused in the femoral region
a- vertebral tuberculous focus;
b- cold abscess fused along the psoas muscle up to the;
c- inguinal region

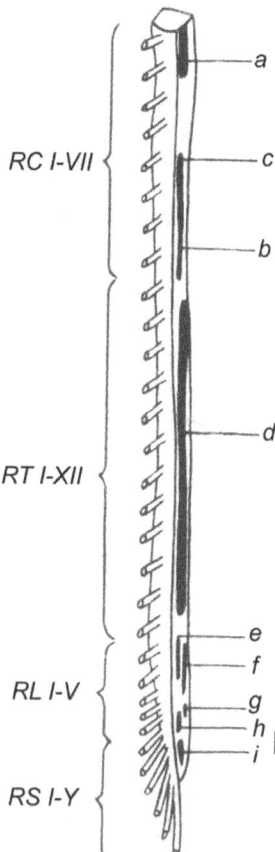

Fig. 432 Epidural or paraspinal abscess, most frequently accompanied by neurological signs

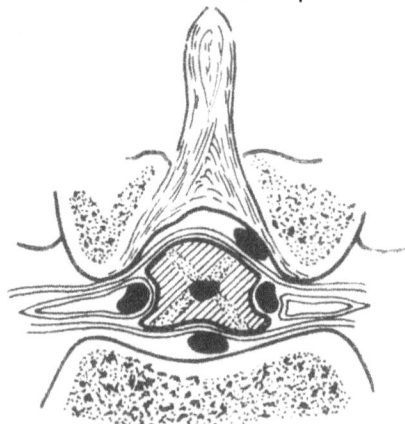

Fig. 434 Site of medullary compression a- vertebral body;
b- spinous process;
c- transverse process;
d- epidural compression;
e- dural compression;
f- extradural compression;
f- radicular compression;
h- medullary compression

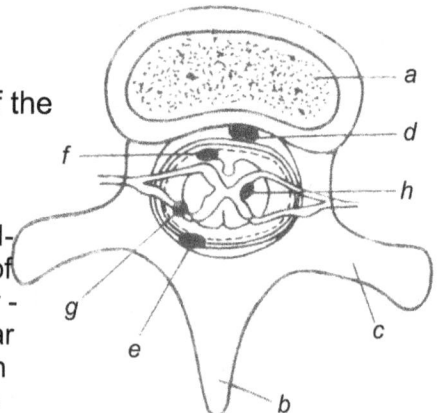

Fig. 433 Location of the tumours of the spinal cord

Fig. 430 Reflexogenic medullary centres -cervical region (I-VIII): a- respiratory centre; b- iliospinal centre; c- centres of cutaneous and tendinous reflexes for the thoracic member - thoracic region (I-XII): centre of abdominal reflex. – Lumbar region (I-V): e- centre of cremasteric reflex, - sacral region (I-V): g- centre of the plantar reflex, centre of the Achilles tendon reflex; h- centre of the anovesical and genital reflexes

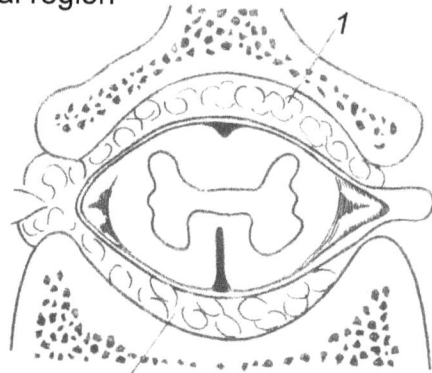

354

Vertebromedullary tumors

They are usually of bony origin and lead to the secondary invasion of the vertebral canal or to medullary compression, followed by severe neurological disorders (motor deficit of a variable intensity). Accorded to the injured histological structures, they may be medullary, meningeal (epidural, dural, extradural) or radicular (Fig.433, 434).

Often a vertebral metastasis of a variable origin is present, which may give rise afterwards to the lysis and even to the fracture of the vertebral body.

The most frequently affected is the thoracic segment, followed by the cervical segment.

The vertebromedullary tumours are grouped into:

- benign tumours, which are well delineated, have a slow course, are non-metastatic and heal after a complete surgical removal;

- malignant tumours, which are infiltrative, restructive, produces metastases and recur even after extensive surgical ablations, leading to the collapse of the vertebral column and appearance of paraplegia.

A significant role in the establishment of the diagnosis plays the scintigraphy.

The tumour may be sited outside the dura (extradural) or inside the dural sac (intra- or subdural; respectively intra- or extramedullar).

Investigation and treatment methods

The lumbar puncture is the method by which cerebrospinal fluid is withdrawn from the subarachnoid space for diagnostic purposes or by which certain substances are introduced into it for therapeutic purposes. The total amount of CSF is 100-150 ml, 25-43 ml are contained in the subarachnoid space at the level of the vertebral canal.

The patient is in a position like on a chair, with the vertebral column in maximum flexion, in order to ensure the enlargement of the intervertebral spaces. The puncture is performed on the median line of the vertebral column, at the crossing with the horizontal line which unites the iliac crests and which corresponds to the L4-L5 space.

After the disinfection of the skin on a surface of 40x60 cm, a needle of 8-10 cm, provided with a mandrin, is introduced sagittally into the intervertebral space, above the thumb which has identified the spinous process.

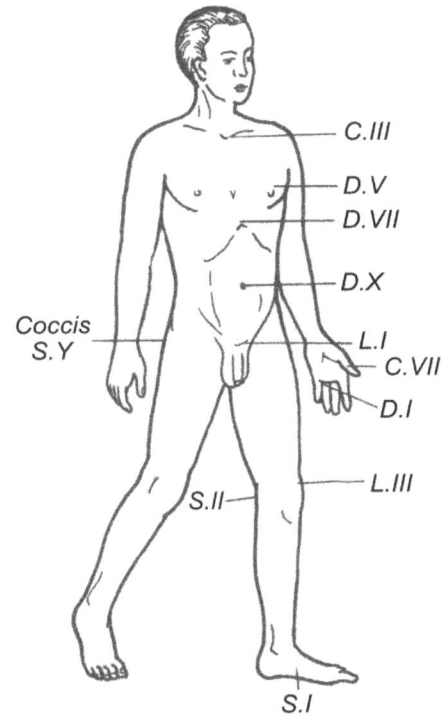

Fig. 435 Orientation landmarks in the choice of the site of spinal anesthesia or peridural anesthesia (after Băjeu)

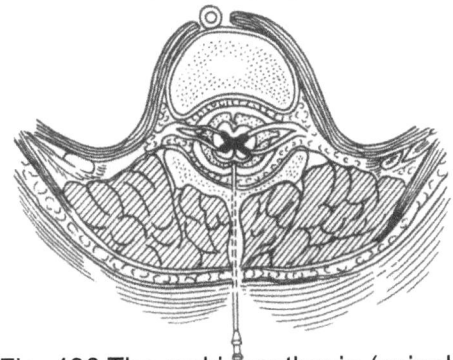

Fig. 436 The rachianesthesia (spinal anesthesia) represents the most frequently used anesthesia method in operation on the abdomen, the pelvis and the lower limb. The anesthetic substance is introduced into the subarachnoid space at the emergence site of the spinal nerves from the spinal cord.

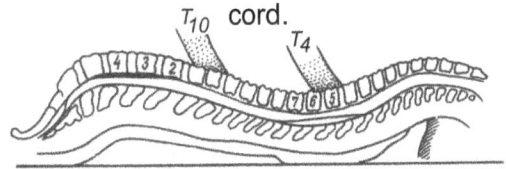

Fig. 437 Declive and proclive areas of the vertebral canal where, the anesthetic, respectively hyperbaric and hypobaric solutions, are sited, when the plane is horizontal: T4-T10 – anesthesia zone obtained in this position with a hyperbaric solution (after Frey)

355

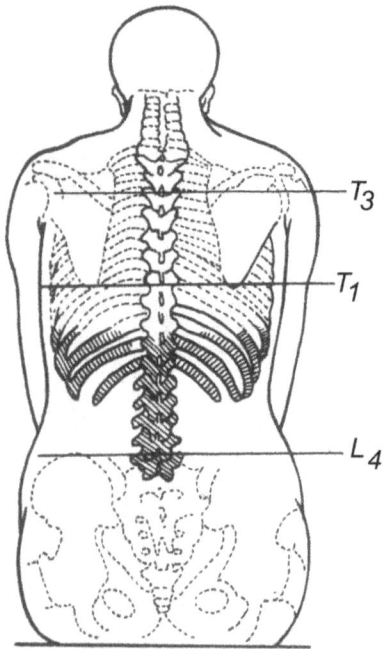

Fig. 438 Subarachnoid space, where the spinal anesthesia is performed. Bony landmarks to delineate the level at which the spinal anesthesia is carried out

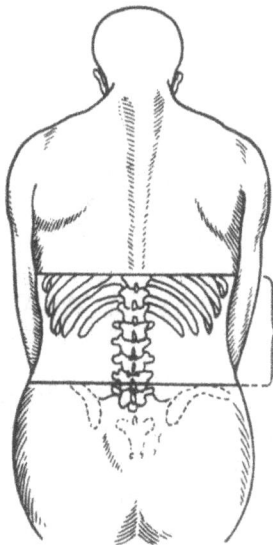

Fig. 439 The zone in which the spinal anesthesia is the least dangerous

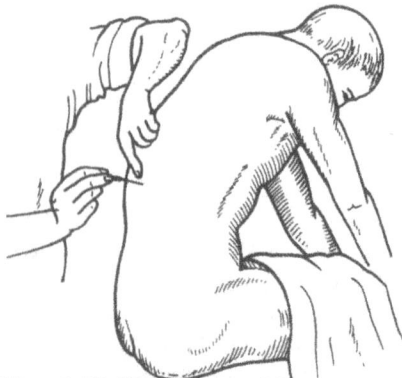

Fig. 440 Sitting position of the patient for the induction of spinal anesthesia

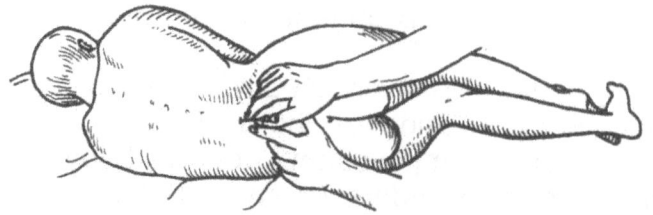

Fig. 441 Lumbar punction in decubitus

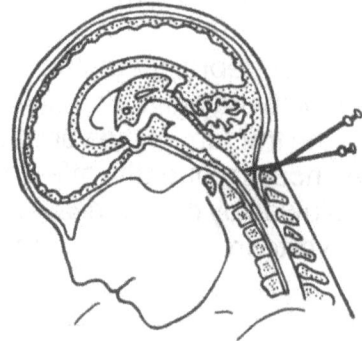

Fig. 442 Suboccipital puncture (stage I)

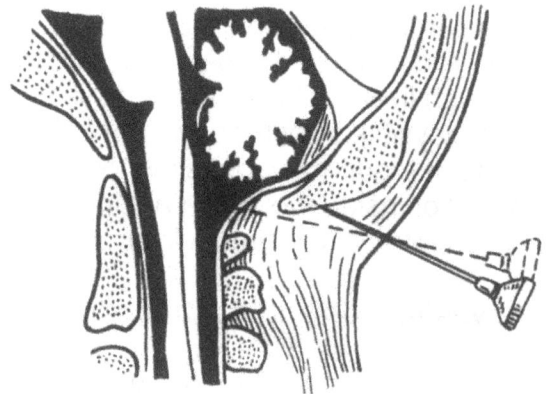

Fig. 443 Suboccipital puncture (stage II)

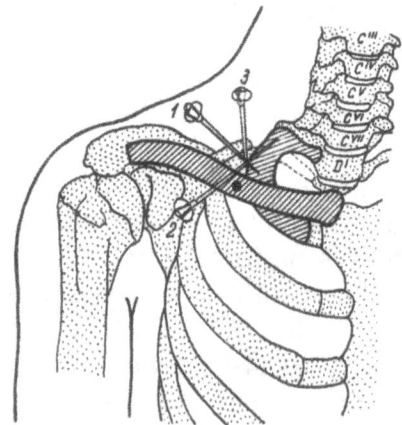

Fig. 444 Anesthesia of the brachial plexus by supraclavicular route. The three positions of the needle according to Labat's technique and the subclavicular topography of the plexus (after Pauchet)

356

The needle penetrates up to the subarachnoid space, traversing the tegument, the subcutaneous fat tissue, the interspinous ligament and then the yellow ligament, which is more resistant. After withdrawing the mandrin, the CSF drips out and either its pressure is measured or various samples are taken or anesthetic substances, air or contrast media (myelography) are introduced.

In the case of intracranial hypertension the puncture is contraindicated, since there is the risk of penetration of the cerebellar tonsils or of the medulla oblongata into the occipital foramen or of the hyppocampal gyrus into the Bichat's fissure.

The rachianesthesia or spinal anesthesia represents a method of anesthesia very frequently used in operations on the abdomen, the pelvis and the lower limbs. Corning performed in 1885 the first epidural anesthesia and in 1894 a spinal anesthesia by introduction of cocaine, after the demonstration by Quincke (1891) of the safety of the subdural puncture and of CSF withdrawal.

The first spinal anesthesia was performed in 1898 by Bier and in our country in 1899 by Severeanu, its development on scientific bases being due to Thoma Ionescu (Fig.435-441).

After carrying out the lumbar puncture, the anesthetic substance is introduced into the subarachnoid space, at the emergence site from the spinal cord of the spinal nerves.

The suboccipital puncture. The patient lies in lateral decubitus or is in the sitting position "like on a chair", with the head in flexion; after the disinfection of the nuchal region, the needle is introduced strictly on the midline, in the space situated immediately above the line which unites the two apexes of the mastoid (Fig.442, 443).

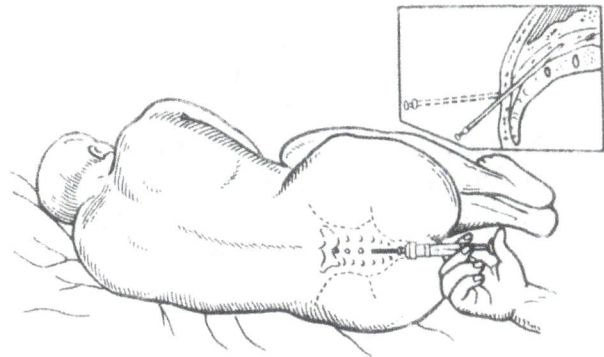

Fig. 445 Low peridural anesthesia
The anesthesia substance is introduced outside the covering of the spinal cord. The anesthesia of the nerves of the cauda equina for operations on the perineum is obtained by this method

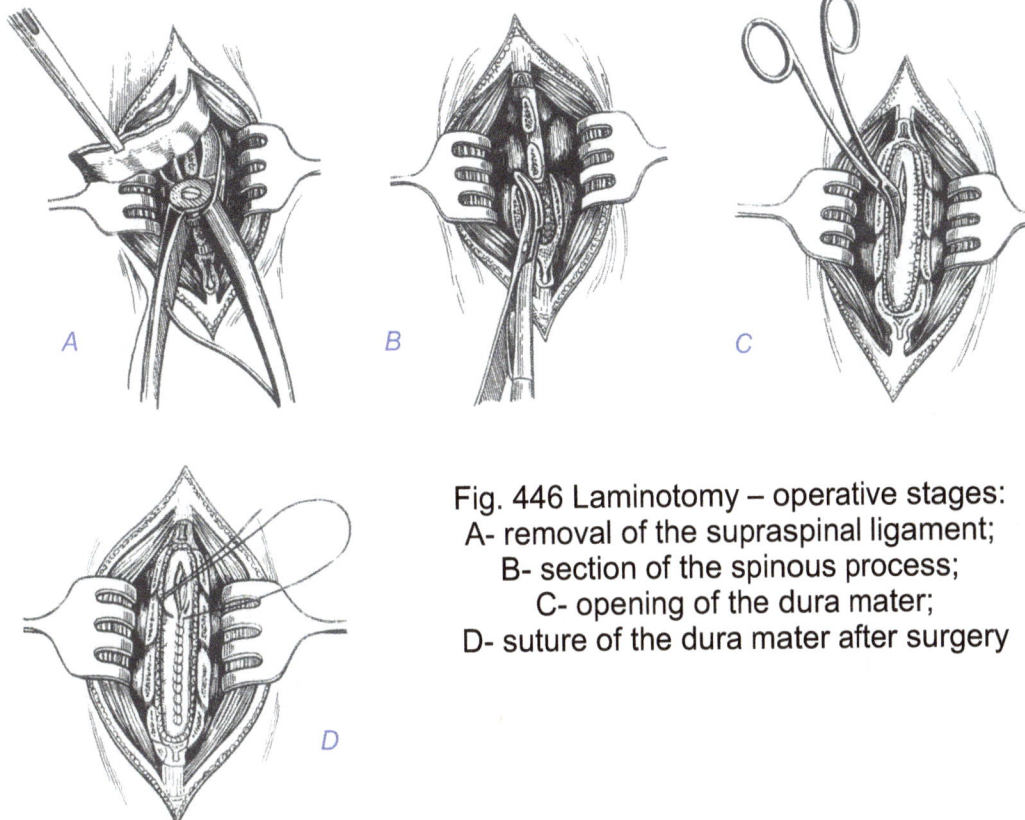

Fig. 446 Laminotomy – operative stages:
A- removal of the supraspinal ligament;
B- section of the spinous process;
C- opening of the dura mater;
D- suture of the dura mater after surgery

The anesthesia of the brachial plexus is performed by four access routes: axillary (Hirschel), subclavicular (Bazy), dorsal (paravertebral – Anglada and Santoni) and supraclavicular (Kulenkampf), the latter being the safest and most frequently used, which we describe below (Fig.444).

The patient lies in dorsal decubitus; above and behind the clavicle, in its middle, are palped the pulsations of the subclavian artery. At first the needle is oriented downwards, inwards and backwards, towards the spinous process T2-T3; after 1.5-3 cm, the needle is stopped by the first rib. It is a little withdrawn and the anesthetic substance is injected. Then the needle is pushed upwards and backwards, the eye of the needle lying nearly horizontally on the clavicle. In a third stage, the needle straightens up nearly vertical, behind the clavicle, for the purpose of identifying the external border of the first rib, and the anesthetic is injected. The appearance of paresthesia or pain is an element which assures that the needle has penetrated the brachial plexus.

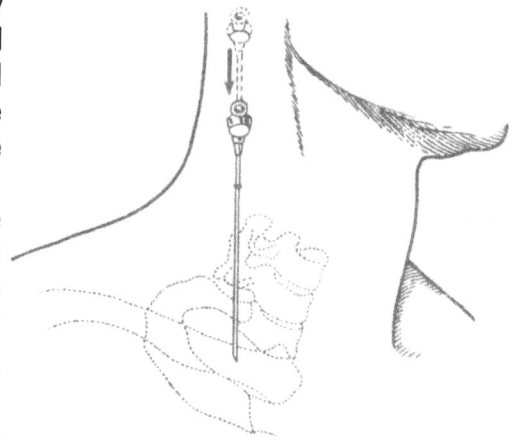

Fig. 447 Blockade of the stellate ganglion by superior external route. The needle advances parallel to the vertebral column

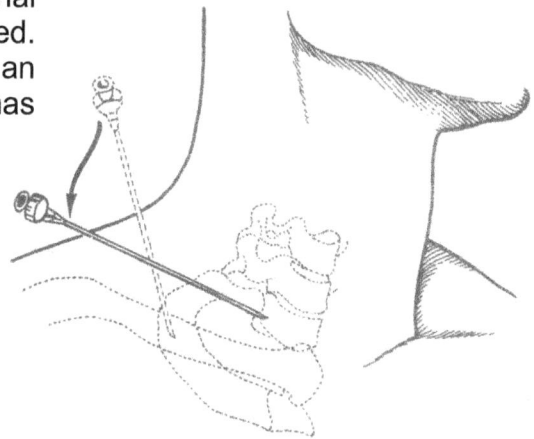

Fig. 448 Infiltration of the stellate ganglion. External route

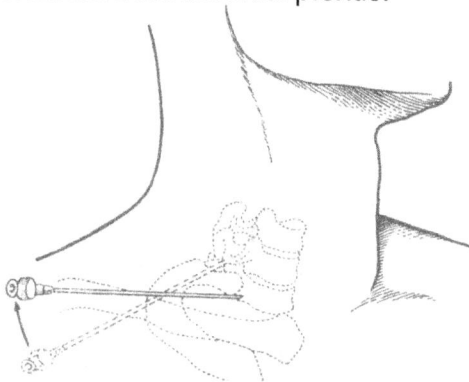

Fig. 449 Infiltration of the stellate ganglion. Anterior route

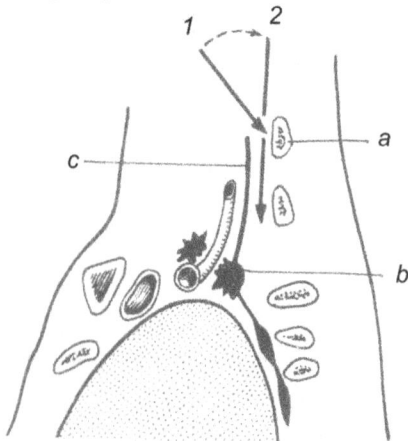

Fig. 451 Infiltration of the stellate ganglion a- transverse process; b- cervical sympathetic chain; c- stellate ganglion: 1- first position of the needle; 2- second position of the needle

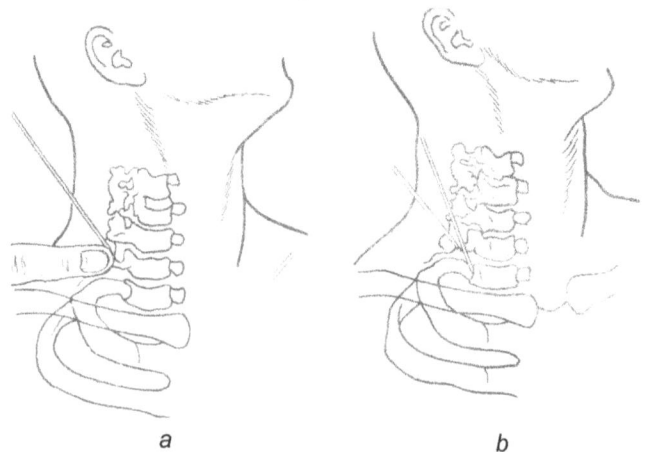

Fig. 450 Blockade of the stellate ganglion a- repair of the transverse process with the index and the direction of the needle to reach the transverse process; b- the needle is directed anteriorly and medially and passes in front of the transverse process, keeping the contact with the latter

The anesthesia of the sacral plexus or *low peridural anesthesia*. The anesthetic substance is introduced outside the covering of the spinal cord, and the anesthesia of the nerves of the cauda equina is thus obtained, for operations on the perineum (Fig.445).

The laminectomy is an emergency operation consisting in the decompression of the spinal cord (Fig.446).

After a previous radiological location and identification of the site of lesion, a longitudinal incision is performed at the level of the spinous processes. The muscle mass is removed and skeletization is performed with a chisel, followed by the section of the spinous processes up to their basis and then by laminectomy. The dura mater is identified and opened, after which the necessary operation is carried aut.

The rachisynthesis consists in the fixation of the posterior spinal column with bone grafts (iliac crest, tibia, rib etc.); it is used in the treatment of unstable fractures, scoliosis, vertebral tumours etc. Good results are obtained with the Rădulescu procedure.

The metal syntheses of the vertebral column are achieved either with Williams' metal plates or with the Skolnikov fixator.

The physiopathic syndromes are due to the vegetative disorders characterized by abolition of sweet secretion, color changes of the skin (violaceus, cyanotic), muscle hypotrophies or even atrophies etc.

The sweat production is investigated by several tests:

- *the pilocarpine test:* 0.001 g pilocarpine are injected subcuteaneously and after 15 minutes the respective zone is painted. The color changes in the presence of sweat production. The test allows to detect the presence of ganglionic lesions or lesions of the postganglionic fibre;

- *the ninhydrin (Moberg) test* is performed in specialized clinics;

- *the Guttmann test:* after powdering the respective zone with a mixture (quinizarine-natrium 2.6 disulfate 35 g, Na zinc carbonate 30 g, starch 60 g), a red-grey color is obtained, which turns to violet in the zones of sweating;

- *the Omer test:* the respective zone is painted with a solution containing 25% cobaltous chloride and 99% alcohol; in the sweating area appear blue spots which can be also photographed.

The cervical sympathetic blockade consists in the introduction of an anesthetic substance so as to surround the stellate ganglion (Fig.447-454).

Various techniques may be used: the Leriche-Fontaine technique, the Goinard technique and the Arnulf technique.

The lumbar sympathetic blockade is used in post-traumatic vasomotor syndromes, arteritides, embolisms, phlebitides, frostbites etc. The blockade is performed by the Leriche technique, the Servelle technique, the Gask and Ross technique, as well as by the Tardieu technique (Fig.455-458).

The lumbar sympathectomy is performed either by transperitoneal route or by subperitoneal route, by means of the Royle-Leriche procedure or by the Diez and Adson technique.

On the right is identified the inferior vena cava and on the left the aorta

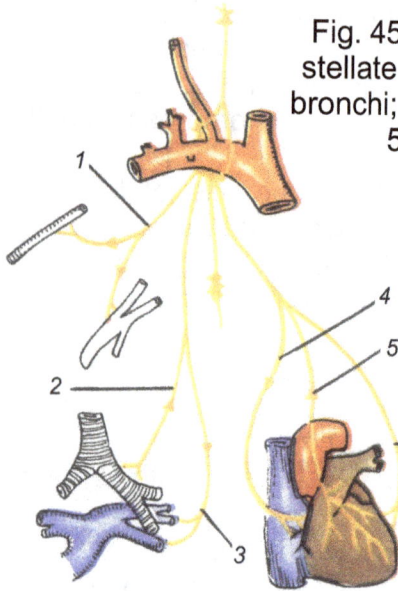

Fig. 452 Scheme of the efferent and of the afferent branches of the stellate ganglion 1- fibres to the upper limb; 2- sensory nerves of the bronchi; 3- pulmonary vasomotor nerves; 4- coronary sensory nerves; 5- cardiac sensory nerves; 6- cardio-accelerator nerves

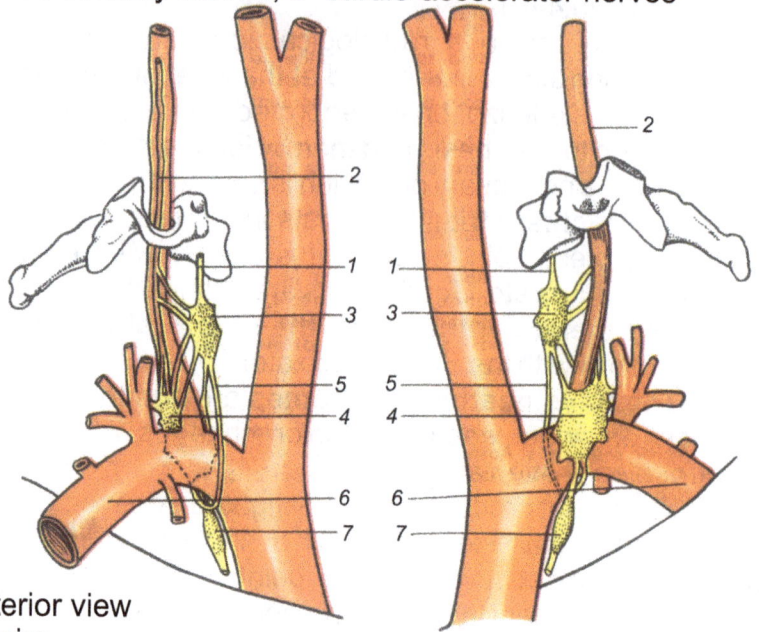

Fig. 453 A- stellate ganglion, anterior view
1- cervical sympathetic chain;
2- vertebral nerve; 3- intermediate ganglion;
4- stellate ganglion; 5- Vieussens' ansa; 6-
subclavian artery; 7- second thoracic
ganglionB- Stellate ganglion, posterior view
1- cervical sympathetic chain; 2- vertebral
artery; 3- intermediate ganglion; 4- stellate
ganglion; 5- Vieussens' ansa; 6- subclavian
artery; 7- second thoracic ganglion

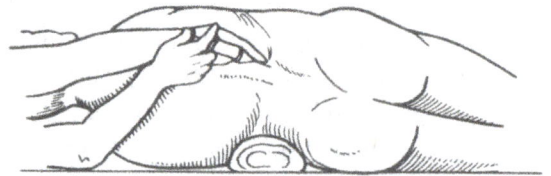

Fig. 455 Novocain lumbar blockade
technique

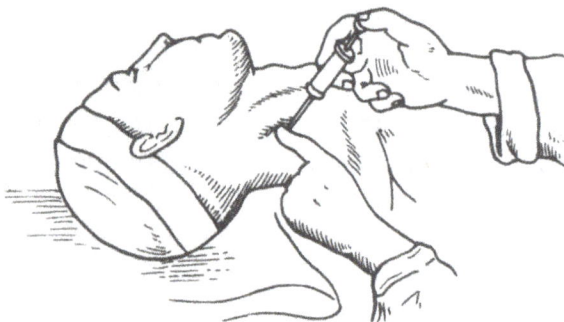

Fig. 454 Novocain cervical vagosympathic
blockade technique: The sinucarotid zone, the
carotid glomus and the vagus nerve with the
sympathetic chain are infiltrated. The puncture
is performed between the posterior margin of
the sternocleidomastoid muscle at the anterior
jugular vein; the needle is introduced in the
depth til it meets the tubercle of the transverse
process, then it is a little withdrawn and 60
cm^3 of 0.5 % xylene are introduced.

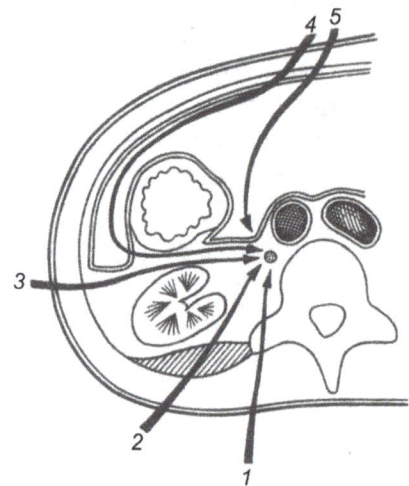

Fig. 456 Access routes to the lumbar
sympathetic plexus
1- lumbar route; 2- posterior
paramedian route; 3- subperitoneal
route; 4-transperitoneal and
laterocolic route; transperitoneal and
transmezocolic route

360

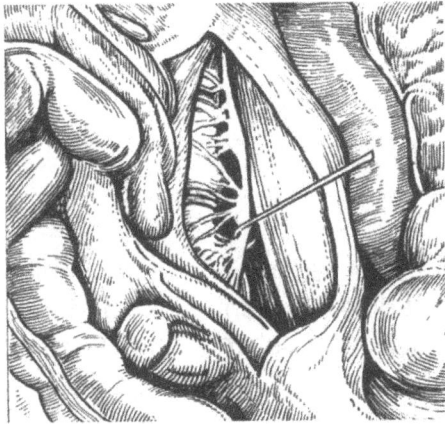

Fig. 457 Laying bare of the right lumbar sympatheticus plexus by transperitoneal route

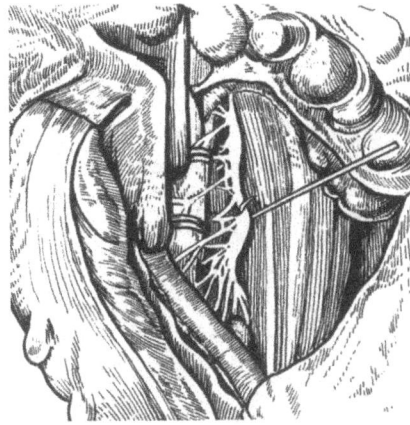

Fig. 458 Laying bare of the left lumbar sympathetic plexus by transperitoneal route

The cranium and the encephalon
The viscerocranium

The functional structure of the visceral cranium is similar to that of the Eiffel tower, but it has the base upwards and is made up of horizontal plates joined by vertical pillars (Rainer). The pressure is exerted on the cranial vault, while the base of the skull is the bony plate through which the pressure is transmitted to the pillars of the face.

As a result, the resistance afforded by the mechanical construction of the skeleton of the face exceeds significantly the pressure exerted by mastication (Triepel).

The pressure exerted by the mandible on the maxilla during mastication is transmitted – according to Blüntschli – through three pillars: nasofrontal, zygomatic, and pterygoid.

Taking into account that the muscles of mastication have a total transverse surface of 40 cm2 and that on 1 cm2 they exert a force of 6-12 kg (usually the value indicated is 10 kgf), it ensues that the total force of these muscles may attain 400 kg (this explains the performances of circus acrobats, who support with their teeth a hanging partner).

Moreover, Rainer has established the role of the nasal septum in the mechanical construction of the visceral cranium (Fig.459-461).

The fractures of the facial bones may involve one or several bones and occur, either as a result of the direct action of the vulnerating agent or by direct impact (fall, hurt etc.).

The reduction and immobilization of these fractures represent an extreme emergency (Fig.462-466).

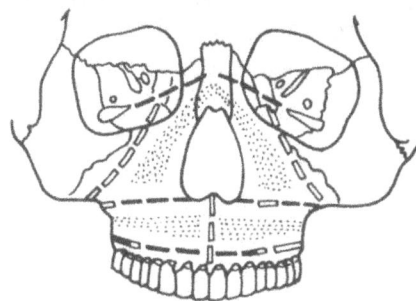

Fig. 459 Points of resistance of the facial structures (anterior view)

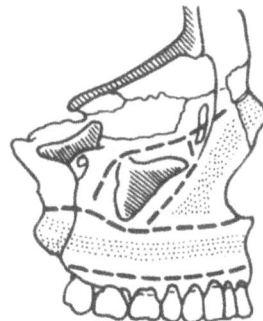

Fig. 460 Points of resistance of the facial structures

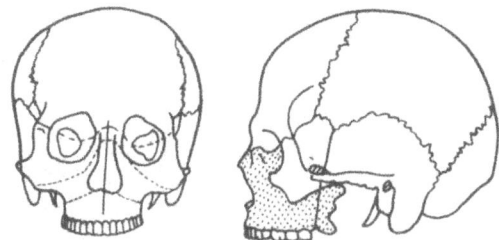

Fig. 461 Zone of low resistance lines (superior, middle, inferior) of the facial structures

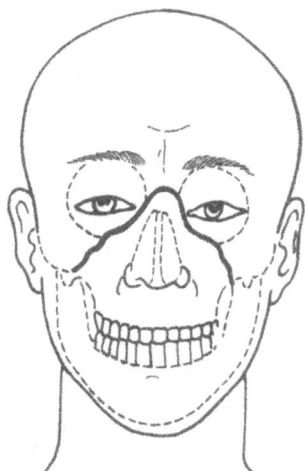

Fig. 462 Fractures of the maxillae, craniofacial disjunction. They are the fractures of the middle region of the face, to which is often associated a bone displacement with consecutive deformation of the face

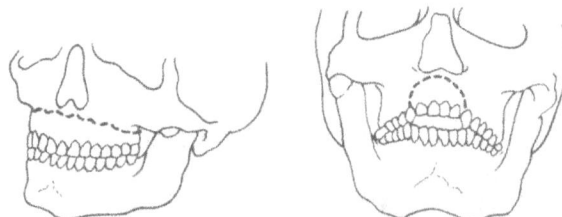

Fig. 463 Type Guérin horizontal fracture without displacement, sinking of the intermaxillary bone between the two maxillae

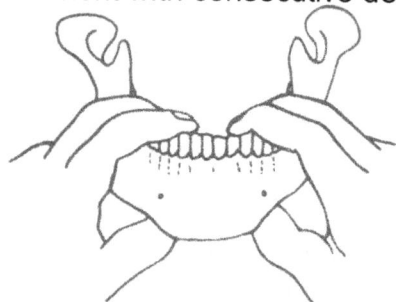

Fig. 464 Investigation technique of the pathological mobility in fractures of the mandible

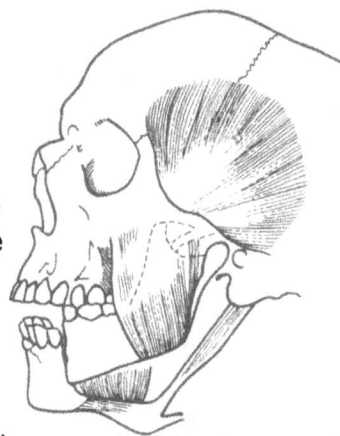

Fig. 465 Displacement of the fragments through the action of the muscular mass in fractures of the horizontal branch of the mandible

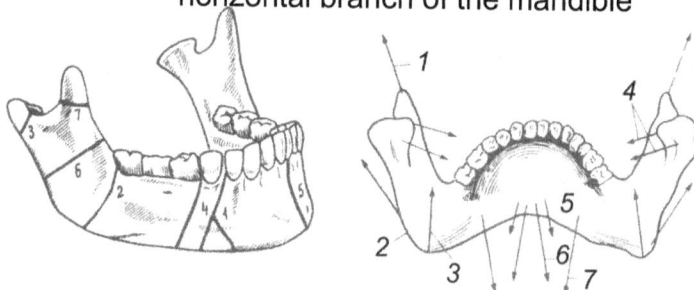

Fig. 467 Lines of least resistance of the mandible (a) and scheme of the action of the mandibular muscles (b)
1- temporal muscle; 2- masseter muscle; 3- internal pterygoid muscle; 4- external pterygoid muscle; 5- digastric muscle; 6- geniohyoid muscle; 7- mylohyoid muscle

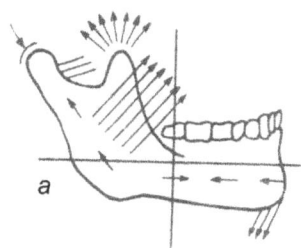

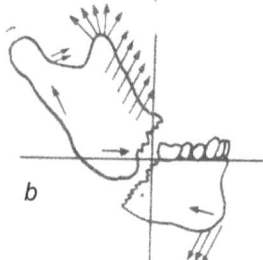

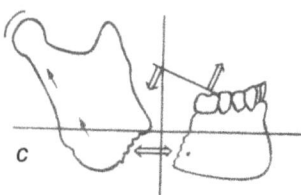

Fig. 466 Scheme representing the action of the mandibular muscles and the displacements of the fragment in fractures

Fig. 468 Reduction manoeuvre of the temporomandibular dislocation. The anterior dislocation is the most frequent form of temporomandibular dislocations: the mandibular condyle leaves the glenoid cavity and lies in front of the temporal condyle. The treatment is usually orthopedic, the reduction is always possible and may be easily performed, if it is carried aut immediately after the accident and by a correct technique

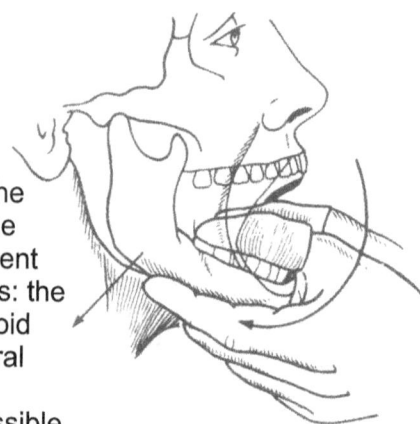

362

The temporomandibular dislocation consists in the displacement of the mandibular condyle, called also condyllar process, out of the glenoid cavity.

The treatment is orthopedic and does not imply difficulties, if it is performed immediately after the accident (Fig.467-468).

Epistaxis is a nasal hemorrhage, in most cases harmless, but often of a great severity, which may occur after nasal injuries or as a consequence of arterial hypertension.

After a previous examination of the nasal septum, gauze dressings impregnated with a 1‰ epinephrine solution, 4% tetracycline and 4-10% cocaine are applied. In cases of a medium severity, local cauterization or silver nitrate is used, while in severe cases posterior tamponment should be performed.

In cases of severe and repeated anterior epistaxis, induced especially by hemorrhagic hereditary diseases (Osler's disease), dermatoplasty or the transantral ligation of the maxillary artery is used.

The rhinoplasty is a technique of the plastique surgery which repairs the significant deformities of the nasal bones.

The neurocranium
Craniocerebral injuries

Cranio-cerebral injuries. The impetuous development of the industry and the intensification of the road traffic led to an increase of the number of craniocerebral injuries, which may assume various clinical aspects (Fig.469).

- *The non-penetrating wounds* involve the scalp, with or without fractures of the braincase, and are of a variable severity, in dependence on the lesion of the brain.

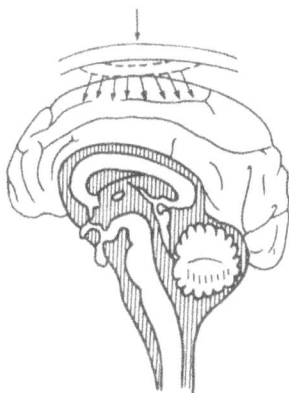

Fig. 469 Production mechanism of fractures of the neurocranium

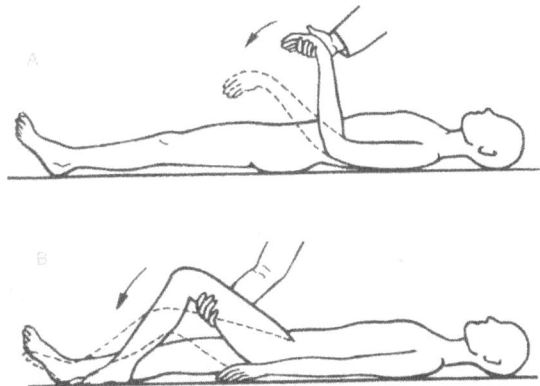

Fig. 470 For the identification of the neurological lesions it is necessary to defect the motor deficit consecutive to the lesion of the pyramidal tracts; this can be obtained by the Barré manoeuvre, either through the "forearm test" or through the "leg test" A-Barré manoeuvre: forearm test – the forearm cannot be maintained in a right angle on the arm and falls inert; B- manoeuvre Barré: leg test – in the damaged side (paresis), the leg cannot be maintained in the flexion position

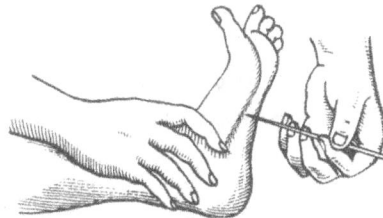

Fig. 471 Signs of cerebral lesion in fractures of neurocranium with intracranial supradural or subdural or/and intracerebral hemorrhage
Unilateral Babinski sign: the plantar region is stimulated with a pointed object

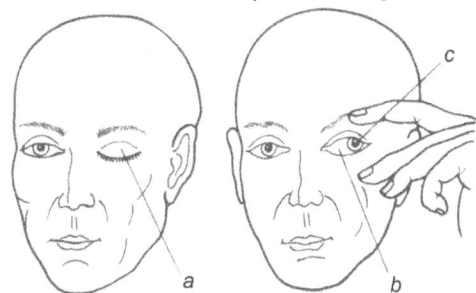

Fig. 472 Oculomotor nerve paralysis, manifested by: a- blepharoptosis (palpebral ptosis); b- divergent strabism; c- mydriasis

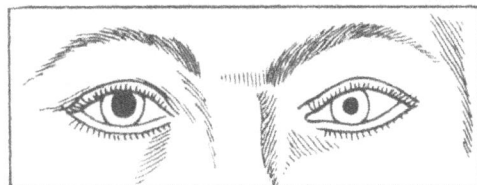

Fig. 473 The dilation of the homolateral pupil may be a useful element for the localization of the nerve lesion

363

According to the conception of C.Arseni, the *"open" craniocerebral injuries* should be treated as emergencies and with competence, as they may lead to various complications, such as osteomyelitis, meningo-encephalitis or cerebral abscesses.

In order to identify the neurological lesions, the investigations are aimed at the detection of the motor deficit consecutive to pyramidal lesions. For this purpose, the Barré manoeuvre through the two techniques, the "forearm test" or the "leg test", or the Babinski's sign is searched (Fig.470-471).

In addition, the oculomotor, abducent, facial and hypoglossal nerves are examined and the presence of mydriasis (dilation of the pupil) is investigated (Fig.472, 473).

The pupillary inequality (unilateral mydriasis) is a useful element for the localization of the nervous lesion (Fig.474, 475).

- *The cranial fractures* occur when the traumatic kinetic energy exceeds the cranial elasticity.

Owing to the existence of the fontanels and to the persistence of the sutures, the elasticity of the calvaria is higher in newborns and children and lower in adults.

At the level of the base of the skull, Félizet has described *the structures of resistance*, both in the midsagittal and in the transverse plane:

- the midfrontal structure, including also, from the anterior region of the base, the vertical ethmoid plate, crista galli;

-the midoccipital structure;

-the basilar area, at the level of the base of the skull;

-the symmetrical spheno-occipital (anterior) and petromastoid (posterior) pillars (Fig.476-479).

The main nervous and vascular formations which could be damaged in the fractures of the base of the skull should be also taken into account (Fig.480).

The longitudinal sinus is the most frequently involved, as it is adherent to the meninges.

Moreover, the hemorrhage is due also to the lesion of the diploic venous lakes and to the thickness difference between the inner and the outer (thicker) table.

The severity of these injuries depends on the velocity of the damaging agent which, in addition, induces also the cerebral lesion.

The clinical signs, according to the injured region of the base of the skull, permit to establish the diagnosis.

- *The fractures of the anterior region* occur after a frontal injury. The fracture line follows the ceiling of the orbit and the optic canal, eventually producing the section of the optic nerve; in other cases the fracture line courses towards the ethmoid and the sphenoid bone producing ecchymoses of the orbit. By nasal route occurs the infection of the meninges (Fig.481, 482), which manifests itself by epistaxis (rhinorrhagia) or CSF discharge from the nasal cavity (rhinorrhea).

- *Fractures of the middle region* occur after a temporal injury and represent the most frequently encountered lesion. The fracture line courses parallel to the axis of the petrous part of the temporal bone, and may involve the cavities of the middle ear. Through the ear the infections are transmitted to the meninges and manifest themselves by otorrhage with or without otorrhea and by retromastoid ecchymosis (Fig.483, 484).

- *Fractures of the posterior region* are accompanied by pharyngeal ecchymosis, nuchal ecchymosis and otorrhagea with or without otorrhea.

The therapeutic indication depends on the general and neurological condition of the injured patient. Surgery is indicated in the case of cerebral compression or meningocerebral dilaceration produced by bone splinters (Fig.485).

- *The craniocerebral wounds* are the apanage of war traumatology and C.Arseni has drawn up their best classification, which we have adopted (Fig.486-490):

- the tangential craniocerebral wounds may involve the scalp or the scalp and the skull. As a rule, the dura mater is dilacerated by a splinter;

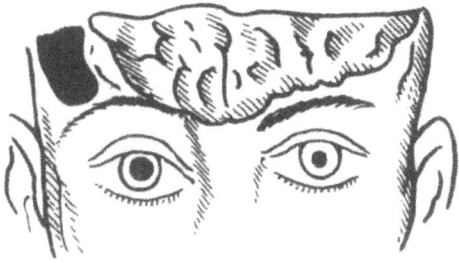

Fig. 474 The pupillary inequality reveals always a volume displacement in the neurocranium. A supra- or subdural intracranial or/and intracerebral hemorrhage displaces the anatomical segments in the neurocranium

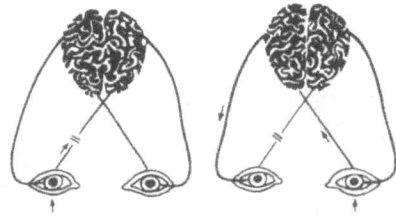

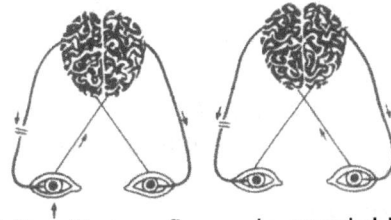

Fig. 475 Pupillary reflexes in cranial lesions

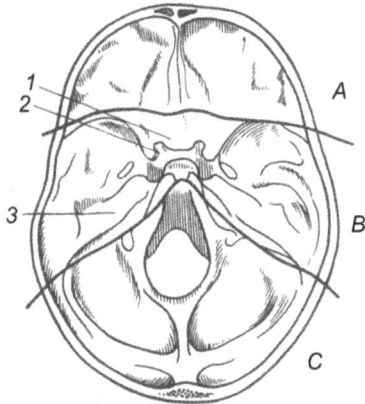

Fig. 476 Regions of the base of the skull
A- anterior region; B- middle region;
C- posterior region: 1- basilar process;
2- lesser wing of the sphenoid bone;
3- petrous portion of the temporal bone

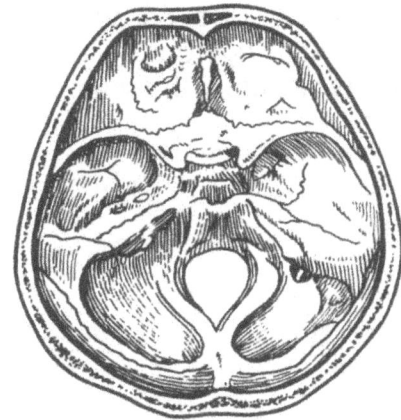

Fig. 477 Fracture of the base of the skull

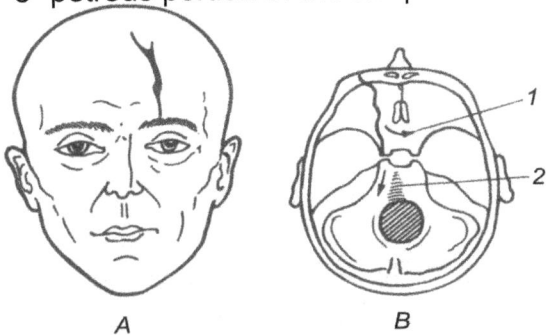

Fig. 478 Fracture of the base of the skull
a- the impact is frontal; b- spreading of the line of fracture from the base of the skull towards the opposite side (b1), the middle region (b2)

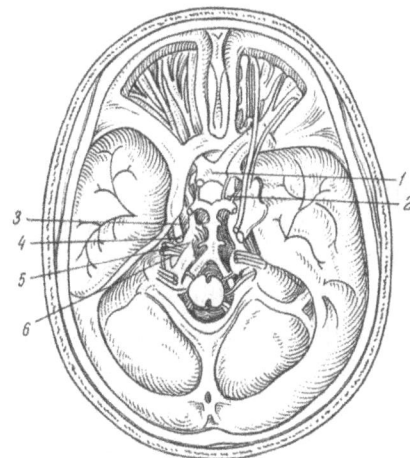

Fig. 480 Main nerve and vascular formations which may be damaged in fractures of the base of the skull 1- optic chiasma; 2- oculomotor nerve; 3- pathetic (trochlear) nerve; 4- external oculomotor nerve; 5- facial nerve; 6- acusticovestibular, glossopharyngeal and spinal nerves

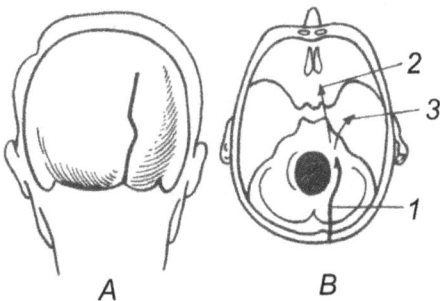

Fig. 479 Fracture of the base of the skull
A- occipital impact; B-1: fracture at the level of the cerebellar fossa; 2- irradiation towards the appex of the petrosa (petrous portion of the temporal bone) and towards the sella turcica; 3- perpendicular irradiation on the petrosa, following the two ruptured holes

- the blind penetrating craniocerebral wounds are the most frequent; they are characterized only by the existence of the penetration orifice of the vulnerating agent. The cerebral component – the cerebral dilaceration – is prevalent, the lesions being more extensive in the superficial areas and more limited in the deep areas;

- the transfixing craniocerebral wounds are varying according to the speed of the projectile and to the distance from which it was fired;

-the craniocerebral wounds by ricocheting are characterized by a cranial fracture at the site of the impact.

The treatment of these lesions is only surgical and consists in the extraction of splinters and of the possible intracerebral foreign bodies, a rigorous hemostasis and the compulsory closure of the scalp.

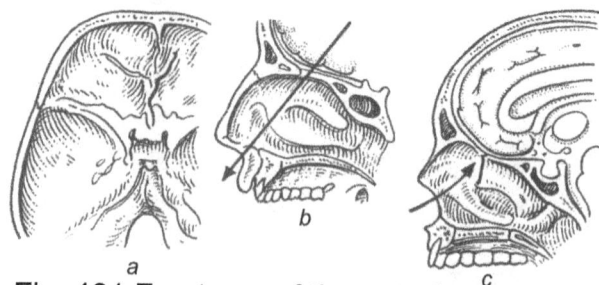

Fig. 481 Fractures of the anterior region of the base of the skull – fracture lines
a- fracture line; b- clinical symptoms: epistaxis or CSF discharge; c- infection of the meninges by nasal route

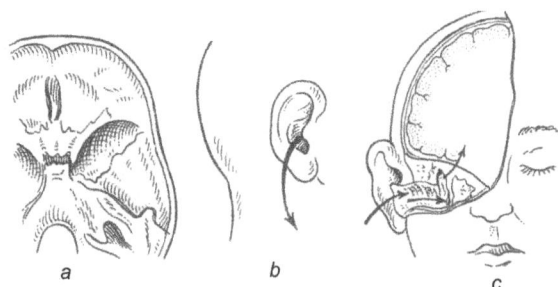

Fig. 482 Fracture of the anterior region – clinical symptoms: a- rhinorrhagia (blood) or/and rhinorrhea (CSF); b- bilateral periorbital ecchymosis

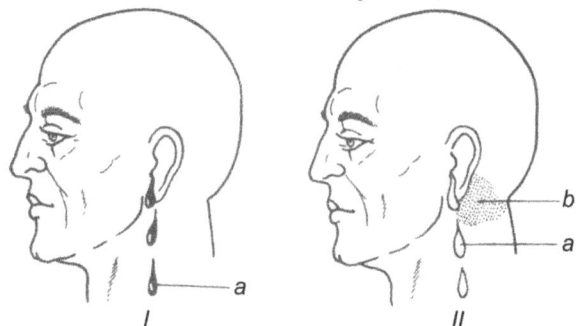

Fig. 483 Fracture of the middle region and of the petrous portion of the temporal bone – fracture lines
a- fracture line; b- clinical symptom: otorrhagia; c- the infection of the meninges occurs through the ear

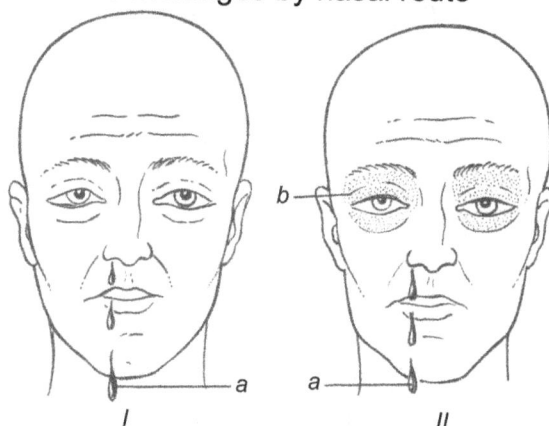

Fig. 484 Fracture of the middle region – clinical signs
a- otorrhagia or/and b- otorrhea; c- retromastoid ecchymosis

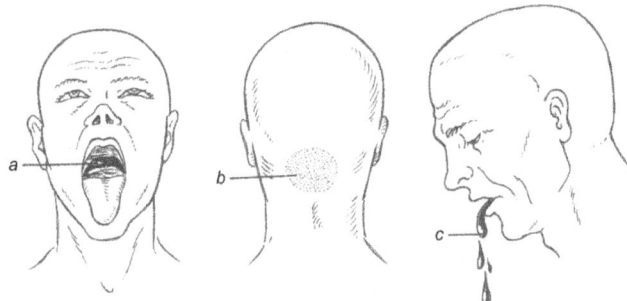

Fig. 485 Fracture of the posterior region – clinical signs
a- pharyngeal ecchymosis; b- nuchal ecchymosis (in the nuchal region); c- bloody vomitus

Cerebral traumatic effects

As a consequence of a severe injury, the patient exhibits lesions of the brainstem, with primary or secondary effects.

According to C.Arseni, any injury of a medium severity may induce vascular reflexes – hence functional disturbances – or anatomopathological lesions exclusively in the brainstem.

The primary traumatic effects assume the following forms:

- *the brain concussion (commotio cerebri)* is characterized by the loss of consciousness for a short period, without vegetative or neural signs, followed by a complete and immediate recovery, without remote consequences (C.Arseni);

- *the brain contusion* is the most frequent lesion following a craniocerebral injury and is characterized by the appearance of small hemorrhages either subdural or subarachnoid, or in the cerebral parenchyma, produced by the impact of the encephalon with the wings of the sphenoid bone, the floor of the orbital fossa and the quadrilateral lamina. It is followed by a posttraumatic cerebral edema with a marked increase of the intracranial pressure. Its manifestations are loss of consciousness for several hours or days, and the presence of convulsive seizures, hemiparesthesias or focal signs (hemiplegia, aphasia, apraxia, agnosia etc.). Most frequently appear contusion lesions at the level of the brainstem.

The minor lesions of the hypothalamohypophyseal complex induce the post traumatic diabetis insipidus syndrome: polyuria (excretion of 6-12 or even 30 l of urine in 24 hours) and intense and continuous polydipsia.

The secondary traumatic effects are of the type of blood effusions (hematomas) and assume various aspects, according to their location: extradural, subdural, subarachnoid and intracerebral hematomas (Fig.491-503).

- *The extradural or epidural hematoma* forms in the virtual extradural spaces between the endocranium and the dura mater and requires emergency surgery.

The source of hemorrhage may be:

- the lesion of the middle meningeal artery or of its branches, which is encountered in over 80% of cases after an injury occurred in the temporal region – the detachable Gérard-Marchand's space. The lesion of the middle meningeal artery has a rapid course; the injury is followed short by a free interval, after which appear rapidly hemiparesis, then hemiplegia, a state of somnolence and coma. In the patient is not operated as an extreme emergency, death may occur. In this injury appears the Battle's sign: ecchymosis of the scalp in the temporal and retroauricular regions;

- the lesions of the longitudinal sinus and of the lateral sinuses, which leads to the posterior fossa syndrome, manifested by respiratory arrest;

- the lesion of the diploic veins, especially in the hemophyliac patients, may lead to the formation of hematomas.

- *The subdural hematoma*, usually unilateral, sited on the left side, is produced by the rupture of the cortical veins afferent to the longitudinal sinus. The treatment is surgical;

- *The intracerebral hematoma* develops within the first three days after the accident, in the cerebral parenchyma. A free interval is possible, after which appear the neurological phenomena and those of intracranial hypertension. The treatment is surgical and requires a rapid and precise diagnosis.

Caput succedaneum is a benign lesion consisting in an oedematous and bloody infiltration of the scalp, which appears in an infant immediately after birth, either as a consequence of a difficult labour or owing to the use of surgical instruments; the swelling may resolve in a few days or after an evacuant puncture.

The cephalhematoma is an effusion formed by the rupture of the epicranial vessels; it is situated between the periosteum and the inner table of the skull, its preferential location being parietal. It resolves slowly or may develop towards calcifications and ossifications, leading to the cranial asimmetry. The intracranial hemorrhagic lesions of the newborn appear after severe and complex fractures of the braincase.

367

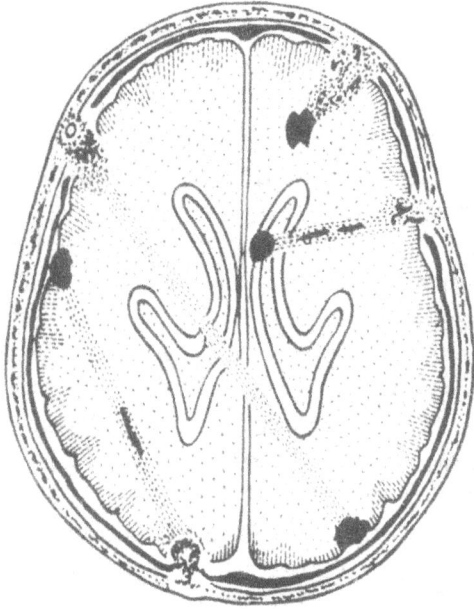

Fig. 486 Section through the brain revealing the possible causes of the craniocerebral wounds

Fig. 487 Non-penetrating tangential gunshot wound, with partial fracture of the calvaria

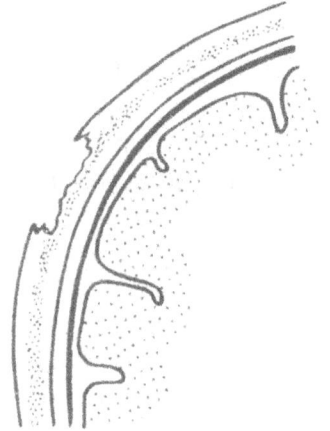

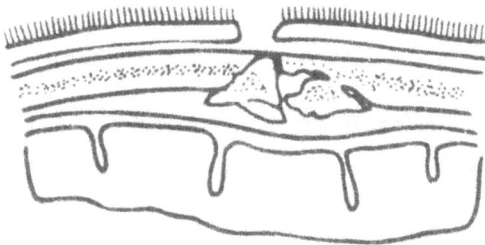

Fig. 488 Open craniocerebral trauma: comssinated fracture

Fig. 489 Perforated craniocerebral wound, with foreign body retention

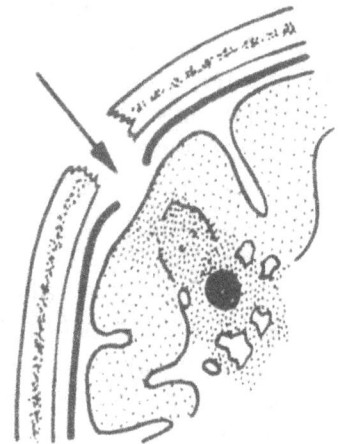

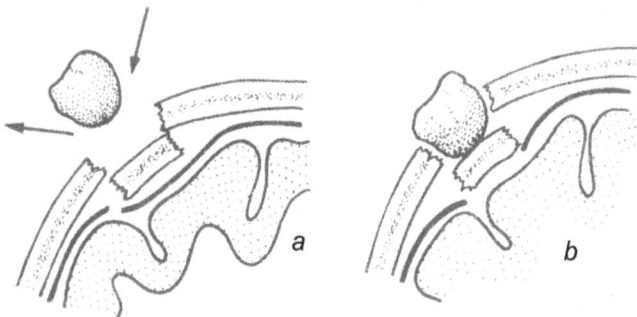

Fig. 490 Penetrating craniocerebral wound
a - penetration of the bony fragment;
b - penetration of the bony fragment with the foreign body;

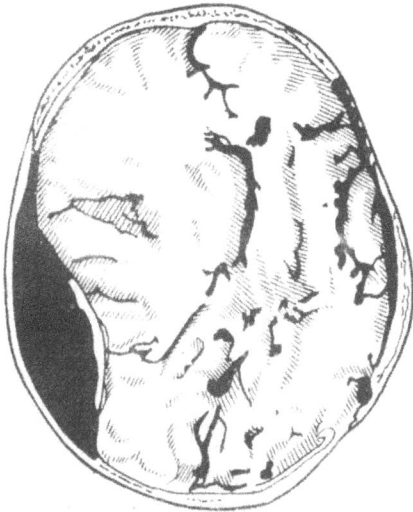

Fig. 491 Extradura hematoma
without the fracture of the skull

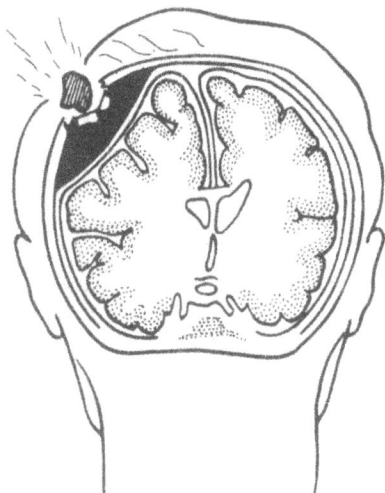

Fig. 492 Epidural or extradural
hematoma accompanied by the fracture
of the bones of the skull

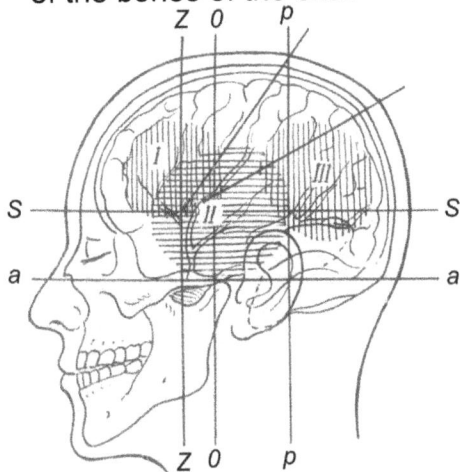

Fig. 493 Location of the cerebral hematoma
I- anterior; II- median; III- posterior; ss.
Supraorbital horizontal line; aa- auriculo-
orbital horizontal line; zz – zygomatic
vertical line; oo – auricular vertical line;
oo – petromastoid vertical line

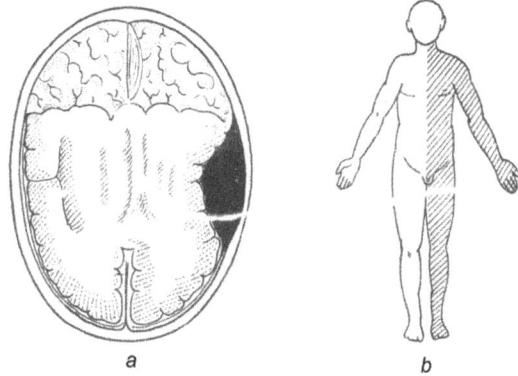

Fig. 494 Hemorrhage due to the lesion of the
middle meningeal artery
a- extradural hematoma in the Marchant's zone
(temporal region); b- contralateral hemiplegia
(hatched zone)

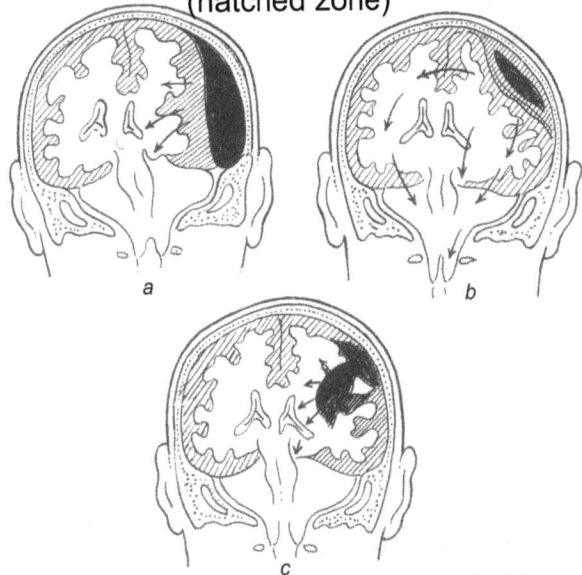

Fig. 495 Hematomas consecutive to
craniocerebral injuries
A- extradural hematoma; B- subdural
hematoma; C- intraparenchimatous hematoma

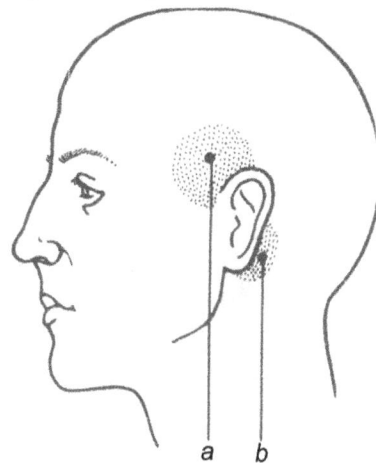

Fig. 496 Extradural hematoma – Battle's sign
a- ecchymosis of the scalp in the temporal
region; b- retroauricular ecchymosis

369

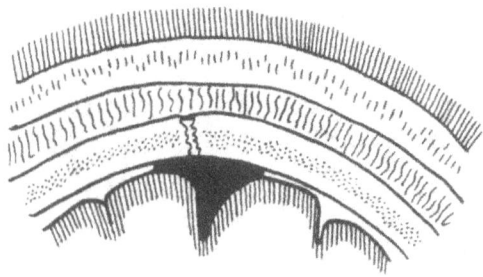

Fig. 497 Hematoma due to the involvement of the longitudinal sinus after a fracture of the skull. The longitudinal sinus is the most frequently involved, at its base, on each side of the adhesion to the sagittal suture, at this level the meninx is very adherent

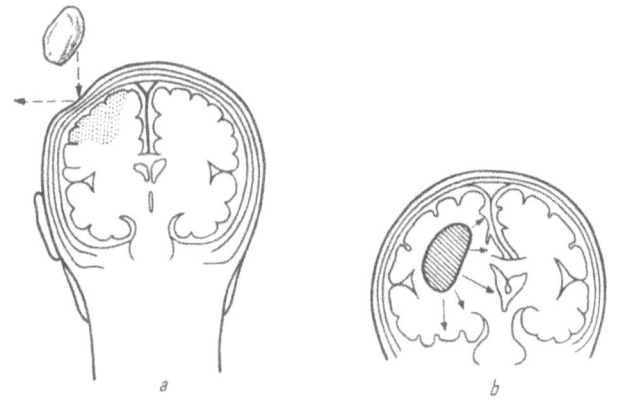

Fig. 500 Interparenchimatous hematoma a- after a comminated fracture; b- with bone integrity. It is a very severe lesion and requires the establishment of a precise and rapid diagnosis to make the correct therapeutic decision

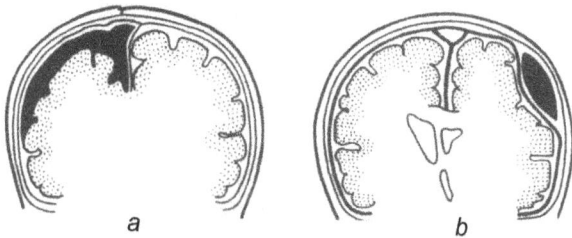

Fig. 498 Subdural hematoma after a commissural fracture: a- diffuse hematoma; b- non-encapsulated hematoma; this is the most frequently encountered hematoma and occurs in over 50 % of the open craniocerebral traumas

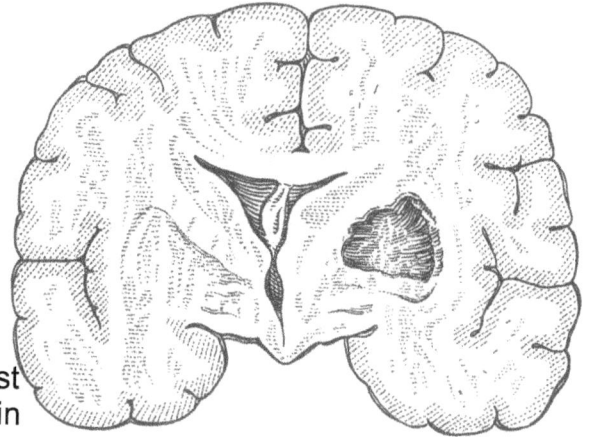

Fig. 501 Intraparenchimatous hematoma

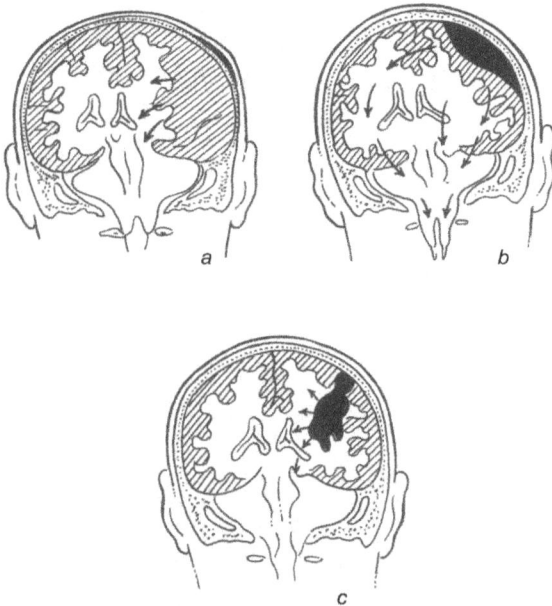

Fig. 499 Craniocerebral injuries
a- extradural hematoma; b- subdural hematoma; c- intracerebral hematoma

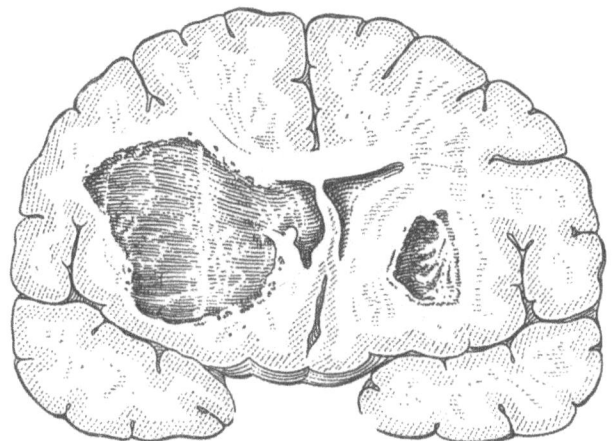

Fig. 502 Cerebral dilaceration

370

Cerebral vascular accidents

They represent, in hierarchical order, the third cause of death, after the cardiovascular diseases and cancer (A.Kreindler, 1972).

Often, no irreversible lesions occur, as the cerebral vascular system is provided with a collateral circulation, which can assure an optimum circulatory output. This is achieved by means of (Fig.504-509):

- *the circle of Willis (circulus arteriosus cerebis)*, which is the reservoir of the cerebral circulation and intervenes by the adaptation of the calibre of the constituent vessels to the necessary blood amount. The existence of arterial hyperplasias and of calibre changes, especially in elder patients, leads to the appearance of severe cerebral impairments;

- the anastomosis, both of the superficial and of the deep branches, assures true substitution relation connections;

- the anastomosis of the internal carotid artery with the external carotid artery, as well as that of the external carotid artery through the occipital artery, provide new pathways able to assure, if necessary, an adequate cerebral irrigation.

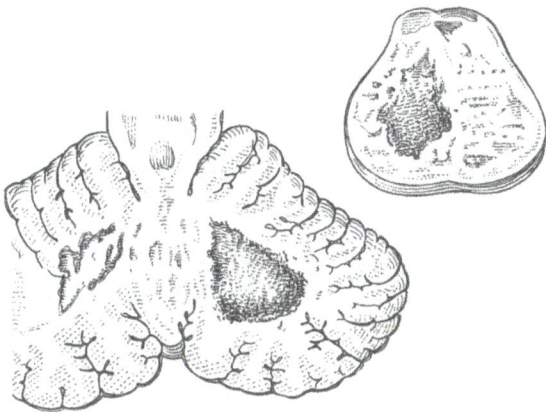

Fig. 503 Hematoma at the level of cerebellum

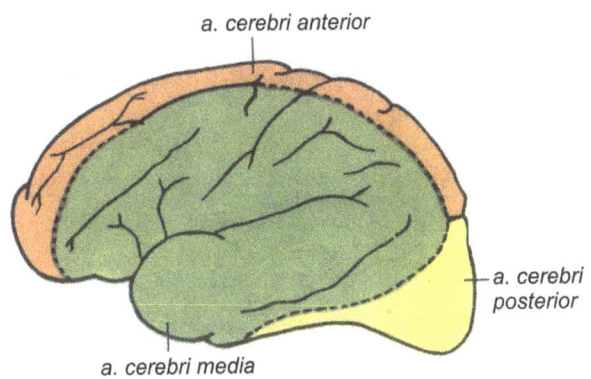

Fig. 504 Vessel supply of the brain (external surface)

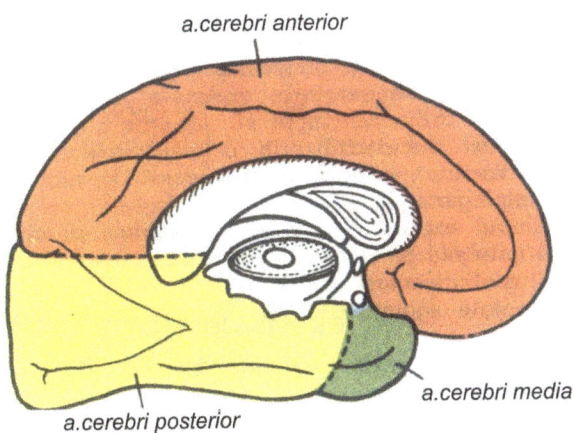

Fig. 505 Vessel supply of the brain (internal surface)

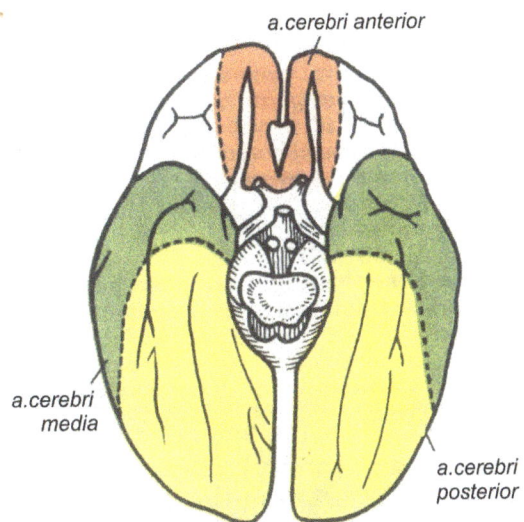

Fig. 506 Vessel supply of the brain (inferior surface)

371

Ischemic cerebral impairments

Ischemia (reduction or suppression of the cerebral circulation in a certain territory of the brain) is induced by general or local causes, in the first place by arteriosclerosis, which may assume various evolutive forms: arterial stenosis, thrombosis and embolisms.

The clinical picture is variable, in dependence on the fact whether these impairments have appeared suddenly or slowly.

The acute ischemic cerebral impairments may be:

- *transient* or *vascular cerebral attacks* manifested by speech, equilibrium and vision impairments up to hemipareses which, if repeated after weeks, months or years, lead to a cerebral infarction with hemiplegia;

- *permanent* or *cerebral infarction*, the onset of which occurs during the night, sometimes (in about 50% of cases) with transient signs (dizziness, diserect cephalalgia, aphasia, paresis, paresthesias), lasting a few hours or days and leading in some cases to the appearance of a superficial comatose state and afterwards to a flaccid or spastic hemiplegia.

The chronic ischemic cerebral impairment may assume one of the following clinical pictures:

- *the neurastheniform syndrome*, characterized by asthenia, cephalalgia, insomnia and emotional lability;

- *status lacunaris (lacunar state)* manifested by discreet aphasia and lacunar gait (small steps); it appears repeatedly and intensifies gradually;

- *the pseudobulbar syndrome*, characterized by mastication, deglutition and speech impairments, spasmodic laughter and crying to minor stimuli.

Cerebral thrombophlebitides

The obstruction of the cerebral veins, either following various infections, or by the direct intervention of alterations at the level of the vascular walls or of blood clotting disorders, is often encountered in the clinic, its site of choice being at the cortical veins and venous sinuses. The most frequent are the post partum cortical and the post abortum cortical thrombophlebitidis and the particularly severe thrombophlebitis of the cavernous sinus, which requires a treatment of extreme emergency.

The cerebral hemorrhage

It is produced by the vascular rupture following a hypertensive outburst at any age, but more frequently in 50-60 year aged patients. The onset is often diurnal, on the occasion of an effort; it is favoured by fatigue, atmospheric pressure or temperature changes, exposure to the sun and alimentary excesses.

It begins, usually, with violent headache, dizziness, neurea and vomiting, leading rapidly to coma, flaccid and spastic hemiplegia.

During expiration, the face sweels on the paralyzed side –"seal sign" or "pipe sign", the head the eyeballs are deviated to the side of hemiplagia; the respiration is noisy (Cheyne-Stokes type respiration).

The intracranial aneurysms

They may be congenital or traumatic, located in 85% of cases at the level of the anterior circulation (in 30% at the anterior cerebral artery, in 30% at the internal carotid and in 25% at the middle cerebral artery) and in 15% at the level of the posterior circulation at the posterior inferior cerebellar artery and in 10% at the basilar artery.

The treatment is surgical.

The spastic cerebral paralysis

Called also *Little's disease*, "it is rather the expression of abnormal posture and movement modalities than a lack of functioning of a neuromuscular activity" (N. Rotanescu); it is due to a non-evolutive lesion of the central nervous system, which is detected during the early development stages of the brain.

Usually the central lesion is the result of a birth trauma or of a neurovascular syndrome occured during or after the engagement of the head in the cervix of the uterus (intra partum). Afterward, cystic type cerebral lesions may appear, involving extensive parts of the encephalon.

Clinically, the presence of the pyramidal (exaggeration of the myotatic reflex), extrapyramidal (involuntary movements) and cerebellar syndrome (tremors, incoordination of movements, gait disorders), to which mintal disorders, ranging from mental debility to idiocy, are associated.

Ocular lesions

Any lesion of the eyes, even the slightest, involves the risk of loss of sight and therefore requires a special care.

The craniocerebral injuries may affect the visual pathways or the oculomotor nerves. However, the severest are the lesions of the orbital region, which appear under the form of penetrating lesions, with or without foreign body retention. Fifteen percent of these accidents lead to blindness.

Investigation and treatment methods

The ventricular puncture is a technique requiring a special equipment and a high competence, therefore it is performed only in neurosurgery departments. After previous trepanation (trephination) of the skull, the needle is introduced at the indicated sites up into the ventricle. In the sucklings, the puncture is performed at the level of the anterior fontanel (Fig.510-512).

The hemostasis of the wounds of the venous sinuses is performed as an emergency, by widening attentively the cranial bone until the bleeding sinus is reached. The bone fragments obstructing the wound are gently removed, after which the plasty of the sinus with the fascia or muscle sutured to the dura mater is performed (Fig.513-515).

The evacuation of the extradural hematoma is urgently necessary and is achieved by 2 or 3 trephine holes performed in the temporal region. The trephine hole permits not only to establish the diagnosis, but also to appreciate the extent of the hematoma. After the outflow of blood and the aspiration of the clots, either the trephine hole is enlarged or a temporoparietal shutter is performed. Hemostasis is carried out without opening the dura mater.

In the coexistence of complications (serous meningitis, cerebral edema, cerebroventricular collapse) is suspected, the dura mater should be opened.

If after the opening of the dura mater, a normal appearance of the brain may be observed through the trephine hole, especially if the injured patient is unconscious and exhibits neurological signs of brain lesion, a trephine hole is performed on the opposite side of the skull during the same operative stage (Fig.516-524).

The evacuation of the subdural hematoma should be performed most rapidly by trephination and large opening of the dura mater. The latter is blue colored, under tension and the brain is no more pulsating. During the first two days after the accident, the subdural hematoma is constituted of pure blood or of an organized clot; later, it is surrounded by a membrane and its content is liquefied. If the condition of the patient does not improve, a trephine hole is performed also on the opposite side after 24 hours (Fig.525-527).

The treatment of penetrating craniocerebral wounds is performed in specialized neurosurgical departments. After the excision of the soft parts surrounding the wound, the injured parts of the bone are removed by pinching the bone with a hooked forceps, the torn dura mater is excized and the accessible splinters, the foreign bodies and the blood clots are removed by means of a slight Ringer serum jet. An accurate hemostasis is performed and the dura mater is closed without drainage, the shutter is again applied and the teguments are sutured (V.N. Shamov). Resuscitation and massive therapy with antibiotics are carried out (Fig. 528-534).

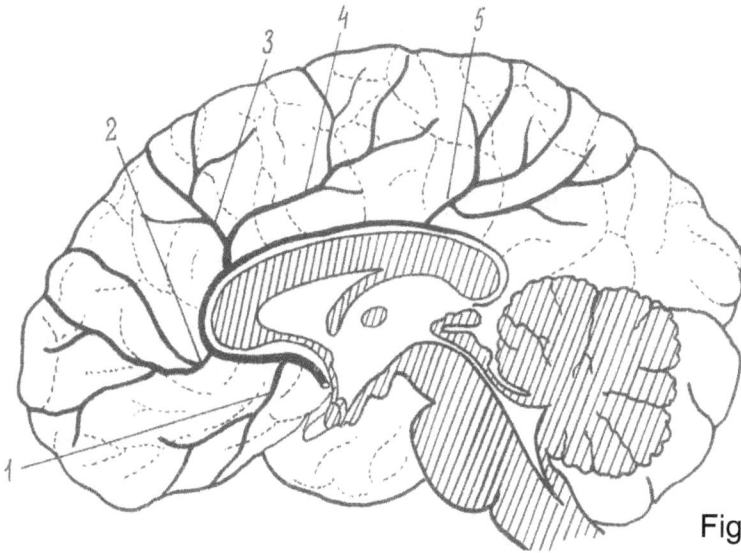

Fig. 507 Cortical distribution of the anterior cerebral artery
1- inferior frontal artery; 2- anterior internal (prefrontal or frontopolar artery); 3- middle internal frontal artery; 4- posterior internal frontal artery; 5- internal parietal (paracentral, precuneal, parieto-occipital) artery; 6- terminal artery

Fig. 508 Cortical distribution of the middle cerebral artery
1- insular artery; 2- orbitofrontal artery; 3- ascending frontal artery; 4- rolandic artery; 5- retrorolandic artery; 6- posterior parietal artery; 7- anterior temporal artery; 8- middle temporal artery; 9- posterior temporal artery; 10- terminal (end) artery

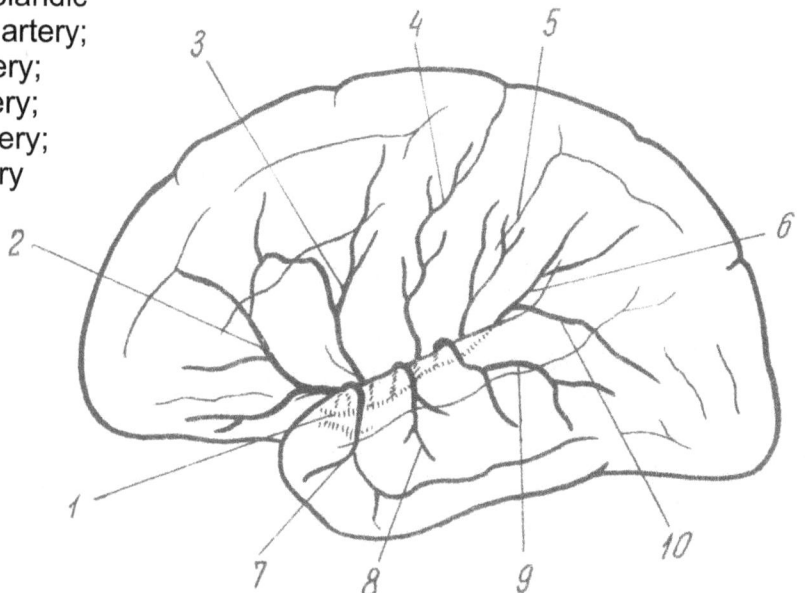

374

a.terminală

a. temporală
anterioară

a. temporală
mijlocie

a. temporo occipit

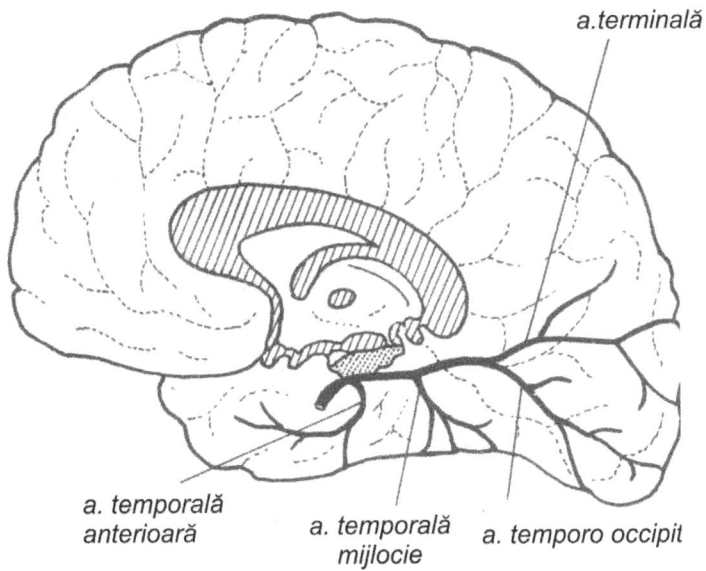

Fig. 509 Distribution of the posterior
cerebral artery

Fig. 510 Hematoma at the level of the
epicranium – evacuant puncture

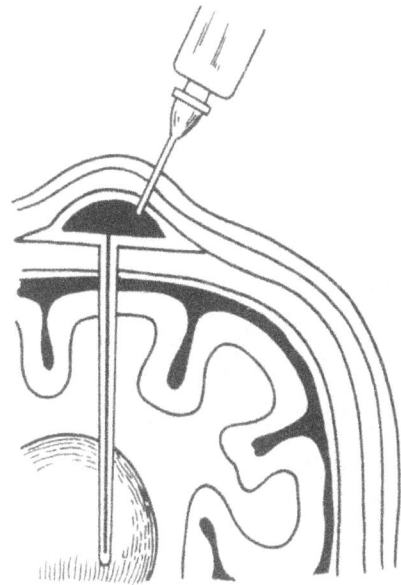

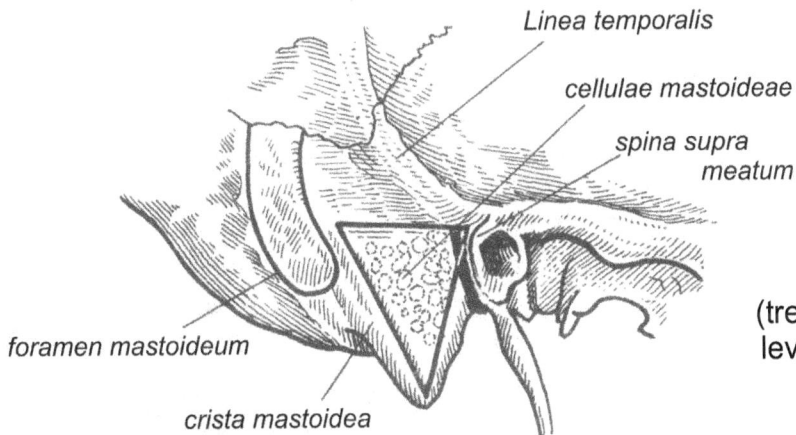

Linea temporalis

cellulae mastoideae

spina supra
meatum

foramen mastoideum

crista mastoidea

Fig. 511 Trephination
(trepanation) triangle at the
level of the mastoid region

Fig. 512 Site of trephine holes for the surgical
exploration of patients with cranial injuries

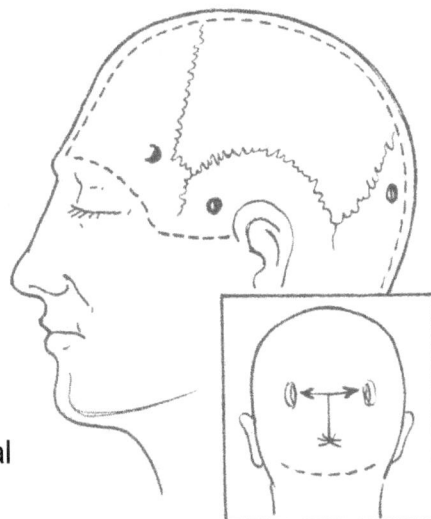

375

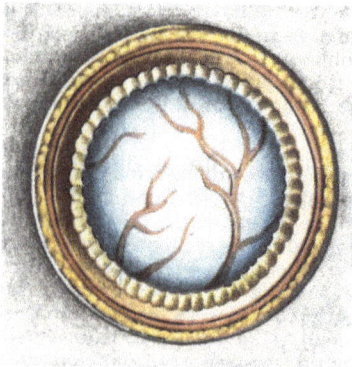
Fig. 513 Normal aspect of the brain before opening the dura (after C. Arseni)

Fig. 514 Normal aspect of the brain before opening the dura. Profile view (after C. Arseni)

Fig. 515 Normal aspect of the brain after opening the dura (after C. Arseni)

Fig. 516 Extradural hematoma (after C. Arseni)

Fig. 517 Extradural hematoma (after C. Arseni)

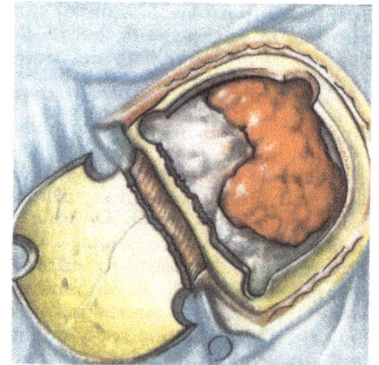
Fig. 518 Extradural cranial hematoma (after C. Arseni)

Fig. 519 Serous meningitis (after C. Arseni)

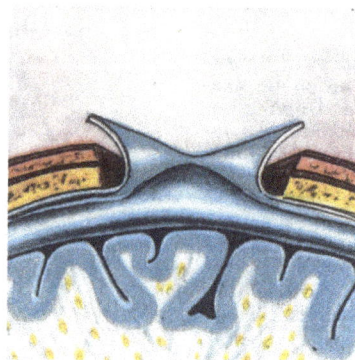
Fig. 520 Serous meningitis (after C. Arseni)

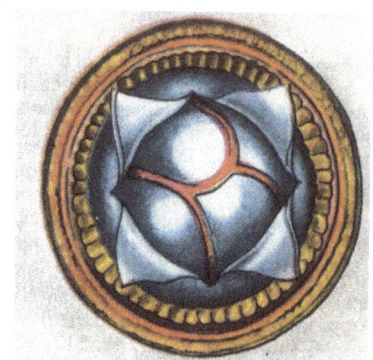
Fig. 521 Cerebral oedema (after C. Arseni)

Fig. 522 Cerebral oedema (after C. Arseni)

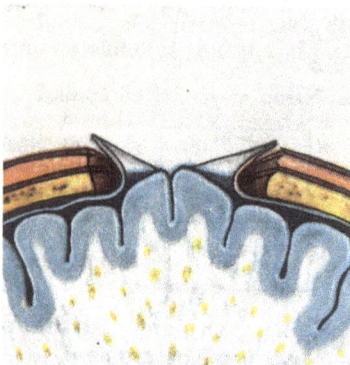

Fig. 523 Cerebroventricular collapse (after C. Arseni)

376

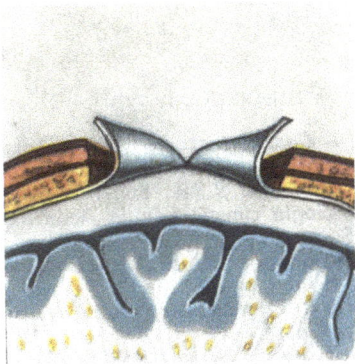
Fig. 524.
Cerebroventricular
collapse. (after C. Arseni)

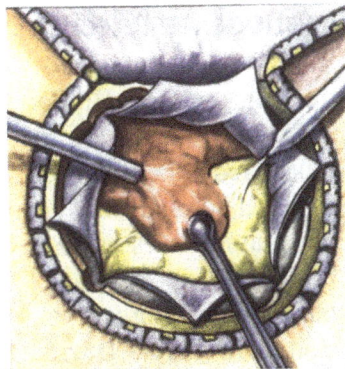
Fig. 525. Subdural cranial
hematoma

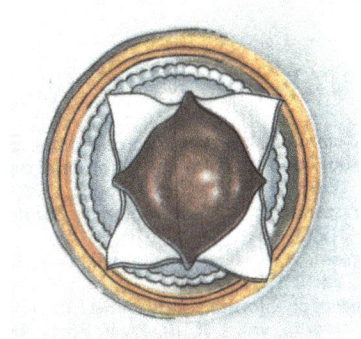
Fig. 526 Chronic subdural
hematoma.
(after C. Arseni)

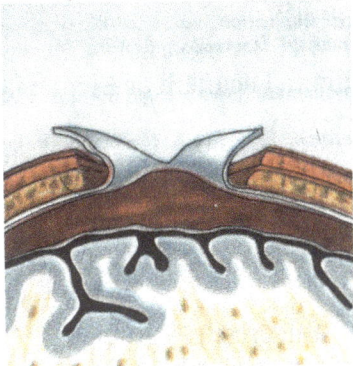
Fig. 527 Chronic subdural
hematoma.
(after C. Arseni)

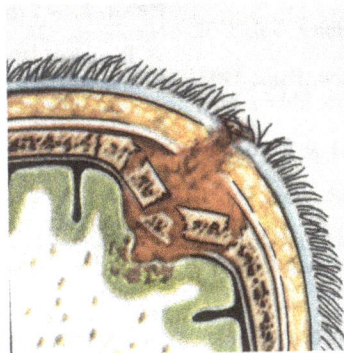
Fig. 528 Fracture of skull
with clogging

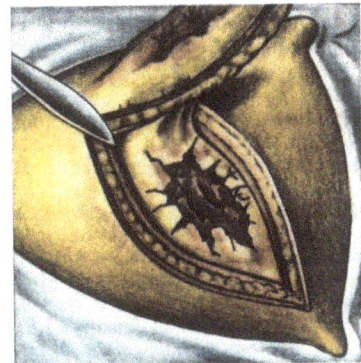
Fig. 529 Excision of soft
tissue around the shot
wound (concussion)

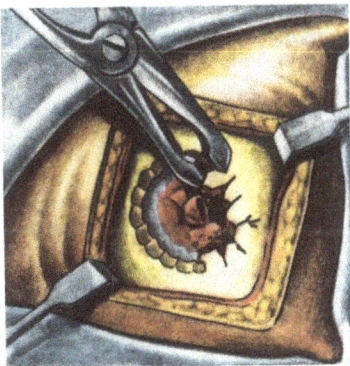
Fig. 530. Bone pinching
(V. N. Şamov)

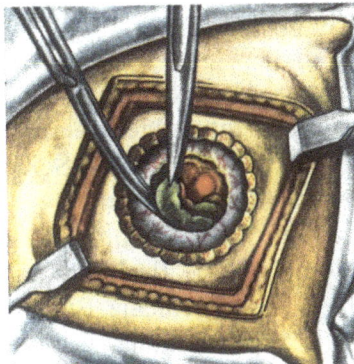
Fig. 531. Excision of the
torn dura mater (V. N.
Şamov)

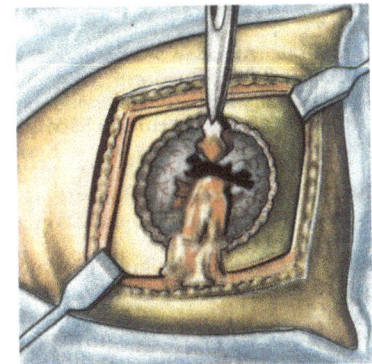
Fig. 532. Removing
affordable bone fragments
(V. N. Şamov)

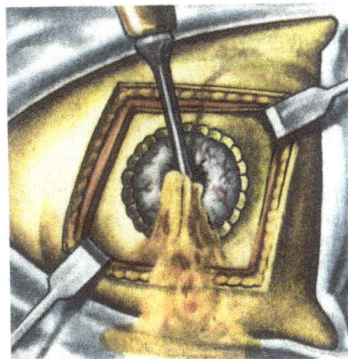
Fig. 533. The wash the wound
tract and the remove of small
bone fragments and blood
clots (V. N. Şamov)

Fig. 534. The cranial dura
suture

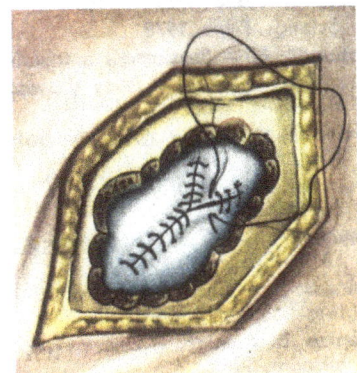

377

SELECTIVE REFERENCES

◆ Albu, Ioan şi colab. - Anatomia omului, curs, vol. I, II, III, Editura I.M.F., Cluj-Napoca, 1980.

◆ Andronescu, A. - Anatomia copilului, Editura Didactică, 1973.

◆ Arseni, A. - Neurologie, vol. I-V, Editura Medicală, Bucureşti, 1981-1983.

◆ Bernard, Claude, Baurgerie, P. - Atlas d' Anatomie Humaine, vol. I, II, III, IV, V, Paris, 1866.

◆ Beuninghoff/Goerttler - Lehrbuch der anatomie des menschen, Ufban-Schwartzenberg, Munchen, 1975.

◆ Boileau, Grant - Atlas of human anatomy, Baltimore, 1972.

◆ Brînzeu, Pius - Tratat de chirurgie, Editura Medicală, Bucureşti, 1966.

◆ Cuningham, I. - Manual of practical Anatomy, vol. I, II, III, London, 1975.

◆ Delmas, A. - Voies et centres nerveux, Masson, Paris, 1983.

◆ Diaconescu, M. - Curs de anatomie, Editura I.M.F., Timişoara, 1980.

◆ Diculescu Mie - Histologie, Editura Didactică şi Pedagogică, Bucureşti, 1983.

◆ Dimulescu, V. şi colab. - Anatomia omului, curs, vol. I, II, III, Editura I.M.F., Timişoara, 1975.

◆ Drăgoi, Gh., Mocanu, Gh. - Anatomia practică a sistemului nervos, Editura Facultăţii de Medicină, Craiova, 1982.

◆ Făgărăşanu, I. - Probleme de chirurgie şi de anatomie clinică, Editura Academiei R.S.R., Bucureşti, 1983.

◆ Firu, P. - Stomatologie infantilă, Editura Didactică şi Pedagogică, Bucureşti, 1984.

◆ Gray, H. - Anatomia descriptivă aplicată, vol. I, II (tradusă şi adaptată de Gr. T. Popa, 1944).

◆ Gregoire, R., Oberlin, S. - Precis d'anatomie, I. B. Bailliere, 1973.

◆ Grigorescu, Aurelian - Receptorii endocrini, Editura Academiei R.S.R., Bucureşti, 1984.

◆ Iagnov Z. şi colab. - Anatomia omului, vol. I, II, III, Editura Medicală, Bucureşti, 1956.

◆ Ianculescu Al. şi Mogaş Ion - Compendiu de Anatomie şi fiziologie, Ediitura Medicală, 1972.

◆ Ifrim, M. şi colab. - Anatomia şi biomecanica educaţiei fizice şi sportului, Editura Didactică şi Pedagogică, Bucureşti, 1978.

◆ Ifrim, M., Căpuşan, Iuliu - Morfologie normală şi patologică a ţesutului conjunctiv, Editura Medicală, Bucureşti, 1983.

◆ Ionescu, Gh. - Chirurgie, Editura Dacia, Cluj-Napoca, 1985.

◆ Juvara, I. - Tratat de tehnici chirurgicale, Editura Medicală, Bucureşti, 1980.

◆ Kahle, W. - Anatomie, tome I, II, III, Flamarion Medicine Science, Paris, 1978.

◆ Kapandji, A. - Physiologie articulaire, Paris, 1968.

◆ Kiss, Szentagothai - Atlas d'anatomie du corp humain, Budapest, 1966.

◆ Kreindler, A., Sager, O. - Neurologie, vol. I, II, Editura Medicală, Bucureşti, 1957.

◆ Kreindler, A. - Neurologie, vol. I, II, Editura Medicală, Bucureşti, 1963.

◆ Menkes, B. - Cercetări de embriologie experimentală, Editura Academiei R.S.R., Bucureşti, 1958.

◆ Menkes, B. şi colab. - Embriologie, curs, Editura I.M.F., Timişoara, 1965.

◆ Menkes, B., Şandor, St. - Cercetări asupra formării somitelor, în Studii şi cercetări de embriologie şi histologie, p. 129-152, .Editura Academiei R.S.R., Bucureşti, 1970.

◆ Milcu, Şt. M. - Endocrinologie clinică, Editura Academiei R.S.R., Bucureşti, 1975.

◆ Milcu, Şt. M. şi colab. - Andrologie clinică, Editura Academiei R.S.R., Bucureşti, 1974.

◆ Milcu, Şt. M. - Introducere în antropologie, Editura Academiei R.S.R., Bucureşti, 1975.

◆ Milcu, Şt. M. şi colab. - Genetică umană. Editura Academiei R.S.R., Bucureşti, 1976.

◆ Moraru, I. - Anatomie patologică, Editura Medicală, Bucureşti, 1981.

◆ Netter, F. - Ciba Medical Ilustration, vol. I, II, III, IV, 1984.

◆ Niculescu, Gh. - Tratamentul fracturilor, Editura Militară, Bucureşti, 1964.

◆ Niculescu, Gh. - Traumatismele, Editura Medicală, Bucureşti, 1975.

◆ Papilian, Victor - Anatomie umană cu aplicaţii practice, vol. I, II, III, Cluj, 1940 (reeditate de Albu I., 1979).

◆ Papilian, V., Victor - Histologie, vol. I, II, Editura Dacia, Cluj-Napoca, 1980.

◆ Paturet, G. - Traite d'anatomie humaine, tome I, II, III, IV, Masson, Paris, 1980.

◆ Rainer, I. Fr. - Oeuvres Scientifique, Editura Academiei Române, Bucureşti, 1945.

◆ Rainer, I. Fr. - Prelegeri de anatomie funcţională, Facultatea de Medicină, Bucureşti, 1929-1944.

◆ Rauber-Kopsch - Lehrbuch und Atlas des Anatomie des Menschen, Georg-Thieme-Verlag, Leipzig, 1940.

◆ Riga, I. Th. - Curs de anatomie topografică şi chirurgie operatorie, vol. I, II, III, IV, Editura Medicală, Bucureşti, 1960.

◆ Robacki, R. şi colab.- Anatomie pe viu, Editura Universităţii Craiova, 1973-1974.

◆ Robacki, R. - Anatomia funcţională a omului, vol. I, II, III, Editura Scrisul Românesc, Craiova.

◆ Robaoki, R. - Dinamica dezvoltării somatice a copiilor între 0 şi 7 ani, în Morfologie normală şi patologică, vol. VIII, nr. 3/1963.

◆ Robacki, R. - Prelegeri de anatomie funcţională, Editura Universităţii Craiova, 1970/76.

◆ Rahen - Yokochi - Photographic Atlas of systematic and regional anatomy, vol. I, II, Schrattauer, 1984.

◆ Rouviere, H. - Anatomie, tome I, II, III, Masson, Paris, 1980.

◆ Rottenberg, N. şi colab. - Anatomie pe viu, Editura Facla, Timişoara, 1982.

◆ Saragea, M. - Fiziopatologie, Editura Academiei R.S.R., Bucureşti, 1983.

◆ Sinelnicov, R. A. - Atlas anatomii celoveca, voi. I, II, III, Moscva, 1972, Modghiz (Izdatelotva mediţina).

◆ Sobota-Becher - Atlas de Anatomie des Menschen, Urban—Schwartzenberg, Wien, 1982, band I, II, III.

◆ Spalteholtz, Werner - Handatlas der Anatomie des Menschen, Leipzig, 1918.

◆ Taşcă, G. şi colab. - Morfopatologia substanţelor de reglare endocrină, Editura Acad. R.S.R., Bucureşti, 1983.

◆ Toldt, W. - Anatomie des Menschen, Leipzig, 1921.

◆ Tonkov, V. - Anatomii celoveca, Medghiz, 1946.

◆ Tuchmann du Plessis - Embriologie, Masson, Paris, 1976.

◆ Vancea, Petre - Curs de oftalmologie, Editura I.M.F., laşi, 1965.

◆ Voiculescu, Vlad - Neurologie, Editura Medicală, Bucureşti, 1975.272

ARA Publisher
American Romanian Academy od Arts and Sciences
University of California Davis
http://www.AmericanRomanianAcademy.org
info@AmericanRomanianAcademy.org
Adresa: P.O. Box 2761
Citrus Heights, CA 95611-2761

www.ingramcontent.com/pod-product-compliance
Lightning Source LLC
Chambersburg PA
CBHW082304210326
41598CB00028B/4439